CW00547949

Practical Reservoir Simulation

Practical Reservoir Simulation

Using, Assessing, and Developing Results

by
M. R. (Mike) Carlson

Disclaimer. The recommendations, advice, descriptions, and the methods in this book are presented solely for educational purposes. The author and publisher assume no liability whatsoever for any loss or damage that results from the use of any of the material in this book. Use of the material in this book is solely at the risk of the user.

PennWell Corporation
1421 South Sheridan Road
Tulsa, Oklahoma 74112-6600 USA

800.752.9764
+1.918.831.9421
sales@pennwell.com
www.pennwellbooks.com
www.pennwell.com

Marketing Manager: Julie Simmons
National Account Executive: Barbara McGee

Director: Mary McGee
Managing Editor: Marla Patterson
Production/Operations Manager: Traci Huntsman
Production Editor: Sue Rhodes Dodd
Book Designer: Alana J. Herron
Cover Designer: Shanon Moore

Library of Congress Cataloging-in-Publication Data

Carlson, M. R.
Practical reservoir simulation : using, assessing, and developing results/M. R. (Mike) Carlson
p. cm.
Includes bibliographical references and index
ISBN 0-87814-803-5
ISBN13 978-0-87814-803-5
1. Oil reservoir engineering—Mathematical models. I. Title.
TN871 .C33 2003
622'.3382'015118—dc21 200301105

Printed in the United States of America

5 6 7 8 9 12 11 10 09

Dedication

This book is dedicated to my family: my parents for their support over many years, my wife for her editorial assistance over the years (not just this book), my children for the lost hours at the computer terminal, and cousin Sandra for her office management skills.

Contents

Chapter 7. Capillary Pressure and Relative Permeability: Data Screening

Chapter 9. Initialization

Chapter 10. Integration and Gridding

Acknowledgments

I would like to acknowledge particularly Eric Denbina, who has provided invaluable editorial assistance, with considerable tact, in putting this book together.

There have been many people along the road who have helped me develop the ideas in this book. They include previous employers, colleagues, supervisors, employees, clients, students, and friends, all of whom deserve a collective thank you. Of course, any mistakes that may exist are my responsibility.

Artwork Acknowledgments

The author thanks these gracious individuals, organizations, and publishers for allowing the use of their materials in this text.

Chapter 2

Figs. 2–4, 2–15, 2–20, 2–25	Society of Petroleum Engineers
Figs. 2–7, 2–11, 2–14	Copyright with permission from Elsevier Science
Figs. 2–8, 2–9, 2–28	Springer-Verlag

Chapter 3

Figs. 3–1, 3–2, 3–3, 3–4, 3–6, 3–7, 3–9, 3–10, 3–30, 3–31, 3–32, 3–33, 3–34, 3–35, 3–37, 3–38, 3–39, 3–40, 3–41, 3–42, 3–43	Reprinted by permission of the AAPG whose permission is required for further use.
Fig. 3–5	Reproduced with permission of the Minister of Public Works, Government Services Canada, 2003, Courtesy of Natural Resources Canada, Geological Survey of Canada
Fig. 3–12	*Netherlands Journal of Geosciences*
Fig. 3–13	New Orleans Geological Society
Figs. 3–16, 3–17, 3–19	Used by permission of Pearson Education Inc., Upper Saddle River, N.J.
Fig. 3–18	*American Journal of Science*
Figs. 3–20	Reprinted by permission of the Canadian Society of Petroleum Geologists
Fig. 3–22	PennWell Publishing
Fig. 3–26	Society of Petroleum Engineers
Figs. 3–47	*Journal of Canadian Petroleum Technology*

Chapter 6

Figs. 6–32, 6–33	American Institute of Chemical Engineers
Fig. 6–9	*Journal of Canadian Petroleum Technology*
Fig. 6–16	Copyright with permission from Elsevier Science
Figs. 6–14, 6–52	Reprinted courtesy of the Gas Processors Suppliers Association
Figs. 6–45, 6–56	Hart Publications
Figs. 6–42	Dr. Karen S. Pederson

Figs. 6–5, 6–10, 6–11, 6–12, 6–13, 6–29, 6–30, 6–51, 6–53, 6–55	Society of Petroleum Engineers

Chapter 7

Table 7–3	Core Laboratories Corporation
Figs. 7–5, 7–6, 7–7, 7–12, 7–13, 7–20, 7–21, 7–27, 7–28, 7–31, 7–35, 7–67, 7–68, 7–69, 7–70, 7–76	Society of Petroleum Engineers
Figs. 7–9, 7–10, 7–11	Core Laboratories Corporation
Figs. 7–14, 7–15, 7–19, 7–54	Marcel Dekker Inc.
Fig. 7–18	Copyright with permission from Elsevier Science
Fig. 7–23	Gas Technology Institute
Fig. 7–38	Reprinted with permission, copyright CRC Press, Boca Raton, Florida

Chapter 8

Figs. 8–1, 8–2, 8–3, 8–4, 8–16, 8–17, 8–18, 8–35, 8–36, 8–37, 8–70	Copyright with permission from Elsevier Science
Figs. 8–8, 8–9, 8–10	Society of Petroleum Engineers

Chapter 9

Figs. 9–5, 9–6	Springer-Verlag

Chapter 10

Fig. 10–7	Society of Petroleum Engineers
Fig. 10–37	*Journal of Canadian Petroleum Technology*

Chapter 11

Fig. 11–4	PennWell Publishing

Chapter 15

Fig. 15–1	Dr. Tarek Ahmed
Fig. 15–3	Reprinted with permission, American Chemical Society
Figs. 15–4, 15–5, 15–6, 15–7, 15–12, 15–13, 15–37	Dr. Karen S. Pederson
Fig. 15–14	Reprinted courtesy of the Gas Processors Suppliers Association

Figs. 15–21, 15–22, 15–23, 15–24, 15–27, 15–28, 15–29, 15–30, 15–31, 15–32, 15–34, 15–35, 15–36	Society of Petroleum Engineers

Chapter 16

Fig. 16–1	PennWell Publishing
Fig. 16–1(b)	Petroway Inc.
Fig. 16–5	Geological Society of America

Chapter 17

Figs. 17–1, 17–2, 17–3	Hart Publications
Figs. 17–4, 17–7, 17–8	Dr. Larry Lake
Figs. 17–5, 17–6, 17–10, 17–11, 17–13	Society of Petroleum Engineers
Fig. 17–12	Mark A. Klins
Fig. 17–15	*Journal of Canadian Petroleum Technology*

Chapter 18

Fig. 18–1	Society of Petroleum Engineers
Fig. 18–4	Copyright Oxford University Press Inc. Used by permission of Oxford University Press Inc.
Figs. 18–19, 18–20	Reprinted by permission of the AAPG, whose permission is required for further use
Fig. 18–28	*World Oil*
Fig. 18–29	*Journal of Canadian Petroleum Technology*
Fig. 18–30	Dr. John Suppe
Figs. 18–31(a–c)	Copyright American Geophysical Union, Reproduced/modified by permission of American Geophysical Union

Chapter 19

Figs. 19–1, 19–2, 19–3, 19–7, 19–12, 19–41, 19–48, 19–52	Gravdrain Inc.

Figs. 19–4, 19–8, 19–9, 19–11, 19–72, 19–73, 19–74, 19–76, 19–100	Permission granted on behalf of Alberta Energy Research Institute, Alberta Science, Research Authority as per Energy Intellectual Property Management Agreement
Fig. 19–5	Dr. K.C. Hong
Figs. 19–6, 19–75, 19–94, 19–95	Dr. Rick Chalaturnyk
Fig. 19–9	Reprinted by permission of the Canadian Society of Petroleum Geologists
Figs. 19–13, 19–30, 19–38, 19–40, 19–53, 19–54, 19–55, 19–56, 19–57, 19–58, 19–59, 19–71, 19–77, 19–78, 19–79, 19–80, 19–82, 19–86, 19–87, 19–91, 19–92, 19–93	*Journal of Canadian Petroleum Technology*
Fig. 19–14	Core Laboratories Corporation
Figs. 19–27, 19–28	Dr. Amin Touhidi-Baghini, Dr. J.D. Scott
Fig. 19–29	Dr. Kaz Oldakowski
Figs. 19–34	Reprinted by permission of John Wiley & Sons, Inc.
Figs. 19–35, 19–36, 19–51, 19–81	Society of Petroleum Engineers
Fig. 19–68	Reprinted by permission of the AAPG, whose permission is required for further use

Chapter 20

Fig. 20–7	Reprinted by permission of the AAPG, whose permission is required for further use

A Note on Hand-Drawn Diagrams

A significant proportion of the diagrams in this text have been hand-drawn. From an efficiency perspective, there is a trade-off on time spent on engineering and time spent on presentation work. Abstractly, there is some ideal balance.

In terms of workflow, most field developments involve the continual addition of new data (from new wells and as lab data is received). The author uses conceptual diagrams "on the fly" to describe what is being done and how approaches are being applied. If a diagram is sufficient to make the point, it has served its purpose. The diagrams have been deliberately left in sketch format to demonstrate this as a realistic and time-efficient working approach. Formal diagrams involve considerable preparation and, in the author's opinion, are best left to final formal presentations and reports.

In actual practice, the ideal balance is not very clear, and diagram usage depends on the culture of the company. The author has worked in companies where hand-drawn and colored working diagrams, sometimes poster-sized, can be used in meetings up to the vice-president level. Such companies are often characterized by a high degree of participation and technical interest by senior management with an emphasis on getting work done.

However, other companies develop a culture in which formal PC graphics are required for any discussion above the group leader level. Such an approach is often preferred where management time is in short supply, management is highly confident in technical staff, the management style is more formal, or where the agenda of the company is driven by financial issues. It is usually left to individuals to figure out the culture and adapt to the style of the organization.

Regardless of the company culture, the author strongly encourages personal working diagrams.

Part 1

Bread and Butter (Black Oil) Simulation

1

Introduction

The Mystique of Reservoir Simulation

During the 1970s and early 1980s, reservoir simulation developed quite a mystique because it was the newest reservoir tool. Much of this has worn off, and reservoir simulation has become quite commonplace. Simulators are very elegant creations. It is easy for someone with a technical inclination to become entranced with using this tool.

Despite the power of reservoir simulation, it can be a dangerous tool. It will calculate meaningless results with incredible precision. Many individuals and companies have been burned when inappropriate use was made of this tool. As a result, some managers became disenchanted with simulation, and they were no longer interested in being fooled by this technique. With significant improvements in simulation techniques, this attitude has diminished gradually. In reality, failures occurred because the people applying the technology either did not understand or properly communicate their assumptions and corresponding limitations of the results. It takes discipline to realize numerical modeling has both strengths and weaknesses and is only one of many tools available.

Screening

Most reservoir engineering textbooks do not have a section dealing with how to screen the input data. There are no examples of bad data. The point is that doing a reservoir simulation probably will put most practicing engineers into new territory.

Breadth of technology

A certain number of students who have attended classes taught by the author—normally those who are just out of university—are quite offended by an apparently established technique being regarded with suspicion. Furthermore, the author regards most data as potentially wrong. Probably the worst offense is the implication that ignorance is a major barrier in reservoir engineering.

With experience, it becomes painfully obvious—many reservoir engineering estimates, plans, and schemes have fundamental flaws. In some cases, it is possible to identify a wrong assumption as the central cause of the problem. Unfortunately, this is not the norm. Most often, it will be attributed to a combination of things; some could never be known, and some could have been better estimated. On a philosophical note, the essence of Buddhism is ignorance is our sole enemy. In simplistic terms, it replaces sin. This general concept is sometimes coined *lifelong learning*.

The truth is that the oil industry encompasses a large span of technology. At present, the Society of Petroleum Engineers (SPE) paper count is more than 75,000 articles and papers, which does not include the works from similar Canadian, British, Australian, and other European countries. Nor does it include the combined works in chemical engineering, mathematics, geological, and geophysical literature. It is difficult, and probably impossible, to know everything contained in the literature. It would take a lifetime to go through all of this material and digest it.

The state of knowledge is not complete. The improvements made in the last 20 years are staggering. A favorite example of this is in pressure transient analysis (PTA). In the past, junior engineers did PTA (Horner plots) by hand, and type curves didn't reveal much information. These days no one does PTA by hand; it is all done with computers and a derivative analysis. There is more certainty in interpretations now; the state of the art has improved vastly.

Our profession involves considerable uncertainty. Consequently, we make educated guesses. Perhaps the most telling indication of this can be found in the salary policies for geologists. Clearly, if certain individuals could be identified as the individuals who could be classified as the *oil or gas finders*, they would attract a huge premium. As yet, no one has been able to make this identification, and the evaluation of geological performance is based on the assessment of supervisors. If this were directly quantifiable, perhaps geologists, and possibly engineers, would work on commission for discoveries.

After performing a number of reservoir simulations and reviewing many more, it is the author's opinion that the proportion of successful simulations could be greatly improved. Although it is impossible to contain the unknowable in our lifetime, significant improvements can be made to the parts individuals can know and control.

Going through life trying to avoid mistakes, i.e., avoiding the negative, is not the best approach. A proactive approach can help, such as reading the literature when one gets into a new area. Learn from the mistakes of other people, and take appropriate steps to avoid them. Take the time to think through the problem, and get help if needed. Join the appropriate technical societies as well.

Value of simulation

In general, reservoir engineering is difficult. Since oil and gas are buried deep underground, it is not possible to see a reservoir or to touch it. Unlike production equipment, such as a gas plant, it is not possible to control physical conditions. A reservoir engineer's job is to understand and predict what cannot be seen or touched. A number of indirect tools and measurements are required.

In this regard, a reservoir simulator is superb. It allows one to test quantitatively how different processes affect production results. It allows complex geometries to be evaluated that cannot be solved easily, if at all, by analytical equations.

Although the simulator allows evaluation of complex problems, it is a passive tool. It can evaluate input, but it is not capable of determining input that should have been entered.

Pattern recognition

It is appropriate to put reservoir simulation in its proper context. The art of reservoir engineering involves pattern recognition. The results from many different techniques must be integrated to develop a correct interpretation of reservoir behavior. Different types of data vary in accuracy. Therefore, input screening is required. Different sources of information can conflict. Frequently, not all of the data desired or required is available. In this case, properties will have to be estimated from other sources. All of this must be weighed mentally, and a best estimate made. To make matters worse, many of the sophisticated techniques used by reservoir engineers provide interpretations that are not unique. Multiple solutions often exist.

As a consequence, the science gets a little fuzzy and becomes an art. Experience counts heavily in reservoir engineering. This leaves the new reservoir engineer in a catch-22 position. He or she needs experience to do the job, and one can't get it without the experience of having done it.

Conceptual model

Certain approaches in different disciplines give a perspective that makes success either easier or more likely. A successful approach to reservoir simulation is to maintain a conceptual model of the reservoir. This must include:

- the shape and location
- the type of reservoir fluid
- fluid flow properties of the rocks
- drive mechanisms
- visualization of flow patterns

Armed with this model, it is possible to proceed, to understand, and to predict performance with reasonable accuracy.

Geological models

First, successful reservoir simulation requires detailed knowledge of geological models. This will immediately put a number of reservoir engineers out of their comfort zone. They will point out they are not geologists. In fact, it is not necessary to develop the model—the job of the geologist—but it is a reservoir engineer's responsibility to understand the model. Geology, to a greater or lesser extent, affects all five of the critical elements of reservoir description listed previously. Two examples are shown in Figures 1–1 and 1–2.

The Keg River Reef is thick and limited in areal extent, somewhat like an office building. Vertical floods work very well in this type of reservoir. The Cardium Bar is long and thin, not very thick, and not very wide. It is normally impossible to implement a pattern waterflood in such a reservoir. Although these are simplified examples, the geological models obviously have immediate implications.

In many respects, a reservoir simulator is nothing more than an ultrasophisticated mental model. The model grid and input will be generated to correspond with this mental model. One of the most powerful aspects of simulation is the ability to propagate these assumptions through the production cycle of a reservoir.

Reservoirs have many features that can be categorized into a number of geological models. An understanding of geological principles is essential to a reservoir simulation engineer. Accordingly, chapter 3 is dedicated to integrating geology.

Model changes with time

As a reservoir matures and more wells are drilled, more pieces of the puzzle are available. The conceptual model of the reservoir will change with time. In most cases, the model will be refined continually. In some instances, accepted ideas will have to be discarded and a newer model developed.

Dangers in application

The consistent use of a mental image, a *reservoir model*, has one serious danger; the model can become reality to the reservoir engineer. Sophisticated reservoir calculations, particularly mathematical models, have an almost hypnotic quality. For this reason, some reservoir engineers earn a reputation for being out of touch. It is necessary to ask if the answers are relevant to developing cash flow for a producing company.

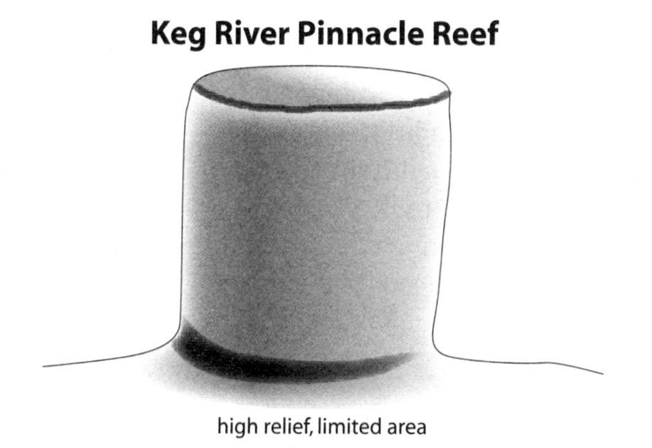

Fig. 1–1 Depiction of Keg River Pinnacle

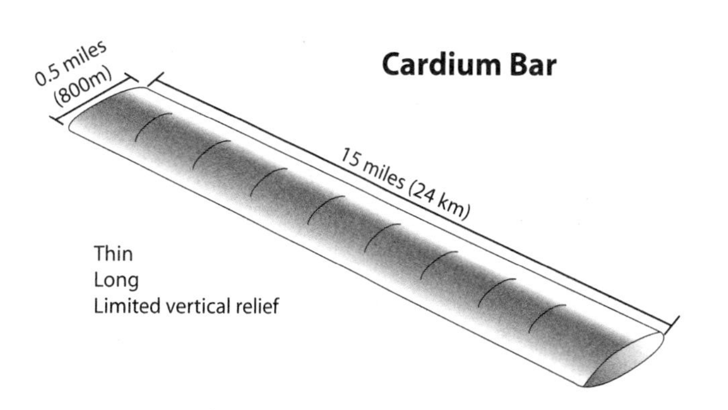

Fig. 1–2 Depiction of Cardium Bar

At the other extreme, some petroleum engineers are convinced reservoir engineering analysis is so theoretical as to be of no practical application. Do not believe this. Most senior engineering managers, vice presidents, and quite a few presidents have a reservoir engineering background.

Background experience

Some engineers, particularly those with an operational background, are intimidated by the high-tech nature of reservoir simulation. They do a disservice to themselves. A narrow simulation specialist is as dangerous, or perhaps more dangerous, than someone who has the experience to tell if results are right or wrong based on gut instinct. One of the most important skills is the ability to recognize good and bad input data. If engineers have operational experience, they have an advantage compared to those who become simulation "experts" via postgraduate studies.

Errors and failures

In some cases, anecdotes are made of errors or failures. This can be a very touchy subject. Some consider it dangerous to the profession to admit error. In the author's opinion, this is rather silly. It is painfully obvious, for example, after a building has fallen down, someone has made an error.

In more philosophical terms, Otto Von Bismarck remarked, "Only fools learn from their mistakes, I prefer to learn from others' mistakes." Although he probably made mistakes, no doubt he avoided a large number of traps using this maxim. A little fear of doing and saying something stupid can be constructive. The process of learning often involves failure. The cases are presented for a positive purpose.

Errors in the oil and gas industry are rarely as dramatic as bridge failures, but the financial implications are huge.

Risk control

Risk control is an important aspect of the oil industry. Most of this analysis concentrates on exploration perspective. This is not the only risk in the oil business. There are significant risks on the technical side. One example is the oil sands. The industry knows where it is. Getting it out economically and without plant failures is a completely different matter. In this case, the issue is operational risk not exploration risk.

One successful operating strategy involves working specific areas of concentration and controlling gas plant facilities. The successful application of this strategy requires a sound knowledge of the distribution of well results, i.e., the number of dry and abandoned wells (D&As), how many high-rate wells, and how many marginal rate wells should be expected. In effect, a statistical profile of well successes can be made and economics run on this distribution. Normally, this is derived empirically from offsets. Companies using this strategy do not do well-by-well economics often. Instead, the staff concentrates on the geological and reservoir engineering input. The mechanics of economics programs and simulation code, while certainly important in some cases, is normally of secondary importance to the input data.

The author has consulted for companies using such an approach. A number of comments have been made to the effect that this strategy is not strong technically. In the author's view, this couldn't be more incorrect. The economics of individual wells are not very accurate in comparison to a well-defined statistical distribution. With this strategy, a higher percentage of time is actually spent on technical issues. The author performed a simulation on a reservoir that did not fit into the empirical experience base. Under these circumstances, the company did not hesitate to use a *technical* approach.

A number of companies have used this approach and have been highly successful. There is an important lesson to derive here. Since input data is not unique in all cases, the answer required may not be a single estimate or economic run but rather a risk-weighted calculation (distribution) of where the correct solution lies.

Industry demographics

The industry is currently in an interesting position. Most operating companies severely curtailed their hiring after the price collapse of 1986. Immediately prior to this, most companies were staffing up for what was anticipated to be a major industry expansion—so much for strategic planning. The net result is a very large group of engineers hired prior to 1985 with almost no one behind them. It appears, in the immediate future, hiring is going to increase dramatically in order to fill junior positions in an industry embarking on an up cycle.

Objectives of the Book

The purpose of this text was originally for a course on reservoir simulation. The objectives are to provide the background necessary to correctly develop, manage, apply, or use simulation results. Simulation techniques are complex and require time to master. This book has been tailored to provide the minimum amount of theory necessary, give pragmatic advice on how to carry out a study, and determine if study results are useful. Understanding of all these techniques in detail takes years of development and experience. No single textbook can teach all of the techniques available. A more realistic objective is to teach how to approach reservoir problems, provide the basic concepts required, and point to more comprehensive references. Case histories and actual data have been provided to give readers a base of experience to build upon. The objective of this book is to outline a system of how to approach the understanding and simulation of a reservoir.

Concrete objectives

- Understand the background skills helpful in performing simulation.
- Recognize the differences in mathematical development commonly used in simulator development.
- Recognize the critical role geology plays in reservoir simulation. It is a reservoir engineer's responsibility to understand the model given (geologists interpret and supply the model).
- Learn data screening techniques for:
    - Capillary pressure
    - Relative permeability
    - Pressure-volume-temperature (PVT) data
- Thoroughly understand pseudo relative permeability.
- Learn how to generate a grid, in particular the use of flow net and streamline visualization.
- Envisage an appropriate approach (game plan) for a complete reservoir study.
- Develop an organized approach to history matching.
- Be familiar with a systematic program for predictions.
- Learn about some advanced topics.
- Be aware of the dangers with gas injection.
- Be able to assess simulation results.
- Learn the realities of managing a study.
- Understand some criteria for evaluating software.

Text outline

The basics will be developed in a logical sequence. Introductory chapters on practicalities are followed by basic data set generation, data screening, geological models, model construction, history matching, predictions, special simulations, and prototypical problem sets. Some of the tools used are presented in an academic style, which is typical of most reservoir engineering textbooks. Understanding the theoretical development is important. It is necessary to use judgment to decide which data is best. The limitations of how data is derived must be understood to do this. Some of the material is presented as stories or case histories. This will provide a base from which to build your own inventory of experience.

This book has been prepared in a conversational style. In many cases, there are a number of observations the author has made and not attempted to prove scientifically. In the author's opinion, a significant aspect of reservoir engineering is the thought process or approach. One of the objectives is to prepare reports to outline this thinking. This has been recognized lately and is often called an expert system, which is coincidentally associated with artificial intelligence (AI). This indicates the thought process has been recognized as a significant determinant of success.

Many engineers find expressing their thought process is difficult. However, the process requires one to "close loops" and clarify one's thinking in a logical manner. Some clients have expressed reluctance when the author takes time to prepare a report, but, once they have it, the feedback is very positive.

There are good reasons for this. It is easy to have confidence if the approach has been outlined thoroughly. Conversely, trying to guess what has been done is unsettling. With the limited time available in today's lean and mean environment, operating company staff are not given time to write any reports. At the same time, relying on contractors and consultants disrupts continuity. A written report is the only way the subsequent contract personnel will ever know what happened previously. Reports are now more important than ever.

Issues

Relatively few books discuss approaches or even issues. An exception is Dake's book, *The Practice of Reservoir Engineering*.[1] He does not like reservoir simulation since he feels it contributes to overlooking the underlying physics. A numerical model is an inanimate object and, in the author's view, cannot be held responsible for what people do or fail to do. Dake definitely has a knack for getting to the nub of a problem. Whether one uses analytical or numerical techniques to get to the bottom line technically is probably not important. The author recommends his book. Dake is a critical thinker, and this is a theme stressed in his book.

Simulation in the Overall Reservoir Life Cycle

Before proceeding to a detailed discussion, an overview of the reservoir life cycle is worthwhile.

Geological review

The first thing required to perform a reservoir simulation study is an understanding of the geology. From an engineering point of view, the key is the physical arrangement of the reservoir and connectivity (or continuity) of the reservoir. The flow properties of different parts of the reservoir will also correspond to parts of the geological model. Normally, petrophysical or log analysis properties are reviewed at this stage.

Reservoir performance review

In the next stage, the reservoir engineer must become familiar with production performance. Normally, this includes such things as water cut, reservoir pressure, and gas-to-oil ratio (GOR) trends. Combined with some PVT data, the potential drive mechanisms should be identified at this stage. The objectives of the study should be determined also. It is important the required results be stated in concrete terms. Often, a material balance study is done. There is a lot of overlap on the input for a material balance study and reservoir simulation input, so this may not involve too much extra time.

Data gathering. Relative permeability and capillary pressure data are normally screened. PVT data is similarly screened. Very often, multiple sets of data must be correlated and differences resolved (if possible). This must be checked for consistency and then put into the data set. This also means digitizing reservoir maps of structure, net pay, porosity, and permeability.

Initialization. The basic data deck is built. (Note this terminology is a remnant of the days when cards were used for reservoir simulation.) An initial run is made in which the model calculates original oil, gas, and water in place (OOIP, OGIP, and OWIP, respectively). These numbers are cross-checked against other results.

History matching. The input to the simulator is tuned. The model is run with the historical base product production specified—usually oil. The idea is to match GORs, water cuts, and pressures predicted by the model to actual performance. This part of the study is usually the most time-consuming, normally averaging one-third of total study time. The permeability × height (kh) or bottomhole pressure (bhp) of the wells is then adjusted so base product production matches actual field performance.

Predictions. At the end of the history-matching phase, the results are stored in a special restart file. With this, it is possible for the reservoir simulation to be continued without rerunning the problem through the history match. After completing the history match, various predictions are made using different production, well, and injection scenarios. The results can be interpreted and ranked according to acceptability.

Reporting/presentation. In the final stage, the results are presented and documented for posterity. Simulation generates an enormous amount of input and output data, so this takes considerable time and ingenuity to appropriately condense the results in a meaningful way (graphs, in many cases). The assumptions on which a simulation is based are extremely important. Therefore, properly documenting the procedures and results is important.

Reservoir life cycle history

The exact details of how much and what kind of detail will be included in a reservoir simulation will depend strongly on where the pool is in its production life cycle. The next section is a quick review of the life of an oil and gas pool as seen through the eyes of a reservoir engineer.

Exploration stage. Most pools are originated as a concept by an exploration geologist. This concept describes a particular size, location, and trapping mechanism for an oil and/or gas pool. The exploration geologist sells his idea or concept to his management. The appropriate land is obtained. Usually, more detailed analysis is done using tools such as seismic to confirm structure.

The next step is to convince management to drill a well and test the concept. Geologists are similar to reservoir engineers in that they must maintain conceptual models of their *plays*. Remember, for every seven or eight exploration wells drilled, only one will be successful on average.

A reservoir engineer normally becomes involved on the minority of successful exploration wells. Typically, this happens after the well has been drilled and logged but not yet tested or cased. At this stage of development, the objective will be to understand as much as possible about the potential new reservoir:

- What does the reservoir look like? Initially, our knowledge of the shape and size of the reservoir is governed by the exploration geologist's exploration concept.
- What are the fluids in the reservoir? One of the reservoir engineer's key objectives is to determine the fluids present. Some information may be known from logs, such as whether it is a gas or oil reservoir.
- What are the fluid flow properties? This involves either specifying or interpreting data from rock samples (core), well tests, and/or logs.
- What are the flow patterns in the reservoir? This information may not be known completely. However, the perforations in some wells need to be located to take advantage of the perceived flow patterns.

Armed with this understanding, management will ask the reservoir engineer to:

- tell them what data needs to be obtained
- locate the initial completion (perforations)
- design appropriate reservoir tests
- assess the significance of the discovery, i.e., reserves

Likely, several wells will be drilled to delineate a field. As each well is drilled, completed, and tested, the process will be repeated.

The analytical approaches and tools used at this stage include test planning and interpretation, core analysis, PVT fluid tests, log calculations, and volumetric reserves determination.

Exploitation stage. Eventually, the field will be turned over to an exploitation team, which normally consists of development geologist(s) and reservoir engineer(s). The basic objective is to finish drilling up the field and completing development. The geological model will become considerably more detailed, since there is much more data now available. Using your understanding of the reservoir, the objectives become:

- To determine the optimum development of the field with regard to:
    - number of wells
    - spacing of wells
    - placement of wells

  This normally involves a great deal of optimization. Each well will likely be justified with detailed economic calculations.
- To identify the best long-term method of depleting the reserves. Is enhanced recovery appropriate? If there is a gas cap, when should it be produced? If there is a nonassociated gas accumulation, is there an underlying aquifer in communication with it that will affect the depletion strategy?
- To determine the amount of oil and gas. This information is normally needed to plan surface facilities such as gas plants, batteries, and offshore platforms.

- To maintain this information for corporate reporting. Management monitors performance and reports the results to shareholders.

The oil industry, at least in Canada, is heavily regulated. The effects of this first become apparent to the reservoir engineer during the exploitation stage. It may be necessary to make an application and get permission from the government for:

- surface facility design
- GOR penalty relief
- water/oil ratio (WOR) penalty relief
- well-spacing designations
- producing an associated gas cap

If the reservoir being developed contains gas, a sale contract will be required. Typically, a gas marketing department gets the contract. However, a reservoir engineer will be called upon to determine reserves and deliverability.

The tools used include PTA test interpretation, economic evaluation packages, reservoir simulation, coning correlations, material balance, and production decline analysis.

Enhanced recovery design. In many cases, this is an extension of the exploitation phase. However, waterflooding may be delayed until more production data is obtained. Enhanced oil recovery (EOR) is often implemented much later than the majority of well development. The calculations involved are different from those used in the other stages, and, therefore, this is covered in a separate section. Tools and concepts used in the implementation of EOR include calculations for:

- *Displacement or Volumetric Efficiency* (E_D): fractional flow, Buckley-Leverett theory
- *Horizontal Sweep Efficiency* (E_A): correlations for various patterns are used
- *Vertical Sweep Efficiency* (E_V): layering calculations include those of Stiles and Dykstra-Parsons
- *Reservoir Simulation*: if properly applied, numerical simulation can take all of the previous factors into account, as well as more detailed accounting for areal variations in reservoir properties
- *Economics*: for these projects, the economics can get quite sophisticated, particularly those involving tertiary schemes such as carbon dioxide (CO_2) and hydrocarbon miscible flooding (HCMF)

Government applications, which are the most demanding applications for a reservoir engineer, usually are made at this time. On some very large projects, such as the Swan Hills, Mitsue, or Judy Creek miscible floods, there were objections from offset operators. In Canada, a quasijudicial process, which involves hearings before one of the provincial government conservation boards or the federal National Energy Board, is used to resolve these disputes. The time requirements for this process can exceed the amount of time spent doing the original technical calculations. Normally, a reservoir engineer is responsible for administering the application process, sending out data packages, answering questions, and appearing before various panels.

Performance monitoring/optimization. Information is normally limited during the planning stages. Very often, therefore, it is necessary to make many assumptions. If these assumptions are correct, then the field should produce as planned. In general, our knowledge is never that good, and some previously unknown aspects of the reservoir will become apparent by the analysis of production performance. The objective is to determine why the reservoir performance is not matching our prediction. This knowledge is used to correct our conceptual model of the reservoir and devise plans to improve reservoir performance. Usually, it is possible to improve performance compared to existing levels, if the field is not performing as expected.

Typically, the following are monitored:

- GORs
- water breakthroughs
- pressure in the reservoir
- production response (hump or no hump)

If everything is as expected, it is likely little else needs to be done. Small, well-defined fields are more likely to perform as expected, but this is rarely the case on large fields.

Some companies formalize this process with a reserves review at regular intervals. Typically, this is on the order of every four years for a large pool. For small pools, which deplete rapidly, this may be required annually.

Enhanced recovery projects require an annual report to the government. This consists of WOR, GOR, and pressure contour maps. A voidage replacement ratio (VRR) calculation is also stipulated. Finally, a summary of workovers for injector/producer conversions is required.

Tools used at this stage include contour plots of GORs, water breakthrough time, and pressures. VRRs are also useful to monitor the effectiveness of injection. For more detailed studies, material balance calculations will be required and reservoir simulations performed.

Typical reservoir simulation study. Most simulation studies follow a similar format and basic procedure. What comprises a typical reservoir simulation study? The process has been summarized in Figure 1–3. The major components are described as follows:

- *Data Gathering.* A reservoir simulation involves a considerable amount of data input. In the first stage, all data are screened for quality. For most simulations, lab data is not available for all of the input. Correlations or data from offsetting or analogous pools must be used. Geological maps of porosity, net pay, and permeability must be developed and translated into the grid format.
- *Initialization.* In this stage, preliminary calculations are made for running the simulator. Grid-block water saturations are calculated based on capillary pressure data. The importance of water saturations is indirect; the correct determination of OOIP is the main objective. Most programs feature data-checking routines at this stage.
- *History Match.* The numerical model is run through time with the base product production (oil or gas) specified in the model input. The idea is to match the rest of the production behavior, such as the GOR, water production, or condensate rates, to the actual behavior that has occurred in the reservoir. History matching consumes roughly one-third of total study time, making it the largest single component.
- *Tuning Phase.* In order to make predictions, the *kh*, or bottomhole pressure, of all wells must be adjusted to match actual production performance. This involves a series of trial and error runs to obtain the correct values.
- *Predictive Stage.* At the end of the tuning phase the model is usually terminated with a restart. This input data file contains all of the information necessary to continue a simulation at a later time. Several different production scenarios or alternatives are run from the same timestep and compared. With different runs, various injector patterns, changes in rates, and producer-injector locations can be studied.
- *Report.* The assumptions on which a simulation is based should be outlined. Simulation also generates an enormous amount of paper output. It takes considerable time and ingenuity to reduce this data to an understandable form from a commonsense perspective.

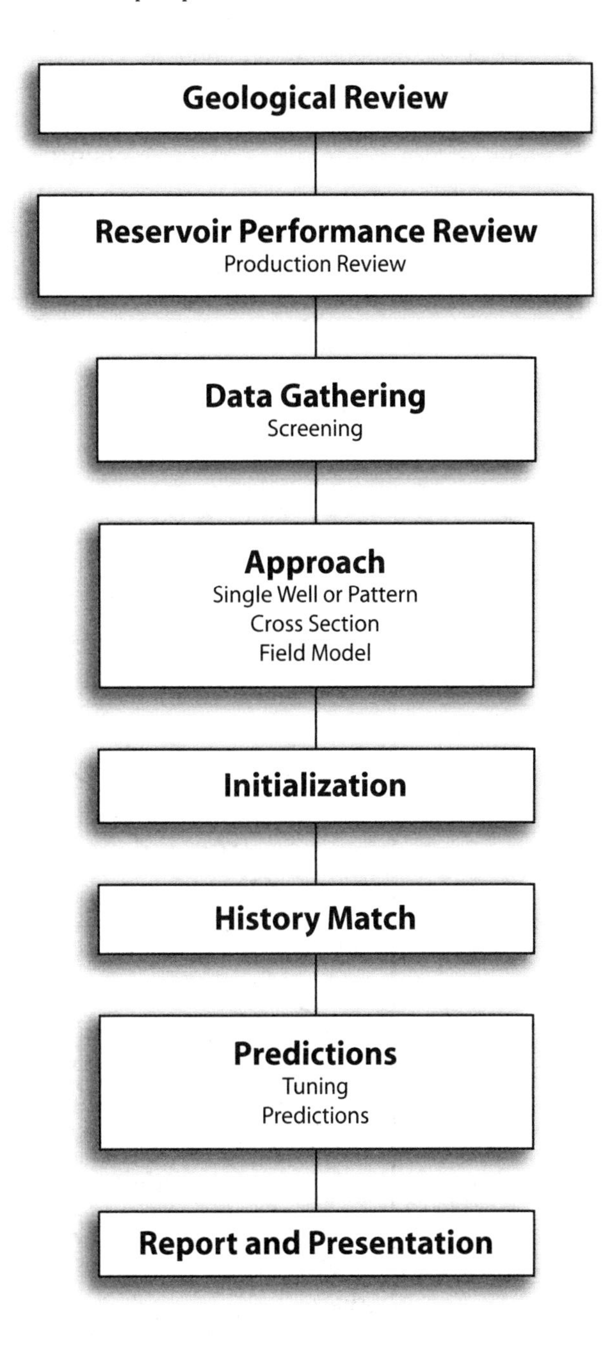

Fig. 1–3 Reservoir Simulation Process

Most simulation reports are written as a variation on the previous outline.

Canadian content

Readers may notice a large number of the examples are taken from the Western Canadian Sedimentary Basin. This reflects a number of things. First, almost all reservoir data is in the public domain in Canada. Second, almost all of the data on overseas reservoirs known to the author is not in the public domain, with some exceptions with respect to securities reporting. In other words, there are few confidentiality concerns with respect to Canadian reservoirs.

The author has spent a great deal of time in Western Canada and has experience on reservoirs in Peru, Bangladesh, India, Australia, Indonesia, and the Former Soviet Union (FSU) as well. The conditions in Western Canada cover a very wide range and are broadly applicable, for the most part. Some Gulf Coast reservoirs in the United States have varying degrees of applicability to Canadian conditions. Unfortunately, it is not possible to cover all of these areas in a single book.

Public data

Some expanded comments on public data are appropriate. Canadian policy in this matter is a significant departure from the practices virtually everywhere else in the world. In Canada, all lab data is, with a limited number of exceptions, in the public domain within one year. This also applies to all production data and well logs. Most recovery schemes require a public application, which includes considerable technical content.

Clearly there are pros and cons to such an approach. Figure 1–4 is taken from *The Dynamics of Oil and Gas Accumulations* by A. Perrodon.[2] The diagram shows Alberta has a relatively high efficiency in exploration. A number of factors, such as basin maturity and geological controls, were considered in preparing this figure; however, it is logical to extrapolate exploration success with increased information. This was—and is—the purpose of making the data public.

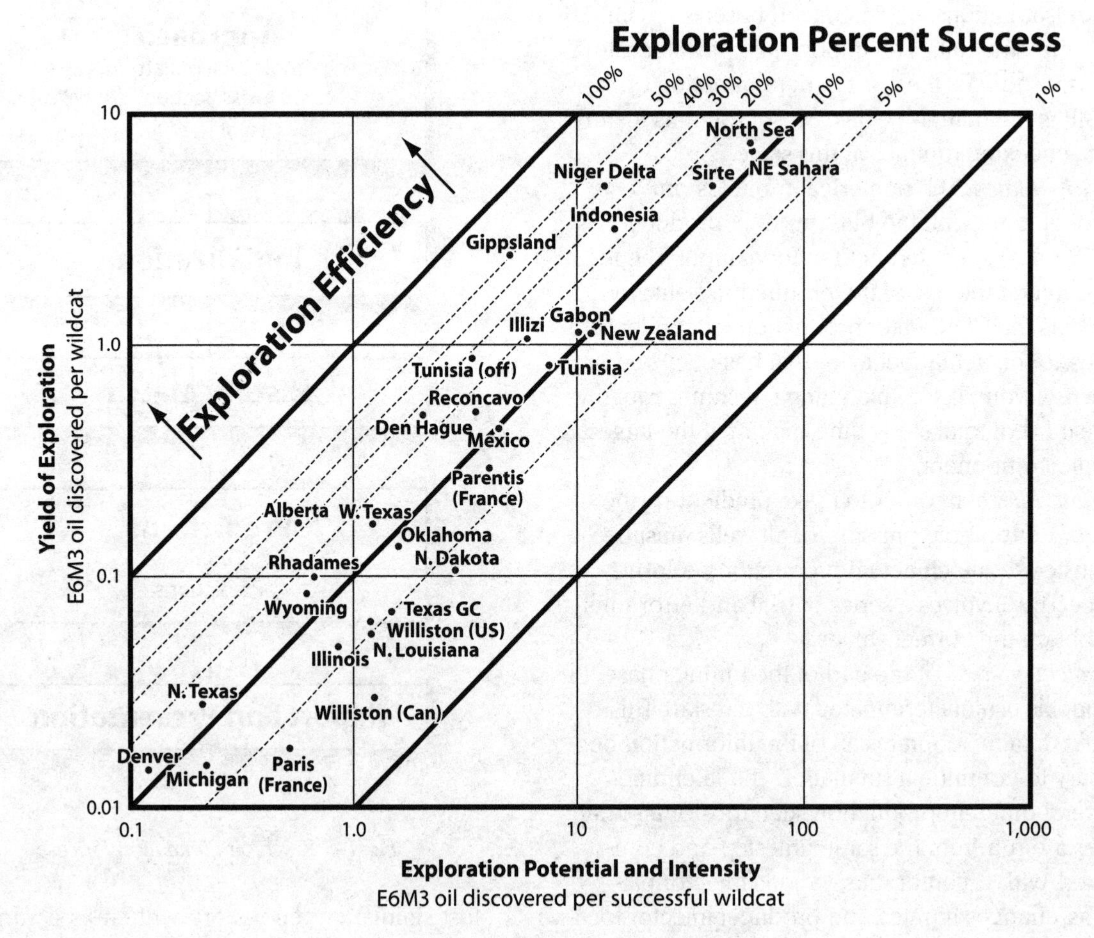

Fig. 1–4 Graph of Exploration Efficiency

There are a number of other implications. From a joint operation perspective, working interest partners with significant interests have and will engineer many aspects of reservoir performance independently. In one, a giant oilfield in Alberta had four partners with approximately the same working interests ranging from 16–17%. In essence, four companies controlled two-thirds of the unit. Three of these companies were major multinational companies. The operator was a senior independent. Needless to say, there were some interesting working interest meetings and substantial technical discussions. The field was sufficiently large to be a major cash flow for all four companies. While it is true some time was wasted in politics, it is fair to point out considerable improvements were made to the plans proposed by the operator and major joint interest owners. Interestingly, many of the smaller companies offered considerable technical insight. Smaller companies have focused objectives and usually have experienced personnel. Frequently, this has been a great surprise to employees of multinational companies.

In general, the Canadian oil industry has run at a higher level of technical expertise because of the availability of raw technical data.

Summary

Correctly applied, reservoir simulation can be a powerful tool. It is extremely dangerous if applied incorrectly. It is a similar to riding a high-performance sailboat in a big wind. If you don't respect it, you will likely get hurt. If you fear it (or are overawed by it), you will not be able to control it, and you will end up in the drink. You can never take your eyes off what is happening and where you are going, because something unpredictable will happen when you do. It takes time and practice to become good. Sailing is not for everyone but is addictive to others.

Each study will be a variation on a common theme. However, the emphasis on the various components in different studies will change with individual reservoirs and the stage of the reservoir in its production life cycle.

References

1 Dake, L.P., *The Practice of Reservoir Engineering, Developments in Petroleum Science 36*, Elsevier, 1994.

2 Perrodon, A., *The Dynamics of Oil and Gas Accumulations,* Elf Aquitane, Pau, France, 1983.

2

Mathematical Considerations

Introduction

An enormous volume of literature exists on different methods of constructing a reservoir simulator. The vast majority of reservoir engineers, however, will use the program provided by their company. Typically, this is one of about a half dozen commercially available programs, which are marketed internationally. In the past, many of the major companies have had their research departments write proprietary programs. With the current downsizing or "right-sizing" trend in the industry, it's likely few companies will continue to maintain their current programs or initiate new ones.

Since readers are more likely to be using commercially available products, an extended discussion on the mathematics of constructing a simulator is, in practical terms, of limited benefit. All that is required of users of this software is:

- sufficient background knowledge to understand the limitations of the simulator
- an understanding of the different simulation solution options
- knowing how to adjust matrix solvers to achieve adequate performance
- knowing that wells and the associated equations have certain limitations, which sometimes require changes in approach
- a minimum understanding of timestep control
- knowing how group and well controls affect simulator performance

The following discussion outlines the basics and concentrates on points where user intervention may be required. This may be regarded as a mixed blessing. The material is presented in a descriptive form. Of course, this means some of the detailed mathematics underlying this material are not shown, although there is worthwhile material in detailed discussions of the mathematics and solutions.

The author participated in a course taught by Dr. A. Settari, which required a simulator be built as a class project. This problem set had a reputation for being long, involved, and difficult. If the reader has career plans for significant exposure to simulation, going through this process is beneficial. It provides a better understanding of how a simulator works, but it requires a major time commitment. Reading pages of differential equations and matrices is not the same as actually programming a simulator. This level of detail would consume another book and would leave much of the practical application unexplained. A number of good textbooks on these issues have been published.

The material in this book has been prepared along a development outline to maintain a logical progression. The purpose of certain points may vary. In some cases, it is designed to explain how things work. In other cases, a description is intended to indicate a limitation or an option or switch commonly available on commercial simulators. Since the purpose of each section is not always obvious, it is indicated in parentheses. The intent is to provide sufficient material to allow the reader an appreciation of critical issues, leaving the detailed mathematical explanation to others.

Detailed derivations (formulation)

The development of the partial differential equations is the same for a reservoir simulator as for pressure transient analysis. The equations are used in both Cartesian and radial coordinates for reservoir simulation. This material is best covered in Aziz and Settari's *Petroleum Reservoir Simulation*, published by the Academic Press.[1] Nevertheless, a quick outline will be made in the following.

The three main elements to the development of the partial differential equation, which govern reservoir simulators, are:

- conservation of mass
- a transport law
- an equation of state

Conservation of mass (basic formulation)

The differential equation in Cartesian coordinates is developed by considering a small block, depicted in Figure 2–1:

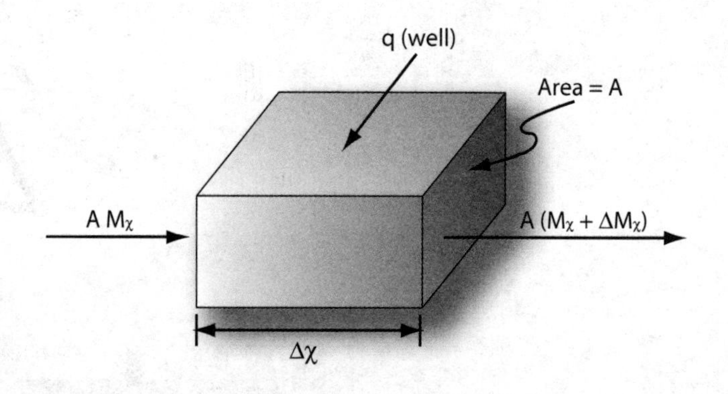

Fig. 2–1 Representative Element

The equation is then developed as follows:

$$\text{In} = A M_x \tag{2.1a}$$

$$\text{Out} = A(M_x + \Delta M_x) \tag{2.1b}$$

$$\text{Storage} = A\,\Delta x\,(\rho_1\phi_1 - \rho_2\phi_2) \tag{2.1c}$$

$$\text{Source/Sink} = q_{\pm} \tag{2.1d}$$

Combining these yields:

$$A\rho\,\frac{(M_x + \Delta M_x - M_x)}{\Delta x} = A\,\frac{\Delta x}{\Delta x}\,\frac{(\rho_1\phi_1 - \rho_2\phi_2)}{\Delta t} + A\,\frac{\Delta x}{\Delta x} \tag{2.2}$$

In the limit as $\Delta x \to 0$ and $\Delta t \to 0$ one derives:

$$\frac{\partial M_x}{\partial x} = \frac{\partial(\rho\phi)}{\partial t} + q \tag{2.3}$$

This conservation of mass is then expanded in two ways. First, a description of the behavior of the fluids at different pressures is required—an equation of state; and, second, an interblock transport relation is required—Darcy's Law.

Equation of state or PVT behavior (formulation)

The material balance in reservoir simulators is usually done on the following basis:

- For gas, the real gas law is used.
- The liquid phase has dissolved gas, which is a linear function of pressure (black oil).
- Water is characterized as a liquid of low compressibility, which is a linear function of pressure.

Transport equation—Darcy's Law (basic formulation)

The current understanding of fluid flow through porous media is derived from work by a French civil engineer named Henri Darcy. He was designing sand filters for a drinking water supply in Lyons. His experiment looked something like the apparatus shown in Figure 2–2.

Apparatus to filter water with sand

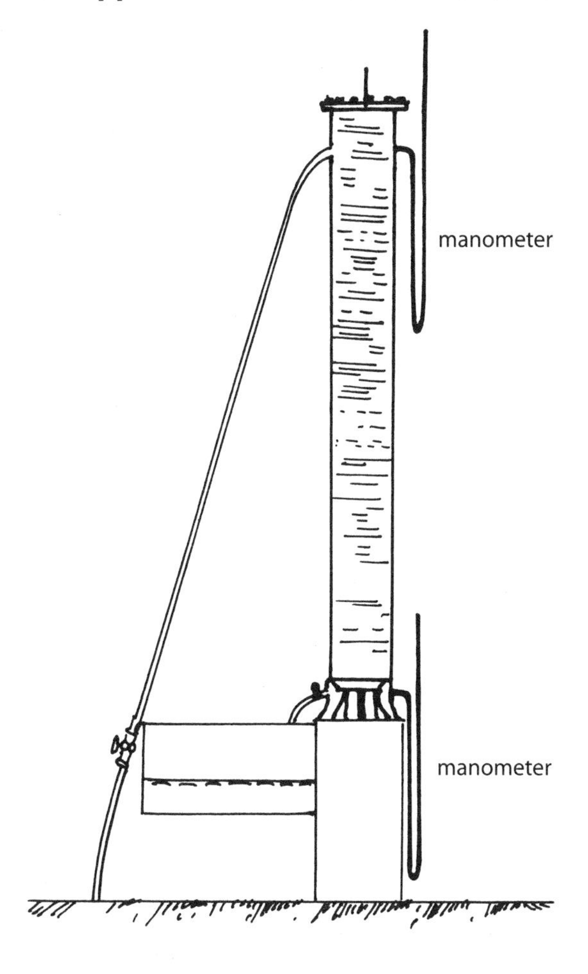

Fig. 2–2 Darcy Apparatus

He found the pressure drop across the sand pack is a linear function of an intrinsic sand property, the applied pressure drop, the length of the sand pack, and the cross-sectional area of flow. It was inversely affected by viscosity. This empirical observation has become known as Darcy's Law and is shown in Equation 2.4.

$$Q = -A \frac{k}{\mu} \frac{\partial \Phi}{\partial x} \tag{2.4}$$

Examples of different materials that have been tested are shown in Figure 2–3. Note the law also applies to the flow of fluid in rocks. The intrinsic property of the rock is termed permeability. Note there is a breakdown at high Reynolds numbers. This is related to turbulence and inertial effects and occurs with high velocity flows.

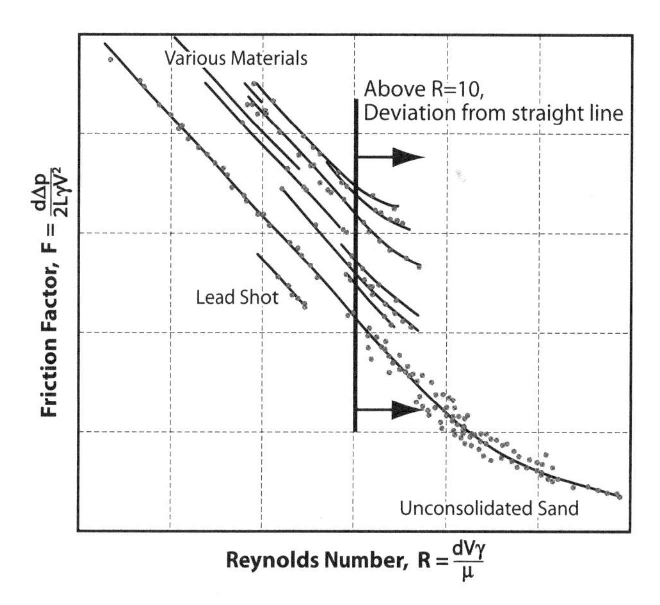

Fig. 2–3 Darcy's Law—Various Porous Media

Concept of a tensor (limitation)

The example shown in Figure 2–3 was for one dimension. In fact, not all materials have perfectly homogeneous permeability, and not all potential gradients are applied along Cartesian axes. The flow in a medium can be calculated exactly under these conditions, since permeability is mathematically a tensor. The permeability in any given direction can be determined by an ellipse based on the minimum and maximum permeability directions. This directional change in permeability is not directly programmed into reservoir simulators, as described in the previous development. This error is inherent in the formulation of simulators, as they are currently built. This level of error is considered tolerable in the majority of cases.

Nine-point implementations (limitation/alternate formulation)

There are certain cases where this error is not acceptable. Probably the most common is in thermal simulation where adverse mobility ratios (i.e., much greater than one) exist in displacements. Gas (steam) displacement of viscous oil—i.e., heavy oil—is a classic example. The converse, where oil displaces gas, does not cause problems.

In this case, the differential equations are formed differently: flow is allowed diagonally between grid blocks, as shown in Figure 2–4.[2]

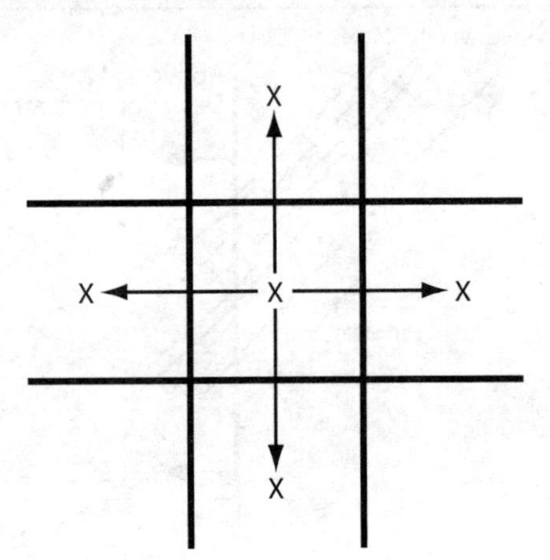

Five-point Formulation

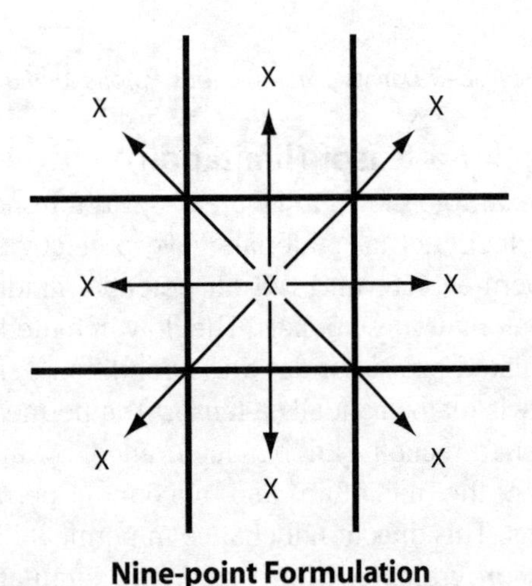

Nine-point Formulation

Fig. 2–4 Computational Molecule—Nine-Point Implementation

Formation of matrix solution (basic formulation)

Returning to the basic differential equation in Equation 2.3, the left side represents spatial terms and includes pressure gradients and permeability. The right side represents change in the material in the block for time and pressure trends. The solution to this equation is not going to be solved exactly. Approximate solutions are going to be developed at certain fixed points in space and, as the simulator marches through certain fixed points, in time. Thus, two discretizations occur, one in space and the other in time.

The discrete points in space are arbitrarily chosen using a grid. Note that two types of grids can be chosen: one is called a block-centered grid and the other a point-centered grid. The boundary conditions for these two types of grids are different; however, all of the commercial simulators use block-centered grids. A block-centered grid is shown in Figure 2–5 and a point-centered grid is shown in Figure 2–6. Note the author's diagram of a point-centered grid differs slightly from those commonly shown in the literature. This is deliberate. The points are intended to be representative of reservoir volumes for both grid systems. The essential difference is at the edge of the grid. With the point-centered grid, the nodes fall on the edge, and, as a result, the boundary conditions are applied differently.

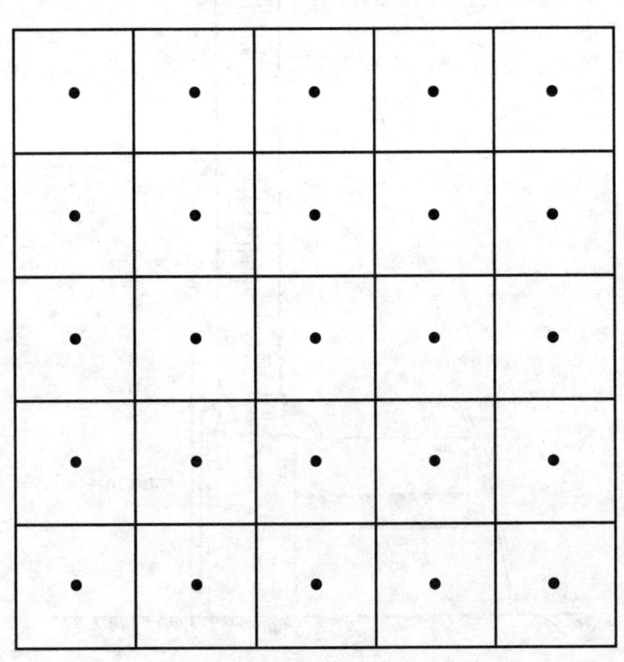

Fig. 2–5 Block-Centered Grid

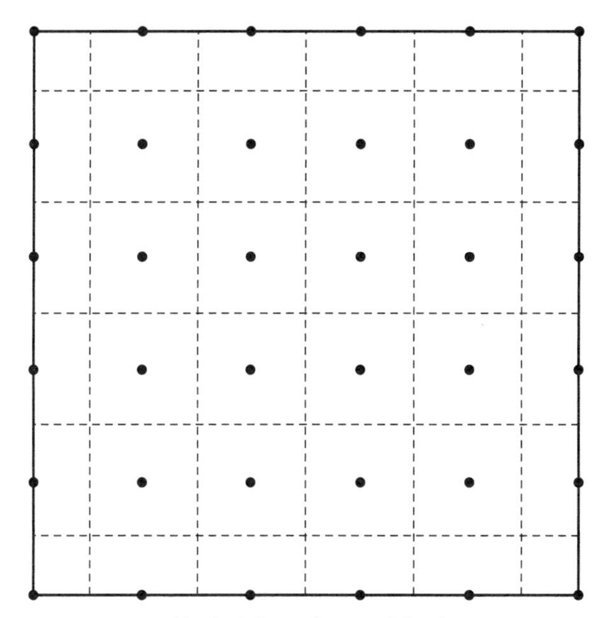

Fig. 2–6 Point-Centered Grid

Time discretization is accomplished with timesteps. These are also arbitrarily chosen.

The accuracy of the solution will be affected by both the size of the grid blocks and the length of the timesteps. Smaller grid blocks and shorter timesteps provide more accurate solutions.

In order to handle the large volumes of computational data, matrices are used to solve these problems. The discretization is accomplished as shown in Equation 2.5.

$$\left(T^{n+1} + \frac{1}{\Delta t}B^n\right)u^{n+1} = \frac{1}{\Delta t}B^n - Q \qquad (2.5)$$

The solution at time $n + 1$ is on the left side. On the far left-hand side are all of the spatial terms, which are made up of permeability and grid-block dimensions. The next term, on the right-hand side, represents the length of the timestep and the previous solution (pressures and grid-block production rates), which completes the information necessary to project the solution at the end of the timestep.

The initial solution or pressures must be set to start the simulation and a series of grid-block values are obtained from the simulator at discrete points in time.

Errors and spatial discretization (limitation)

Using the element described with a Taylor series expansion, it is possible to show the error involved in discretization is related to the spacing of the grid. For a regular grid, the error is proportional to the spacing squared. For an irregular grid, the error is directly related to the spacing.

It is not possible to determine the absolute level of error from this technique, only how it varies with spacing. Therefore, to determine the absolute level of error, it is necessary to compare a computer simulation with an analytical solution. In many cases, numerical techniques are used to solve functions (such as on a computer), and thus, there is normally some numerical error associated with a solution. Nevertheless, the latter error can be quite small, if implemented correctly. These analytical solutions are only available for certain geometries, and, as a result, the accuracy of most real simulations cannot be determined absolutely. However, multiple runs made with different grid spacing can be compared to each other, which indicates convergence to a consistent solution is rapid and changes in solutions are of a small magnitude. This is called a grid sensitivity test.

Grid sensitivity tests (practical knowledge)

Therefore, it is practical to test the accuracy of a grid using multiple simulation runs. These sensitivities need not be performed on the entire grid. It is generally sufficient to make a grid for a small group of wells, which feature the important mechanisms in the reservoir. This allows detailed (small dimension) grid blocks to be explored with realistic simulation execution times.

Errors and temporal discretization, i.e., timesteps (limitation)

The errors associated with timesteps are directly related to the length of the timestep. This means timestep lengths usually vary in a simulation. Specific controls governing timestep sizes will be discussed later.

Turbulent flow (limitation/alternate formulation)

As shown in Figure 2–3 and mentioned earlier, there are deviations from Darcy's Law at high flow rates. Four significant situations where this deviation from Darcy's Law is commonly encountered are:

- high-rate gas wells
- propped hydraulic fractures
- natural fractures in fractured reservoirs
- steam injection in heavy oil

Turbulent flow is described by the Forcheimer equation, which is shown as Equation 2.6.

$$\frac{dp}{dr} = \frac{\mu}{k}u + \beta\rho u^2 \quad (2.6)$$

This has an additional term, which is rate dependent. Beta, the Forcheimer coefficient, can be correlated with permeability as shown in Figure 2–7.

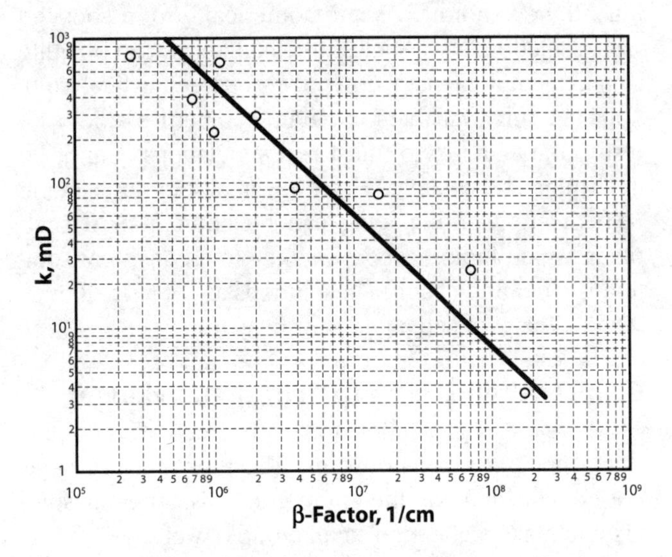

Fig. 2–7 Forcheimer Beta Factor

This rate-dependent permeability can be programmed in a reservoir simulator, although it has been implemented in relatively few. In these instances, conventional reservoir simulators may not be capable of matching field results.

Pseudo pressures (limitation/alternate formulation)

The base differential equations for pressure transient analysis and reservoir simulation are the same, although in different coordinate forms:

$$\frac{\partial^2 p}{\partial x^2} = \frac{\partial p}{\partial t}\frac{\phi\mu c}{k} \quad (2.7)$$

$$\frac{\partial^2 p}{\partial r^2} + \frac{1}{r}\frac{\partial p}{\partial r} = \frac{\varphi\mu c}{k}\frac{\partial p}{\partial t} \quad (2.8)$$

It is natural to ask is why reservoir simulators are not formulated based on pseudo pressures. The answer is they can. It is even possible to assign a pseudo pressure to oil. The use of pseudo pressure linearizes the $1/\mu z$ when dealing with gas. In a reservoir simulation, these factors are all calculated for time as a function of pressure. The use of pseudo pressure is a limited improvement for a reservoir simulator. This is because the values of μ and z are already pressure dependent from PVT tables. Note also a performance prediction from a pressure transient package will be different from a simulator prediction, even if pseudo pressure is used. The simulator is more accurate. This is due to the right-side $\varphi\mu c/z$ term. The use of pseudo time (Agarwal) is intended to linearize the term on the right side of the differential equation.[3]

Settari and others have built a reservoir simulator specifically for pressure transient analysis, which utilizes pseudo pressure for both oil and gas.

Matrix structure (formulation/practical knowledge)

By systematic ordering of the equations from the Cartesian equation described previously for each grid block, it is possible to create a matrix having a particular shape or character. This is shown in Figure 2–8 and is typical of reservoir simulation problems. Sometimes this is called D2 ordering.

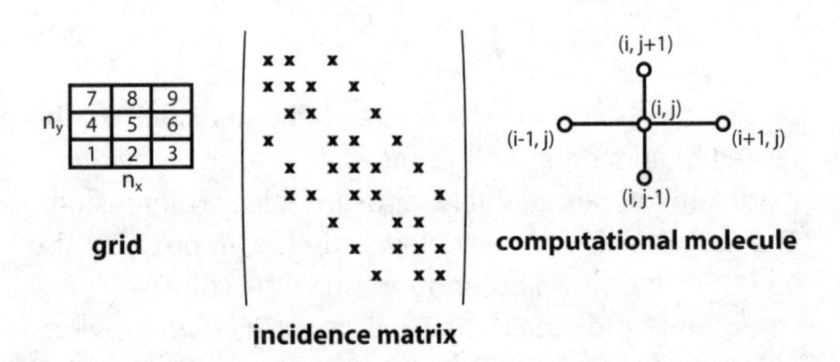

Fig. 2–8 Example of Matrix Structure (D2 Ordering)

There is also another method, which was first applied to reservoir simulation by Coats.[4] It is called red-black ordering or D4 ordering. This results in some increase in efficiency in some cases and is shown in Figure 2–9.

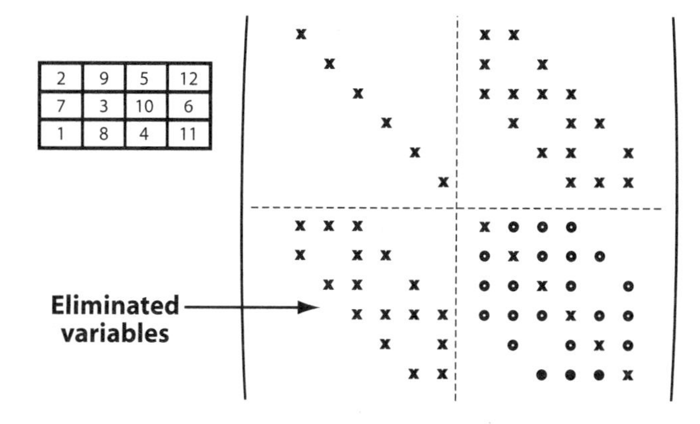

Fig. 2–9 Matrix Structure (D4 Ordering)

Since D4 ordering does not always decrease run times (or increase solver efficiency), a number of simulators allow you to specify the ordering used. This is specified with a line and keyword in a numerical section such as *ORDERING = D4* or *ORDERING = D2*. Most simulators seem to default to D2 ordering at present.

The terms are concentrated along a diagonal in the matrix. For implementation in a simulator, it is possible to take advantage of the bandwidth and use a minimum of storage, because calculations are required in active areas of the matrix only. Matrix solution is faster for narrower bandwidths.

These are not the only ordering schemes available. A reverse Cuthill-McKee ordering is found on some simulators. Other orderings may also be developed.

Since it is not always possible to tell which ordering scheme will work best, this has to be determined by trial and error. It becomes an issue when you are dealing with large simulations, where execution time is a significant issue.

Solution of the matrices will be discussed later, since further background is required related to multiphase flow and displacement.

Stable matrix terms (formulation)

The terms in the matrix on the left side, which at this point have been formulated for single-phase flow, are quite stable. These are calculated once, which is done before the simulator proceeds with calculations through time.

Both compressibility of the fluid and porosity of the grid block are normally pressure dependent.

Implicit versus explicit solutions (formulation/practical knowledge)

Using a simplified single-phase example can be solved in two ways. The first assumes the pressure gradients in space don't change much between timesteps. Therefore, the solution at time $n + 1$ can be estimated from existing values of pressure. *This is known as an explicit solution.*

After a few experiments, the prediction of the solution at time $n + 1$ is improved if one takes the solution from time $n + 1$ and averages it with the original solution at time n, yielding an average effective pressure of $(P_{n+1} + P_n)/2$ for the iteration. This updated average pressure is then used to reestimate the pressure at time $n + 1$. The variation in each successive estimate of the solution at time $n + 1$ is observable using the average pressure across the timestep. The difference between successive iterations can be set at a tolerance defining convergence, and a form of control is introduced. *This is known as an implicit solution.*

The difference between these two types of calculations can be summarized as follows. Explicit solutions are based on the solution at the last timestep, and only one leap forward in time is made. Implicit solutions are based on both the previous and current solution and involve multiple estimates of the solution at time $n + 1$.

Clearly, the second method involves more calculations—i.e., more work—on behalf of a computer. The second method is more accurate and allows longer timesteps to be taken. So there is a trade-off in accuracy, calculations required, and the length of the timesteps. The two methods will approach exact analytical solutions from different sides of the actual solution.

What is not immediately apparent is that both explicit and implicit methods have limits on the maximum length of timesteps, with the explicit limit being much shorter than the implicit limit. Explicit solutions can become unstable much more easily, in which case the accuracy of the solution can diverge strongly. This is depicted in cartoon format in Figure 2–10.

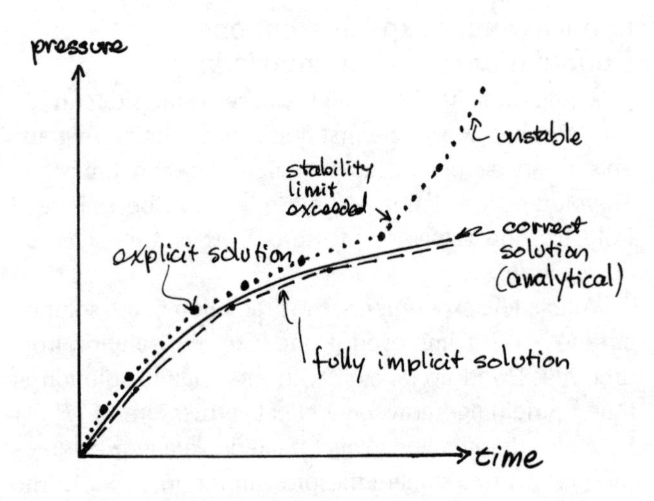

Fig. 2–10 Divergence of Implicit and Explicit Formulations

All modern reservoir simulators use an implicit formulation for determining pressures.

Inner and outer loops (formulation)

One other subtlety should be pointed out. In the explicit formulation, only one matrix solution is required. When an implicit formulation is used, the matrix is solved multiple times with different estimates of the average pressure for the timestep. Changing the pressure estimate is known as the outer loop, while the matrix solution is known as the inner loop.

Multiphase development (formulation)

When multiple phases are introduced, the equations are written for each individual component separately, which usually comprise oil, gas, and water. The saturations of all three phases, expressed as a fraction of pore volume, all have to add up to one, which gives another equation for each grid block to solve the problem. The solution vector now includes pressures and saturations for each phase. This is not difficult to implement.

The PVT description is not difficult. Simple table lookups are used for the oil, and the real gas law is implemented with a table lookup to account for variations in real gas compressibility or z factor with pressure. Water compressibility is handled with an analytical equation.

The messy part of multiple phase simulation relates to the transfer between grid blocks or Darcy's Law. The absolute permeability must be multiplied by the respective phase relative permeability, which is a function of saturations. This introduces a strong nonlinearity. The relative permeabilities, as with the PVT data, are usually determined by a table lookup function.

Accounting for changes in relative permeability (formulation)

This type of problem is approached in a number of ways. First, one can assume the relative permeability, and, hence, saturation does not change much during a timestep. In this case, the relative permeabilities are determined at the start of the timestep and are not updated. This is known as an IMPES formulation for implicit pressure explicit saturation.

The second way to do this is to use the saturations from the last timestep to estimate the average saturation for the first iteration, in essence an IMPES timestep. At the end of the first timestep, the average saturation for each grid block is updated using the saturations at the end of the timestep to calculate an average saturation and average relative permeability. A control check is then made to see how much difference there is between the first estimated solution (saturation) and the second estimated solution. When the difference between successive iterations becomes smaller than a specified accuracy, the latest solution is adopted as the solution for time $n + 1$. This is really an implicit pressure and implicit saturation formulation (IMPIMS). A consistent terminology such as IMPIMS has not caught on as an acronym; this formulation has been called *fully implicit*. Later in this chapter, it will be shown this last term is actually quite misleading.

Finally, there are some intermediate methods, which solve first for pressure implicitly and then update saturations in a multistep manner. The latter methods (generally termed semiimplicit) are now rarely used, although they were common in a number of in-house and commercial programs at one time. These equations are no longer in widespread use due to mathematical problems related to the conservation of mass. This is only an issue for older software.

Newton-Raphson technique (formulation)

The nonlinearity associated with saturation changes is quite strong. The process of averaging the saturation for the timestep can lead to slow convergence on the outer iterations. In addition, based on the author's experience, the convergence of the pressure solution is not always fast if there are steep pressure gradients.

There is a way to speed this up—use an acceleration technique. The one used is the classic Newton-Raphson technique. This requires partial derivatives of the transmissibility matrix be calculated. A derivative matrix is called a Jacobian, which is a rather funky name often referred to in software documentation. The basic Newton-Raphson technique involves recasting the equation into a root form as shown in Equation 2.9. Successive estimates are calculated as shown in Equation 2.10. These are shown for a regular algebraic equation as an example.

$$\left(T^{n+1}+\frac{1}{\Delta t}B^{n+1}\right)u^{n+1}-\frac{1}{\Delta t}B^{n+1}u^{n}+Q=0 \qquad (2.9)$$

$$x^{n+1}=x^{n}-\frac{f(x^{n})}{f'(x^{n})} \qquad (2.10)$$

For the square root of 2, and an initial guess of 1.2:

$$0=x^{2}-c \qquad (2.10a)$$

$$x^{n+1}=x^{n}-\frac{x^{2}-c}{2x} \qquad (2.10b)$$

$$x^{1}=1.2-\frac{(1.2^{2}-2)}{2*1.2}=1.43333 \qquad (2.10c)$$

$$x^{1}=1.4333-\frac{(1.4333^{2}-2)}{2*1.4333}=1.4143 \qquad (2.10d)$$

This process replaces calculating a simple average pressure and saturation across the timestep and results in much faster convergence.When this technique is used (which is universally the case), the *outer iterations* are called *Newtonian iterations*.The two terms are synonymous.

Material balance (practical knowledge)

Since the methods described previously are numerical, it is useful to have an indication the solution is converging to a correct solution. The material balance is a tool developed to monitor this. The idea is not difficult; the total mass of oil, gas, and water within the reservoir is calculated by the simulator, as well as the total mass of the produced fluids, less the mass of injected fluids. The amount of mass missing or added represents an indication of the quality or accuracy of the solution. This calculation is made on a global basis.

This tool is not perfect, since different parts of the grid may have errors that cancel out. In practice, it is a useful tool.

To add some confusion, there are two common ways of calculating the material balance. The first is for the cumulative timesteps to date, and the other method involves calculating the material balance for each individual timestep. The latter is a much smaller representation of the total solution and normally would have a smaller tolerance. Having said this, the timestep version often gives insight to where and at what time problems are occurring in the simulation. It is possible to rectify these by using shorter timesteps or modifying other input.

Some simulators, particularly thermal simulators, allow you to use material balance for a single timestep as a control variable. If the material balance error for a timestep is too large, then the timestep is cut. This variable is usually implemented in the numerical controls section of a data set.

Since most simulations involve comparing relatively small differences in recovery between competing depletion plans, material balance errors should be kept to low values. This is a relative guideline. Investigating a difference in recovery between depletion plans of 2% is common. On this basis, material balance should be kept to less than 0.1%.

Such accuracy is hard to achieve on thermal simulations. Economics depend heavily on rate. Changes in economics will occur with 5% variations in rates. Actually, this is good accuracy for rate predictions. Therefore, for overall project economics, higher levels of error are acceptable.

Frontal displacement (limitation)

Frontal displacement may seem like an odd topic for a book on reservoir simulation; however, the reader should be aware of this area of simulator weakness. In fact, there are some fundamental areas in reservoir simulation to be resolved. This is one of the most important.

Buckley-Leverett theory (formulation)

The Buckley-Leverett theory is explained well in a number of places. The SPE monograph on waterflooding edited by Craig, Dake's *Fundamentals of Reservoir Engineering*, and Lake's *Enhanced Oil Recovery* are just a few.[5,6,7] These excellent explanations will not be repeated in this text. The main point is that this theory predicts an abrupt flood front, such as the one shown in Figure 2–11.

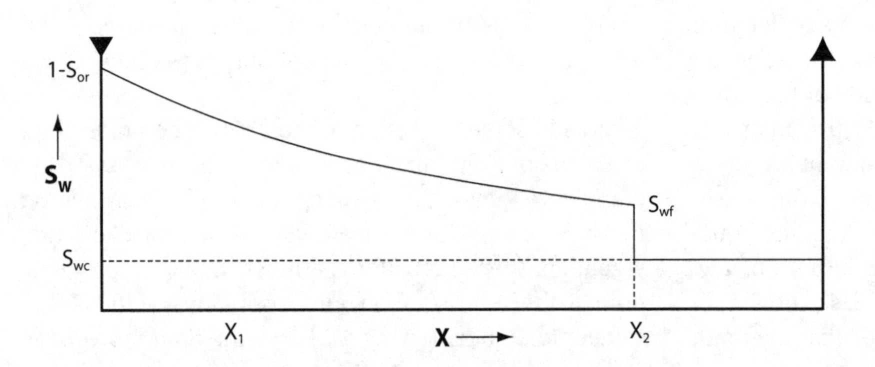

Fig. 2–11 Buckley-Leverett Immiscible Displacement Profile

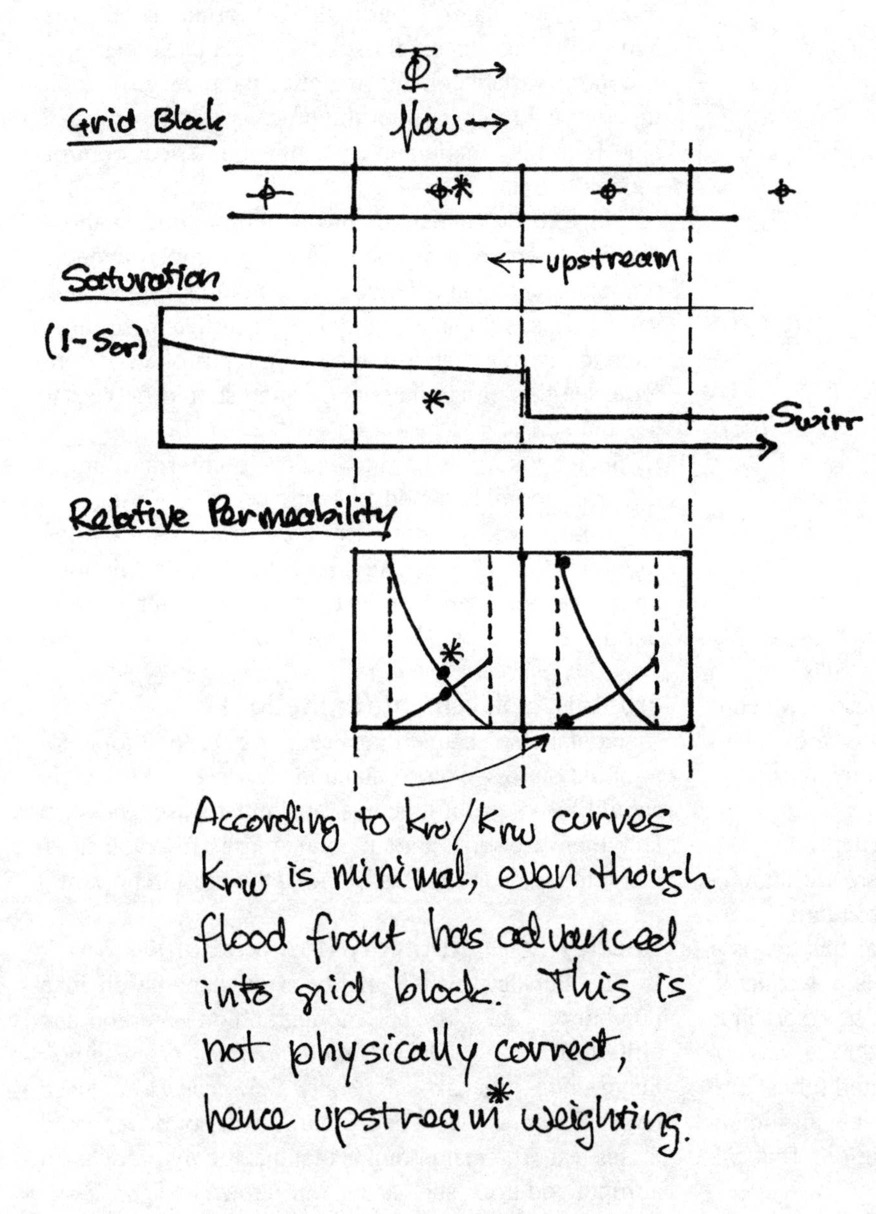

Fig. 2–12 Idea of Upstream Weighting

Weighting of transmissibilities (formulation)

Displacement processes have caused problems in generating the basic formulation for reservoir simulators. The general idea, depicted in Figure 2–12, shows a simplified one-dimensional (1-D) grid. As injected water displaces oil in a grid block during a timestep, what relative permeability should be used? Should it be the relative permeability from the upstream side of the block, which has a high water saturation? Alternatively, would the downstream side of the block be preferable, where the relative permeability to water is low due to the low water saturation? An average is required, reflecting the relative permeabilities of the flood front upstream and downstream, as well as the proportion of the grid block that has been traversed.

Effects of different formulations—upstream weighting (limitation/formulation)

Without going into detail, many mathematical approaches do not produce stable results, as shown in Figures 2–13(a) and 2–13(b).[8]

The only method producing stable results is the use of upstream weighting. In the injected water problem discussed previously, the water relative permeability for the grid block is determined by the water saturation in the block upstream of the grid block being calculated (note this is defined by the potential gradient).

At first, this may cause consternation. A block at irreducible water saturation has a water relative permeability of zero. If downstream weighting is used, water should not be able to move. However, this analysis overlooks the direction of saturation changes. We determine the irreducible water saturation by flooding a

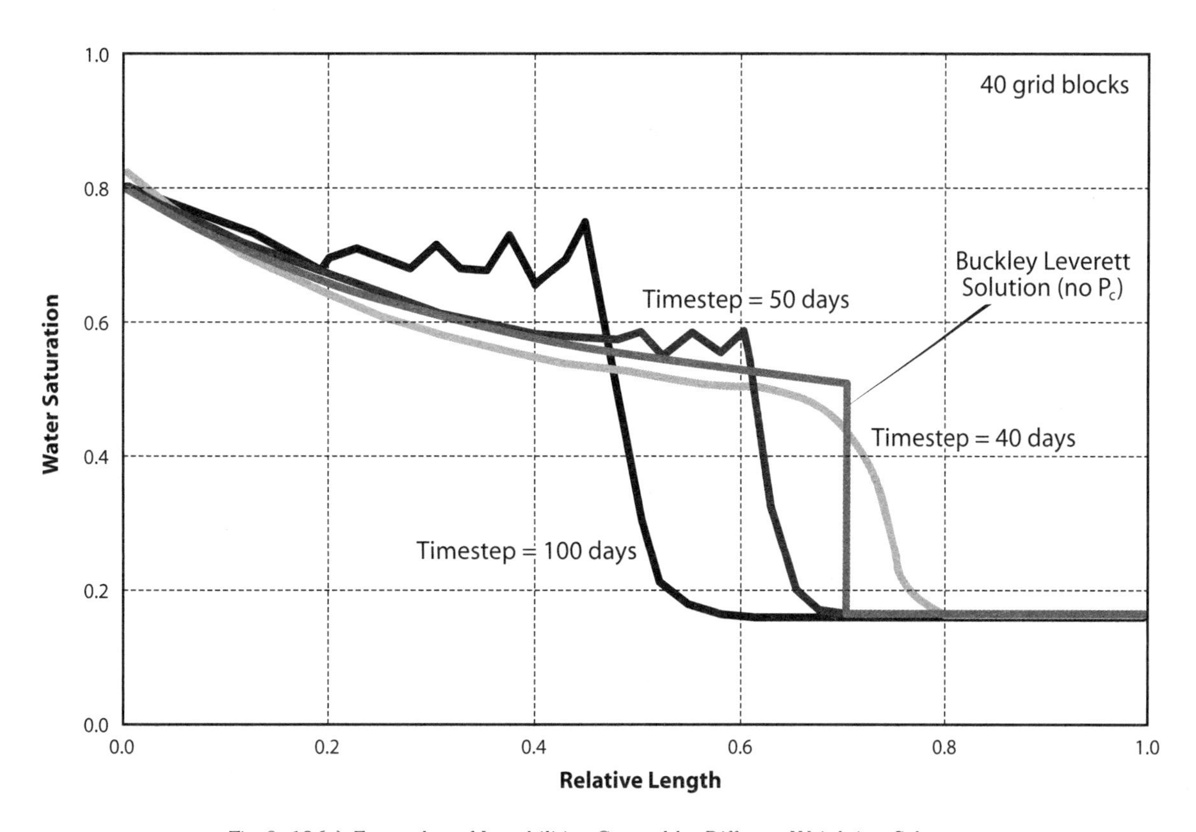

Fig. 2–13(a) Examples of Instabilities Caused by Different Weighting Schemes

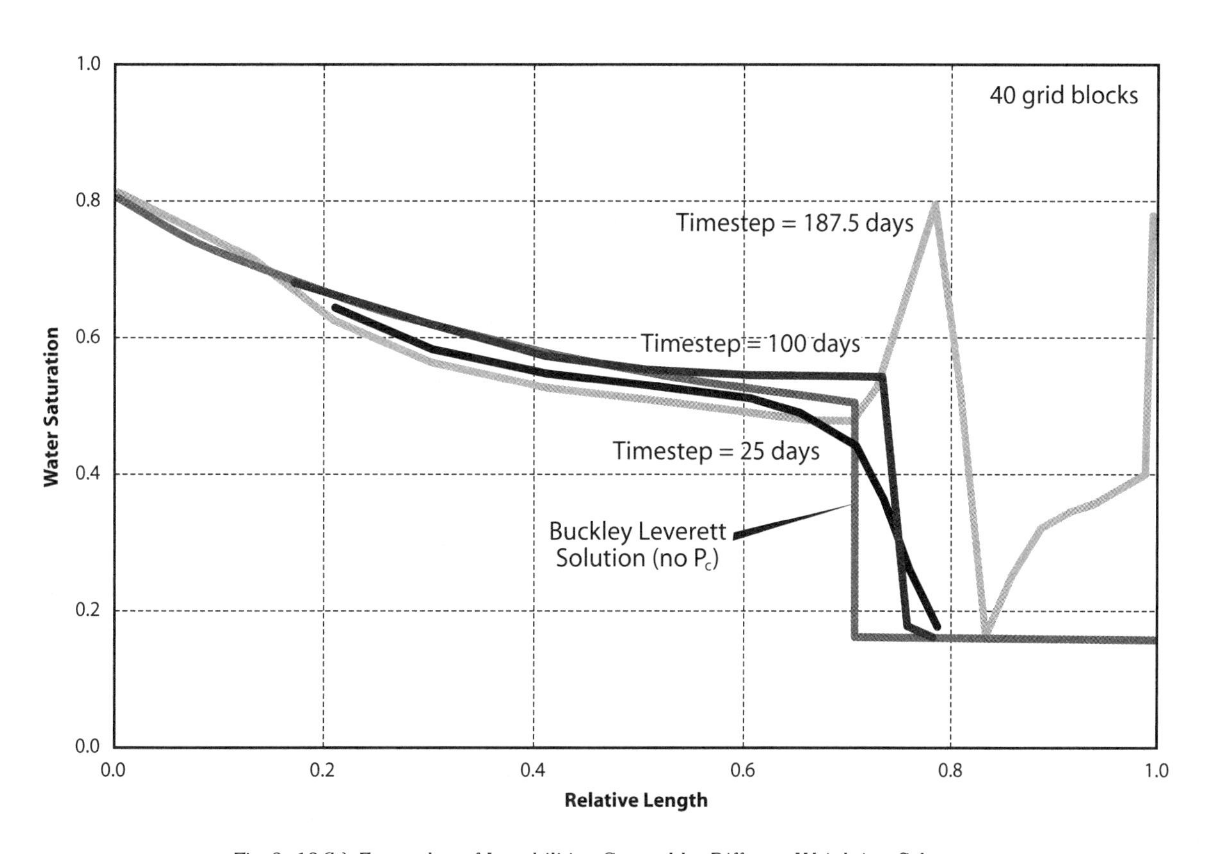

Fig. 2–13(b) Examples of Instabilities Caused by Different Weighting Schemes

100% water-saturated core with oil until we cannot get the water saturation to go any lower in a reasonable test period. In this case, the water saturation is decreasing, and, since it is the wetting phase, it is a drainage process.

Commonly, water is injected into samples, which is our so-called waterflood test. Since the wetting phase is increasing, the process becomes an imbibition displacement process. This is not a tricky operation and, in fact, supports the assertion that some adjustment must be made to calculate relative permeabilities when displacement is occurring.

Capillary pressure (limitation)

The Buckley-Leverett theory is deficient in one major respect—accounting for capillary pressure. The relation used is shown in Equation 2.11. An equation has been constructed with this effect for radial flow; however, no one has solved this equation directly for use in commercial simulators. Chin has solved it for radial flow for drilling invasion profiles.

The problem term is the calculation of the partial derivative of P_c with respect to x. We do not know this directly, but it would be possible to calculate it with Equation 2.12:

$$f_w(S_w) = \frac{1 + \frac{k_0 A}{\mu_o q_t}\left(\frac{\partial P_c}{\partial x} + \frac{\Delta \rho g \sin\alpha}{1.0133E6}\right)}{\left(1 + \frac{\mu_w}{k_{rw}}\frac{k_{ro}}{\mu_o}\right)} \quad (2.11)$$

(field units)

$$\frac{\partial P_c}{\partial x} = \frac{d P_c}{d S_w}\frac{\partial S_w}{\partial x} \quad (2.12)$$

Qualitatively, we know what this looks like, as shown in Figure 2–14(a) and (b):

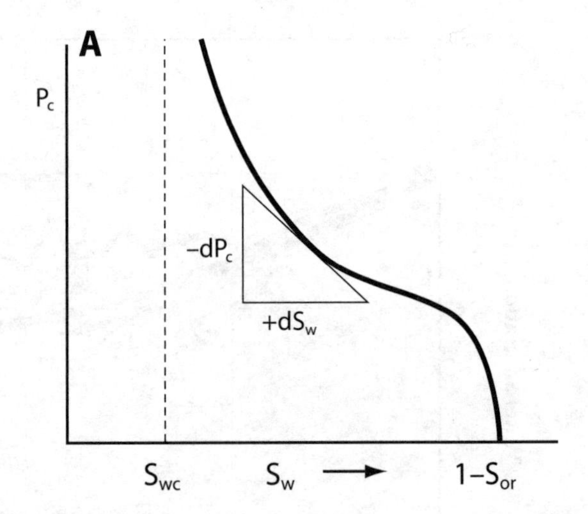

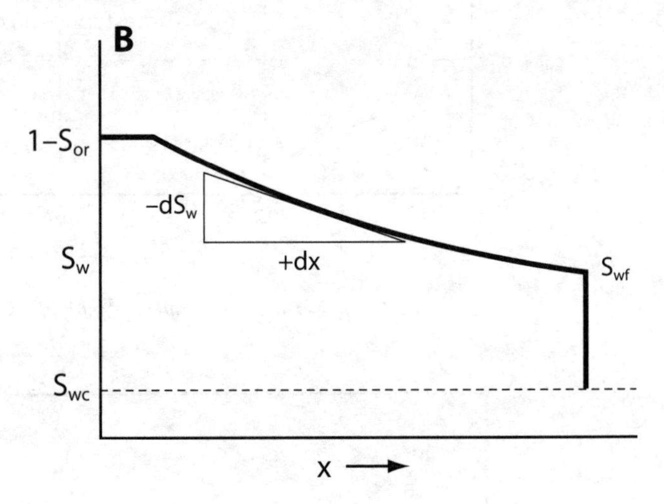

Fig. 2–14 Derivative of P_c with Saturation and S_w with Distance

From this, it is obvious both (dP_c/dS_w) and $(\partial S_w/\partial x)$ are negative. If the terms are multiplied together, they are positive. Therefore, $\partial P_c/\partial x$ is positive and, referring to the previous equation, the fractional flow of water should always increase. Buckley-Leverett theory suggests a discontinuity at the shock front; however, the capillary pressure will round the sharp points and cause the singularity in the slope to disappear, resulting in a continuous curve (i.e., one with a finite slope). The effects of this are shown in Figures 2–15(a) and Figure 2–15(b).[9]

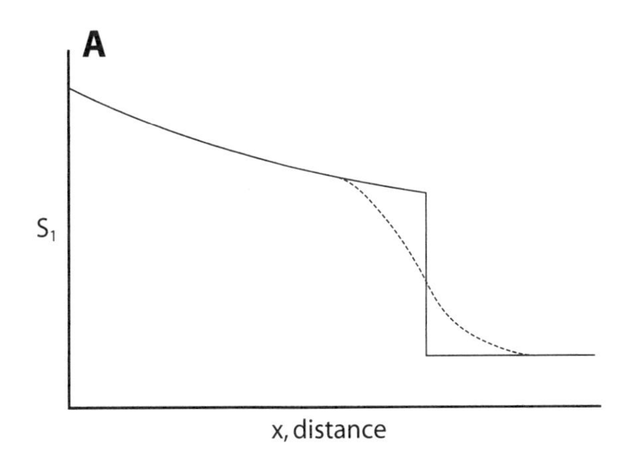

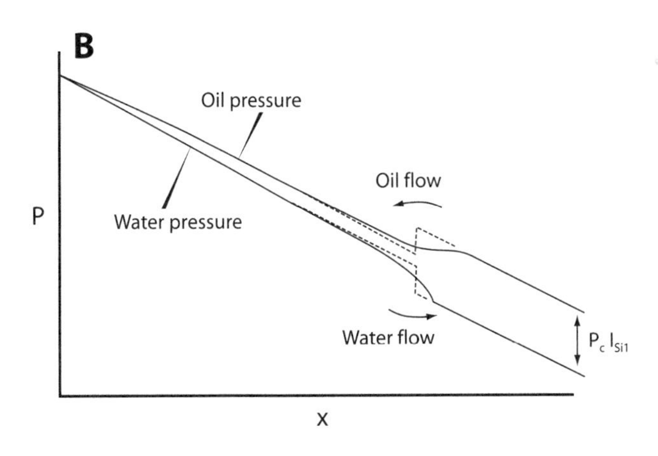

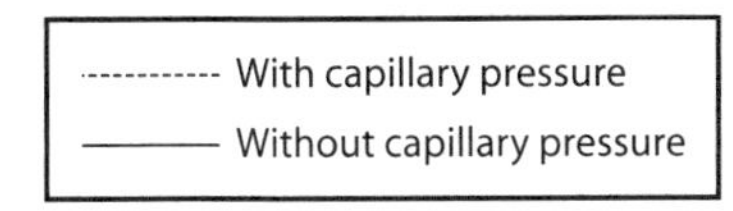

Fig. 2–15a P_c *and Water Movements with Immiscible Displacement*
Fig. 2–15b P_c *and Water Movements with Immiscible Displacement*

It is appropriate to point out that the P_c curve is oversimplified when real displacement is considered. We use drainage curves for doing our initial saturations. This will be discussed later. In most waterflood calculations, our wetting phase is increasing, which requires an imbibition curve. Thus, a case can be made for having different capillary pressure data for initialization and for displacement calculations.

Fundamental problem (limitation)

Such fundamental problems in displacement theory have been noticed. A considerable amount of detail on this exact topic is found in Dullien's *Multiphase Flow in Porous Media* and in Springer Verlag's *Springer-Verlag Lecture Notes in Engineering*, by M. B. Allen III, G. A. Behie, and J. A. Trangenstein.[10,11] Wilson C. Chin deals with these issues in a number of his books, which include *Formation Invasion, Wave Propagation in Petroleum Engineering*, and *Modern Reservoir Flow and Well Test Analysis*. [12,13] All of the latter are from Gulf Publishing.

One-dimensional IMPES displacement stability (limitation/practical knowledge)

Some simple 1-D tests with a simulator show the nature of errors caused by multiphase flow. First, there is the nature of the numerical error, which is a smearing of the shock front. This numerical dispersion effect resembles the smearing caused by capillary pressure, but it is not the same. The two may be additive. Second, the use of timesteps that are too large may introduce instabilities. A simple 1-D situation has been modeled using two different commercial simulators. In Figure 2–16, the high values of water saturation represent a numerical stability problem, which occurs with IMPES-type simulators.

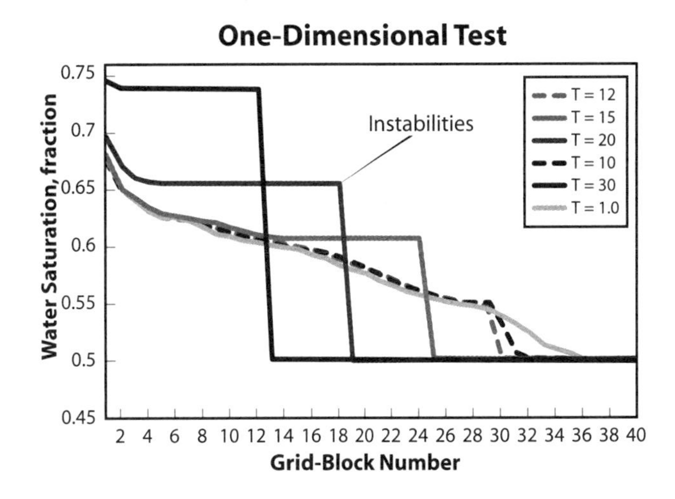

Fig. 2–16 Water Saturations When IMPES Stability Limit Is Exceeded

This problem is more difficult to spot in an areal (2-D or 3-D) simulation. It can be identified by looking for values of S_w higher than $(1 - S_{or})$. They will most likely occur in well blocks. This will identify the worst errors. Look for a stability calculation in the simulation report. These are actually rare but should instill confidence if included. The flood front should not advance more than one grid block per timestep.

Contrary to popular belief, these instability errors will not show up as a material balance error. They will overestimate waterflood recovery if the history match is based on primary production. It will make history matching difficult if a waterflood history is being matched.

One-dimensional dispersion (limitation/practical knowledge)

Dispersion, another numerical error, is shown in Figure 2–17. A smearing of the waterflood front occurs in simulation. A small tongue of water (saturation) will sneak ahead of the flood front predicted by Buckley-Leverett theory, which is shown on the subject diagram.

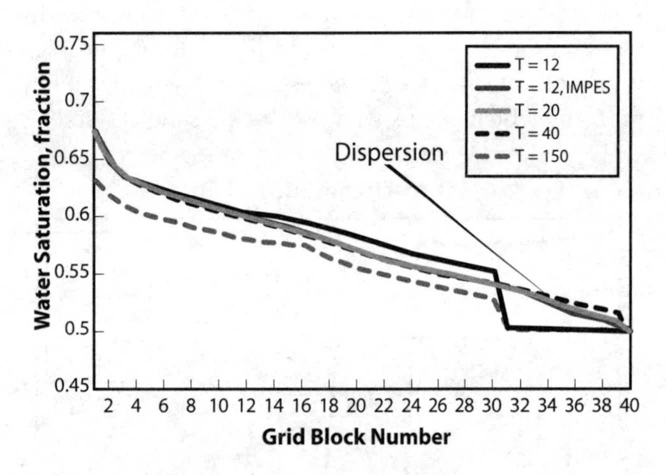

Fig. 2–17 Water Profile from Simulator Including P_c *and Dispersion*

In order to avoid matching water production resulting from numerical dispersion, it has been common to make this problem go away by using a well pseudo relative permeability curve. Some simulation engineers will try to match breakthroughs at the 20% water cut to ensure the proper breakthrough is matched. These are matters of personal style.

Maximum accuracy (practical knowledge)

The numerical dispersion is higher on the newer fully implicit simulators than on the old IMPES-type simulators. The IMPES simulator more closely matches the predictions of Buckley-Leverett theory. Therefore, a newer (implicit) simulator does not necessarily give better results in all respects.

Adaptive implicit method (practical knowledge)

A frequent compromise is to use a combination of the two techniques. This is known as an adaptive implicit method (AIM). It can result in increases in speed. This method requires some overhead to change the grid blocks from one formulation to the other. There are also various ways of controlling this. Some controls utilize a saturation jump limit and some utilize a stability calculation.

It is possible the overhead can exceed the gains in speed. If correctly implemented, the use of a saturation change control will be faster. Stability calculations are more rigorous but slower.

Overall simulator flow diagram (formulation)

The overall flow of a simulator program is shown in Figure 2–18. The major steps are as follows:

1. Stage 1 during *Input Data* is data scanning. The simulator will evaluate all of the input information. Normally, this involves some consistency checks to evaluate if the input data is correct. At this time, the information is stored in lookup tables.
2. Stage 2 during *Input Data* is to develop all of the matrices. Most of the terms in the matrices are fixed and only need to be calculated once.
3. The initialization is made. Recall that an initial solution is required to propagate the solution through time. In a modern simulator, the grid-block saturations may be calculated by a number of different user-defined methods. See chapter 9 for more information on initialization.
4. Stage 3 in *Read Run Data* is the timestep loop. During this stage, any production information (or well rates) and well controls are read by the simulator and added to the matrix. This is followed by *Timestep Control*. The timesteps must be smaller than the next time the well rates and/or well controls are specified.

a. Once the time-dependent information has been read, the *Outer* or Newtonian iteration control occurs. This checks the differences between successive solutions. A minimum of two solutions is required: the initial guess (the solution from the last timestep) and at least one Newton Raphson updated solution.
   i. Within the *Outer Iterations* are successive *Inner Iterations*. It is possible to solve the matrices in one loop using a direct (Gaussian) matrix solution. In this case, the number of outer and inner loops for a timestep will be the same. However, all simulators have iterative solvers that usually provide a faster solution by trial and error methods. Using iterative solvers, the number of inner iterations will usually be larger than the number of outer loop iterations. Solvers will be discussed later.

b. At the end of each set of inner iterations, it may be necessary to shorten the timestep. This control occurs at the outer iteration level.

5. At the end of well rate and/or well control information, the simulator is instructed to stop, and the simulation is complete.

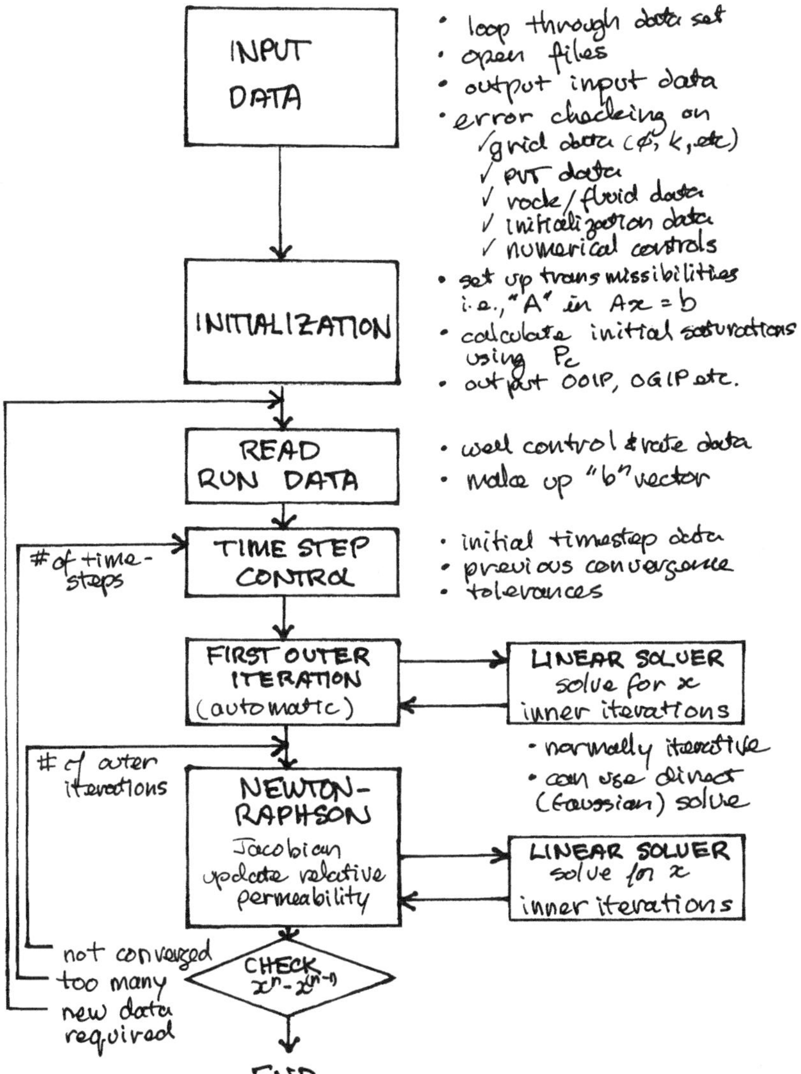

INPUT DATA
- loop through data set
- open files
- output input data
- error checking on
  - ✓ grid data (ϕ, k, etc)
  - ✓ PVT data
  - ✓ rock/fluid data
  - ✓ initialization data
  - ✓ numerical controls

INITIALIZATION
- set up transmissibilities i.e., "A" in $Ax = b$
- calculate initial saturations using P_c
- output OOIP, OGIP etc.

READ RUN DATA
- well control & rate data
- make up "b" vector

TIME STEP CONTROL (# of time-steps)
- initial timestep data
- previous convergence
- tolerances

FIRST OUTER ITERATION (automatic) → LINEAR SOLVER solve for x inner iterations
- normally iterative
- can use direct (Gaussian) solve

NEWTON-RAPHSON (# of outer iterations) Jacobian update relative permeability → LINEAR SOLVER solve for x inner iterations

CHECK $x^n - x^{(n-1)}$
- not converged
- too many
- new data required

END

Fig. 2–18 Overall Flowchart for Reservoir Simulation Program

The overall outline of a simulator is relatively simple. The first part of the simulator is not repeated and is usually referred to generically as the *initialization* section of the simulator. The second part, with the timestep control comprising the outer and inner loops, is usually referred to generically as the *run* section of the simulation. The data decks and execution of the code are also usually split into these two logical divisions.

Timestep control (practical knowledge)

Initially, the first timestep is specified in the input data. Normally, there is also a default. For field simulations, a day is usually a good starting point. If a fine grid is being used and near-wellbore phenomena are being investigated, 0.01 days seems to be a good initial timestep size. This comes with a bit of experience, but, usually, the issue is limited to fine grids.

The end of a timestep occurs when a particular threshold of accuracy has been reached. There are a number of ways of doing this. It can be expressed as an arbitrary pressure and/or saturation change. Alternatively, it can be expressed as a percentage or fraction. In addition, average errors and maximum errors for individual lines in the matrix can be specified. In general, the simulator will have reasonable defaults built in or described in the documentation.

Timestep optimization (practical knowledge)

The simulation has an absolute maximum level of error set by the timestep control. Optimally, better accuracy can be obtained and a second set of tolerances can be set to, in effect, indicate *try for this accuracy*. The fewer the timesteps there are in a simulation, the faster the work is done, which allows for more engineering or more rapid work. With timestep optimization, the objective is to aggressively maximize timestep size and still meet the optimal level of accuracy.

This is a difficult control to design. Most simulators will try to double the size of each timestep compared to the previous one. However, this can be either too aggressive or not aggressive enough. This has become a user-adjustable input, and it is specified as a multiplier of the previous timestep length.

If the timestep is too long, then the timestep length will have to be cut. There is a corresponding timestep divider, which is the opposite of the acceleration factor.

Overall, the defaults seem to work quite well, but this is not always the case. Timestep lengths can sawtooth, which is characterized by several rapid increases, followed by a drastic cut in timestep length. This should be avoided because it slows down execution.

Finally, there are practical limits for timestep lengths. Both absolute minimums and absolute maximums can be specified as user input.

Analytical inflow equations (formulation)

Wells become active during the run portion of the simulation. This is, therefore, an appropriate point to discuss how wells work in the simulator.

For the majority of areal simulations, an analytical inflow equation is used within the model. Implementing an analytical equation presents a few problems. A typical grid block is shown in Figure 2–19.

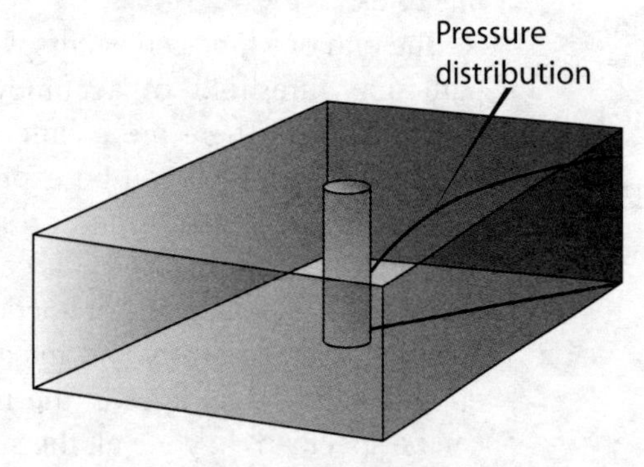

Fig. 2–19 Pressure Distribution in Grid Block from Well Production

The first problem is to relate the well pressure to the average block pressure. There are a number of choices of analytical equations that can be used, which utilize different boundary conditions. These include no-flow boundaries as well as constant pressure boundaries. In truth, a reservoir simulator grid block falls between these two extremes. Flow does enter from the next block as a function of the difference in pressure between the two blocks. In reality, neither equation is directly applicable. Equation 2.13 is the steady state (not transient) equation for a no-flow boundary:

$$q = (0.001127)\,\frac{2\pi\, k_{ro}\, kh}{\mu\, B_o}\,\frac{(p_e - p_{wf})}{\left[\ln\left(\dfrac{r_e}{r_{wf}}\right) + s - 0.5\right]} \qquad (2.13)$$

The intent of developing an inflow equation for the well is to allow it to be solved along with the other reservoir equations. This is done by adding it to the matrix developed for the grid as a whole. Note that well equations are usually added to the matrix in a separate section at the bottom.

Peaceman's equation (formulation/limitation)

The solution to this problem was developed by Peaceman.[14] In essence, he solved a steady state inflow of a single-phase fluid using numerical methods. The grid he used is shown in Figure 2–20 and is intended to represent a repetitive five-spot injection pattern.

10 x 10 Computing Grid for Repeated Five-Spot Pattern

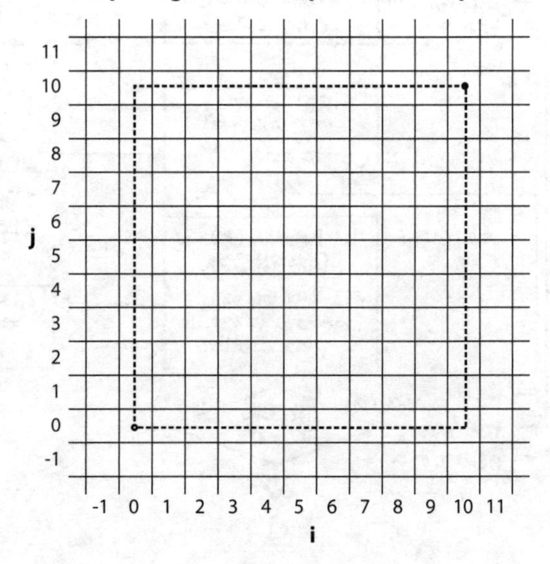

Fig. 2–20 Steady State Grid Used by Peaceman

The ultimate outcome of this simulation was an equivalent grid-block radius. The equation for this is given as Equation 2.14:

$$r_e = 0.2\sqrt{\Delta x \Delta y} \qquad (2.14)$$

This effective grid-block radius, r_e, is input into a discretized equation for solving well rates, which is shown in Equation 2.15.

$$q_o = (0.001127)\frac{2\pi k_{ro}kh}{\left[\ln\left(\frac{r_e}{r_w}\right) + s - 0.5\right]}\sum_{i=1}^{i=n}(p_{bi} - p_w) \qquad (2.15)$$

It is important to recognize this was developed based on steady state flow conditions in which rates and pressure profiles are stabilized. The pressure profile is depicted in Figure 2–21.

This steady state flow assumption would not be correct for a single well located in the middle of an infinite aquifer. Most wells are turned on and off on a number of occasions during a producing month, and, therefore, they are rarely in steady state conditions. This level of detail cannot be captured unless short timesteps are used. In practice, this may or may not be a significant issue.

Another way to look at this is the production performance of low-permeability reservoirs. There are numerous examples of this in the Western Canadian Sedimentary Basin. Most low-permeability Deep Basin gas reservoirs show a one- or two-year flush production. This transient effect cannot be described with a steady state equation. The deviation from the idealized conditions is more severe for low-permeability reservoirs because the transient effects take longer to die out.

Implicit wells versus explicit wells (formulation/practical knowledge)

Just like the development of the grid, wells can be solved either in an explicit fashion or in an implicit fashion. In the earlier days, wells were not always solved implicitly. Recent observations indicate all simulators now include implicit wells. In this case, the relative permeability for the well and the pressure in the well are updated during outer iterations.

The relative permeability is controlled by the saturation in the well grid block. As with displacement, this situation requires upstream weighting; otherwise, when water injection is attempted, relative permeabilities of zero will stop injection. A specialized form of upstream weighting is utilized on injection well relative permeabilities to allow water injection to occur—just as it does in reality. Some simulators do not have this feature built in. It is necessary to mislead the simulator by arbitrarily including water saturations in injection well grid blocks that are above the irreducible water saturation during initialization. This requirement is rare, but such simulators are used.

Pressure buildups (limitation)

It is important to realize, since the well inflow equation was based on steady state flow, the simulator cannot predict a pressure buildup unless special fine gridding is used.

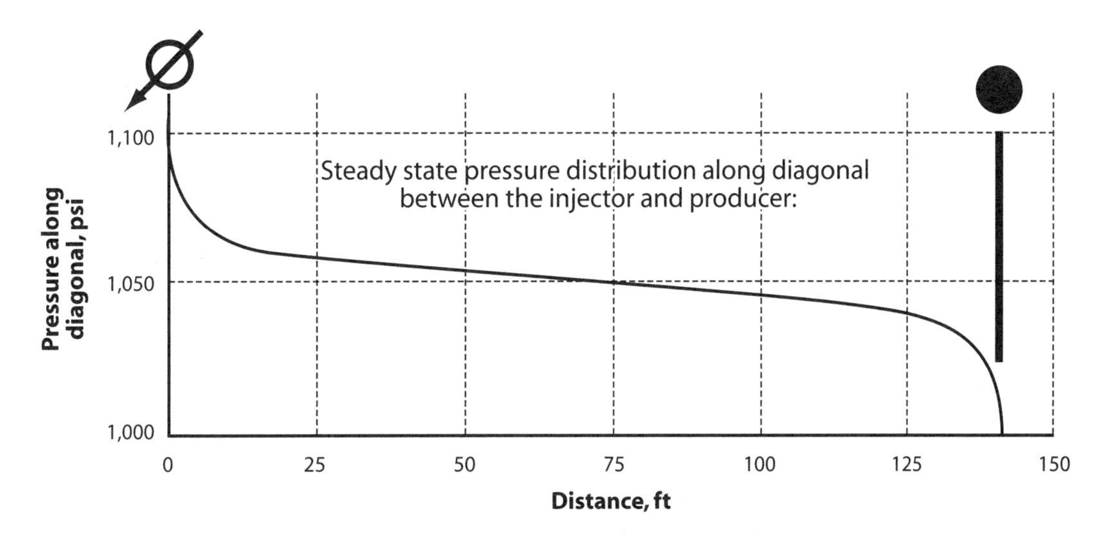

Fig. 2–21 Pressure Profile between Steady State Water Injector and Producer

Anisotropy

If the reservoir permeability is anisotropic, then the equation can be modified as follows:

$$r_o = \frac{\left[\left(\frac{k_y}{k_x}\right)\Delta x^2 + \left(\frac{k_x}{k_y}\right)\Delta y^2\right]^{0.5}}{\left(\frac{k_y}{k_x}\right)^{0.25} + \left(\frac{k_x}{k_y}\right)^{0.25}} \tag{2.16}$$

Skin factor limitations (limitation/practical knowledge)

The discretized well equation has a skin factor. In severe cases, the skin becomes so large the denominator in the inflow equation can be less than zero. The magnitude of $\ln(r_e/r_w)$ for most well equations is normally about seven or eight. With smaller grid blocks, the effective grid-block radius can drop to values of skins typical of fracturing treatments (4 or 5). Since s is subtracted from the $\ln(r_e/r_w)$ term, the denominator in the discretized well equation can be less than zero. However, the log of a negative number is not defined. Further mathematical trickery is needed as defined by Odeh.[15] R_s is the radius of the improved zone. The $k \times h$ is adjusted upwards as follows:

$$k_s = \frac{k_{avg} \ln\left(\frac{R_s}{r_w}\right)}{s_A \left(\frac{\sum_{i=1}^{i=n} \Delta z_i k_i}{(k \times h)_{avg}}\right) + \ln\left(\frac{R_s}{r_w}\right)} \tag{2.17}$$

Partial completion (limitation/practical knowledge)

The analytical inflow equation also assumes the entire grid-block height is perforated. Frequently, the extent of perforations is limited. In such cases, it is normal to adjust the skin factor via the method of Brons and Marting.[16]

Multiphase flow in wells (limitation)

The development of the Peaceman equation was only for a single-phase fluid (water). Although the wellbore equation is implemented in simulators by weighting the individual phase components (water, gas, and oil), the well equation was not developed and tested under these conditions. This situation happens frequently in reservoir simulation, as shown in Figure 2–22.

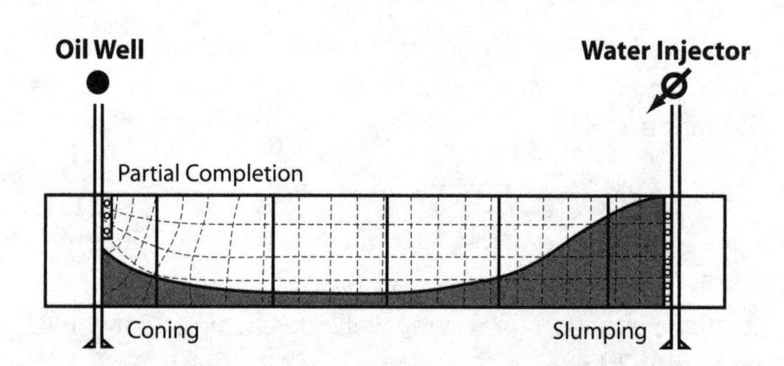

Fig. 2–22 Diagram Illustrating Slumping and Coning

Analytical coning calculations indicate we should not expect a single-phase approach to provide accurate results. Therefore, when multiple saturations exist, it is necessary to model the situation with a detailed grid. This leads naturally to the next topic.

Handling multiphase wells in reservoir simulation (practical knowledge)

Wells with multiphase flow and near-wellbore pressure transients present special difficulties in reservoir simulation. A number of different techniques have been developed to handle them. The appropriate technique depends on the application. The various techniques are outlined:

- use of radial grids
- local grid refinement
- analytical coning equations
- composite (multigrid) reservoir simulators
- well pseudo relative permeabilities

These techniques will be explained in the following section.

In some cases, the entire objective of simulation will be to determine well performance. Under these circumstances, the model will be restricted to a single well. Single-well studies are discussed in the latter part of this chapter.

Detailed well studies (practical knowledge)

For detailed well studies, a modification on the basic simulator is used. Radial coordinates are used in favor of Cartesian coordinates. A typical radial grid is shown in Figure 2–23, which represents a single *slice of a pie*. In most cases, the pie will extend around all 360 degrees if properties do not vary radially.

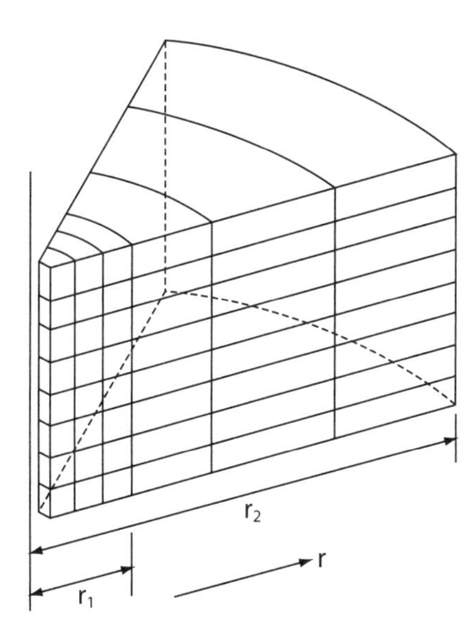

Fig. 2–23 Example of Radial Grid

Such a model is capable of handling multiple phases correctly. Note the size of the grid blocks is now much smaller than the size of the earlier flow net. The inflow into the wellbore is still calculated using the Peaceman steady state equation. Theoretically, this is a farce since there is nothing really steady state about the flow. The error is not serious though, because the volumes immediately surrounding the wellbore are so small. The pressure transient when the well is put on production is propagated through interblock flow. A simulation using this technique will start with very small timesteps.

Alternatives to reduce the number of grid blocks (practical knowledge)

From the previous discussion, it could be concluded that handling wells can require a large number of grid blocks. This can exceed the capacity of large, modern computers.

For an areal model with multiple wells and fluid contacts, more sophisticated techniques have been developed to keep grids to reasonable dimensions:

1. A separate detailed coning study is performed to determine well behavior. Provided the saturations change in one direction consistently, the radial inflow equation may be fooled with a well pseudo relative permeability curve. Pseudo relative permeability and upscaling are discussed in chapter 8. The proportion of oil, gas, and water produced because of coning can be accurately replicated this way. This is shown in Figure 2–24.

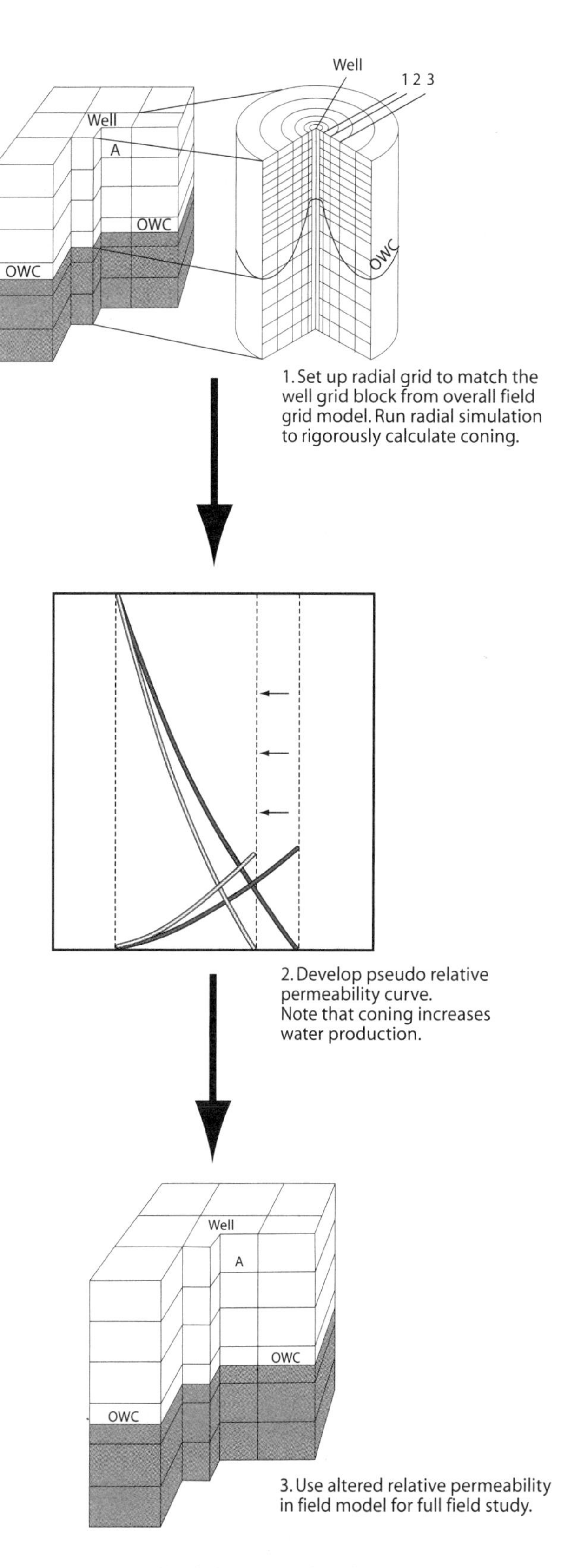

Fig. 2–24 Radial Grid Study

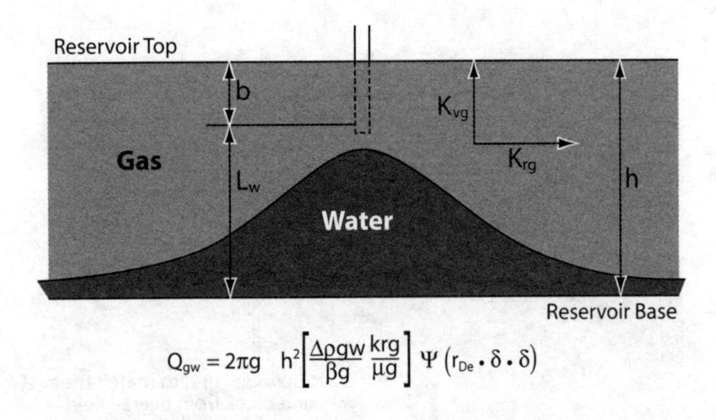

Fig. 2–25 Example of Water Cone and Analytical Equation

2. An analytical equation may sometimes be used to calculate the proportion of water or gas resulting from coning. This involves some custom modifications to the simulator. A number of correlations can be used. Ordinarily, historical match tuning to the equation will be required to get good results.[17] Such a situation is depicted in Figure 2–25.
3. A special areal model can be developed that has a special coning model routine built in, i.e., a simulator within a simulator. This requires a custom model. The general idea is outlined in Figure 2–26.
4. Recently, models have been built that allow subgrids or local grid refinement. This effectively handles the coning phenomena. Typical subgrids used are shown in Figure 2–27.

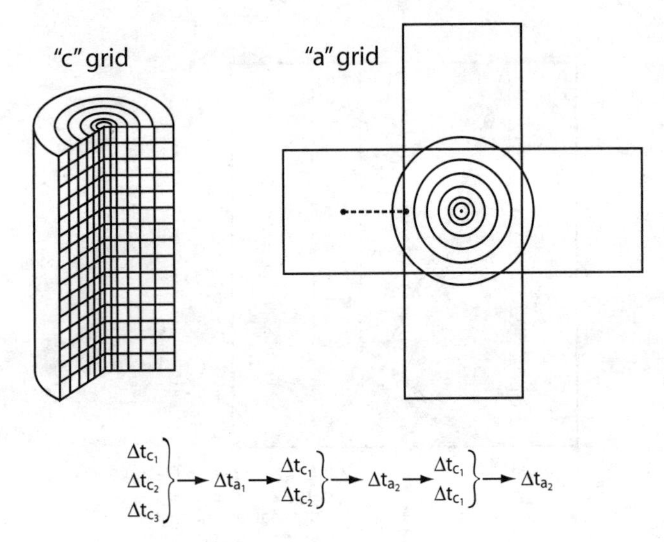

Fig. 2–26 Combination Simulator: Radial Subgrid Inclusion Advancing through Time

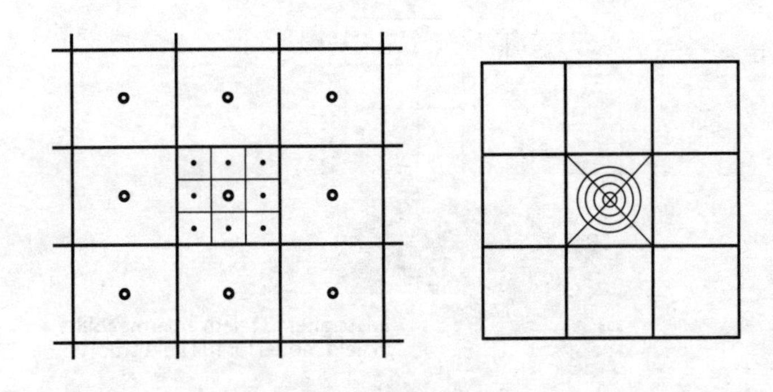

Fig. 2–27 Two Local Grid Refinement Methods

The number of local grid refinements allowed in a model may be limited. Therefore, in large studies, the older techniques described in items 1 through 3 may still be used.

Well controls (formulation/practical knowledge)

The next logical topic is well controls. These are mostly used in predictive reservoir simulation, i.e., after history matching. These controls allow maximum and minimum well rates, maximum water cuts, and GORs, as well as allowing similar controls on groups of wells.

These controls are added as equations after the well equations, which connect the wells to the grid. Well controls are not a trivial problem mathematically. They can produce what mathematicians refer to as stiff matrices. A slight shift in productivity will trigger additional wells being put on production or other operating changes. It is easy for the simulator to cycle across one of these set points and become unstable. Frequently, the solution to this is the use of dampening, which is applied in the matrix solver.

Matrix solvers—numerical (formulation)

Two basic types of solvers are used in reservoir simulation. The first is direct solution, which is done normally by Gaussian elimination. The second type is an iterative solution. There are three major issues in solver selection. The first is storage, which is limited normally by computer memory. The second is the amount of work involved. The third is the amount of overhead required to implement a system.

There have been a number of significant changes during the last few years in matrix solvers. For this reason, a number of the standard works on reservoir simulation, such as Aziz and Settari's *Petroleum Reservoir Simulation*, while correct on formulation, are no longer up-to-date on modern solvers. For the most part, solvers have become much stronger, more robust, and generally easier to use. A brief discussion of the two types of solvers is outlined in the following.

Direct elimination (formulation)

This method is similar to the method first taught for solving equations in university and high school math. This process should look familiar:

$$\begin{aligned} 3x + 4y + z &= 1 \\ 2x + 3y &= 0 \\ 4x + 3y - z &= -2 \end{aligned}$$

$$\begin{bmatrix} 3 & 4 & 1 \\ 2 & 3 & 0 \\ 4 & 3 & -1 \end{bmatrix} \begin{bmatrix} x \\ y \\ z \end{bmatrix} = \begin{bmatrix} 1 \\ 0 \\ -2 \end{bmatrix} \tag{2.18}$$

This can be translated into an augmented matrix and solved by row operations:

$$\left[\begin{array}{ccc|c} 3 & 4 & 1 & 1 \\ 2 & 3 & 0 & 0 \\ 4 & 3 & -1 & -2 \end{array}\right]$$

Row 1 - Row 2

$$\left[\begin{array}{ccc|c} 1 & 1 & 1 & 1 \\ 2 & 3 & 0 & 0 \\ 4 & 3 & -1 & -2 \end{array}\right]$$

Row 2 - (2 × Row 1) and Row 3 - (4 × Row 1)

$$\left[\begin{array}{ccc|c} 1 & 1 & 1 & 1 \\ 0 & 1 & -2 & -2 \\ 0 & -1 & -5 & -6 \end{array}\right]$$

(Row 1 - Row 2) and (Row 3 + Row 2)

$$\left[\begin{array}{ccc|c} 1 & 0 & 3 & 3 \\ 0 & 1 & -2 & -2 \\ 0 & 0 & -7 & -8 \end{array}\right]$$

Multiply Row 3 by -1/7

$$\left[\begin{array}{ccc|c} 1 & 0 & 3 & 3 \\ 0 & 1 & -2 & -2 \\ 0 & 0 & 1 & 8/7 \end{array}\right]$$

By back substitution, it is possible to calculate

$$x = -\tfrac{3}{7}$$

$$y = \tfrac{2}{7}$$

$$z = \tfrac{8}{7}$$

A systematic operation has been used. The first row was normalized to a value of one on the diagonal. Then, the lower triangle of the matrix was systematically eliminated. Back substitution allowed direct calculation of z, y, and x by working from the bottom to the top.

For small problems where the amount of work and storage is not a major concern, this is the preferred method of solving equations. Most simulation documentation sets an upper limit of the order of 500 to 1,000 grid blocks for Gaussian elimination. Gaussian elimination sometimes is as efficient on larger problems as the suggested maximum. For problems of as many as 3,000 grid blocks, not much harm is done by running a problem for two or three timesteps with the various solvers available. Normally, it is easy to determine which one works best by using trial and error.

The previous scheme is usually broken down and described by mathematicians as *LU Decomposition*. The implementation to minimize storage is normally via a series of vectors, such as those shown in Figure 2–28.

Second Degree ILU (Natural Ordering)

Third Degree ILU (Natural Ordering)

Fig. 2–28 LU Factorization for Preconditioning

The author used one of the solvers in the back of Aziz and Settari's book for a university course and messed up the vector numbering convention. Reasoning there must be a bug, the solver was rewritten with different indices. Zero terms have to be handled differently and each matrix row operation has many index manipulations. This experience demonstrated to the author how matrix solutions involve a number of integer operations for each floating point calculation.

Iterative solvers (formulation)

Larger simulations will result in matrices more efficiently solved by iterative methods. These methods require less storage. The number of iterations must be sufficiently low to produce a faster solution.

Traditionally, this has involved a number of successive overrelaxation (SOR) techniques. An acceleration factor, normally designated with the small Greek letter omega (ω), is used in this method. Tuning was normally required on the acceleration factor by trial and error. This system has been virtually replaced by accelerated preconditioned Orthomin solvers.[18] Thus, a discussion of SOR techniques is considered to be of historical rather than practical interest.

In iterative solutions, the matrix is not solved directly. In essence, a guess is made as to what the solution will be (perhaps the saturations and pressures from the previous timestep), and then a method is used that will correct subsequent guesses. An arbitrary level of accuracy is normally specified and iterations are stopped when this is achieved.

In more mathematical terms, the general solution of a matrix is given as:

$$Ax = b \tag{2.19}$$

In the case of an iterative solver, A is split as follows:

$$A = C - R \tag{2.20}$$

where:

C = the approximate coefficient matrix
R = the residual matrix, representing the error in C

The iterative method is then defined as:

$$C_x = R_x + b \tag{2.21}$$

Or, in residual notation:

$$x^{(n+1)} = x^{(n)} + C^{-1} r^{(n+1)} \tag{2.22}$$

where:

$$r^{(n)} = b - Ax^{(n)} \tag{2.23}$$

The size of r is used to determine if convergence has been achieved. Recall that both A and b are known, therefore the size of the error can be calculated at each iteration. The previous equation takes the current guess and updates it to the next one. Note the size of the error is used for correction.

A variety of different methods can be used by changing the definition of C:

- If $C = A$ and $R = 0$, then the method becomes Gaussian elimination.
- If $C = I$, the solution becomes the Richardson method.
- If $C = D$, solution is via Jacobi method, where D is the diagonal part of A.
- If $C = 1/\omega \times D - G_L$, it gives an SOR solution where $(A = D - G_L - G_U)$.
- If $C = LU$, then the method becomes incomplete factorization.

The rate of convergence will depend on the method used and how well it applies to the particular problem. When the residual reaches zero, one has an exact solution. This also means the solver can be controlled by setting a tolerance on the residual vector. This will be a matrix solution control.

Acceleration (formulation)

The techniques listed previously are not the full story on a modern matrix solver. This methodology can be upgraded by using an acceleration technique. The entire matrix solution process then becomes a two-stage process, as outlined in Figure 2–29.

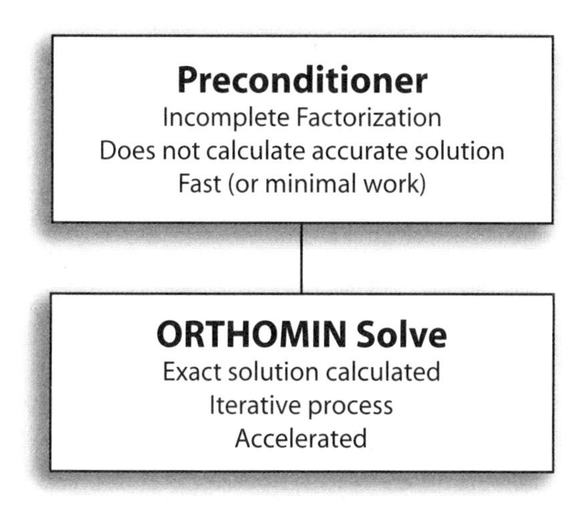

Fig. 2–29 Iterative Solver Flow Chart

Acceleration is the second part of an iterative solver. The method tries to find which direction causes the residual to change the most. Recall that in 3-D space (in fact, *n*-dimensional space), a plane can be represented by a single vector orthogonal to the surface. The term Orthomin is taken from orthogonal and minimization. The method also determines a magnitude to the orthogonal vector, which is designed to rapidly determine a closer solution. Note the iterative solution is applied to the original *A* matrix—this part is not an approximation.

With regard to the flow diagram described previously, it may be concluded that the closer your initial estimate is to the correct solution, the faster you will get your solution. The hitch here is more accurate estimates require more sophisticated methods. A variety of solutions can be used as initial guesses. These were outlined earlier as traditional techniques, which include successive overrelaxation, partial Gaussian eliminations, or incomplete factorization methods. This is a trade-off, and getting the right degree and type of preconditioning will depend on individual problems.

In the end, the Orthomin pointers direct successive iterations to the right answer. The various correction vectors are also stored, and, therefore, the system has a memory. However, if early estimates are off, the previous corrections may not contribute much toward pointing to the correct solution.

User control of solvers (practical knowledge)

The number of Orthomin solutions to be remembered can be specified. Depending on the problem, the solution speed could be increased or impaired. This is, in fact, a control in reservoir simulator solvers and is known as the number of Orthomin vectors (usually called NORTH). The most common number of maximum solutions remembered is between 10 and 25.

Some simulators allow one to specify the preconditioning technique. Some keywords are quite specific about the mathematical technique to be used. Others hide the details and simply specify a level of difficulty. Under these circumstances, one may be able to test the combination providing optimal solution time. Many simulators do not allow this flexibility.

When working on difficult problems and having trouble getting adequate (i.e., accurate and stable) solutions, the author has obtained best results by tightening the residual tolerances on the solver (inner loop). To date, these problems have been insensitive to the number of Orthomin vectors (solutions) used. However, it is worth trying various changes if the defaults do not work satisfactorily. The storage of these vectors can be significant, and one program allows the user to store this information on the hard disk instead of in memory.

Pivoting (formulation and practical knowledge)

This is a standard technique in the solution of matrices. It is used in linear programming and is used as a starting point for finding an optimal solution. There are really two objectives in modern simulation programming. The first objective is to speed up solution. The second objective is to reduce numerical errors.

Pivoting is desirable whenever there are large contrasts in the matrices. Where would this occur in numerical simulation? There are a number of cases. Layering can cause severe changes in transmissibilities. A directional permeability trend will have the same effect. The most common example the author has used concerns modeling hydraulic fractures directly in models.

A pivoting option in a simulator will reorder the equations so the high values are centrally located about the diagonal. This requires overhead to test the matrix but can mean the difference in whether or not a solution is obtained. Since simulators generate banded matrices, the reordering of the equations can cause an increase in bandwidth, which in turn makes solution longer. However, this will be offset somewhat because the ease with which a solution is obtained depends on the size of the matrix elements along the diagonal. After the matrix is solved, *deordering* is required between the main part of the simulator and the matrix solver.

For those simulators using a level of difficulty, pivoting may be implemented, although this will be transparent to the user. Other simulators will have a specific option such as *PIVOT = ON* or *OFF*. The default is normally *OFF*. Some popular simulators don't have this option at all. This limits their capability.

Explanation of dampening

One way to control the cycling problem associated with well controls is to use dampening. Simulators use acceleration in the form of the Newton-Raphson technique on the outer iterations and, in the form of Orthomin, on the inner iterations. Maximum group rate controls often result in step changes, such as a well becoming shut-in or put on production. The acceleration techniques used often overshoot the actual solution. This is a bit like projecting a time of arrival on the highway when the driver in front crashes into the car ahead. The estimate based on derivatives (velocity) drastically overshoots the future projected position.

The mathematical solution to this is equivalent to determining where the crash will occur (or where the maximum rate is encountered). In essence, an arbitrary *fudge factor* is applied to the well control or well equations to stop overshooting the target solution.

Since *fudge factor* is somewhat frowned upon as a technical term, this has been renamed *dampening factor*. It must be less than one (one being no dampening), greater than zero, and normally must exceed some minimum. The correct factor is determined by trial and error, also know as *guessing*, another nontechnical term that is frowned upon.

Summary

Buckley-Leverett theory, which is the foundation of simulation calculations, has some significant simplifications. Reservoir simulators partially consider capillary pressure and therefore represent a somewhat improved calculation compared to pure Buckley-Leverett theory. The effects of capillary pressure have not been solved with rigorous analysis. The general impact of these effects can be deduced. Numerical dispersion and potential stability effects offset the improved formulation. Therefore, some fundamental mathematical issues in reservoir simulation have not been resolved yet.

Research is being conducted on this problem and involves alternate mathematical techniques, such as the methods of characteristics and finite element analysis. Simulator technology is not nearly as mature as most petroleum engineers would believe. Based on Chin's results, fundamental improvements in the near future are quite possible.

The manner in which wells are handled in a simulator is based on restrictive assumptions. The equation used is based on homogeneous properties, a five spot injection pattern, and a single phase (water). Frequently, this is not particularly representative. The assumption can work adequately in many cases. At the same time, it is easily abused and used in inappropriate situations. Regrettably, no error messages result. The best advice is *Caveat emptor* (Buyer beware). Simulators can calculate meaningless results with incredible precision.

Solvers have improved greatly during the years and improvements are still to come in this field. Expect some changes in this regard.

At present, this text contains relatively little detail on mathematical development. As stated in chapter 1, this can be found readily in other references on the subject. To truly understand this material, a course on how to build a simulator is highly recommended.

References

1 Aziz, K., and A. Settari, *Petroleum Reservoir Simulation*, Applied Science Publishers, 1979.

2 Yanosik, J.L., and T.A. McCracken, "A Nine-Point, Finite-Difference Reservoir Simulator for Realistic Prediction of Adverse Mobility Ratio Displacements," SPE 5734, *Society of Petroleum Engineers Journal*, pp. 253–262, Aug. 1979.

3 Agarwal, R.G., "Real Gas Pseudo-Time—A New Function for Pressure Buildup Analysis of MHF Gas Wells," paper 8279, presented at the 1979 SPE ATCE, Dallas.

4 Price, H.S., and K.H. Coats, "Direct Methods in Reservoir Simulation," *Society of Petroleum Engineers Journal*, 1974.

5 Craig, F.F. Jr., *The Reservoir Engineering Aspects of Waterflooding*, SPE Monograph Vol. 3 of the Henry L. Doherty Series, 1971.

6 Dake, L.P., *Fundamentals of Reservoir Engineering*, Developments in Petroleum Science 8, Elsevier, 1978.

7 Lake, L.W., *Enhanced Oil Recovery*, Prentice Hall, 1989.

8 Settari, A., and K. Aziz, "A Generalization of the Additive Correction Methods for the Iterative Solution of Matrix Equations," *SIAM Journal of Numerical Analysis*, 10, No. 3, pp. 506–521.

9 Yokoyama, Y., and L. Lake, "The Effect of Capillary Pressure on Immiscible Displacements in Stratified Porous Media," SPE 10109, 56th ATM of the SPE, San Antonio Texas, 1981.

10 Dullien, F.A., *Porous Media: Fluid Transport and Pore Structure*, 2nd ed., Academic Press, 1992.

11 Allen, M.B. III, G.A. Behie, and J.A. Trangenstein, *Multiphase Flow in Porous Media: Mechanics, Mathematics and Numerics*, Springer-Verlag Lecture Notes in Engineering, New York, 1991.

12 Chin, W.C., *Modern Reservoir Flow and Well Transient Analysis*, Gulf Publishing Company, 1993.

13 Chin's Wave Propagation; Chin, W.C., *Wave Propagation in Petroleum Engineering*, Gulf Publishing Company, 1994.

14 Peaceman, D.W., "Interpretation of Well Block Pressures in Numerical Simulation," *Society of Petroleum Engineers Journal*, June 1978, pp. 183–194, *Transactions of AIME* 265.

15 Odeh, A.S., "The Proper Interpretation of Field Determined Buildup Pressure and Skin Values for Simulator Use," *Society of Petroleum Engineers Journal*, Feb. 1985.

16 Brons and Marting, "The Effect of Restricted Fluid Entry on Well Productivity," *Journal Petroleum Technology*, Feb. 1961, pp. 172–174.

17 Chierici, L.G., and G.M. Ciucci, "A Systematic Study of Gas and Water Coning by Potentiometric Models," *Journal of Petroleum Technology*, Aug. 1984.

18 Vinsome, P.K.W., "Orthomin, an Iterative Method for Solving Sparse Banded Sets of Simultaneous Linear Equations," SPE 5729, presented at the SPE Symposium on Numerical Simulation of Reservoir Performance, Los Angeles, Feb. 1976.

3

Geology, Petrophysics, and Layering

Introduction

This chapter on geology is based on a theme similar to that proposed in the first chapter—the use of conceptual models. The only difference is the models in this chapter are geological in nature. Although each reservoir is individually unique, there are great similarities between certain formations and fields around the world. The most effective way of understanding these is to group them according to basic common characteristics. Examples from the Western Canadian basin will be shown.

To effectively discuss geological models, it is necessary to understand some basic geological concepts and terminology. This is presented first in the following sections.

Uniformity

The foundation of modern geology is the assumption that the same laws of physics and chemistry applied in ancient geological times apply today. By studying modern processes and modern environments, it is possible to develop a better understanding of what has occurred in the past.

Geological models are based on studies of modern reefs, beaches, rivers, and lakes. This is why the major oil companies send their geologists to Florida and the Bahamas. They don't get to surf on the beach; they are expected to go snorkeling in the various reef-building environments to observe sedimentation processes first hand. If they're not so lucky, they go to Texas or North Carolina. These areas have excellent modern examples of beach and bar environments.

Sources of oil and gas

The origin of oil will be discussed briefly. A number of theories account for this, ranging from frozen hydrocarbons (comets) landing on earth, or volcanic activity, to the most commonly accepted theory of thermally altered organic debris.

In most cases, oil and gas are thought to have migrated into what are current traps. No exact mechanisms have been widely accepted. Considering the recent advances being made in hydrogeology, the author predicts considerable progress will be made during the next decade in this matter. This will have serious impacts on reservoir engineering analysis and will become the hottest technique in exploration.

One of the pieces of evidence that most strongly supports migration theories is the systematic change in fluid properties. This is of use to a reservoir engineer since it makes it easier to remember the characteristics of different pools and can aid in correlating petroleum properties.

Lower Cretaceous oil and gas in Western Canada

This section has been summarized from an article written by John Masters. The original paper is contained in AAPG Memoir No. 38, which is titled "Elmworth, Case Study of a Deep Basin Gas Field."[1] This discussion is limited to the Cretaceous; however, there are similar trends in the Paleozoic.

The source of oil and gas is considered the Lower Cretaceous shales. This is supported by total carbon contents from shale samples and the thickness of the shale deposits. Figure 3–1 shows a simplified view of the Alberta geosyncline. Note the disturbed belt is not shown. The organic content of the shale is converted into oil and gas by thermal alteration under pressure. Temperature and pressure in the subsurface are generally related to burial depth. The conditions thought to be conducive for generating oil and gas are different. These two windows are often referred to as the *gas window* and the *oil window*. These are shown in Figures 3–2 and 3–3.

The oil and gas migrate updip due to buoyancy. Figure 3–4 is a structure map of the Precambrian basement. There are a number of important features. The predominant flow path is updip, parallel to the basement trend. To the south is the Sweetgrass Arch, which is expressed in the Precambrian basement. Well above the basement, the Lower Cretaceous sediments have an anticlinal feature called the Athabasca Arch, which extends off the end of the Sweetgrass Arch and then along the Alberta-Saskatchewan border. This acts as a huge updip seal for the majority of the Lower Cretaceous.

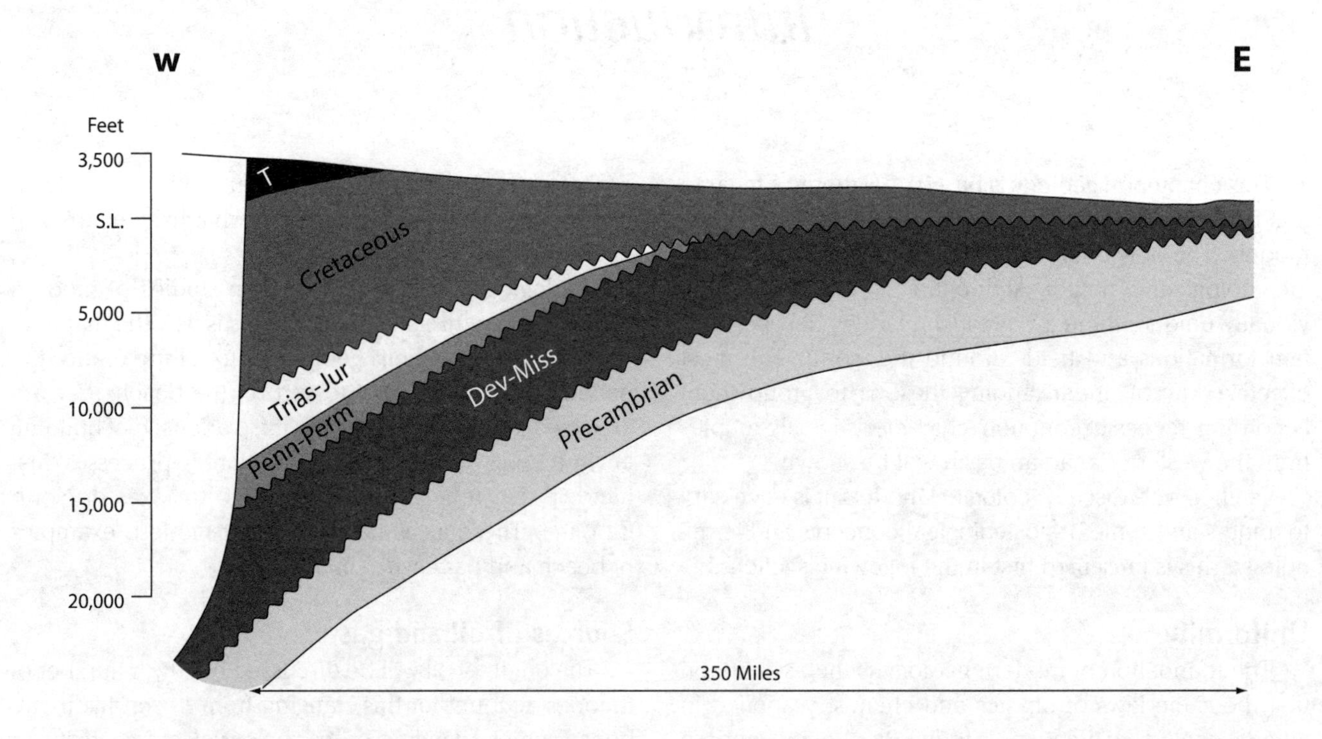

Fig. 3–1 Cross Section of General Western Canadian Sedimentary Basin Stratigraphy

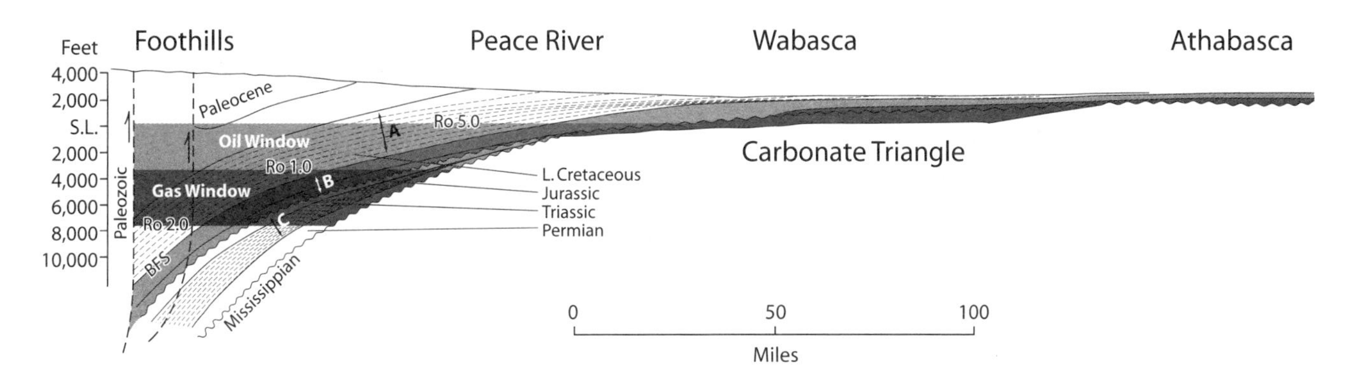

Fig. 3–2 Cross Section of Oil and Gas Generating Windows, Western Canadian Sedimentary Basin

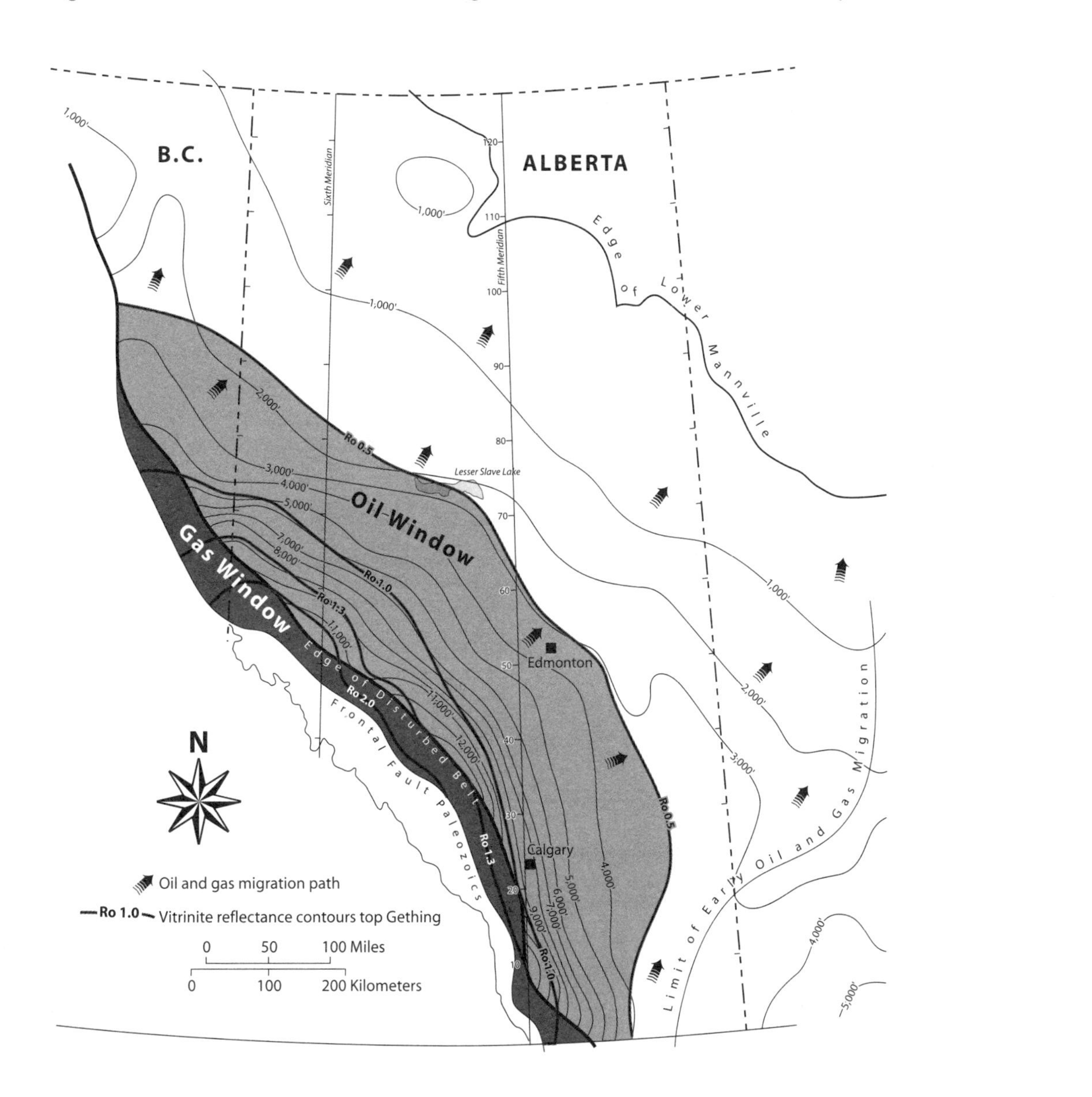

Fig. 3–3 Map of Oil and Gas Generating Windows, Western Canadian Sedimentary Basin

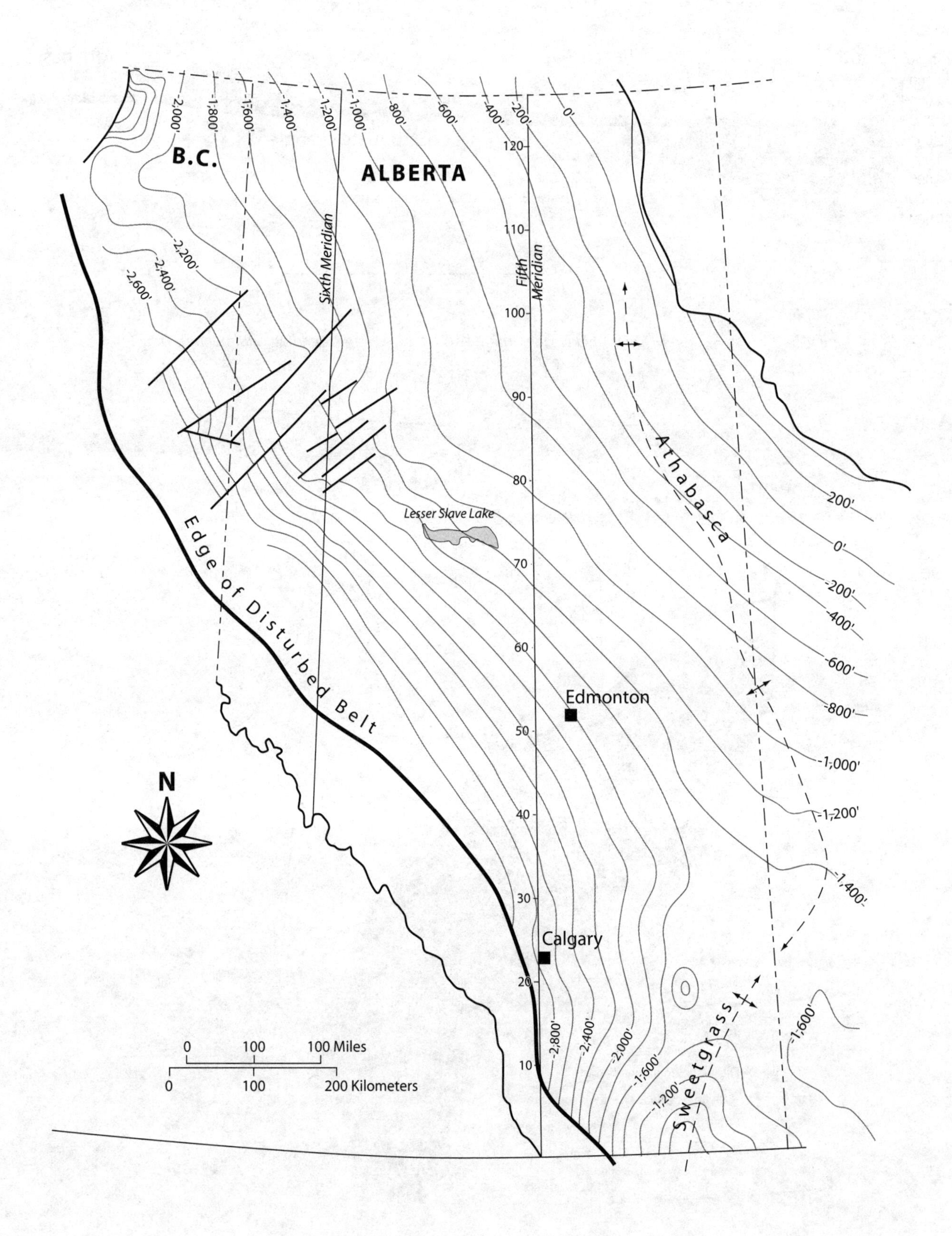

Fig. 3–4 Basement Contours, Western Canadian Sedimentary Basin

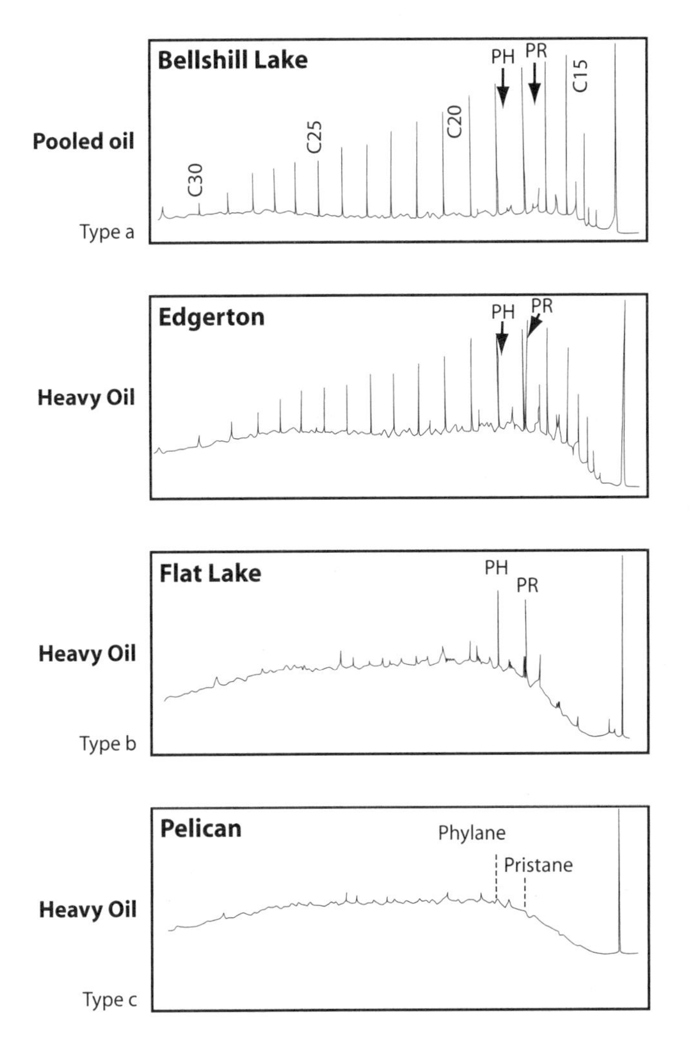

Fig. 3–5 Variation of Gas Chromatograms, Western Canadian Sedimentary Basin

Most of the hydrocarbons trapped along the Athabasca Arch are better known as the tar sands. This tar was once conventional liquid crude oil. Studies have determined this oil was altered by biodegradation, water washing, and, possibly, some inorganic oxidation. This data was determined by analyzing gas chromatographs from pools across the province. An example of these changes is shown in Figure 3–5.[2]

The migration follows the path of least resistance or the beds with the best permeability. Figure 3–6 shows a cross section through the Lower Cretaceous, which shows the major reservoir rocks. These trends are shown areally in Figure 3–7. Oil and gas migrated updip towards the east. In this figure, it is easy to see the reservoir fluids that will be closely related.

This trend also exists in the deeper Paleozoic as well. Trends in reefs, which also correspond to migration routes, are shown in Figure 3–8.

In summary, the properties of reservoir fluids in the Western Canadian Sedimentary Basin change systematically across the province. Although this supports migration theories, it also has direct application to reservoir engineering. In some cases, a PVT fluid analysis is not available. This forces the use of correlations for properties, as well as looking at sample analyses from nearby pools. If fluid migrations are known, this gives a rational basis with which to extrapolate where the fluid properties should lie.

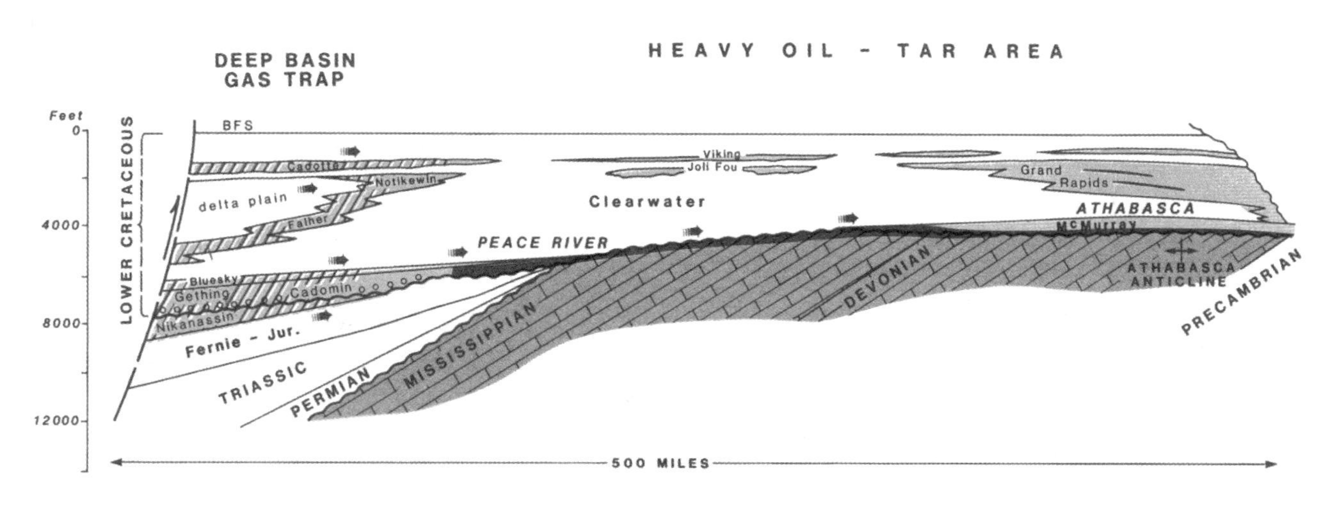

Fig. 3–6 Cross Section of Cretaceous Facies, Migration, and Accumulations

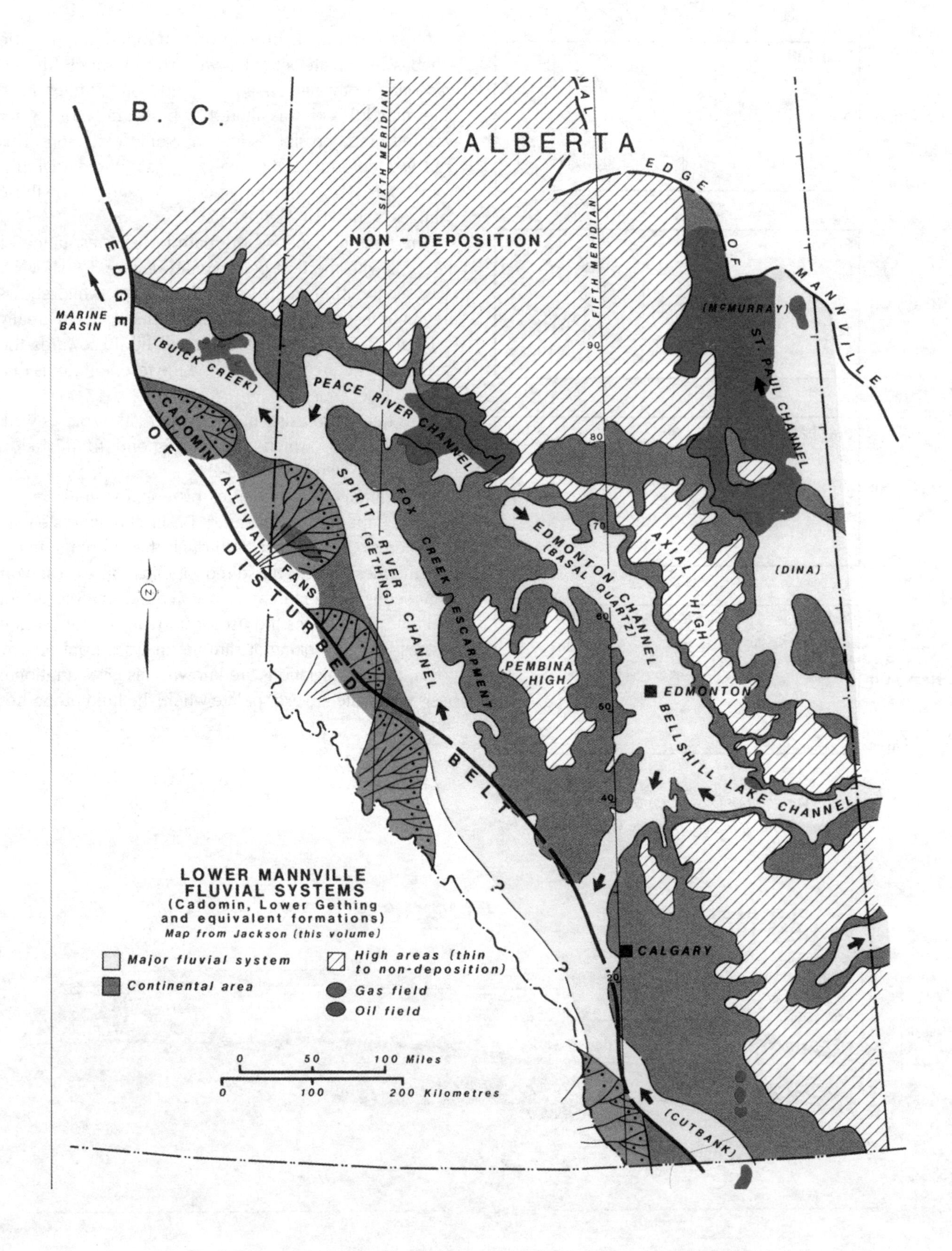

Fig. 3–7 Map of Cretaceous Fluvial Systems from Which Oil Trends May Be Inferred

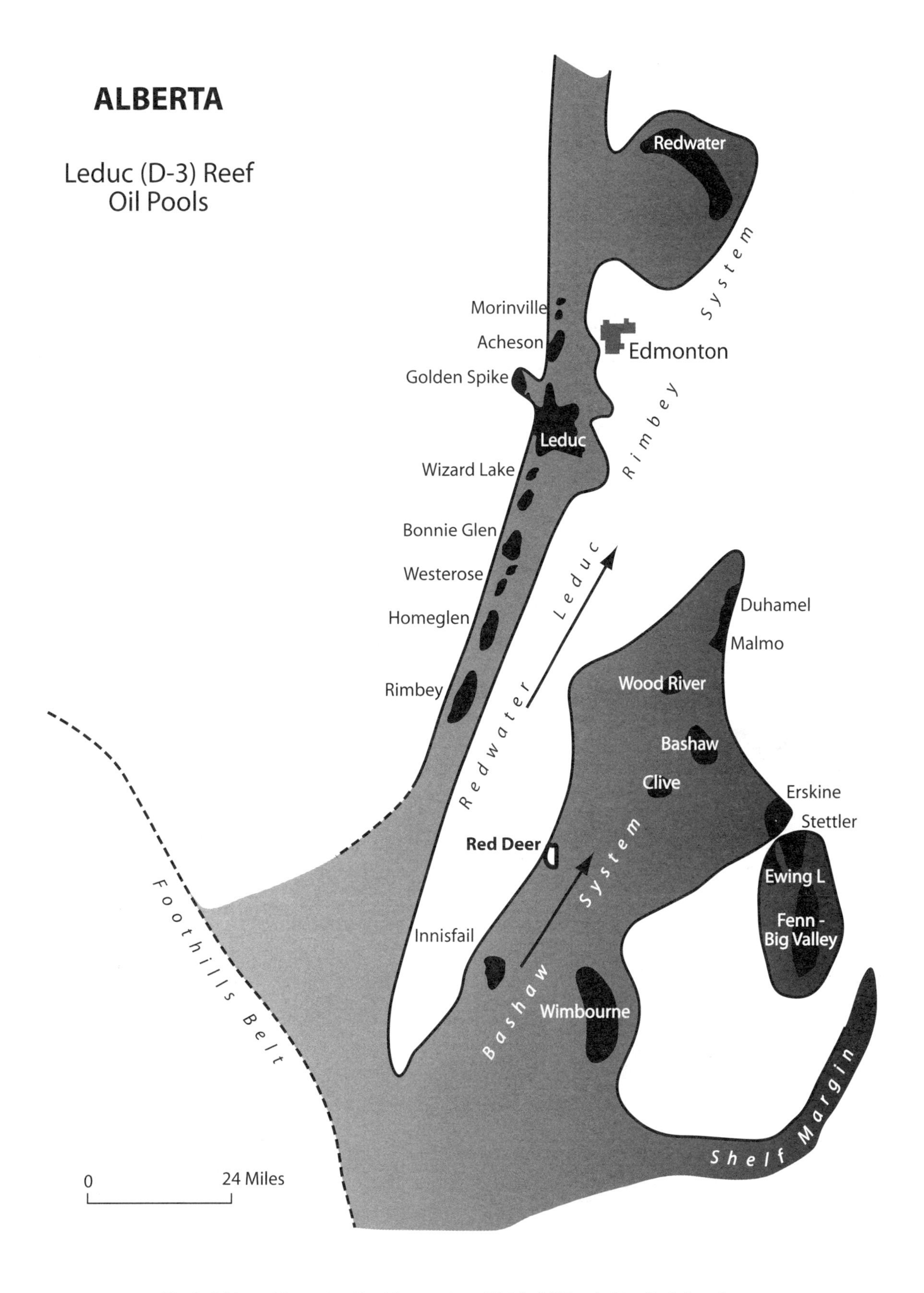

Fig. 3–8 Map of Devonian Reef System from Which Oil Trends May Be Inferred

Geological Models

One of the best ways to demonstrate the application of a geological model is by an example. The giant Hoadley gas field of South Central Alberta is an excellent case study. This example is particularly good because the ancient geomorphology has been compared explicitly to the modern environment. For instance, the areal maps of the modern environment are based on aerial photographs. This section was also based on a paper contained in AAPG Memoir No. 38 written by Kam K. Chiang.

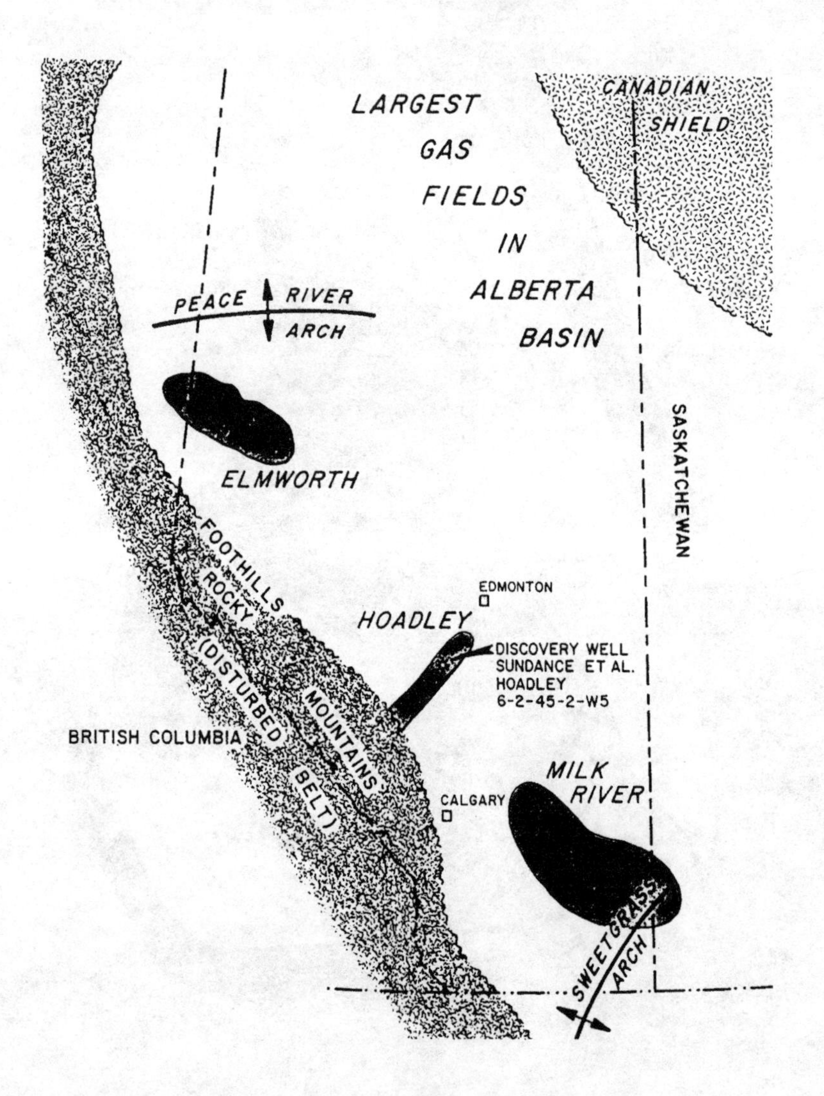

Fig. 3–9 General Location of Hoadley Barrier Bar System

Hoadley barrier bar

The Hoadley barrier bar is a relatively recent discovery. The first well was drilled in 1977. It ranks as the third-largest gas field in Canada, behind the Elmworth and Medicine Hat gas fields. It is rich in natural gas liquids (NGL), which also makes it a significant liquids discovery.

The general location of the bar is shown in Figure 3–9. A detailed map of the different rock types, known as facies, is shown in Figure 3–10. The bar extends a considerable distance of more than 130 miles. Downdip the bar disappears into the disturbed belt, while updip the bar has eroded away. To the northwest is open ocean, with land located to the southeast.

The principal facies include:

- Eolian Sand Dunes
- Tidal Channel
- Levee
- Interbar Lagoon
- Back Bar Washovers

Immediately behind the barrier bar is the Medicine River Delta Complex. In this area the following facies may be found:

- Distributary Channel
- Abandoned Channel
- Deltaic Deposits

The major significance of these facies to a reservoir engineer is that each rock type has different grain-size distributions, mineral content, and particle size. This is a direct result of the depositional process. The effects of these factors are discussed in more detail in the following chapter. At this time, it is sufficient to point out the different types of rock have different flow properties. This ultimately determines the pattern of flow in the reservoir and recovery. The different properties of the various facies are discussed in the following sections.

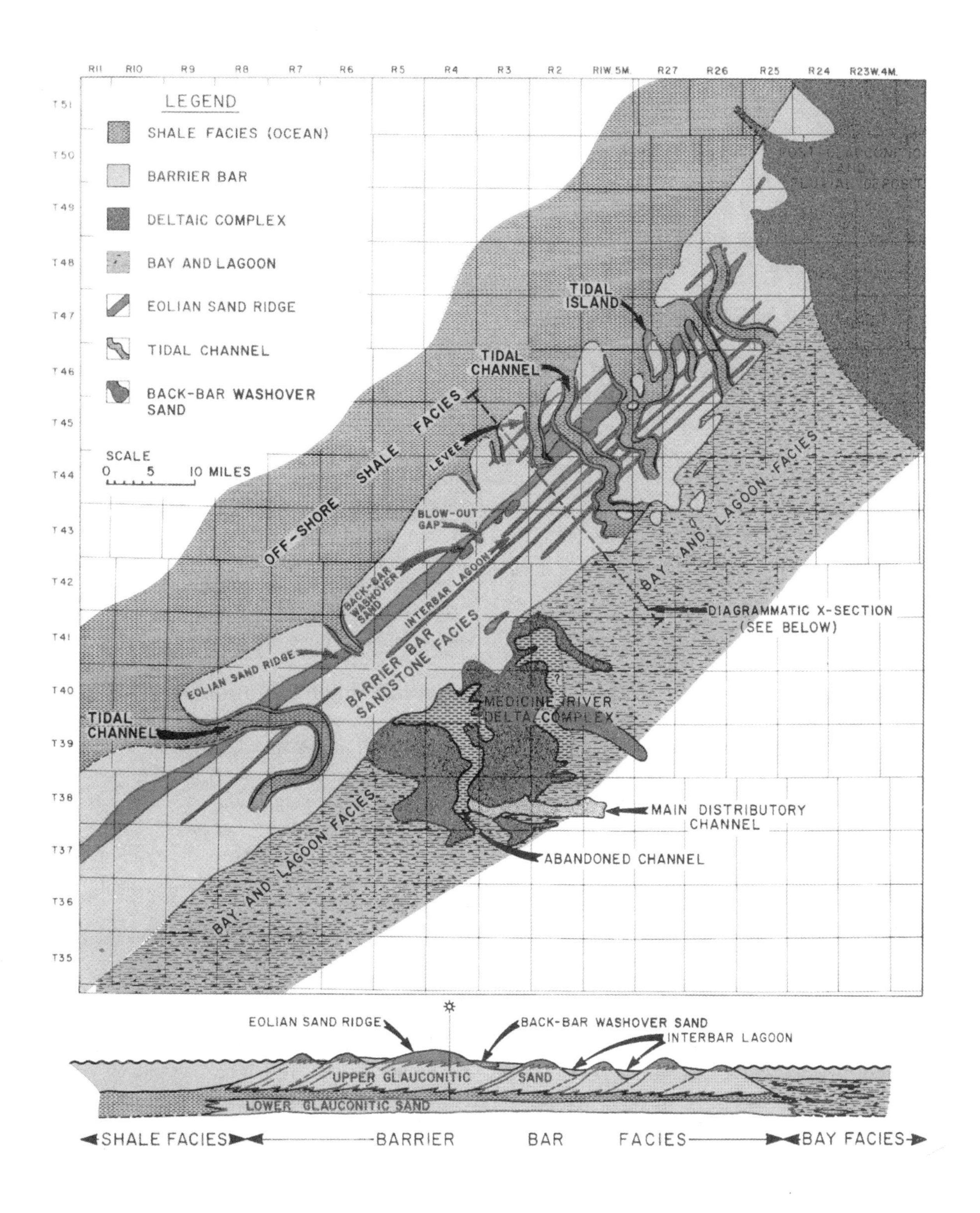

Fig. 3–10 Map of Hoadley Barrier Bar

Marine shale facies. The ocean offshore from the barrier bar deposition consists of fine-grained material. The total thickness of this unit is thinner than the barrier bar. Deposition is slower offshore, and fine sediments normally show more compaction. The shale is of very low permeability and forms part of the trapping mechanism.

Bay facies. In the bay behind the bar, a mixture of mud, seaweed, plant debris, and minor amounts of sand are deposited. This becomes coal and siltstone after burial and has a low permeability due to the fine grain size. It acts as a reservoir seal in a number of places.

Eolian sand facies. The wind-blown sand ridges feature fine- to medium-grained quartz and chert sediments. There is very little clay in this rock. This rock features above-average permeabilities and porosities. The presence of these facies can improve well deliverability enough to make a substantial difference in the profitability of a well.

Interbar lagoon facies. Interbar lagoons exist between eolian sand ridges. These low areas are good for vegetation growth. This material becomes coal after burial.

Back-bar washover sands. These sands lie immediately behind the eolian sand ridge in the interbar lagoon areas. These sands are built up as the result of large storms. Sand is washed across a low point on the bar into the lagoon area. These facies can be recognized by thin porous sand, which occurs between two coal seams or which is underlain by carbonaceous sediment. This is significant from a reservoir engineering perspective. Wells in this facies will normally produce at high rates initially; however, they are limited in areal extent and will deplete quickly.

Tidal channel facies. Major tidal channels will cut transversely through barrier bars connecting the ocean and the bay. This will effectively divide a bar into separate reservoirs. Smaller channels may not bisect the bar but produce barriers to flow. These channels are infilled with shale and do not normally feature carbonaceous material.

Levee facies. Levee deposits occur along the edges of tidal channels. In the Hoadley barrier bar, this sand is fine grained and has low permeability. Levee sand can be identified on logs by subdued gamma ray response and no indication of upward sorting or cleaning.

Deltaic sand facies. A large oilfield is located in the Medicine River Delta. Overall, the reservoir quality is poor. This is a thin sheet of mixed fluvial-, wave-, and tide-dominated sediments. This type of reservoir often shows poor continuity, because there are often abandoned channels. There is an associated gas cap on the updip side of the field. Virtually no pressure decrease was in the gas cap after 16 years of primary oil production, demonstrating very low permeabilities.

Distributary channel. This channel contains the Sylvan Lake gas field. It is about one mile wide and extends for about 15 miles. The facies have the distinctive bell-shaped gamma ray log response of a channel, which indicates a gradual decrease in porosity and permeability upwards.

Abandoned channel facies. When a distributary in the delta changes course, the abandoned spillway fills with fine-grained materials. These areas are barriers to flow and will often divide delta deposits into separate reservoirs.

Reservoir characterization

The advent of geostatistics and fractal mathematics, coupled with increased use of reservoir simulation, has greatly increased the demand and the techniques available to quantify reservoir characteristics. This becomes particularly important in EOR, which is very sensitive to reservoir layering. Effective understanding of reservoir behavior demands different lithologies be correlated separately.

For instance, use of core data from the Sylvan Lake distributary bar glauconite to create permeability versus permeability relationship in the Medicine River deltaic glauconite would result in gross errors. On the other hand, the permeability versus porosity relationships for glauconite distributary channels virtually overlay each other as shown on Figure 3–11. This data was from a number of channels separated by large distances.

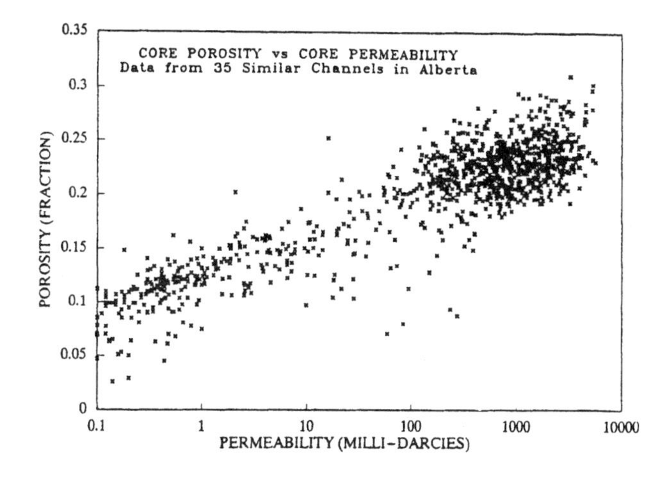

Fig. 3–11 Striking Similarity of Glauconite Channels from Geographically Dispersed Areas

Visualization tools

Most fields a reservoir engineer will oversee are not likely to cover an area as large as the Hoadley barrier bar. As described later in this text, correlating the continuity of porosity and permeability on a fine scale is important in determining reservoir performance. There won't be a paper available to spell out exactly how reservoir properties vary areally within the field. It will therefore be up to the engineer and a geologist to determine this information.

Contour maps are the traditional method of recording this data. These are powerful tools. The following sections describe two other visual approach tools.

Stratigraphic cross sections. This technique is commonly used by geologists to correlate data laterally. For a reservoir model, this often means a series of cross sections covering each row of the well pattern, as shown in Figure 3–12.[3] Geologists frequently build the cross sections, utilize log analysis, and then build a series of maps. Normally, geologists will describe differences in lithology or distinctive character on the gamma ray log to the various layers.

Structural cross sections. Logs from adjacent wells are pasted or taped to a large sheet of paper utilizing a depth datum. This technique is great for identifying how fluid contacts will affect production. It also shows where secondary gas caps form. An example of a structural cross section is shown in Figure 3–13.[4]

The following is a story demonstrating how a structural cross section can provide practical insight.

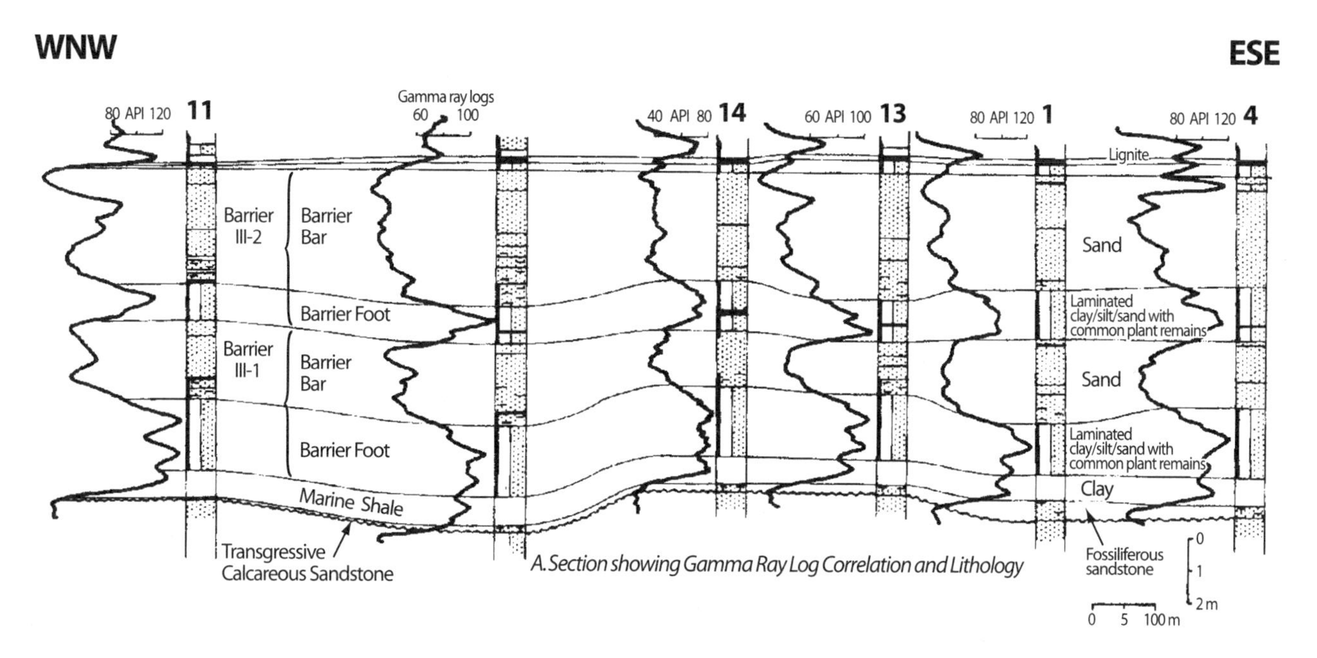

Fig. 3–12 Example of Stratigraphic Cross Section

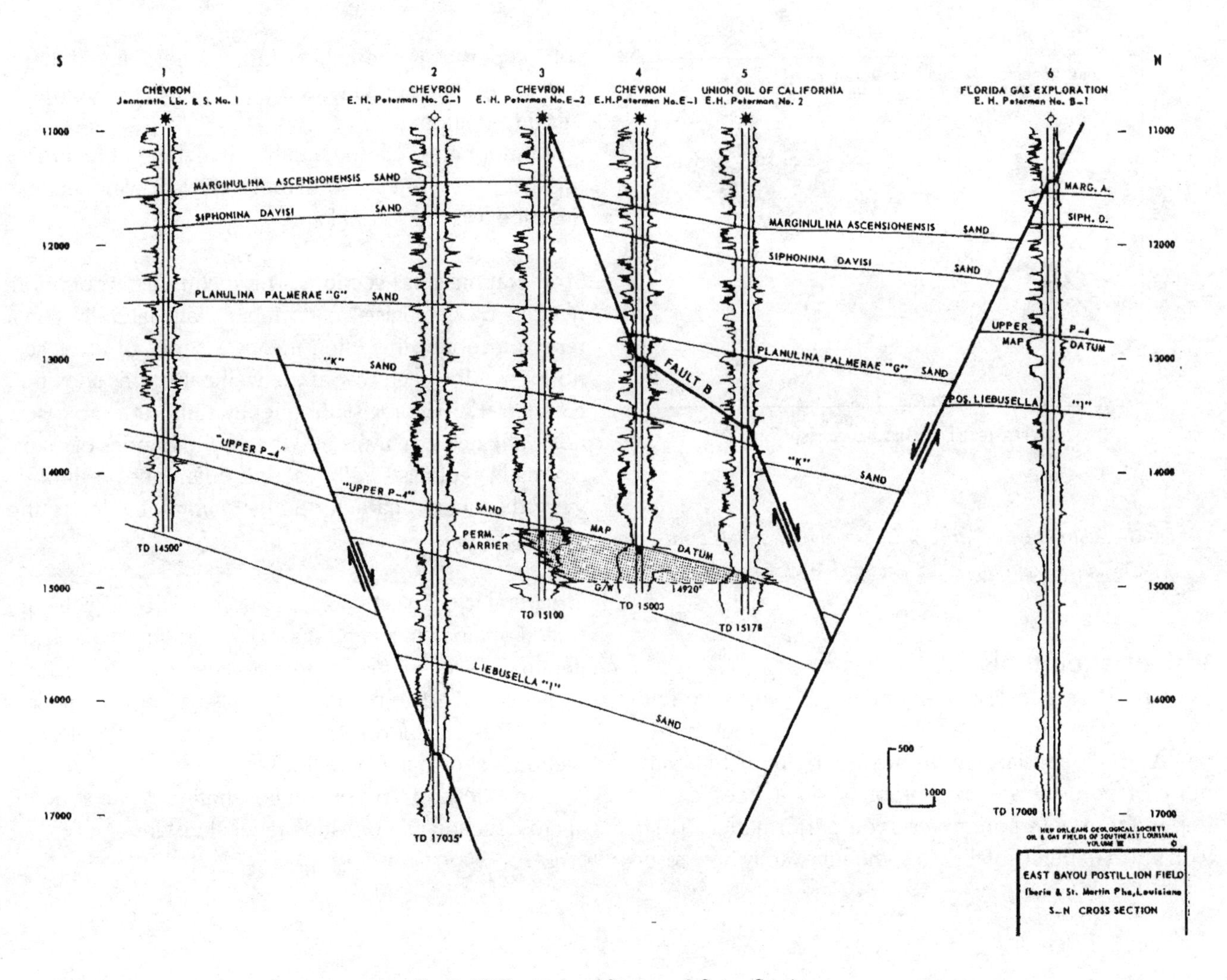

Fig. 3–13 Example of Structural Cross Section

Cretaceous sandstone completions. The author was the reservoir engineer for a gas field located in Northern Alberta that had been on production for an extended period. This sandstone was located immediately above a Paleozoic carbonate subcrop. Gas was contained in the Cretaceous sandstone and the underling Devonian formation. The lower zone had a gas-water contact and was highly permeable. The two were separated by thin shale, which may or may not have been a barrier. The degree of communication seemed to be controlled both by rock quality and drilling-completion practices.

During the mid-1980s, it was anticipated more deliverability would be required to meet contractual obligations. A number of infill wells were drilled to increase production capability. A preliminary completions program had been developed by one of the producing departments. The well logs looked something like the one shown in Figure 3–14.

Based on the log response, the completions engineer had assumed there was bottom water present. The gradual decrease in resistivity was attributed to a capillary pressure transition zone. The proposed completion called for only the top 30% of the zone to be perforated and no acid stimulation.

A reservoir engineer was asked to approve the completions program prior to implementing the workover. Fortunately, in the process of familiarizing himself with the pool, the reservoir engineer had made a structural cross section as shown in Figure 3–15.

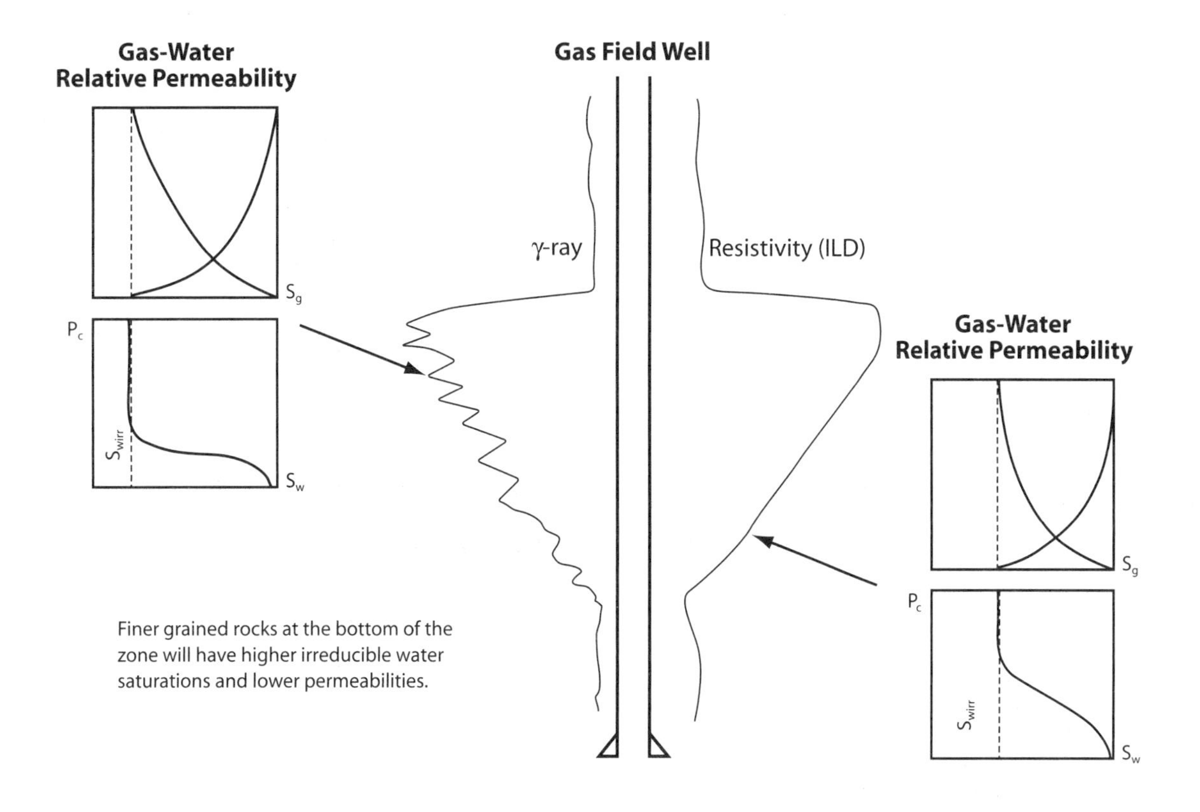

Fig. 3–14 Gamma Ray and Resistivity Logs, Northern Alberta Reservoir

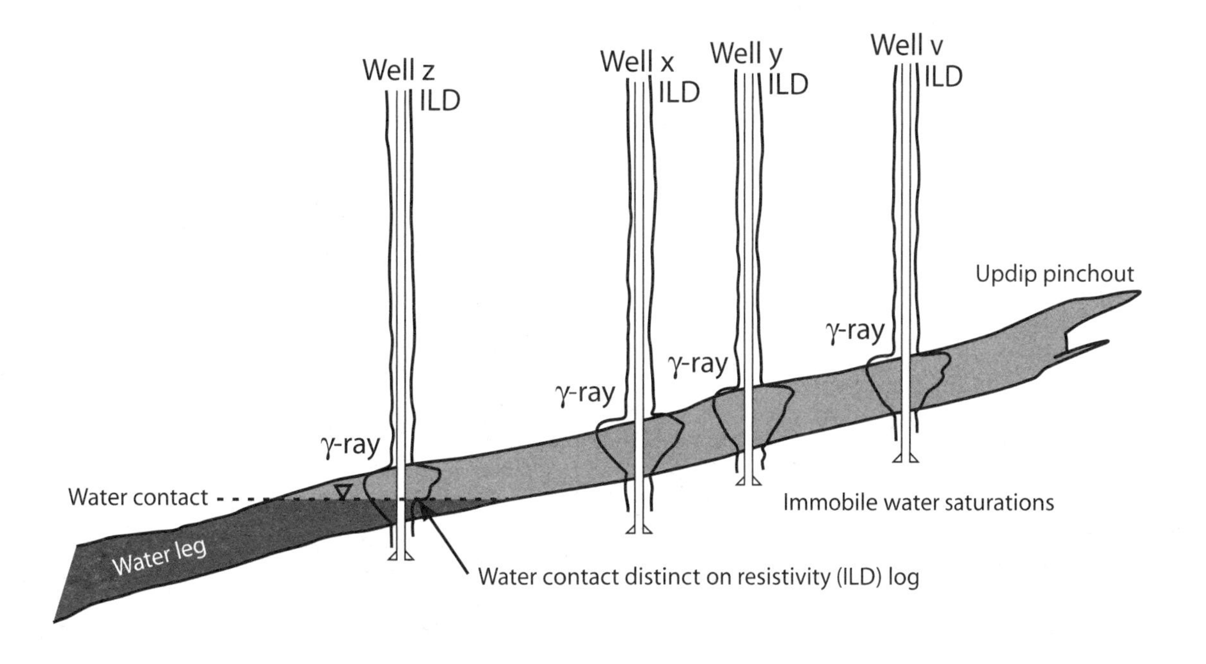

Fig. 3–15 Cross Section Resolving Actual Water Contact Location

The field had a downdip edge water contact. The decreased resistivity is due to finer grain size and the presence of clay. All moveable water had long since rolled down hill into the water leg.

The reservoir engineer realized edge water encroachment was a possibility. The material balance, which is discussed later in this text, did not show any water influx. To make doubly sure there was no water encroaching, the production histories of all wells in the field were reviewed. Downdip water encroachment would cause wells to progressively water out, starting from the downdip edge.

The review of production histories confirmed there was no water production and no edge water encroachment. Based on the previous information, plus core analysis, the reservoir engineer was able to identify that completing the entire interval would result in at least a 20% increase in well deliverability, with no danger of water production.

Engineering cross sections. If cross sections have not been provided, the author suggests reservoir and simulation engineers make their own. Further, if a set of cross sections is prepared by a geologist, study them carefully. In one case, the layers in a carbonate reservoir did not correlate. This wasn't completely the fault of the geologist. There was no correlation, yet his instructions were to develop a map with layers for the engineers. The layers were abandoned, and the simulation worked fine. In another case, there were two possible interpretations. Ordinarily, the geologist will pick one, which may or may not be the correct one. There was roughly a fifty-fifty chance of either. With better communication, there is another possible approach. The geologist or engineer draws up both geological interpretations, and the simulator is used to see how much, if any, difference there is between the production performances of the two interpretations. The simulator can provide a guide, which is supported by performance.

Another possibility exists: production history is insufficient for accurate interpretation. In this case, the best approach may be to run both cases and present the client or management with a range of answers.

Fence diagrams. The author favors fence diagrams in many cases. This provides a better visualization than cross sections. Frequently, history-match adjustments are made based on a feel for what is connected and what isn't. This can be particularly helpful in identifying shale lags. As with cross sections, a correlation of some log-observed property is normally made. An example of a fence diagram is shown in Figure 3–16.

Structural geology

From the foregoing, it is clear structure can be important in producing oil and gas. In addition to depositional models, a number of structural models reoccur in similar geological settings throughout the world. The basis of these models is the theory of plate tectonics.

The outer portion of the earth is regarded as a series of rigid segments floating on an interior core of liquid. The composition and structural properties of plates vary, depending on how they were formed and their location.

Continental plates are normally considered to be fairly thick and of relatively lower density. Deformation within continental plates is normally considered limited. Oceanic plates are denser, thinner, and continuously growing. Deformation within oceanic plates is also considered limited.

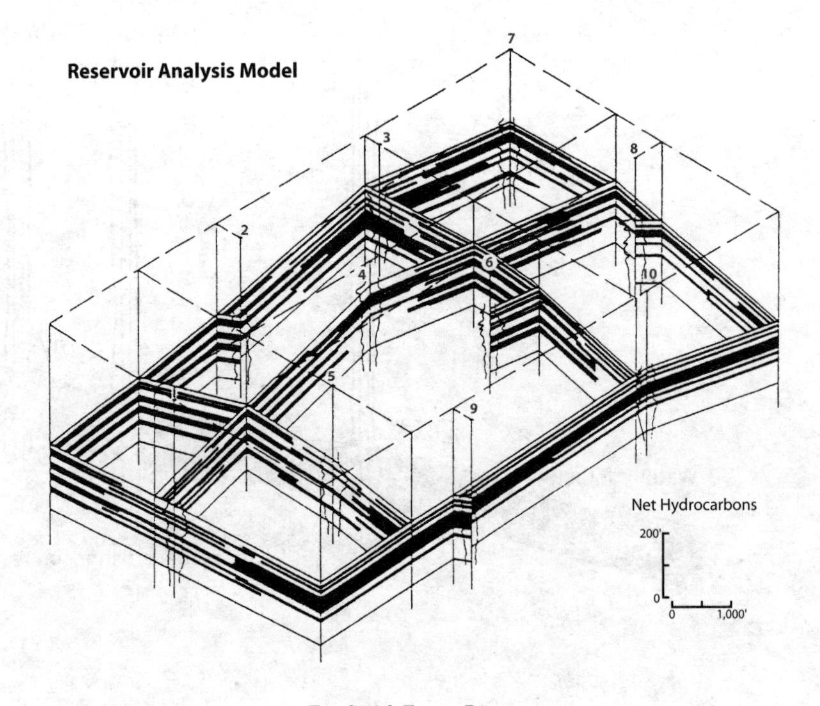

Fig. 3–16 Fence Diagram

These plates are not fixed, but rather are moving. The forces behind this movement are not well understood. Wherever these plates meet, a lot happens from a structural geological point of view. Some major boundary types are discussed in the following sections.

Divergent plate boundaries. These occur in the middle of the oceans. The ocean floors are continuously spreading. The areas where this spreading occurs manifest as mid-oceanic ridges.

Convergent plate boundaries. When continental plates converge, there is normally strong orogeny (mountain building). This is the root cause of the Himalayas. When oceanic plates crash into continental plates, the denser oceanic plate is shoved below the lighter continental plate. This results in an oceanic trench in the ocean. Mountains also result on the continent due to the strong compressive forces induced.

Transform fault boundaries. At a transform plate boundary, there is a lateral movement or sideswipe instead of straight-on collision. In reality, few plate boundaries consist 100% of either type of collision. Therefore, this reference refers to the dominant mechanism that takes place. This is the type of movement occurring in the San Andreas Fault off California. Earthquakes from this type of motion tend to be shallow in origin and severe.

A map of the boundaries of world plates is shown in Figure 3–17.[5] The edges have been defined by earthquake activity.

Plate tectonics are the source of most structural events in petroleum geology. Some of the more common environments and features associated with oil and gas reservoirs are outlined in James D. Lowell's *Structural Styles in Petroleum Exploration*.[6]

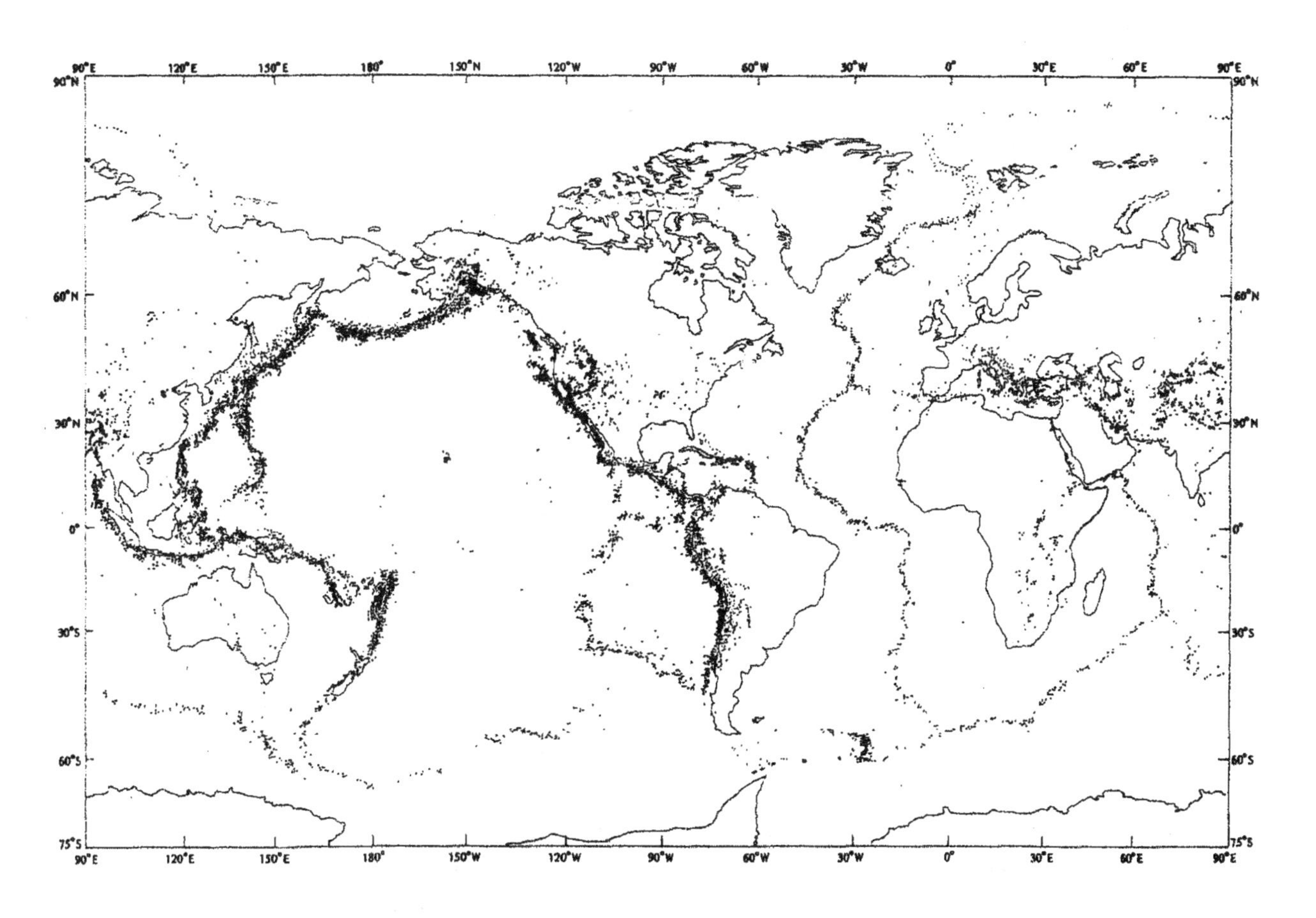

Fig. 3–17 Map of Worldwide Seismic Events

Western Canadian Sedimentary Basin

The majority of the Alberta basin is not strongly affected by structural geology. This is because most of the basin is located on a stable continental plate. Detailed discussion may seem academic; however, Calgary-based companies have been moving into international exploration programs. In addition, having said that the majority of the Western Canadian Sedimentary Basin is unaffected, the area near the Rocky Mountains is at the other extreme. The foothills of Alberta are described in the following sections.

Thrust faulting. The basic mechanism involved in the formation of the Western Canadian disturbed belt is thrust faulting. Failure occurs diagonally across competent beds with displacement occurring along a line of weakness. This line of weakness is called a detachment or a decollement. The basic mechanism is shown in Figure 3–18.[7]

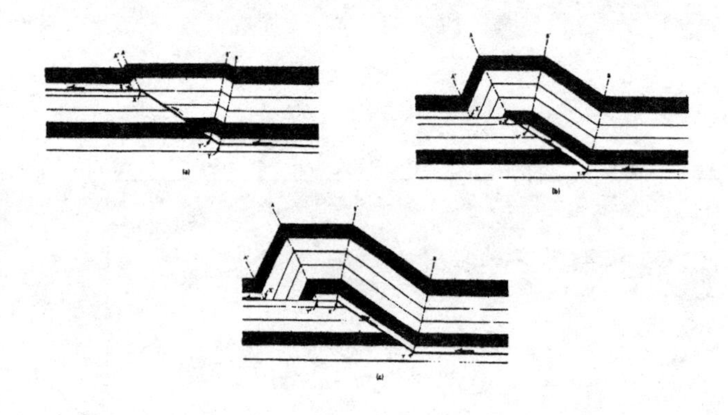

Fig. 3–18 Generation of Thrust Fault Structure

The basic mechanism can produce quite complex structures via various episodes of failure. Some of these are depicted in Figure 3–19.[8]

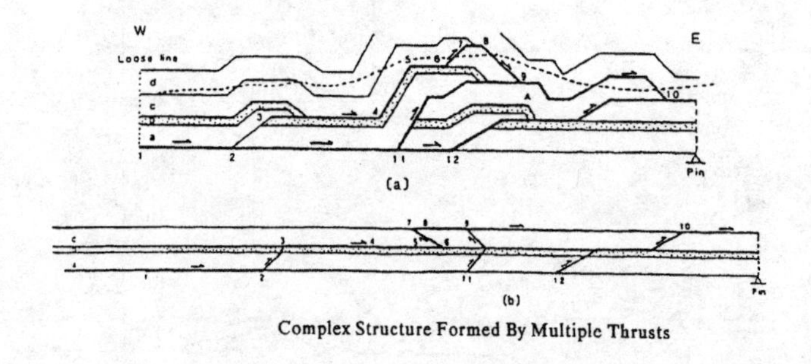

Fig. 3–19 More Complex Imbricate Structure and Palinspastic Restoration

Canadian Rockies. The following has been summarized from John Suppe's *Principles of Structural Geology.*[9] The southern Canadian Rockies are considered a textbook example of a *fold and thrust* belt.

The Canadian Rockies can be subdivided into a series of belts running parallel to the front of the mountain ranges. Each is characterized by a distinctive structural style, stratigraphy, and topography.

The easternmost belt is the Alberta Basin. This is the undeformed foredeep where up to 4,000 m of Upper Jurassic through Lower Tertiary clastic sediments have accumulated from the developing cordillera to the west. The timing of mountain building is not the same as found in most of the United States. Both were active in the late Jurassic and the Cretaceous; however, the United States activity ceased in the late Cretaceous, whereas in southern Canada, this extended into the more recent Paleocene and Eocene.

In the foothills belt, strong deformation is exposed at the surface, which displays steep dips, abundant fault repetitions of the stratigraphy, and tight folding. This does not extend to great depth. At between 3 and 5 km, the Paleozoic cratonic sediments are undeformed and overlie the Precambrian crystalline basement. Most of the faults flatten to detachment horizons in the Mesozoic or, in some places, the upper Paleozoic. A horizontal shortening to 50% of the original strata's length takes place along the detachment plane.

The western limit of the foothills belt is marked by the topographically abrupt front ranges, where the first Paleozoic rocks rise above surface. The Cambrian and younger carbonates form cliff faces, since they are more resistant to erosion than the Cretaceous sandstones and shales. Some of the best-known thrusts include the Lewis in Glacier-Waterton National Park, the McConnell in Banff National Park, and the Boule thrust in Jasper National Park.

The major thrust sheets of the front ranges are more widely spaced than in the foothills—about every 5 km instead of a spacing of 1 to 2 km. The difference reflects the much larger steps in detachment, which exist in the geosynclinal section of the front ranges. The thrusts generally step from low in the Cambrian to the Upper Paleozoic or Lower Mesozoic in a single step. Each of the mountain ranges or ranges running parallel to strike in the front ranges is generally a single thrust sheet. Hence, as one travels along the Trans-Canada highway, one may observe the same Paleozoic stratigraphic section repeated in each mountain range, with the strata always dipping to the west. The shortening in this section is not as strong, at about 25%.

Two changes occur in the main ranges, which are further to the west. The first change is structural. The detachment becomes deeper, originating from a thick sequence of Precambrian clastic sediments. The second change is stratigraphic. In the east, the Cambrian comprises thick, massive, carbonate sequences typical of the Paleozoic carbonate bank. In the west, the carbonate bank is terminated and is replaced by a deepwater shale basin. The basin contains carbonate submarine landslide breccias derived from the bank to the east. This new sequence is much less resistant to buckling than the stiffer bank carbonates. Very tight chevron folding, therefore, dominates the faces of the mountains in the main ranges. The rocks in this sequence have reached only low green schist facies of regional metamorphism. The shales have been deformed into slates.

Within each of the main belts described in the preceding, there are unique features. Figure 3–20 shows a number of these features found in different areas along the foothills.[10] Additionally, many of the common trap types are depicted. Identifying the style relevant to a field helps with visualization.

Structural analysis. Analysis of the foothills structures requires the use of specialized techniques. The building block for this analysis is the structural cross section, which is oriented at right angles to the strike of the thrust faults. These cross sections must be done to the same scale vertically and horizontally. The existing structural cross sections have been prepared on this basis.

Typically, the data used to interpret these cross sections consists of surface geology (Geological Survey of Canada or GSC), well formation tops, dip meter data, and seismic data. This information is plotted on a cross section, and an interpretation is made.

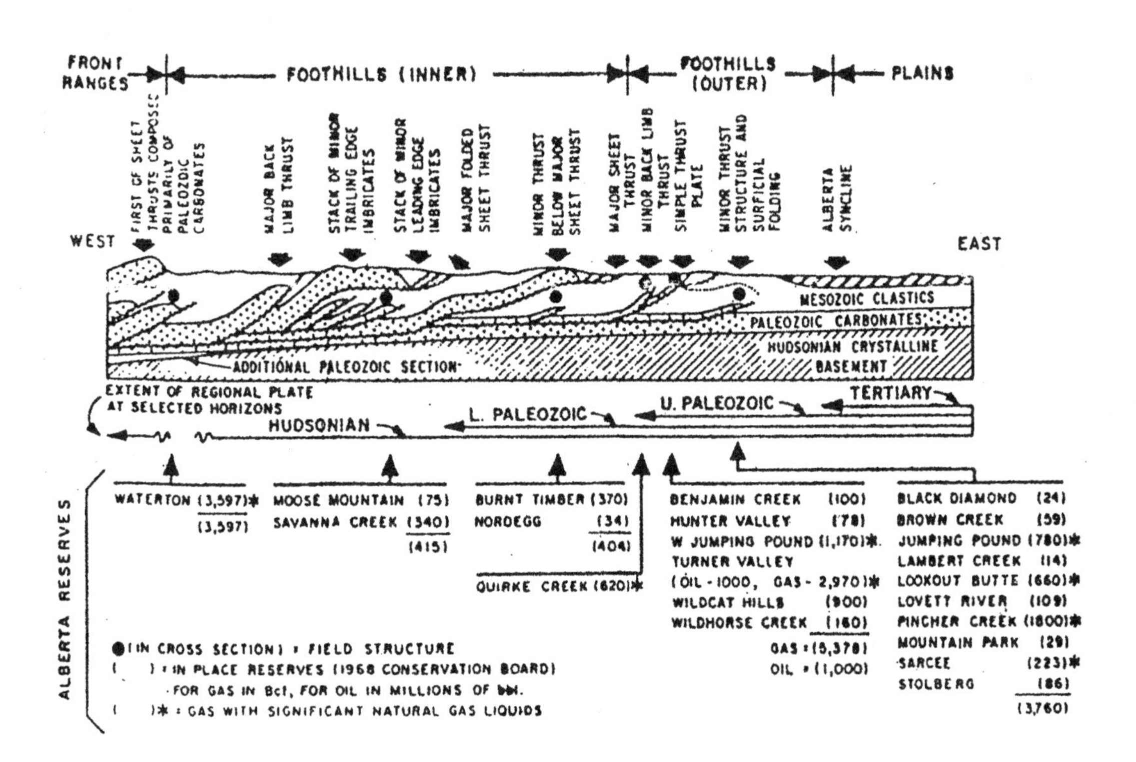

(Dahlstrom, 1970)—Schematic cross section showing zonation of the Canadian Rocky foothills with common structures, field types and reserves. Imbricate thrust zone-foreland detached zone break of figure 6-38 would be placed at the Front Ranges-Inner Foothills boundary.

Fig. 3–20 Foothills, Reservoirs, and Characteristic Structures, Western Canadian Sedimentary Basin

A more powerful version of the basic structural cross section is the balanced cross section. The use of a balanced cross section is predicated on a number of assumptions:

- The Precambrian basement extends, unbroken, beneath the foothills structure.
- The rocks east of the foothills are essentially undeformed.
- The foothills structures were formed in the later stages of the Laramide Orogeny, i.e., long after most of the rocks now preserved were deposited.
- There is no thinning or thickening of beds by flow, and, therefore, folding is concentric.

This single postdepositional (and lithification) deformation is important since it simplifies structural interpretation. The following may be excluded:

- pronounced unconformities
- structures growing during deformation
- substantial compaction after deformation

Therefore, conservation of volume can be used. From a cross-sectional perspective, area is conserved. A further simplification may also be made. Since there is no thickening, thinning, or plastic deformation, the original bed length will be conserved. The major part of the structure will occur through brittle failure.

The most rigorous method of testing the cross section is by palinspastic restoration. This is done by doing an area balance on the entire cross section for each zone or stratigraphic segment. The geometric acceptability of a cross section can be determined by *undoing* the deformation. Note this method will only generate physically possible cross sections, not necessarily the correct one or a unique solution. If tied into surface and well data, such a cross section will allow the structure to be more accurately resolved.

Structural geology is a separate specialty within the science of geology. Often, exploration in the foothills will be carried out by a separate group of exploration geologists who have specialized in structural geology at the postgraduate level.

Foothills carbonate reservoir. The applications of these techniques can have an important impact on reservoir understanding. In one instance, the author worked on a foothills carbonate reservoir similar to the diagram shown in Figure 3–21.

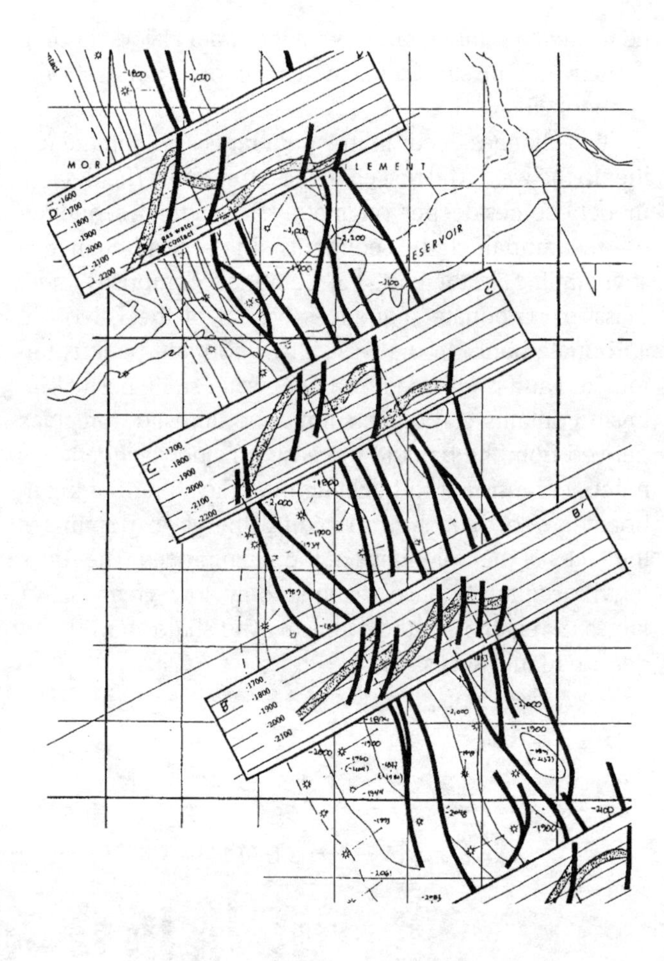

Fig. 3–21 Contour Map with Representative Cross Sections

In this case, miniature cross sections were placed on an areal map. From this plot, it is obvious that communication throughout the field is discontinuous due to the large displacements of the thrust sheets.

This diagram was prepared for an economic evaluation. Because of this interpretation, production decline reserves were used in favor of a material balance calculation (*P/Z* plot). A previous economic evaluation, prepared by another company, had used material balance. The other company was forced to reevaluate their reserves estimate upwards after only a few years. This was despite more than 20 years of production history and a large number of wells.

By assuming a material balance, they implicitly assumed communication across the entire field. However, since the reservoir is really a number of discrete sheets, undrained sections may be found and drained. The other company's evaluation failed to recognize considerable upside potential. Correct structural interpretation will often dictate the most appropriate technique of reservoir analysis.

Familiarization

Some engineers are reluctant to do cross sections, since they feel geologists should do this. In the author's experience, this attitude can lead to poor results, similar to what almost happened in the previous anecdotes. Here are some good reasons for engineers to do their own cross sections:

- On smaller pools, generating a request for a cross section to the geology department can be more work than actually doing it.
- If the geology department is busy, they may not be able to accommodate timetable requirements.
- Doing a cross section, armed with colored marker or colored pencils, is not only an enjoyable break but also a necessary process to absorb large amounts of data.
- With improved familiarization, the reservoir engineer will be in a better position to understand the same work done by a more qualified geologist. This may lead to some animated discussion, but it can also result in better understanding by both parties.

In the author's experience, geologists don't usually object to someone making his own cross sections. The author's intent in making cross sections is primarily as a visualization tool. The tricky part, which is done by a geologist, is adding the interpretation of environment. Best results occur when the geologist and engineer work well together, which is one of the reasons team building is emphasized as part of the new management concepts.

Geological and Engineering Perspectives

Reservoir engineers will spend a great deal of time interfacing with geologists. Knowing a little about geology will help immensely since it allows one to understand the language of geology. This will expedite communication.

However, the reservoir engineer and the geologist have different priorities. Some judicious screening of the information they provide is necessary. Typically, geologists are primarily concerned with building an accurate conceptual exploration model. Their style will strike most engineers as an extremely broad-brush approach. Fluid flow properties will be of secondary importance to the geologist. In contrast, for an engineer, the facies are academic. Engineers are most interested in flow properties, and this can lead to miscommunication.

A reservoir engineer should be aware of the different specialties within geology. Exploration geologists typically are least familiar with flow properties. At the other extreme, there are development geologists who are accustomed to making detailed continuity correlations between wells and are extremely knowledgeable about reservoir engineering. They will likely be experts at reading logs, and some can analyze pressure transients.

Often, it is necessary to educate professional colleagues about specific requirements. Exploration geologists normally have more questions than development geologists have. Don't be reluctant to ask them for an explanation of their jargon. Similarly, they will be looking to the engineer for answers outside their main expertise. This typically involves gathering data, interpreting tests, predicting performance, making economic analyses, and continually updating reserves.

As outlined earlier, exploration involves overcoming a lot of failure. It is a psychological necessity in the exploration game to be optimistic. To engineers, this can sometimes appear to border on outright fantasy. In the earlier stages of development, particularly in calculating reserves, geologists exhibit human traits and will have highs and lows. A little tact will be necessary as these hopes are realized or, more often, destroyed. In the early stages of a field's development, it will be necessary to use reserves guidelines for areal assignments, etc. to maintain peace and order until more information is gathered.

Geologists deserve a lot of respect. They have to know a considerable amount about conceptual models, understand regional geology, and be able to apply this information to their working project. Many people say a good geologist requires imagination. This is required for effective pattern recognition; it is not a matter of fantasy.

Geophysics

As a rule, reservoir engineers normally have much less to do with geophysicists than geologists. Most exploration departments usually pair geologists and geophysicists according to area. Most communication tends to be directed through the geologist.

The author found geophysics is far less precise than anticipated. The geophysicist's job is interpretative. Sections are correlated by finding distinctive responses on seismic line trace maps. Different seismic lines are then correlated where they cross. This is analogous to closing a loop on an engineering level survey. In general, geological tops do not correspond directly to distinctive seismic reflections.

The seismic line trace maps are based on time in the vertical scale. This does not correlate directly to vertical distance, since average velocities vary above the horizons chosen for analysis. Structure maps prepared by total travel times are not generally accurate. To achieve greater accuracy requires two things:

- Synthetic seismic responses. These are made from sonic logs and are intended to reproduce the characteristic pattern seen on seismic line traces.
- Velocity surveys. These are taken by measuring the travel time from surface to a recorder run on an openhole wireline.

With this information, a graph correlating travel time versus depth can be made.

It is more common to see isochron maps. This measures the difference in time between two horizons. It is assumed one of the horizons is level. For instance, if the upper layer is assumed level, shorter times indicate pull up or structural highs. This technique is a relative one and can be more accurate.

A minimum thickness of formation can be seen, which is related to the wavelength of the seismic signal. Often in Western Canada, it is impossible to distinguish the specific production horizon.

Make sure the expectations from geophysics are reasonable. When dealing with geophysical data, ask the geophysicist how accurate the data is.

Petrophysics

Simulation input often depends heavily on log analysis input. In some cases, one will have a recent and complete computerized log analysis for the field or project. Reality is less than this ideal, and data is frequently not ready for the simulator.

In the author's view, everyone running a simulator should read logs and be able to do quantitative calculations. This is not quite the same thing as being an expert log analyst, but it is important to have a feeling or sense of what data is right and what degree of certainty should be attached to the data.

Log analysis

The author was fortunate to be involved in a major gas contract reserve negotiation. He supported the geologists in the log analysis. Not only does such an experience provide a lot of practice, but also the process of arguing the points on a large variety of wells and zones makes one sharper and more confident.

Later, while working for a large international consulting company, the author had the opportunity to learn computerized log analysis. This was a sophisticated program and required a thorough knowledge of many different calculation techniques. Quantitative log calculations involve many judgments and a good knowledge of individual tools from the various manufacturers. The point is log analysis involves a number of judgment calls, which can have large influences on the calculated results. Just as in reservoir simulation, some people are good at twiddling the right input parameters to come up with realistic solutions.

A good log analysis report will indicate the technique used and why. It will also indicate where the calculations could be wrong or what level of probable error exists. Not all reports are this detailed.

Data required

The data normally expected from a log study includes:

- Net Pay
- Porosity
- Water Saturation
- Permeability
- Lithology
- Capillary Transition Zone
- Water-Oil Contact
- Gas-Oil Contact
- Gas-Water Contact

This is fundamental input. It frequently controls the pore volume in conjunction with the mapping. Lithology is often determined from log responses as well as gas-oil, oil-water, and gas-water contacts.

Special core analysis

Typically, the better log analysts will obtain capillary pressure data and will use this for tuning their log calculations. In the author's experience, they will not screen relative permeability data and will not normally go through this data in tuning their log calculations (the analysis does have end-point or connate water saturation information). Log water saturations usually should be matched to the end-point for relative permeability data. In some cases, this will be somewhat different from the limited number of special cores analyzed. Further, if the reservoir has multiple facies, one can apply limited relative permeability data to other facies in the reservoir, which have different porosities, permeabilities, and water saturations.

A number of common log relations can be used and may be helpful. The first is a Buckles Number, which is Porosity × S_w. In certain reservoirs, such as the Keg River reefs of northwestern Alberta, connate water saturations will follow this curve as shown in Figure 3–22.[11]

The data below 200 will be water-free production and data exceeding 400 will typically produce water. This technique can be used to calculate water saturations or identify where water production will occur.

Sometimes it is possible to make other correlations. Figure 3–23 shows a relation between S_{wc} and permeability. Sometimes there is no correlation at all, such as shown in this figure. The correlation shown, in addition to Buckles Numbers, is formation and area sensitive, and one has to use trial and error to find out where it applies.

Screening on special core data will often tell if these are typical samples or whether there is a systematic bias in the data. Such an example is shown in Figure 3–24.

Log analysts will often use conventional core data, so one may be able to obtain this data from them. In addition, they may have compared the computed porosity histogram with the log-calculated histogram. This type of analysis, shown in Figure 3–25, will show skews in data and is a good indication of quality control.

Difficult analyses

In some areas, log analysis is inherently difficult. The author performed an extensive (log) study in Bangladesh in one such situation. The area featured a number of heavy minerals and generally immature sediments. Due to micas and potassium feldspars, the logs were quite radioactive, and the sands would not appear on the gamma ray as normal sand or limestone would. The heavy minerals masked gas crossover effects.

The formation waters were very fresh, and there was not much resistivity contrast with the gas. There was a reasonable amount of core data; however, the matrix

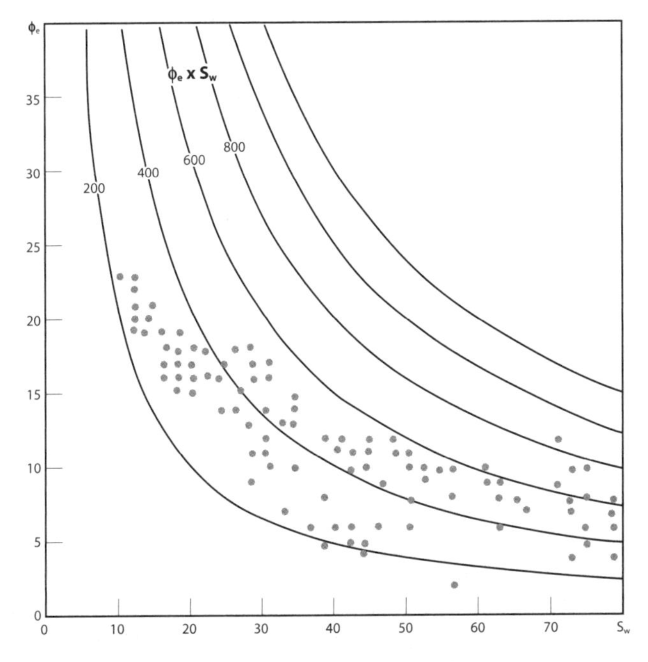

Fig. 3–22 Buckles Number Plot

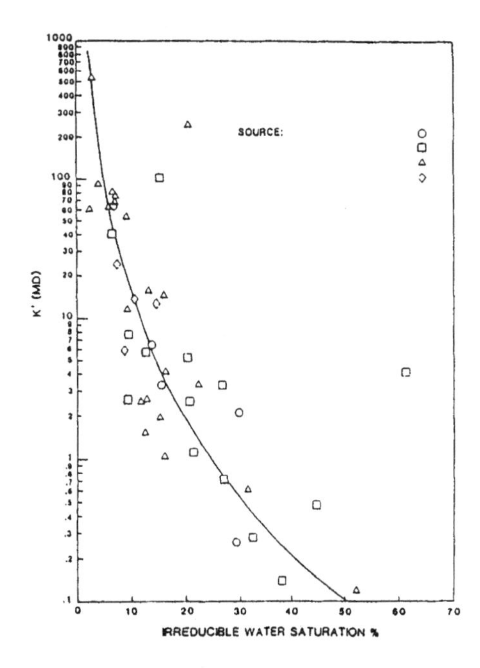

Fig. 3–23 Correlation of S_{wir} and Permeability

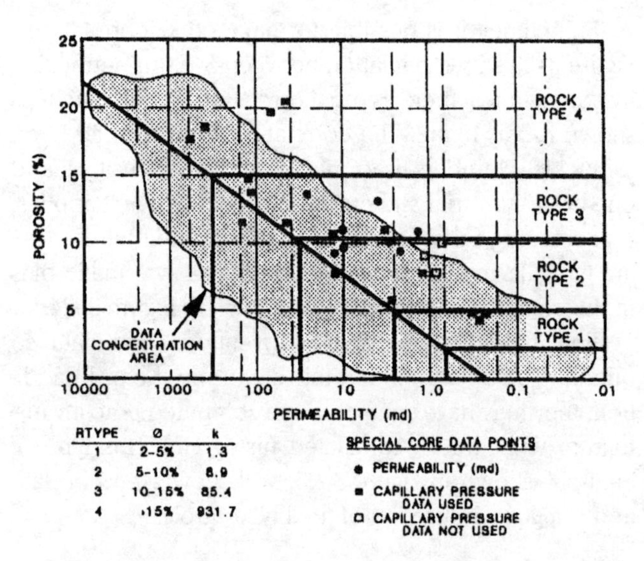

Fig. 3–24 Use of Permeability vs. Porosity Cross from Conventional Core to Screen Special Core Data Samples

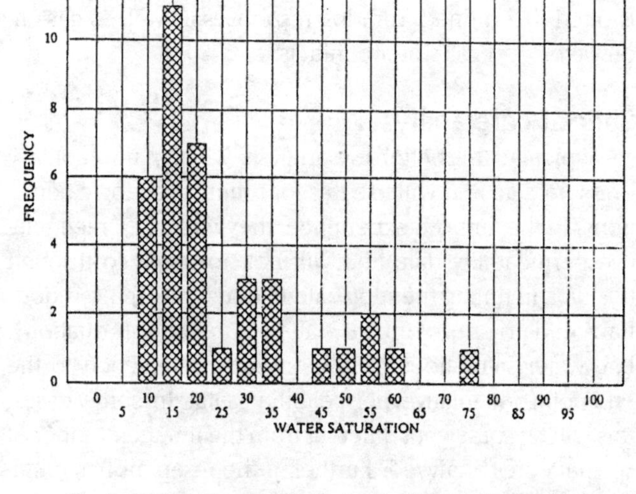

Fig. 3–25 Bar Graph (Histogram) of Water Saturations from Relative Permeability Tests

density of the rock was quite variable. Since the calculated water saturation depends on the porosity, this made both the output porosity and water saturations quite variable. In a situation like this, no amount of analysis will correct these problems. It is necessary to work knowing the variability exists.

Qualitative indications

Much of what can be gained from log analysis is qualitative in nature rather than quantitative. On one reservoir where the author worked, they had trouble matching reservoir performance. Looking through the logs, a number of apparent high permeability streaks were discovered that occurred at varying points through a carbonate zone. The logs were run in 1967 and 1968 and were done with early sonic logs. These tools predated borehole compensation and were prone to cycle skipping. These sonic skips can sometimes be useful; they are often indicative of fracturing. Modern digital electronics rarely miss the first arrival, and this is now uncommon. The geologist who had worked on the maps was not familiar with logging tools, and, not recognizing the cycle skips for what they were, had mapped high porosity and permeability streaks. Therefore, the porosities mapped could not be correlated with any certainty.

Permeability transforms

One area can be quite misleading—the use of permeability transforms. In one case, well test data did not support the permeabilities calculated by transforms for wells in a particular area. The problem wasn't really a log analysis problem *per se*. Rather, the geological model was not correct. The permeability-porosity relations generated were based on erroneous lithology correlations. Grouping data this way did not indicate any unique relations. Had the proper rock lithologies been input, then different rock properties would have resulted. In this sense, log analysis programs are like a reservoir simulator. They will only test the scenarios input and are quite incapable of independent intuition.

Gas-oil and gas-water contacts

Contacts are often picked off logs. Gas-oil contacts can be particularly difficult. The crossover effect on the logs can be quite different depending on the invasion profile of the mud filtrate. This is because the depths of investigation of the neutron and density tool are different. The neutron tool is a pad device and has the shorter depth of investigation. Note this problem typically occurs in lower permeability formations. The higher permeability formations develop a filter cake more quickly and normally have lower filtrate losses. There is only a slight resistivity contrast between oil and gas, which is not always discernible. Therefore, contact depths will have some variation.

One unusual occurrence the author has found with difficult log picks is some geologists will often go through all the data and make their best pick. Having done this, the matter is cast in stone. It becomes *the* gas-oil contact. Normally this will move a meter or two in history matching; however, there is no point discussing the matter once the contact has been set. Geologists use a different thought process from engineers, and debating the unknowable will simply annoy them.

This attitude is necessary in their work. Geological supervisors are looking for geologists who are confident in their interpretations. They have a different starting point from engineers. Since geological work is almost all interpretational, they assume uncertainty. Their certainty is understood to include *subject to other data being subsequently obtained.* Admitting doubt indicates to their bosses they knew they had not bothered to do as complete as job as they could. Geologists are often happy to agree to variations in interpretation without any ill feeling towards each other. Engineers want to be more precise and are much more likely to take offense if the other person can't agree.

Gas-water is similar. With saline formation waters, the contrast between gas and water is quite high. However, as described earlier, fresh water may make the contacts difficult to calculate. In this case, the location of the contact may have some uncertainty, and varying invasion profiles may obscure the location of the gas-water contact.

Small-scale geology: layering

Simulation is often used for waterflood design. It has been long recognized that layering has a strong effect on waterflood performance. Layering frequently has an effect on reservoir performance. An example of this is a high-permeability conglomerate overlying tighter sandstone. In the past, it has been suggested gas (or perhaps oil) will feed vertically into the higher permeability sand. Another example is basal chert conglomerate, which conducts the oil to the wellbores and greatly enhances productivity.

Why does this chapter address so much geology when engineers do the majority of simulations? In the author's experience, engineers have developed a much better sense of flow than geologists have. There are, of course, some outstanding exceptions. Nonetheless, the importance of these matters is often overlooked in constructing reservoir models. Part of the purpose of this book is to bridge the gap between geology and reservoir engineering application.

Thus, layering presents a number of problems:

- How is it determined?
- Is there a procedure to be followed?
- How does one decide the number of layers required?
- When does one acknowledge the layering cannot be determined?

The author has a number of stories on the mischief layering can cause.

Vertical permeability is often a difficult decision and the use of k_v/k_h ratios can cause some serious problems. A methodology is suggested. Shale lenses are also relatively common, and this affects vertical permeability.

Finally, faults are a common geological occurrence. It is a virtual certainty an engineer will encounter them in reservoir simulation. What effect will they have?

Determination of layering

In the preceding chapter, a case was made for engineers becoming familiar with log analysis. This is the key starting point. In the broader picture, it is necessary to go beyond the well level and examine the reservoir as a whole. A number of techniques can be used that are discussed in the following sections.

Dykstra-Parsons plots. Applied Reservoir Engineering & Evaluation Ltd. (ARE) made it a policy to process all core data (core permeability versus core porosity) on its reservoir studies. The data is plotted on a group as well as on an individual well basis. In addition, Dyskstra-Parsons plots were made on a group and individual well basis.[12] Figure 3–26(a) shows a typical well plot. Figure 3–26(b) shows the typical range of Dykstra-Parsons ratios (V).[13] The data was tabulated as shown in Table 3–1, to indicate the range and average of V.

The purpose behind this process was to spot areas where there may be a difference in lithology not obvious from examining logs or core descriptions. In Western Canada, all of this data is available digitally, and, with the use of a standardized spreadsheet, this is not an onerous amount of work. Strangely, engineers normally make these diagrams. This is in contrast to cross sections.

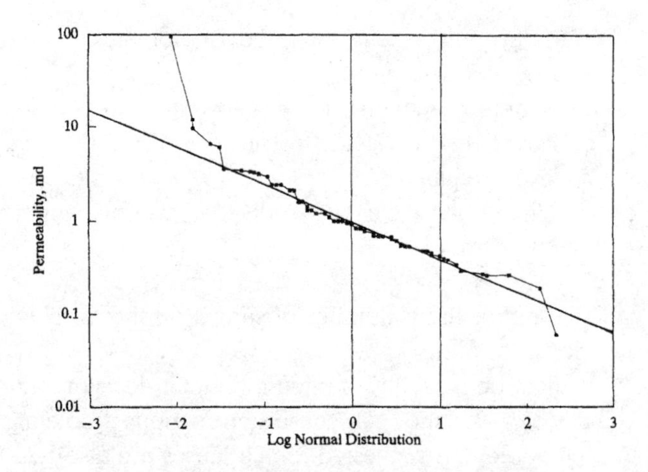

Fig. 3–26a Dykstra-Parsons Plot of Well Data and Typical Range of Data

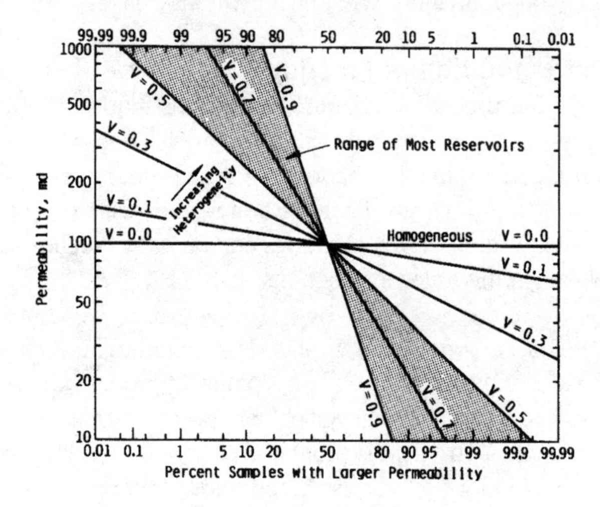

Fig. 3–26b Dykstra-Parsons Plot of Well Data and Typical Range of Data

Table 3–1 Common Environments and Features

Well	V
Well 1	0.70
Well 2	0.72
———	
Well n	0.75
Average	0.73

Mission: layering. One of the worst problems in the author's experience is the geologist who is given the mission to layer the reservoir. The conceived mission was to provide layers for the cake. To make matters worse, the geologist was given an insufficient amount of time to do the layering, and, of course, the resultant layers were total nonsense. This is an important time for critical thinking. More importantly, are the layers provided consistent with the geological model?

In the end, the rushed geology produced a failed model. Due to a limited production history of only two years, the problems with the model never become apparent during the history match. From a commonsense perspective, it was obvious the final predictions were not consistent with expected well deliverability. At this stage, the model had to be largely redone. The geology needed to be done again, along with the history matching, tuning, and the predictions. Some items did not need to be repeated, such as the PVT properties. Regardless, improper geological preparation can waste enormous amounts of engineering time.

Permeability. One of the difficulties one now faces is changes in lithology do not necessarily correlate directly with permeability. One can have variations in permeability within a single lithological unit. Conversely, different lithologies can sometimes have the same flow properties.

This may seem ridiculous; however, the author has seen separate reservoirs mapped based solely on changes in lithology. One case in point related to an unconformity in which porous and permeable Cretaceous sandstone overlay a porous and permeable Mississippian carbonate. Of course, both zones were in direct communication. Nevertheless, the company was about to make an application to the government based on two separate reservoirs. There are some direct methods to approach this.

Layering technique of Testerman. This technique has been around for quite a long time. It was originally developed in about 1962. The paper was published in the Transactions of the SPE of AIME.[14] The method takes a column of horizontal permeabilities at varying depths and determines at which (single) point there is a statistically significant difference in reservoir properties. To develop a series of layers, the method must be applied to the preceding splits. Therefore, the technique will not provide one with a complete layering sequence. Despite the unusual way in which it is implemented, ARE has used this technique with good success as shown in Figure 3–27.

The example in Figure 3–27 is from a single (channel) lithology and demonstrates the point regarding variations in permeability within a lithological unit.

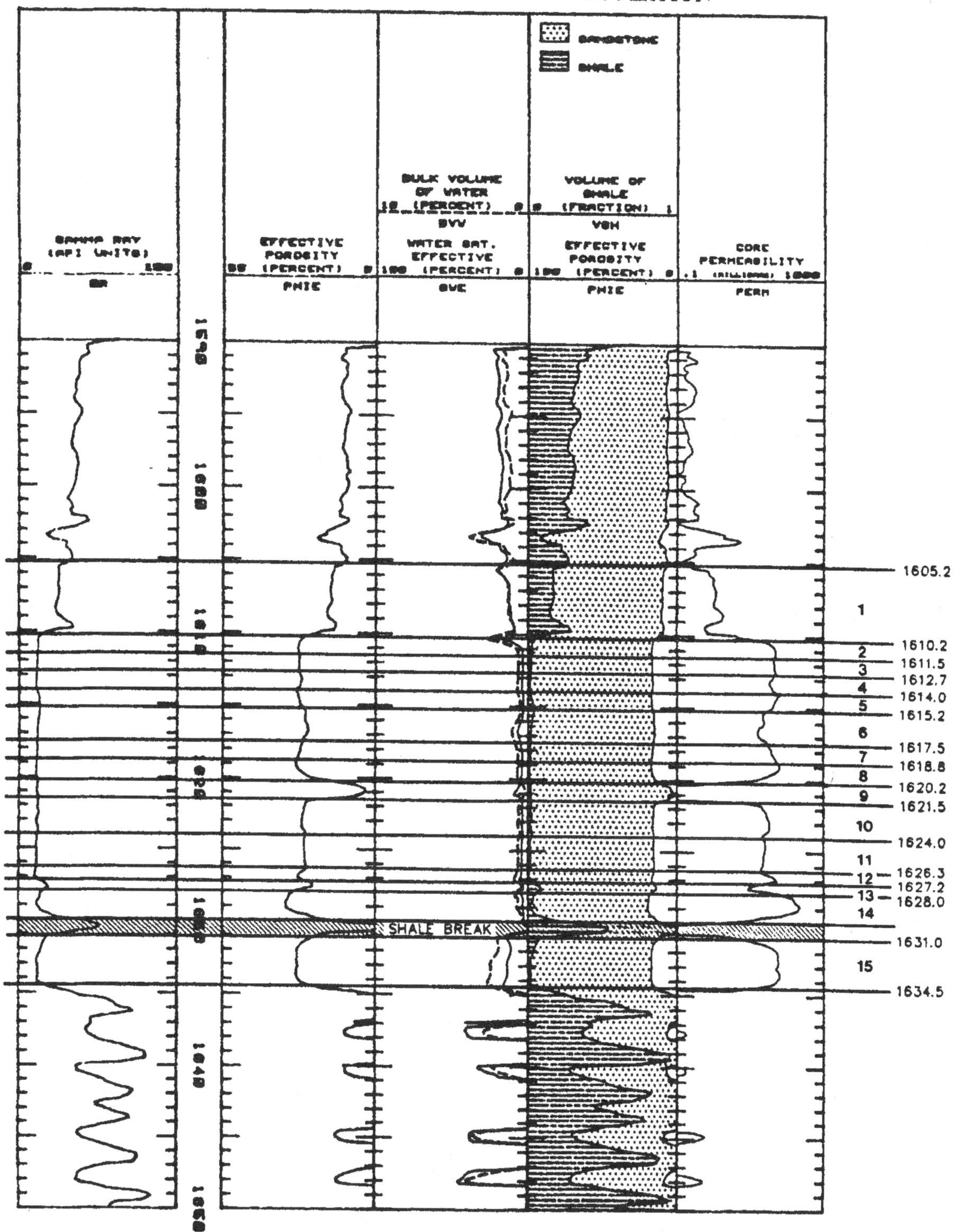

Fig. 3–27 Layering Developed Using Testerman Technique

View core. One of the best things an engineer can do is to look at the core. On one study, I invited a geologist to take me up to the government core storage center and view the core. When we got there, he did not know his way around the building, and I realized the layering could not be correct. The reservoir in question was a Devonian carbonate reef. These reservoirs have tremendous variability in lithology, which cannot be determined from logs. He eventually admitted he had prepared the layering and maps without looking at the core. He had relied, to some extent, on notes written by another geologist. In truth, this can be a reasonable approach, although most people would be innately curious and want to see for themselves.

Many companies use a particular consultant in Calgary who has evaluated almost every one of these Devonian reefs (although no consulting report had been prepared in the case of this unit).[15] At times, an individual simply must have the requisite experience that cannot be obtained without considerable research and development. In the end, the reservoir was too complicated, and detailed input of lithological variation proved impossible; however, the simulation was successful in spite of these problems.

Reservoirs that defy description

In the final analysis, a reservoir may defy detailed description. This can represent:

- extreme complexity
- lack of core data
- limitations of log analysis
- random processes such as dolomitization
- low well density
- incomplete penetrations

This can be frustrating. A search of the literature yields little guidance on what to do. This is inherent in the paper selection process. Someone finds a "neat" correlation that works and reports it in a paper, and, because it's "neat" and worked, it is published. The petroleum literature is short on describing failures. This is quite different from other industries, such as mining, where a mine failure almost guarantees a paper about the cause.

In these cases, an engineer may be forced to do some gross averaging. In the final analysis, this will reduce the chances of success in modeling; however, it doesn't eliminate the chances either. ARE has completed a number of projects where bulk properties were estimated.

Where does this leave the engineer? In the author's opinion, it means providing a realistic appraisal of what may be expected. As an employee and as a consultant, we are obligated to let the boss/client know the risks in completing a study. Management is not receptive to "maybe." Yet, simulation is no different from drilling wells and is much cheaper. Not every well works out. Overall, the margins on individual successful wells are so large that a percentage of failure is acceptable.

Some reservoirs, which defy deterministic description, are good candidates for geostatistical analysis. This is discussed later in the text.

Vertical permeability

Typical core samples indicate a range of permeability 0.1 to 0.3 times horizontal permeability. An example of how such ratios are estimated is shown in Figure 3–28. Such ratios, however, typically result in overstating the actual larger scale effective vertical permeabilities. The reason is related to permeability averaging. An arithmetic average should be used to determine well productivity, since layers flow in parallel. Vertical permeability should be averaged in series. The appropriate averaging formulas are shown in Figure 3–29. The series method results in a lower average vertical permeability.

Layers tend to stabilize vertical flood fronts and destabilize horizontal flood fronts. Hence, the effective average k_v/k_h will be lower than indicated by individual point measurement of k_v/k_h. Accurate averaging of the vertical permeability should be done with detailed core analysis. Sometimes there is no core data available from the well being modeled. In some cases, there is also relatively little data from the nearby offsets. In this case, a multiple can be estimated from a distant, offsetting core. Although this may sound weak, it is better than simply guessing. As a rule, realistic multiples using the complete core will be lower than the commonly assumed 0.1.

Incidentally, coning simulations have a reputation for overestimating the amount of water coning and/or gas coning. In the author's opinion, small-scale stabilization is the cause of these overestimates. To include such small-scale detail means returning to small-scale modeling. Local grid refinement does not give enough detail to detect this without some vertical permeability averaging.

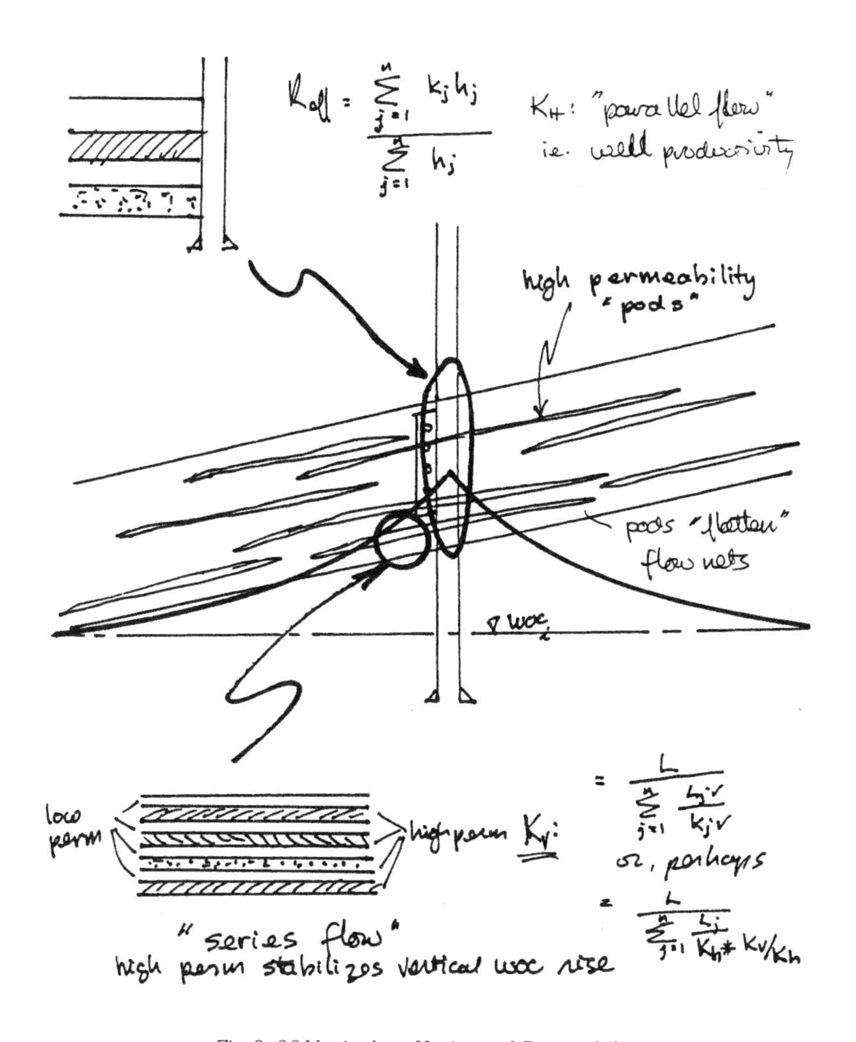

Fig. 3–28 Vertical vs. Horizontal Permeability

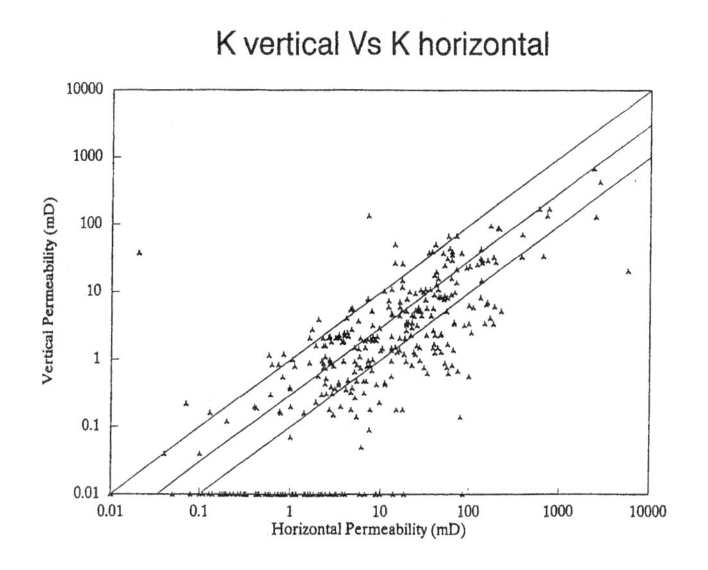

Fig. 3–29 Methods of Averaging Permeability

Faults

Faults are quite common in some environments. This is not typically true of Western Canada, however, since most of the prairies are relatively stable geologically. The question is whether they are sealing or nonsealing. An informal debate on this topic accomplished little except a rumor that Chevron had done some large study and concluded 50% of the time they were and 50% of the time they were not. The classic reservoir engineering texts are almost silent on this issue. Material exists in the geological literature; however, this was relatively scant until recently. From a geologist's perspective, their primary interest relates to trapping.

Trapping mechanisms. The majority of the geological material relating to the issue pertains to whether there is a trapping mechanism for oil or gas. Reservoir seals are shown, in general, in Figure 3–30.[16] This diagram shows that reservoirs can be sealed laterally, at the bottom and at the top. This diagram does not show fault seals. In a variety of situations, faults may or may not be sealing, as shown in Figure 3–31.[17]

Fault traps. Following is material extracted from Marlan W. Downey's *Evaluating Seals for Hydrocarbon Accumulations* under the heading "Assessment of Seal Risk in Exploration, Fault Traps."

> *Faulted Structures:—The term "fault trap" is somewhat of a misnomer; faults do not trap. Faults can place porous reservoirs adjacent to seals and form traps. This trivial sounding distinction deserves attention as the general tendency*

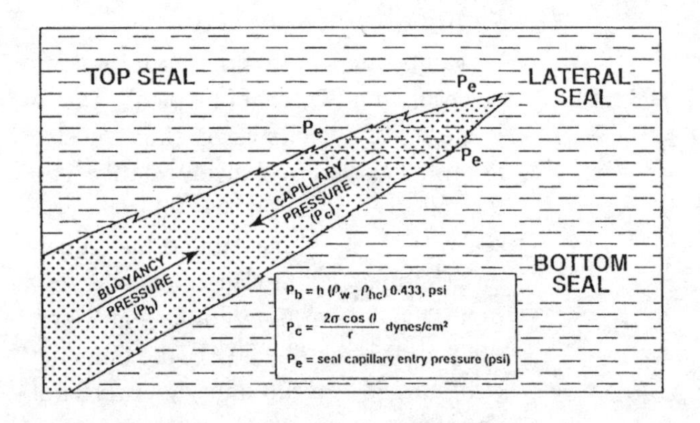

Fig. 3–30 Types of Seals

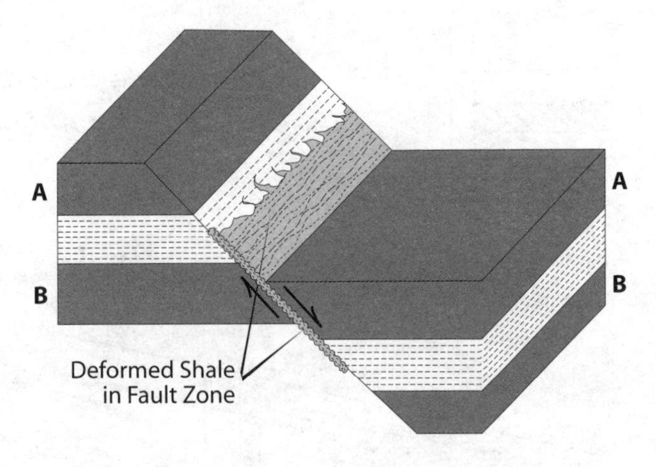

Dragging of undercompacted clays into fault plane can locally emplace scaling material along a fault. Modified from Smith (1980).

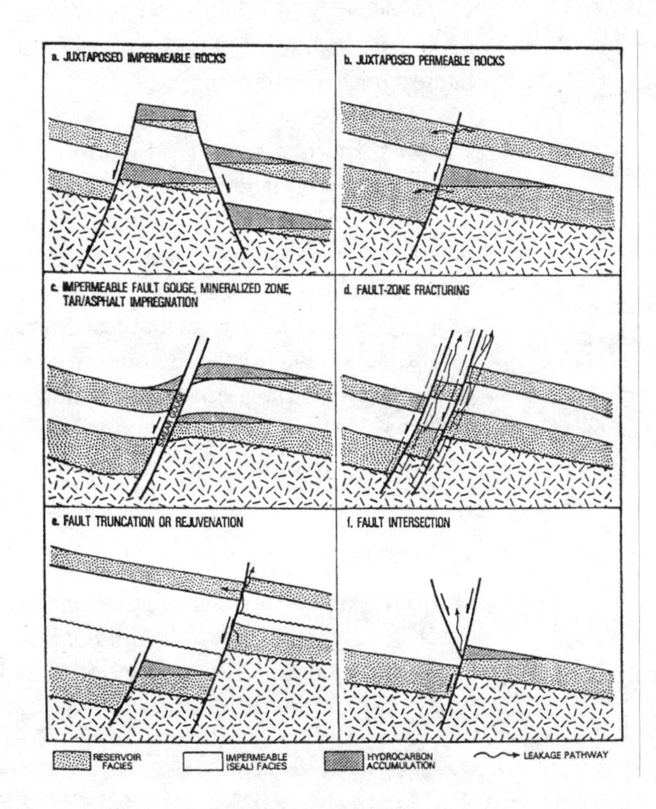

Fig. 3–31 Faults and Sealing

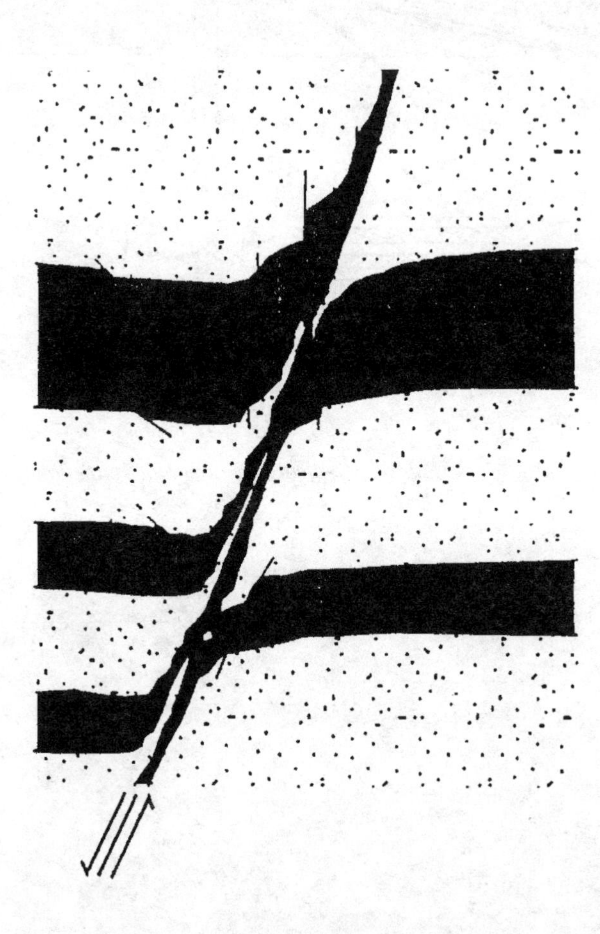

Fig. 3–32 Fault Gouge (Smear) May Provide Seal

> *of explorationists to consider that the mapping of a fault trace is sufficient to define a sealing surface.*
>
> *Entirely different seal risk is attached to the portion of an exploration prospect with simple domal closure (requiring only a top seal) versus fault closure (dependent on seal juxtaposing laterally to reservoir). The portion of the structure depending on fault closure has two seal requirements: (1) top seal and (2) presence of a lateral seal across the fault from the reservoir.* [18]

Later in the book, the topic of fault seals is discussed in more detail.

Juxtaposition. Fault planes are normally inconsequential to migrating fluids and generally are of significance as sealing surfaces only because they may juxtapose rocks of differing capillarity (sic) properties and fluid pressures.[19]

Flow along fault plane. The fault plane itself offers open passage to migrating fluids under special circumstances. The most significant of those special conditions is the circumstance of shallow near-surface faulting in an overall tensional regional stress field. In such cases, field observations and theory (Secor, 1965) agree the fault plane may act as an open transmissive fracture.[9] Faulted shallow structures have substantial risk the fault plane may leak.

Clay smears. In special circumstances, where thick, undercompacted clay shales are interspersed between reservoirs, clay smears can be emplaced along a fault plane as shown in Figure 3–32.[20] Such fault plane shale smears are common small-scale phenomena where thick, soft shales are present in the stratigraphic section. These clay smears have been reported in East Texas outcrops (Smith, 1980), have been studied in coal mines in Germany, and have been inferred from log interpretation of fault zones in Nigeria.[21] Thin clay smears have been invoked to prevent fluid migration between some adjoining sandstone reservoirs in the Gulf Coast Tertiary (Smith, 1980).

The special circumstances governing the emplacement of these fault plane clay smears indicate they are relatively rare in providing traps for significant hydrocarbon accumulations. Important exceptions are found in the Tertiary sediments of the Mississippi and Niger deltas. Note this paper predated North Sea research. There are also sealing faults in sandstones in the North Sea.

Faults *per se* are not generally considered to be sealing.

Accumulations can indicate if faults are sealing. Combinations of faults are shown in Figure 3–33. A number of different combinations can occur. Sealing faults can normally be identified by the arrangement of hydrocarbon accumulations.

Fault seals. If the sealing capability of a fault is not obvious from well data, determination of fault properties is more difficult. A methodical procedure for evaluating whether faults are sealing or nonsealing has been developed.

Figure 3–34 is a flow chart of reservoir seal analysis.[22] There are a number of key points related to seal analysis:

- *Lithology/Stratigraphy.* The first item of importance is lithology and stratigraphy. Thus, it is expected dolomites will behave differently than sandstones or limestones, etc.
- *Structure Map Quality.* The quality of the structure map is important.
- *Tests, Hydrocarbon Contacts, Pressure.* If there are logically different pressures at adjacent locations across a fault, then it is most likely sealing. Similarly, it would be reasonable to expect there would be different gas-water contacts if the faults were truly sealing.
- *ID Critical Seal Aspects by Basin.* In a number of basins, the sealing tendencies of faults are well-known. For instance, the Niger Delta and Gulf Coast areas are known to have many sealing faults. There are reasons for this, which will be discussed later quantitatively. This also underscores the potential importance of local area knowledge.

Sealing and Nonsealing Faults, Gulf Coast

Hypothetical Situation	Analysis of Fault Seal	
	Vertical Migration	Lateral Migration
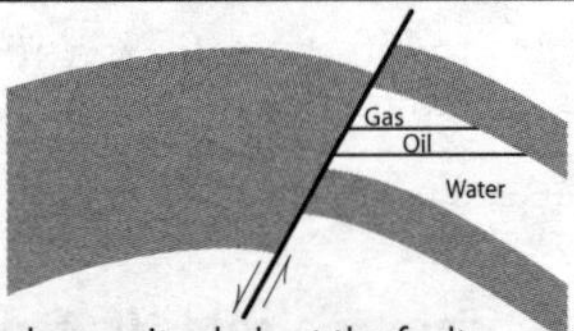 a) Sand opposite shale at the fault. Hydrocarbons juxtaposed with shale.	Sealing	SEALING Reservoir boundary material may be the shale formation or fault zone material.
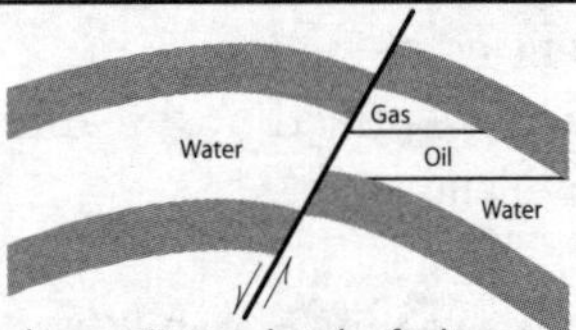 b) Sand opposite sand at the fault. Hydrocarbons juxtaposed with water.	Sealing	SEALING Seal may be due to a difference in displacement pressures of the sands or to fault zone material with a displacement pressure greater than that of the sands.
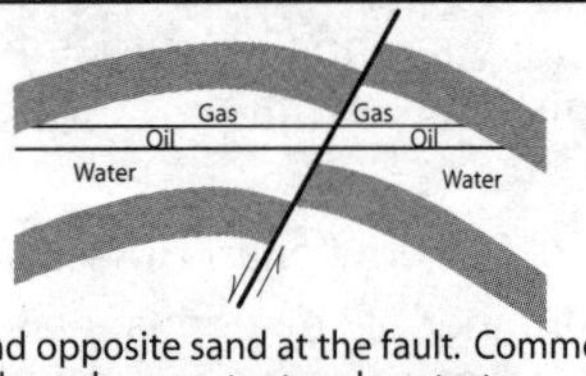 c) Sand opposite sand at the fault. Common hydrocarbon content and contacts.	Sealing	NONSEALING Possibility is remote that fault is sealing and the reservoirs of different capacity have been filled to exactly the same level by migrating hydrocarbons.
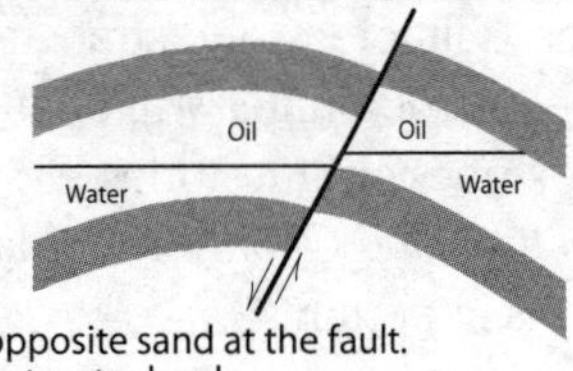 d) Sand opposite sand at the fault. Different water levels.	Sealing	UNKNOWN Nonsealing if water level difference is due to differences in capillary properties of the juxtaposed sands. Sealing if water level difference is not due to differences in capillary properties of the juxtaposed sands.
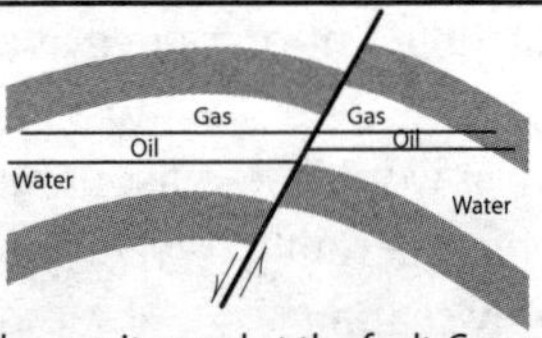 e) Sand opposite sand at the fault. Common gas-oil contact, different oil-water contact.	Sealing	NONSEALING Possibility is remote that fault is sealing and migrating gas has filled the reservoirs of different capacity to exactly the same level.
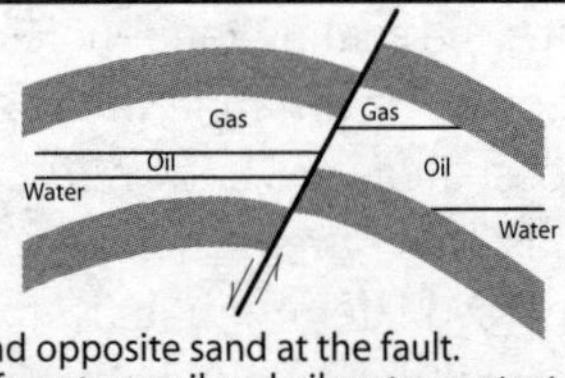 f) Sand opposite sand at the fault. Different gas-oil and oil-water contacts.	Sealing	SEALING A difference in both gas-oil contact and oil-water contact infers the presence of boundary fault zone material along the fault.
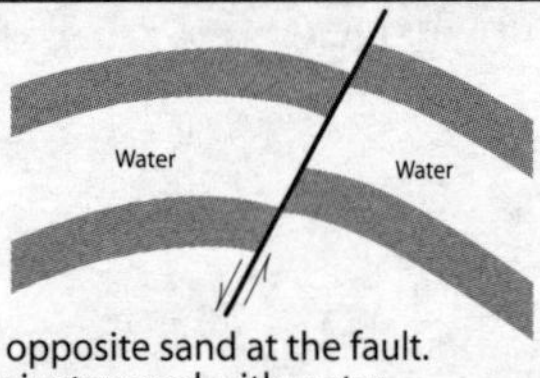 g) Sand opposite sand at the fault. Water juxtaposed with water.	Sealing	UNKNOWN

Fig. 3–33 Sealing and Nonsealing Faults, Gulf Coast

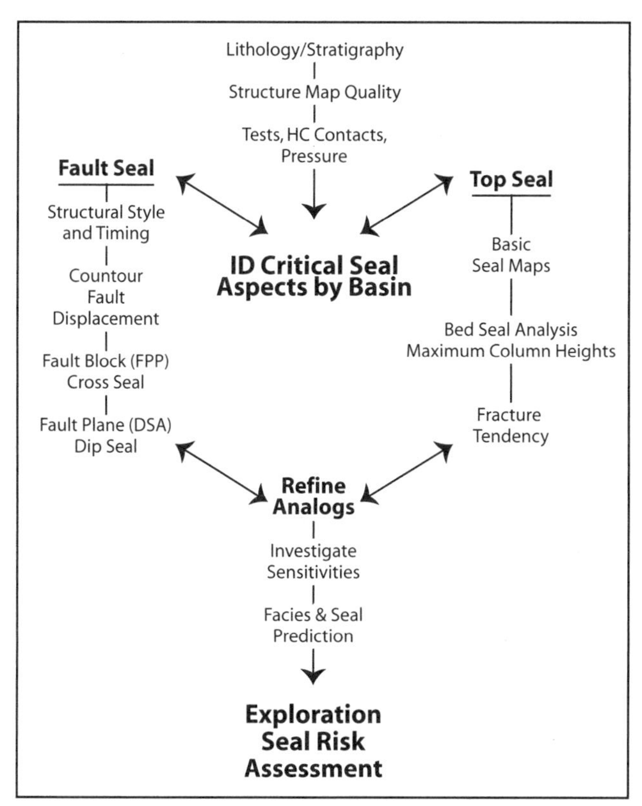

A seal analysis method focused on improving prediction of seal integrity in an exploration project.

Fig. 3–34 Seal Analysis Method

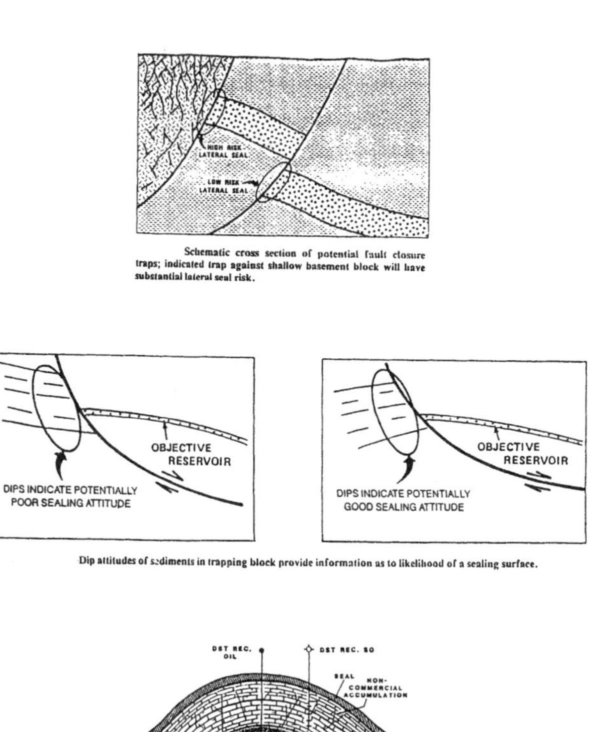

Fig. 3–35 Effects of Fractures on Fault Seals

Top seal analysis. At this point, the analysis branches into two components. Starting with the top seal analysis, basic seal maps are normally structured as top or net pay top maps. Bed seal analysis is done via capillary pressure, which will be discussed in more detail later. Note the top seal can be limited by fracture tendency. Brittle rocks, when deformed, will fracture. This is shown in Figure 3–35. In addition, the dip of a potentially sealing rock can affect sealing tendency. The permeability of shale is much lower across bedding than parallel to bedding. If the lithology of the cap rock is prone to fracturing, then the seal can be breached (and no oil and gas field will remain in the trap).

Evaluation of fault seals. Following is a brief review of the fault seal:

- *Structural Style and Timing.* The degree of fracturing depends on structural position. Normally the footwall is not as good a location for fractures as the top of the hanging wall.
- *Contour Fault Displacement.* This is an exercise in geological engineering geometry, an example of which is shown in Figure 3–36. From this, it is clear where juxtaposition will or will not occur. These plots are sometimes called *Allen plots*.
- *Fault Block Cross Seal.* This will be discussed in more detail later. A key item to note is communications across the fault and along the fault are different but related issues. This refers to communication across a fault.
- *Fault Plane Dip Seal.* This refers to communication along the fault plane.

General seal analysis. Other steps in evaluation include:

- *Refine Analogues.* The first major step is to refine analogues. Comparing performance to offsetting reservoirs is a normal process in evaluating fault seal potential.
- *Investigate Sensitivities.* Important parameters in determining seal can be evaluated by using sensitivities.
- *Facies and Seal Prediction.* This topic will be explored in detail in the next section.
- *Exploration Seal Risk Assessment.* This is the final result or summary step. For an exploration geologist, this can result in a decision to abandon a play or field development. In many cases, this cannot be determined with certainty and a relative weighting will be estimated.

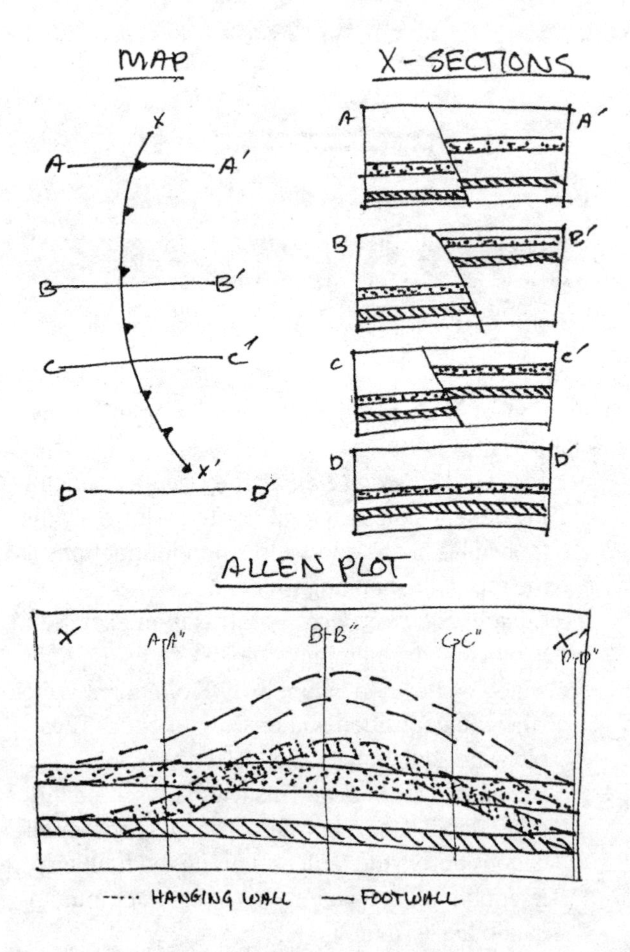

Fig. 3–36 Allen Plots and Contact Areas

Fault gouge analysis. Fault gouge is one mechanism causing faults to be sealing. Dolomite lithology is not prone to gouge production. Conversely, clay smears are common on the Gulf Coast, Nigeria, and, to some extent, in the North Sea. In fact, the clay must be located between layers of reservoir rock. A soft clay layer would have to exist in the faulted stratigraphic sequence.

The topic is covered in more detail later in this chapter, after the discussion of reservoir seals.

Fractured reservoir analysis. Fractured reservoirs are often associated with faults. This topic will be discussed in more detail later in this book. However, a few comments are in order in this chapter. For a fractured reservoir to be productive, there must be fracture conductivity. The important point is rocks prone to sealing faults have the opposite properties of potentially fractured reservoirs. Fractured reservoirs are most prevalent in brittle rock facies, which do not produce low-permeability gouge. Hence, the faults associated with the fractures will be conductive. Conversely, formations comprised of plastic rocks will not tend to be good fractured reservoirs.

Capillary pressure trapping. Virtually all rocks (except perhaps some naturally occurring glasses) have some porosity and can transmit fluids. The ability to transmit fluid is empirically described with Darcy's Law. For some materials, such as clay and shale, this permeability can be very low. However, there is a distinction between no permeability and not very much. The oil industry also uses a number of specialized terms rather loosely. For instance, we call a well D&A for dry and abandoned, even though it may produce copious amounts of salt water. It would be more correct to say no economic hydrocarbons and abandoned (NEH&A). Of course, NEH&A doesn't roll off the tongue as quickly as D&A. This is yet another example of misleading terminology in our industry.

When multiple phases are present in the pore space of rock, such as oil, gas, and water, understanding porous media becomes more complex. Oil and gas are sealed in place by capillary imbibition barrier pressures. This is shown diagrammatically in Figure 3–37, which has been extracted from "Evaluating Seals for Hydrocarbon Accumulations" by Marlan W. Downey.[18]

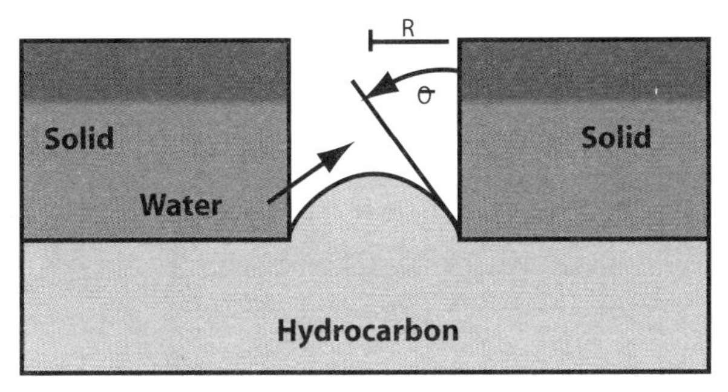

Capillary displacement is dependent on three parameters: radius of largest connected pore throats (R), wettability (theta) and interfacial tension (gamma).

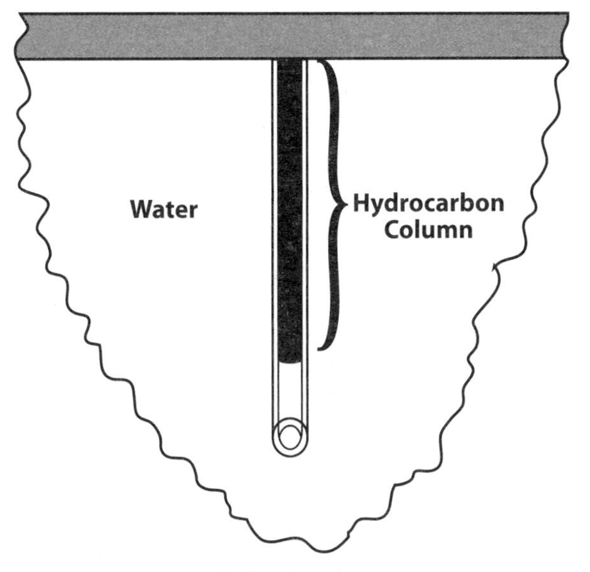

Buoyancy pressure of hydrocarbon column acts to attempt to force hydrocarbon through larger pores in seal.

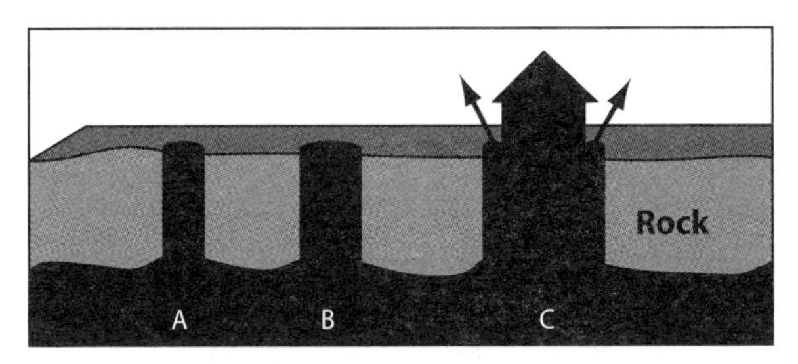

Rock pore throats A and B are constricted (have high entry pressures); pore throat C is sufficiently large that buoyancy pressure of hydrocarbon column can displace pore water and pass through pore throat.

Fig. 3–37 Capillary Pressure Imbibition Threshold Pressure

Barrier pressures. Barrier pressures are a feature of imbibition capillary pressure. Recall most rocks are laid down in a marine (submerged) environment and uplifted onto the continents later. Due to rainfall and water movement, all but a thin layer of the subsurface is water filled. Therefore, when hydrocarbons move, they must *force* entry to move into new rock.

The required *force* or pressure differential can be estimated from laboratory data, which is typically obtained by injecting mercury into a rock. An example of this is shown in Figure 3–38.[23] The barrier pressure is strongly related to the size of the pores. A general correlation is available as shown in Figure 3–39.[24] This figure cannot be relied upon to predict specific examples; however, it shows the general trend. The quality of barriers has also been classified, as shown in Figure 3–40.[25]

Implications of barrier pressures. The air-mercury imbibition capillary pressure can be converted to apply to gas-water, gas-oil, and water-oil systems. Significant implications to this data include:

- Some seals can only hold a certain pressure differential. Thus, the size of a trap can be limited. An example of this is shown in Figure 3–41.[26]
- In some cases, traps will preferentially catch oil or gas.
- In some cases, a seal may exist initially. However, the pressure barrier may be broken by increased pressure differentials that result from continued production. In such cases, seals will break down during the producing life of the reservoir.

Evaluating seals is a complex process.

Fault gouge. In the case of a fault, the seal is provided by the fine material ground up when the two rock faces scrape together on either side of the fault. Such a situation is depicted in Figure 3–42.[27]

The amount of gouge produced is also related to fault displacement. In the case of minor (extensional) faults, the amount generated will not be large. As described in "Numerical Simulation of Jointed and Faulted Rock Behavior for Petroleum Reservoirs," fracture conductivity is also dependent on shear movement history.[28] In particular, a reversal of direction decreases a joint's ability to transmit fluid.

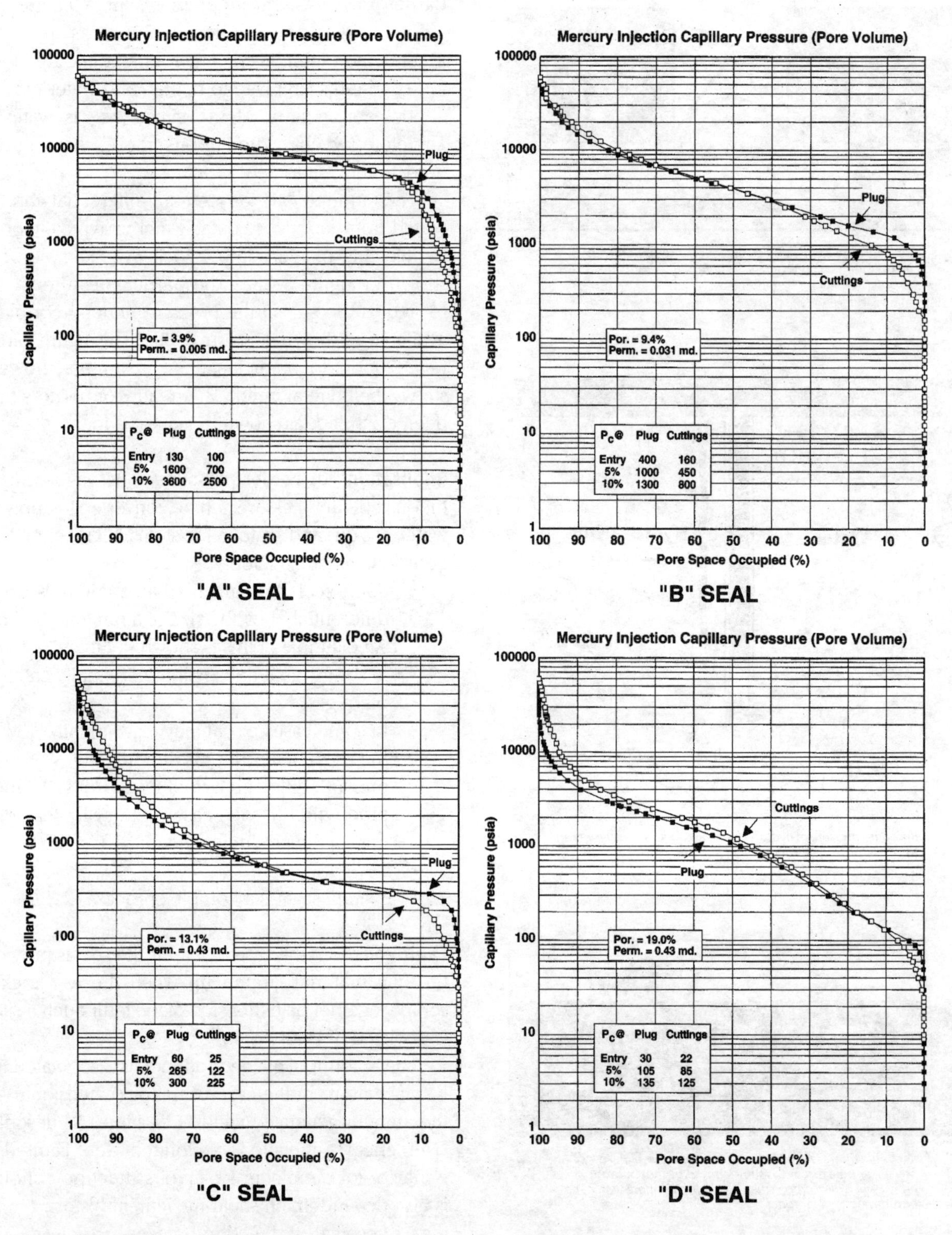

Examples of high-pressure mercury/air injection curves for seal types "A", "B", "C", and "D". Curves for the vertical plug are designated with black squares; the cuttings curves (open squares) are of simulated cuttings from rock adjacent to the vertical plug.

Fig. 3–38 Classification of Seals via Mercury Imbibition Pressure Thresholds

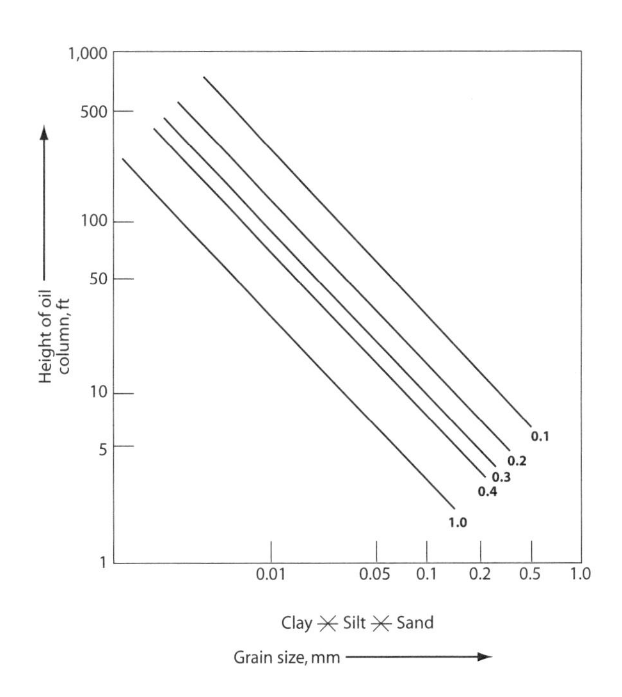

Fig. 3–39 Generalized Indication of Seal Capacity

Reservoir-Seal Couplet

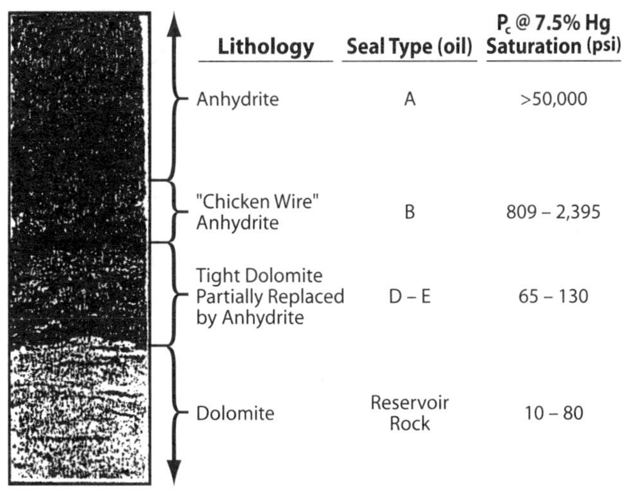

Lithology	Seal Type (oil)	P_c @ 7.5% Hg Saturation (psi)
Anhydrite	A	>50,000
"Chicken Wire" Anhydrite	B	809 – 2,395
Tight Dolomite Partially Replaced by Anhydrite	D – E	65 – 130
Dolomite	Reservoir Rock	10 – 80

Core of an anhydrite top seal on a dolomite reservoir, San Andres Formation, New Mexico

Fig. 3–40 Seal Classifications

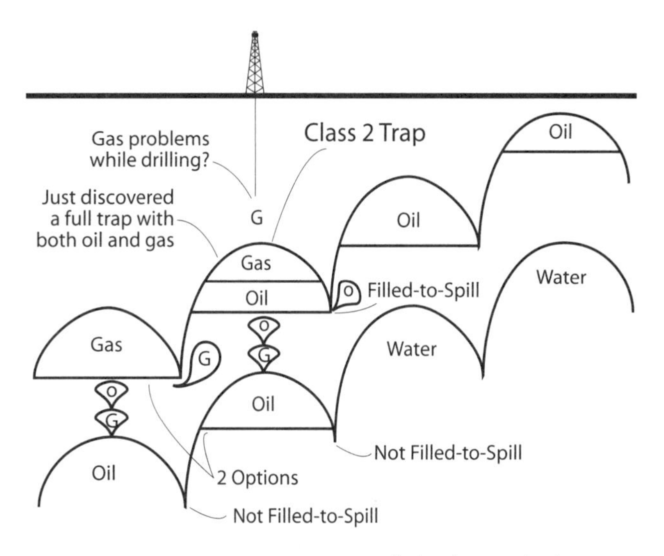

Fig. 3–41 Classes of Traps for Oil and Gas

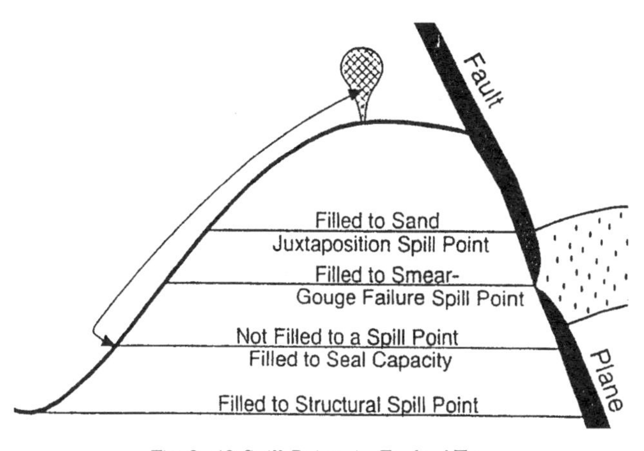

Fig. 3–42 Spill Points in Faulted Trap

It is possible to obtain samples of the gouge and perform air mercury imbibition capillary pressure laboratory data. With this data, and a defined procedure, it has recently become possible to predict sealing pressure differentials. A software company in London, England, has developed a program to predict sealing pressure differentials.

The software company provides information on its website. A paper on this procedure, "Quantitative Fault Seal Prediction," is quite recent. [29]

Figure 3–43 shows the fault throw and presence of clay (or shale) material affects the predicted fault seal. Figure 3–44 shows the seal strength (from capillary pressure data) for gouge obtained by clay smearing, cataclasis, and reservoir rock properties.[30]

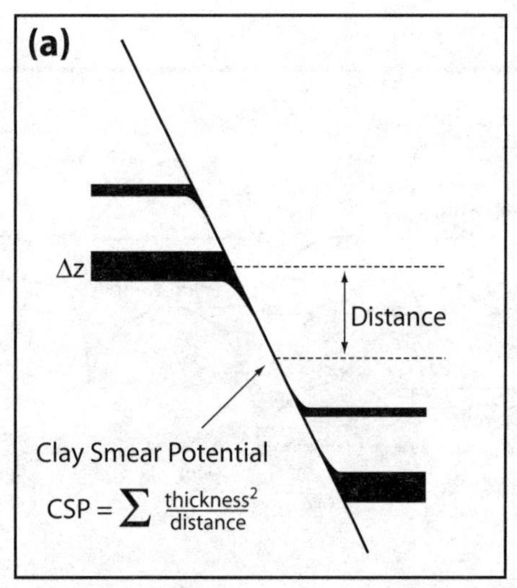

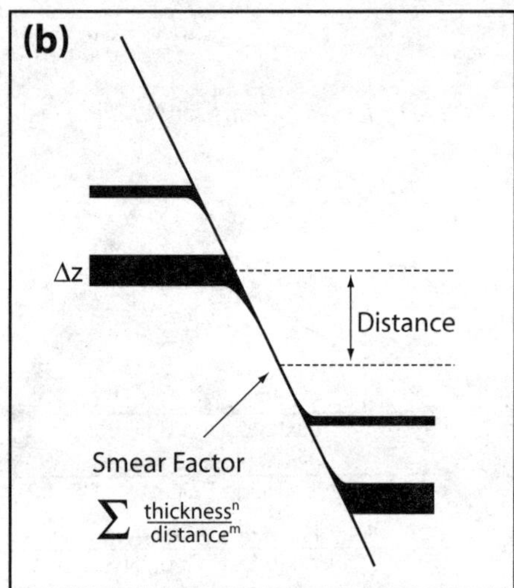

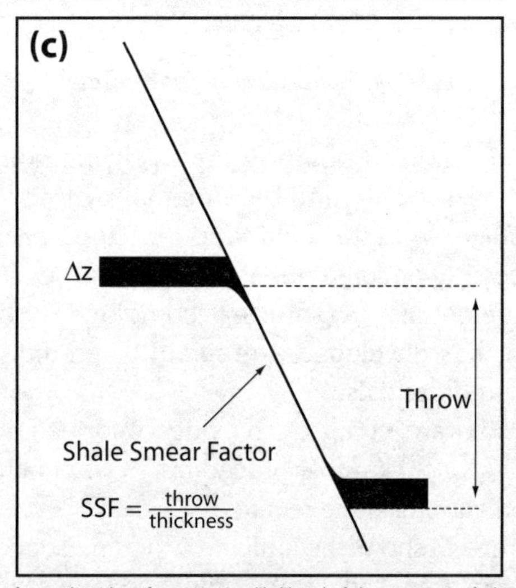

Smear factor algorithms for estimating likelihood of clay smear on a fault plane. (a) Clay smear potential (CSP) (Bouvier et al., 1989; Fulljames et al., 1996) given by the square of source-bed thickness divided by smear distance; (b) generalized smear factor, given by source-bed thickness divided by smear distance, with variable exponents; (c) shale smear factor (SSF) (Lindsay et al., 1993) given by fault throw divided by source-bed thickness. Methods (a) and (b) model the distance-tapering of shear-type smears, whereas method (c) models the form of abrasion smears.

Fig. 3–43 Quantitative Criteria for Fault Seal Analysis

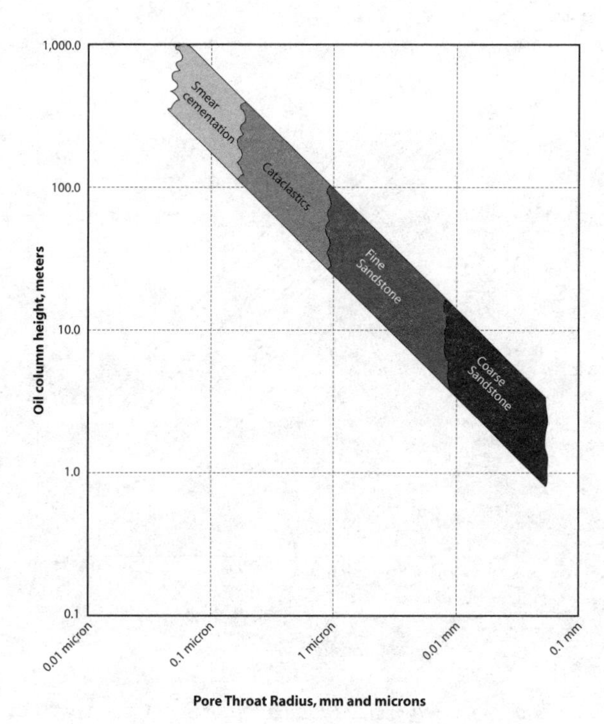

Fig. 3–44 Generalized Fault Seal Trapping Potential

Shales

In a number of cases, shale barriers become significant:

- Two zones are separated by shale. The question is, How thick does the shale have to be to act as a reservoir barrier?
- There is a series of shale lenses. Often shales are seen in a lithological unit at varying locations. This can represent, in part, changes in deposition or an annual cycle. The questions are: Is the shale continuous? How much will it affect vertical permeability?
- There are lag shales. In fact, these are hard to distinguish from shale lenses or variable sand thickness. The question is whether to model the reservoir as a layer cake or compartmentalize the reservoir.

Shale barriers. Based on the author's experience, three to four meters of shale are required to form a barrier. Having said this, the author has not been sufficiently systematic to record this and cannot provide a list of proofs where this "rule" applies and where it does not. Users beware.

Shale lenses. Shale lenses appear on logs and are of limited areal extent. Therefore, they do not constitute total barriers from one well to the next. They do have an effect on vertical permeability. A number of different approaches to this problem exist. The first approach was based on streamlines, and a diagram is shown in Figure 3–45.[31] This covers only single-phase flow. The calculations are, in ARE's view, somewhat simplified. No allowance was made for varying k_v and k_h, which is a remnant of deposition. This is shown in Figure 3–46. The effects of shale breaks can also be determined by electrical analogues.

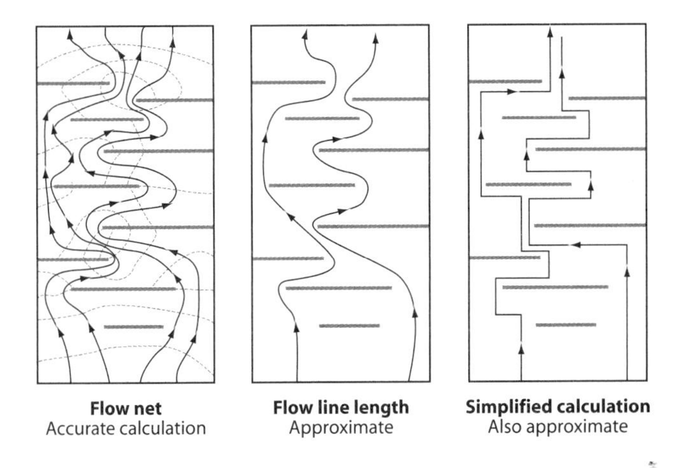

Fig. 3–45 Effect of Shale Intercalations

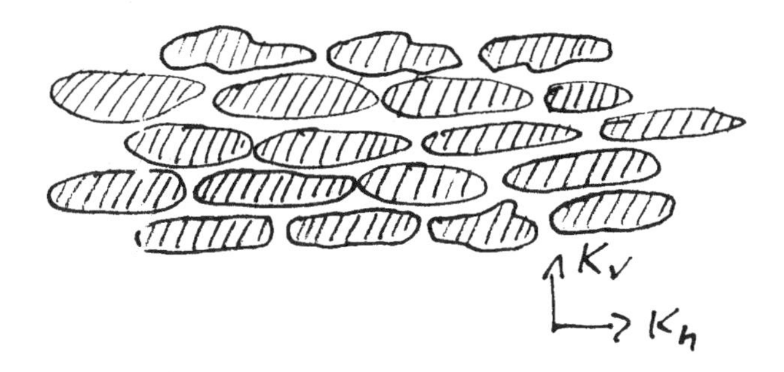

Fig. 3–46 Effect of Grain Orientation and Vertical vs. Horizontal Permeability

There is also a method by Deutsch. A more sophisticated approach is to use statistical shales. An example of this is shown in Figure 3–47. A correlation has been developed by Bora, which can be used to estimate the effective vertical permeability.[32] Of course, knowing the dimensions of the shale lenses is not easy. It must be estimated from modern environments or from outcrops.

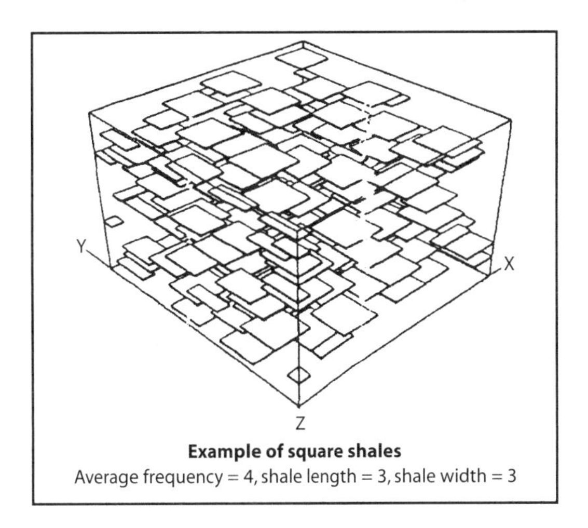

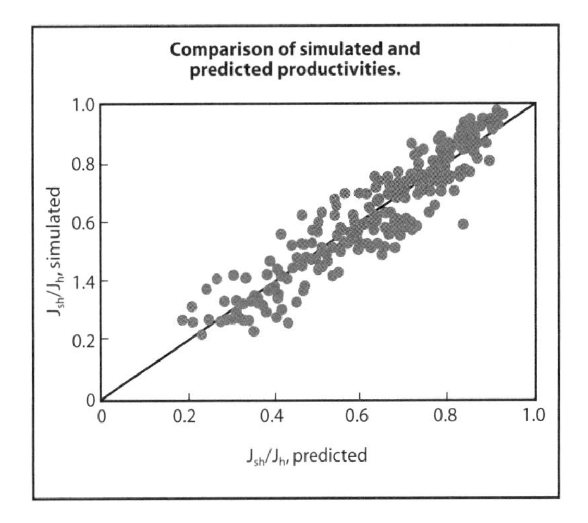

Fig. 3–47 Stochastic Shale Distributions and Correlation of Effective Vertical Permeabilities from Stochastic Realizations

Lag shales

The author has seen definite examples of lag shales affecting communication in a shoreface deposit reservoir. Ironically, it was mapped as an inferred fault based on pressure information. The fault was thought to have a sufficiently small throw that it could not be interpreted on seismic data. The lag shale was put into ARE's simulation based on the geological model and a fence diagram. The reservoir was then modeled as overlapping sands rather than in a layer. This is notoriously difficult to determine from correlation, as shown Figure 3–48.

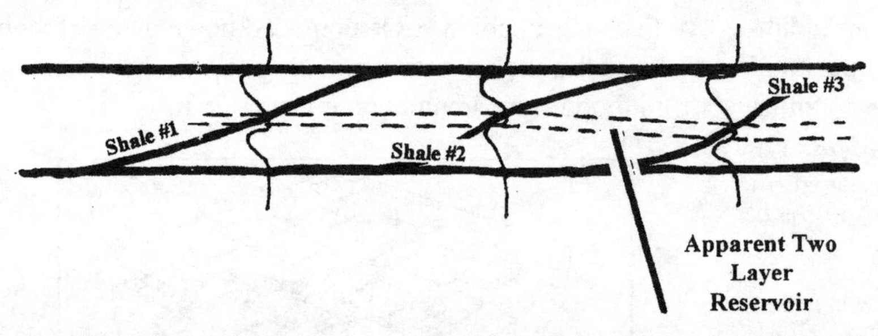

This reservoir is actually considerably more heterogenous than early "simple" interpretations would indicate.

Fig. 3–48 Shales Are Easily Misinterpreted and Cause Mischief

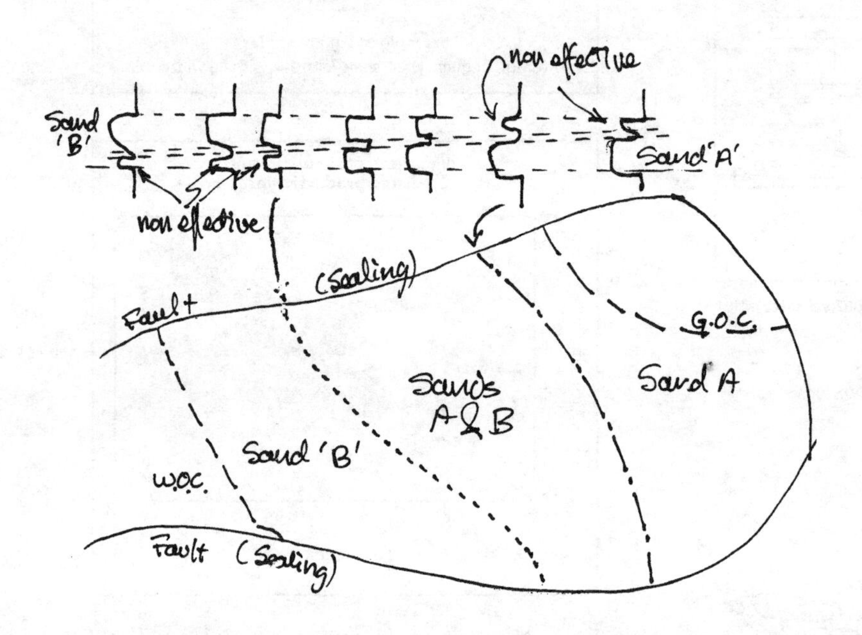

Fig. 3–49 Example of History Match Affected by Lag Shales

Figure 3–49 shows a map developed to model a reservoir. History matching identified a number of lag shales.

Summary

When a new field is encountered, the first objective should be to learn the geological model for the reservoir. Query the geologist assigned to the field for as much information as possible. Do whatever is necessary get a good mental picture of the reservoir. Coloring crayons, some log strips, and a large piece of paper are nearly unbeatable for developing mental imagery.

There is obviously not enough space in a reservoir simulation text to write extensively on log analysis. This subject would comprise a book in itself. At the same time, the input data for a simulation depends heavily on this data. The objective of this chapter is to identify the need to thoroughly understand the topic. It is hoped one or all of the following occurs:

- One can use this chapter to justify the expense of a course in log analysis to the boss. It is important to understand where these numbers are derived.
- One asks the log analyst not only what his answer is but also why he used the approach he did, and what uncertainties exist.
- One takes the opportunity to go through the calculations in a number of wells.
- One feels more comfortable making changes to pore volume or contact depths for gas-oil, gas-water, or oil-water based on a better understanding of data accuracy.

The results from log analysis, although quantitative, have a significant amount of uncertainty associated with them.

If anything, the author hopes this chapter has reiterated that reservoir description is a shared responsibility between the geologist and the simulation engineer. The use of visualization tools such as cross sections, fence diagrams, and Allen plots (for faults) is heavily emphasized. Sometimes there is no clear-cut answer and one must average properties or guess. This means making realistic decisions as well as understanding and communicating the risks involved.

Choosing layering is a difficult art and requires hard work and analysis.

Faults often provide a trapping mechanism by juxtaposition. They are rarely sealing faults of their own accord. Examples where they are sealing are described and may be predicted.

Shale barriers are difficult to assess. A general rule of thumb has been proposed—with no justification. The influence and placement of shales are strongly controlled by environment, and the effects on the numerical model should be consistent with the conceptual geological model wherever possible.

Adjustments to vertical permeability can be used to adjust for the presence of shale lenses. This is still difficult to implement since the dimensions and distributions of the shale lenses are rarely known. Analogues must be used as an estimate. More realistically, if it is known they are there, this can justify a history-match change. Better still, include a rough estimate during the construction of the geological model and fine-tune the estimate while history matching.

Lag shales are notoriously difficult to correlate, although the author has seen it done via some high-quality geological work. The key to including these phenomena is to recognize this as a part of the geological model. It is not easy to implement geometrically. The author has interpreted a number of lag shales during the history-match process.

References

1 Masters, J.A., "Elmworth, Case Study of a Deep Basin Gas Field," *AAPG Memoir No. 38*, 1984.

2 Deroo, G., et al., "The Origin and Migration of Oil in the Western Canadian Sedimentary Basin—A Geochemical and Thermal Maturation Study," *Geological Survey of Canada*, Bulletin 262, 136 pages.

3 Weber, K.J., "Sedimentological Aspects of the Niger Delta," Geologie en Mijnbouw, Vol. 50, pp. 559–576.

4 NOGS, *Oil and Gas Fields of Louisiana*, Vol. 3, New Orleans Geological Society, 1983.

5 Sawkins, F.J., et al., *The Evolving Earth*, Macmillan Publishing Inc., 1974.

6 Lowell, J.D., "Structural Styles in Petroleum Exploration," *Oil and Gas Consultants International*, 1997.

7 Suppe, J., "Geometry and Dinematics of Fault Bend Folding," *American Journal of Science*, Vol. 283, pp. 648–721, 1983.

8 Marshak, S., and G. Mitra, *Basic Methods of Structural Geology*, Prentice-Hall, 1988.

9 Suppe, J., *Principles of Structural Geology*, Prentice-Hall, 1985.

10 Dahlstrom, C.D.A., "Structural Geology in the Eastern Margin of the Canadian Rocky Mountains," *Bulletin of Canadian Petroleum Geology*, Vol. 18, no. 3, pp. 332–406, 1970.

11 Crain, E.R., *The Log Analysis Handbook, Volume 1, Quantitative Log Analysis Methods*, PennWell Publishing Co., 1986.

12 Dyskstra, H., and R. L. Parsons, *Secondary Recovery of Oil in the United States, Principles and Practice: The Prediction of Oil Recovery by Waterflood*, 2nd ed., American Petroleum Institute, 1950.

13 Willhite, G. Paul, *Waterflooding*, SPE Textbook Series Vol. 3, 1986.

14 Testerman, J. D., "A Statistical Reservoir-Zonation Technique", *Journal of Petroleum Technology*, 1962.

15 Stoakes, F.A., "Nature and Control of Shale Basin Fill and its Effect on Reef Growth and Termination: Upper Devonian Duvernay and Ireton Formations of Alberta, Canada," *Bulletin of Canadian Petroleum Geology*, 1980.

16 Sneider, R.M., et al., "Comparison of Seal Capacity Determinations: Conventional Cores vs. Cuttings," *AAPG Memoir* 67, pp. 1–12. 1987.

17 Harding, T.P., and A.C. Tuminas, "Structural Interpretation of Hydrocarbon Traps Sealed by Basement Normal Faults at Stable Flank of Foredeep Basins and at Rift Basins," *AAPG Bulletin*, Vol. 73, no. 7, pp. 812–840, July 1989.

18 Downey, M.W., "Evaluating Seals for Hydrocarbon Accumulations" under the heading "Assessment of Seal Risk in Exploration, Fault Traps," *AAPG Bulletin*, 1984.

19 Smith, D.A., "Theoretical Consideration of Sealing and Nonsealing Faults," *AAPG Bulletin*, 1966.

20 Harding and Tuminas, 1989.

21 Weber, J.K., "Simulation of Water Injection in a Barrier-Bar-Type, Oil-Rim Reservoir in Nigeria," SPE 6702, Society of Petroleum Engineers, 1978.

22 Allard, D.M., "Fault Leak Controlled Trap Fill: Rift Basin Examples," *AAPG Memoir 67*, pp. 135–42, 1987.

23 Sneider, Sneider, Bolber, and Neasham, 1987.

24 Berg, R.R., "Capillary Pressures in Stratigraphic Traps," *AAPG Bulletin*, Vol. 99, pp. 939–956, 1975.

25 Sneider, et al., 1987.

26 Sales, J.K., "Seal Strength vs. Trap Closure—A Fundamental Control on the Distribution of Oil and Gas," *AAPG Memoir 67*, pp. 57–83, 1987.

27 Ibid.

28 Koestler, A. G., et al., "Numerical Simulation of Jointed and Faulted Rock Behavior for Petroleum Reservoirs," *Structural and Tectonic Modelling and its Application to Petroleum Geology, Norwegian Petroleum Society Special Publication 1*, Elsevier, 1992.

29 Yielding, G., B. Freeman, and D.T. Needham, "Quantitative Fault Seal Prediction," *AAPG Bulletin*, 1997.

30 After Knipe, R.J., "Faulting Processes and Fault Seal," *Structural and Tectonic Modelling and its Application to Petroleum Geology*, edited by R.M. Larsen, H. Bekke, B.T. Larsen, and E. Talleraas, Elsevier, Norwegian Petroleum Society Special Publication No. 1, 1992.

31 After Haldorsen, H.H. and D.M. Chang, "Notes on Stochastic Shales; From Outcrop to Simulation Model," *Reservoir Classification* (I), Academic Press Inc., 1986.

32 Belgrave, J. D. M., and R. Bora, "On the Performance of Horizontal Wells in Reservoirs Containing Discontinuous Shales," *Journal of Canadian Petroleum Technology* (May): 38–44, 1996.

4

Multidisciplinary Integration and Geostatistics

Introduction

Many people would consider multidisciplinary integration and geostatistics to be advanced topics. On this basis, perhaps this chapter should have been put in the second half of the book; however, it is naturally related to the topic of geology. This is where multidisciplinary integration is key, and the subjects are presented here to provide background for the remainder of the text.

Both of these topics have become popular in the literature. To some degree, multidisciplinary integration and reservoir management have been practiced on a *de facto* basis by many companies for many years as a regular part of reservoir engineering. However, this has been transformed of late into a new discipline that seems to be most often called one of *integrated studies* or *reservoir management*. Overall, good practice has become somewhat more formalized and, in the process, more concretely identifiable.

There have been a number of books written on reservoir management, which touch on a variety of geological, geophysical, petrophysical, and reservoir engineering disciplines from an overview basis.

Some of these include books by Ganesh Thakur, such as *Integrated Waterflood Asset Management and Integrated Reservoir Management*.[1] More recently, Luca Cosentino wrote *Integrated Reservoir Studies*.[2] These books are very general and are a useful overview.

As reservoir simulation has developed, the shift has been away from the problems of getting the numerical program to work properly toward developing as accurate a representation as possible. Initially this was called reservoir characterization. Since reservoir characterization is so strongly dependent on geology and petrophysics, this is linked to multidisciplinary studies. One of the more important developments in the area of reservoir characterization is geostatistics.

The purpose behind geostatistics is to remove bias in analyzing data. Geostatistical tools are designed to provide estimates and bounds on various reservoir properties. However, the author has found geostatistics difficult to follow from the literature. Many of the papers rapidly launch into statistical notation. This is not something with which most engineers have maintained an intimate familiarity. There is nothing intrinsically wrong with this technical approach; however, as practicing engineers, our primary interest is less oriented to the detailed mathematical development of the technique and more oriented to the details of application.

Background

Most of the textbooks on geostatistics concentrate on the statistical derivation rather than a practical application point of view. The American Association of Petroleum Geologists (AAPG) has produced an excellent memoir. It has a good introductory chapter and describes a large number of case histories. The Society of Petroleum Engineers (SPE) has a two-book summary of reservoir characterization papers, and Academic Press has made three quite nice books on reservoir characterization. In fact, these three books are really the proceedings from three separate conferences on this topic. Recently, Clayton Deutsch prepared a book titled *Geostatistical Reservoir Modeling*.

Incidentally, the development of geostatistics came from the mining industry. Its primary application was for mapping ore bodies. Similarly, the objective was to remove bias in the preparation of reserves estimates for ore bodies.

Large software packages

Geostatistics and multidisciplinary integration are commonly associated with one of the large software packages, which have a robust geological model and a reservoir simulator. This is related to large studies that have been documented in the literature for some very large fields.

A first look at GSLIB, the geostatistical software package prepared at Stanford University, can be anticlimactic. All this software package contains is a collection of Fortran programs that manipulate ASCII data files. Significantly, GSLIB was used as the basis for a considerable number of the larger integrated geological and geostatistical packages. With a little bit of practice, these apparently simple programs are rapid and powerful. The developer of the programs does not seem to rely on an integrated package; rather, he stays closer to the data and uses ASCII flat files.

The large integrated packages are really more of a data handling and graphics tool than a geostatistical program *per se*. This is, of course, extremely valuable. This discussion differentiates among these software characteristics: data management, graphics, and geostatistics. It is easier to understand what is required if the process is broken up.

Study integration/team approaches

This type of approach has also been reinforced by recent management techniques, such as the increasing use of teams assigned to specific projects. The significant amount of material written on this concept and its implementation is not addressed in this book. The structure of these teams is quite flexible, and the distribution of labor will often depend on individuals' personalities and fields of expertise. At the same time, it is useful to have a general notion of what is expected of the simulation or reservoir engineer.

Overall responsibility for large-scale development projects seems to fall somewhere between geology and engineering. The fundamental assessment of the resource generally falls within the domain of the geology. For the most part, developing the project and evaluating economics falls within the domain of engineering. Of course, there cannot be one without the other, so both are equally important and critical. Overall, management tends to be split between individuals with these two backgrounds.

Geostatistical work

Most geostatistical work falls into the domain of geologists. Certainly, the development of the basic geological model is the primary function of the geologists. They usually obtain a considerable amount of input from geophysicists. This is dependent on the situation. Geophysics is often used for structure, gross thickness, and/or edge definition. More recently, the amplitude of the geophysical traces is used as a relative indicator of porosity. This often involves specialized processing, which is done by geophysical specialists. Although these stages involve both the geologist and the geophysicist, the greater part of this work seems to be coordinated by a geologist.

Most large studies will also involve a detailed petrophysical study. In such studies, permeability is normally generated from core porosity versus core permeability relationships. The output from the log analysis is then included in the map prepared by the geologist. Interestingly, the background of quantitative log analysts is split between engineers and geologists. The split is not exact, however. In the author's observation, there are slightly more log analysts with engineering than geological backgrounds. This is probably because a strong background in logging tool operation and function is helpful in quantitative analysis. Logging tools (the sondes) are very sophisticated, and a detailed understanding of their responses is technically complex. This is often obtained while working for one of the major logging companies, and the latter appear to hire more engineers than geologists.

Reservoir and simulation work

Identification of reservoir rock properties seems to fall on the dividing line of work between geologists and engineers. Traditionally, in less-sophisticated studies, averaging permeabilities and core permeability versus core porosity usually fell in the engineers' domain. Capillary pressure data is normally utilized by good log analysts. It is likely the reservoir engineer will either use this same data or adapt it for use. This is one area where some overlap in work may occur with the log analyst. Relative permeability will normally fall in the reservoir engineers' domain of responsibilities, as will PVT data and the majority of input data covered by the other chapters in this text.

One additional point: the choice of grid at the front-end of the geological description is critical for the reservoir engineer. Ideally, the geological grid is the same size as the simulation grid blocks used by the reservoir engineer. Failing this, an odd multiple is very convenient. It is easy to add three grid blocks (or points) together or, alternatively, use the center block to represent reservoir properties.

If the spacing of the reservoir model grid and the geostatistical grid are irregular, it is necessary to interpolate the reservoir grid block values. The process of resampling degrades much of the sophisticated mathematics from the geostatistics, and this should be avoided, if possible.

If the reservoir grid and the geological grid do not coincide, it is advantageous if the geological grid is finer, rather than coarser, than the reservoir grid.

One of the earlier steps should be a grid sensitivity study, which was discussed in chapter 2. Finalize the grid at a later stage, when an evaluation of all of the data is available. Indeed, depending on the course of a study, the grid may need to be reset later.

Stepwise approach

The process ARE has used relies less on an integrated package but is conducted in a stepwise fashion using ASCII files. ARE has been using geostatistics for many years; however, most of this work has been with PC-based kriging programs or spreadsheets. Our experience has shown kriging gives much more realistic contours and grid-block values than most other interpolation programs. The application of kriging normally occurs far more often than simulation.

The following addresses geostatistics—with some simplifications. There are two different types of analysis and two fundamental processes.

Fundamental approach—objects versus distribution

Of the two fundamental approaches, one deals with *objects* and the other with the *distribution* of properties. An object refers to a reservoir, a channel, a lithology type, or some other physical entity. Within an object, there can be a distribution of properties.

On a recent project in the northeastern part of Alberta in the McMurray heavy oil trend, the first job was to identify the object. In the McMurray case, the bitumen is found in channels, and these channels are the fundamental object. The first problems are related to this fundamental object, that is, where it is located and its physical properties. This could include:

- width
- height
- cross-sectional shape
- length

Layering, permeability, porosity, and water saturations are within the objects.

This can also be looked upon as a variation of scale.

Fundamental process—averaging versus projecting

The distribution of properties in geostatistics can act in two ways. Kriging, which is a sophisticated method of averaging data, tends to smear or smooth details. The opposite of this is the use of simulation, which is designed to re-create realistic heterogeneity.

Therefore, part of the time, geostatistics attempts to add order or simplify a complex data set, and, at other times, it is used to add realistic detail or re-create a random element to properties.

In general, kriging (averaging) is used far more than simulation (randomizing).

Statistics

Having made these broad clarifications, there are a number of misconceptions with regard to statistics. First, statistics uses mathematics; however, statistics is not entirely mathematically based. The laws of mathematics can be rigorously tested and derived from formal proofs and logic. This is not true of all statistical techniques.

Arbitrary development

For example, it is common to use least squares regressions. In most cases, we assume a relation—for instance, a straight line or a polynomial curve. This is arbitrary. It may be true, or it may not be true. The statistical technique does not require a formal mathematical proof the relation must be linear, although there are arguably some cases where this is known to be the case.

Continuing with a least squares regression development, the technique of taking the difference between the actual data point and the regression point is arbitrary. Even more arbitrary is squaring the difference. It is certainly convenient in that we always get a positive number using a square. It may also be desirable that it weights points further away from the trend more heavily. It certainly makes differentiation (for minimization of error) convenient. However, no formal mathematical proof can be derived that shows squaring the difference is *the* one and only correct method. This is an arbitrary choice made by statisticians, read *error-prone humans*, in developing the technique.

Clearly, many of the arbitrary choices in statistics are made based on the ease with which the requisite mathematics can be implemented. In a large number of cases, this is precisely why certain assumptions are made, such as the use of least squares regression or Gaussian probability distributions. The latter is commonly used in geostatistics. However, this is very different from an absolute proof or logical derivation. This topic is discussed later.

Obviously, linear regressions are extremely common and almost all readers of this text will have used them extensively. Statistics can provide a useful guideline, but not an absolute answer, in dealing with various forms of risk and measurement errors.

Independence of variables

The second biggest problem with statistical techniques is that independent variables are usually assumed in the development of these techniques. This overlooks two things, which can be illustrated with a simple core permeability versus core porosity graph:

1. All the variation is assumed to be in the permeability, and the porosity measurement is assumed to be correct. In practice, the porosity measurement is probably more accurate; however, it is not strictly true that the precise porosity is known. There are many elements of uncertainty associated with a measurement of porosity.
2. More significantly, permeability and porosity are not strictly independent. In general, we can describe the trend between porosity and permeability. Permeability generally increases with porosity, but we could have three different rocks:
   - a conglomerate with low porosity and high permeability
   - sandstone with intermediate porosity and permeability
   - chalk with high porosity and low permeability

   An analysis of representative points for these rocks would show the permeability decreasing with porosity. Hence, although there is usually a relationship between permeability and porosity, it depends to some extent on other factors—in this case, lithology.

Like so many other sophisticated tools, some assumptions are not strictly correct. Consequently, quantitative results from geostatistics are guidelines and do not constitute comprehensive results.

Spatial relations

The concept of geostatistics could be described better as spatial statistics. *Geo* is an accident of history; the geological sciences are where spatial data was first used. Such analysis would have application in many non-geological fields, such as city planning.

Part of the following discussion will deal with straight spatial relations, and the other part will deal with how this is applied to modeling.

Spatial variables. Two types of variables are commonly used in spatial descriptions. The first relates to continuous variables, such as porosity and permeability, and the other relates to properties normally referred to as indicators. This might be a lithology or facies. Examples would include channel sand, beach sand, limestone, or sandstone.

For indicator variables, some predetermined geometry can be developed from assumed data. For instance, channels in a turbidite system might be represented in cross section by a half ellipse. Depending on the environment, we know roughly what kind of bends will exist in a fluvial channel or river and the approximate dimensions of those bends. Delta fans may be represented with a triangular shape. These are generally known as objects. Within these objects, we may have spatial patterns of certain geological features such as facies. The statistics comes in when we randomly place these objects and when we allow the features to vary in size. The objects are classified by integer numbers, i.e., 1, 2, 3, … etc.

This type of variable is normally used to assign unique values. For instance, a geologist may create a finite number of lithologies, such as fine sandstone, conglomerate, and coarse sandstone. Each section of a well comprising part of the reservoir has one of these values associated with it.

Continuous variables. Continuous variables include properties such as porosity or permeability. These properties can have any value, generally within some known range. These values are represented by real numbers.

Describing spatial variation. The method by which space is taken into account for continuous variables is with the use of a semivariogram. At its simplest level, which is in 2-D space, the general idea is to take all points and correlate the difference in properties with distance. This is shown in Figure 4–1.

Since the use of statistics is required with large volumes of data, the calculation of semivariograms is best accomplished using a computer. A number of programs do this. Some of these programs are subsets of the larger integrated petroleum packages, as a subset of specialty geostatistical packages (such as ISIS) or, in a simpler form, as one of the programs in GSLIB.

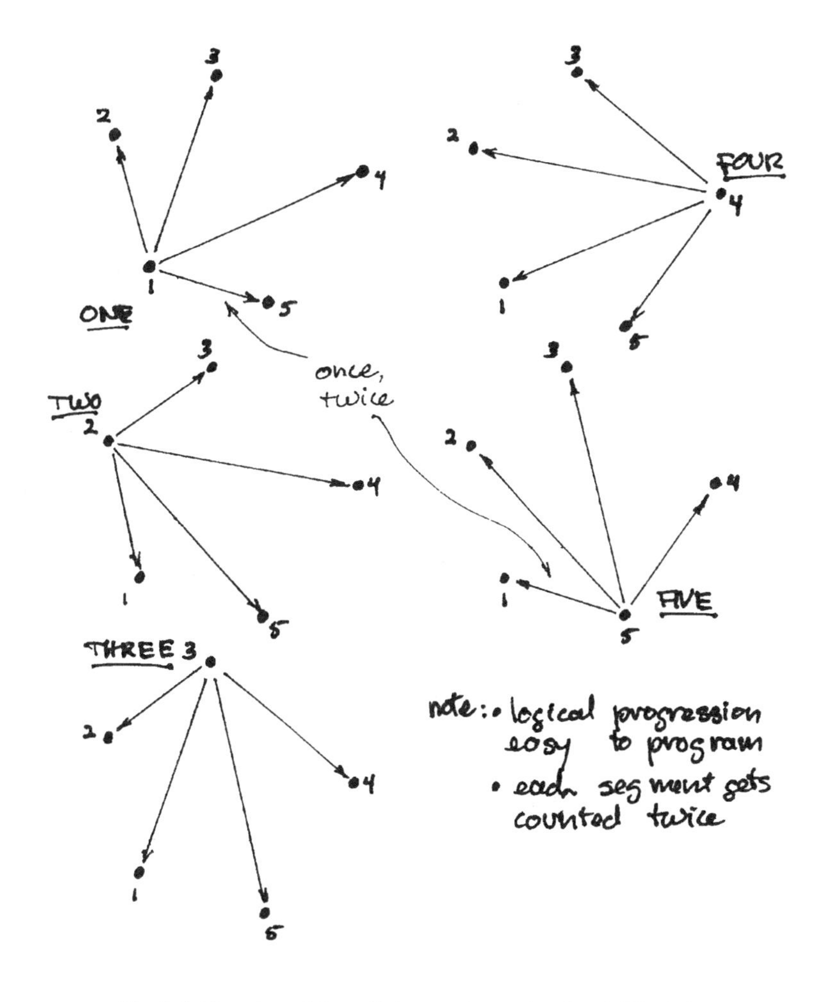

Fig. 4–1 Mapping of Data Points to Determine Spatial Variation

With a little development and the influence of conventional statistics, this data can be shown graphically, as illustrated in Figure 4–2. The ordinate is scaled in variance (standard deviations) and the abscissa in distance. This type of display is known as a semivariogram.

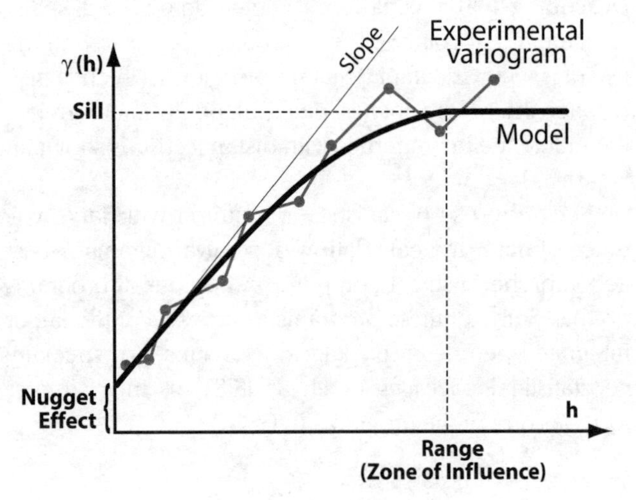

Fig. 4–2 Semivariogram

Interpretation of a semivariogram. The data on this example chart has some physical interpretation. The sill represents the total standard deviation of the sampled points. If sufficiently far from a known data point, there is no spatial information, and this distance is known as the range. Measurement error or background randomness is known as the nugget effect.

Different models—Gaussian, spherical, linear, etc. The semivariogram is used a number of times within geostatistical software. For this reason, most software has been designed to use an analytical equation to represent the semivariogram information. This choice, like many statistical decisions, is often arbitrary. The choice of analytical equation is usually made on a graph of the data. The actual data will rarely fit one of the analytical equations exactly. The idea is to find an equation that is a reasonably close match for the data.

In the past, the wording of some of the papers on geostatistics implied the data somehow was controlled by an underlying physics to a particular relation. A typical phrase might be, *the data fits the spherical model.* This has nothing to do with the actual physics and simply represents replacing a noisy, slow-to-calculate table of data with a smooth, quick-to-calculate analytical equation.

Fundamental concept of kriging

Kriging is first based on using a grid. Geostatistical grids are normally grid centered and the *output results* are located at points. Input data can have any *x*- and *y*-value and need not correspond to a grid point. Each grid point does need an *x*- and *y*-location. Geostatistical grids, unlike reservoir grids, are generally of constant cell sizes. Note it is common practice to put actual data at points. This will not cause a problem.

The idea is to set up a series of equations for each point on the output grid such that all of the actual input data is weighted according to distance using the semivariogram. This is only calculated once. This means every solution point will affect all of the others.

The implications to this are that output grid points existing at a distance greater than the range of the semivariogram will be related by the standard deviation. For a grid point with an actual input data value, the equation for this point is weighted completely on the value. Therefore, actual data is honored completely.

A matrix solution is made to optimize all of the point values to produce a minimum expected error. The error is described in a least squares fashion. The most common equation is shown in Equation 4.1.

$$z^*(u) = \sum_{\alpha=1}^{n} \lambda_\alpha(u) Z(u_\alpha) + \left[1 - \sum_{\alpha=1}^{n} \lambda_\alpha(u_\alpha)\right] m \qquad (4.1)$$

Modifications to this function can be used to create different variations in kriging. This is the key to kriging; it does not give an interpretation but rather an optimized interpretation. Variations can be made in terms of anisotropy and the inclusion of trends.

Simple experimentation will show the output results can be affected by grid spacing for the geological grid. Because of the error minimization, kriging reduces data detail and is an effective averaging technique.

Practical examples. Figures 4–3, 4–4, and 4–5 show three examples of reservoir descriptions created with kriging. For the most part, this work was done with Surfer, a product of Golden Software, Inc.

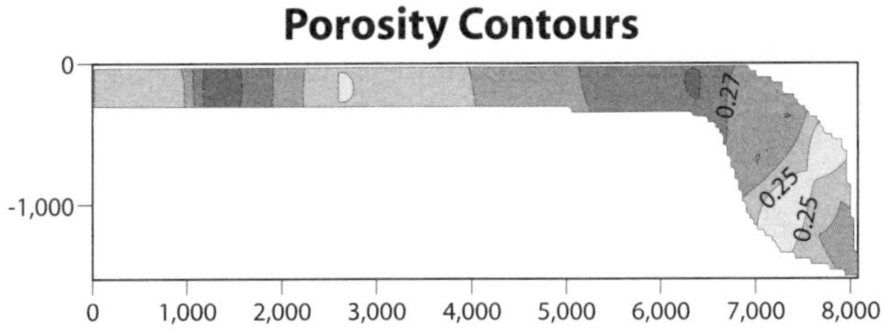

Porosity Contours

Fig. 4–3 Example of Reservoir Description—
Kriged Porosity

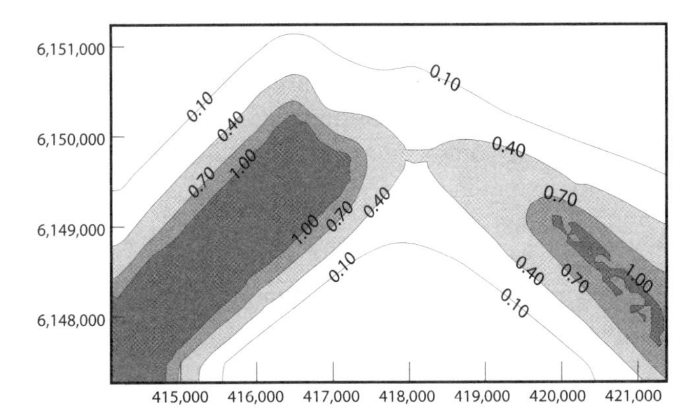

Fig. 4–4 Example of Reservoir Description—
Kriged Permeability

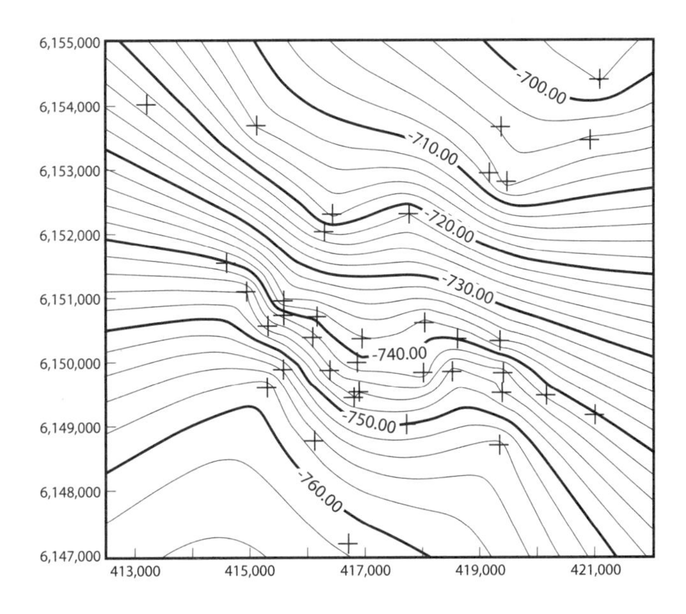

Fig. 4–5 Example of Reservoir Description—
Kriged Structural Top

Fundamental concept of simulation

The idea of simulation is similar to kriging in that an output grid is specified. A simulation also makes use of a semivariogram. In this process, a starting point and location have to be chosen on a statistical or random basis. The rest of the grid points are filled in using one of several techniques and results in the creation of a *realization* of the reservoir. In this manner, a possible picture of the reservoir is generated.

These realizations are not reality and are termed *equiprobable*. There are two implications to this:

- It suggests that, if some of the equiprobable solutions correspond to reality via history matching, it might be possible to infer some physical meaning to the realization.
- It suggests that the equiprobable realizations, through predictions, could be averaged to develop a possible distribution of production responses. It might also be possible to select some of these as more probable based on to-date history matching.

Figure 4–6 shows a cartoon drawn by the author for the cover of the *Journal of Canadian Petroleum Technology* that alludes to these possibilities.

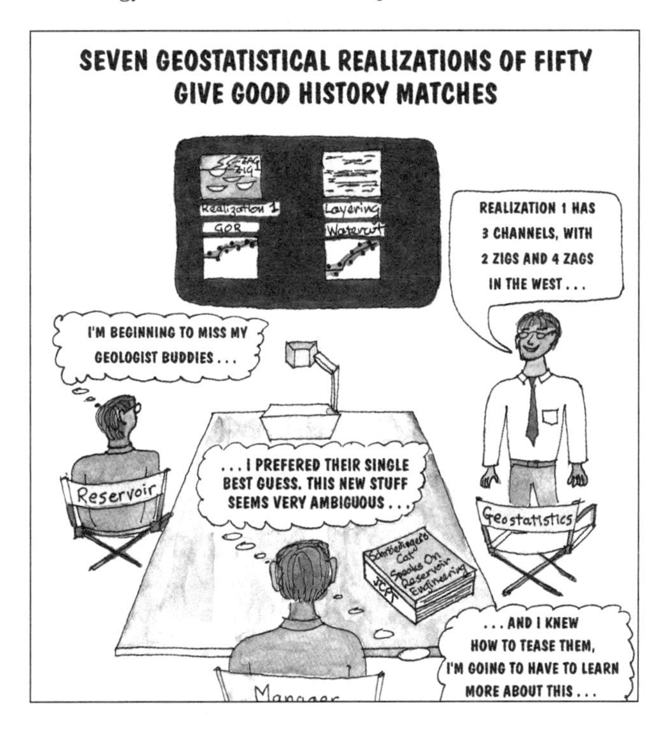

Fig. 4–6 Cartoon of Equiprobable Concepts

Annealing

Another possible method exists for taking reservoir information into account. An example of this is the results from a well test. The well's test permeability is largely affected by the permeability of the wellbore area. Of course, this data is not measured directly, and its exact distribution cannot be known.

However, if we take a segment of the reservoir and develop an average permeability, we can test the model. Alternatively, we can take a simulation that meets an overall distribution and average and swap some of the grid cells to match the well test data in the area surrounding the well. This general process is known as *annealing*. It offers the possibility to alter our equiprobable models based on well test information.

Workflow process

In this discussion, a general workflow has been developed for an integrated/multidisciplinary study.

- *Geological Model.* In order to evaluate rock properties, the various lithologies must be known. Geological models are always ascertained deterministically. There will always be some element of analogy, i.e., by a geological interpretation as outlined in the previous chapter. Clearly, in order to work with a semivariogram for example, the *bins* (normally, some type of facies) must be known.
- *Structural Model.* If the reservoir has been faulted or changed in shape by structural processes, the original spatial relationship in which geological relationships and properties were brought about will also be disrupted. In the previous chapter, the use of structural models and how to re-create the original bedding was described. These tools are used to undo the structure for geostatistical analysis and then reapply the structure to develop the final simulation grid. For correlating fracture properties, which can also be analyzed using geostatistics, the present structure would be more appropriate.
- *Petrophysics.* With the geological model complete, the log analyst will normally use the lithologies developed to prepare a foot-by-foot log analysis. The different lithologies are set with flags, and this allows grouping and analysis that varies with these groups.
- *Rock Property Generation.* Following the process described previously, the simulation engineer will use the lithologies, available core, and special core analyses to develop rock properties for each lithology. Note this is an iterative process. The analysis of the core data might help differentiate geological properties, which may result in the geologist revising or altering the interpretation.
- *Reservoir Model.* The next stage is reservoir simulation, which is described in detail in this text. This starts with understanding the previous work and screening the input data.
- *Economic Evaluation.* The results from the simulation will be used to forecast production, reserves, and costs and to determine discounted cash flows. In some companies, this is done by the reservoir engineer. This is typical for most Canadian companies working in the Western Canadian Sedimentary Basin. In a significant number of companies, economic evaluation is done by specialist groups.

Case history—McMurray formation in Northeastern Alberta

Potentially large reserves of heavy oil are present in Northeastern Alberta. This topic is discussed in more detail in chapter 19. One of the major issues is the size and distribution of shales within the McMurray formation. The reservoir is located in unconsolidated channel sands. Economic development is most attractive when a number of these channels intersect vertically, providing large pays. The channels are of fluvial origin.

From a development perspective, geostatistics offers a number of potentially valuable tools. First, it would be useful to determine the exact location of these channels. This cannot be determined seismically since there is relatively little contrast between the velocities in the shale and the sand. Furthermore, it is the author's understanding that obtaining 3-D seismic at the relatively shallow depths of the McMurray requires a dense shot point array. This type of program is expensive to obtain.

Once the location of the channels has been determined, a technique for placing shale intercalations and mud-filled channels would be useful in evaluating how steam would rise vertically in the steam-assisted gravity drainage (SAGD) process.

A number of major issues arose as follows:

- There are no statistics on how wide the channels are, where they are located, or what profile they have. Miall, in his book *The Geology of Fluvial Deposits*, gives a thorough description of the different fluvial environments.[3] There is significant variation in this type of environment. At present, none of the analogues suggested have complete sets of statistical information.
- Typically, the development of the early SAGD projects utilized quite dense drilling on 16 ha (40 acres). Even with such close spacing, correlation from one well to the other can be quite difficult due to the rapid changes in the depositional environment. There is no data for the intervening area.
- It is known that the channels overlap along trends, and it is clear there are some physical controls on where they are located. Although not directly determinable, it is clearly not entirely random either. There are mathematical fixes, such as the use of attractors, but this does not rely on fundamental physics. It may still work provided the right value can be found.
- The channels are not straight, and an indication of the path of the channel would be helpful.
- Some of the shales, actually mudstones, are from channel infills, and some are contained within the channels as a type of lag mud. This suggests the mud-filled channels could be modeled as objects. The intercalations or lag muds could be modeled with porosity-permeability distributions within the channels.

Some attempts have been made to use geostatistics in the public domain. This material was filed as part of a government hearing. The early work was based on results from fields not having a considerable volume of statistical study, e.g., Prudhoe Bay. This submission was made based only on the information the submitting company could find. In the end, the approach was voluntarily withdrawn when further work was unable to resolve a number of issues.

The problem can likely be solved. Potential sources of detailed geological data are available that include a number of open pit mines and numerous river outcrops. This is a major work commitment the author believes will ultimately be completed. Such results will not be available in sufficient quantity for probably 5–10 years. It is possible other analogues could be used; for instance, Miall has tabulated deposits with similar characteristics. These may eventually be completed, if proper surface outcrops can be found. Geology is becoming more quantitatively oriented in these fields.

Such studies may help with cross-sectional dimensions and detailed description of the permeability within the channels. In terms of calculating channel trends, this will likely be developed based on other techniques. Typical channel radii were shown in chapter 3. The early work on this type of analysis was done by Leopold and Maddock as well as Leopold and Wolman.[4, 5] The latter was also evaluated as part of the early geostatistical work on the McMurray formation. However, this work has progressed greatly since 1957. Such work may be useful in helping to develop channel paths and the underlying physical controls.

Limitations of geostatistics

In broad terms, geostatistics has a number of limitations:

- The development of geostatistics has provided powerful tools. Yet, much of the development is based on arbitrary decisions on how to weight data. The use of mathematics often gives the appearance of underlying physical relations, which, in fact, do not exist.
- Most data in geology is not independent, such as porosity and permeability. This is a fundamental assumption in the development of much of the mathematics. This does not mean statistics are not useful; it does mean the results are not concrete but provide a basis for interpretation.
- Considerable interpretation is made in setting up the appropriate categories for data analysis. Restated, the technique really depends on geological interpretation.
- Finding the appropriate analogue field with the detailed data required for geostatistical analysis and geostatistical simulation can be time-consuming and difficult. There may not be sufficient data to use the technique. In this case, it is necessary to use traditional techniques.

Summary

In the modern context, a reservoir or simulation engineer will most likely spend considerable time in a multidisciplinary development where the team's key objective is accurate reservoir characterization. This leads to division of labor, but it also identifies certain specialized areas where individuals will have particular responsibilities. Simultaneously, the work done by individuals will affect others, and it is important to understand the implications of these interactions. The author has outlined his view of where simulation work fits in a broader context.

Reservoir characterization is technically complex. This book is about the simulation elements and does not address all of the broader issues. One area that overlaps simulation extensively is the generation of input permeability, porosity, and net pay with the use of geostatistics. Some basic comments on these techniques are outlined. Again, there is considerably more detail available in other references.

One area that concerns the author is the gridding options available in geostatistical software. This is one area where geologists (who typically perform geostatistics) need to be cross-trained in reservoir simulation. Alternatively, the work needs to be done in concert between the geologist and simulation engineer. There are many misconceptions about gridding, which are discussed later.

Geostatistics is a useful tool, but it does have some limitations. It is likely kriging will become a regular part of a simulation engineer's work. Geostatistical simulation will likely be less common, but this usage should increase with time. The use of geostatistical tools requires a significant investment in engineering and geological time and expense. A large proportion of work, particularly on smaller onshore oil and gas fields, is conducted using traditional deterministic techniques. In many cases, geological interpretation and judgment can achieve the necessary results without the extensive use of statistics. There are still many cases where insufficient data will limit the depth of geostatistical analysis.

References

1 Thakur, G., et al., *Integrated Waterflood Asset Management and Integrated Reservoir Management*, PennWell Publishing Co., 1998.

2 Cosentino, L., *Integrated Reservoir Studies*, IFP and Editions TECHNIP, 2001.

3 Miall, A.D., *The Geology of Fluvial Deposits,* Springer Verlag, 1996.

4 Leopold, L.B., and T. Maddock Jr., "The Hydraulic Geometry of Stream Channels and Some Physiographic Implications," Paper No. 252, U.S. Geological Survey, 1953.

5 Leopold, L.B., and M.G. Wolman, "River Floodplains: Some Observations on Their Formation," Paper No. 282-C, USGS, 1957.

5

Production Performance Analysis

Introduction

Reviewing the production performance of the reservoir is an important part of constructing a simulator for two principal reasons. First, it will help determine the correct input data required. Second, it will give direct clues as to the depletion processes, i.e., mechanisms occurring in the reservoir. In a number of examples, this process could have avoided problems, and in some cases, production performance data was useful in setting the scope of a simulation project.

Conceptual model—reports

This is a process of putting together a consistent mental model. In the cases mentioned previously, some of the inputs were simply wrong. The only way to tell is to look at all aspects of the problem and ask if all the information adds up. This skill is essentially one of pattern recognition. Production performance data can be used positively. To ensure this isn't overlooked, include a production performance evaluation as part of the simulation reports.

In the next two sections, two situations involving misconceptions about reservoir mechanisms are outlined. In the first case, a mistake was made in PVT data interpretation, and in the second case conclusions were made about a reservoir based on early and incomplete data. The significance of the latter case is previous interpretations often need to be scrutinized when subsequent work is done. The initial interpretations were likely the best interpretations available at the time they were made using the limited data available. The mistake is to continue these interpretations when they become inconsistent with observed performance.

Examples of interpretation errors

The author reviewed a material balance study based on production occurring above the bubble point. However, a review of production plots showed increasing GORs on a number of wells. The calculation overestimated the OOIP based on an assumed undersaturated reservoir. The engineer was tricked by PVT data, which indicated the bubble point was lower than the original reservoir pressure. As will be discussed in chapter 6, this is a frequent occurrence.

Another field was assumed to have a major gas cap. The original conceptual model for the reservoir, as near as anyone could tell, came from a drillstem test (DST) interpretation. In this case, the objective was a material balance calculation. The production plots indicated no increasing GORs despite a 1,000 psi pressure drop corresponding to more than 30 years of production.

Practical experience indicated it was inconceivable *no* gas coning occurred (see the course problem set). This is another example of where one set of PVT data (out of five) and some preconceived notions led a very large number of people in the wrong direction.

A well-established author had published a material balance study during the early stages of field production. In the process, the gas cap interpretation had become gospel, despite production data to the contrary.

An offsetting operator, who also had significant production in the same field, attempted a reservoir simulation assuming both a gas cap and vertical equilibrium (VE), explained later. It was a simulation failure because there was no primary gas cap. Today, few engineers would consider using this technique, since the field could now be modeled with a 3-D grid using substantially more powerful computers.

Drive mechanisms

During initial training in reservoir engineering, many drive mechanisms are outlined, such as solution gas drive, water drive, and gas cap expansion drive. From an elementary text on reservoir engineering using solution gas drive as an example, three main plots are explained. The plots are used to describe reservoir performance (and a fourth will be noted here):

1. A plot of reservoir pressure. Above the bubble point, the pressure drops rapidly with cumulative production, followed by a decrease in pressure decline below the bubble point.
2. A plot of production. As time goes on, the production rate declines. In fact, for solution gas drives, production rate normally shows as a relatively straight line on a semilog plot of production.
3. A plot of GORs. This is constant or slightly decreasing at first and then rises with time.
4. A plot of water cuts. This normally doesn't increase for solution gas drives, ignoring potential coning. However, for waterfloods or water drives, water breakthroughs will occur.

Points 3 and 4 are normally plotted on a single plot using different colors, usually green for oil, red for gas, blue for water, and black for operating hours or well count. Most reservoir engineers use a semilog scale.

Production performance analysis

In large part, production performance analysis consists of the reverse of this process. We plot graphs of reservoir pressure versus time, or cumulative production, and production. From this we can interpret the reservoir mechanism.

On the surface, this is a simple process, and this is true in some cases. However, in the majority of cases, it is not nearly so simple, and experience helps. It is common to plot both the reservoir as a whole, groups of wells that may be isolated (such as a fault block), and individual wells.

Two major topics will be discussed in this chapter. The first is production plots, and the second is building a plot of reservoir pressure versus time or cumulative production. The first topic is easier to address and will be discussed first. The second topic can involve significant interpretation using fairly sophisticated techniques.

The analysis of production performance data should lead to increased chances of success in interpreting the drive mechanism(s) in the reservoir and in the quality-applicability-correctness of the simulation study subsequently carried out.

Production plots

Organization. As with any other data, organization helps. Typically, production plots often come from the computer based on legal subdivision (LSD). This, of course, does not always correspond to the most systematic pattern in the reservoir. Therefore, the author normally reorders the plots more logically, using 11 × 17 in. size paper and inserting them in an Acco-Press binder. The larger plots allow for annotations. Included in the binder is a location map inside the cover and on the backside of the previous plot. Note the computer plots come out single sided. The well is highlighted for the production plot on the opposite page. The location of offset injectors is normally also highlighted in a different color. If the wells are grouped, the total reservoir goes on the first page, and then each individual well is found after the groupings.

Part of the reason for doing this is purely technical; it helps spot patterns. The second reason is it makes presentation of the material much easier.

Raw data. Later, it is necessary to input production data in the simulator. Although this topic will be addressed later, bear in mind this data will be used again. Therefore, it is advisable to keep track of the data in tabular format.

If a company does not have computer records of production, it is possible to buy shrink-wrapped production accounting/decline analysis packages off the shelf. These allow different pools to be defined, different well groupings to be defined, etc. and can be very useful for manipulating, analyzing, and printing production data. At least one ARE client company uses this type of software on their international operations.

Overcoming data problems. Despite extensive government regulations and inspections, data problems with production still occur in some instances. In southern Saskatchewan, a large field of relatively shallow wells produced from the Viking formation. These wells were in advanced stages of decline.

In these southern Saskatchewan fields, the original measurement facilities were designed and approved by the government. For this particular oilfield, this consists of flowlines, satellites, and test and group separators followed by treaters. An important design criterion is whether the test separator is measuring flowline fill or actual well production. In essence, the volume from the flowline has to be a sufficiently small part of the total fluid coming from the well. While this is true during the early part of the life of the well, it is not always true later when rates drop considerably.

Wells do not flow, or pump for that matter, consistently. This is due mostly to wellbore effects where liquid often builds up and slugs. Pumps will go down with power shutdowns, maintenance, and, depending on the pump sizing, the well may be pumped intermittently—known as a pump off controller.

The field in southern Saskatchewan was on waterflood. Depending on what stage of breakthrough exists on individual wells, the amount of oil will vary, even if the well itself is running relatively consistently.

While the reasons are varied, the net result is the production reports from individual wells show a considerable amount of noise. For other reasons discussed later, this reservoir was best suited to simulation analysis using a single type well. Variable production data makes picking and modeling an individual well difficult.

For this particular field, a relatively easy solution is to use data from multiple wells. The production data is normalized to a common start date. An average production profile can be created by adding all the production and dividing by the number of wells.

Other reservoirs, particularly those in advanced stages of depletion, have similar issues that can often be resolved by this approach.

Production data in general. Sometimes it seems as if every reservoir has production data problems. Knowing the level of inspection occurring in the Province of Alberta, this is actually rare in the jurisdiction. Much of the time, this represents production data difficult to explain from a reservoir engineering perspective. It is difficult to explain the production data using the physical distribution of assumed properties, i.e., production data does not fit with the geological interpretation assumed.

Most surface facilities are fairly simple from a flowchart perspective. The industry also has extensive experience with flow meters. With relatively rare exceptions, field personnel are quite conscientious regardless of the country or area where they work.

While data problems can and do occur, this may be an indication our understanding of the reservoir is not correct. In the author's experience with a number of metering problems encountered in the field, the probability of metering problems depends on the level of inspection, but it is just as likely, if not more so, the reservoir description is incorrect.

Decline patterns. Very often, well declines will show a distinctive pattern as shown in Figure 5–1. This plot is from a gas condensate reservoir. The reservoir has low permeability, so there is an element of flush production. There is also condensate dropout and corresponding productivity impairment.

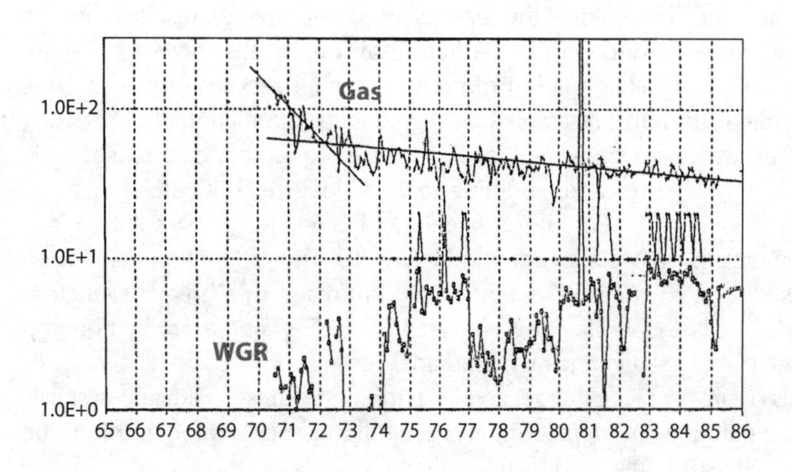

Fig. 5–1 Distinctive Production Profile—Gas Condensate Reservoir

It is an example of the type of production profile an engineer wants to generate using single well type models. As indicated previously, production analysis can suggest the type of model to work towards using single well models.

Contouring. Much can be said for contouring water breakthroughs and other data. An example is shown in Figures 5–2, 5–3, 5–4, 5–5, 5–6, 5–7, and 5–8. This pool is hydrodynamically trapped. This can be seen in the cross section on Figure 5–2. Note the dip of the water contact is quite subtle—about 0.23 degrees from the horizontal.

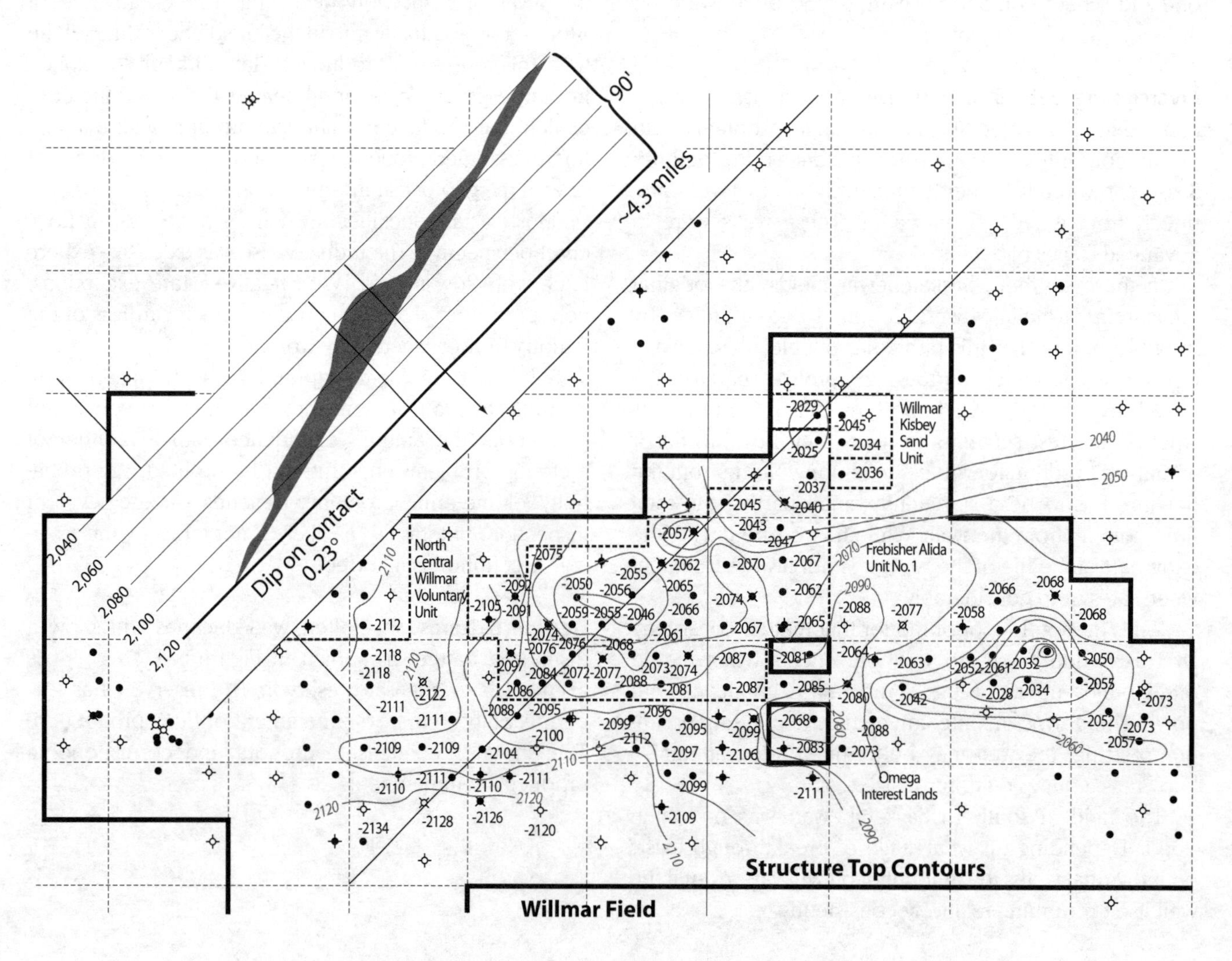

Fig. 5–2 Production Analysis Plots, Contours/Structure

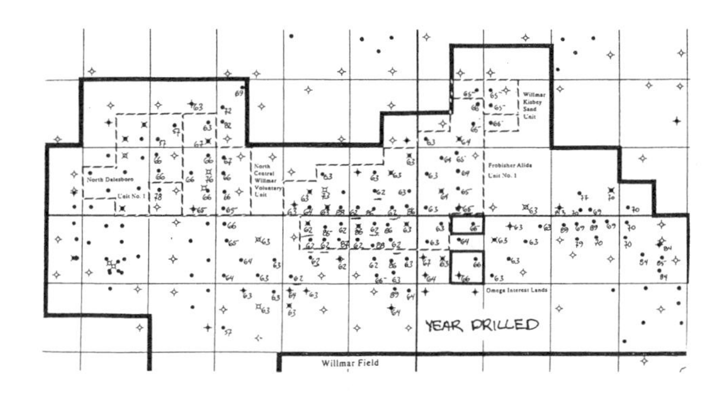

Fig. 5–3 Production Analysis Plots, Contours/ Year Well Drilled

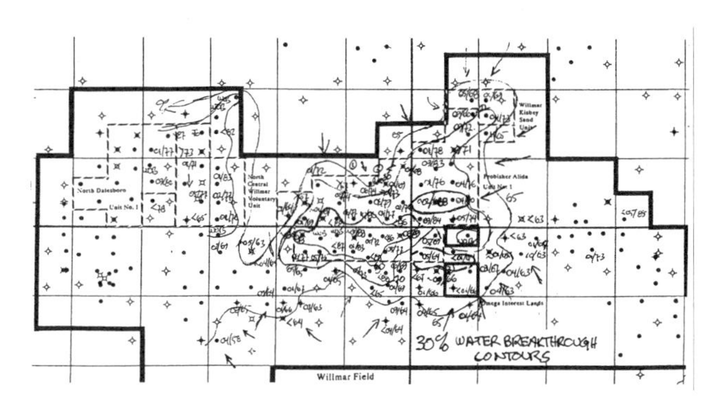

Fig. 5–6 Production Analysis Plots, Contours/ Breakthrough Time—30% Water Cut

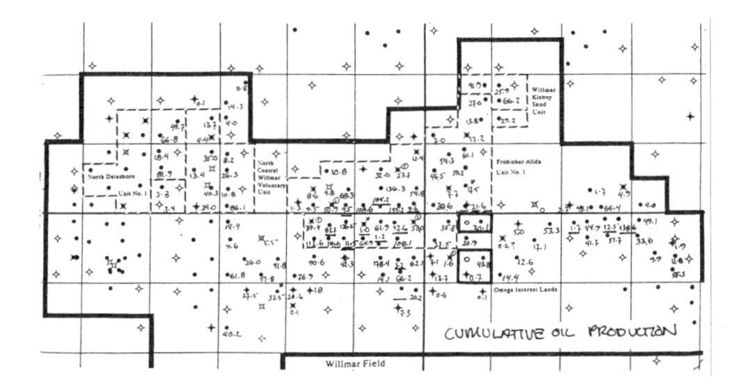

Fig. 5–4 Production Analysis Plots, Contours/ Cumulative Oil Production

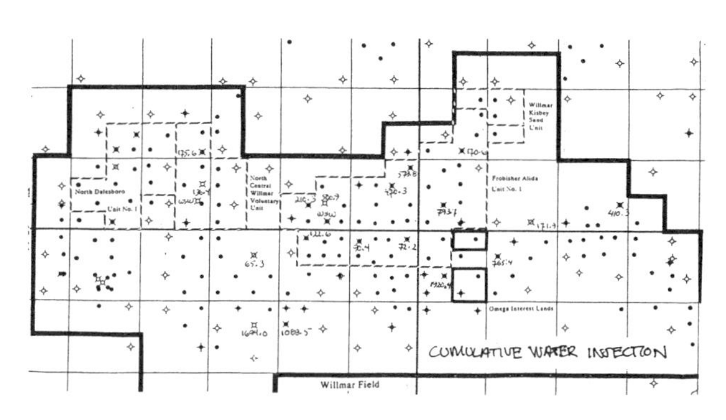

Fig. 5–7 Production Analysis Plots, Contours/ Cumulative Water Injection

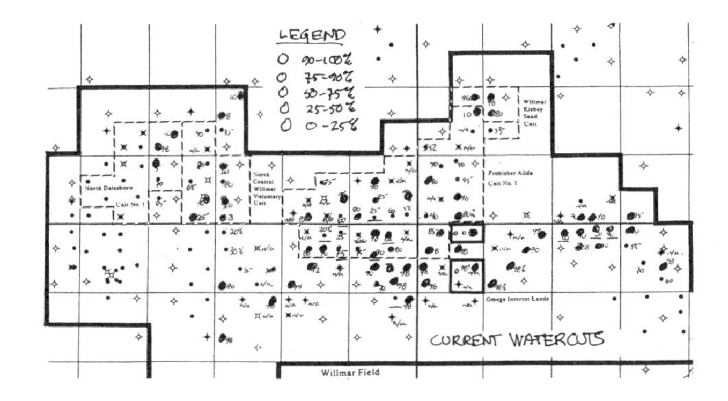

Fig. 5–5 Production Analysis Plots, Contours/ Current Water Cuts

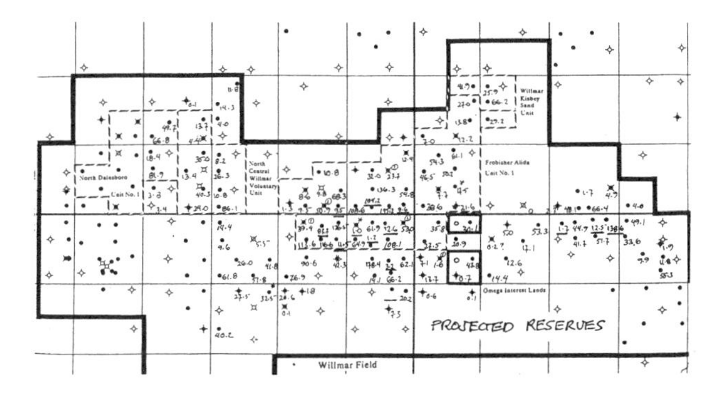

Fig. 5–8 Production Analysis Plots, Contours/ Projected Reserves

The most interesting plot is the progressive influx of water. A water cut of 30% has been arbitrarily chosen as water breakthrough. Note the contours gradually progress to the middle of the reservoir, where the pay is thickest but not structurally highest.

Other plots in the suite include drilling rig release date, cumulative production, extrapolated reserves from production decline plots, as well as current water cuts. This type of analysis takes a few days to perform and can be done from well summary forms and production plots.

Some water injection was implemented. Water production does take place throughout the reservoir. The central parts where the water converges are excellent locations for horizontal wells.

Systematic water encroachment. An example of this relates to a reservoir ARE worked on in the foothills of northeastern British Columbia. Geologically it looked like Figure 5–9.

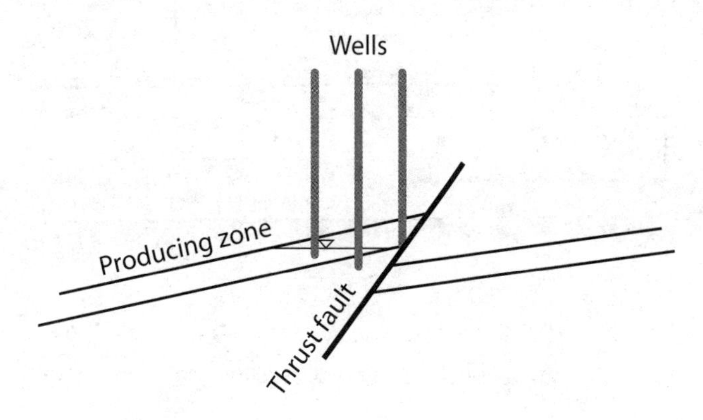

Fig. 5–9 Geological Cross Section—British Columbia Reservoir

Production performance data indicated the updip wells produced water at the same time or earlier than the downdip wells. Perhaps the water moved up by fingering; however, this would require the fingers to wriggle around the downdip wells. While abstractly possible, this did not seem to be likely, which indicated something else must have been happening rather than updip migration. Two possibilities were envisaged. Either the water was moving up the fault plane at the updip extreme of the thrust fault, or water was coning up from below as shown in Figure 5–10.

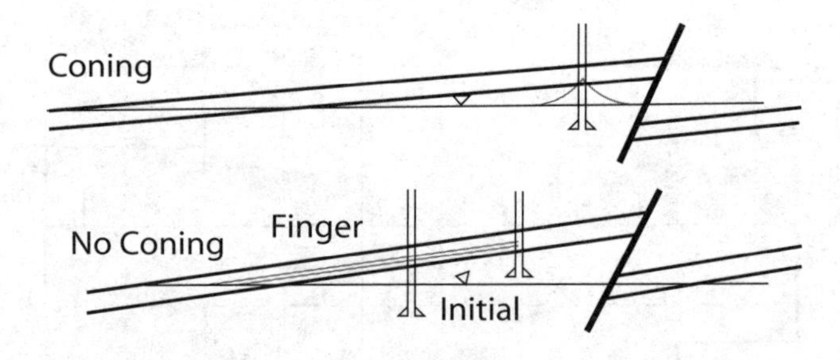

Fig. 5–10 Possible Sources of Water—British Columbia Reservoir

The latter point is quite subtle. The underlying formation was incapable of economic production due to its low permeability. To detect this requires a good understanding of flow, which the author learned from a friend in the hydrogeology business. Vertical flows in the subsurface are very significant. Vertical flow of water actually takes place across shales because vertical distances are very short, i.e., the shales are relatively thin. Large flow areas are involved in the vertical direction, and pressure gradients are much larger than typically occur horizontally. Horizontal flows are limited by low-height and low-pressure gradients but are enhanced by larger permeabilities. The model shown in Figure 5–11 was constructed and tested.

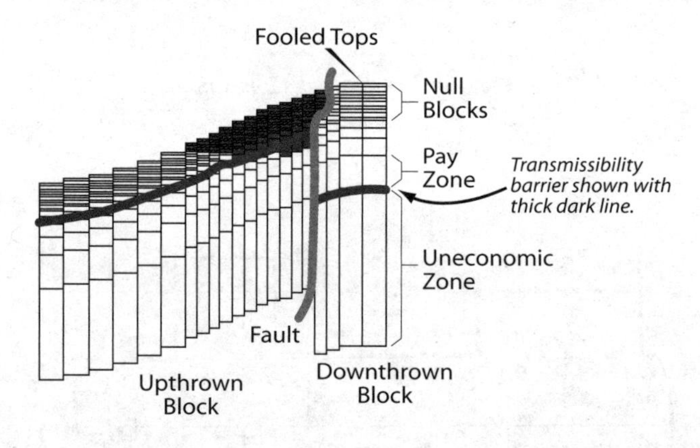

Fig. 5–11 Model Used to Test Actual Source of Water

Testing either method is easy because the simulator allows one to stop flow by specifying a transmissibility barrier. The results conclusively showed upward migration, which was enhanced near the leading edge of the thrust sheet.

Pressure plots

Types of raw data. Typical raw data consists of a mixture of static gradients and pressure buildup analyses. Ideally, the static gradients are taken after sufficient time so the wellbore pressure represents the local reservoir pressure. Sometimes this is true, and sometimes it is not. Lower permeability reservoirs take longer to build up than high-permeability reservoirs.

Many pressure tests will have complete pressure buildup analyses. Pressure tests offer the possibility of providing effective average reservoir properties as well as a pressure. This potential is frequently not realized and will be discussed in more detail later. For the time being, it is sufficient to assume the pressures from such tests are correct.

Gathering pressure data. Obtaining all of the pressure data is a difficult process. In general, this demands each well file be reviewed in detail. Pressure data can also be obtained from government sources. Needless to say, not all pressure data actually makes it to the government coffers, even when this is a legal requirement. More surprisingly, the company subsequently loses much of the data sent to the government, and the government records need to be reviewed to make sure all of the data available is gathered. It is ironic that much of the record keeping by the government seems to protect oil companies from oversights. Sometimes pressure data will be available from the pressure transient analysis specialist in the company. Again, serendipity often seems to play a hand here, and, although most data falls into the well file, there are exceptions.

The downsizing process has really hurt the quality of files in operating companies. Very frequently, data is contained in individual engineers' offices. With less time available, the housekeeping functions are not as tightly followed, and a tremendous amount of data gets lost when someone is terminated. No one has time to either clean their material or know where things should be forwarded. Some companies now retain the files of the departed; however, they often end up in a filing cabinet or another engineer's desk. This can involve some detective work to determine who the past engineer(s) was and where the material is located.

Plots of raw data. Figures 5–12, 5–13, 5–14, and 5–15 are four examples of simple pressure plots made with raw data. In many cases, this will be sufficient to indicate what is happening in the reservoir. The following is a brief discussion of these plots:

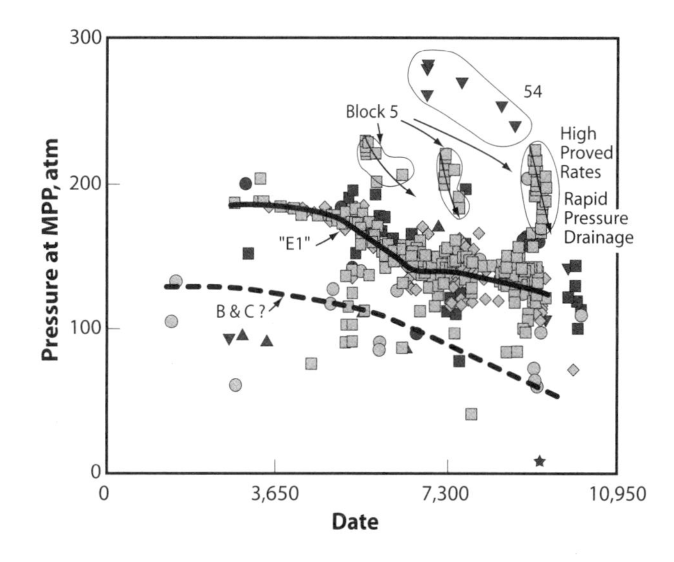

Fig. 5–12 Pressure vs. Time Plot—International Reservoir

- In the example illustrated in Figure 5–12, it was anticipated the reservoir would be separated by lateral shear faults running at right angles to the front of the thrust sheets. In this case, all of the anticipated well groupings proved to be incorrect. The kink in the pressure trend represents a bubblepoint. This is from an international reservoir.

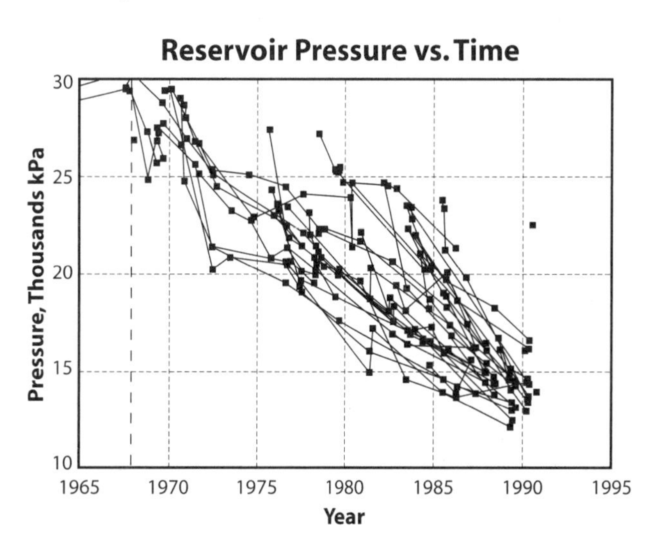

Fig. 5–13 Pressure vs. Time Plot—Thrust-Faulted Reservoir

- The graph in Figure 5–13 is also a reservoir in a thrust-faulted environment. Here the field is not really a single reservoir but rather many different smaller pools. Note this is a gas reservoir.

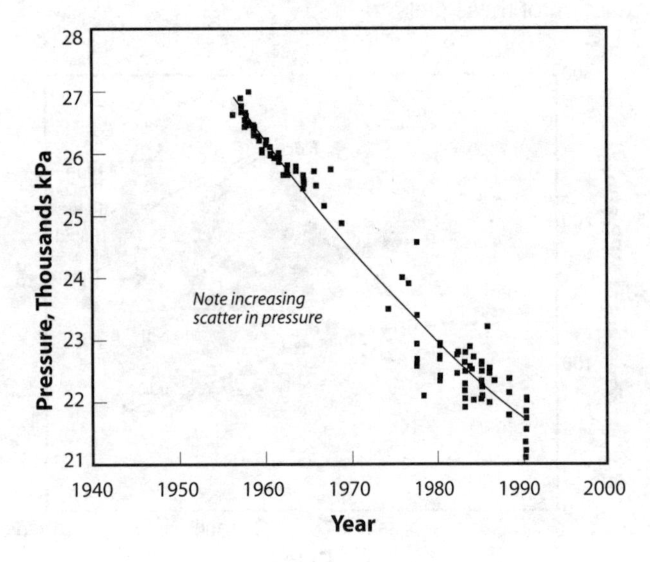

Fig. 5–14 Pressure vs. Time Plot—Undersaturated Reservoir

- In Figure 5–14, the reservoir was thought to have a gas cap. The reservoir has partial water drive. The fact the GOR never increased was positive proof that no gas cap existed. The reservoir had not dropped below the saturation pressure. The aquifer has high permeability but was of a finite size. The size of the aquifer corresponded to the wet areas of a very large carbonate reef.

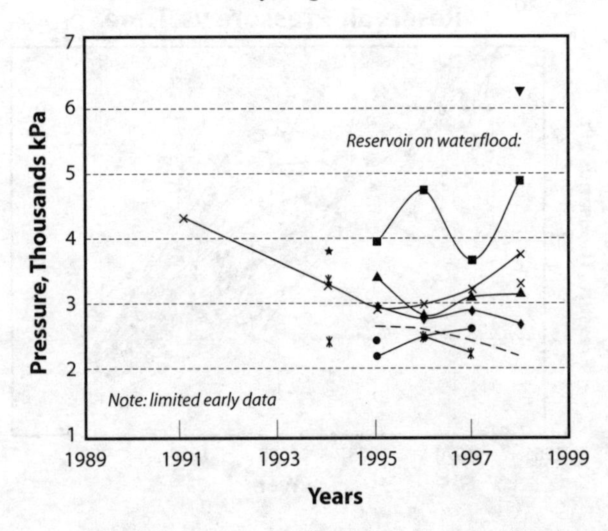

Fig. 5–15 Pressure vs. Time Plot—Saturated Reservoir

- Figure 5–15 shows a plot of a relatively standard, saturated reservoir. Note the scatter of data. From this plot, we may conclude the scatter of data often tells as much about what is happening as the trend itself. Unless the reservoir is of very high permeability, such scatter is to be expected. Sometimes there will be more than one discernible trend, in which multiple pools can be defined.

Contour plots. As with production data, contour plots can be very useful. The plots shown in Figures 5–16 and 5–17 were created using the kriging option in Surfer, a product of Golden Software. Areal variations in pressure are quite common and can often indicate the parts of the reservoir that either have better permeability or may be separated by barriers.

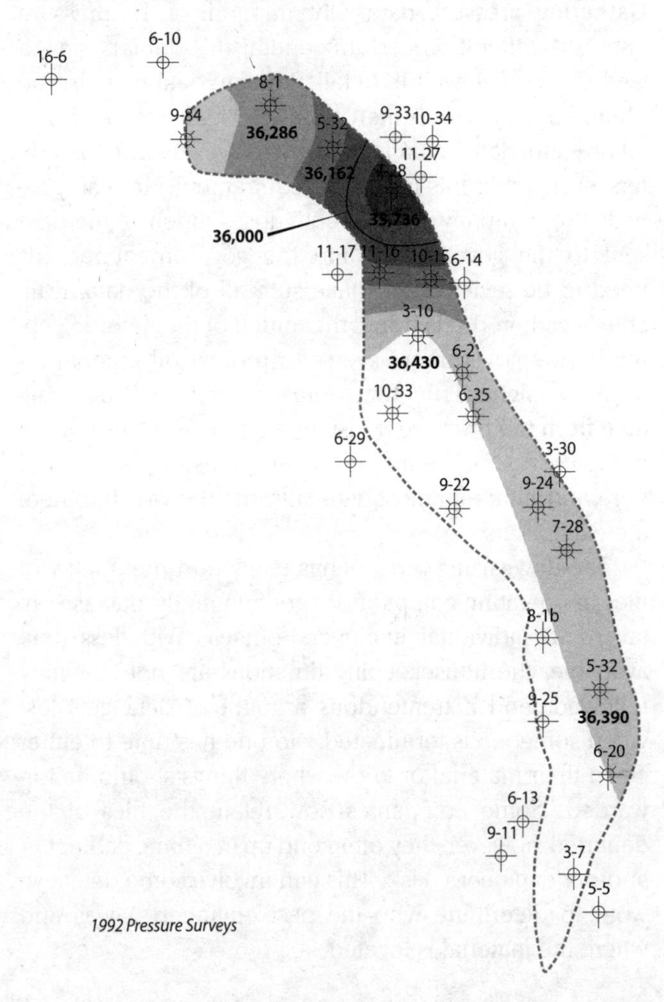

Fig. 5–16 Reservoir Pressure Contours—Gas Reservoir, 1992

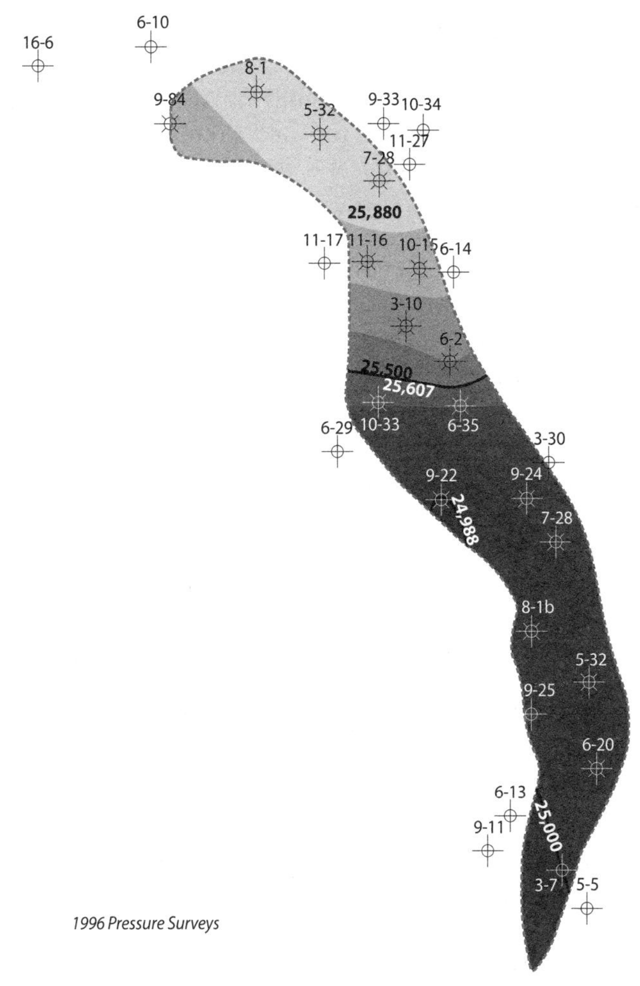

Fig. 5–17 Reservoir Pressure Contours—Gas Reservoir, 1996

Field practices. Bottomhole pressure recorders are typically run in pairs. This is colloquially referred to as *tandem recorders*. They are, in fact, physically separate devices and can be run independently. Although this practice was started with the older Amerada gauges, failure frequency is still sufficiently high among the modern gauges such that this remains a prudent practice. It also provides a built-in quality control.

Note the two pressures will not be identical. The recorders are separated vertically in the well. The two pressure readings differ by the pressure gradient of the well fluids in the interval times the difference in depth.

Types of bottomhole pressure recorders. A number of basic types of bottomhole pressure recorders are available. Historically, the Amerada gauge was the most popular pressure recorder used in the industry. This gauge used a Bourdon tube for sensing the pressure and a mechanical clock that moved a recording medium (metal card) during the test. Note a Bourdon tube measures pressure differences, and the measurements obtained from these instruments are relative to the atmospheric pressure contained in the tool. They are, therefore, measurements of gauge rather than of absolute pressure.

In the early 1980s, quartz pressure gauges were developed. The frequency at which quartz crystals vibrate changes with pressure, and this signal can be calibrated. Quartz gauges inherently measure absolute pressure, which controls the frequency at which they vibrate. The vibration frequency of crystals varies with temperature. Hence, quartz gauge readings are different from Amerada-type gauges. Also, each crystal must be calibrated—all quartz pressure gauges do not have the exact pressure reading or frequency of vibration.

The accuracy and precision of these gauges provided a dramatic improvement compared to the existing Amerada/Bourdon tube gauges. This resulted in a number of new effects seen on pressure gauges, such as the effects of tides in large high-permeability reservoirs.

Recently, further changes in pressure gauge technology have taken place. The electronics in the quartz crystal gauges are delicate, and measurements drift with extended exposure to temperature and vibration. Vibrating wire gauges have proven to be more accurate and more durable in hostile environments.

Finally, there are gauges utilizing mechanical strain gauges. The strain gauge responds to changes in resistivity as the pressure changes.

Temperature stabilization. The temperature within the earth increases with depth in oil and gas wells. Therefore, temperatures are always changing as a gauge is lowered into the well. Since the readings vary with temperature, it is important to allow sufficient time for the gauge assembly and recorders to equilibrate. Specified accuracies assume temperature equilibration has taken place.

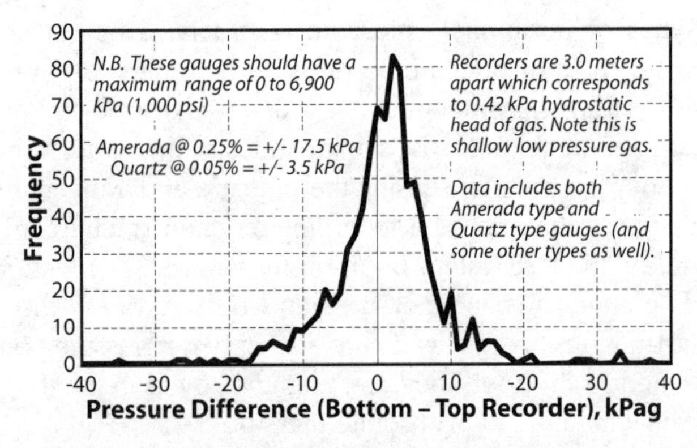

Fig. 5–18 Differences in Pressure Top and Bottom Recorders at Surface (in Lubricator)

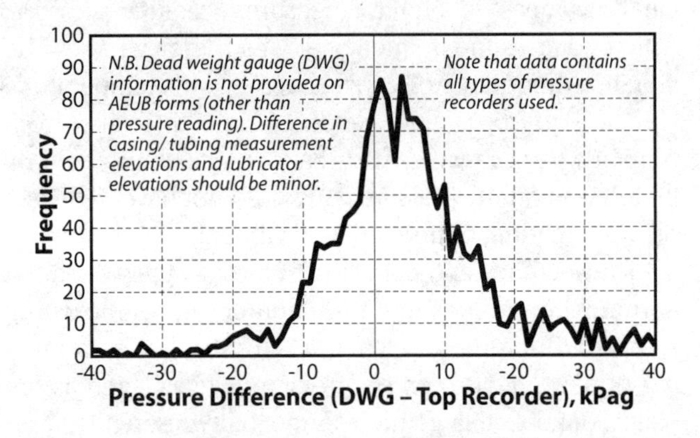

Fig. 5–19 Differences in Pressure Bottom Recorders and Casing Drilling with Gas at Surface

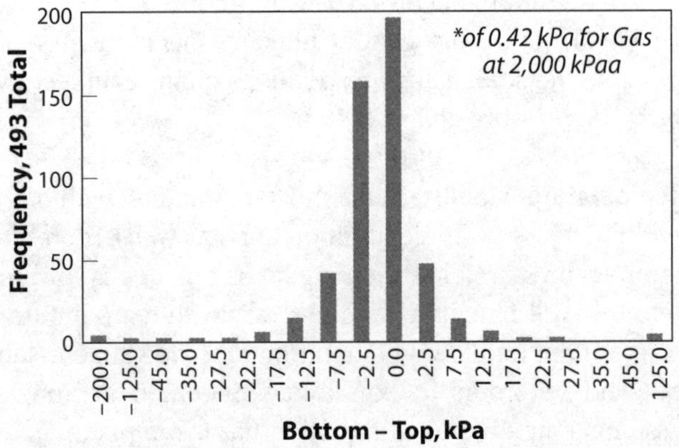

Fig. 5–20 Empirical Accuracy of a Large Volume of Data—Recent (Post-1985) Surveys

Empirical data accuracy. Figure 5–18 shows the distribution plot of the bottomhole recorder error taken from 856 static gradients from a study by ARE. In this figure, the difference between top and bottom recorders is plotted. This graph indicates the bottomhole pressure recorders are not as accurate in the field as the published specifications.

The dots in Figure 5–19 represent the distribution of the difference in the reported pressure between the surface deadweight gauge reading and the bottomhole recorder while at the initial surface stop. The data set includes combinations of electronic, mechanical, deadweight, and bottomhole recorders.

Figure 5–20 shows the difference in pressures between top and bottom recorders for recent (post-1985) data. This data indicates improved accuracy.

Calibration of bottomhole pressure recorders. Bottomhole recorders are calibrated at a constant temperature inside an oven where they are steady and not subject to vibration or movement. Measurements are required with both different temperatures and pressures.

On the older mechanical gauges, corrections were performed manually or with a computer program. The newer gauges often use a table of corrections loaded into the pressure gauge memory. Extracting the pressure measurements from the newer pressure gauges requires the use of a computer link to a personal computer.

The frequency of calibrations varies somewhat with the type of gauge. In Alberta, mechanical gauges must be calibrated every three months, with a bench test once a year. There are no strict requirements yet for electronic gauges.

Actual measurement conditions. In the field, pressure recorders are transported across dirt roads in the backs of trucks, where they are subject to considerable vibration and shock. Often the recorder is taken from –30° C to +15° C and back to –30° C, sometimes in less than a one-hour period, for instance, during a static gradient test.

While it might seem that quartz gauges would be the most stable, having no moving parts, this has not proven to be the case. Most ultra-reliable installations use vibrating wire gauges. In time, both the electronics and the actual quartz crystals can drift, in particular due to the effects of high temperature. The crystals are also subject to physical degradation.

Field calibration. It is normal to double-check all of the gauges in the field by taking pressures in both the tubing and casing. Historically, this was done with a portable deadweight gauge. The deadweight piston gauge is used to measure pressure in terms of fundamental units—force and area. A piston is inserted into a close-fitting cylinder. Weights are placed on one end of the piston and are supported by pressure applied to the other end on a smaller piston. For absolute pressure measurements, the assembly can be placed inside an evacuated container. The accuracy is consistent throughout the full pressure range. Note deadweight testers also come in more accurate large-scale bench models.

In recent years, these mechanical devices have been replaced. Electronic deadweight gauges are now used. These have built-in compensation for temperature and altitude changes. The following is the type of accuracy generally specified for an electronic deadweight gauge. FS indicates Full Scale (the maximum pressure reading):

Gauge range	5516 kPaa	
Resolution	0.0003 % FS	0.01655 kPa
Accuracy	0.024 % FS	1.3 kPa

Considerable difference exists in the quality of hand-held electronic deadweight pressure testers available. This can be demonstrated on a portable unit by closing off the external valve, manually increasing the pressure inside the unit, and then releasing the valve. Often, poorer quality gauges will not return to the original measurement. This lack of repeatability can exceed the specifications of the pressure gauge; therefore, quality gauges are important.

Electronic deadweight gauges have proven to be significantly more accurate than the traditional deadweight gauge, which is still specified as the standard by Measurement Canada (federal government). Recently the government has been reevaluating its standards.

Corrections required. The mechanical or electronic deadweight gauge is the most accurate pressure measurement device on the wellsite and, barring operator error, should be accurate to within 7 kPa, i.e., within one psi.

By calibrating the bottomhole recorders to the surface deadweight gauge readings, it is possible to spot gauge errors. Failing this, the downhole readings can be corrected if a calibration difference can be measured. This assumes the error can be represented as a simple difference, which may not be the case.

ARE has done large areal evaluations of pressure, and experience has shown that some bottomhole pressure recorders are seriously miscalibrated. It is possible to track the serial numbers and observe this effect in many different wells.

Changing recorders. Common practice during an absolute open flow test for gas wells is to use a different set of bottomhole recorders for the initial static gradient, flow and buildup, and the final static gradient. By calibrating the bottomhole recorders to the surface deadweight gauges, the pressures obtained from the initial static gradient, buildup extrapolation, and final static gradient can be consistently compared. If the pressure determined from all three tests match, then the value can be considered reliable.

Absolute pressure conversion. Industry practice is to correct gauge pressures to absolute pressures by adding the barometric pressure. Unfortunately, it seems the corrections used are not very consistent. Values of barometric pressure conversion have been observed ranging from 85 to 101.35 kPa within Alberta, Canada.

This results in the introduction of unnecessary additional error into pressure measurements. In the case of high pressures, this error will be relatively small. In fact, these changes are sufficiently minor that omitting the correction is hardly serious. However, there are cases involving lower pressures, in particular with shallow gas reservoirs, where this error can be significant. Large reserves of shallow gas are present in Western Canada.

The simplest method to minimize the error introduced by the barometric conversion is to ensure a consistent value is used. It is not always possible to determine what was used in some test reports. Pressures are frequently identified as absolute, without specifying what atmospheric pressure was assumed. In fact, a good number of reports are silent on the type of pressure utilized or reported.

The pressure at the land surface decreases with increasing elevation. At 1,000 meters elevation, the average atmospheric pressure is 89.9 kPaa, down from the average sea level value of 101.3 kPaa according to the International Civil Aviation Organization. Barometric pressure corrections could therefore differ by 11.4 kPa, depending on what value is assumed. Recall also, weather causes surface pressure variations. Normal atmospheric variations of ± 2 kPa can be expected. Wellsite barometric pressures are rarely recorded or known. Hence, to adjust to the absolute pressure, it is simplest to estimate the barometric pressure based on the surface elevation.

Calculated adjusted pressure. For static gradients, both recorders are used to calculate a bottomhole pressure. When a recorder has malfunctioned or is erratic, then only one recorder in used. The surface tubing-casing pressure is taken, and the average pressure value of the recorders at surface is subtracted from it. This is added to the average of the pressure calculated to the midpoint of perforations.

Program to screen data. Pressure data screening can be reduced to a systematic program. Especially when large volumes of data are required, it is helpful to tabulate the data. This can be done with a spreadsheet. Table 5–1 shows a table for a very large area used for pressure analysis of gas wells. This was a huge area, and only one page of approximately 100 pages of data is shown.

The use of such a data table helps screen out unusual pressures. In fact, a surprising number of data points have significant problems with them. Sometimes this is all we have to work with and the data can't be totally ignored, so some judicious interpretation (really a guess) is needed. It is also a good idea to keep the base analysis and well logs in a set of binders.

The database has been arranged in columns. It is a duplicate of the system from an actual report. Note there are a number of data cross-checking steps. The number of glitches one can encounter in practice is truly amazing. Buyer beware.

- Well location.
- The top and bottom of the pay zone was determined from well logs. This data was also recorded for gas and bitumen zones. Note the gas-water contact was not recorded since multiple gas zones exist in many of the wells.
- The Kelly bushing (KB) and casing flange (CF) elevations were obtained from well tickets and checked against values recorded on log headers. The well logs were examined to ensure the perforation interval corresponded to a consistent interval naming convention. These steps are designed to eliminate KB to CF conversion errors, as well as to confirm the tested interval was from the pay zone. All depth values were reported relative to KB to facilitate a quick comparison to the well logs.
- The midpoint of the perforation interval was usually calculated as being halfway between the top and bottom perforations. In some cases, modifications were made for DSTs. The top and bottom packers are normally set in nonpermeable and porous zones, and the area midway between packers is often not representative of the zone.
- The number of gas zones perforated within the pay zone was determined from the logs and recorded in the database. In general, the lower gas

Table 5–1 Spreadsheet

Well Location (UWID)	KB ASL (meters)	CF ASL (meters)	McMurray/ Wabiskaw (meters)	Gas (meters)	Bit (meters)	Perf Top (meters)	Perf Bottom (meters)	MPP (meters)	Completion Equipment (meters)	Gas Zones	Prod Date	Test Date	Test Type	Shut-in Time (hours)	Shut-in Pressure (kPaa)
N.B. all depths are TVD															
100/08-09	623.1	620	371	375	395	369	374	371.5		1			DST #1		
102/08-09	623.5	619.2	373	376	374.5					1	Jan-95		AWT/DWG	88	1,376
100/13-11	616.6	612.8	364	370	384	364.8	370.2	367.5	328.1	1	None		StGr	?	875
													AOF/BU	47	
													StGr	48	
100/10-14	616	612.9	361	367	380	368	365.5	364.3		1			None		
100/02-15	624.3	620.7	367	379	386	366.5	373.9	370.2	?	1	Mar-93	Jul-98	StGr	46	689
100/07-15	624.8	621	367	378	385	369.1	377.1	373.1		1	Apr-94		None		
100/01-21	623.4	619.4	369	374	392	370.5	373.5	372	359.6	1	May-91	Jan-90	StGr	48	2,103
												Jan-90	AOF1/BU	43	
												Jan-90	StGr	44	
												Sep-95	AWT/DWG	71	1,317
100/08-25	607.8	604.1	372	372	413	402	419	410.5		0		Jan-90	D&A		
100/06-27	633.1	629.7	376	389	403	378	384	381	365.8	2	May-91	Jan-90	StGr	18	2,085
												Jan-90	AOF/BU	99	
												Jan-90	SDtGr	100	
												Sep-95	AWT/DWG	71	1,152
100/06-28	628.4	625	374	382	393	377	378.5	377.8	372	1	May-91	Jan-90	StGr	18	2,076
												Jan-90	AOF1/BU	99	

zones have a higher pressure than the upper gas zones. As demonstrated in later sections, the pressure in different sands may have different gas-water contacts or be subject to varying degrees of drainage. In this case, there can be substantial cross-flows.

- The month the well was put on continuous production was recorded. This typically would not include short-term well tests.
- Test date was the month in which the pressure test was done. In the case of an initial absolute open flow (AOF) test conducted on a well, the initial static and final static gradients are given the same date. In some cases, tests straddle two months.
- The type of test was also recorded in the database as follows:

a) Gauge	Surface gauge reading
b) DWG	Surface deadweight gauge
c) StGr	Static gradient
d) AOF1/BU	Single-point absolute open flow test and buildup
e) AOF4/BU	Four-point absolute open flow test and buildup
f) DST#1	Drill stem test number 1
g) Ac/DWG	Acoustic well test using a deadweight gauge
h) BU	Buildup only

- Recorder depths on the static gradients are generally reported relative to CF. These values were adjusted for the database by adding the CF to KB interval. Hence, recorder positions can be easily compared to the log, the perforated interval, and the midpoint of perforations.
- The shut-in period before the pressure reading was taken was recorded. Whenever possible, the shut-in time for the initial static gradient was estimated using the well's completion-perforation date.
- The final pressure value at the midpoint of perforations is the best estimate based on the information available. When an accurate deadweight gauge reading was available, this value is considered to be accurate to ± 6 kPa. This can be further improved to ± 3 kPa by accounting for the surface barometric variation.

Pressure transient analysis. The basic idea behind pressure transient analysis is to determine the reservoir properties around a well by manipulating well rates and measuring the pressure response in the well. These properties may be determined by the use of sophisticated mathematical analysis.

In more specific terms, the following may be determined:

- an estimate of reservoir permeability
- an estimate of reservoir pressure
- an estimate of efficiency of the completion
- an inference about reservoir shape, size, and flow patterns
- an estimate of future well performance

Pressure transient analysis can provide very useful data about reservoirs and how they behave. Very often a considerable amount of time and effort will be expended on gathering this data and analyzing it. Unfortunately, a significant number of tests will not result in obtaining much useful data. The key to making sure useable data is obtained is pretest planning.

Basic pressure drawdowns and buildups. Before proceeding to the mathematical aspects of pressure transient analysis, a qualitative description will be made of what a pressure test looks like. The starting situation looks like Figure 5–21.

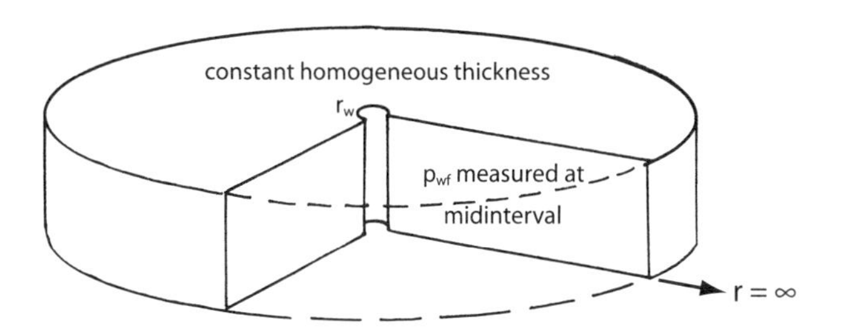

Fig. 5–21 General Pressure Test Assumptions

The idealized reservoir depicted has never been produced before, and the pressure is constant throughout. The reservoir is infinite in all directions. The next step is to start producing the well at a constant rate. As a result, the pressure in the well starts to decline or show drawdown, as shown in Figure 5–22.

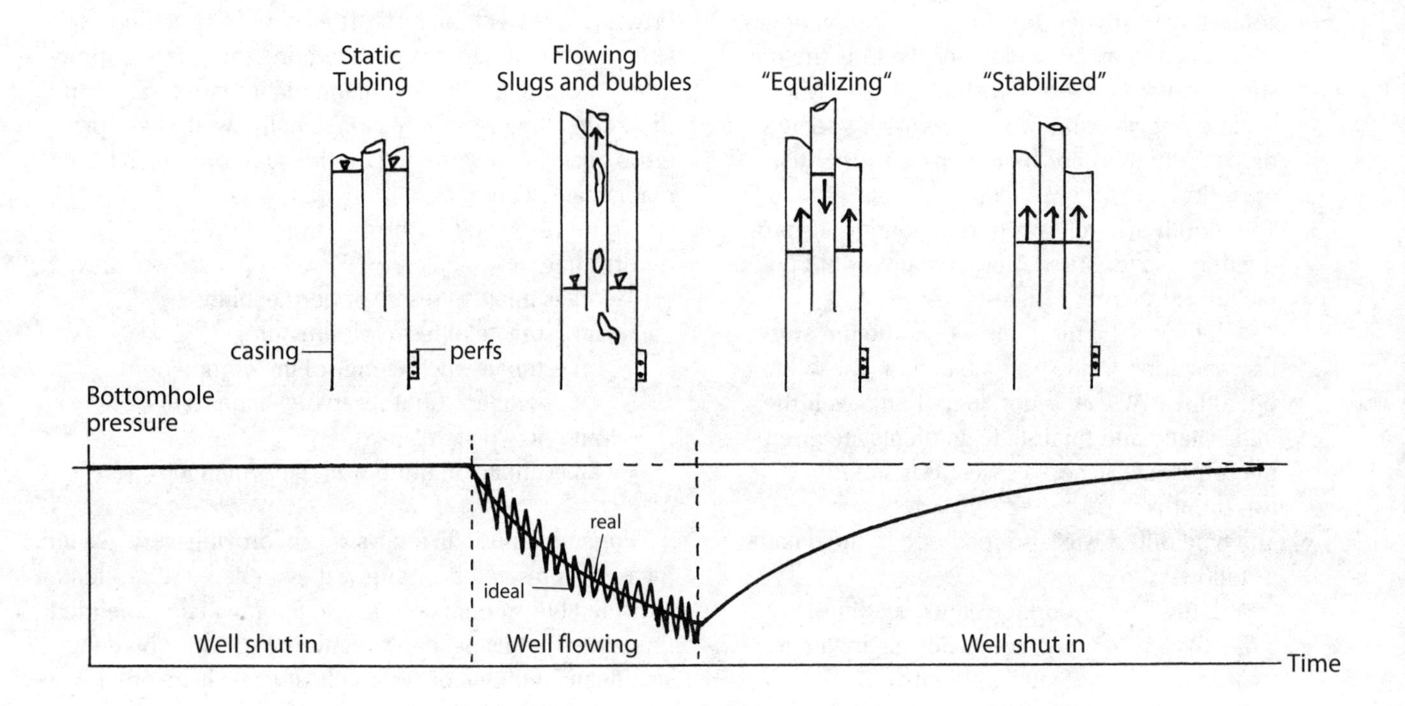

Fig. 5–22 Drawdown and Buildup Well Pressures

The amount of drawdown will depend on the permeability of the formation. The rate at which the pressure reduces can be monitored. Such a test is called a pressure drawdown test. When a well is flowed, there can be a considerable amount of slugging, gas liberation or NGL condensation, and turbulent flow effects. For these reasons, drawdown pressure measurements have a lot of noise in them. In truth, this noise obscures the reservoir test, and drawdown data rarely can be analyzed.

After the well has been flowed, it is possible to shut in the well and watch how fast the well builds up. This is called a pressure buildup test. These tests look something like the one shown in Figure 5–22.

Ordinarily, once the well is shut in, the fluids have a tendency to settle, and the well becomes quiet as it reaches static equilibrium. For this reason, the vast majority of well test analyses are done on pressure buildup tests.

Armed with this bit of information, it is possible to proceed through more practical comments on test analysis.

Pressure transient analysis failures. There are a number of reasons for the poor success rate of pressure transient analyses. In ARE's view, the major cause is most training on this subject concentrates on very sophisticated analysis techniques, which reservoir engineers apply after the fact, i.e., after the test has already been done. As indicated previously, the real trick to obtaining good pressure transient analysis results is pretest planning. No amount of analysis can correct useless data.

Problem test. This point is illustrated in the following story, which relates to a West Pembina high-productivity pinnacle reef typical of the Nisku formation encountered in the area. This reef covered about a quarter of a section. The vertical relief was quite high so there were considerable reserves. The subject pool was in the early stages of production and had two wells. A new battery had been built to process production. Immediately to the north of the pool there was another, similar pool found by another operator.

The problem started with a pressure test done on one of the producing wells. The test had been recorded for a period of 180 hours, about a week, with an Amerada-type pressure recorder. The test results are depicted in Figure 5–23.

No obvious buildup had been recorded in the test. When the pressure recordings were examined closely, the pressure had dropped about 8 kPa throughout the test. During this period, both producing wells had been shut in. Someone in the reservoir engineering department had extrapolated the pressure drop and, knowing the other well in the pool was shut in, had concluded the pool was

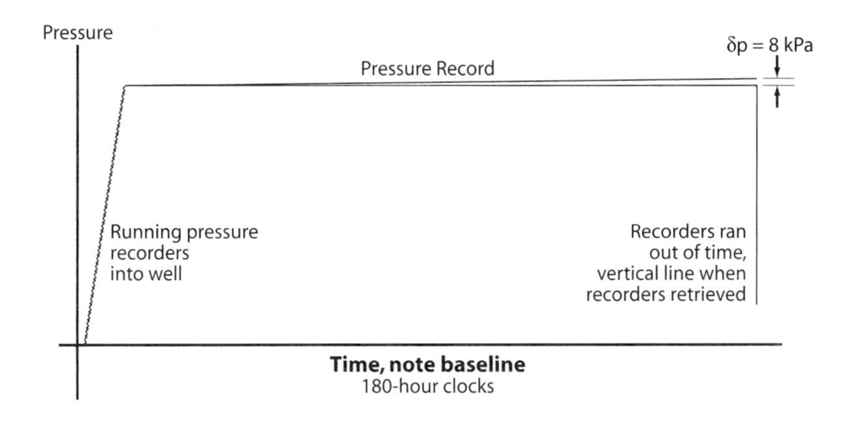

Fig. 5–23 Pressure Card—West Pembina Reservoir

being pressure depleted by the offset operator through a connecting aquifer—i.e., this interpretation implied oil was being pulled downward into the virgin aquifer zone, thereby losing some reserves in the process of leaving a residual oil saturation. In addition to this, the GORs in the field had been increasing faster than was anticipated based on pressure and PVT data. Management had been alerted that offset operators were rapidly draining this major find. Naturally they wanted to ascertain, as quickly as possible, how serious the drainage was.

The operations engineer received a high-priority memo from the reservoir group. An immediate new pressure test to monitor the pressure drop was requested. Senior management was waiting for the answer. Now, all of this would have seemed fairly reasonable, unless you know a bit about Amerada pressure gauges—some basic explanation was provided earlier.

The accuracy and precision specifications for an Amerada pressure gauge are a precision or sensitivity of 0.05% of full deflection and an accuracy of 0.2%. This level of performance is truly impressive for a mechanical device. However, the precision on a 21,000 kPa gauge amounts to 10 kPa, which happened to be the size of the pressure drop. In other words, the supposed pressure drop was below the stated measuring capability of the gauge.

The request had been generated about a week after the first test. Following this line of thought, the operations engineer realized that, since the accuracy was only 40 kPa (0.2%), it would be necessary to wait four times longer than the last test to prove there had been a pressure drop. This was a period of four weeks; therefore, the new pressure test would have to be delayed in order to obtain a reading of any meaning.

As fortune would have it, the operations engineer had been responsible for the implementation of a number of pressure buildup and falloff tests. The previous test results bothered him since he could not see any buildup. The next step was to discuss the proposed test with the reservoir engineer. In the course of the discussion, the core data was reviewed. The well had something like 10 meters of vuggy dolomite with air permeabilities as high as 2,600 mD. The problem with the lack of buildup was immediately obvious. Using the stabilized rate calculation, the operations engineer estimated this well would produce 1,000 bbl/day, with about a 20 psi drawdown. High-permeability formations result in fast buildup times, and a buildup for this well would probably be complete in less than a few hours.

Once again, it helps to know some more about pressure gauges. Amerada gauges come with a number of different clocks. The standard static gradient is done with a 3-hour clock. The normal pressure buildup is done with a 180-hour clock. However, they are available in increments of 3-, 6-, 12-, 24-, 72-, 180-, and 360-hour clocks. The actual recording card is exactly the same; it moves past the recording stylus at a faster speed with the shorter clock times. Armed with this information, a 6-hour clock seemed to make more sense.

The test program was reviewed with the reservoir engineering department. After reviewing the information on the pressure gauges and the proposed program, it was agreed the test would be given to the operations department, who took responsibility for obtaining results from the test.

Two other changes were made to the test. Since the buildup was only going to be 20 psi, and the precision of the gauge was slightly more than 1 psi, 10 kPa, a more accurate pressure gauge would be required to record an accurate buildup shape. Quartz pressure gauges had recently become available featuring a fantastic accuracy of 0.01 psi. The test was also run with the pressure gauge on a wireline. With this, it would be possible to plot the data to ensure useable data had been obtained before leaving the lease.

When the fateful day arrived, the pressure drawdown amounted to 13 psi. The buildup lasted less than 18 minutes. The results looked somewhat like those shown in Figure 5–24.

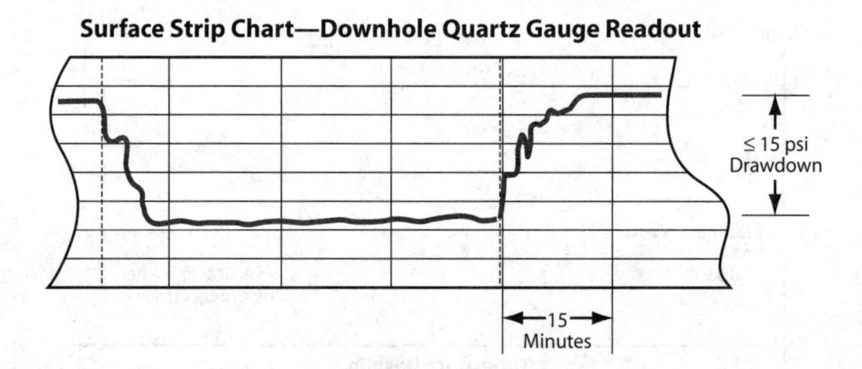

Fig. 5–24 Downhole Electronic Measurements of Pressure

This data was completely useless for analysis, because segregation was occurring in the wellbore, which resulted in the erratic response. Such behavior is common in very high permeability wells. In fact, no matter how well designed the test, no data could have been analyzed from this well. With a 180-hour clock, this entire buildup would appear as a vertical straight line.

Obtaining accurate pressure transient analysis is somewhat academic in this well. It is easy to achieve a stabilized reservoir pressure by waiting an hour. The pressure drawdown is so small, a stimulation would barely produce a measurable difference. The well's production was limited by allowables, so no incremental production could be obtained. Further data simply wasn't required.

Surface rates. Most tests are intended to be run with more or less constant rates.Although it is possible to enter a rate history and use superposition,a finite number of rates are used. More particularly, rates are determined by subtracting cumulative productions normally measured in hourly intervals. This immediately introduces a practical problem: How does one control a well so it will flow at a relatively constant rate? A little field experience goes a long way to understanding the problem.

Maintaining a constant rate can be very difficult. Ignoring pumping wells for the moment, this requires adjusting a choke continuously and monitoring the production. For a gas well, this means doing a gas meter calculation. These calculations take a little time and a calculator. For an oil well, it is normal to gauge the tank every hour and then calculate the rate. This is hardly what one can call instant feedback.

Perhaps we could put a positive displacement meter on the oil well. Even this will not work since the gas coming out of solution over spins a positive displacement meter. Putting the meter downstream of a separator will not work either, because separators use dump valves.

Very few oil wells in Alberta flow naturally. Of the few that do, most need to be swabbed in. This again disrupts the requirement for a constant rate. Unless one is dealing with a rare overpressured reservoir, flowing wells cannot achieve a full drawdown and lift all of the oil and gas.

Sometimes it is possible to get an initial flow rate; however, the rate on the well reduces to zero and kills itself in 10 or 12 hours. Under these circumstances, a constant flow rate test cannot be achieved.

The solution to this problem is to put the well on pump. This introduces yet another problem: how to measure the pressure. On the surface this is simple enough; use a pressure gauge at the wellhead.

The problem with a surface measurement is you don't know the exact gradient in the well. Normally, there is a mixture of oil, water, and gas in the well. The density of the oil and gas mix is not consistent. If the well is shut in, it is impossible to calculate exactly the location of the fluid level. Pressures must be taken as close to the perforations as possible. This is normally done with a slick line and pressure bombs.

The pumping rods block the way for installing a pressure gauge downhole. The conclusion would be to pump the well, shut it in, pull the rods, and run the bombs. However, to pull the rods the well must be killed by pumping an overbalance of oil in the hole. The pressure in the well goes up according to what is put in the tubing from the surface instead of based on what the formation can provide. The net result: no data.

Perhaps it would be possible to run a pressure bomb underneath the pump, produce the well by pumping, shut it in, and then measure the pressure buildup. The results of this type of test look something like Figure 5–25, but the pumping action rattles the sense out of these mechanically sensitive pressure gauges.

There is a way to accomplish the goal—use a surface pressure gauge and a sonolog. This procedure also has problems. Foam develops on the top of the oil and can obscure the true fluid level. The use of chemicals helps control this. Sonolog measurements reduce the available

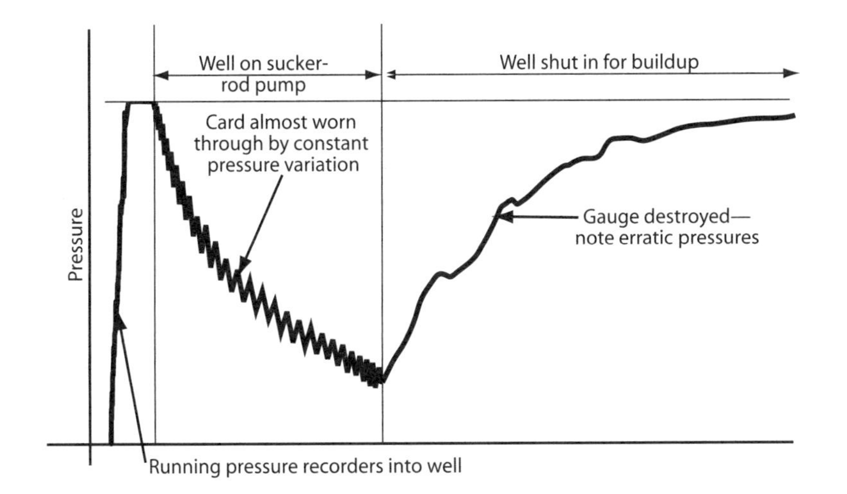

Fig. 5–25 Effect of Sucker-Rod Pump on Pressure Recorder

accuracy, which means one may not be able to use the latest and more sophisticated derivative techniques discussed later in this chapter.

Nevertheless, it is possible to get useable data by this process. The author used this in the Davey Belly River oil pool and obtained good-looking results. Permeabilities calculated correlated well with core data. During the last 10 years, others have made significant progress with this technique. Recently, automated equipment has been developed and is making the process more economical than before.

In truth, an engineer may not be able to use the latest wonder techniques promoted in the literature on the majority of wells. Most of the times when suitable data is acquired, it will relate to a gas well because data is much easier to obtain from flowing wells.

Wellbore storage effects. The next step to study is what actually happens when a well is shut in. The following discussion is related to the diagram of a well shown in Figure 5–26. Since this is a gas well, a packer has been installed. The real problems begin when one has condensate production.

The well is shut in at the surface using a wing valve on the wellhead, but the well doesn't stop flowing immediately at the bottom. First, the NGLs, which are condensing on the way up the tubing, start to fall down the hole. The gas also continues to come into the well from the formation, compressing the gas in the wellbore. As the pressure builds, the gas in the wellbore compresses progressively more. This still allows gas to enter the wellbore, although at increasingly lower rates as time elapses.

There is sufficient compressibility in a wellbore full of water on an injection well to seriously affect results. It has been possible to modify the inflow equations to adjust for compressibility effects. However, no one has yet determined how to handle segregation effects. During the wellbore storage phase, relatively little information is obtained about the

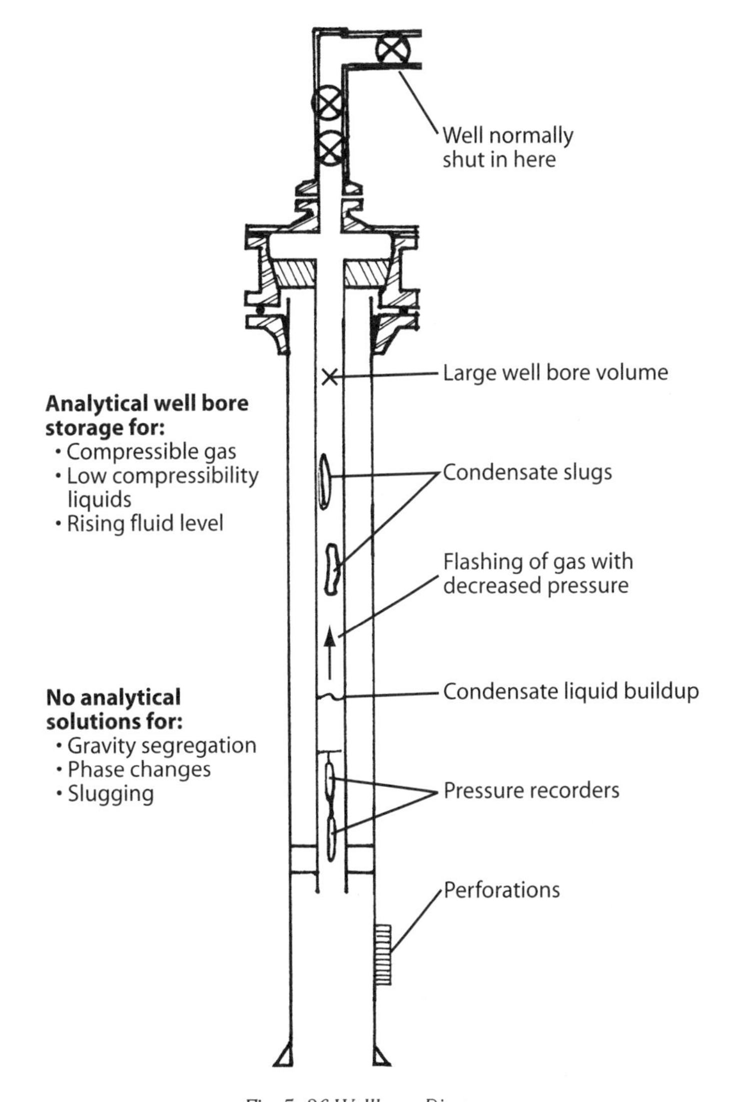

Fig. 5–26 Wellbore Diagram

formation. Quite often, wellbore storage effects are referred to as after flow, in recognition of the fact that a shut-in is far from instantaneous.

Some methods can provide an estimate of the transmissibility and an indication of the damage-stimulation in the near-wellbore region, as well as a rough estimate of the bulk formation transmissibility, specifically the McKinley afterflow analysis type curves. Although approximate, these appear to provide good commonsense results.

If doing a buildup on a high-permeability well where the buildup occurs very quickly, the entire pressure transient may be obscured by wellbore effects, as mentioned in the Nisku pinnacle reef problem well case described earlier. No interpretation can be made.

Downhole shut-in. There is a way around this problem—shutting in the well downhole. This requires the use of DST tools. A DST tool has a valve located at the bottom of the well, which can be manipulated by twisting the tubing. This adds considerably to the cost of the test. It is necessary to have a service rig on the hole while the test is being conducted. However, substantially better data is obtained in this manner.

Lately tools designed to run a slickline have become available.

Clean up. Most wells are flowed immediately after they are perforated. This does two things. First, it sweeps all the debris from perforating, drilling fluid leak off, and drilling fines out of the well. Second, a stimulation may not be done automatically. If the rates are too low, the well will be acidized and/or fraced. Once a treatment has been done, it is necessary to recover spent acid and/or broken frac gel. All of this occurs before pressure bombs are run in and the production test started. If the reservoir is extremely permeable, like the Nisku pinnacle reef described earlier, this rarely matters. The well will have stabilized before one set of equipment is torn out and the testing equipment rigged up. On the other hand, if it is a tight formation—i.e., the kind usually stimulated—it could be a week, perhaps a month, before the well is stabilized.

Field personnel hate sitting on wells and doing nothing. It is against their work ethic, which is really a good thing. Drilling and completion superintendents are also under considerable pressure to keep well completion and testing costs to a minimum. The net result is that sometimes the test starts whether the well is ready or not. This can completely corrupt a beautiful mathematical analysis later on. It may be necessary to send the well crew home for a week, or to another job, until things have settled down.

Operational difficulties

The real world often intrudes on the beauty of mathematical analysis:

- Something breaks toward the end of the flow period. The well has to be shut in immediately to make repairs. This can be a stuck dump valve, a slug that overflows the separator, the pilot in the line heater can get blown out, or the flare stack can get blown down in a high windstorm. Just about anything imaginable has happened at one time or another—the Murphy's Law corollary applicable to pressure transient analysis.
- A frequent problem concerns hydrates. Water combined with natural gas produces some very interesting compounds called natural gas hydrates. This stuff looks like an Italian ice dessert with a slight yellowish tinge. Remember not to eat the yellow snow. If you manage to get some of this out of a surface flowline or on a wireline tool, hold it in your hand. It will degrade under atmospheric conditions. It is like soda water and makes a snap, crackle, and pop. There will never be any left to show the head office; however, it can completely obstruct tubing. More often than not, it will simply cause erratic rates on your test.
- One way to elicit strong emotions and rude comments is to give the testers a hard time about this. This problem normally only occurs between 2:00 and 5:00 a.m., when it is coldest. It is possible to control it by injecting methanol or glycol. It can take a bit of experimenting to get the injection rigged up and set to the right volumes. Strong responses are virtually assured.
- Downhole tools can leak. This is most significant when you have two zones in a single well. It can be spotted by monitoring the annulus and tubing. In rare instances, this will also show on the casing vent. Management is never happy to hear about this since it involves capital expense to repair.

Sometimes the field personnel will know how to work around these problems. More often than not, it will be up to the reservoir engineer to figure out how long the flow time should be extended.

Sitting tests

Every engineer should spend some time sitting in on tests and completions. The testing crew and well site supervisor usually work much harder when they know the results are important enough to send out someone from head office. More importantly, getting to know these people is a great source of information on how things really work. Very often when something does go wrong, the engineer will be the only one who can say how much longer to flow the well. It makes interpreting testing reports much easier when an engineer appreciates the difficulties in establishing good data.

On extremely high-profile wells, such as East Coast and Arctic offshore wells, it is quite common to send out a reservoir engineer on the test armed with a computer. With the computer and a pressure transient analysis program, the test can be simulated and analyzed on the spot. This is an excellent opportunity. Anyone who gets the chance should take it.

Problems with interpretation

Having a pressure buildup and interpretation is generally better than a static pressure gradient. However, even if the test gets past the operational problems, an engineer may still have considerable problems deciding on an absolutely correct pressure:

- Horner plot extrapolations of pressure may actually overestimate the reservoir pressure. There are methods to account for finite drainage areas; however, real reservoirs never follow these patterns exactly.
- The engineer needs to be able to relate the buildup pressure to the grid blocks in the simulator. This item is discussed somewhat in chapter 4. The grid blocks are of a finite size; therefore, grid-block pressure will not necessarily correspond to the average pressure surrounding the well.
- There may be multiple completions, and the zones are crossflowing.
- There will be multiple interpretations possible, such as dual permeability, i.e., two layers of different permeability, or dual porosity, to be discussed later.
- Many tests will be older, and they may lack modern interpretations.
- Analytical pressure buildup solutions are based on single-phase flow. On some high-GOR wells, there may actually be more gas flow than oil flow in reservoir cubic meters (or barrels) per day. In this case, the solution form can be changed from an oil analysis to a gas analysis.
- Frequently, it is hard to get pressure recordings at very late times, which is when true reservoir properties are observed. This requires a long flow and buildup time for lower permeability formations. This is particularly true for hydraulically fractured wells and horizontal wells.

Potential problems are extensive. Getting good pressure transient results, although easy to outline in theory, is actually quite difficult in practice.

Organized test procedures

Once again, the secret to well tests is proper planning. Many situations do not lend themselves to either drawdown or buildup analysis. If this is the case, look for operational opportunities to run static gradients.

It is imperative to model tests before running them. This can be done with a simulator, or better yet it can be done with pressure transient software. Often, there is no point in waiting for late data on very high permeability reservoirs. Low-permeability reservoirs can require very long buildups. It may be better to have one or two very long tests than a series of intermediate length tests that don't reach pseudo steady state or radial flow.

For the most part, when an engineer gets to simulation, he may have to accept the available data. However, some cases allow sufficient time to ask for tests before getting to the history-match stage. In these circumstances, it may be a high priority to get good data as far in advance as possible.

Perspectives on pressure transient analysis

Talking about PTA could easily dissolve into a book in itself, which is not consistent with the basic objectives of this text. At the same time, many larger companies have a PTA specialist who may do all of the evaluations in the company. It is also quite common to send this type of analysis to a specialty consultant.

This is one of those areas that justify having all reservoir engineering staff gain experience in well test analysis. This is a somewhat painful process since there is roughly one week required to learn a basic well test or PTA program. There is also a reasonable investment in learning and understanding the mathematical models. Two one-week courses on each of these topics would be extremely valuable for those involved in reservoir simulation. With practical experience, one develops a feel for the accuracy of pressure transient analysis.

Another potential problem with sending all tests to internal or external consultants is they do not get a complete background in the geology or the other reservoir data. Since multiple interpretations are frequently possible, they often cannot *a priori* determine the correct model to use. In most cases, they will pick a model with the least error; however, this does not necessarily correspond to either the correct or best model.

As with many other aspects of life, balance is important. The specialists develop extensive experience that helps resolves difficult problems. A minimal level of experience provides enough expertise to know the issues.

Reinterpretation of tests

In a number of cases, older pressure data is perfectly suitable for analysis. It helps to reanalyze the data using a modern well test package. The new type curve analysis with derivatives may resolve many issues that previously could not be interpreted, but getting data from the old 5.25-in. floppy disks can be a problem. Not many computers available today can still read this format.

Even with modern analyses, some reinterpretation may be required. If the model of the reservoir has changed, then the model used in a recent analysis might not be usable. This is normally easier to correct than the older tests, where the pressure data must be input by hand.

Pattern recognition—paste board

In a number of studies, the author has arranged the pressure transient responses on the wall, with the plots arranged in the same geometric pattern as they would appear on a field map. In many cases, there is a pattern of pressure responses that has dominated. An example of this is shown in chapter 18 for a simple thrust-faulted reservoir.

Using simulator for buildups

The layering in the field does not always have an analytical model available. This can include the taper in fractures, a layer of high permeability of limited extent, and fracture patterns. In these complex situations, the simulator can be used to model a buildup.

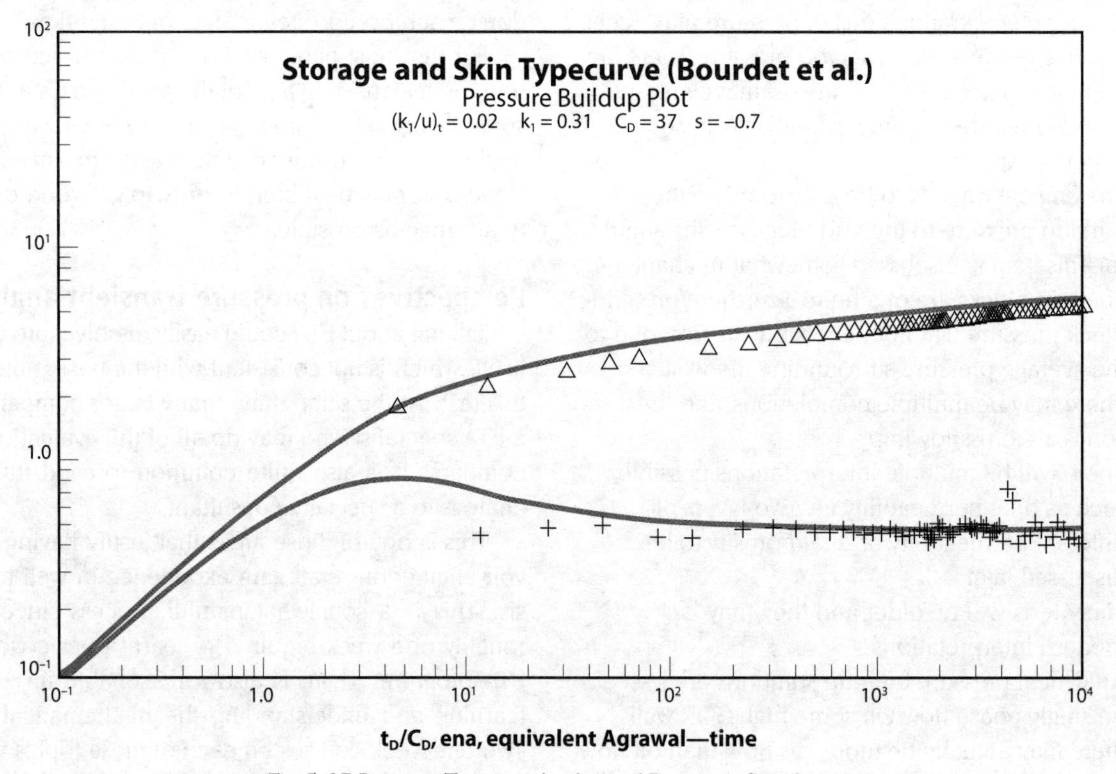

Fig. 5–27 Pressure Transient Analysis of Reservoir Simulation

In these circumstances, a very fine grid must be used, particularly around the well. This topic will be discussed in more detail later in this text. In general, the author has not directly modeled wellbore storage; however, this is not a major problem.

Pressure versus time can be exported in a tabular format from the postprocessing package, and this can be imported in ASCII file format into most pressure transient programs. The simulator results can be analyzed for accuracy and used to evaluate the shape of the type curve and derivative. Reservoir simulation provides very analyzable data, i.e., it's free of noise. An example is shown in Figure 5–27.

ARE has used this as a separate study phase in a number of circumstances. These include matching buildups of single-well models with observed pressures, including tapers in hydraulic fractures—a feature not available in analytical models, and modeling high-permeability pods, which are of limited areal extent and for which there is no analytical model.

Pressure buildup programs

A number of pressure transient programs can analyze simulator results and actual well test results. It is not economical to analyze this data by hand. The author has used almost all of the major commercial programs. In general, they give similar results and have similar capabilities. There can be some big variations in price; in particular one program charged considerably more for the inclusion of some analytical models. This program is worth the cost *only* if you require these models, which are not commonly required.

Various programs come from Canada, the United States, France, Norway, and the United Kingdom. Test them and decide. One of the larger differences is in the error minimization options and how they work. Some include specific reporting requirements for Canadian government bodies. A number of the international programs have this capability built in. This market is surprisingly competitive, and the quality of the software is remarkably good.

Presenting pressure data in history-match files

Many of the post processors require a pressure be given at fixed time increments corresponding to production data. Naturally, pressure data is much more difficult to obtain, and many records have no data. These empty spots will often default to zero. When the history-match plots are made, one has ugly dots along the time axis. One way to improve the appearance of a history-match plot is to set null data points to a negative number and restrict the plot scale to positive numbers.

Summary

Production performance data is key to creating a realistic model. In the number of examples given in this chapter, simplistic models were discarded before modeling was begun. Any one or all of these could have been simulation failures.

Simulation provided a method of verifying whether the new conceptual models were physically possible. It is important to point out that, had simplistic models been used and failed, the modeling technique would not have indicated the right geometry or production mechanisms. The simulator will evaluate what is input, but the simulator does not have the intuition required to get the right scenarios to test.

Pressure data, like production performance data, is an important part of reservoir screening. The author prefers to see data analyzed for consistency and in particular, the type or style of response. If the analyses are old or inconsistent, it is probably worthwhile to redo the entire suite on a consistent basis. If the response of pressure tests is anomalous, it is probable the reservoir is not sufficiently understood to be modeled. Many types of tests do not lend themselves well to analytical techniques. A reservoir simulator can be used very effectively under these circumstances.

Collecting and analyzing this data can provide very valuable insight or simply waste a lot of time with very little return. This data provides enough key interpretations that it needs to be done for every study. Many practical problems can occur that are not immediately obvious. Although detailed review of this data might appear to be rather tedious in some circumstances, doing this job correctly actually takes considerable expertise. As with other fields of endeavor, preplanning can make a huge difference in success.

In both of these areas, part of the secret to successful modeling relates to general reservoir engineering knowledge. Both production data review and pressure test analysis are important tools in this process.

6

Reservoir Fluid PVT: Data Screening

Introduction

Initially, this chapter will deal with traditional PVT relationships for oil reservoirs. Quite often, PVT data is not understood well. This is despite the fact it is an area that has existed and been researched for a long time. This may be part of the problem—it has become so well-accepted and taken for granted that the basic principles and underlying relationships have been forgotten. Helpful information is found in some very old material, specifically, M. B. Standing's *Volumetric and Phase Behavior of Oil Field Hydrocarbons Systems.*[1] This is a tremendous book, especially considering it was written in 1951. Standing was a fluids researcher for Chevron in California, and his material is mostly presented from the perspective of the reservoirs encountered in Chevron's operations there.

Handling PVT data is difficult when using what are considered accepted methodologies. For example, negative solution gas oil ratios are extrapolated when correcting for flash conditions. In the author's experience, the corrections to PVT lab data proposed traditionally are normally of no help in applying actual field conditions. Part of this problem is related to matching field and lab separator conditions. In this chapter, the author offers a solution that appears to work within close tolerances most of the time.

The use of correlations of oil properties is discussed, as well as the limitations of the correlations. Volatile oils and typical PVT relations are also discussed.

Gas PVT is reviewed briefly, followed by the standard tests for gas condensate systems and volatile oil reservoirs.

Obtaining representative oil samples and the correct design of test procedures is discussed. A test is outlined to check whether separator samples are in equilibrium. Some common problems are outlined.

A number of examples have been included of what are, in the author's experience, typical PVT lab properties, as well as normally expected ranges of consistency. Some typical input requirements for black oil input are also outlined.

Flash liberation

In a flash liberation process, all of the gas evolved from a reduction in pressure remains in contact with the liquid phase. A typical example of this is a surface separator, as shown in Figure 6–1. In this particular case, the oil and gas are kept in the vessel sufficiently long to approximate equilibrium for the pressure and temperature in the vessel.

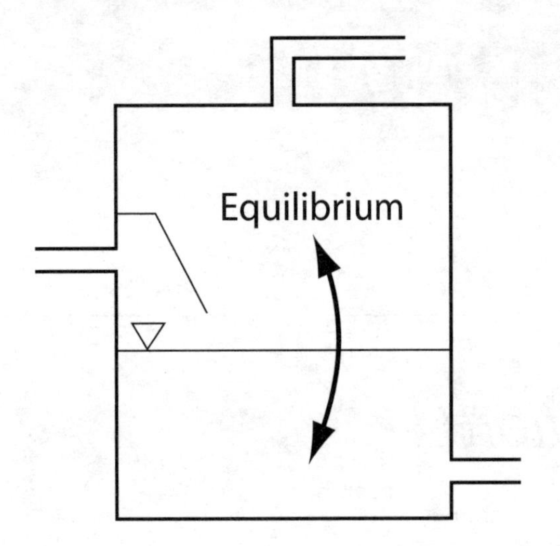

Fig. 6–1 Flash Equilibrium in Separator

This is not the only flash process. A lab test called a constant composition expansion (CCE) or constant mass expansion (CME) is available. The lab setup for this is shown in Figure 6–2.

This is done for oil samples, and the inflection point corresponds to the bubblepoint. In addition, it is possible to measure the volume of liquid and gas if a windowed cell is used. This is not commonly done for oil but is used frequently for gas condensates.

Normally this test is run first, and then differential liberation is done. The two types of flash processes can lead to miscommunications.

Differential liberation

In the differential process, the gas evolved during pressure reduction is withdrawn. In fact, this is a stepwise and path-dependent process, which means slightly different results would be obtained if different pressure decrements were utilized. The process is shown in Figure 6–3.

The differential liberation process is more difficult to run than a constant composition expansion process and is more prone to experimental error. This is because the stages must be allowed to come to equilibrium, and this dynamic process typically does not occur rapidly.

Actual liberation processes

Neither of the previous processes is completely representative of what happens in the reservoir. In Figure 6–4, at point A, the reservoir gas saturation is below the critical gas saturation S_{gc}. Hence, gas, which is immobile, is evolved in a flash (CCE) process. At point B, gas is flowing freely in the reservoir. The gas passing by point C is derived from oil further away from the well, at higher pressure, which will have a slightly different gas gravity than is currently being evolved at C. This differs from the lab processes described previously.

Historically, there was some debate on these issues, and it seems commonly accepted that the differential process is more representative of the actual reservoir process. It is fair to point out the assertion made until recently—gas-condensate systems were thought to be adequately represented by constant volume depletion (CVD) experiments. (CVD experiments and corresponding figures are presented later in this chapter.) Compositional equations of state (EOS) simulation indicates this may not be the case. This figure also shows wellbore tubulars and surface separation equipment. A flash process occurs in these after the reservoir process.

Separator tests

Approximately 50% of PVT studies include a separator test. Significant production optimization can be achieved by cascading separators on light crude oils. This is outlined in Craft and Hawkins and is not detailed here, since it does not relate directly to simulation. Strangely, the effect of this separator is often missed. Production data represents oil that has passed through a separator, and adjustment is required from the differential data to back calculate what has actually been withdrawn from the reservoir.

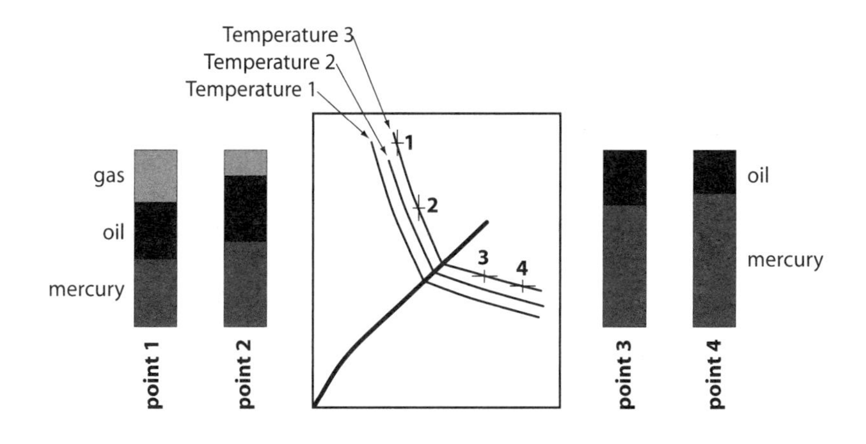

Fig. 6–2 Stages in Constant Mass Expansion

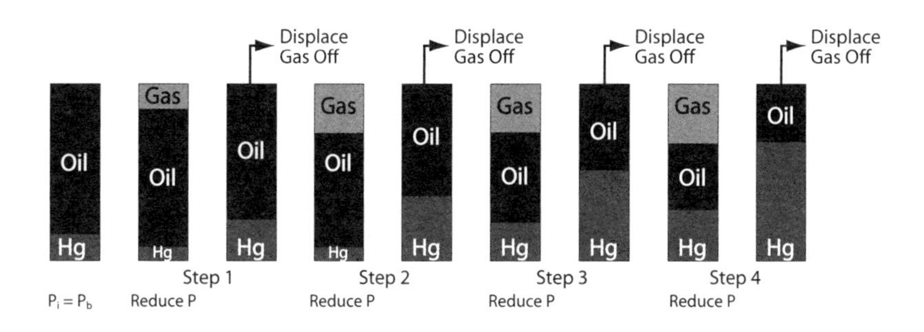

Fig. 6–3 Stages in Differential Expansion

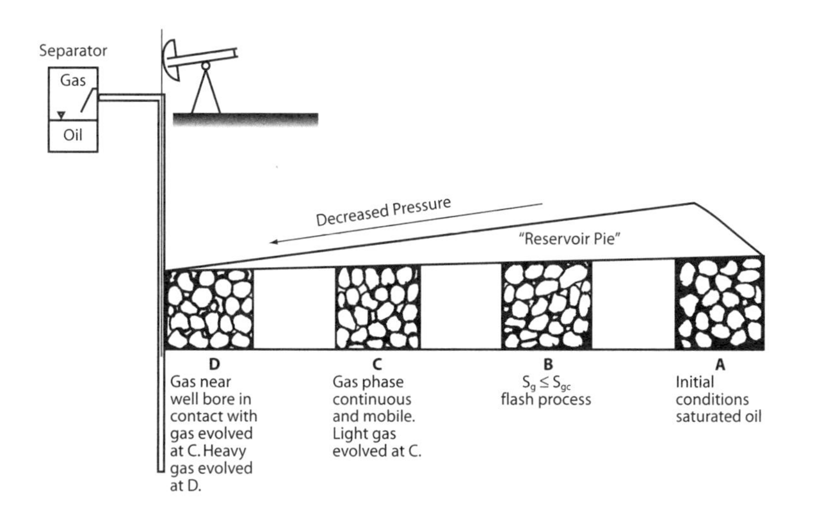

Fig. 6–4 Reservoir Pie Diagram—Gas Condensate System

Expected adjustments

It is extremely rare that PVT data can be used without some adjustment. The following is normally required:

- The data must be smoothed. If multiple data sets are available, it is necessary to evaluate the quality and select the best data or data to be averaged.
- Depending on the sample and how it was obtained, the bubblepoint must be adjusted to fit observed field conditions. Samples tend to systematically underestimate the saturation pressure in the reservoir.

More recently, some labs have taken to smoothing data for the customer. This is a mixed blessing. Data smoothing precludes the reservoir engineer from observing the actual data character, which really gives an indication of accuracy. Conversely, many reservoir engineers are not comfortable modifying data they did not prepare and have difficulty assessing the true precision of the figures supplied. Both perspectives are fair.

In the final analysis, it is preferable to know what has been done to the data. This preference is for the precision of the results to be reported as well as detailed commentary on how the tests were performed. In the author's experience, most lab companies respond well to an expression of interest in the details. Detailed write-ups add to expense, and lab personnel have indicated the majority of customers are simply not interested in the gory details.

PVT adjustment for separator conditions

The accepted procedure for adjusting PVT data is shown in Equations 6.1 and 6.2.

$$B_o = B_{od} \frac{B_{ofb}}{B_{odb}} \tag{6.1}$$

$$R_s = R_{sfb} - (R_{sfb} - R_{sd}) \frac{B_{ofb}}{B_{odb}} \tag{6.2}$$

The author has found this difficult to implement. The calculation is always predicated on having separator data from the PVT analysis. In practice, a significant proportion of oil PVT analysis has no or limited separator tests. Invariably, the author has also found field implementation of separator temperature and pressure does not match the PVT implementation. In many cases, operating conditions are not accurately recorded throughout the producing life of a property.

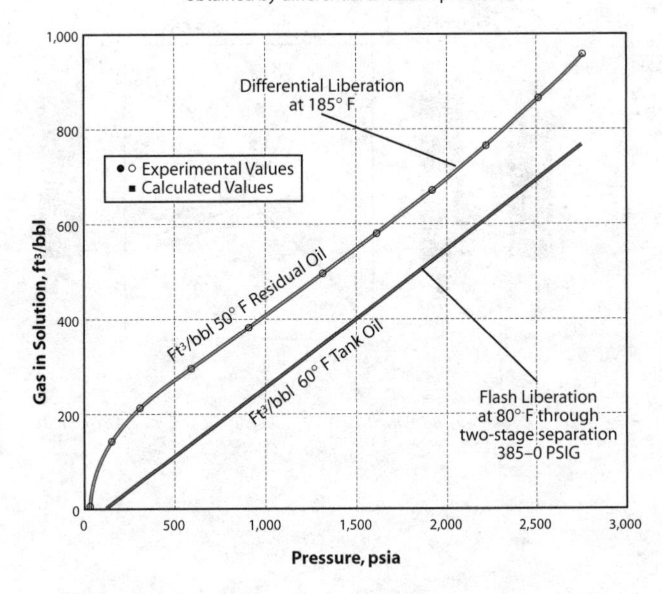

Fig. 6–5 Negative R_s from Classic Adjustment of Differential to Flash Process after Dodson

Negative solution GORs

One aspect of the existing methodology is quite unattractive. It frequently extrapolates to negative values of solution gas at low pressures. Figure 6–5 was extracted from Standing's Douglas field example. The intent of this was clearly to adjust for variations in evolved gas gravity; however, it results in what is obviously a nonphysical extrapolation. This observation is not new; it is highlighted for instance in OGCI's reservoir course notes and Amyx, Bass, and Whiting (1960) as shown in Figure 6–6. Reservoirs do not normally reach such low average pressures and, in practice, it was often possible to complete a material balance calculation without creating a noticeable problem. In the author's opinion, a reasonable adjustment in Standing's calculation was mistaken to be a general rule.

The author highly recommends reading Standing's material. It is interesting because he outlines the lab techniques in detail. A material balance calculation is presented in his book on the Douglas field. A number of subsequent changes have occurred since this material was prepared, such as Tracy's method and the Havlena and Odeh technique. Almost all material balance calculations are now done with the latter, and the Standing material balance is not consistent with the modern style. The format in which PVT data is presented by commercial labs has changed and been improved.

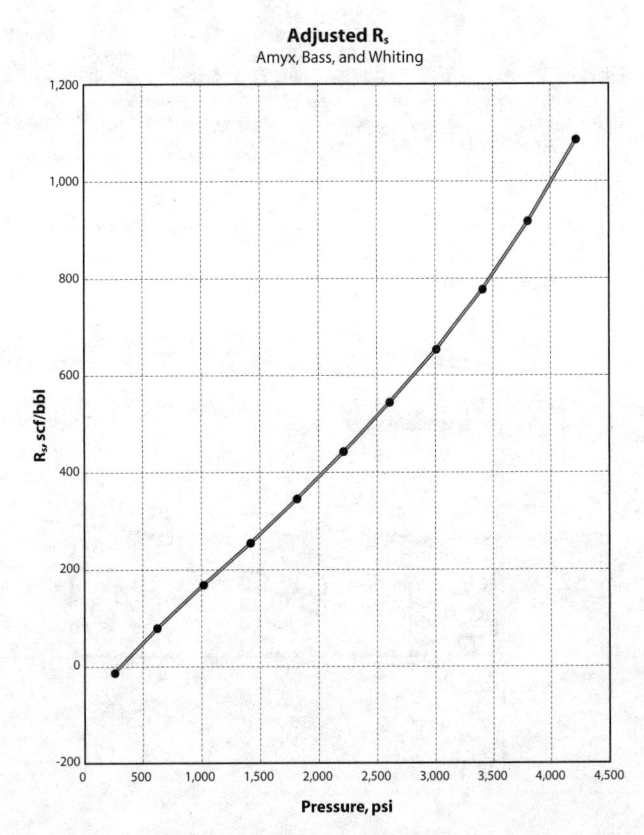

Fig. 6–6 Negative R_s from Classic Adjustment of Differential to Flash Process after Amyx, Bass, and Whiting

Recent material

Moses of Core Labs prepared a more recent paper on these adjustments.[1] However, one element of this paper is not to the author's liking, i.e., the sample used is not representative of most oils. The sample has an unusually high proportion of intermediates (C2 though C4). Consequently, it does not extrapolate to negative solution gas oil ratios; therefore, it hides a serious problem instead of dealing with it. In the author's experience working on many reservoirs and adjusting PVT data, all but one reservoir produced this problem. Moreover, this one looks much like the sample Moses presents. He clearly states the methodology is an approximation.

Treaters

Continuing with the review of PVT reports, the author has never found the inclusion of such common equipment as a treater in laboratory PVT data. Treaters are run at higher temperatures to improve separation of the oil and generate higher GORs.

Material balance calculation experience

The PVT adjustment techniques were originally developed for material balance calculations and were adapted for use in black oil reservoir simulation. The problems with GOR mismatches cause problems. The author has persevered in material balance calculations with the best separator data available, only to find instances where negative underground withdrawals could occur. This was caused by mismatches in field initial GORs and the lab-predicted GORs.

In the material balance calculation, rarely does the average reservoir pressure become low enough to enter the range of negative solution GORs. The problem becomes more severe in reservoir simulations because (well) grid blocks reach low pressures where the adjusted PVT data is clearly in error.

It is necessary to go back and look at the fundamental assumptions in black oil simulation. Two hydrocarbon fluids have distinct and separate properties, i.e., changes in gravity or composition are not tracked. Therefore, adjusting the separator conditions using the commonly accepted practice can cause a contradiction in the most basic assumption.

Note also it is common practice to inject gas in black oil simulations, ignoring quite large differences in gas gravity between the solution and injected gas.

Empirical observation

This author's solution has been to modify the B_o and R_s curves based on field-observed GORs. The differential data is adjusted to field data based on the ratio of field-observed GOR divided by lab differential GOR as outlined in Equations 6.3 and 6.4.

$$B_o = 1 + (B_{od} - 1)\frac{R_{si,\,produced}}{R_{sdb}} \tag{6.3}$$

$$R_s = R_{sd}\frac{R_{si,\,produced}}{R_{sdb}} \tag{6.4}$$

The adjustment ratio is normally a number less than 1, although the author can recall one exception, which was slightly more than 1. Usually, a production plot with a reliable estimate of initial GOR can be found. This method is simple, always applies to the facilities actually installed, and conserves mass within our fundamental assumption.

Interestingly enough, when the author has applied the ratio of lab separator GOR divided by differential GOR to the lab differential oil formation volume factor, the estimated flash B_{oi} is normally only 2–3% different from measured values. Since production data is accurate to ±2% under ideal conditions, the author considers this an acceptable level of error. Normally, this is a built-in accuracy check, even when a single separator test has been run. An example with a higher-than-normal level of error is shown in Table 6–1.

Recently, the author encountered another set of PVT data where this rule was off by about 5.0% based on actual separator tests. It was a 36° API oil, and had a high GOR. Still, a 5.0% error is much smaller than the error associated with Standing's correlation.

Ironically, justification for this methodology can be derived from Standing's correlation. All of the data he presents is flash data. If the nomograph from low pressures is used and a PVT data set is built, there are no negative values. An example is shown in Figure 6–7.

Equation of state (EOS) matching

If the previous simplification does not seem justified, try using an EOS match of separator data to match field operating conditions. This was not an option historically. Compositional simulation may also be carried out. However, in view of the accuracy of other input data, the use of black oil PVT in simulations and material balance, in the author's opinion, has proven to be a more-than-adequate technique in practice.

Table 6–1 Separator Tests of Reservoir Fluid Sample

Separator Tests of Reservoir Fluid Sample										
Separator Pressure (psig)	**Separator Temperature (°F)**	**GOR Differential**	**GOR Flash**	**Stock Tank Gravity (API)**	**B_o**	**Separator B_o**	**G of Flashed Gas**	**Author's Technique For Separator Adjustment**	**Estimated B_o**	**Error in B_o**
50	75	715	737			1.031	0.84			
to										
0	75	41	41	40.5	**1.481**	1.007	1.338			
			778					1+778/854* (1.6-1.0)	**1.547**	**4.40%**
100	75	637	676			1.062	0.786			
to										
0	75	91	92	40.7	**1.474**	1.007	1.363			
			768					1+768/854* (1.6-1.0)	**1.54**	**4.40%**
200	75	542	602			1.112	0.732			
to										
0	75	177	178	40.4	**1.483**	1.007	1.329			
			780					1+780/854* (1.6-1.0)	**1.548**	**4.40%**
300	75	478	549			1.148	0.704			
to										
0	75	245	246	40.1	**1.495**	1.007	1.286			
			795					1+795/854* (1.6-1.0)	**1.559**	**4.30%**

**N.B. Differential P_b was 2,630 psig, R_{si} was 854 scf/bbl, B_o was 1.600*

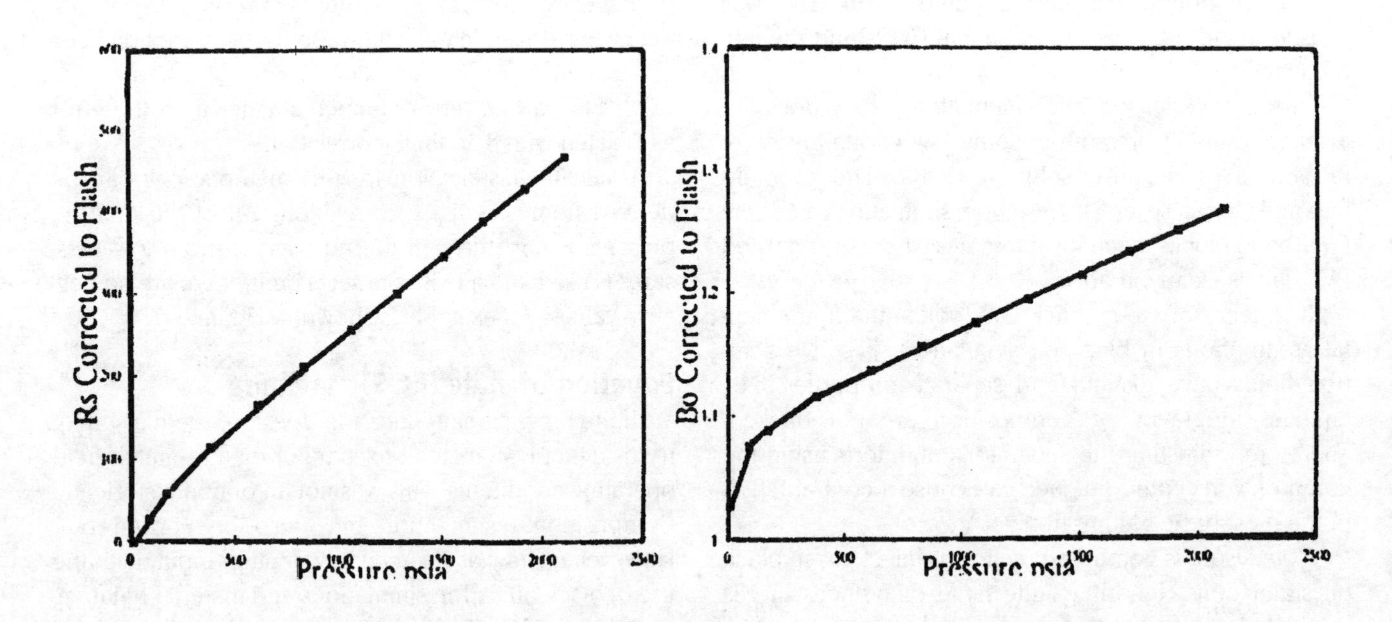

Fig. 6–7 Oil PVT Data Generated Using Standing Correlation

Returning to the main point, if it is assumed both separator and stock tank are operated at consistent temperatures and pressures, the stock tank GOR and formation volume factor should be relatively constant. Of course, this is an unlikely assumption given the range of ambient conditions in Canadian fields.

Realistically, we rely on representative averages. One does not have to look at many production plots to realize GOR data has considerable variation from month to month. Sometimes, exceptions occur, as outlined in the next section.

Composite liberation

Dodson et al. suggested an improvement on the standard oil PVT test by developing a composite liberation.[2] An example of this is shown in Figure 6–8. Oil from differential liberations is flashed through a separator system to derive GOR and B_o values. This requires large sample sizes and dramatically increases the cost of the experiment. The author has yet to see a PVT analysis of this type available on an actual project. This raises a number of questions as to why the methodology has not been widely used. The author has found oil companies are often inordinately sensitive to lab costs, despite the obviously large potential financial impacts. It may be the extra cost was not viewed as justified. The differential lab process was regarded as only an approximation of the reservoir process and, perhaps, the improvement was not regarded as adequately significant.

The computed separator flash is well below the observed separator flash. This would be more consistent with the correction suggested by the author earlier. It is also possible to calculate a composite liberation using an EOS software package.

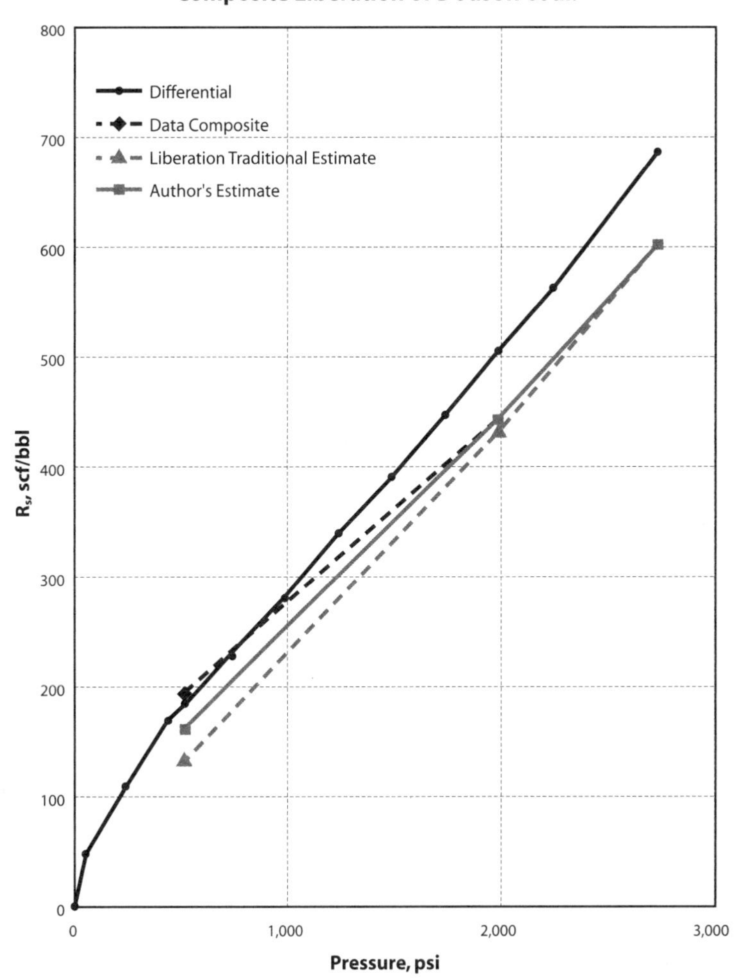

Fig. 6–8 Adjustments from Dodson's Composite Liberation (after Amyx, Bass, and Whiting)

Built-in composite liberation

Historically, some black oil simulators allowed changes in PVT tables with time to take into account different surface operating conditions. This is a potentially useful feature. After waterfloods are installed, it is common to use treaters, which, among other things, typically heat the oil. This changes the surface conditions and, with moderately high gravities, can obviously change GORs. Figure 6–9 shows an example of a simulation on a reservoir with a 39° API oil.

In this figure, the GOR increases immediately after water breakthrough. The reservoir pressure history indicates a highly undersaturated reservoir. This reservoir was produced well above the saturation pressure for almost all of its life. Thus, there was not a significant change in reservoir fluid properties during the producing life of the field. The simulator used could not account for changing PVT data with time. Overall, the simulation was still successful.

Apparently, a black oil reservoir simulator is on the market that uses an EOS to adjust for composite liberation and separator temperature and pressure. In the author's experience, EOS requires tuning,

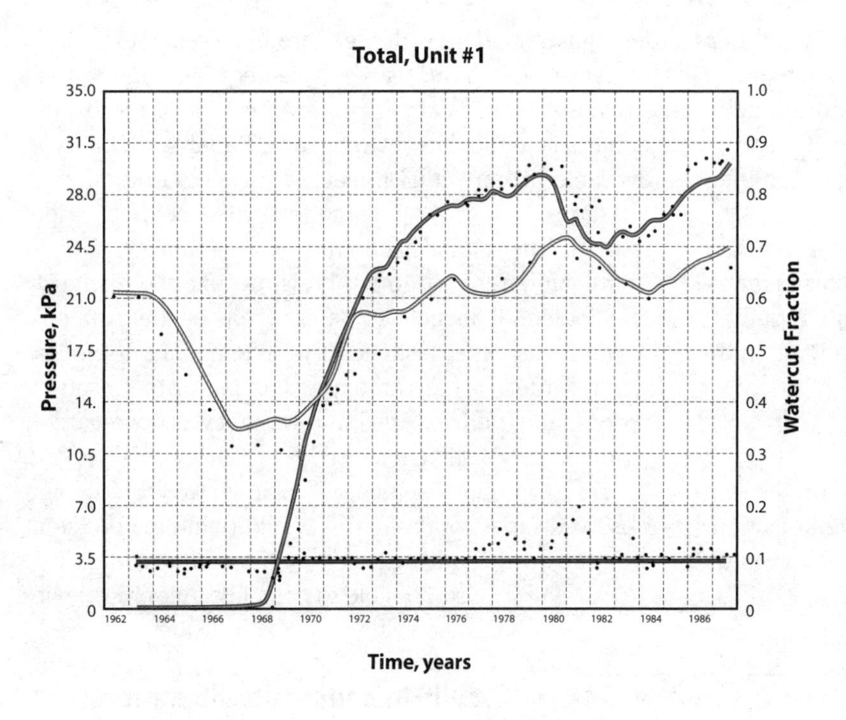

Fig. 6–9 Effect of Treater on GOR Trend in a History Match

and it's uncertain whether incorporating this, with the attendant increase in run time, is an enormous advantage. Still, there are real situations where PVT adjustments change with the passage of time and can give rise to noticeable effects.

Validity of using adjusted data

Before completing this discussion of separator adjustments, it is necessary to return to the fundamental theoretical concept, which has been discussed. When the author first learned about this subject, production volumes clearly needed to be adjusted for separator conditions. At the time, it seemed equally clear the fluid in the reservoir, which expands with further withdrawals, was not going through a separator at all (nor could one even place a separator in the reservoir). To correctly derive the OOIP, one must correctly describe the expansion in the reservoir, which clearly did not require adjustment. Conversion to stock tank volumes would be a last step to meet the convention of expressing OOIP in stock tank barrels. Actually, this is an issue in compositional simulation.

Carrying this argument further, we expend much effort getting a sample representative of the original fluid in place. If we succeed at this, and if one believes the differential liberation process most closely resembles what happens in the reservoir or reservoir model grid blocks, we should use a differential PVT table in the simulator for the reservoir grid blocks. Recall the comments on varying gas gravities and the implied flash equilibrium effects, which show differential liberation is not the absolute truth.

In further support of this, it can be argued, since oil recovery is usually less than about one-third of OOIP, it would be more important to accurately represent the oil properties with a differential PVT table. Withdrawal should be of secondary importance since it is the minority of the volume of oil originally in place.

Although the argument in favor of a pure unadjusted differential PVT table has some merit, this overlooks the main drive mechanism in most reservoirs, which is normally the evolution of solution gas. Recovery of solution gas typically amounts to more than 65%. Since most of this gas is recovered through a flash process, this tips the scale in favor of adjusted data, dependent on the reservoir recovery pressures.

Some people have said the use of adjusted data is self-evident, but the author does not agree, because the use of a black oil description is an approximation, no more, no less.

At this point, an engineer has two choices. One could alter black oil formulations with the use of multiple PVT relations, or one could proceed directly to compositional simulation. The idea is not new. People have found, for oils characterized as black oil, the comparison with compositional simulations results in essentially similar results. With further increases in available computer power, the extra time to run and tune a compositional model may be trivial, and this could become the accepted practice. Clearly, with current technology, black oil simulation is much more readily accomplished and provides an acceptable level of accuracy.

In the past, the author has done simulations without making separator adjustments. The attached problem sets in the final chapter are such examples. The examples in this book are not intended to represent a real reservoir. Nonetheless, the results are often quite good, and the trends are not significantly affected. In view of the difficulties with separator adjustments outlined previously, and given the amount of time spent sorting out these issues, the author suggests ignoring them in the same applications is not unreasonable.

Oil property correlations

In some cases, no data is available for the reservoir. In fact, if engineers are in a completely unexplored area, they may have to rely entirely on correlations or offsets. A number of correlations are available. For the most part, the accuracy of the different correlations is quite good. Standing's correlation, which was prepared using what is really a limited data set, is truly remarkable.[3] The correlations have been reproduced in Figures 6–10, 6–11, and 6–12 (Standing's correlation).

It is important to realize Standing's correlation represents flash-corrected data. This correlation can therefore be used directly for data input.

Standing's correlation is by no means the only one. More recent correlations have been made by Lasater and Glaso representing improvements under most conditions.[4,5]

Correlation accuracy

Abdul-Majeed and Salman compare a number of different solution GOR correlations to actual PVT lab data. Their results are summarized in Table 6–2.[6]

Table 6–2 Solution Gas-Oil Ratios

Correlation	Average Percent Error	Average Absolute Percent Error	Standard Deviation
Standing	-5.4	15.78	31.76
Lasater	-6.2	13.79	28.34
Vasquez-Beggs	-6.7	12.45	23.25

Approximately 66% of the data should fall within plus or minus one standard deviation, assuming a normal distribution. For the Standing correlation, this is ±31.8%. To fall above the upper extreme, it is probable that one-half this, approximately 17%, fell outside the expected range. The probability of this data occurring was one in six. In the author's opinion, knowing the accuracy of these correlations can be extremely valuable. Abdul-Majeed and Salman's paper is very valuable. They have applied all of the common correlations to a large sample set, which was completely different from the data with which the correlations were derived.

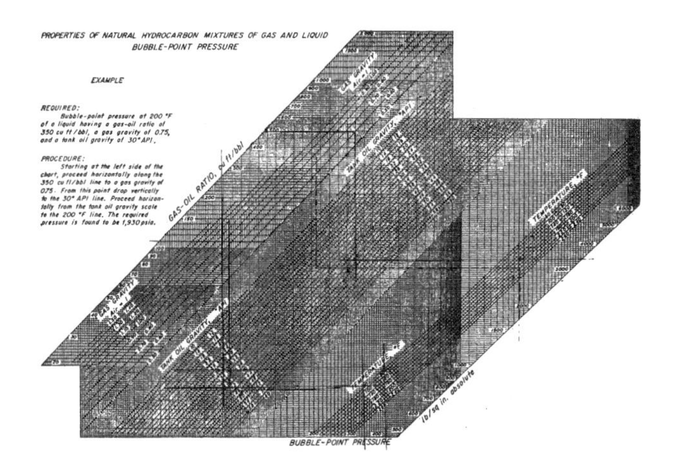

Fig. 6–10 Standing Correlation GOR vs. P_b

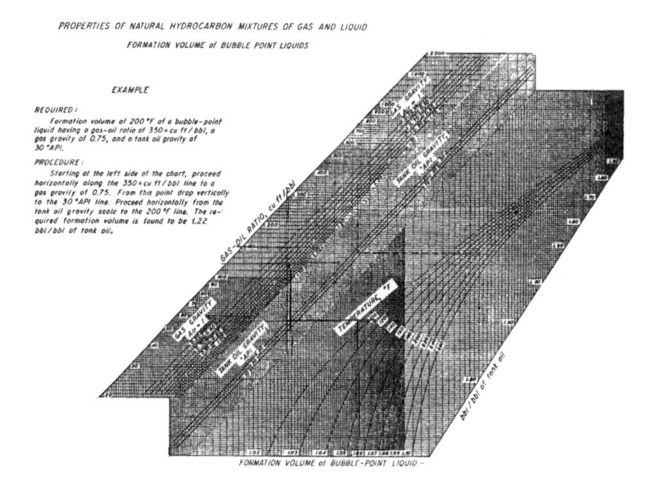

Fig. 6–11 Standing Correlation GOR vs. B_o

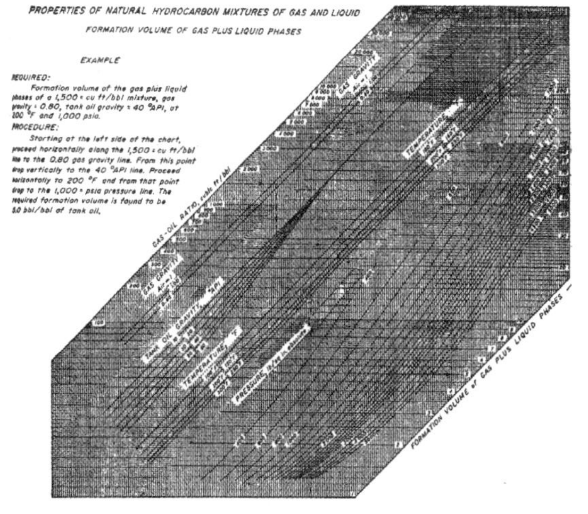

Fig. 6–12 Standing Correlation GOR vs. Two-Phase Formation Volume Factor

Limitation of black oil usage

There is a limit to where black oil descriptions can be used. The normal limit corresponds to a B_o of 2.0. Beyond this, more sophisticated methods should be used. These are most likely a compositional simulator with an EOS or a volatile oil option. The volatile oil option allows liquid to be entrained in the gas. The liquid precipitates as the pressure drops below the dew point.

Shape of PVT curves

The shape of PVT curves can reveal information about oil composition. Figure 6–13 has been extracted from a paper by Chapman Cronquist.[7] The PVT curves have been normalized so the bubblepoint or saturation pressure is equal to one. Similarly, the formation volume factor or B_o has been scaled to one.

The volatile oils have an S-like shape, which is characteristic of this type of sample. In these samples, the intermediate components have a significant effect on PVT properties.

Gas PVT properties

Different people prefer to deal with different formats. It is a good way to trick the unwary. Three different methodologies are common:

- specifying Z factors
- specifying B_g in units of rb/mscf or rm^3/sm^3
- specifying E_g or b_g in units of rmscf/rb or sm^3/rm^3

As an aside, the author was trained on the more consistent metric units and finds the field units of gas properties quite unattractive.

Gas formation volume factors are calculated using the real gas law, $pV = ZnRT$. The most difficult part of this is calculating the Z factor. The basis for Z is the theorem of corresponding states. In essence, the deviation from the ideal gas law is scaled to a proportion of the critical pressure and critical temperature of the gas. This is accomplished with what are called pseudo reduced pressure and pseudo reduced temperature:

$$pT_r = \frac{reservoir\ pressure}{pT_c} \tag{6.5}$$

$$pP_r = \frac{reservoir\ pressure}{pP_c} \tag{6.6}$$

It is then possible to use a single scaled graph to determine the Z factor as shown in Figure 6–14[8]. In common practice, this is recorded to three decimal places. However, the actual accuracy of this graph is approximately ±3%. Although it is possible to use computer routines fitted to this graph, they are no more accurate than the original graph. The actual level of significance is still only slightly less than two decimal points.

This is a phenomenally convenient tool. The methodology for calculating the critical point of mixtures is to arithmetically add the critical temperature and pressure of each component, weighed by its molar concentration expressed as a fraction. This is demonstrated in Table 6–3.

Gas condensate systems

Gas condensate systems and, frequently, volatile oil systems are normally tested using a different process. This type of test is known as a constant volume depletion or CVD test. The process used is quite different from a differential test and is shown in Figure 6–15. A PVT cell is loaded with gas, and the pressure is reduced in increments. In this process, the gas drawn off the top of the cell and any liquid (condensate) that precipitates, due to pressure decline below the dew point, is free to accumulate by gravity in the bottom of the cell. In most tests, the gas extracted from the top is analyzed for composition. This is one area in which it is possible to (foolishly) reduce costs. Typical results are shown in Figure 6–16.

Gas condensate systems seem to frequently have a tail that extends up the pressure axis to the dew point. This feature is common and should not cause alarm. There has been some speculation on this issue. Danesh indicates this may be a time-dependent effect in the PVT cells.[9] It is also speculated this may have to do with sampling problems, as outlined in the following.

"Red" fluids

There is rumor in the Canadian oil patch, which the author has heard from a number of different sources. A sample was obtained from a gas condensate system that had a peculiar red coloring in the PVT cell. The lab CVD dropout apparently showed a very pronounced tail as well.

Further investigation indicates the red coloring and peculiar tail were the result of some exotic (yes, it was red) pipe dope being used in the area. The heavy hydrocarbons in the pipe dope were apparently captured in the sample and had a pronounced effect on the PVT samples.

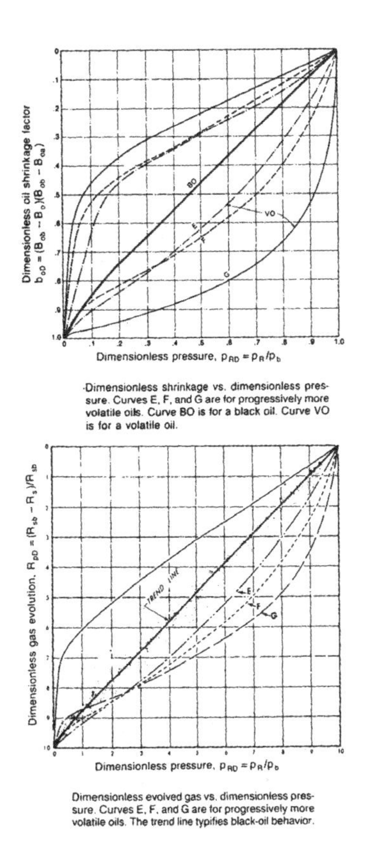

Fig. 6–13 Shape of Oil PVT Data (after Chronquist)

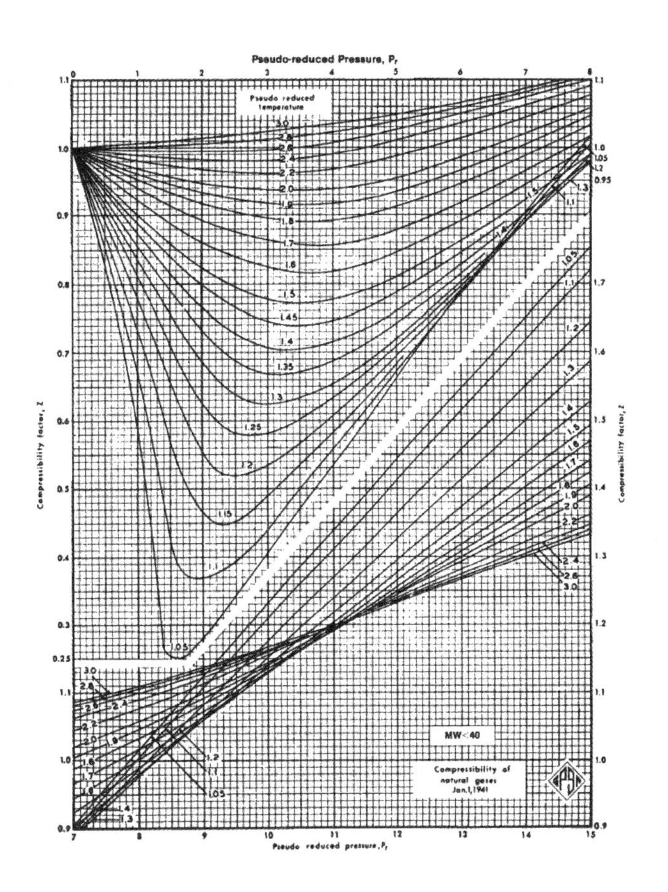

Fig. 6–14 Standing and Katz Z Factor Chart

Table 6–3 Calculating the Critical Point of Mixtures

	M mol mass	P_c Kpa	T_c deg. K	Total Mole Frac.	X_i*M_i mol mass	X_i*P_c kPa	X_i*T_c deg. K
H_2	2.016	3399	33.2	0.0001	0	0.3	0
He	4.003	22118	5.2	0.0015	0.006	33.1	0
N_2	28.013	5081	126.1	0.074	2.074	376.1	9.3
CO_2	44.01	3499	304.19	0.0191	0.84	66.8	5.8
H_2S	34.076	9005	373.5	0.0019	0.065	17.1	0.7
C1	16.043	4604	190.55	0.8453	13.561	3891.6	161.1
C2	30.07	4880	305.43	0.0352	1.058	171.7	10.7
C3	44.097	4249	389.82	0.0152	0.671	64.6	5.9
iC4	58.124	3648	408.13	0.0028	0.163	10.3	1.1
nC4	58.124	3797	425.16	0.0026	0.152	9.9	1.1
iC5	72.151	3381	460.39	0.0008	0.059	2.8	0.4
nC5	72.151	3369	469.6	0.0006	0.045	2.1	0.3
C6	86.178	3012	507.4	0.0003	0.029	1	0.2
C7	100.205	2736	540.2	0	0	0	0
C8	114.232	2486	568.76	0.0005	0.061	1.3	0.3
C9	128.259	2288	594.56	0	0	0	0
C10	142.286	2099	617.4	0	0	0	0
				1	18.783	4648.8	197
				G =	0.6485		

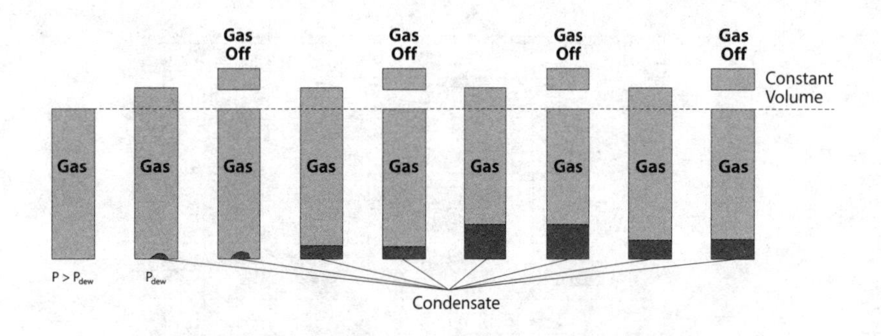

Fig. 6–15 Constant Volume Depletion Stages

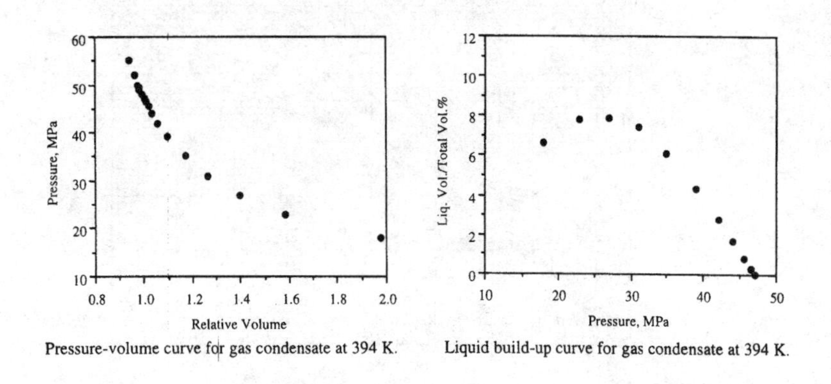

Fig. 6–16 PVT Curves for Gas Condensates

What are the morals of the story? First, it is wise to be more sparing with the pipe dope. Drilling engineers are generally very conscientious, but PVT sampling is a little out of their normal expertise. This suggests at least occasional interdepartmental communication. It makes one wonder if the tails regularly seen on CVD samples might not be affected by smaller amounts of colorless pipe dope. Finally, the consistent use of bright red pipe dope is recommended, since other colors have not yet been proven in PVT experiments.

Representative samples

Reservoir engineers are primarily concerned with what is happening in the reservoir. It is highly desirable to have a sample closely resembling the original reservoir fluid as much as possible. The following example is for an oil reservoir. Condensate reservoirs are discussed in Part II of this book, and this specific issue is addressed.

Referring to Figure 6–4, it illustrates an oil reservoir producing just below the bubblepoint pressure. At this stage, gas is coming out of solution in the reservoir. Due to relative permeability effects and the low viscosity of gas, the solution gas is more mobile than the oil. Therefore, any sample we take of total reservoir effluent will have an excess of gas.

In order to get as representative a sample as possible, it would be better to have a sample of the reservoir oil. Ideally, we can also have a sample before the pressure around the wellbore dropped below the bubblepoint. Hence, we would sample an oil reservoir above the bubblepoint with the pressure drawdown sufficiently low such that no solution gas is evolved.

Since some reservoirs are initially saturated, we do not always have this opportunity, and we must sample the reservoir as early as possible and with as low a drawdown as possible. This process of producing the reservoir at low pressure drawdown, or low rates, is known as *well conditioning*. It is also discussed in detail in the American Petroleum Institute (API) recommended practice RP-45.

The author has encountered other interesting twists on older (circa 1957) PVT sampling programs.

Historical thinking suggested that, if one could shut in the well, the pressure surrounding the well would increase, and the afterflow or wellbore storage, which causes so many problems in pressure transient analysis, could be minimized and the well sampled with what might represent an accurate representation of the original reservoir fluid.

To the author's knowledge, the accuracy of this process has not been studied in detail. In the author's opinion, evolved solution gas does not redissolve quickly, since pore level effects indicate contact areas may not be as large as one might expect. The use of afterflow is somewhat tricky, since one must be directly opposite the perforations as the fluid is entering the reservoir. It is obviously dependent on timing. It would also depend on how fast equilibrium occurs with the fluids in the wellbore, which is the subject of the next topic.

Wellbore samples

For the most part, wellbore samples are a disaster. It seems reservoirs do not flow consistently due to wellbore effects. This includes liquid holdup, bubble flow, slugging, and other rapid PVT effects occurring as the fluids travel up the wellbore. Restated, bottomhole pressures are not, in fact, constant. Since slugging occurs when the valve on the wellhead is closed, the total ratio of gas and liquids in the reservoir will generally not be representative of the material exiting the reservoir. The wellbore will eventually segregate by gravity, and there will be compositional change within the wellbore dependent on the speed with which equilibrium is established. As discussed previously, there will also be afterflow into the wellbore.

These problems have been widely recognized for a considerable time. Standing outlines the issues in his 1951 book.

Taking samples from the wellhead is also equally difficult. It is generally impossible to get the correct mix of liquid and gas due to the chaotic flow in tubulars and slugging effects.

Sampling options

Sampling options fall into two specific categories. The first option now is to obtain a sample of the virgin fluid before the reservoir is produced from an uncased wellbore. The second option is to obtain a bottomhole sample. This is best done on a flowing well producing downhole at above the bubblepoint pressure. The next best option is to use samples taken from a separator and recombined according to a measured GOR. Finally, one can take the afterflow option and take a downhole sample from a recently shut-in saturated reservoir. The following discusses these briefly.

Open-hole reservoir samplers. During the past 15 years, newer tools have been developed derived from the repeat formation tester (RFT) tool. The latter was designed to take point pressure measurements in exploration wells. After this, tools such as the MDT tool were developed. In essence, one of the pressure points, taken with a probe against the wellbore side, is directed into a sample holder. Ideally, the reservoir is undersaturated, as this will make gathering a single-phase sample possible. A bit of mud filtrate and mud cake is also recovered with this tool. Hence, such samples were generally dirty.

The tool has an additional feature, which is an optical sensor designed to distinguish between oil and gas. Although a good concept, the times the author needed this information were on near-critical fluids. Unfortunately, the tool results were not sufficiently accurate in these circumstances to differentiate near-critical fluids. Part of this problem was samples were sucked into an evacuated cell, which caused immediate phase separation.

Recently, these tools are said to have been considerably refined. The initial sample, which contains the majority of contaminants, is directed to a small dummy sample catcher. The main sample is no longer an evacuated container. Rather, the sample is drawn from the formation with a motor-driven piston, which minimizes both drawdown and phase flashing effects.

While the author has not worked with the new improvements, there are many positive indications the historical problems with this type of tool are likely solved.

Bottomhole samplers. Most bottomhole samplers are supplied by laboratories. The sample chamber is triggered by an electrical switch, which opens a valve to an evacuated sample chamber. The tool is generally opened for a fixed time and is automatically closed.

Surface sampling—recombination. The other alternative is to sample at surface. This is the standard procedure when saturated reservoir fluids are being sampled. There are also some tricks to taking proper samples in the field. Liquids can be carried into the gas stream of the separators. Therefore, the test separator must be sized correctly. To help control this liquids carryover problem, a steel wool demister is installed on separators. Demisters can and do deteriorate, often by corrosion. In rare cases, they may have been removed. Check to assure these are installed and in good condition. Do not hesitate to check the size of the separator (see production operations textbook), and discuss this issue with the testing company.

Dump valves on condensates will often spin out due to gas breakout while passing through a positive displacement meter. Therefore, samples should be taken upstream of the dump valve. The author has seen a meter run improperly mounted and full of condensate. It may be inferred the gas volumes measured weren't correct.

In addition, the gas must be measured when conditions are stabilized to obtain the correct gas density. Similarly, samples need to be taken near the end of the flow cycle with stabilized rates. One other point: it is possible to take samples at a number of times. Multiple samples are cheap insurance when compared to the cost of redoing the test or lab testing program. Consistent sample compositions also give much more confidence to lab results. The correct GOR is critical to correct recombination of the reservoir fluid.

Proper labeling is, surprisingly, a real issue. This sounds dreadfully simple, but, as one moves from one sample point to another to get samples as closely as possible to the same time, it is actually easy to mislabel things. One of the biggest problems is water. Cardboard tags and markers deteriorate to a pulp with just the water that collects in the back of a pickup truck. Reinterpreting these spoiled labels is virtually impossible. Tyvek (nylon paper) labels and permanent markers are suggested.

Murphy's Law seems to apply widely to field operations and can be a source of problems. A well laid-out written program is recommended for the field staff. It is highly appropriate the reservoir staff witness well tests. Finally, although good experienced testing staff may cost a little more, they are worth the added expense.

Leases are potentially dangerous places and the field foremen are responsible for everyone's safety, particularly with sour gas wells. Get the proper training and safety equipment. This will mean some lead time before you can go out in the field. Most field staff are very happy when someone from the head office pays a visit to see how things actually run. Engineers can learn a great deal from these people. It is their turf, however, and respect for their expertise is important.

Afterflow sampling

Recently ARE worked on a study featuring an afterflow type test. The well had been shut in for a long period of time—68 hours. The author immediately pronounced this to be a wellbore sample and of no validity whatsoever, pursuant to the previous explanations. One of our project engineers felt the test was representative. There was, of course, an intense debate. We spent some time evaluating a number of offsets, which were, in this case, not particularly helpful. The test was consistent with oil correlations and, in the end, the sample results did prove to be consistent with volumetrics and observed GORs. This was positive proof the methodology can work (and the author can and will likely be wrong again).

Nonetheless, the author still holds reservations about wellbore samples. The warnings in Standing's material indicate wellbore slugging generally, but not absolutely, results in unrepresentative samples. The author would be receptive to simulator and wellbore modeling that would indicate how much fluid is actively moving into the wellbore across from the perforations. Given the author's experiences with PVT cells, the timing of equilibrium in the wellbore is still troubling.

Lab checks of downhole samples

The sample is taken into the lab, heated with an electric tape, and then the contents displaced with mercury to an intermediate storage vessel. This allows rapid return of the downhole tool to the service company. The sample is then tested for composition, a saturation pressure is likely determined, and the contents are transferred to a PVT cell.

Lab checks of separator samples

The samples are received in the lab. A bubblepoint is determined for the liquid sample and checked against the reported operating conditions of the separator. Compositions of the liquid and gas are taken and processed. A recombination procedure is implemented to combine the two fluids. This requires a number of steps to ensure the accurate amounts of each are combined. A sample is then taken and checked against a mathematical recombination of the liquid and gas. At this point, the material is transferred to a PVT cell.

Composition

One check recommended by ARE is to compare compositions across a number of samples. Heavy ends have a disproportionate effect on phase behavior based purely

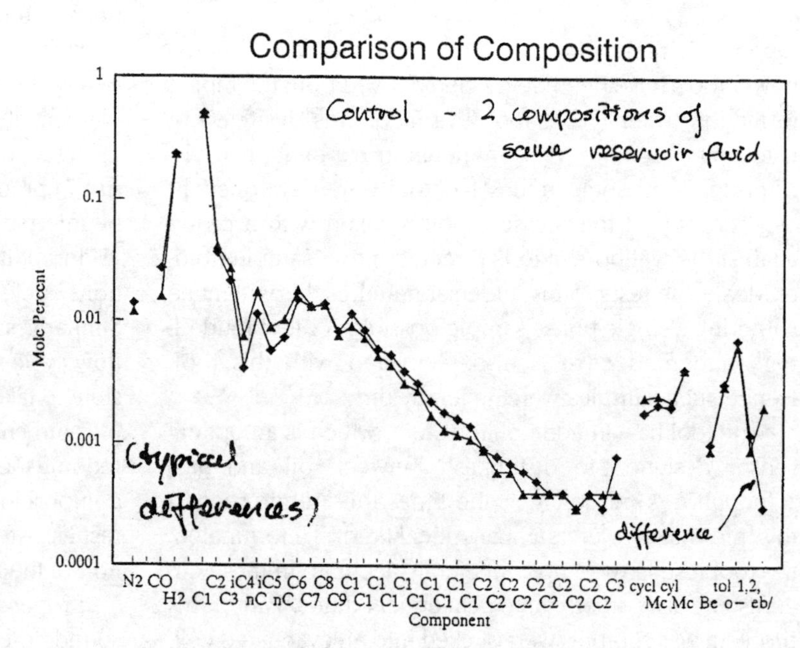

Fig. 6–17 Comparison of Compositions

on mole fraction. On the other hand, they are quite large compared to the smaller and lighter species. Therefore, plotting on a log scale against carbon number is recommended. An example of this is shown in Figure 6–17.

Differences are easy to spot on a proportional basis. Cyclic and aromatic components are separated on the right side of the graph. In addition, condensates and near-critical fluids will show as a broader distribution than gases.

Equilibrium

It is possible to check to determine if the samples chosen were in equilibrium. This technique is empirical and is a remnant of the old flash calculation, which is comforting to those of us advancing in age. There will be some deviation on carbon dioxide and hydrogen sulphide. Figure 6–18 shows a good sample, and Figure 6–19 shows a problem sample.

Common sense would also tell us a recombined sample from an undersaturated reservoir should have a pressure somewhat less than the reservoir pressure. This drop in dew point or bubblepoint pressure is systematic and is discussed in chapter 16. If a recombined sample has a saturation pressure above or equal to the reservoir pressure, it would cause suspicion.

Some laboratory problems

As outlined previously, a remarkable number of problems can occur in the sampling process. Many examples in the following section on screening PVT data show problems that have occurred in the lab, but we will jump ahead slightly and first describe what happens in the lab.

Not all lab data is created equal. One has to be alert to pick up problems. A trip to the lab is recommended while the experiments are being done. Again, there are safety issues in visiting the lab. These must be respected. The best lab personnel are highly responsive to interest from the reservoir engineer. As a rule, most people like to tell about the good job they are doing. Talking with lab personnel reveals information that would not be known otherwise.

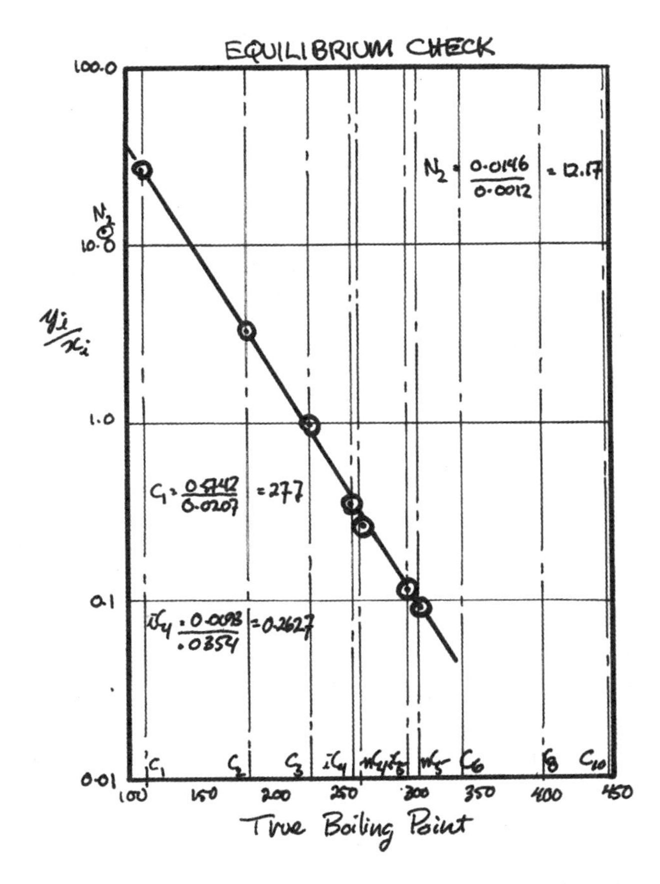

Fig. 6–18 Use of log K vs. True Boiling Point to Screen Samples (Good Data)

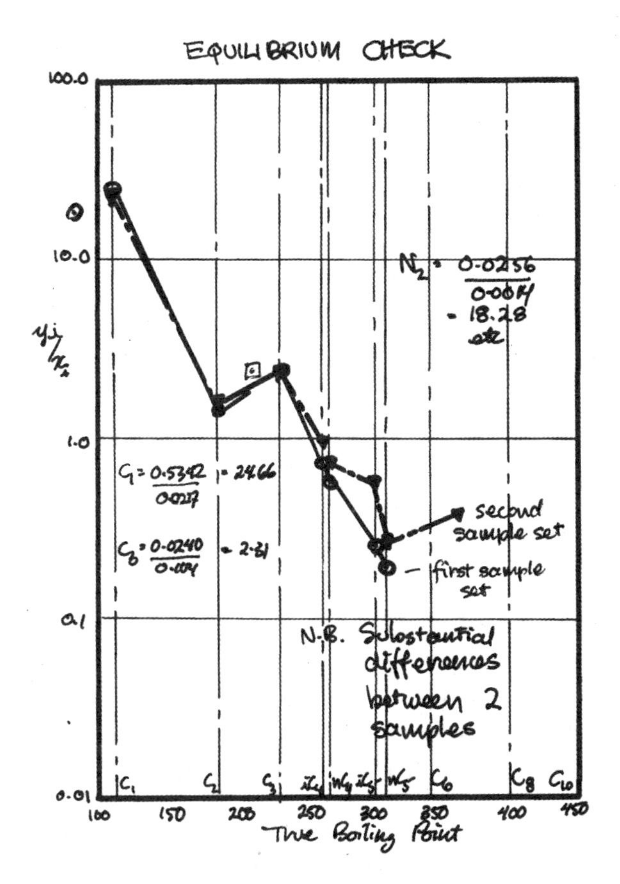

Fig. 6–19 Use of log K vs. True Boiling Point to Screen Samples (Bad Data)

In the lab procedures described, with the exception of the CCE or CME, all involve taking gas off the top of the cell. Right off the top, this would seem like a straightforward process (pun intended). So, to start off, a certain amount of skill is required to get the liquid to "kiss" the top of the cell, exactly displacing all of the gas and not carrying over any of the liquid. Oh yes, there will be a meniscus on the liquid. A light touch and eagle eyes are therefore a job requirement.

In order to maintain the pressure at a constant level, mercury must be pumped simultaneously into the cell as the gas is withdrawn.

The reality is, however, the windows on some cells do not go right to the top. Therefore, for these cells, it is necessary to estimate when the liquid will reach the top of the cell.

After witnessing PVT tests, it becomes clear the assumption of chemical equilibrium is not quite as simple as it might seem. It is possible to back off the mercury in the cell and, with a gas condensate, observe relatively little change. This is true until the side of the cell is tapped or hit, at which point a rapid phase change occurs. Liquid magically appears at the bottom of the cell, and the pressure gauge drops like a rock and comes to a rapid stop accompanied by intense vibrations of the pressure gauge needle, which quickly die out. It is actually necessary to agitate or rock the samples to ensure chemical equilibrium is actually happening.

This entails several competing requirements: a rock steady interface, constant agitation, constant pressure, and deadly accurate eyesight or volume estimation. This may not be so easy after all.

Laboratory equipment

Sometimes it is difficult to read very low levels of condensate dropout. It is generally considered unsafe to look through gas windows on high-pressure cells. Although windows rarely break, it can and does happen. In one case, a window went in a Calgary lab with devastating results to the solid brick wall behind the cell, but no one was reported as injured. For safety, the contents of the cell are observed through a mirror on the wall using a telescope. This also solves the problem of parallax in making readings. Interestingly, the spyglass is similar to ones the pirates used and makes everything appear upside down.

There are solutions to a number of these problems. For instance, some labs have developed clear cells, which allow observation—at extra expense. You can make use of a capillary pressure line, i.e., very small diameter piping similar to capillary imbibition trapping that reduces the vision and light touch requirements to exactly displace a cell. The latter, due to the small diameter of the gas off-take, requires considerable time to drain—at extra expense. It is also possible to use a pressure regulator to keep the pressure constant in the cell while the gas is taken off, which requires an automatic feedback servo—at extra expense. To make measurements of small volumes of liquid easier, it is possible to use a tapered cylinder instead of mercury. This amplifies the changes in fluid level measured and increases accuracy—at extra expense.

It should be apparent the lab equipment is very important. Not all labs use these various features. Coincidentally, there seems to be a widespread attitude that lab testing is ridiculously expensive. The more the author has learned about this, the more he is convinced lab data is probably a better bargain than many people realize. As always, buyer beware.

Replication of reservoir conditions

All PVT samples are tested at reservoir temperature and pressure conditions. In order to achieve this, the high-pressure containers must be heated, and there are a number of ways of doing this. First, vessels can be wrapped in heating tapes and then wrapped in insulation. Alternatively, tests can be run in an enclosed oven. The trouble with an oven is it cannot be opened, and the equipment cannot be handled directly. Air does not have a high heat capacity, and any leaks could result in changes of the temperature around the cell.

PVT behavior is sensitive to temperature. Hence, the temperature must be carefully controlled. Any deviations may cause serious errors. Does a temperature chart in a lab report show the temperature remained constant?

Most lab cells use mercury to displace the reservoir fluids. The mercury in the cell is at reservoir temperature, but the pump used to manipulate is at room temperature. Mercury expands significantly with temperature, and appropriate calculations must be made to match cell displacement with the displacement on the pump on the bench.

The laboratory cells and lines into the cell are made of metal and expand with increased temperature; therefore, the volume of the cell and the lines must be carefully measured, and the temperature must be adjusted.

Material balance

One other calculation must be done to evaluate the integrity of lab data, particularly for condensate systems. That is, each step should be checked for molar or mass balance. One should be able to account for all of the moles, from the start of the test to the last step. It is surprising how many lab tests do not achieve material balance. If there is a substantial material balance error, the test is useless.

The example outlined in Figures 6–20 and 6–21 as well as Tables 6–4 and 6–5 show the differential liberation from different labs do not match. This is a sweet but volatile oil. Expect more problems for volatile oil systems. The traditional process used for evaluating gas condensates is also the one most prone to material balance problems. The Australians have an excellent paper by on this topic. CVD material balance problems are exacerbated on sour retrograde gas condensate systems.

In the calculation of the material balance, a unit volume of stock tank oil has been assumed. A single barrel would be acceptable when using field units. The laboratory samples are, of course, determined with much smaller volumes. The size of the sample is almost never reported. It is customary to make predictions of reservoir performance based on extremely small test volumes, in comparison to the size of a reservoir.

It is natural to ask how such an error occurred. Some of the items outlined previously are possible explanations. However, in this case the lab report mentioned a mist in the gas taken off the top of the cell. Apparently, the gas withdrawn from the cell was transferred directly to the gas measurement station. After consultation with one of the experts in this field, the following was learned:

1. Gas volumes are measured with a gas-o-meter. This is a clear Plexiglas cylinder, which has a nylon puck on top and uses water as a seal. The puck is not heavy; however, the gas volume is adjusted for the slight pressure exerted. It is at room temperature. Some adjustments are made for the line fill from the cell to the gas-o-meter, although this volume is small.

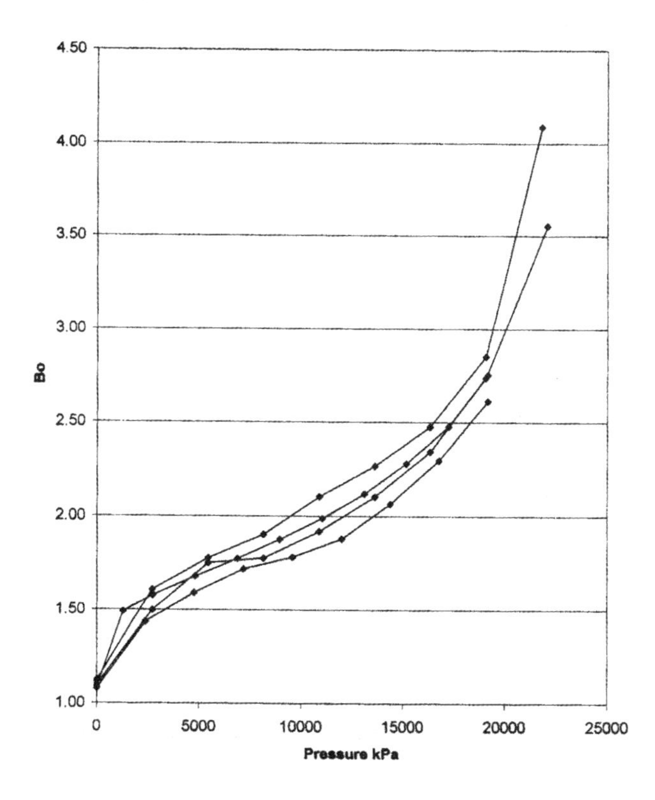

Fig. 6–20 Comparisons of Various B_o readings—Volatile Oil

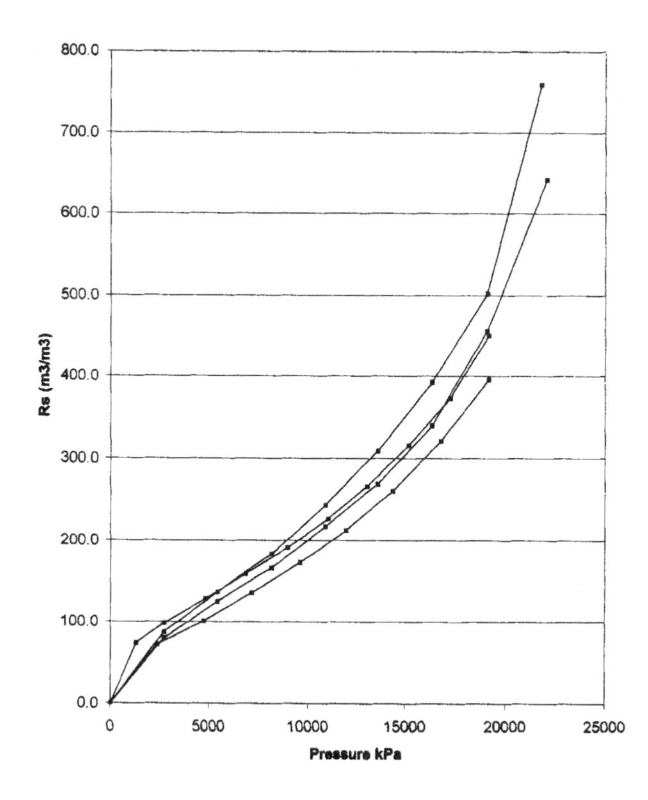

Fig. 6–21 Comparisons of Various R_s readings—Volatile Oil

Table 6–4 Material Balance Laboratory Sample A—with Acceptable Accuracy

Temp. (C)	Pressure (kPag)	ROV (m^3/std m^3)	Density Oil (kg/m^3)	Mass Oil (kg)	Incr. GOR (m^3/m^3)	Lab Gas Gravity	MW Gas Air = 28.964	Mass Gas (kg)	Mass Gas + Oil (kg)	% error in cum mass
99	19,167	2.757	511.0	1,408.8	0.0	-	-	-	-	-
99	17,237	2.480	532.0	1,319.4	77.4	0.942	27.284	89.3	1,408.6	0.01
99	15,168	2.283	549.6	1,254.7	57.2	0.917	26.560	64.2	1,408.2	0.04
99	13,100	2.118	566.2	1,199.2	49.9	0.905	26.212	55.3	1,408.0	0.06
99	11,032	1.990	580.6	1,155.4	39.5	0.892	25.836	43.1	1,407.4	0.10
99	8,963	1.878	594.6	1,116.7	35.4	0.906	26.241	39.3	1,407.9	0.07
99	6,895	1.775	608.0	1,079.2	32.4	0.927	26.850	36.8	1,407.2	0.11
99	4,826	1.681	620.8	1,043.6	30.1	0.967	28.008	35.6	1,407.2	0.11
99	2,758	1.580	634.5	1,002.5	30.4	1.056	30.586	39.4	1,405.6	0.23
99	1,310	1.492	647.5	966.1	24.8	1.238	35.857	37.6	1,406.7	0.15
99	0	1.088	720.5	783.9	73.6	2.012	58.276	181.5	1,406.0	0.20
									AAD >>	**0.11**

Table 6–5 Material Balance Laboratory Sample B—Unacceptable Accuracy

Temp. (C)	Pressure (kPag)	ROV (m^3/std m^3)	Density Oil (kg/m^3)	Mass Oil (kg)	Incr. GOR (m^3/m^3)	Lab Gas Gravity	MW Gas Air = 28.964	Mass Gas (kg)	Mass Gas + Oil (kg)	% error in cum mass
99	19,167	2.757	511.0	1,408.80	0.0	-	-	-	-	-
99	17,237	2.480	532.0	1,319.40	77.4	0.942	27.284	89.3	1,408.6	0.01
99	15,168	2.283	549.6	1,254.70	57.2	0.917	26.560	64.2	1,408.2	0.04
99	13,100	2.118	566.2	1,199.20	49.9	0.905	26.212	55.3	1,408.0	0.06
99	11,032	1.990	580.6	1,155.40	39.5	0.892	25.836	43.1	1,407.4	0.10
99	8,963	1.878	594.6	1,116.70	35.4	0.906	26.241	39.3	1,407.9	0.07
99	6,895	1.775	608.0	1,079.20	32.4	0.927	26.850	36.8	1,407.2	0.11
99	4,826	1.681	620.8	1,043.60	30.1	0.967	28.008	35.6	1,407.2	0.11
99	2,758	1.580	634.5	1,002.50	30.4	1.056	30.586	39.4	1,405.6	0.23
99	1,310	1.492	647.5	966.10	24.8	1.238	35.857	37.6	1,406.7	0.15
99	0	1.088	720.5	783.90	73.6	2.012	58.276	181.5	1,406.0	0.20
									AAD >>	**0.11**

2. The gravity of the gas is determined by using an evacuated glass bulb and directly weighing the gas-filled bulb. The volume of the bulb is known, as well as its weight. Therefore, the gas gravities reported on lab tests are directly measured.
3. It is common practice to run the line from the top of the cell through a trap, which is filled with ice or dry ice. The liquid volume can be measured, weighed, and then converted into a gas equivalent volume.

Step 3 was not done, ergo the mist in the cell.

All gases coming off the top of a cell should be saturated with liquid, and it is logical to assume some liquid would always be recovered. More liquid will be recovered with volatile systems, and, apparently, this error is smaller with less-volatile black oil systems. The suggested methodology is shown in Figure 6–22.

For this particular study, there were other samples of the reservoir fluid that all featured significant material balance errors. A number of corrections via equation of state modeling were attempted to resolve errors due to condensation. This was unsuccessful and suggested there were other undetermined errors.

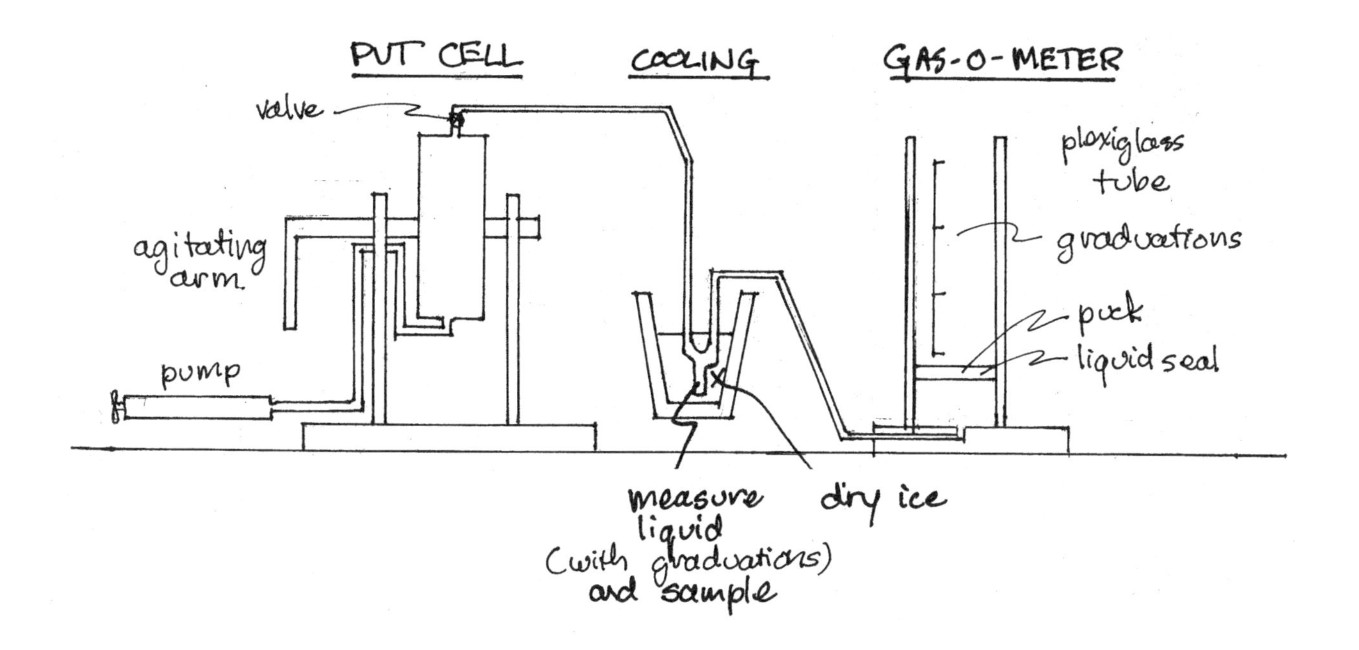

Fig. 6–22 Resolving PVT Bugs

A number of efforts to find the errors were made. This was done on only one of the tests, which had a 5% material balance error. In general, there seemed to be some discrepancies with gas densities as well as gas volume. Comparisons were made on composition to ensure uniformity. The attempted corrections are shown in Figures 6–23, 6–24, 6–25, and 6–26.

Lab tests from reservoirs that were not highly volatile also have had substantial material balance problems similar to the ones described previously. With respect to the plots of B_o versus pressure and R_s versus pressure, there is nothing to indicate the laboratory tests would not achieve material balance. Two earlier studies were carried out on this reservoir, which included compositional modeling and extensive EOS characterization based on the good and bad data. Buyer beware.

Use of laboratory gases

The author does not recommend using laboratory gases for solution gas.

It is true the lab gases are likely quite pure and gas chromatography is well-established, but this is not the end of the story. Although most of the author's experience on this topic is derived from sour gases, which is a special case, there are similar issues applying to sweet fluids. The testing done with gas chromatography only measures what the tests are designed to measure. The ionizing detectors used do not reliably detect sulphur compounds accurately, including H_2S. Mercaptans and thiols also exist in sour gases and, for the same reason, are not measured, except in rare cases. On the same line of reasoning, the amounts of structural isomers in the heptanes, septanes, and octanes are heavier components and are not distinguished with typical GC testing. Most GC tests are not set up to look at the heavier napthenes, alkenes, and aromatics. It is widely recognized the heavier components have a major effect on PVT behavior. So why deal with the uncertainty, and why take the risk of compromised PVT results?

At least one major oil company insists on sampled gases as an operating policy.

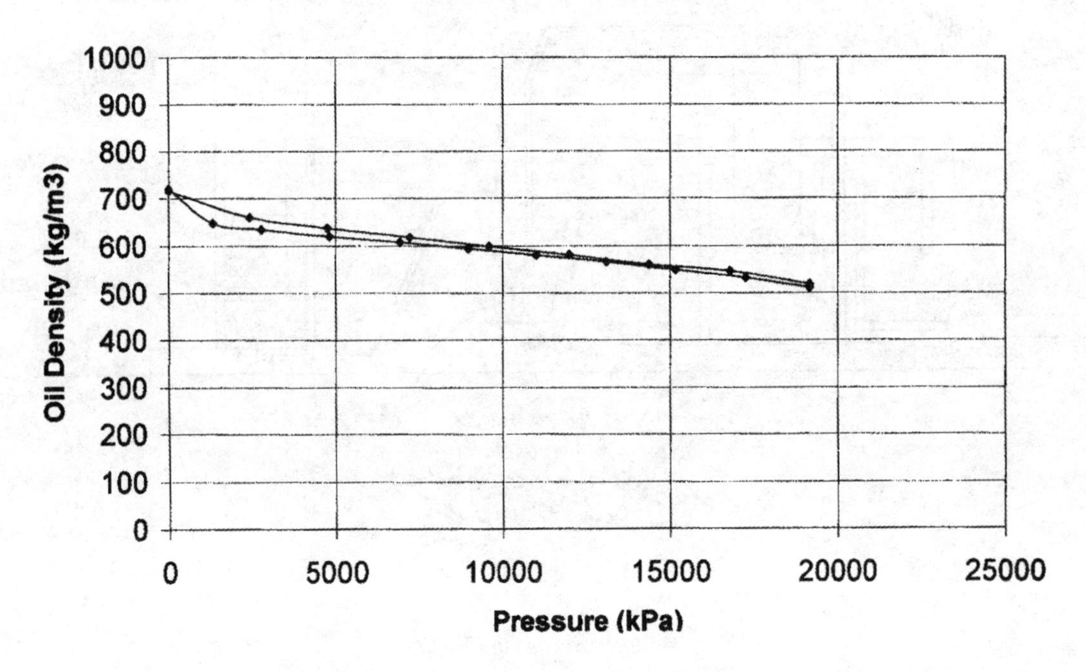

Fig. 6–23 Comparison of Oil Densities

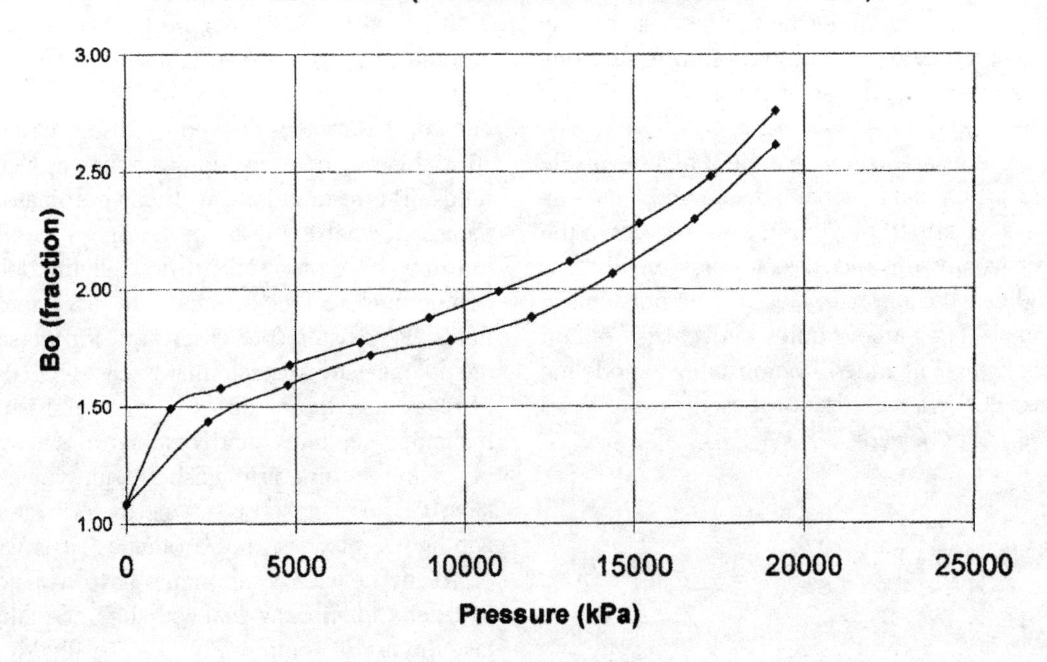

Fig. 6–24 Comparison of B_o Readings

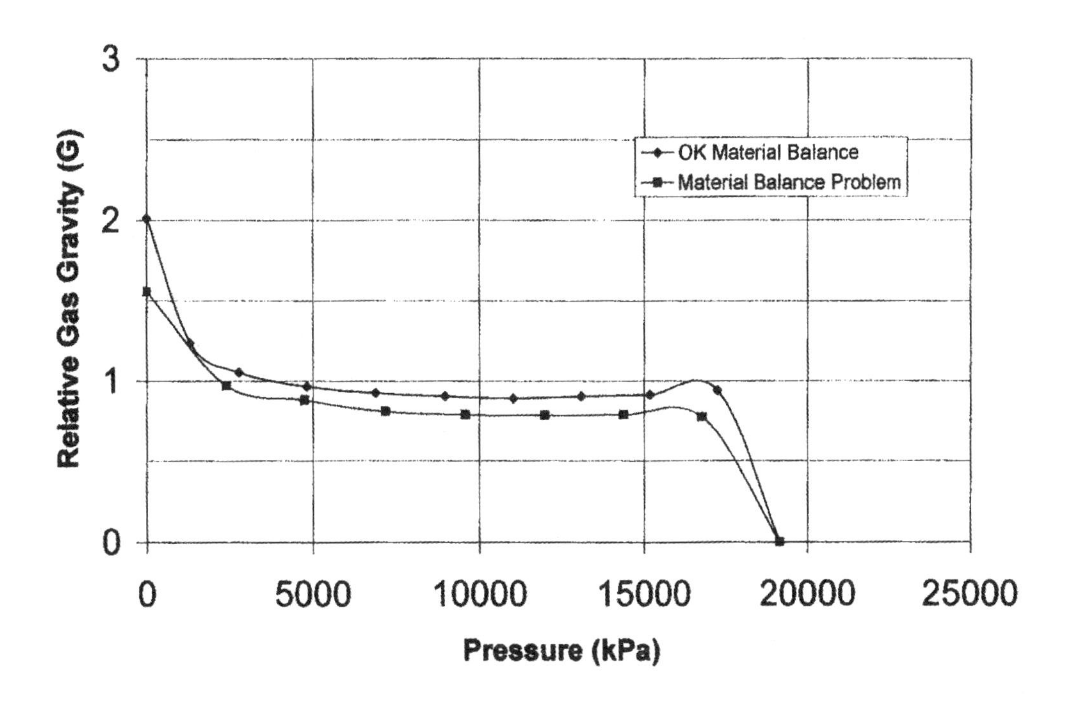

Fig. 6–25 Comparison of Gas Gravities

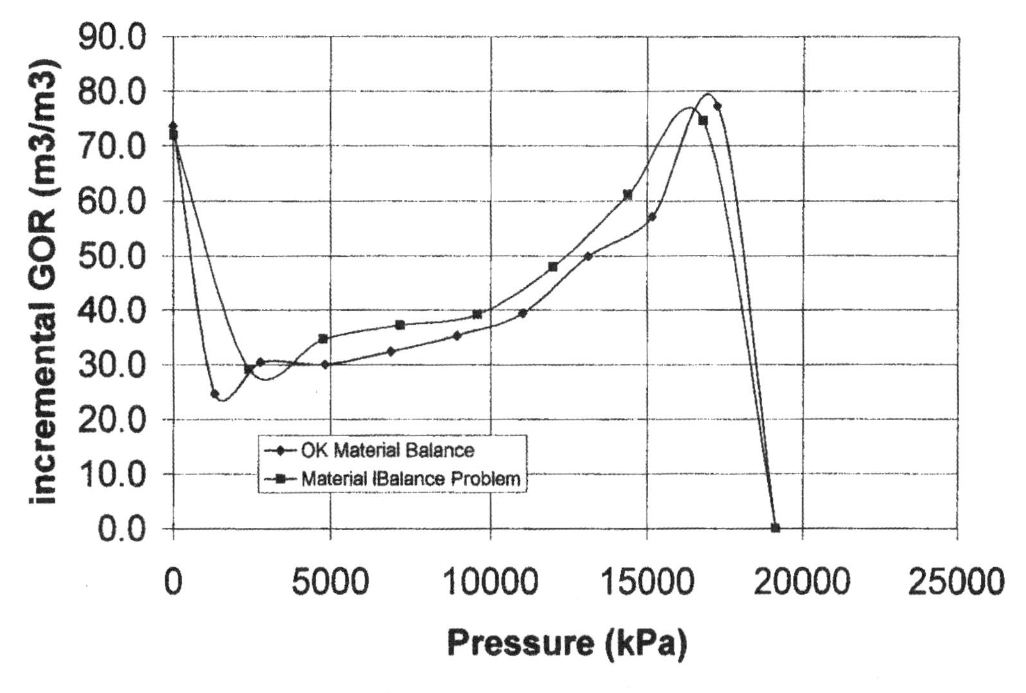

Fig. 6–26 Comparison of Incremental GORs

Laboratory testing program

A good lab program can make a big difference. In ARE's opinion, this includes cross-checks, for example, when comparing results from a CCE visual cell and the CVD. When dealing with sour fluids, specialized equipment is required. It is also ARE's experience some labs produce more consistent results than others. One company actually sent quality control samples to different labs. There was astonishing variability. Gas chromatographs are tricky to run and require frequent calibration.

A knowledgeable lab will normally be glad to help design a program. Relative volume does not provide much additional resolution on a PVT study. On the other hand, some equilibrium ratio work will help significantly in EOS characterization. The author recommends expenditures on the latter rather than relative volume.

The accuracy of EOS tuning is highly dependent on the completeness of the data, as shown earlier. It is also necessary to put PVT data in perspective. Expensive drilling and development plans depend on accurate PVT data. Some managers are reluctant to spend as little as $20,000 on lab work but think nothing of investing $10,000,000 on a well. This can be difficult for technical people, since the benefit seems obvious. Make the justifications obvious in dollar terms. It helps to show management what kind of differences and problems can occur when laboratory testing is minimized or not done at all. In this regard, it is hoped this book may be of assistance. Do not hesitate to make a distinct list of problems that can and will occur: sampling, variations in PVT properties, potential lab problems, and proper well conditioning.

Someone and no one

Regarding the potential impact, here is another interesting story. A gas plant was built south of town about 15 years ago. The production engineers dutifully supplied gas analyses that were perfectly correct from some wells.

It became painfully obvious during plant start up, *No one* failed to ask *Someone* to determine and provide the expected *average* gas plant input composition. *No one* and *Someone* are noted for their problems with communication.

The plant engineers and the construction company were paid to build most of the same plant twice, a stroke of financial good luck. The latter party was smart enough to have documented the gas compositions provided. The fates of *No one* and *Someone* are not known, although apparently they have been responsible for such problems in a number of companies and on a number of plants. It seems *Someone* and *No one* are either not clearly identified or change identities, and so history repeats itself. It is very bad to be associated with these two characters on your *curriculum vitae*.

It is a good idea to determine the purpose of requests for analyses. If a plant is being designed, then an average composition might include many different reservoirs with different compositions. On the other hand, the request for an analysis may be for production accounting purposes on an individual well. This is required on a periodic basis and is usually used for calculating gas rates. In this case, the individual well gas gravity may be the desired information. Individual well compositions are also used to prorate NGL production from a plant to individual wells and, hence, to reservoirs.

Compositional variations

In most reservoir engineering calculations, it is assumed the reservoir fluid is constant in composition. For practical purposes, this is true in the vast majority of cases. However, a number of situations arise:

- The engineer wants to know if sample variations are measurement errors.
- The engineer wants to know why and how to account for significant sample compositional variation problems.
- The engineer may wonder if the changes are sufficiently large that there must be different reservoirs.

The fundamental purpose in this section is to rationalize the observed variations in composition.

In some reservoirs, it is necessary to account for gravity and the interface curvature between phases. It is also customary to assume the system is in a static equilibrium. However, due to the cooling of the earth, there is a net flux of heat through reservoirs. In the following, a quick summary is made of the thermodynamics. This material has been shortened from *Thermodynamics of Hydrocarbon Reservoirs* by Abbas Firoozabadi.[10] More detailed and accurate descriptions may be found in this reference. The following will be discussed:

- Gravity equilibrium
- The effect of a curved interface
- Irreversible thermodynamics and steady state
- The effect of temperature gradient, without convection
- The effect of temperature gradient, with convection

Influence of gravity on equilibria

The basis of the calculation starts with the amount of work done by a fluid when it expands or contracts. This is described in Equation 6.7.

$$dW = -PdV + mgdz \tag{6.7}$$

The expression of dU for a closed system is shown by Equation 6.8.

$$dU = TdS - Pdv + mgdz \tag{6.8}$$

It is also possible to derive expressions for dH, dA and dG as shown in Equations 6.9 through 6.11.

$$dH = TdS + VdP + mgdz \tag{6.9}$$

$$dA = -SdT - PdV + mgdz \tag{6.10}$$

$$dG = -SdT + VdP + mgdz \tag{6.11}$$

It is assumed for equilibrium that the Gibbs criterion of equilibrium must vanish at equilibrium. Hence, Equations 6.12 and 6.13 apply.

$$dT = 0 \tag{6.12}$$

$$VdP + mgdz = 0 \tag{6.13}$$

Restated, the temperature must be the same everywhere in the system. Since $\rho = m/V$, then Equation 6.14 can be used.

$$dP = -\rho gdz \tag{6.14}$$

This can then be expanded to multiphase systems as follows, in Equation 6.15, for component 1 to c:

$$dW = \sum_{i=1}^{c} (\mu_i + M_i zg)dn_i \tag{6.15}$$

Similarly, dU may be calculated as shown in Equation 6.16.

$$dU = TdS - PdV + mgdz + \sum_{i=1}^{c} (\mu_i + M_i zg)dn_i \tag{6.16}$$

where μ_i is the molar free energy of component i, in the fluid.

Then dG can be expressed as shown in Equation 6.17.

$$dG = -SdT + VdP + mgdz + \sum_{i=1}^{c} (\mu_i + M_i zg)dn_i \tag{6.17}$$

The conditions for equilibrium, in the same vein as the earlier development for a single component system, are outlined in Equations 6.18 through 6.21.

$$dT = 0 \tag{6.18}$$

$$\mu_i + M_i gdz = 0 \tag{6.19}$$

$$VdP + mgdz = 0 \tag{6.20}$$

$$(d\mu_i = -M_i gdz)_T \tag{6.21}$$

Equation 6.22 is known as the Gibbs sedimentation rate.

$$(RT \ln fi = -M_i gdz)_T \qquad (6.22)$$

From $(d\mu_i = RT\, d \ln f_i)_T$ and the immediately preceding equation, one may derive Equation 6.23.

$$f_i = f_i^{\,o} \exp\left[-\frac{M_i}{RT} gz\right] \qquad (6.23)$$

Then, integrating from a reference depth of zero to z, Equation 6.24 is derived.

$$F_i = f_i(T, P, y_i, \ldots, y_{c-1}) - f_i^{\,o}(T, P^o, y_i^o, \ldots, y_{c-1}^o) \times \exp\left[-\frac{M_i}{RT} gz\right] \qquad (6.24)$$

where

$$i = 1, \ldots, c$$

$$\sum_{i=1}^{c} y_i = 1$$

The Newton-Raphson technique can then be used to calculate both the composition and pressure as a function of depth. The sum of all of the components must equal 1.0.

Conditions for pronounced compositional variation

Taking the previous analysis and using some matrix techniques provides Equation 6.25.

$$\left[d\mu_i = \left(\frac{\partial \mu_i}{\partial P}\right)_y dP + \sum_{j=1}^{c-1} \left(\frac{\partial \mu_i}{\partial y_j}\right)_{P, y_j} dy_j \right]_T \qquad (6.25)$$

Using $(d\mu_i = V_i dp)_{T,y}$, it is possible to derive Equation 6.26.

$$\sum_{j=1}^{c-1} \left(\frac{\partial \mu_j}{\partial y_j}\right)_{P,T,y_j} \left(\frac{dy_j}{dz}\right) = (\rho \overline{V}_i - M_{i)})g \qquad (6.26)$$

This may be translated into matrix form as shown in Equation 6.27.

$$\begin{bmatrix} \frac{\partial \mu_1}{\partial y_1} & \frac{\partial \mu_{i1}}{\partial y_2} & \cdots & \frac{\partial \mu_1}{\partial y_{c-1}} \\ \frac{\partial \mu_2}{\partial y_1} & \cdots & \cdots & \cdots \\ \cdots & \cdots & \cdots & \cdots \\ \frac{\partial \mu_{c-1}}{\partial y_{c-1}} & \cdots & \cdots & \frac{\partial \mu_{c-1}}{\partial y_{c-1}} \end{bmatrix} \begin{bmatrix} \frac{dy_1}{dz} \\ \frac{dy_2}{dz} \\ \cdots \\ \frac{dy_{c-1}}{dz} \end{bmatrix} = g \begin{bmatrix} \rho \overline{V}_1 - M_1 \\ \rho \overline{V}_2 - M_2 \\ \cdots \\ \rho \overline{V}_{c-1} - M_{c-1} \end{bmatrix} \qquad (6.27)$$

The variation in composition will be large when the term $(\rho V_i - M_i)$ is large or when the determinant of the matrix is small. This occurs for two conditions:

- the presence of ashphaltene micelles in oil
- near the critical point

Figures 6–27 and 6–28 show the effects of compositional gradients in an oil reservoir and a near-critical gas condensate reservoir.

Systems with strong compositional gradients

It is interesting to look at systems having compositional gradients. Some examples of this are the East Painter reservoir as shown in Figures 6–29 and 6–30.[11] The mole percent methane varies from 55–70%, which is a substantial range of 15%. The predictions using an analysis similar to the preceding closely matched observed compositional variations.

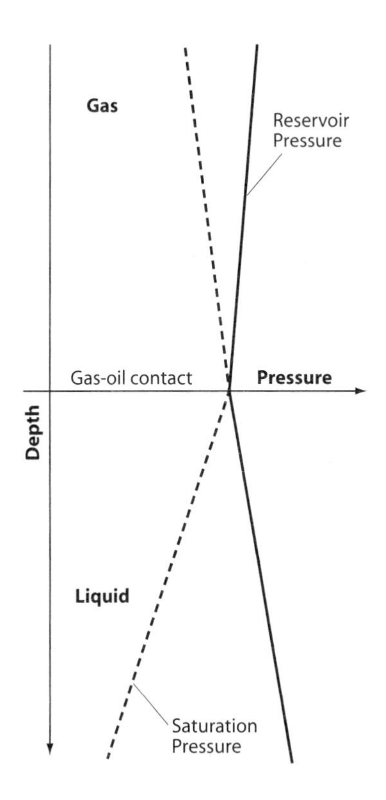

Fig. 6–27 Saturation Pressure vs. Depth—Gas-Oil System with Sharp GOC

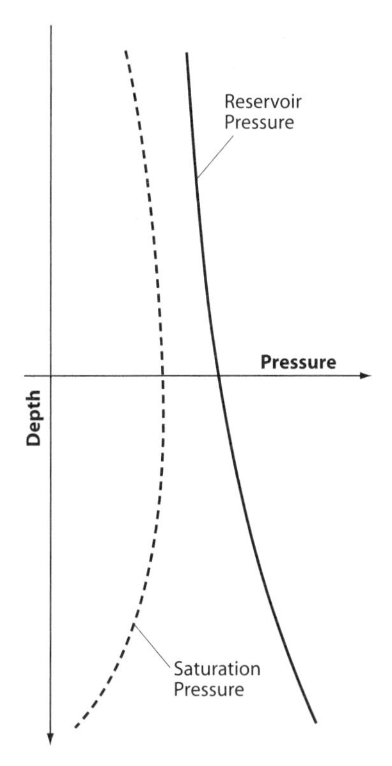

Fig. 6–28 Saturation Pressure vs. Depth—Near-Critical Fluid, No Contact

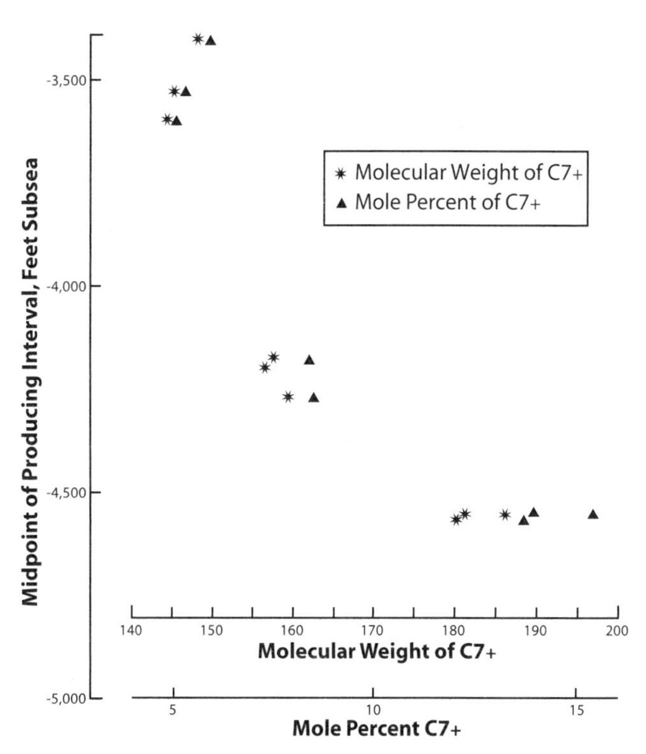

Fig. 6–29 Variation in Composition with Depth—Mole Percent and Molecular Weight of C7+ with Depth

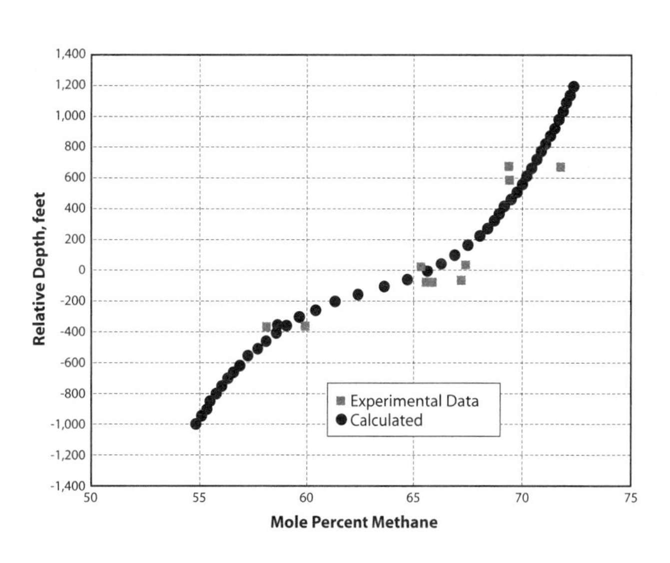

Fig. 6–30 Variation in Composition with Depth—Mole Percent C1 with Depth

Effects of curved interfaces

No theoretical development is presented in this case. However, the effects of a curved interface can be shown via thermodynamics. Figure 6–31 shows some experimental data, which was prepared by two Russians, Trebin and Zadora.[12] The dew point varies depending on pore size, which was modified using packs of silica sand of different grain sizes.

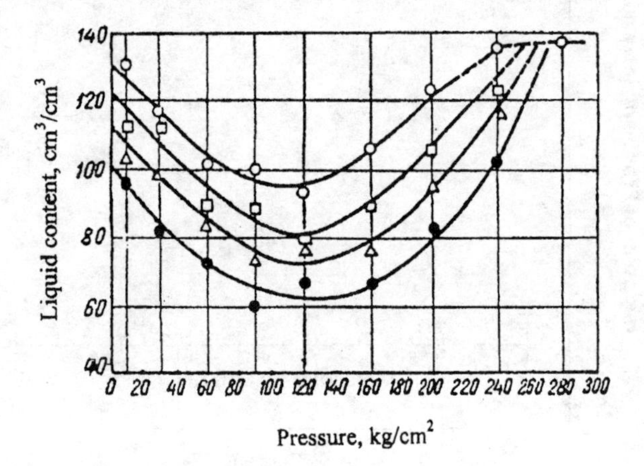

Fig. 6–31 Effect of Surface Area on Condensation

Restated, dew point pressures measured in the reservoir may be somewhat different from those measured in a laboratory cell. The actual PVT properties in the reservoir will be a function of the precise conditions, such as rock type and pore system.

Systems with thermal diffusion

It is interesting to look at a number of solutions generated by Firoozabadi for a number of systems that include a temperature gradient. In these calculations, Gibbs' free energy minimization cannot be used. Figure 6–32 shows a system of k = 0.2 mD.[13] In this system, the C1 concentration varies from 17.35% to 22.86%. This is a range of 5.5%. This should be realistic, as real reservoirs will have a temperature gradient, even if it is not obvious on temperature measurements, due to data scatter. Higher permeabilities are shown in Figure 6–33.

This analysis is somewhat simplified. Only two components are used. However, it goes a long way toward showing what range in compositions one might expect from a reservoir under normal circumstances.

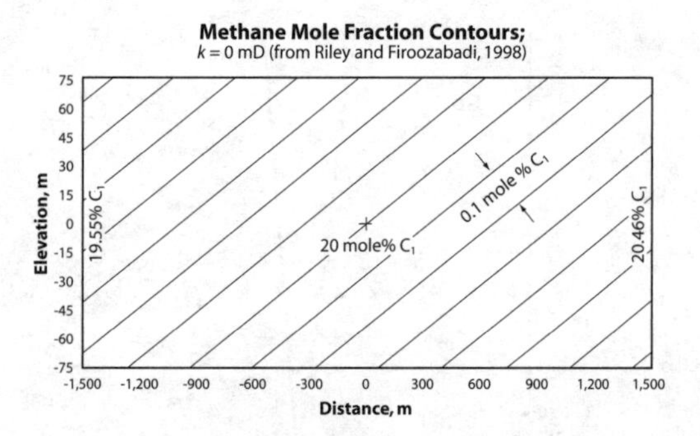

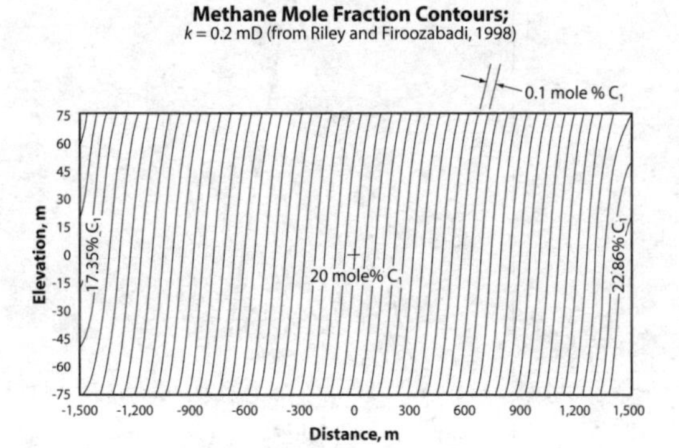

Fig. 6–32 Compositional Variation with Thermal Flux for Low Permeabilities

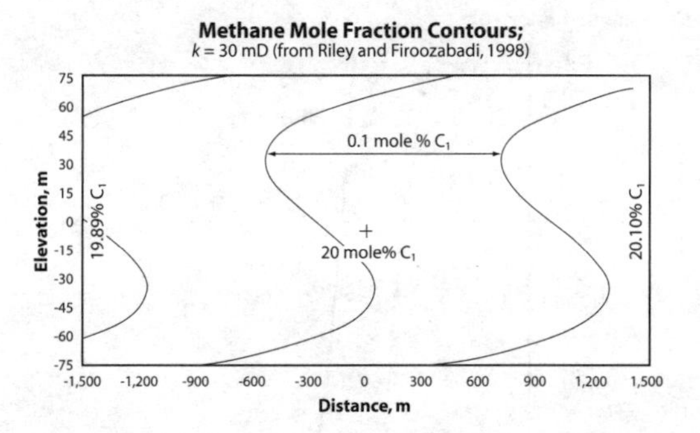

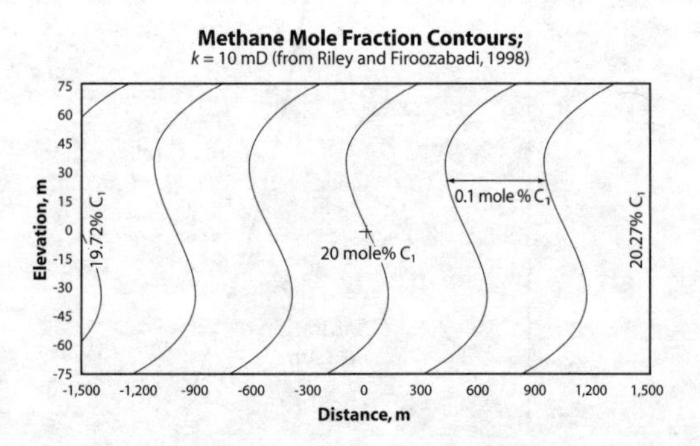

Fig. 6–33 Compositional Variation with Thermal Flux for Moderate Permeabilities

Bubblepoint variations with depth

In a number of reservoirs, the saturation pressure varies with either pressure, or, if you wish, depth. The author has only worked on one such reservoir and the reservoir falls into one of the classes where compositional variations are expected to exist, i.e., heavy oils or near-critical systems. The following is an example of the former.

Analysis of extensive wellhead samples on this reservoir indicated there was significant variation in the oil produced from this field ranging in API gravity from 9° to 12°. It was possible, by analyzing well completion data and well pressure histories, to correlate bubblepoint pressure with depth. The laboratory saturation pressures were also compared on this basis and a correlation made. The correlation, shown in Figure 6–34, is remarkably consistent given the spatial variations in oil properties. The outstanding feature of the data on this graph is the good match with individual well pressure production histories that it yielded. Such an example is shown in Figure 6–35.

One immediate implication of these variations in bubblepoint pressures is that individual wells may be affected in the history matching and prediction processes. In truth, it is likely, without extensive areal correlation, only the general trends will be included in most reservoir studies.

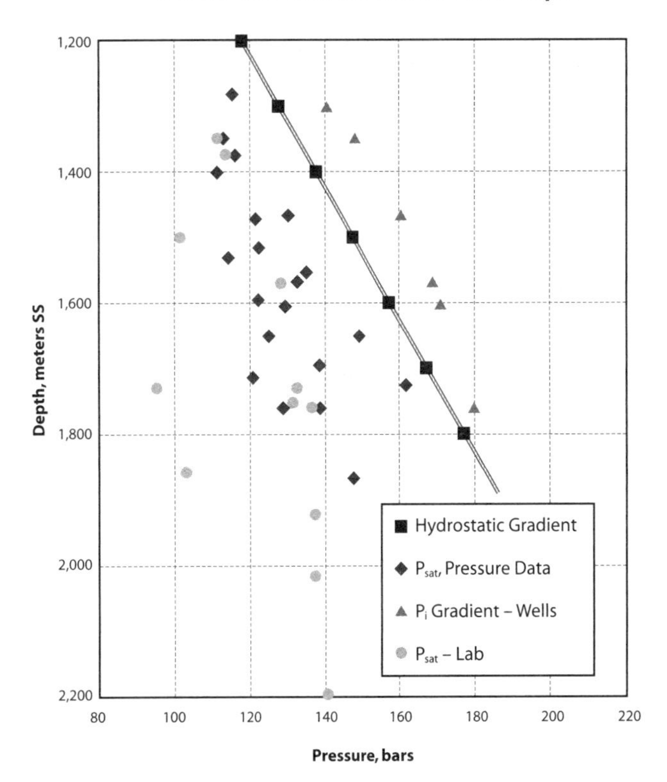

Fig. 6–34 Field Example of Saturation Pressure with Depth

Screening table

Whenever multiple sets of PVT data exist, the author prepares a table of all PVT data. Such a table is shown in Table 6–6. This general type of process will be repeated for other data types such as capillary pressure and relative permeability.

Such a table makes noticing unusual sampling and other anomalous data easier.

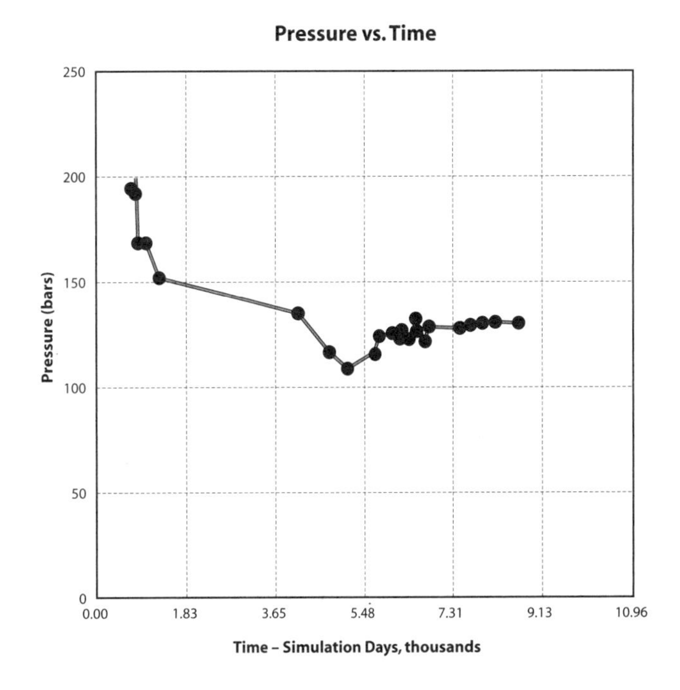

Fig. 6–35 Discontinuity of Pressure Response with Time due to Saturation Pressure

Table 6–6 PVT data

Date	Wells	Lab	Sample Type	P_b (psi)	B_{oi} (frac)	R_{si} (scf/bbl)	μ_o (cp)	T Res. (°F)	API Gravity
Apr. 10, 1968	14-Apr 14-Jul 14-Oct	Core Labs Houston	Recombined	798	1.058	146	7.2 4	82	
May 03,1968	15-Oct	Core Labs Dallas	Bottomhole	776	1.063	145	8.2 4.1	93	32.4
Oct. 15,1953	Jul-35	Chemical Casper	Bottomhole	195	1.034	60	9.3	86	32.5
Dec. 18, 1953	26-Jul	Chemical Calgary	Bottomhole	265	1.042	70	8.8	85	33.9
Dec. 03, 1981	14-Oct	Core Dallas	Recombined	933	1.074	152	6.9 2.8	89	
Apr. 28,1969	20-Dec	Core Dallas	Bottomhole	748	1.06	140	9.2 5.4	86	
Apr. 28, 1969	17-Dec	Core Labs Dallas	Bottomhole	793	1.073	155	8.3 4.6	86	33.5
Apr. 30, 1969	17-Apr	Core Labs Dallas	Bottomhole	850	1.072	165	7.6 4.4	86	33.6
Mar. 28, 1961	24-Jul	Core Labs Dallas	Recombined	750	1.061	132	7.2 4	90	33.8

Oil screening graphs

The following assumes the reader has more than one PVT sample available for a reservoir. This is quite often the case. The first step is therefore to graph all of the data on the same series of graphs:

- Differential B_o versus Pressure
- Differential R_s versus Pressure
- Oil Viscosity versus Pressure
- Evolved Z versus Pressure

In the author's experience, the slope of black oil PVT data (B_o and R_s) is remarkably consistent. This is shown in the PVT data from the Provost field of Alberta in Figure 6–36.

Another example of relative consistency is shown in Figure 6–37. This data comes from the Blueberry Debolt field in northeastern British Columbia.

There are, of course, exceptions where there are problems with the data. One of these is shown in Figure 6–38.

The two odd samples stand out clearly. One sample was taken from the wellhead, which is extremely unusual and which the author cannot explain. The second odd sample was a recombination, whereas all of the other data was from bottomhole samples.

Gas property screening graphs

A plot of gas formation volume factor B_g is often provided in simulation reports. Because it is quite close to an inverse pressure relation, the values at high pressures are hard to read, as shown in Figure 6–39. If you use a semi-log scale, Figure 6–40 shows it will be possible to read even the small values. Kinks are also revealed more clearly since there is not as much curvature.

Since oil PVT data is normally close to a straight line, it is possible to input this data with sparse spacing. This is not true of gas formation volume factor, which follows $1/p$. Data spacing should be governed by the character of the gas formation volume factor data.

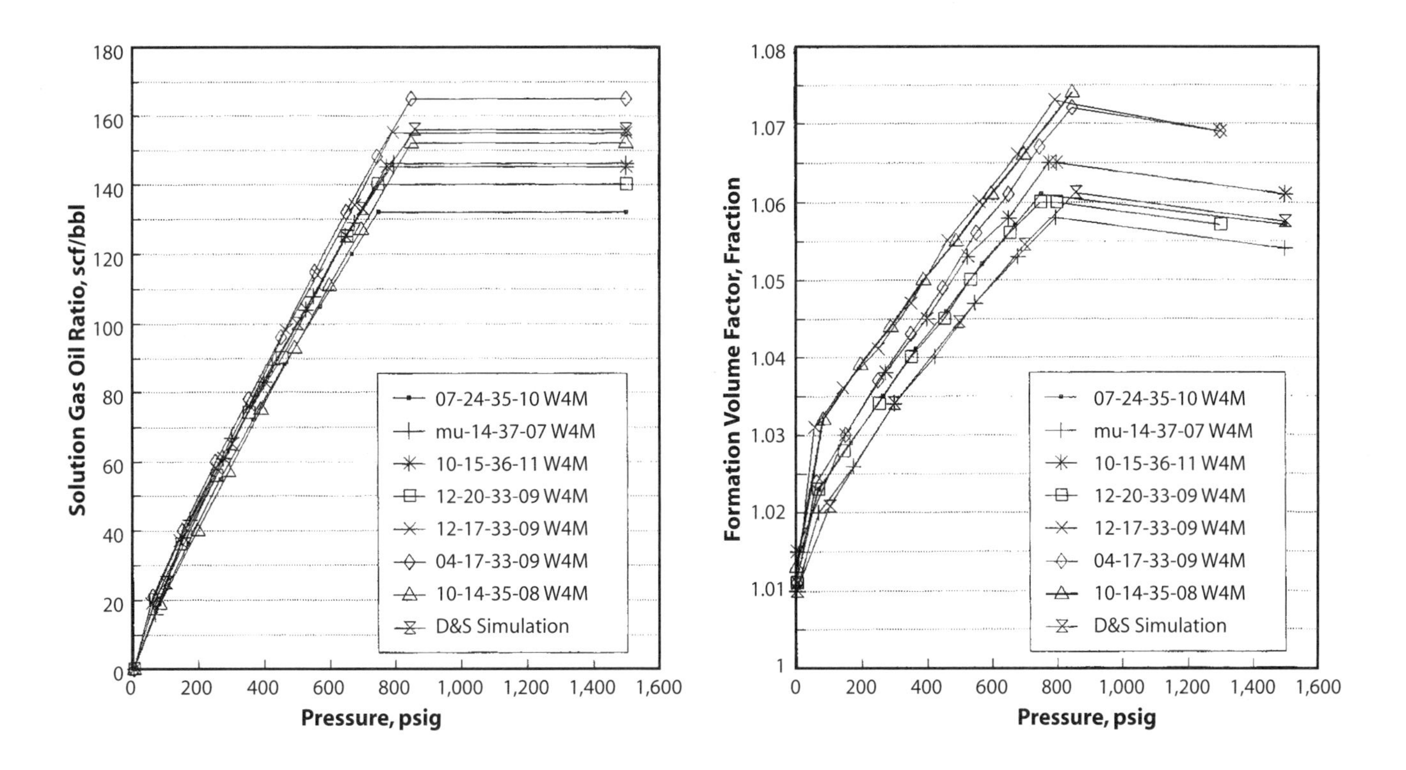

Fig. 6–36 Field Example of Variation in Oil Properties with Geographic Location

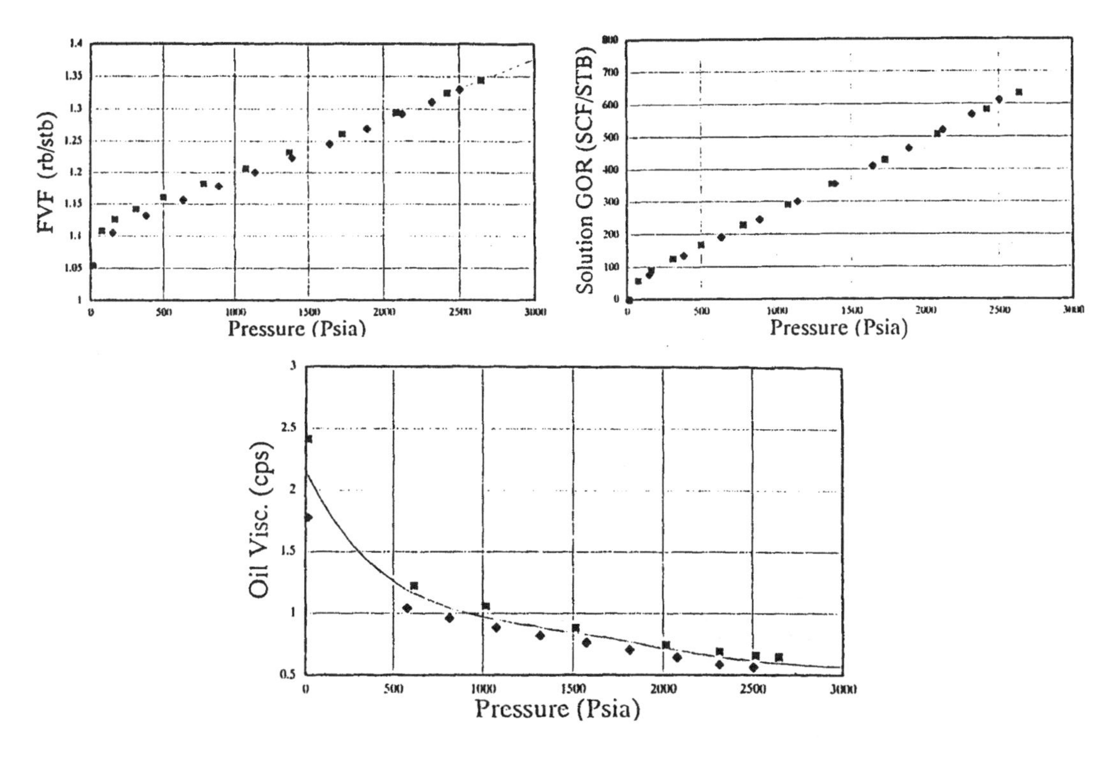

Fig. 6–37 Another Field Example of Variation in Oil Properties with Pressure

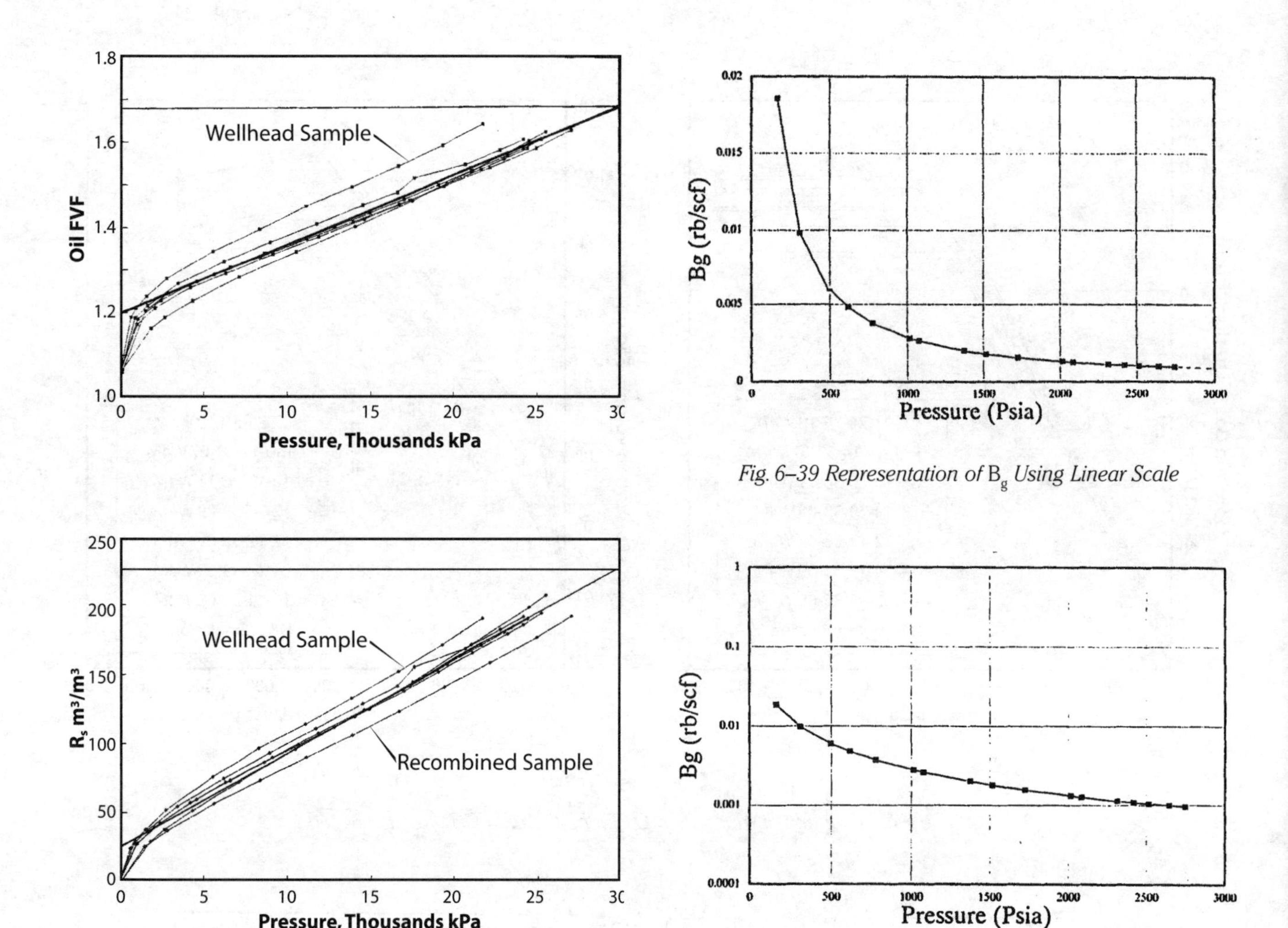

Fig. 6–39 Representation of B_g *Using Linear Scale*

Fig. 6–38 Carbonate Reef Field Example of Variation in Oil Properties—Problem Data

Fig. 6–40 Representation of B_g *Using Semilog Scale*

Gas condensate screening

Normally, the main graph used for gas condensates is a plot of the CVD results. If multiple samples are available, then it is a good idea to overlay the results as shown in Figure 6–41.[14]

The example shown is not an easy one to deal with. This sample involves extremely sour samples that are difficult and dangerous to work with. This is not common worldwide but is quite common in certain areas.

Offset comparison

In Canada, all PVT data is sent to the government, and it is easy to check against other data sets for an area. This is not true everywhere. However, if you work for a major company, they often have a large inventory of internal data that can be used. Cross-checking is a good way to establish trends. The collection of samples shown in Figure 6–42 shows a clustering around one set of data. This is a sour fluid. Note the anomalous results.

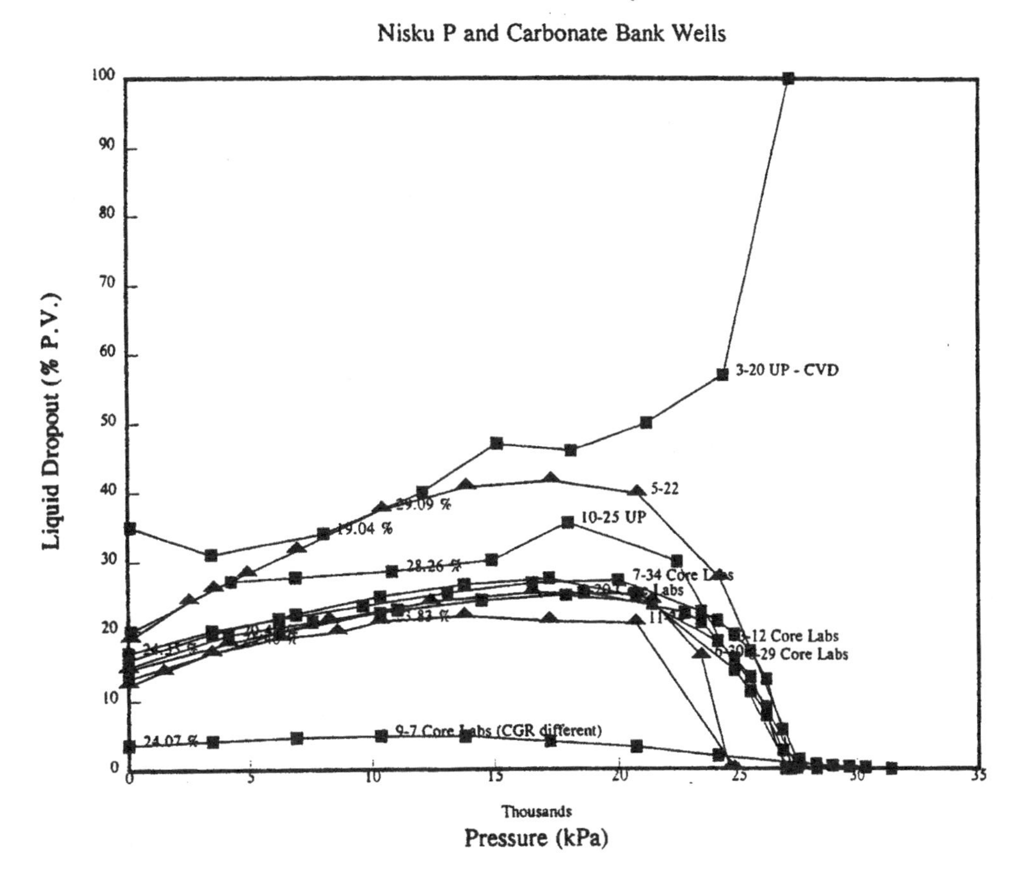

Fig. 6–41 Constant Volume Depletion Tests—Sour Retrograde Condensate Field

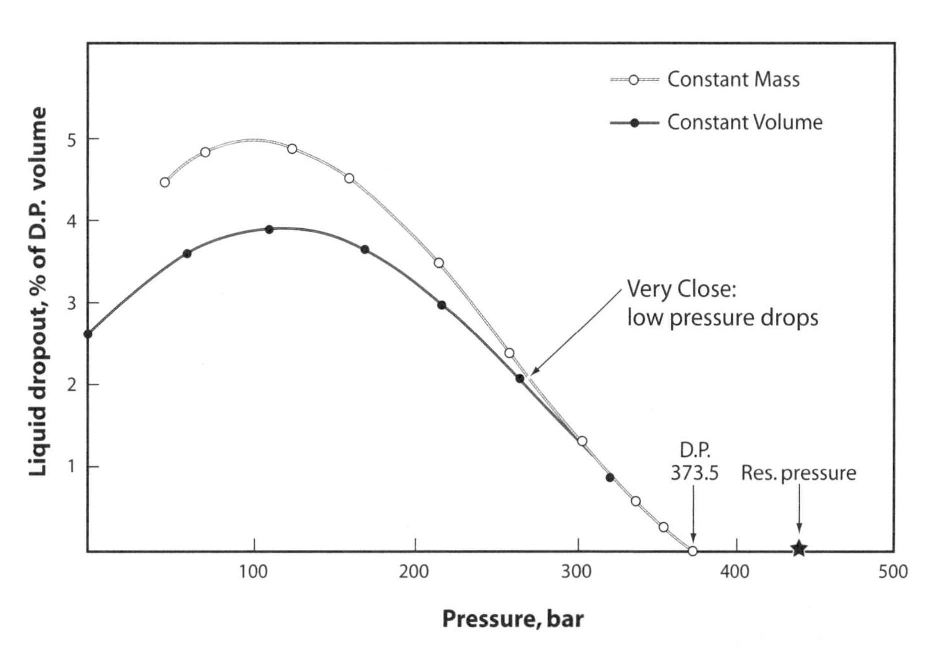

Fig. 6–42 Comparison of CCE vs. CVD Depletion Liquid Volumes—Lab Data

Data consistency

One way to screen gas condensate data is to overlay differential liberations and CVD tests with CCE liquid volumes. Pederson et al. shows these two curves generally overlay for the first 25% of pressure depletion on CVD tests. Their example does not represent a highly volatile fluid. Figure 6–43 shows their data, and some data for a near-critical fluid is shown in Figure 6–44. The EOS checking described earlier was used to demonstrate the general applicability of this rule to a high dropout liquid.

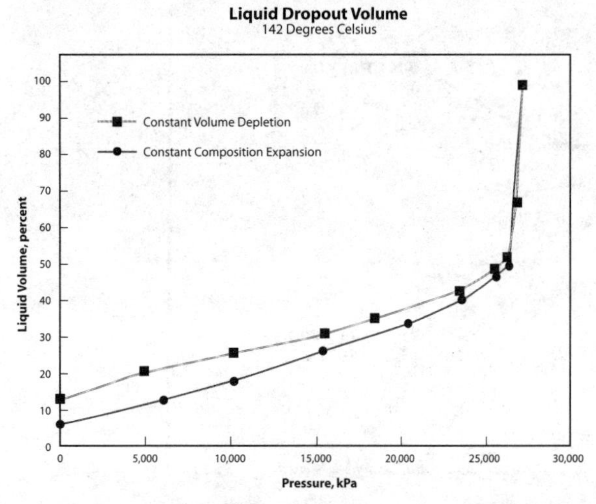

Fig. 6–43 Comparison of CVD vs. CCE for Oil—Modeled with EOS

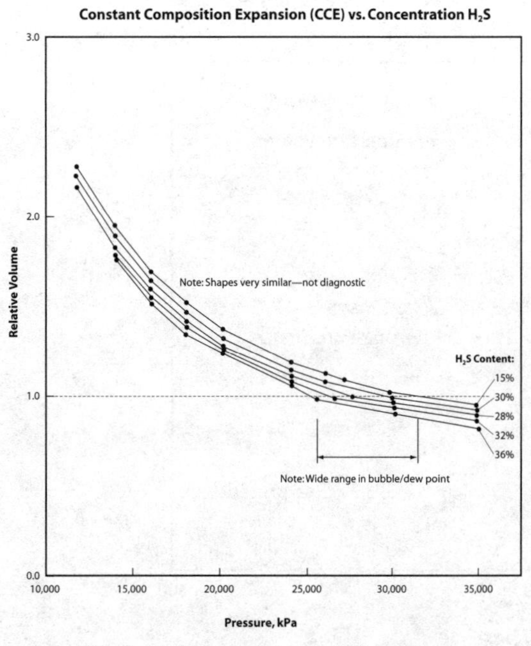

Fig. 6–44 Comparison of CCE Relation vs. Hydrogen Sulfide Concentration

Advantage of CCE-type experiments

A CCE can offer some significant advantages. It is much simpler to implement than a CVD. The gas in the cell is not drawn off in stages; hence, this test is much less susceptible to errors. The liquid volume observed offers good quality control on CVD work done later. It is also correspondingly less expensive. Not all CCE tests are conducted in windowed cells. Based on this project, the author found CCE data to be of little use unless liquid dropout volumes are reported, which requires a windowed cell.

ARE has found, via EOS sensitivities, CCE data, in particular the relative volume, proved to be of little use in distinguishing between high dropout dew point and volatile oils systems. Curves from both systems will virtually overlay because there is no abrupt transition from low liquid compressibilities to higher gas compressibilities. This is shown in Figure 6–45.[15] This is also relevant to the discussion of near-critical fluids later in this book.

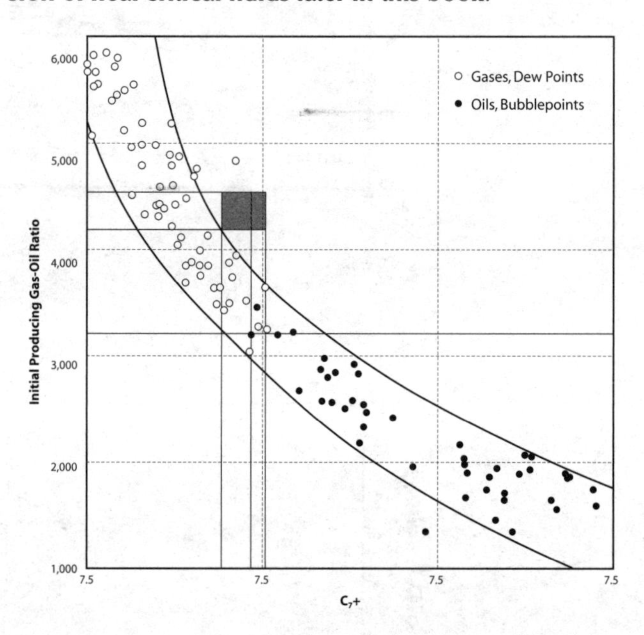

Fig. 6–45 Plot of Saturation Pressure and C7+ Content Sour Sample Also Shown

Differentiating volatile oils and gas condensate systems

A rule of thumb for differentiating volatile oils from retrograde condensate systems was recommended to the author on a number of occasions. The rule is as follows:

> *Laboratory-determined compositions of volatile oils will have 12.5% to 30% heptanes plus. The dividing line between volatile oils and retrograde condensate gases of 12.5 mole percent heptanes plus is fairly definite. When the heptanes plus concentration is greater than 12.5 mole percent, the reservoir fluid is almost always a liquid and exhibits a bubblepoint. When the heptanes plus concentration is less than 12.5 mole percent, the reservoir fluid is almost always a gas and exhibits a dew point. Any exceptions to this rule normally do not meet the rules of thumb with regard to stock tank oil gravity and color.*

The data upon which the rule of thumb appears to have been drawn is from McCain and Bridges.[15] A graph is shown in Figure 6–45. The author's interpretation of the data is slightly different. If the heptanes plus content is between 12 and 13.5 mole percent, it can go either way.

The author has plotted a number of sour systems that fall close to the dividing line on the previous figure. The data plotted includes the heptanes plus mole percent data from an MDT test on a dew point system and two laboratory volatile oil analyses. The data clearly falls off the trend. Thus, in the author's opinion, this conventional rule of thumb should not be used for sour systems. There is no indication McCain et al. intended their rule of thumb would ever be applied to very sour systems. If so, a generally usable rule of thumb could easily be misapplied in these, albeit limited, circumstances.

Which is correct?

The question then becomes, *Which sets or set of data is correct?* Knowing that PVT data for oils is normally consistent, the author would drop the odd remaining samples. It is also possible to use Standing's or similar correlation(s). In the case of production data, it is often possible to check GORs and condensate production rates against the laboratory data.

Oil PVT input

There are two common types of reservoir input styles. The difference concerns the manner in which data above the bubblepoint is handled. The older style is shown in Figure 6–46. Observe the B_o values decrease with increasing pressure above the bubblepoint. The more modern method, shown in Figure 6–47, extends the R_s and B_o curves well above the reservoir bubblepoint. Concurrently, an undersaturated oil compressibility is specified with a keyword and is not included in the PVT table. Both are still available.

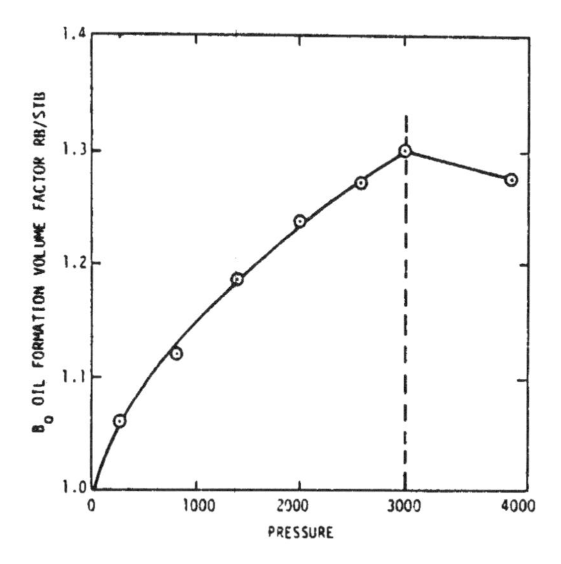

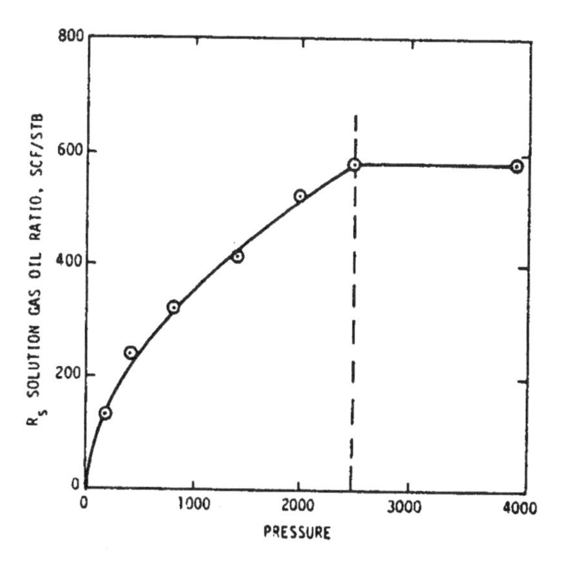

Fig. 6–46 First Method of Inputting PVT Data for Oil Formation Volume Factor (B_o)

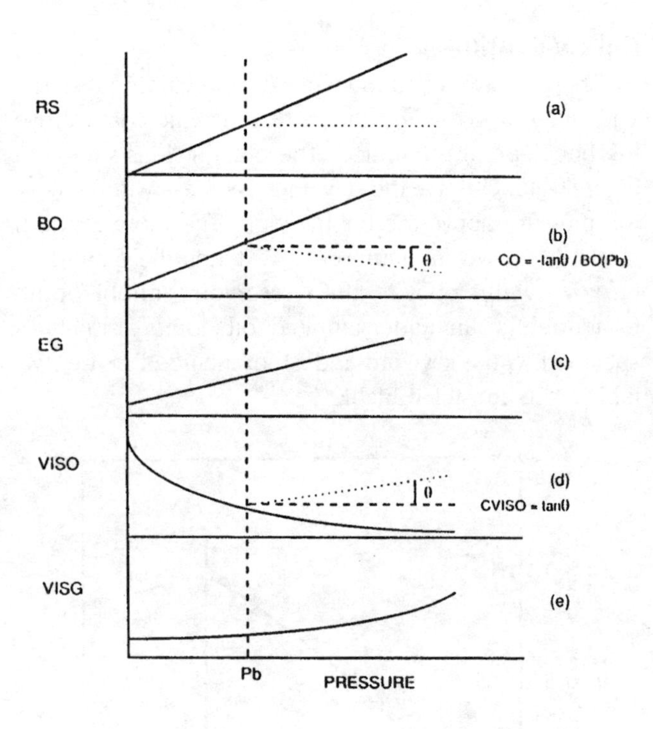

Fig. 6–47 Second Method of Inputting PVT Data for Oil Formation Volume Factor (B_o)

The author prefers the type that extends to high pressures. If gas injection is implemented, then more gas should be soluble in the oil. Some people feel this does not actually occur. The latter is not related to the PVT data *per se*. Rather, gas fingers and does not physically contact all of the oil that would absorb the gas. The latter is better dealt with by including reservoir heterogeneities and other improvements rather than in the PVT table. The idea is easy and practical, however. For more detail, see chapter 14.

Oil input data consistency requirement

It is a mathematical necessity in reservoir simulation that the total compressibility of an oil system remain positive. This condition is:

$$\frac{\partial B_o}{\partial p} - Bg\frac{\partial R_s}{\partial p} \succ 0 \qquad (6.28)$$

Most simulators will confirm this as part of the automatic data checking. That is, provided you have not turned it off. Since error checking is not always performed automatically, it might not be prudent to rely on the simulator to do this checking.

Variable bubblepoint pressure input

The input of variable depth bubblepoints is quite direct. A simple table of depth with saturation pressure or pressure with saturation pressure is made. The author prefers the former. The simulator interpolates automatically.

Gas PVT input

As outlined earlier, different people prefer to deal with different formats of gas PVT input. Simulators cater to these preferences by having different input options. Three different user options are common:

- specifying Z factors
- specifying B_g
- specifying E_g or b_g

Normally the statements used to indicate these options are only slightly different:

```
*PVTG *ZG
*PVTG *BG
*PVTG *EG
```

These different options are a good way to trick the unwary. With only one minor letter error in the input specification, one can waste hours rechecking calculations to find a "mistake" made in calculations that were actually perfectly correct. If Z is specified, the reservoir temperature must be specified also. The simulator will then calculate the gas formation volume factor from the Real Gas Law.

Extend for water injection

One trick that comes with experience is to extend the PVT table to high pressures if injection predictions will be analyzed. This is hardly a fatal error; however, it is much easier to extend the data to bottomhole injection pressure while the PVT data is available and being worked on.

If this is forgotten, the waterflood prediction will stop on an *out of PVT table range* error.

Kinks in reservoir data

If there are kinks in input PVT data, then, as is the case with relative permeability data, it is possible to induce convergence problems for the simulator. Therefore, it is necessary to plot your data and check for changes in slope. The data shown in Figure 6–48 will definitely cause difficulties.

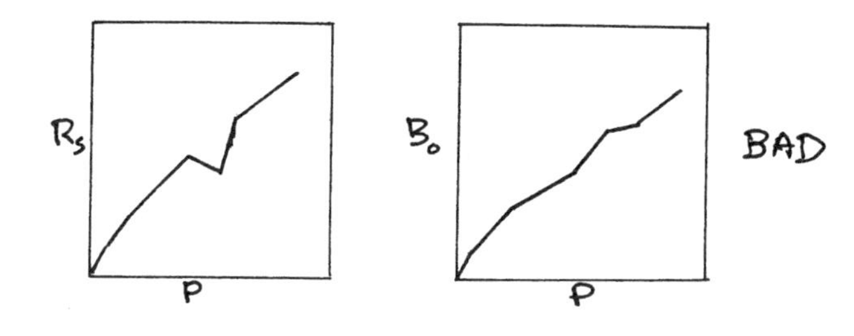

Fig. 6–48 Kinked Data Input (Bad)

Oil densities

Oil density is not normally input in a pressure table, although a surface density or API gravity will be required. The simulator will automatically calculate liquid densities. The relative volume is known, as well as the mass of the gas plus stock tank oil.

Oil viscosity

Most oil PVT analyses also measure oil viscosity. An example of oil viscosity data from a set of lab data is shown in Figure 6–49. Oil viscosity drops with increasing solution gas content. Viscosity will usually rise again slowly above the bubblepoint.

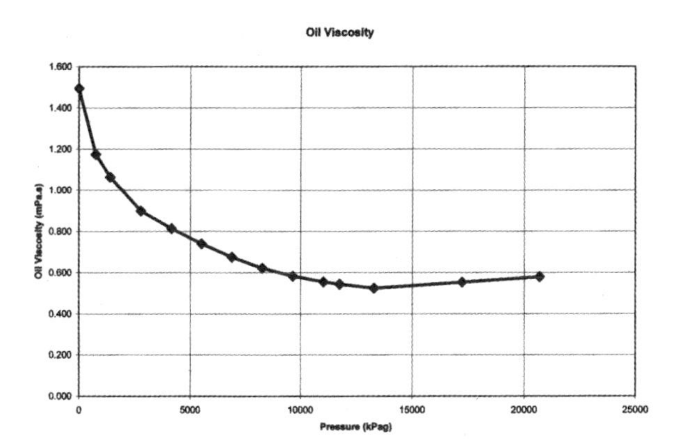

Fig. 6–49 Oil Viscosity—Lab Data

Failing lab data, there are also correlations available to estimate viscosity. This is a two-stage process. First, the surface viscosity is correlated; then adjustment is made for the solution gas contained in the oil. Correlations are shown in Figures 6–50 and 6–51.[16,17]

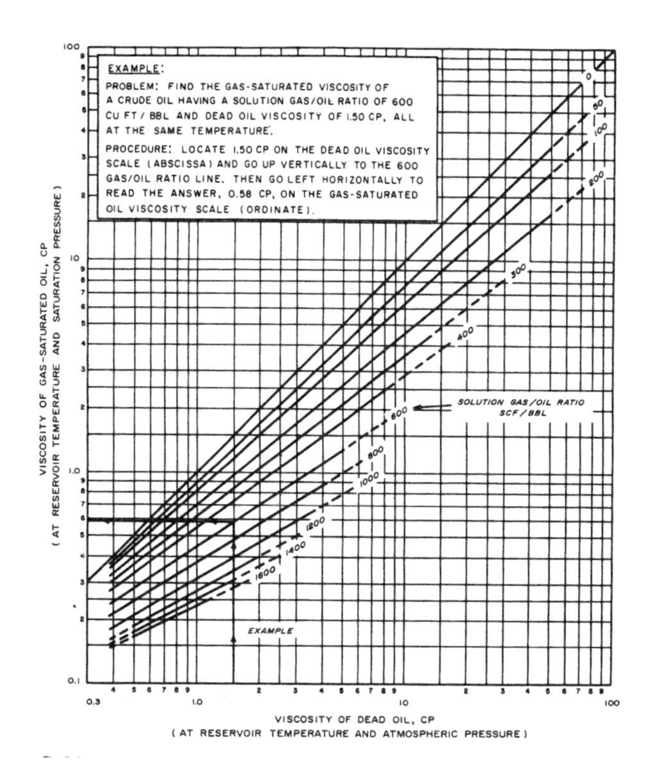

Fig. 6–50 Oil Viscosity Correlation—Live Oil

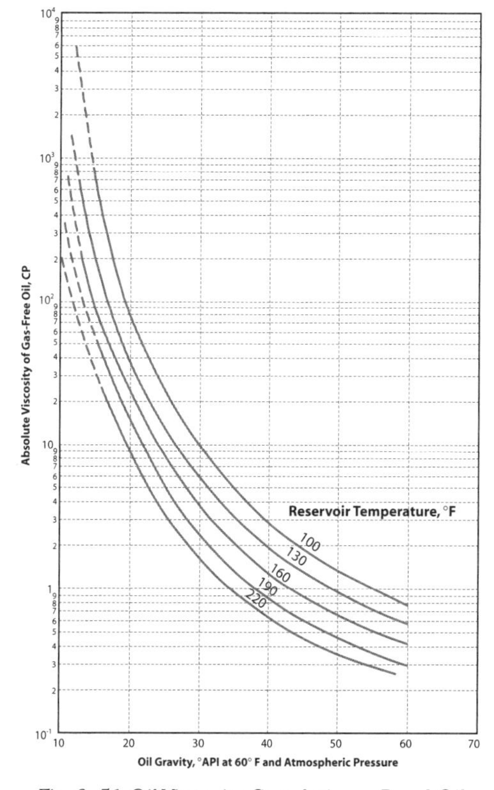

Fig. 6–51 Oil Viscosity Correlation—Dead Oil

Gas viscosity

Gas viscosity is almost never measured. Even if the data is presented in a lab report, it will normally have been calculated from a correlation. The most commonly used correlation is contained in the Gas Processors Suppliers Association (GPSA) data book. A copy is shown for reference in Figure 6–52.[18]

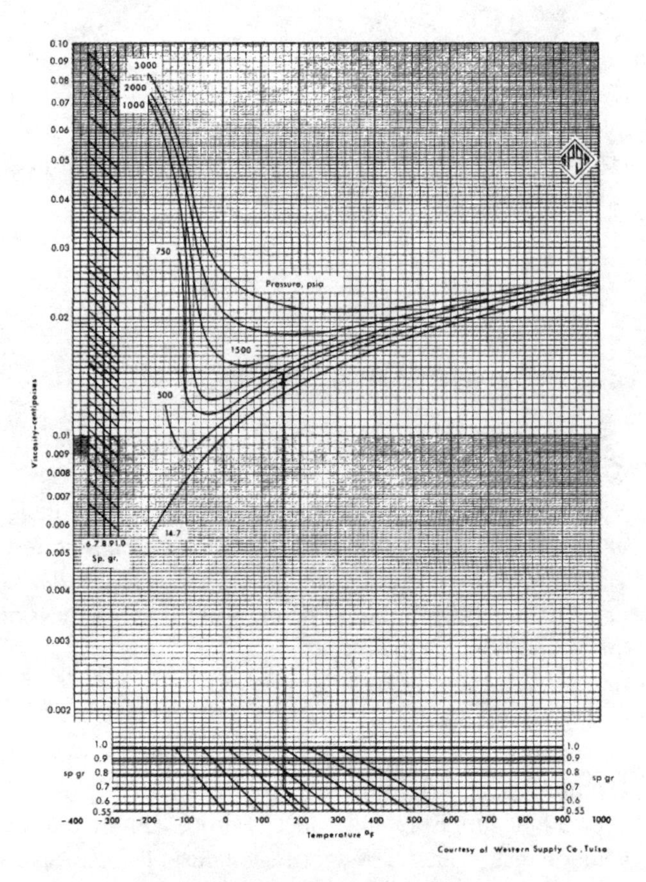

Fig. 6–52 Gas Viscosity Correlation

Water formation volume factor

Most simulators require water properties be input. This is normally done with a standard water correlation, which is shown in Figure 6–53. The latter is found in a number of reservoir engineering textbooks and the SPE monograph, *Advances in Well Test Analysis*. The author normally puts these figures in reports and shows the data points. Frequently, the statement, "Water properties were determined using standard correlations" is found in simulation reports.[19]

Water formation volume factors are adjusted for dissolved solution gas and temperature effects, as shown in Figure 6–54.[20] Usually, water formation volume factors are slightly more than one and represent the thermal expansion of water at high temperatures plus some dissolved gas.

If H_2S is in the water, there can be a lot more gas in the water. At times, the H_2S contents of solution gas increases as the wells mature and produce more water. There is limited hydrocarbon solubility in water.

Most simulators only allow you to specify one B_w. If a waterflood is installed, there will be a discrepancy between the B_w for the formation water and injected water. There do not appear to be any tricks to get around this. If minor water production exists, use the gas free B_w. Most waterfloods will cycle water to some extent.

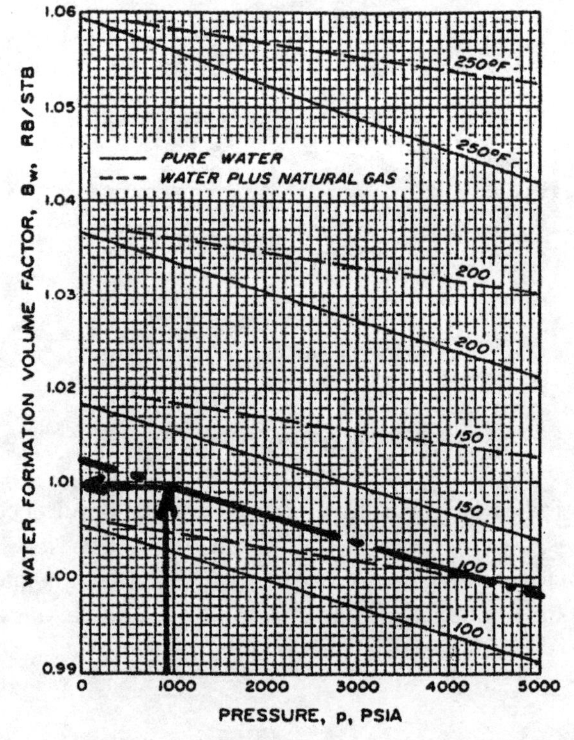

Fig. 6–53 Formation Volume Factor of Water and a Mixture of Natural Gas in Water

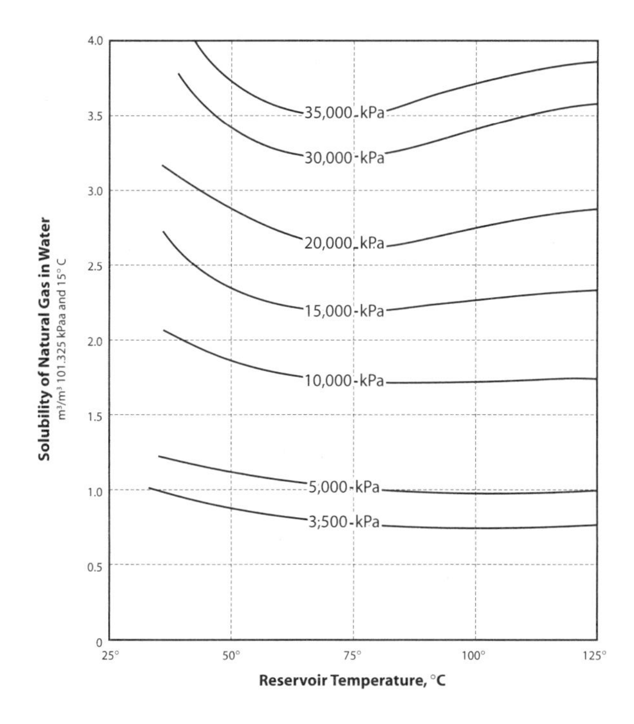

Correction Factor for Salinity

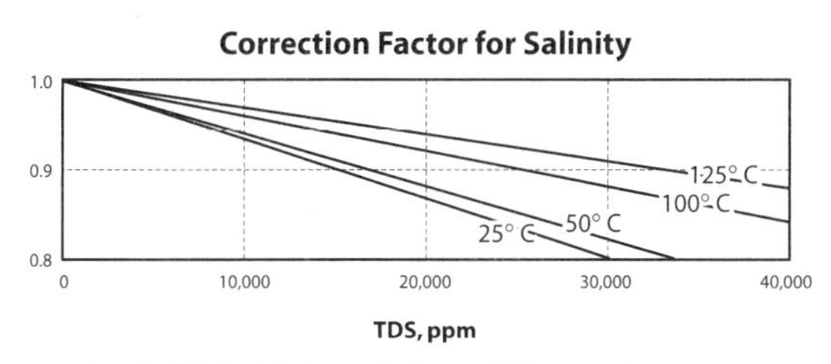

Fig. 6–54 Solubility of Natural Gas in Water

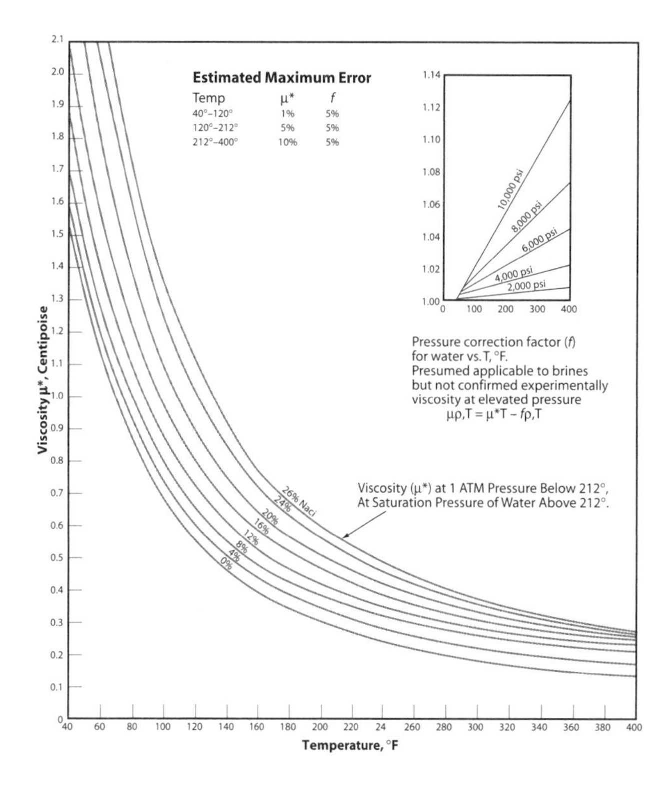

Fig. 6–55 Water Viscosity vs. Temperature and Salinity

Water viscosity

Water viscosity is controlled by temperature and salinity. The standard correlation most commonly used is shown in Figure 6–55.[21]

The same comments on water formation volume factor apply to reservoirs with both produced and injected water. The fresh water viscosity more correctly represents fractional flow.

Compositional simulators will allow variations in water properties such as calculations from Bingham's correlation (Perry et al., 1977), which is shown in Equation 6.29.

$$\mu = \frac{100}{2.4182\left[T + (8078.4 + T)^{0.5}\right] - 120} \qquad (6.29)$$

The author has found this gives slightly different values than the most common graphical correlation. The difference is not large, but it could affect displacement efficiency (fractional flow curve) and will change water injectivity.

Water compressibility

Water compressibility is usually calculated using the data of Long and Chierici with the chart shown in Figure 6–56.[22]

It is quite common to ignore the complex calculations outlined previously and simply use 3.0E-6 psi^{-1} or 0.437E-6 kPa^{-1}. Water compressibility is normally not important except in undersaturated oil reservoirs.

Average Compressibility
of distilled water
P=369.8 PSI
1422.3
2844.7
4267.0
5689.4
6756.1
3.60
3.40
3.20
3.00
2.80
2.60
$C_w \times 10^6$, PSI^{-1}
40 60 80 100 120 140 160 180 200 220
Temperature, °F

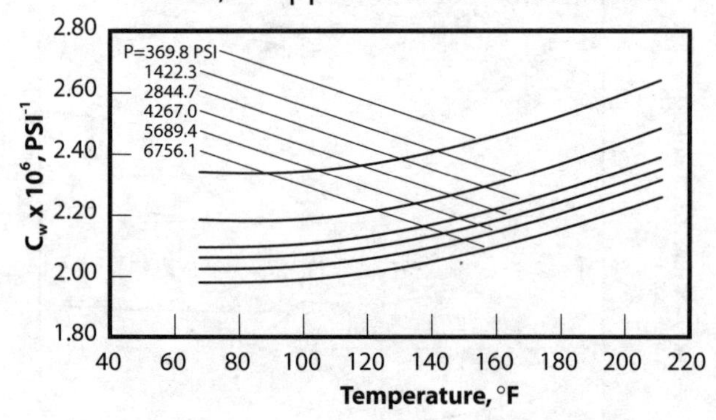

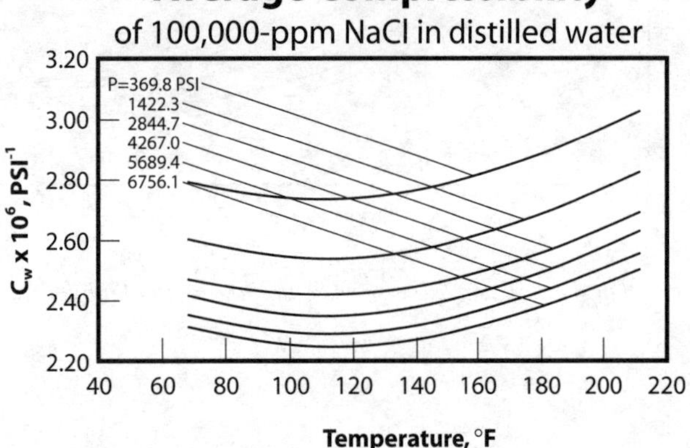

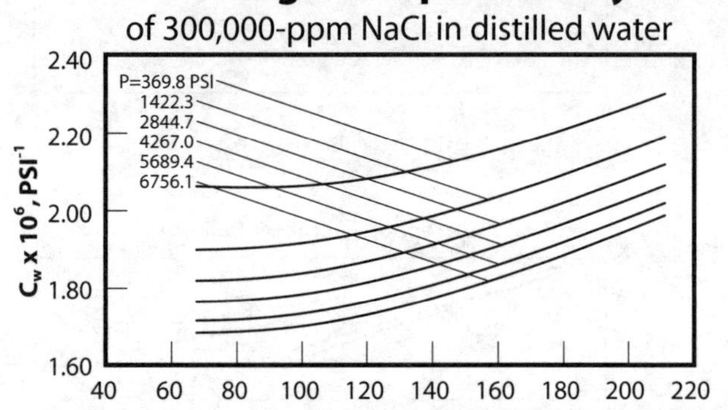

After Long and Chierici.

Fig. 6–56 Compressibility of Water

Summary

Adjusting PVT data can be somewhat frustrating. The existing guidelines are almost impossible to implement. The author has suggested an alternate methodology, which can always be implemented. Some adjustments, such as the linear extrapolations to the bubblepoint, are actually easy. Most PVT data is remarkably consistent. Screening includes checks against correlations and ensuring PVT data is consistent with both production and reservoir pressure performance. Gas formation volume factors govern the number of PVT table entries. Water properties are usually input based on correlations.

A number of factors cause variation in reservoir fluid sample compositions. Issues include measurement of compositions, separator conditions, variations in well rates, and thermal fluxes. It is possible to quantify these, although some sophisticated investigation may be required.

A number of standard correlations are used for water properties. These have been outlined briefly.

References

1 Moses, P.L., "Engineering Applications of Phase Behavior of Crude Oil and Condensate Systems," SPE 15835, *Journal of Petroleum Technology*, p. 715, 1996.

2 Dodson, C.R., D. Goodwill, and E.H. Mayer, "Application of Laboratory PVT Data to Reservoir Engineering Problems," *AIME Petroleum Transactions*, Vol. 198, 1953.

3 Standing, M.B., Volumetric and Phase Behavior of Oil Field Hydrocarbons Systems, 7th ed., Society of Petroleum Engineers, 1977.

4 Lasater, J.A., "Bubble-Point Pressure Correlation," *Tranactions of AIME*, Vol. 213, 1958.

5 Glaso, O., "Generalized Pressure-Volume-Temperature Correlations," *Journal of Petroleum Technology*, 1980.

6 Abdul-Majeed, G.H., and M.A.-S. Salman, "Evaluation of PVT Correlations," SPE 14478, Society of Petroleum Engineers, 1985.

7 Cronquist, C., "Dimensionless PVT Behavior of Gulf Coast Reservoir Oils," SPE 4100, *Journal of Petroleum Technology*, 1973.

8 Standing M.B., and D.L. Katz, "Density of Natural Gases," *Transactions of AIME*, No. 146, pp. 140–149, 1942.

9 Danesh, A., *PVT and Phase Behaviour of Petroleum Reservoir Fluids, Developments in Petroleum Science*, Elsevier, 1998.

10 Firoozabadi, A., *Thermodynamics of Hydrocarbon Reservoirs*, McGraw-Hill, 1999.

11 Creek, J.L., and M.L. Schrader, "East Painter Reservoir: An Example of a Compositional Gradient From a Gravitational Field," SPE 14411, 60th ATCE, Las Vegas, Nevada, Sept. 22–25, 1985.

12 Trebin, F.A., and G.I. Zadora, "Experimental Study of the Effect of Porous Media on Phase Changes in Gas Condensate Systems," *Neft'i Gas*, Vol. 81, p. 37, 1968.

13 Riley, M., and A. Firoozabadi, "Compositional Variation in Hydrocarbon Reservoirs with Natural Convection and Diffusion," *American Institute of Chemical Engineers Journal*, p. 452, Feb. 1998.

14 Pedersen, K.S., A. Fredenslund, and P. Thomassen, *Properties of Oils and Natural Gases*, Gulf Publishing Company, 1989.

15 McCain, W.D. Jr., and B. Bridges, "Volatile Oils and Retrograde Gases—What's the Difference?" *Petroleum Engineer International*, Jan. 1994.

16 Chew, J.-N. and C.A. Connally, Jr., "A Viscosity Correlation for Gas-Saturated Crude Oils," *Trans. of AIME*, No. 216, pp. 23–25, 1959.

17 Beal, C., "The Viscosity of Air, Water, Natural Gas, Crude Oil and Its Associated Gases at Oil-Field Temperatures and Pressures," *Trans. of AIME*, No. 165, pp. 94–115, 1946.

18 Gas Processors Suppliers Association, *Engineering Data Book*, Tulsa, OK: PennWell Publishing Co., 1998.

19 Earlougher, E.C. Jr., "Advances in Well Test Analysis," SPE Monograph No. 5, Henry L. Doherty Series, 1977.

20 After Dodson, C.R. and M.B. Standing, "Pressure-Volume-Temperature and Solubility Relations for Natural Gas Mixtures," Drilling and Production Practice, API, 1944.

21 Matthews, C.S, and D.G. Russell, "Pressure Buildup and Flow Tests in Wells," SPE Monograph No. 1, Henry L. Doherty Series, 1967.

22 Long, G., and G. Chierici, "Salt Content Changes Compressibility of Reservoir Brines," *Petroleum Engineer*, B-25 to B-31, July 1961.

7 Capillary Pressure and Relative Permeability: Data Screening

Introduction

After geology, one of the biggest problem areas in simulation is acquiring and inputting correct reservoir data. After entering data in the first few simulation models, the author wondered, *How do I know the data is correct?* When the question was reviewed with the bosses and other engineers, most answers were, "It looks OK to me," or "Is this an oil-wet or water-wet rock?" With regard to the latter, it seemed none of the reservoirs had wettability tests performed. In other words, the question was answered with the classic Zen-like circular question. To add to the frustration, many reservoirs are defined as oil wet or water wet solely on the saturation where the relative permeability curves cross.

The answer to the question is not simple. Many times the author has relied on experience, which could not be uniquely identified. Ergo, the *Looks OK to me* answer often prevails. In this sense, perhaps the answer was fair, but it also leaves one with the feeling a certain *something* is missing.

What is missing is a considerable amount of subtle details. Although most reservoir engineers understand the general concepts behind capillary pressure, relative permeability, and PVT data, there is very little call on a daily basis to understand the nuts and bolts of how the tests are run. In fact, it is rare for the typical reservoir engineer, who generally works in exploitation, to do detailed reservoir calculations. Here then is a quick rundown of this data usage, ignoring simulation for the time being:

- Capillary Pressure. Usually used by log analysts to confirm connate water saturations.
- Relative Permeability. Usually used in waterflood calculations, which have been replaced by simulation for the most part.
- PVT Data. Usually used in a material balance study—most reservoir engineers might do one to five studies in a career, i.e., perhaps no more often than once every five years.

If the boss has been in management for a considerable period, he or she may have even less direct experience. Restated, simulation is actually the technique making the most use of this data.

The literature does cover a great deal of the background material for these tests. However, this is spread throughout a significant number of papers, most of which are quite old. Although most labs have proprietary manuals on how they perform these tests, they do not hand out these procedures, since this is often their competitive advantage. Restated, this material does not make it into the public domain.

The apparently circuitous answers one often gets about data quality and interpretation reflect genuine unfamiliarity with lab testing. Very few summary articles exist in the petroleum literature dealing with this subject.

Commercial lab results

Some have said they think the major commercial lab results are manipulated, since they are noticeably more consistent than those obtained from other sources. In the author's experience, the level of training and experience is actually much better in commercial labs. More importantly, the procedures and data screening they utilize are, at least in the cases evaluated, much more sophisticated. The commercial labs also tend to utilize more chemists than engineers in their operations. Analytical chemists have a very different approach and a literature of their own. Therefore, what really happens in the commercial labs is strongly influenced by factors not usually discussed in the petroleum engineering literature.

The quality of testing does vary considerably from one lab to the next, to a much higher degree than one expects. Buyer beware.

Some companies send out samples to a number of different commercial labs as a quality control test. The companies do their own internal tests prior to sending them out. Overall, this is a good idea. It is recommended the labs be informed about what the company is doing. In one case, a lab stood by their test, which provided different results from the base testing. Regardless of how the matter was resolved, it is fair to say the base testing has to be done very carefully. It is also surprisingly difficult to get uniform samples. This requires careful planning and sometimes the use of artificial materials.

Capillary pressure

Capillary pressure is a key phenomenon in understanding reservoir behavior. To completely discuss this topic would require several detailed chapters in a book. However, this book is predicated on a basic understanding of reservoir engineering, so this level of detail will not be discussed. Some review comments will be made to get into the discussion of simulator input.

The largest effect of capillary pressure is during initialization when the original saturations are determined by calculations that assume gravity equilibrium. As outlined earlier, capillary pressure also affects frontal displacement.

Surface chemistry

At the interface between two immiscible liquids or between two different phases, there will be a force imbalance. For instance, in a gas, the molecules have a very random motion, and the distances between molecules are very high. For this reason, if you release a gas into a container, it will eventually fill all of the space. Intermolecular forces are very weak. In a liquid, the molecules are semi-ordered. Thus, there are significant intermolecular forces. When water is poured into a glass, it stays together and aggregates at the bottom rather than filling the entire container. The force imbalance between the air and the water creates surface tension.

Similarly, liquids and solids will exhibit a surface force demonstrated by water beading on the surface of a waxed car hood.

Rise of fluid in capillaries

The situation illustrated by the glass is quite sensitive to physical geometry. As the dimensions of the glass become smaller, the curvature of the surface becomes more significant. This is shown for a small diameter glass straw in Figure 7–1.

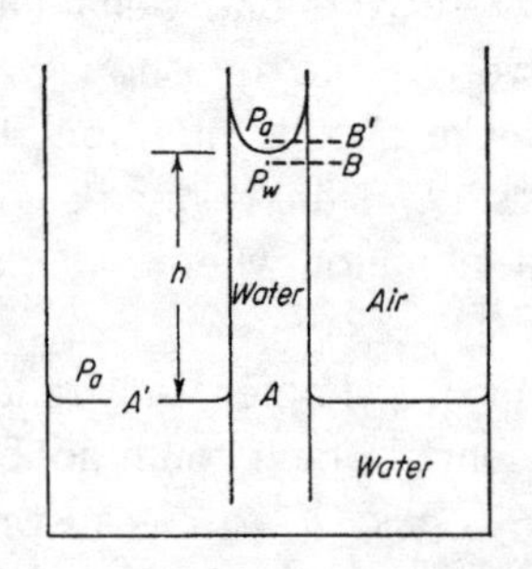

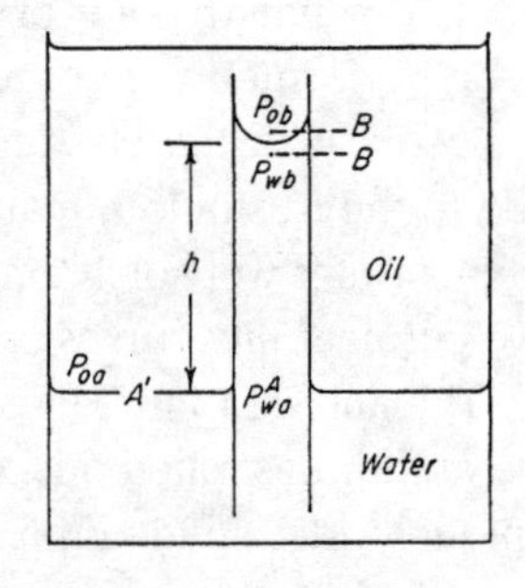

Fig. 7–1 Basic Example of Capillary Pressure—Vertical Orientation

In the capillary tube, there is a solid-liquid interface. The liquid meets the glass straw at a diagnostic angle, depending on the type of liquid and the composition of the glass. Similarly, there is a force imbalance between the gas and the liquid. The net result is the liquid creeps up the straw until hydrostatic and surface pressures are equalized. One can analytically calculate the rise in the straw knowing the contact angle and the surface tension between the oil and the gas.

Horizontal movement—soldering

Capillary pressure occurs horizontally. A practical example is soldering. If the pieces of copper are close together, joints can be soldered, and the solder is distributed evenly by capillary forces. This is shown in Figure 7–2.

The displacement is amazingly fast. It is very difficult to see the solder move. Once it melts, the solder moves faster than the eye can follow.

Directional sensitivity (drainage versus imbibition)

As shown by the irregularly shaped vertical capillaries in Figure 7–3, the capillary pressure rise is affected by the direction of water movement.[1] In this case, the example on the left is a drainage sample, and the one on the right is an imbibition sample. These terms imply a direction of saturation change.

The terms are an important convention and are summarized as follows:

- Imbibition — wetting saturation increases
- Drainage — wetting saturation decreases

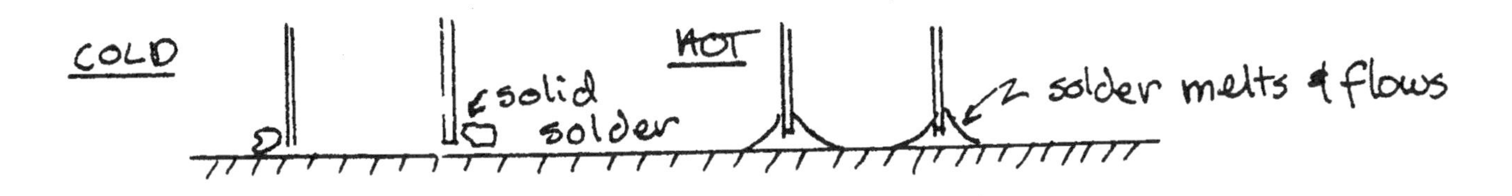

Fig. 7–2 Horizontal Example of Capillary Pressure

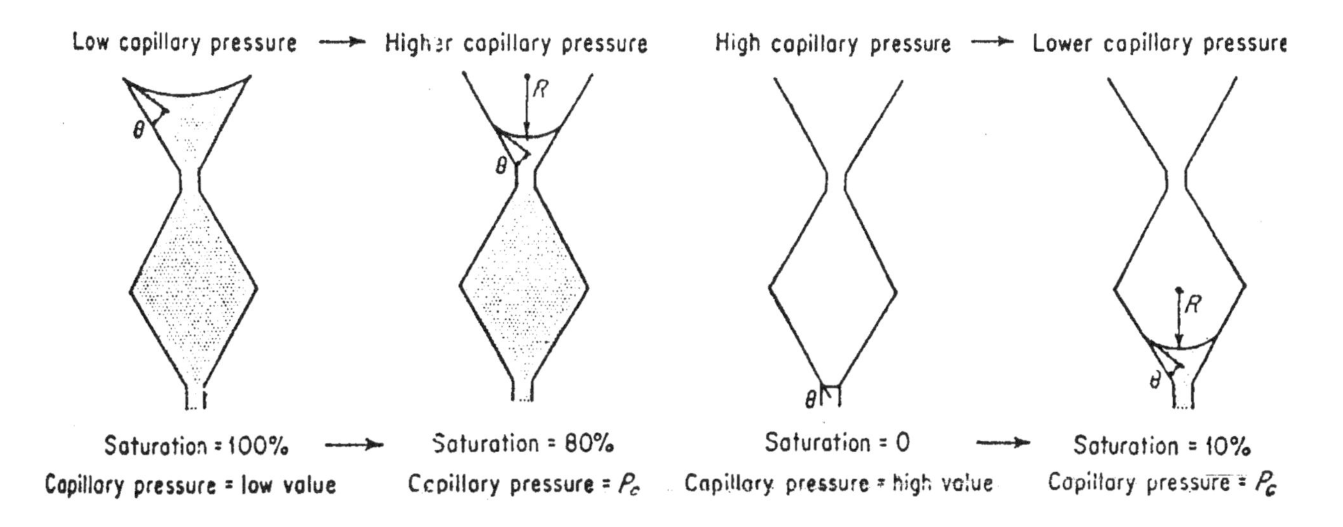

Fig. 7–3 Irregular Pore Shapes Cause Directional Difference in Saturation

Reservoir rocks

The pore spaces inside rocks are the size of capillaries. Therefore, capillary forces have a significant effect. However, rock pores do not have the convenient geometric consistency of the irregular straws shown previously. This is not the end of the problems. The sides of the pores are made up of mineral grains, which inconveniently have a different contact angle as shown in Figure 7–4.[2]

Adjustments

Obviously, the information gathered in the laboratory is not gathered at the same conditions existing at the reservoir. Adjustments must be made for the following:

- surface tensions at reservoir conditions
- contact angles at reservoir conditions

The adjustment is normally done as follows:

$$P_c = P_c' \frac{T\cos\theta}{T\cos\theta'} \tag{7.1}$$

The prime indicates lab conditions. Core Labs provides the interfacial tension (IFT) and contact angles as illustrated in Table 7–1.

Table 7–1 IFT and Contact Angles

System	Contact Angle θ	Cos θ	T IFT	T Cos θ
Laboratory				
Air-water	0°	1	72	72
Oil-water	30°	0.866	48	42
Air–mercury	140°	0.765	480	367
Air-oil	0°	1	24	24
Reservoir				
Water-oil	30°	0.866	30	26
Water-gas	0°	1	50*	50

**Pressure and temperature dependent, Value reasonable to 5,000 ft.*

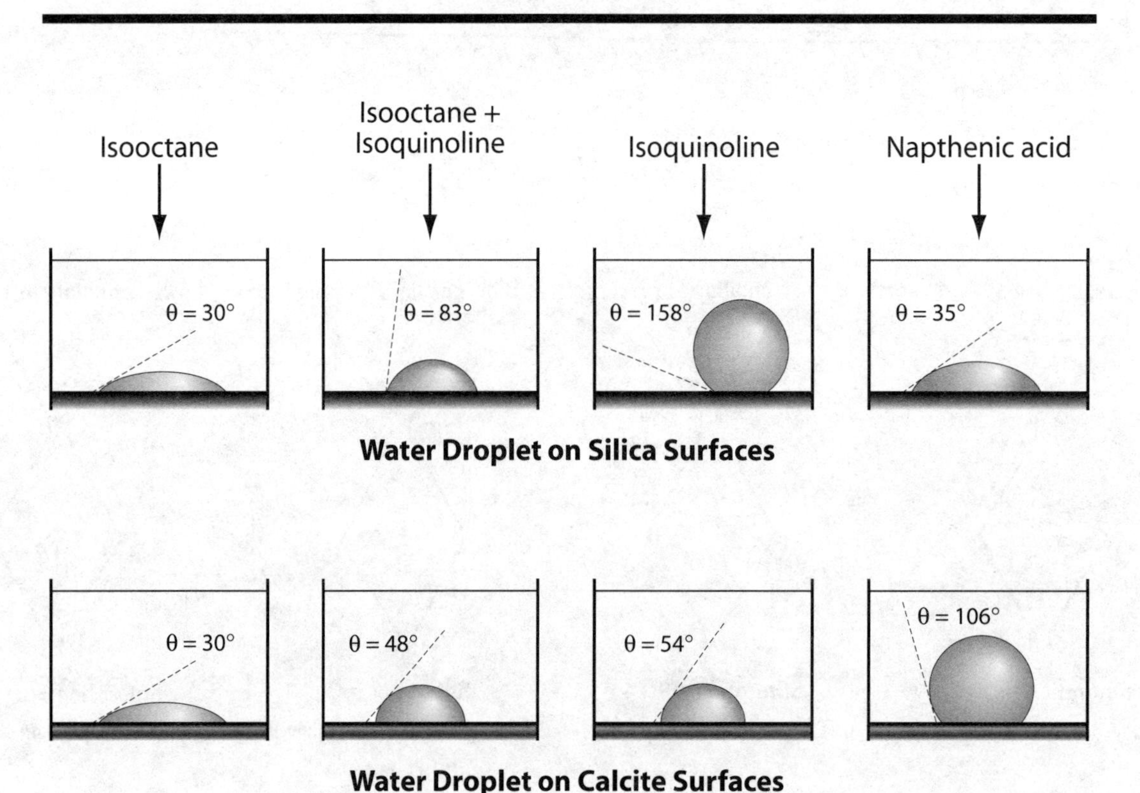

Fig. 7–4 Surface Force Effects with Different Fluids and Surfaces

Caveats on capillary pressure adjustments

Although the technique is commonly used, a number of comments are in order:

- These adjustments are done as if the capillary system were uniform straws. Given the complexity of real pore structures and mineralogy, this is a huge leap of faith.
- More than likely the contact angles and interfacial tensions between the oil, gas, water, and the minerals in the reservoir will not have been measured at reservoir temperature and pressure. Adjustments will be made on general correlations.
- Most data is simple drainage data. In rare circumstances, imbibition data is available. However, none of this data conforms to the real processes in the reservoir, which are countercurrent drainage for initialization and countercurrent imbibition for frontal displacement.

Therefore, the data from the lab must be adjusted, and the adjustments are not rigorous.

Figure 7–5 shows some data on the interfacial tension between water and gas.[3] This data is rather rare and not very consistent. Gas-water contacts tend to be more difficult to match with lab data.

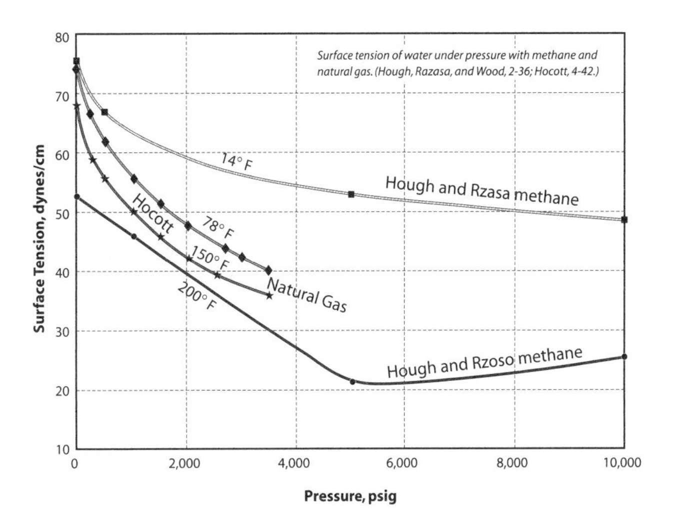

Fig. 7–5 Variation of Surface Tension of Gas and Water with Pressure and Temperature

Oil-water surface tension

The interfacial tension between oil and water also varies considerably. A major factor is temperature, which is shown in Figure 7–6.[4]

The saturation pressure (bubblepoint) of the oil also has an effect as shown in Figure 7–7.[5]

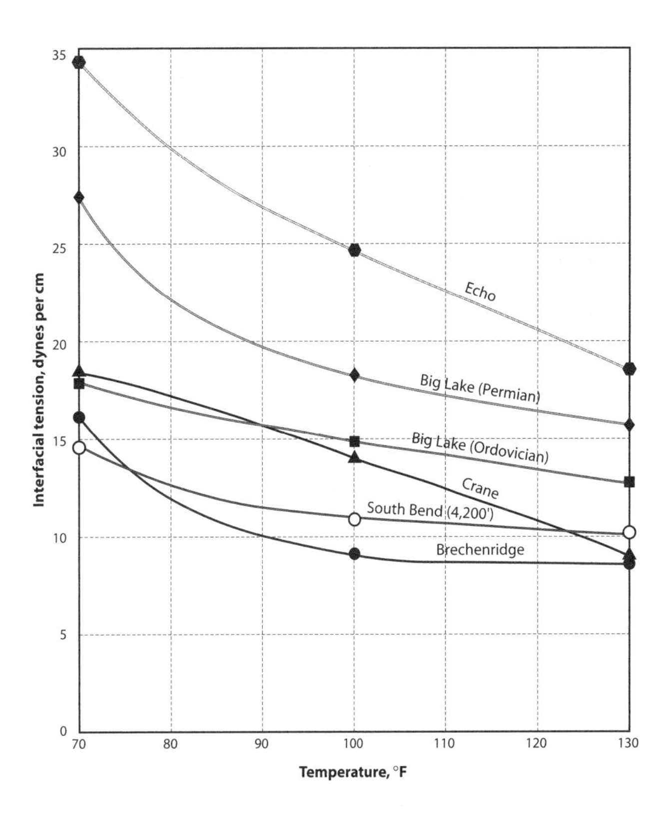

Fig. 7–6 Variation of Surface Tension of Oil and Water with Temperature for Different Oils

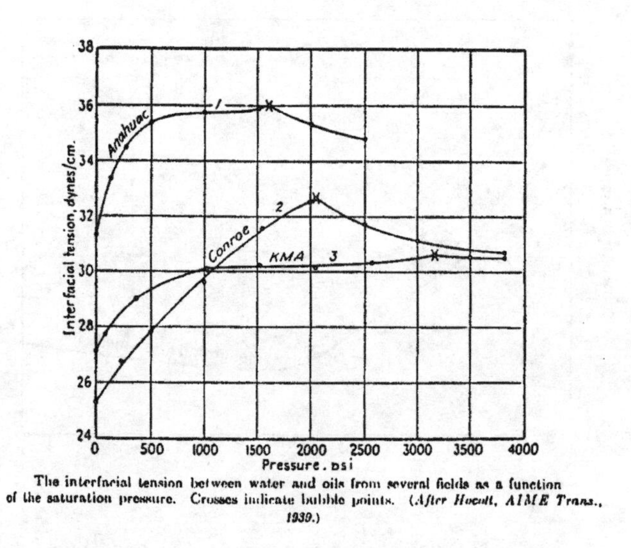

Fig. 7–7 Variation of Interfacial Tension of Oil and Water for Different Oil Saturation Pressures

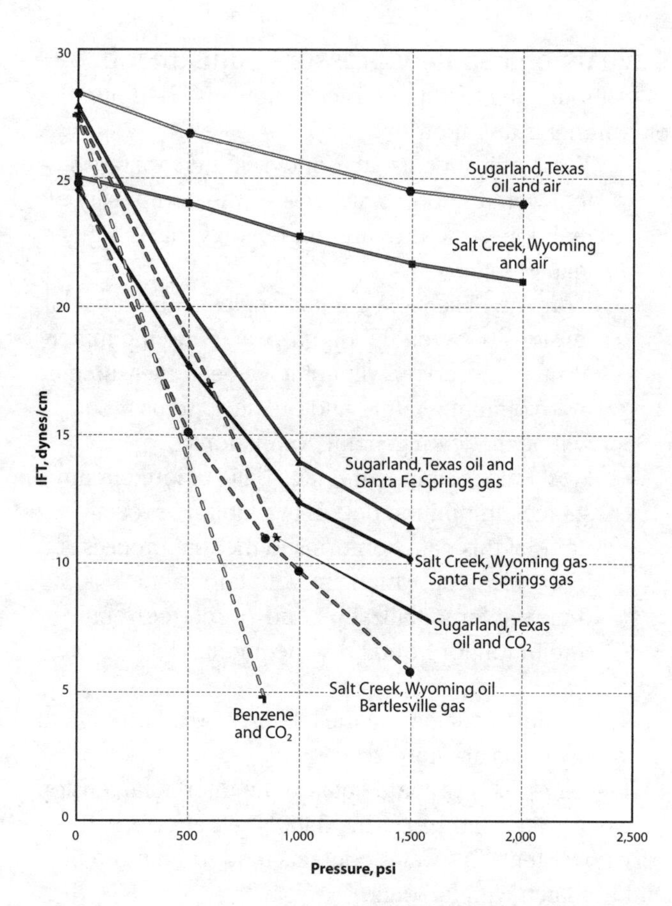

Fig. 7–8 Variation of IFT of Various Oils with Benzene for Different Oil Saturation Pressures

Gas-oil surface tension

Considerable differences exist in the interfacial tension between oil, or benzene, and various gases, as shown in Figure 7–8.[6]

Measuring capillary pressure

There are four different methods of measuring capillary pressure curves:

- Porous plate
- Centrifuge
- Air mercury injection test
- Water vapor desorption

The most common data is the air-brine (porous plate) test and the air mercury injection test. Frequently, both types of data will be available. Diagrams of those two types of tests as well as centrifuge (less common) are shown in Figures 7–9, 7–10, and 7–11.

Frequently, data from multiple types of tests is present. Presumably, all of the methods will give the same results. Early work from Purcell comparing the former two would indicate this is the case as shown in Figure 7–12.[7]

Based on Core Labs Special Core Analysis notes (recommended reading), Air-Hg P_c can be adjusted to air-brine P_c by dividing by 5.1. However, the notes also demonstrate that data from same samples often requires division by 7.4 for limestones and 7.2 for sandstones. Given the simplistic basis of the adjustment, this is not surprising. *Unless you have tests conducted on the same*

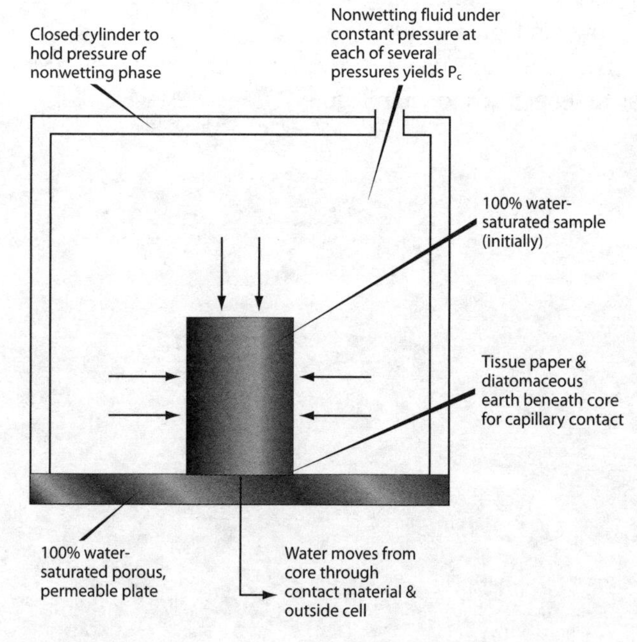

Fig. 7–9 Restored State (Porous Disk) Cell Measurement of Capillary Pressure

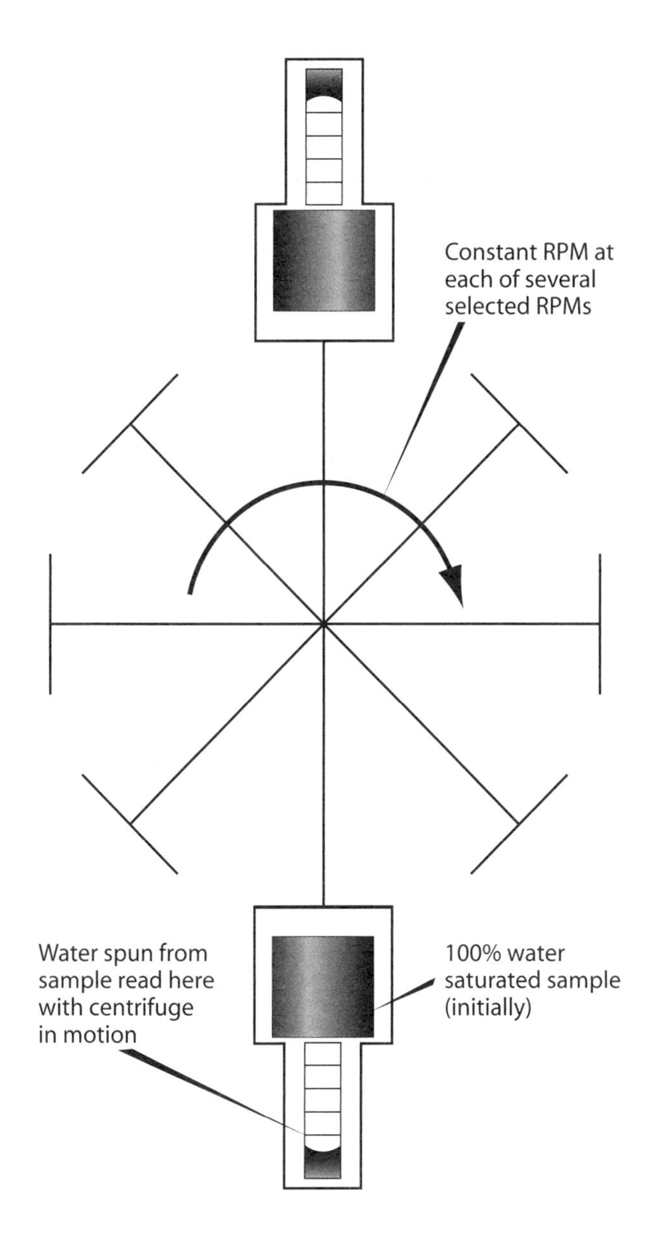

Fig. 7–10 Centrifuge Measurement of Capillary Pressure

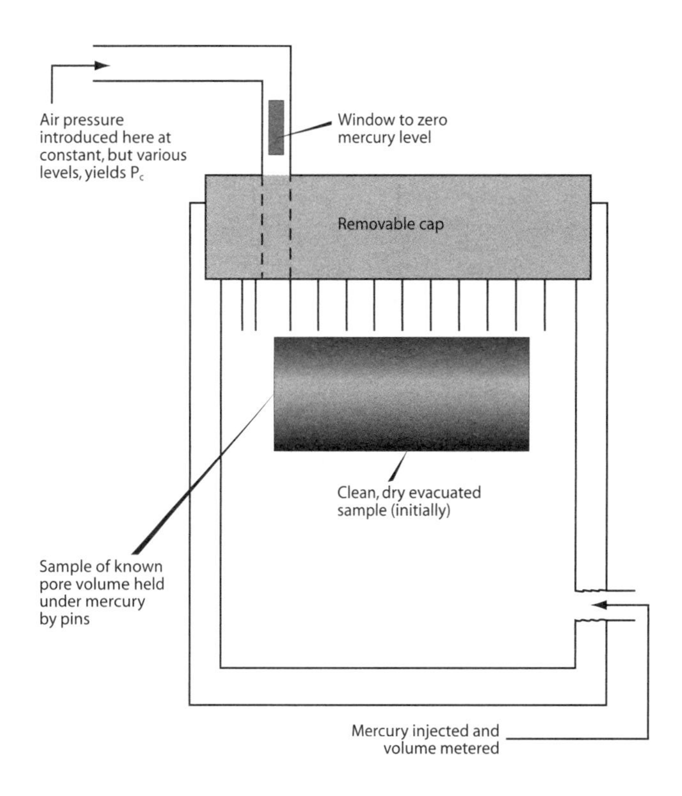

Fig. 7–11 Mercury Injection Measurement of Capillary Pressure

sample with multiple techniques, the data from multiple test types cannot be compared directly. Regrettably, in practice, this is almost never available.

Core Labs does not purport unique correctness to the figures they provide. In truth, the figures they provide are better than nothing. The figures indeed seem to be of a reasonable range.

Nonetheless, it is frustrating when one is attempting to average data impartially. The revised correction factors were calculated from samples analyzed with both techniques. Invariably, many practical applications involve different samples, which were obtained with different methods, and the engineer will not know the real correction factor. Bluntly stated, the accepted methodology does not work too well here. The author normally takes one set of data and ignores the other.

Many engineers arbitrarily use air-brine data in favor of mercury data. There are reasons for this, which will be explained later.

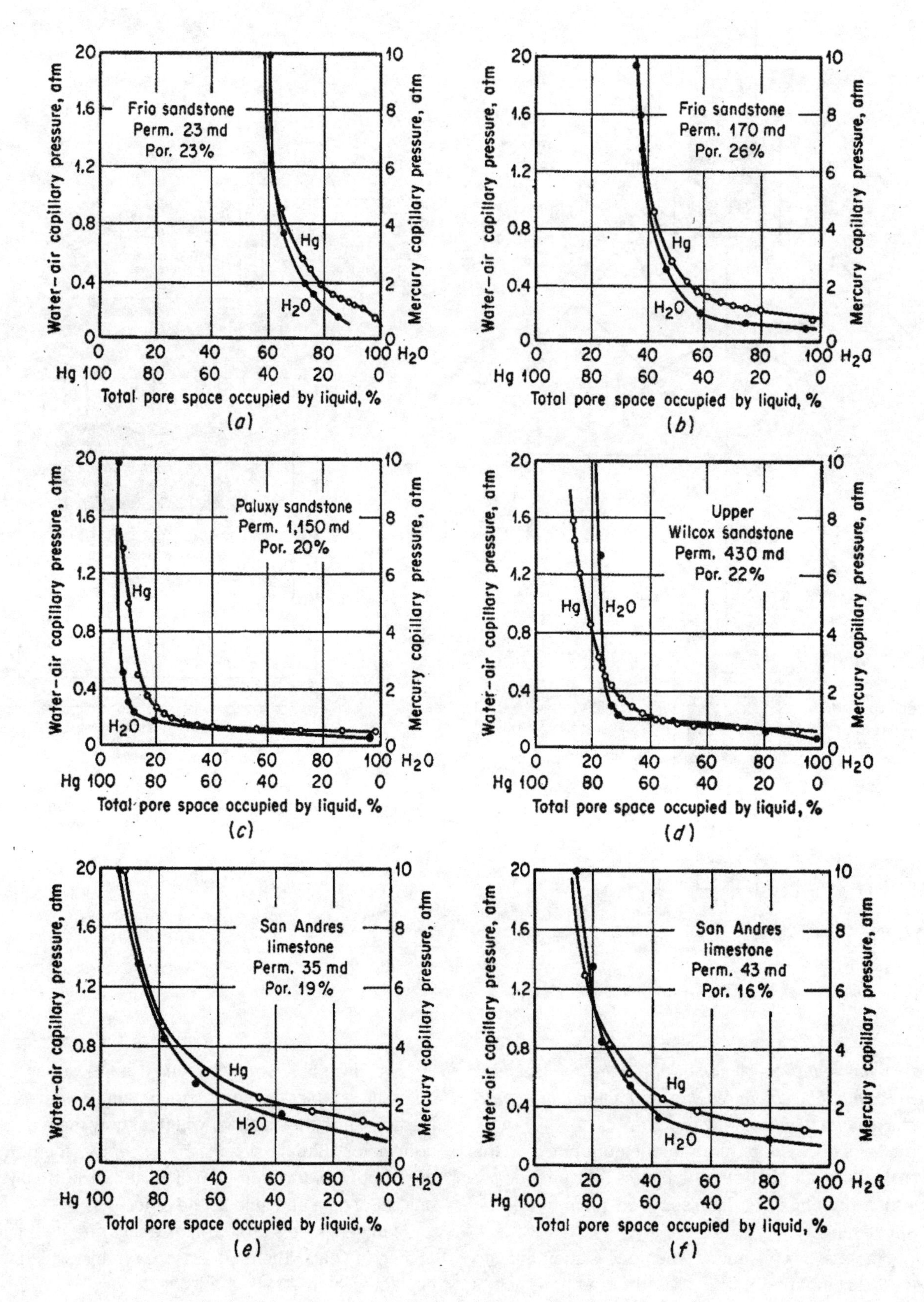

Fig. 7–12 Comparison of Air Brine and Mercury Capillary Pressure Data

Capillarity and connate water saturations

This one area has caused some noticeable disparities between theory and practice. The most comprehensive description of the problem may be found in *Interfacial Phenomena in Petroleum Recovery*.[8]

In the book, Morrow and Melrose make the point that, in bead or sand packs, very low levels of water saturation should be achieved. The source of irreducible water saturations is pendular rings, and this is supported by laboratory data. Figure 7–13 is from Leverett's original paper.[9]

Time is required for a film of water on the surface of the beads to drain. As a result, data obtained from the lab can represent nonequilibrium data. The effects of time are shown in Figure 7–14.

Time requirements for film flow are not obvious. For a 1 cp oil film, it would take about one day to drain, and for 3.1 cp oil film, drainage takes approximately three days.

For factual tests, this apparently small change will extend the total time significantly. For *w* points, a 10-day test becomes a 31-day test. It can be shown that the various methods of measuring capillary pressure, when carefully conducted, produce identical results as shown in Figure 7–15.

The air-Hg data is modified by dividing by 5.8 and not 5.1, 6.4, or 7.2. The irreducible water saturation is trapped in pendular rings. This is a very low level of water saturation, in the range of 1.5–2.0%. The pendular ring calculation is not discussed in detail, since it can be found in a number of sources.

This issue is not new; it is discussed thoroughly by Muskat, Leverett, and quite a few others. The bottom line is *good capillary pressure data will always yield very low irreducible water saturations.*

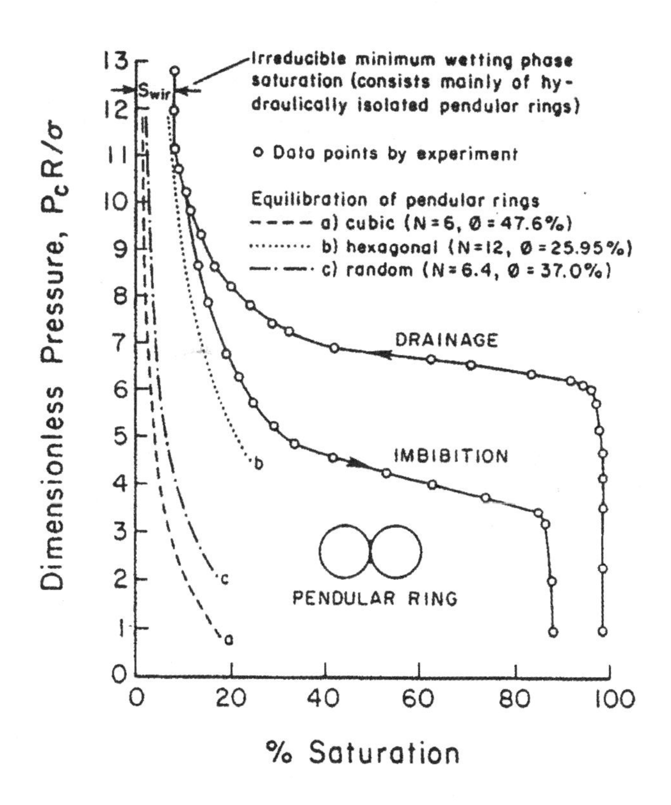

Fig. 7–13 Classic Capillary Pressure Relation from Leverett

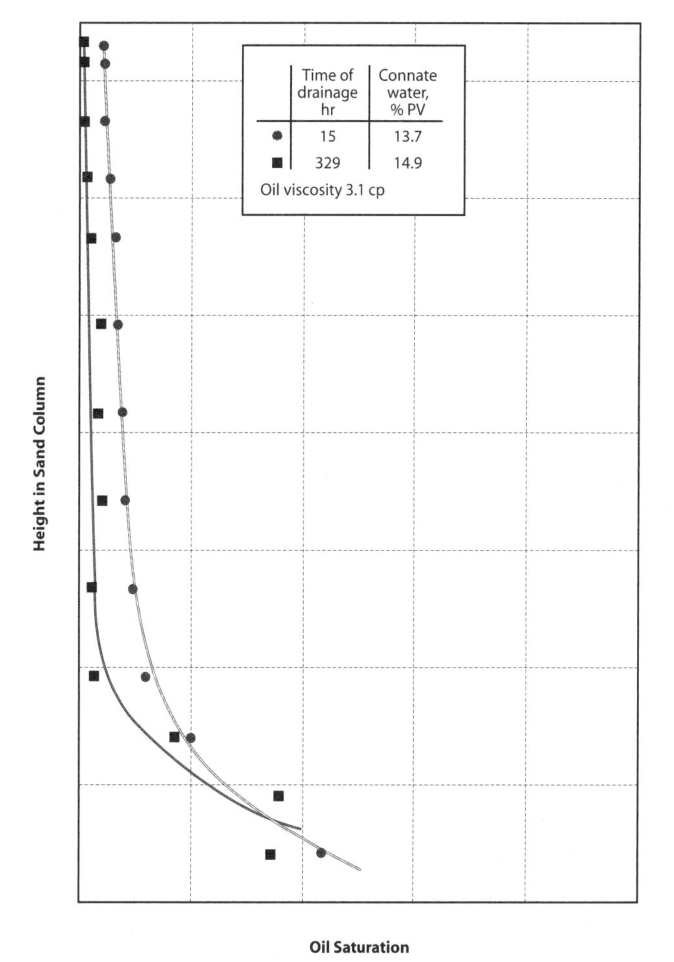

Fig. 7–14 Effect of Time on Saturations in Capillary Pressure Lab Test

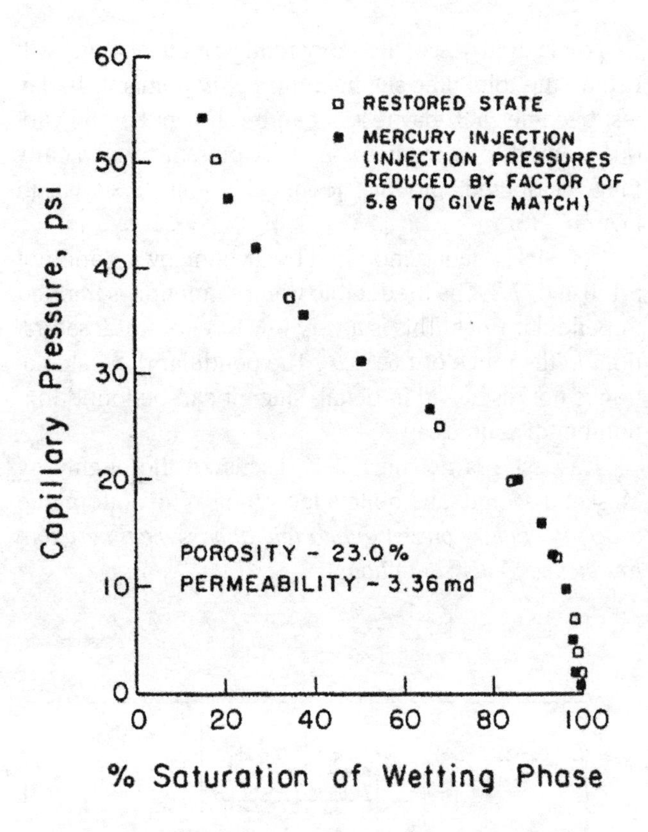

Fig. 7–15 Example of Match between Restored State (Air-Brine) and Mercury Injection Capillary Pressure

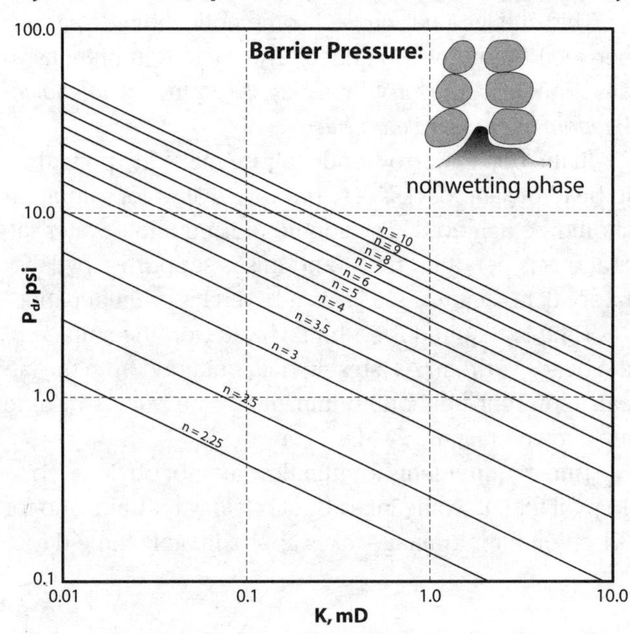

Relation between displacement pressure (P_d) and rock permeability (k) for various values of n (poresize distribution index).

In general, the larger values of n are found in rocks having high porosities. Denser rocks usually have a wide range of grain sizes, lower porosity, and probably a wider range of pore sizes, which results in a smaller value of n. A value of n of about eight might be considered as representing an average distribution of pore sizes.

Fig. 7–16 General Correlation of Imbibition Capillary Pressure Thresholds

Field observation

When log data is reviewed, the low saturations predicted by static capillary pressure lab data are not observed. Further, there is a strong correlation between residual water saturations derived from oil-based cores, where no water will invade the core, and nonequilibrium lab capillary pressure data. It would seem logical that, across geological time, the equilibrium curves would be achieved.

Hydrogeology

In fact, this last statement is completely wrong. Hydrogeology shows us that water is moving pervasively in the subsurface. The shale on top of the reservoir rock is, in fact, a barrier to oil and gas because of imbibition capillary pressure. Fine-grained materials have a high capillary pressure imbibition threshold. This was introduced in chapter 3.

The water, which is the wetting phase to the clays in shale, is actually free to flow vertically—either up or down. Although static conditions can be achieved in the lab, they do not occur strictly in the field. Since the water is always moving, the apparent connate water saturation is really more of an end-point on a relative permeability curve. More on hydrogeology is outlined in chapter 9. Since the oil and the gas are stuck and immobile, the vertical water flow is free to take place, and the physics change immediately.

The static model has another discrepancy. It is natural to view the water, or wetting phase, in an air-brine test as moving down. The air, or nonwetting phase, is open to the chamber above. However, in the ground, the nonwetting phase is moving in the opposite direction, as it pushes up against the seal via buoyancy.

There is no requirement for large amounts of flowing water. Most seals have permeability, but the permeability is low. Figure 7–16 correlates imbibition threshold pressure with permeability. This is from the groundwater literature.

Flow through shale

Many readers will immediately question the amount of fluid flowing across a shale. The amount of fluid is much larger than training has led us to believe. This can be calculated using Darcy's Law. An example calculation follows:

Darcy's Law

$$Q = \frac{k}{\mu_w} A \frac{\partial p_{atm}}{\partial x}$$

(from chapter 2, equation 2.4)

Input Data:

Area = 1 section = 5280' * 5280'
k_{shale} = 2.5 μD
h = 40'
μ_w = 1.0 cp
head difference between sands = 80' = 2.4 atm

$$Q = \frac{(2.5E\text{-}6)\ \{(5280)*(12)*(2.43)\}^2\ (2.40)}{(1.0)\ \{(40)*(12)*(2.54)\}}$$

$$= 127.5 \text{ cc/sec}$$

$$= \frac{(127.5)*(6.291)*\{(60)(60)(24)\}}{(1{,}000{,}000)}$$

$$= 69.3 \text{ bbl/day} = 25{,}294 \text{ bbl/year}$$

Although the permeability of shales is indeed low, in hydrogeological terms, there are large potential head differences across relatively short distances resulting in high potential gradients vertically. With the cross-sectional area of flow substantially larger in the vertical direction than in the horizontal direction, significant fluid flow can indeed occur vertically across shales.

An interesting article by Parke A. Dickey regarding water flows in the subsurface of Oman is recommended reading.[10] The answer to the apparent black hole problem is substantial vertical movements of water.

Pore scale continuity of water saturations

At very low water saturations and high capillary pressures, the water phase is not continuous as shown in Figure 7–17.

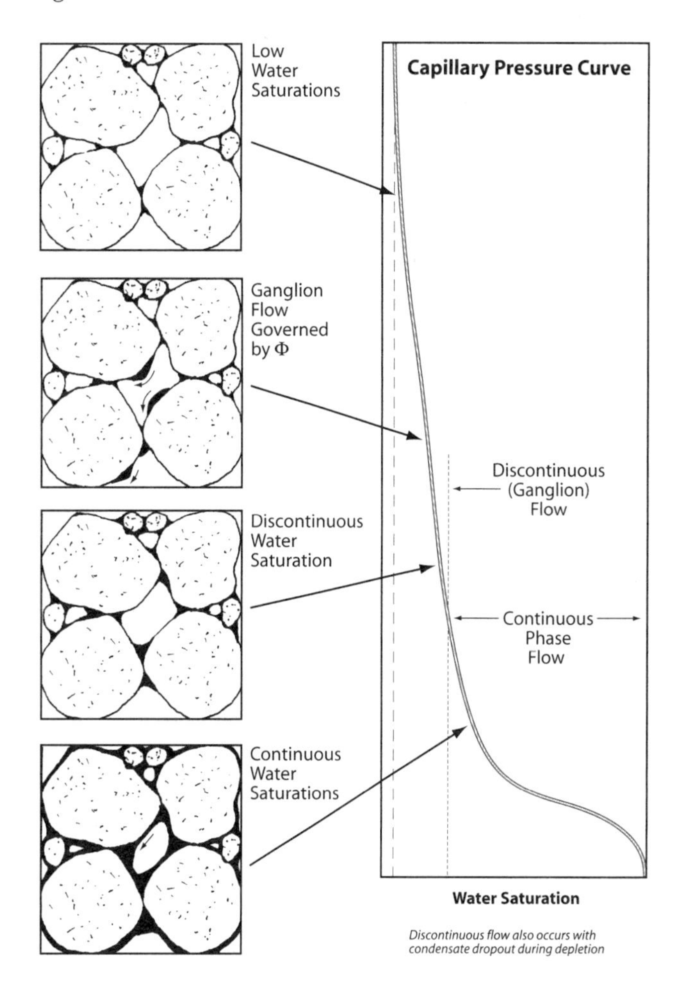

Fig. 7–17 Diagram Outlining Continuity of Wetting Phase

Since water is actually free to flow in the subsurface, the author suggests there will be continuous water saturations. If this is the case, the *in situ* water saturations observed on logs and in oil-based cores will be close to the transition point from a discontinuous wetting phase to a continuous wetting phase. This would also correspond closely to the end-point of a relative permeability curve.

Resolving the discrepancy between P_c data and field observations

The assumption of static equilibrium, which has been misapplied to the reservoir, has caused misinterpretations of the lab data. For instance, Figure 7–18 has been taken from a modern reservoir engineering text. It shows a limiting, connate or irreducible, water saturation greater than pendular ring saturations and implies an unlimited slope.

This does not contradict field observations, although the infinite slope is troubling theoretically.

Capillary pressure data errors

Experimental drainage capillary pressure data, which is very typical of the commercial data reviewed, is shown in Figure 7–19.

This data appears to support the incorrect model shown previously. There are a number of factors in play here. Most porous plate data is limited to 35 psi, and this is not very high up the P_c curve. The noncontinuous wetting phase data is normally missing. Some authors have speculated there are experimental errors with maintaining fluid continuity with the porous plate. Contact is normally maintained with a piece of Kleenex or tissue paper—literally. Higher readings will be obtained if equilibrium has not been maintained. Rapid data acquisition may not allow film flow to occur.

Practical comments

The previous incorrect interpretation will actually be close to a realistic distribution of fluids in the reservoir. Naturally, the next question is, *What is the problem if everything works out in the wash?* The real problem occurs when one has *good* data. In particular, mercury data seems to yield water saturations that are hard to correlate with log calculations and centrifuge and/or air-brine tests.

The real data one wants to correlate for determining initial saturations is the transition from continuous wetting phases to discontinuous wetting phases or film flow.

Saturation history

This is not the end of the fundamental problems with capillary pressure. Earlier the difference between drainage and imbibition curves was outlined. Almost all rocks were deposited in a marine environment and were initially water wet. Oil and gas are emplaced or trapped after deposition. Because the connate water is already attached to the mineral grains, it is not easily displaced.

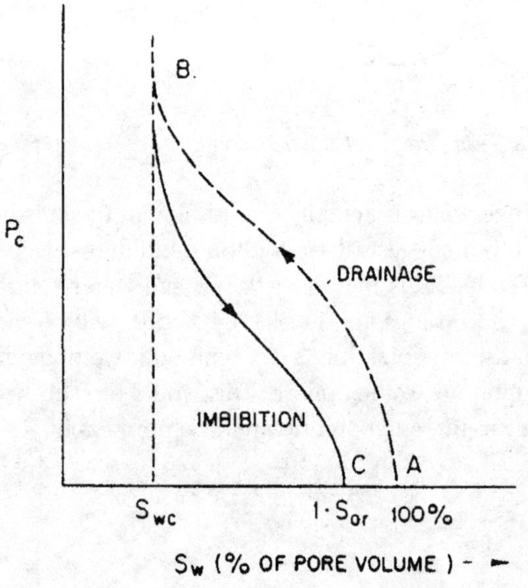

Fig. 7–18 Classic Schematic of Connate Water Saturation

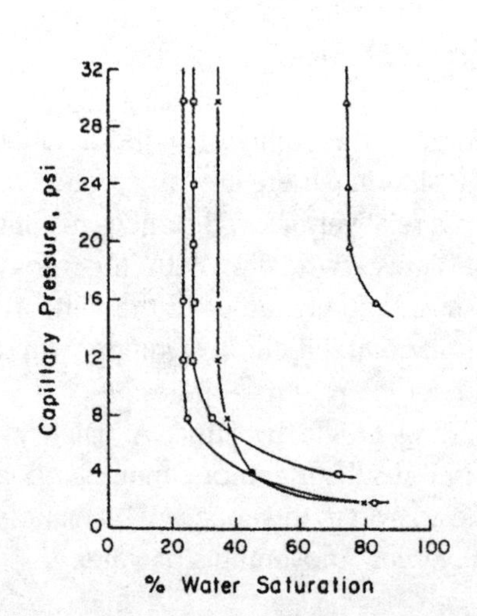

Examples of restored-state capillary pressure curves frequently described as typical. The well-defined irreducible saturations are the result of nonequilibrium, but this is not usually recognized in analysis and application of results to the reservoir.

Fig. 7–19 Examples of Lab Data That Have Not Reached Equilibrium

Countercurrent drainage

Oil and gas emplacement in the reservoir is likely a countercurrent process, i.e., the oil and gas rise at the same time the water is moving downward. Similarly, in a waterflood displacement, some water has to be sneaking or shooting ahead of the main water saturation increase.

This problem also occurs in fractured reservoir engineering, where the oil is expelled from the block and the water imbibes into the reservoir rock. In this case, a number of experiments have been designed to determine the imbibition nature.

The only reference found by the author on this topic is in F. A. Dullien's *Porous Media, Fluid Transport and Pore Structure* (see page 412).[11] Lenormand (1981) used capillary micro models to demonstrate countercurrent imbibition does not take place if all pores have the same size.[12] A pore-sized distribution is required to obtain countercurrent imbibition. It is inferred that water imbibes into relatively fine pores, and oil is expelled through relatively large pores.

Dullien describes experiments with Berea core to compare cocurrent and countercurrent imbibition. More interestingly, results from Bourbiaux and Kalaykjian are displayed in Figures 7–20[10] and 7–21.[13]

The curves on the left are for upward, cocurrent, displacement of oil by water; the ones on the left display results where the bottom was sealed off. In the latter, the oil leaks out the top, and the water must be absorbed downwards, which is countercurrent imbibition. The saturation profiles were determined by X-ray absorption.

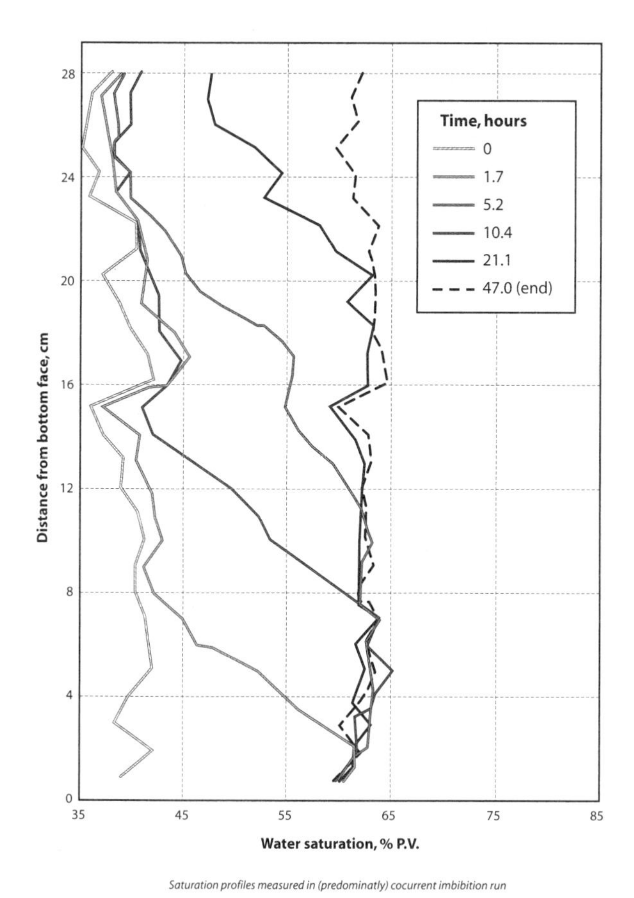

Fig. 7–20 Countercurrent vs. Cocurrent Imbibition Profiles—Saturation Profiles Measured in Cocurrent Imbibition Run

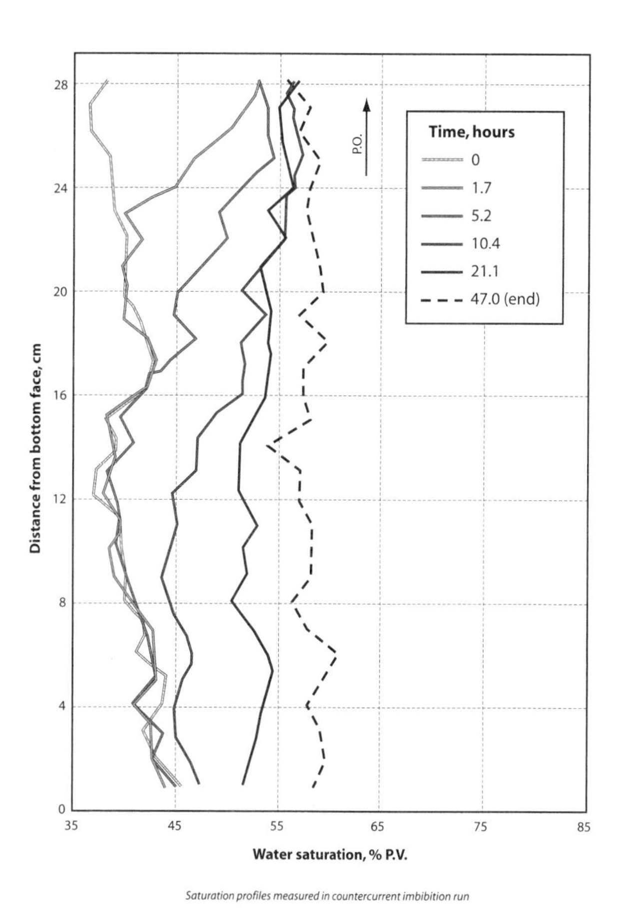

Fig. 7–21 Countercurrent vs. Cocurrent Imbibition Profiles—Saturation Profiles Measured in Countercurrent Imbibition Run

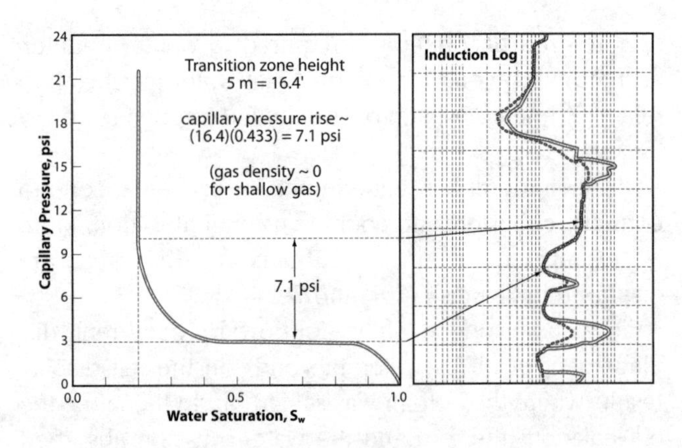

Fig. 7–22 Use of Log Capillary Pressure Transition Zones (Uniform Reservoir)

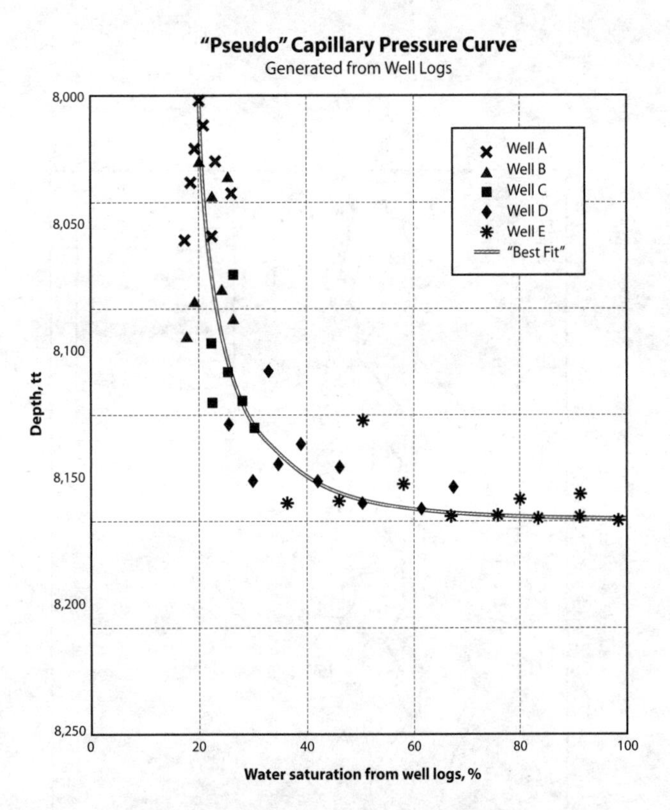

Fig. 7–23 Use of Log Capillary Pressure Transition Zones Using Averaging

Drainage capillary pressure curves don't really represent the process occurring in the reservoir at all. Is there proof? The author believes the answer is yes. As a rule, lab data does not match reservoir transition zones at all. Drainage capillary pressure curves overestimate the thickness or height of capillary pressure zones.

We are always dealing with uncertainty in sampling with special core data, and this makes identification difficult.

Log transition zones

Actually, the best data one can get on the transition zone is from logs. This will always be somewhat difficult, since the porosity and permeability vary somewhat vertically. Figure 7–22 is an example where the transition zone was in abnormally uniform reservoir.

In most studies, the reservoir properties will not be as uniform as in the preceding example. In such cases, if one plots up a number of transition zones, one can get a nice average. The Gas Recovery Institute (GRI) suggests the use of logs if lab data is not available. They present a graph, in Figure 7–23, which features averaging from a number of wells.[14]

The GRI commented the log data will often mimic the lab data. Given the shaky adjustments on which we have to rely, it is actually the other way around. Lab data sometimes mimics what the reservoir *in situ* conditions are really like.

Zero capillary pressure point

If you adjust the P_c curve data to log data, you are not matching $P_c = 0$ but rather the inflection point in all probability. This is shown in Figure 7–24. There will be a finite amount of additional capillary pressure from the inflection point and the true point for $P_c = 0$. The best estimate for this is probably the drainage lab data. One of the commercial models allows you to adjust for this point.

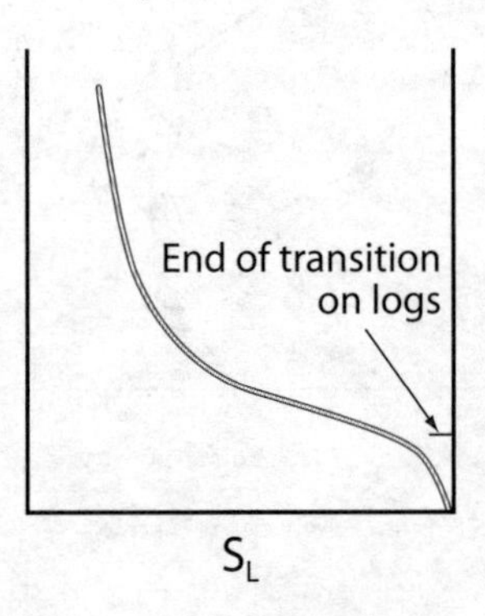

Fig. 7–24 End of Transition Zones and What Can Be Distinguished on Logs

Observations on lab data applicability

Process all of the lab data thoroughly. If it does not match the logs, scale the lab data to match the real transition zones, which are shown on the logs. In general, lab data overestimates the height of the transition zone.

The data is worse for gas than for oil. This is likely because gas-water surface tension is not well-known. The author has come across only one graph dealing with this issue, and the results hardly qualify as consistent. This data was shown previously in Figure 7–5.

Data available

In Canada, most data is available from the government at minimal cost. Here is how to get the data:

Alberta: The Alberta Energy and Utilities Board (AEUB) makes a convenient summary of all special core data and PVT analyses in a Cerlox bound book entitled, *PVT and Core Studies, Index, Guide 14*. This has almost every study done in Alberta. Order by phone, or visit the AEUB's publication desk.[15]

British Columbia: There is not as much data in British Columbia, and the structure is not as formalized as Alberta. A coordinator at the BCPRD has compiled a list of all data available up until late 1994, which will be faxed to you at nominal cost if requested.

Manitoba: Although working in Manitoba is rare, nonetheless they have all the data Saskatchewan and British Columbia have.

Saskatchewan: Phone Saskatchewan Energy and Mines (SEM). Traditionally the SEM has had a very stable and experienced staff, and the author has found them to be a very valuable source of information on applications and data.

In other jurisdictions, the author has always been provided with lab data as part of the study. When some of the author's students have been queried about central records, some indicate oil companies have these records centrally located; however, this is not true for the majority of companies. Comparing various PVT studies is helpful, and copies of all PVT data in a central library would be useful.

Special core analysis

The following example represents an ARE project for which there was a considerable amount of data. This is from the Brazeau River and Pembina areas and consists of data publicly available from the AEUB. Most of this data represents offsetting well information, since relatively little data was available on the reservoir.

This question naturally arises, *How representative is this offsetting data?* In order to evaluate this, all of the core data available from the immediately offsetting wells was plotted on a core permeability vs. core porosity plot. The graph showed a reasonable straightline relation. The data has some spread in it, and this has been represented with the two lines drawn to represent a fairway.

To this graph, special core analysis data from the Nisku S pool was added as shown in Figure 7–25. With the exception of a few points, the data from these wells appears to fall within the fairway. Although this approach is hardly exhaustive, it does indicate all of the data from the area could reasonably be used to estimate capillary pressure and relative permeability. This process is quite work intensive and was conducted in a stepwise fashion.

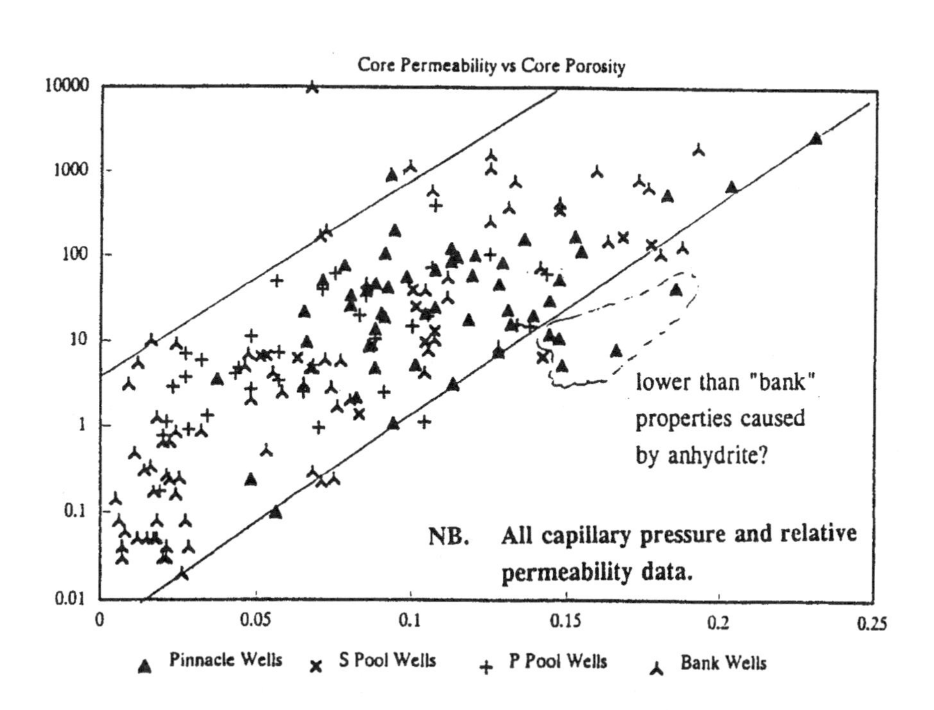

Fig. 7–25 Use of Core Permeability vs. Core Porosity to Screen Special Core Data

Tabulating capillary pressure data

Capillary pressure data was gathered from the special core analysis and was summarized as shown in Table 7–2.

The data was sorted by permeability and by porosity as shown in Figures 7–26(a) and (b).

The data indicates a long transition zone. The majority of the P_c curves start to straighten at about 800 kPa on the mercury data. The bottomhole density of the oil-gas is approximately 0.5 g/cm^3. Adjusting for density and the known gradient of water yields a transition zone thickness of 9.13 meters (30.5 feet).

Inspection of the log analysis showed the indicated transition zone is no more than four meters thick. Such a discrepancy is common. Although drainage data is the most commonly used form of capillary pressure test, it does not represent the countercurrent flow in which gas or oil become separated in a reservoir trap.

Table 7–2 Capillary Pressure Data

Lab	Location	Sample	k_{air}	Phi	Date	Type
Core Labs	14-32-49-11W5M	47	220.00	0.065	79-03-28	Hg
Core Labs	7-30-49-10W5M	65	6.36	0.065	79-03-28	Hg
Core Labs	14-32-49-11W5M	90	13.80	0.088	79-03-28	Hg
Core Labs	14-32-49-11W5M	133	909.00	0.093	79-03-28	Hg
Core Labs	14-32-49-11W5M	163	5.29	0.101	79-03-28	Hg
Core Labs	14-32-49-11W5M	184	0.24	0.048	79-03-28	Hg
Core Labs	9-1-48-14W5M	199	68.40	0.107	79-03-28	Hg
Core Labs	14-32-49-11W5M	211	3.56	0.037	79-03-28	Hg
Core Labs	14-32-49-11W5M	287	199.00	0.094	79-03-28	Hg
Core Labs	9-1-48-14W5M	314	122.00	0.112	79-03-28	Hg
Chevron	1-9-50-12W5M	1	115.00	0.154	79-05-18	Centrifugal
Chevron	1-9-50-12W5M	2	7.60	0.128	79-05-18	Centrifugal
Chevron	1-9-50-12W5M	3	19.00	0.091	79-05-18	Centrifugal
Chevron	1-9-50-12W5M	4	18.00	0.118	79-05-18	Centrifugal
Core Labs	15-9-47-14W5M	29A/15-9	9.79	0.104	81-12-21	Brine
Core Labs	15-9-47-14W5M	45B/15-9	25.40	0.101	81-12-21	Brine
Core Labs	15-9-47-14W5M	81A/15-9	39.90	0.100	81-12-21	Brine
Core Labs	15-9-47-14W5M	91A/15-9	6.80	0.053	81-12-21	Hg
Core Labs	9-1-48-14W5M	58/9-1	1.41	0.083	81-12-21	Hg
Core Labs	9-1-48-14W5M	31B/9-1	142.00	0.177	81-12-21	Hg
Core Labs	14-13-49-10W5M	1	-	0.053	89-07-07	Hg, fractured
Core Labs	14-13-49-10W5M	8	-	0.059	89-07-07	Hg, fractured
Core Labs	14-13-49-10W5M	17	16.20	0.087	89-07-07	Hg

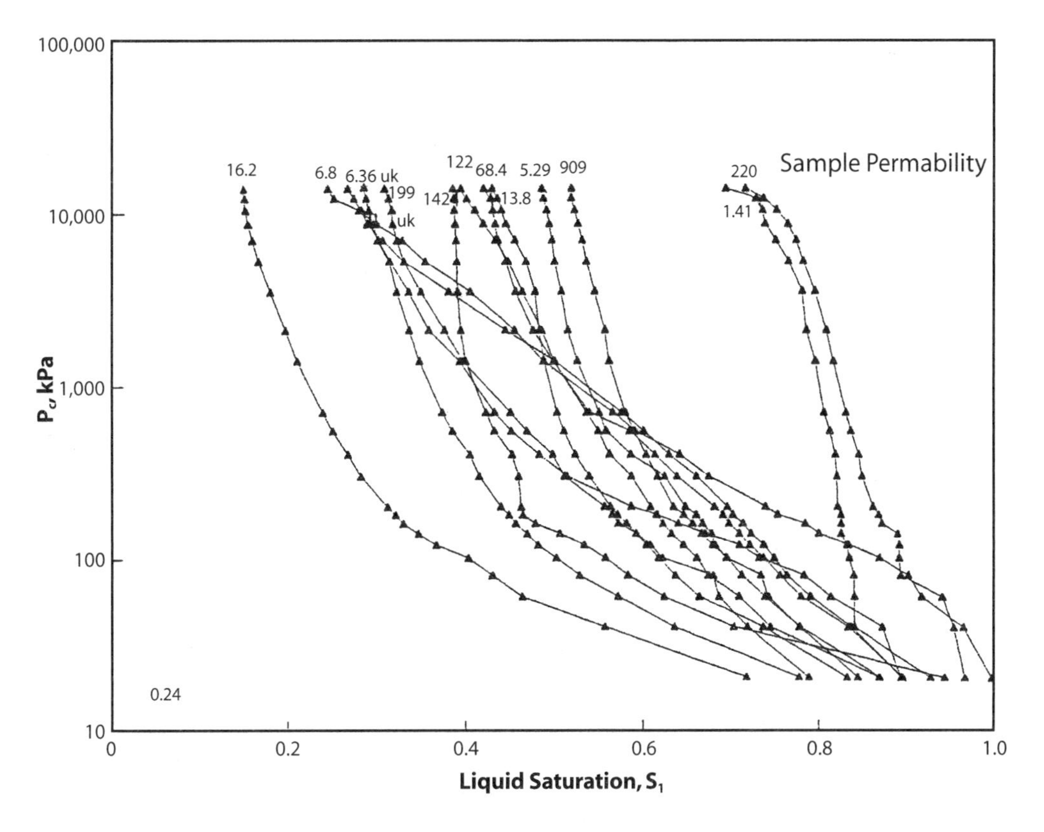

Fig. 7–26a Example of Field Capillary Pressure Data Correlated with Permeability and Porosity

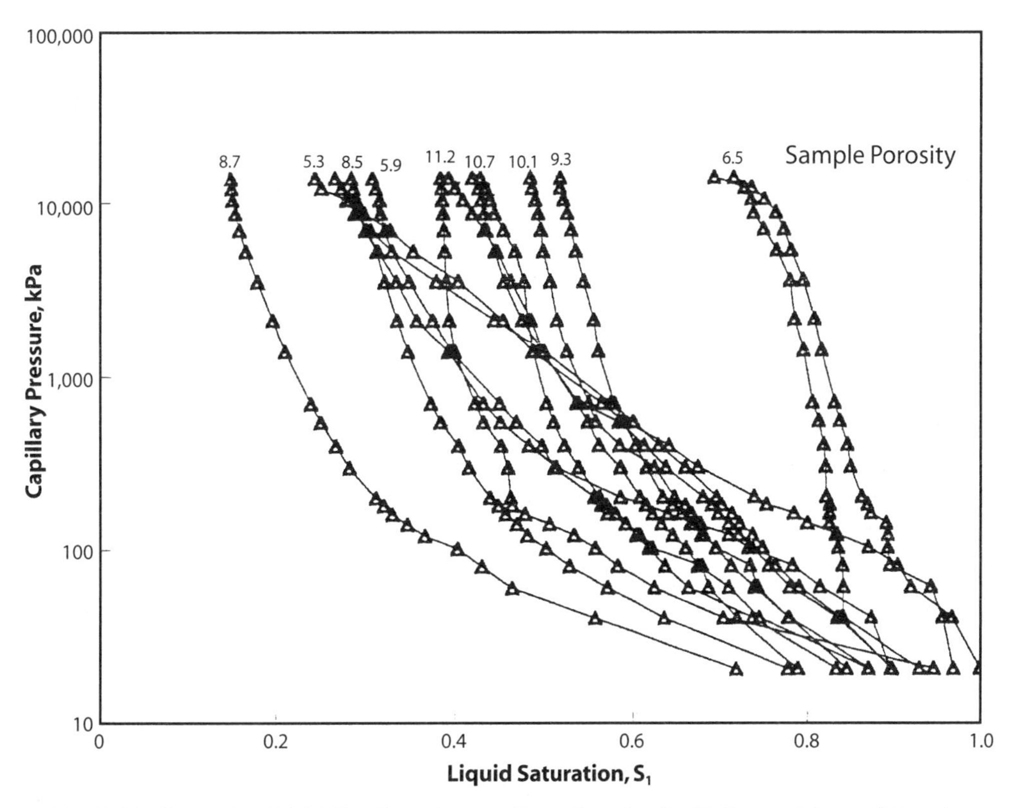

Fig. 7–26b Example of Field Capillary Pressure Data Correlated with Permeability and Porosity

The systematic overestimation of capillary pressure transition zones is another factor causing well-coning simulations to overestimate water production. This is known from practical experience. Often, it is not clear from well logs how much of the transition zone is due to changes in lithology.

Leverett J-function

One method proposed for averaging capillary pressure is due to Leverett. The expression is shown as follows:

$$J(S_w) = \frac{P_c}{\sigma \cos\theta}\sqrt{\frac{k}{\phi}} \tag{7.2}$$

In the author's experience, it did not work at all satisfactorily. However, there are reservoirs where it does work, as evidenced by the material from Amyx, Bass, and Whiting (after Brown) in Figure 7–27.

Imperial Oil in Canada has used this in the Redwater area. Having shown it can work, the author has yet to work on a reservoir directly where this correlation was usable.

Effect of permeability

Capillary transition zones are affected mostly by the permeability of the reservoir. This is due to the strong link between pore size and permeability. It is not as strong a function of porosity. This is shown in the classic correlation of Wright and Woody, Figure 7–28.[16]

The author has attempted to average the lab data based on the porosity and permeability he assigns for rock types.

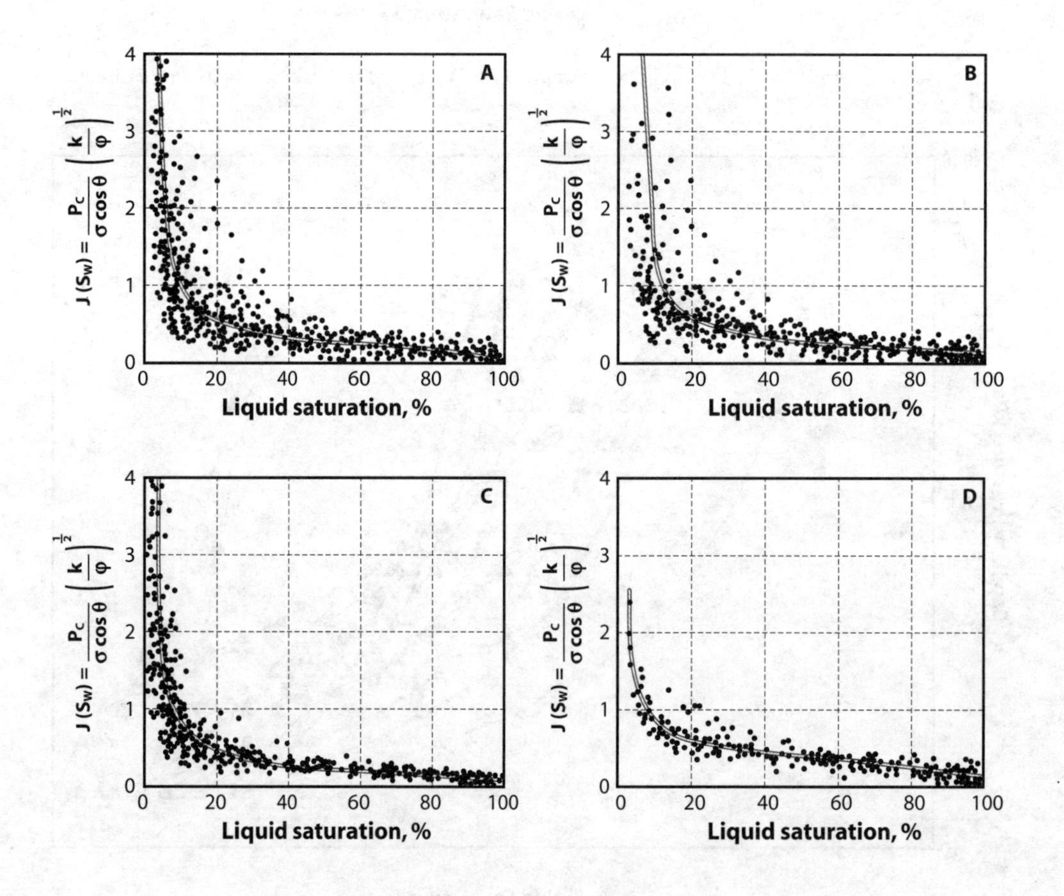

Fig. 7–27 Leverett J-Function Works in Some Circumstances

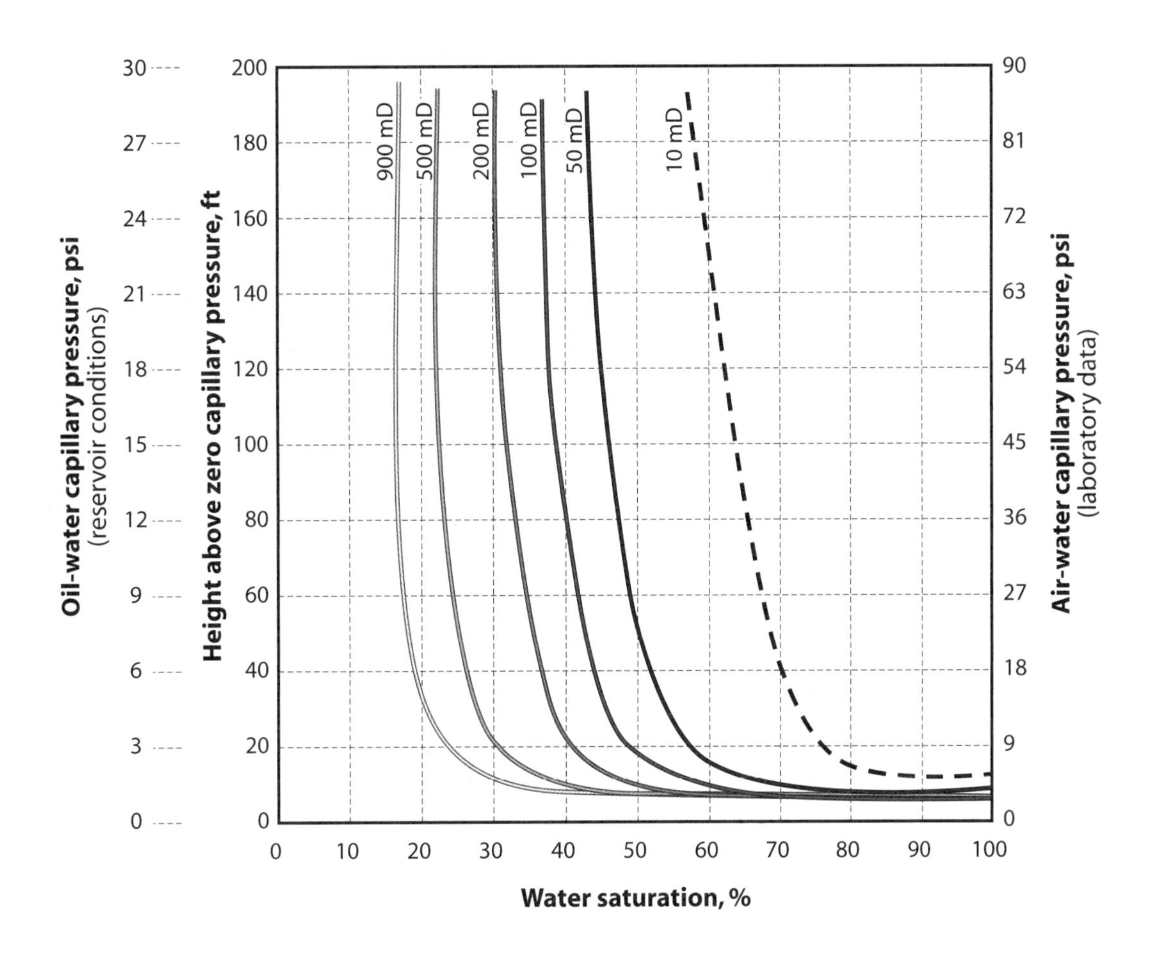

Fig. 7–28 General Correlation of P_c *with Permeability*

Relative permeability

To talk about everything with respect to relative permeability would require a book. In fact, there is such a book titled *Relative Permeability of Petroleum Reservoirs* by Mehdi Honarpour, Leonard Koederitz, and Herbert Harvey. Although very extensive, it does not cover everything engineers need to know for reservoir simulation. It would not be effective to reproduce this and other excellent references. What this chapter will attempt to do is give some advice on how to input the data into a simulator.

Sample preparation

The core has to be prepared before the test can begin. This reflects the manner in which core is obtained. Refer to Figure 7–29 where the core is subject to a number of physical changes:

- drilling filtrate invasion
- physical damage as the core barrel rotates
- mechanical capture of the core as it is drilled over
- solution gas evolves and gas expands as core is taken to surface
- the core is relieved of stress as it is taken to surface
- the surface of the core is plugged and covered with mud cake

Therefore, unless extremely special precautions are taken, core undergoes changes as it is obtained. The saturations in the recovered core are not representative of what was in the ground. These changes are summarized in Figure 7–30.

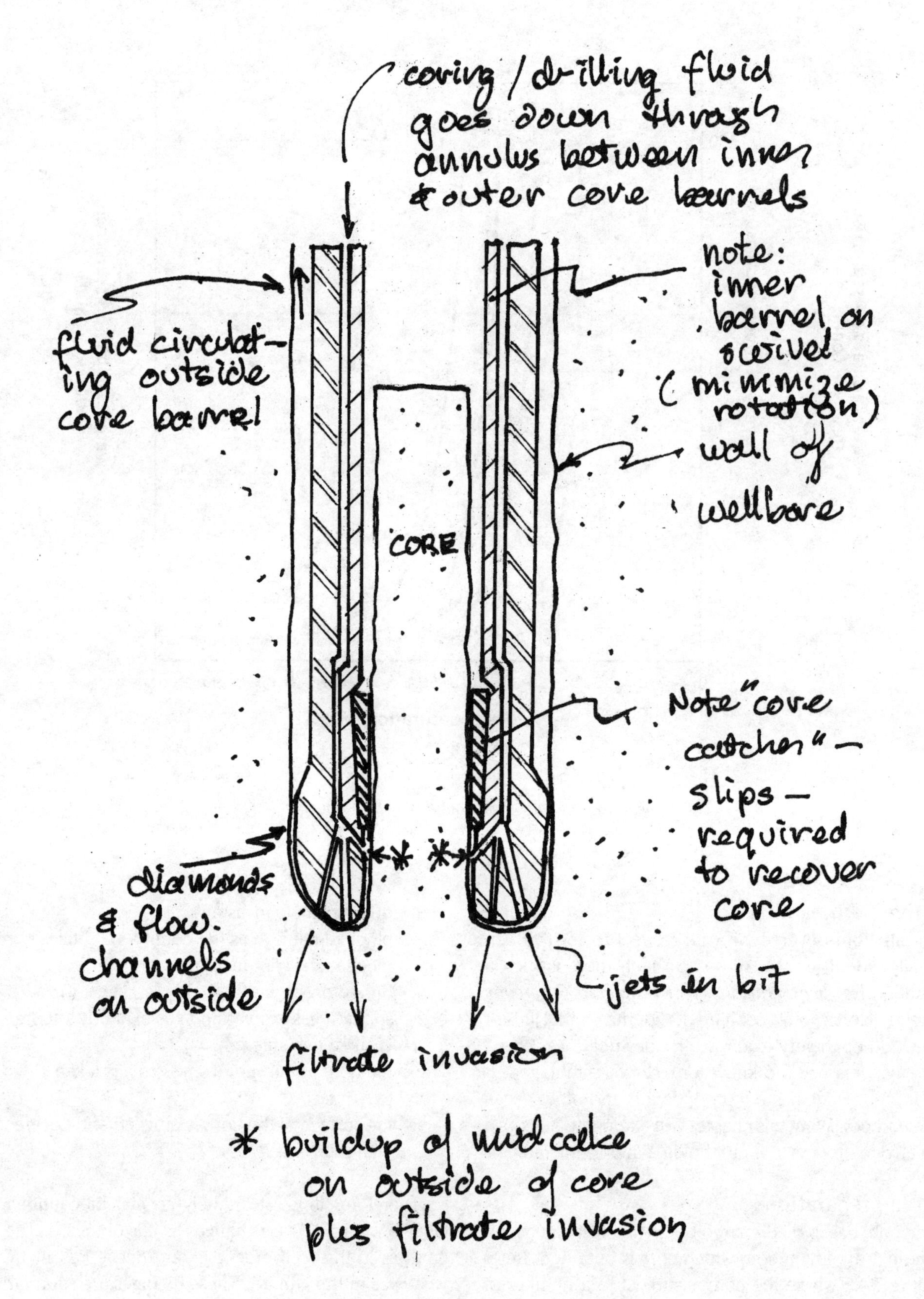

Fig. 7–29 Influences on Core during Coring Operations

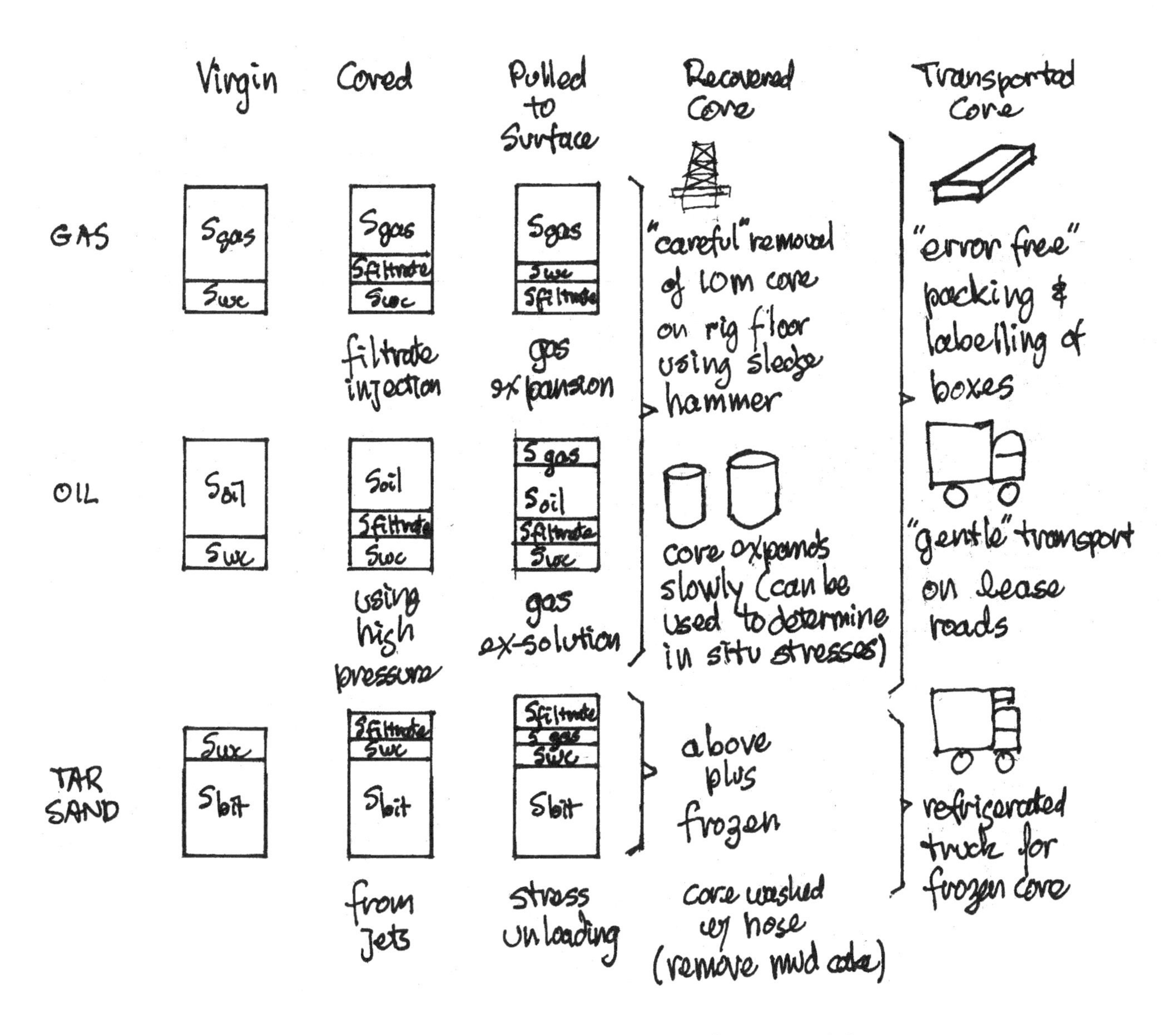

Fig. 7–30 Summary of Changes to Core from Virgin Conditions to Laboratory

Measuring relative permeability

Of the number of ways to measure relative permeability, the steady state method will be discussed next. The apparatus used is shown in Figure 7–31.[17]

As discussed earlier, formations are normally saturated with water before oil and gas migrate into a trap. Core obtained from oil-producing formations is normally cleaned with a mixture of toluene and carbon dioxide. Therefore, the core is saturated with water first and then displaced with oil prior to starting the relative permeability test.

The idea is simple. Oil and water are injected in one side of a cylindrical core and the pressure drop across the core is measured. This is done for a series of water and oil flow rates as shown in Figure 7–32.

Knowing the length and cross-sectional area of the core sample, one can calculate the effective permeability for both phases. The saturations in the core are determined via electrical resistivity measurements.

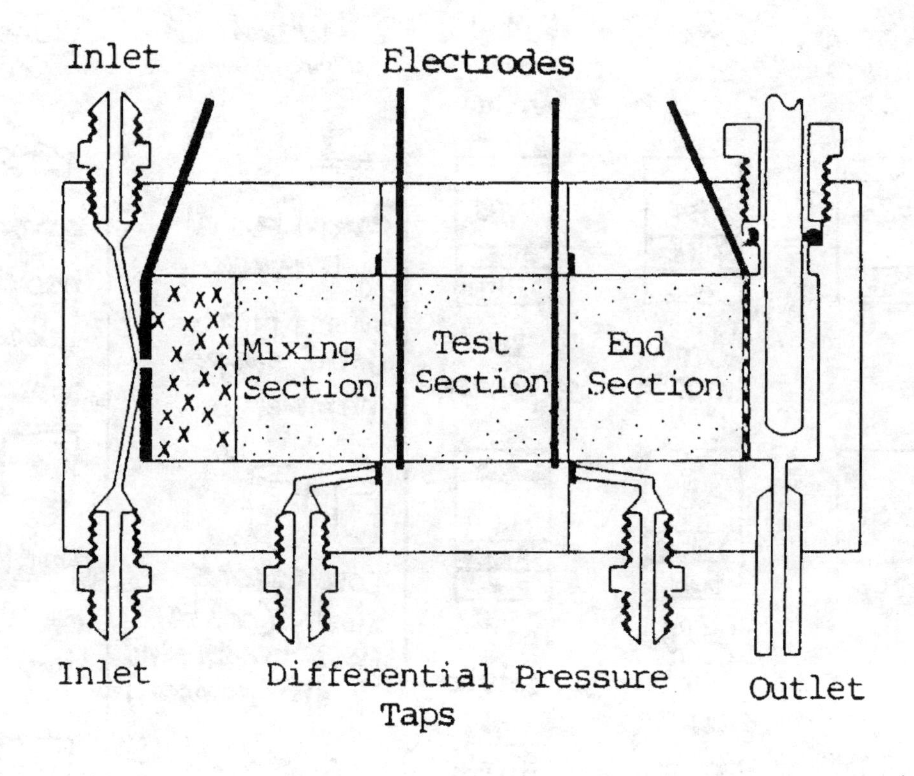

Fig. 7–31 Test Cell for Relative Permeability

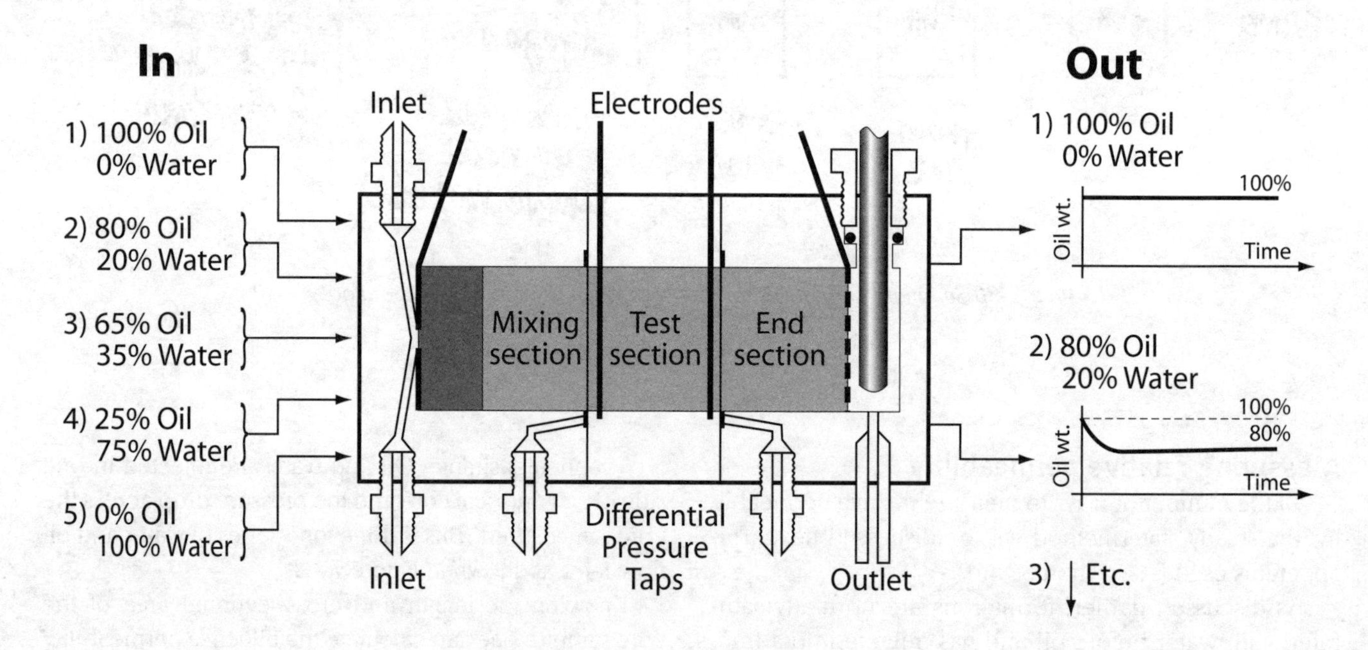

Fig. 7–32 Steps for Steady State Test

Effect of connate water saturation

The test starts with injection of 100% oil. The measured permeability obtained will be different from the absolute permeability of the sample with only single-phase fluid. The reason for this is shown in Figure 7–33.

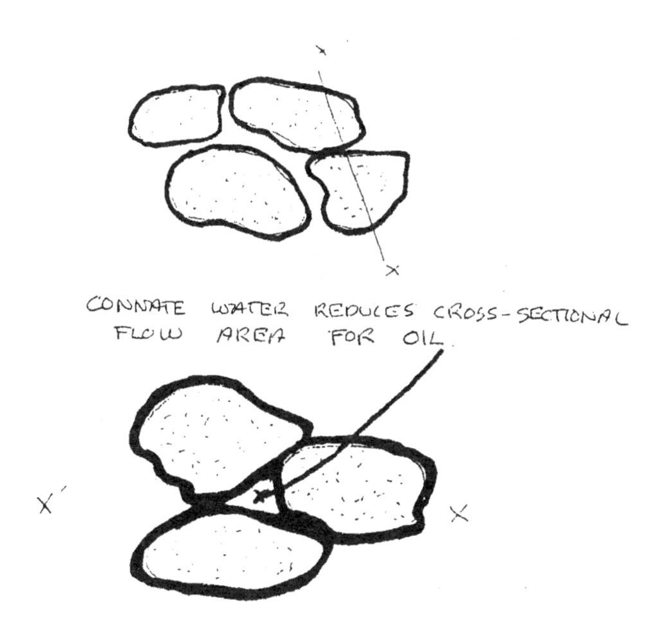

Fig. 7–33 Effect of Connate Water Saturation on Flow Area

The connate water saturation, which is attached to the rock grains via wetting, obscures part of the port throats. Therefore, there is less cross-sectional area for oil flow.

Definition of relative permeability

For the time being, the relative permeability is defined as shown in Equation 7.3.

$$k_{ro} = -\frac{Q_{oil}\mu_{oil}}{k_{absolute}A\dfrac{\partial p_{oil}}{\partial x}} \tag{7.3}$$

This is used in Darcy's Law, which is then applied to each phase separately.

Oil:

$$Q_{oil} = -\frac{k_{absolute}k_{ro}A}{\mu_{oil}}\frac{\partial p_{oil}}{\partial x} \tag{7.4}$$

Water:

$$Q_{water} = -\frac{k_{absolute}k_{rw}A}{\mu_{water}}\frac{\partial p_{water}}{\partial x} \tag{7.5}$$

Gas:

$$Q_{gas} = -\frac{k_{absolute}k_{rg}A}{\mu_{gas}}\frac{\partial p_{gas}}{\partial x} \tag{7.6}$$

Total flow can then be calculated by adding the flow from all three phases.

Development of relative permeability curve

Returning to the experiment being described, the results can then be plotted using the relative permeability defined previously and shown in Figure 7–34. The total flow of water and oil results in a reduction in total relative permeability. Restated, multiple phases interfere with each other and less flow occurs than if only one phase were flowing.

A number of key points can be derived from this graph. The total permeability, k_{ro} and k_{rw}, sum to a total permeability less than one at the end of the test—on the left side.

Residual oil saturation

Oil relative permeability starts at the connate water saturation. The k_{ro} decreases with increased water saturation. The k_{ro} curve stops at the residual oil saturation, which is known as S_{or}. For a water-wet rock, this represents discontinuous drops of oil trapped in the middle of pores. These are actually trapped by capillary pressures. This is shown in Figure 7–35.[18]

Because of the drops of oil trapped in the middle of the pores, the relative permeability to water at the endpoint is considerably less than the permeability to single-phase oil or water.

Movable oil saturation

The movable oil saturation is the difference between the residual oil saturation and the oil saturation at initial conditions. The original oil saturation is $1 - S_{wc}$. Hence, the movable oil saturation is then $1 - S_{wc} - S_{or}$. This is a commonly used term when discussing waterflooding, which is discussed later in the chapter.

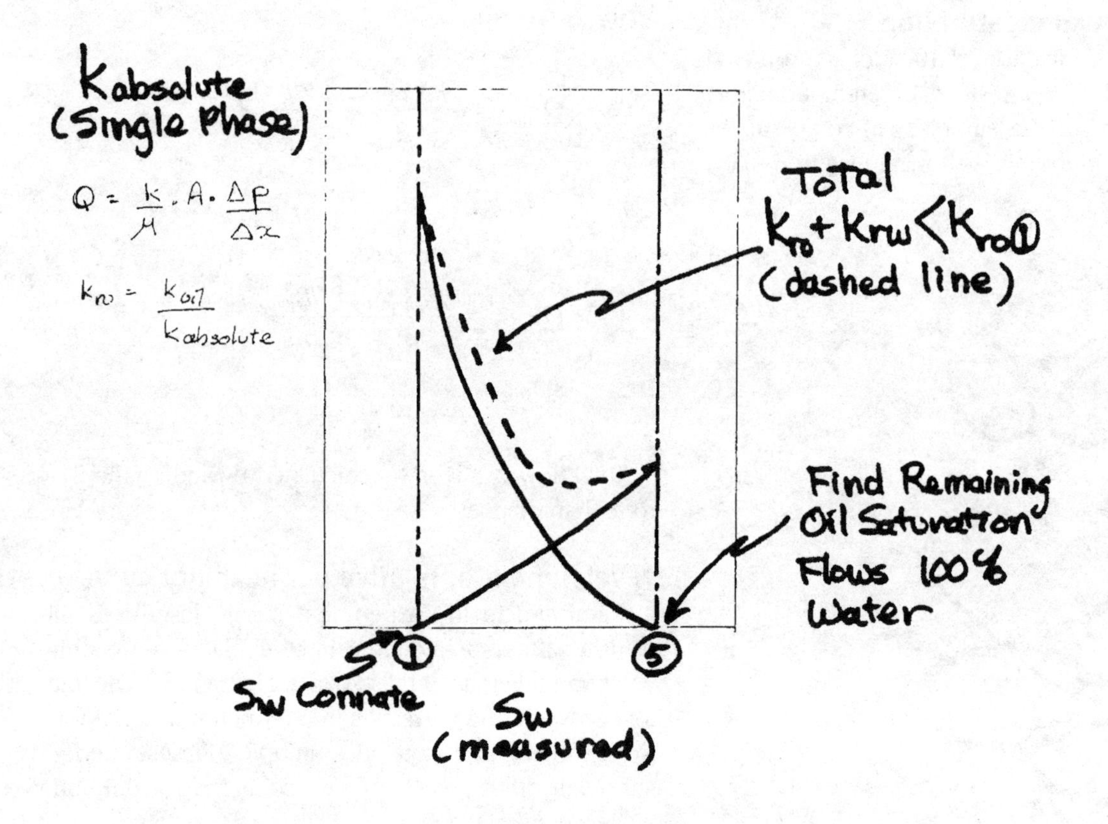

Fig. 7–34 Generalized Results of Relative Permeability Test

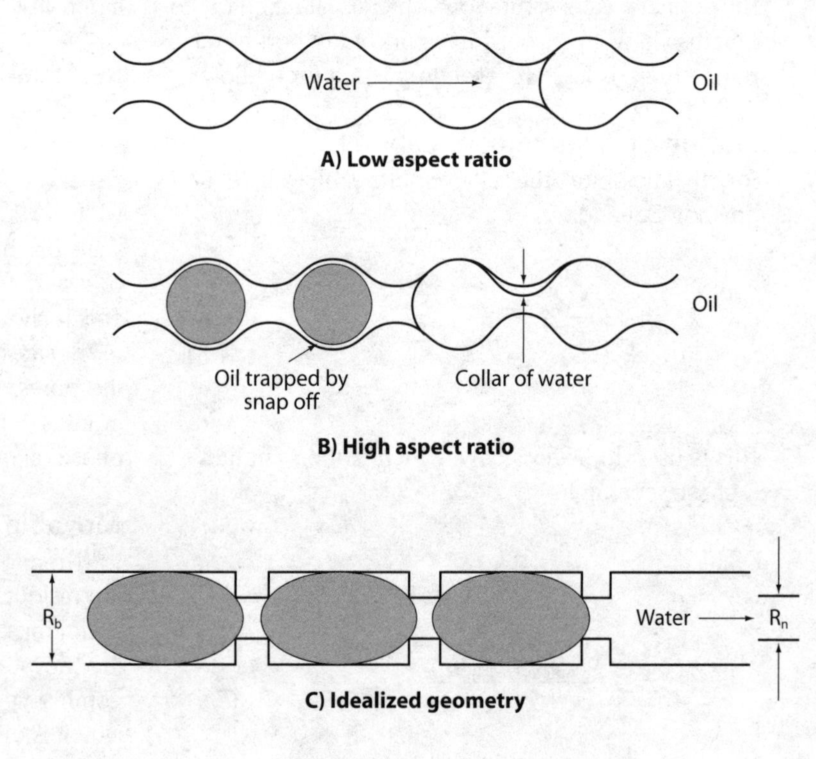

Fig. 7–35 Trapped Oil Saturations

Relative permeability convention

Two relative permeability conventions are important. The curves can be presented in two styles, which are shown in Figures 7–36 and 7–37.

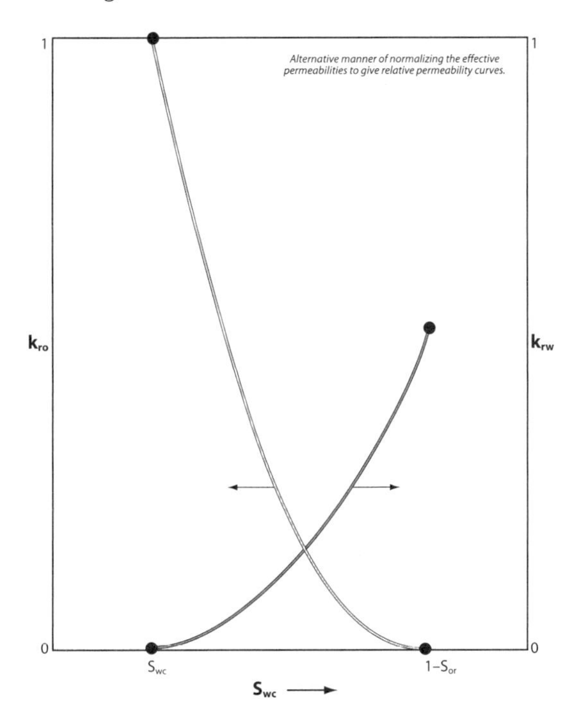

Fig. 7–36 Relative Permeability Convention I

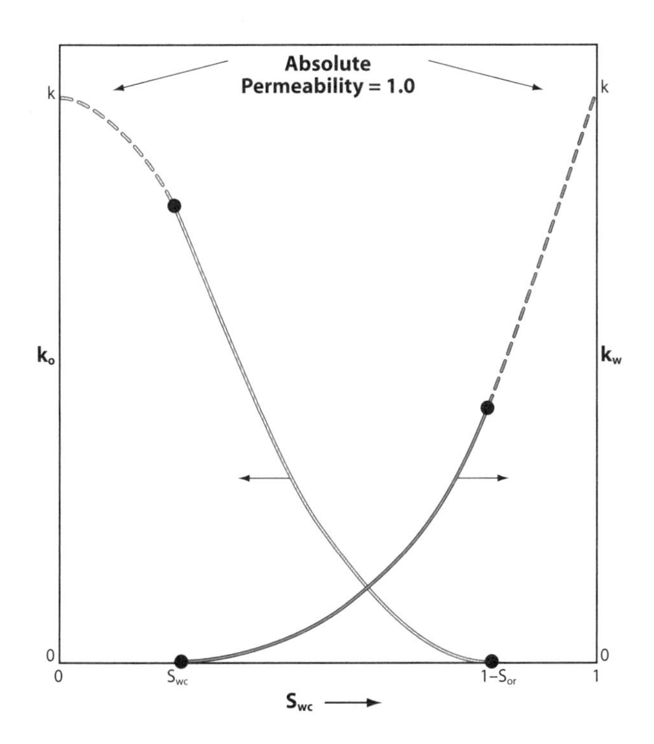

Fig. 7–37 Relative Permeability Convention II

The first convention is the one favored by most commercial labs. The second is the style favored by some operating company labs. In previous sections of this chapter, it was mentioned there needs to be a correction for the presence of the connate water saturation on the edge of the pore spaces. The net result is that the cross-sectional area, at which the first relative permeability to oil is measured, is less than that available to flow in the rock with only one phase. This correction is normally about 20%.

Why do the major commercial labs present the data using one as the end-point? It makes using the data in calculations easier, such as fractional flow curves and three-phase relative permeability. Although the correction for the effect of connate water saturation is useful information, most permeability maps are generated from air permeabilities. The end-point correction is small in comparison to the liquid-air adjustment, which is seldom measured. In other words, reservoir calculations are usually made on air permeability data, which is corrected approximately. Little accuracy is gained by precise knowledge of the end-point adjustment.

Water-wet and oil-wet rocks

Experience has shown water-oil relative permeability data can be broadly classified into two types, which are shown in Figure 7–38.

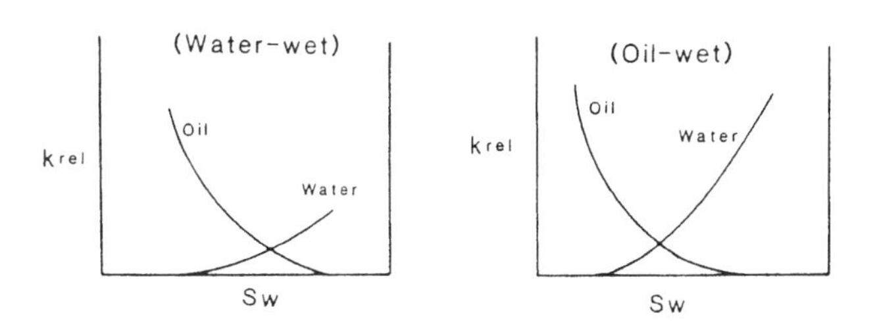

Fig. 7–38 Water-Wet vs. Oil-Wet Relative Permeability Relations

The two types of curves are caused by the wetting characteristics of the oil and the rock minerals. This depends not only on the mineralogy but also upon the characteristics of the oil. In broad terms, the two curves (k_{ro} and k_{rw}) will cross at less than a 50% water saturation (S_w) for an oil-wet system and at more than 50% S_w for a water-wet system. The end value of the water relative permeability is much higher for oil-wet systems. The dividing line is a k_{ro} of approximately 0.3 to 0.4. Rocks do not have to be either one or the other. There is a range of characteristics. Relative permeability data between these two extremes is often referred to as having intermediate or mixed wettability.

Terminology convention

As with capillary pressure, a convention has been developed to describe the direction in which saturation changes are occurring. The convention is as follows:

Imbibition: wetting saturation increases
Drainage: wetting saturation decreases

Due to the shape of pore spaces, different relative permeability curves are obtained if the direction of fluid movement changes.

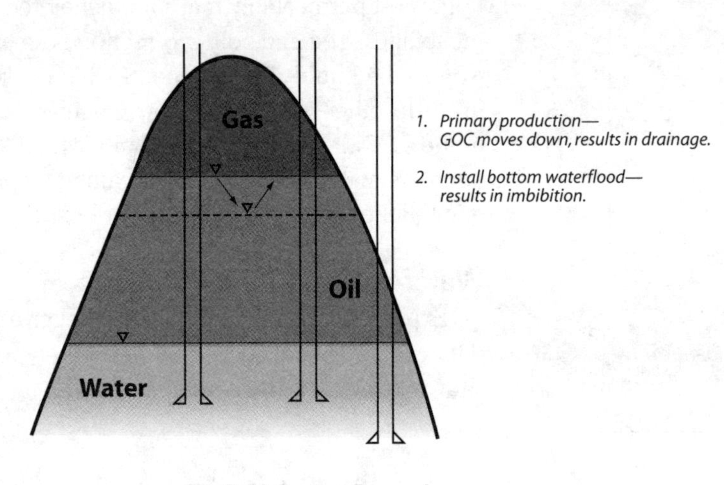

Fig. 7–39 Contact Reversal

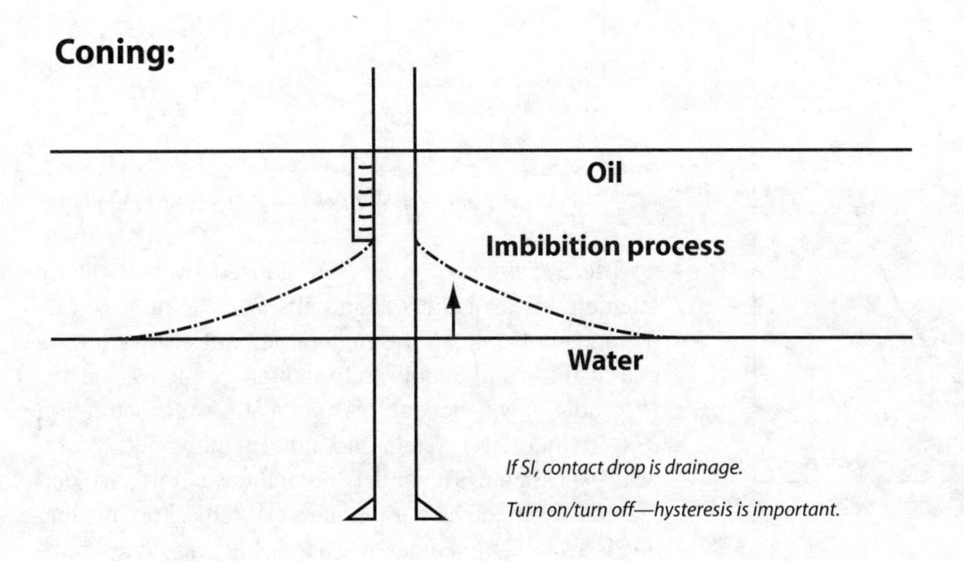

Fig. 7–40 Coning with Varying Rates

Hysteresis

Very often, the direction of saturation changes is not the same in the reservoir. There are also a number of occurrences when the direction of saturation changes reverses. Two such situations are shown in Figures 7–39 and 7–40.

Gas storage reservoirs

The effect on a gas storage reservoir is shown in Figure 7–41.[19] The lower axis is the amount of gas in the reservoir, and the left axis is the surface shut-in pressure. One of the major design concerns for gas storage is how much gas is available and for how long. Gas storage schemes are usually designed to provide short-term peak capacity, which a long pipeline may not provide. The use of storage can also result in cost savings. The storage reservoir is filled in the summer, when gas prices are historically low, and then produced during the winter when gas prices are traditionally higher. The difference in acquisition costs provides the cash to run the storage project. The variation in pressure represents changes in the water level of an underlying aquifer and the varying saturations occurring in the transition zone. As may be observed, the reservoir pressure varies considerably depending on the size and number of the production-injection cycles.

Scanning curves—accounting for hysteresis

It is possible to adjust for these effects by using scanning curves, as shown in Figure 7–42. These are usually part of a simulation input. Analytical techniques generally do not lend themselves to such adjustments.

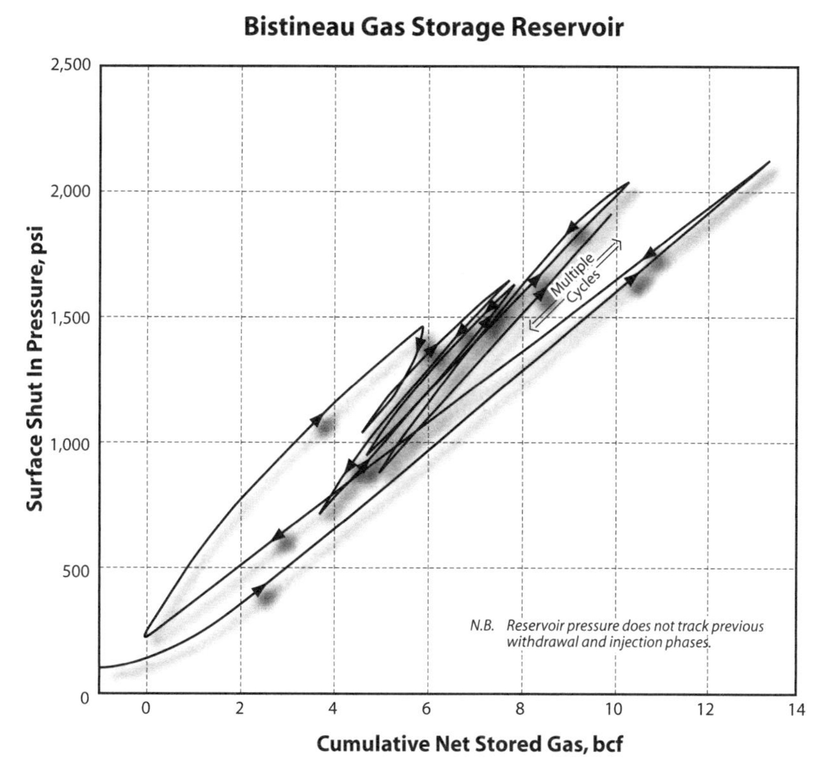

Fig. 7–41 Gas Storage Scheme

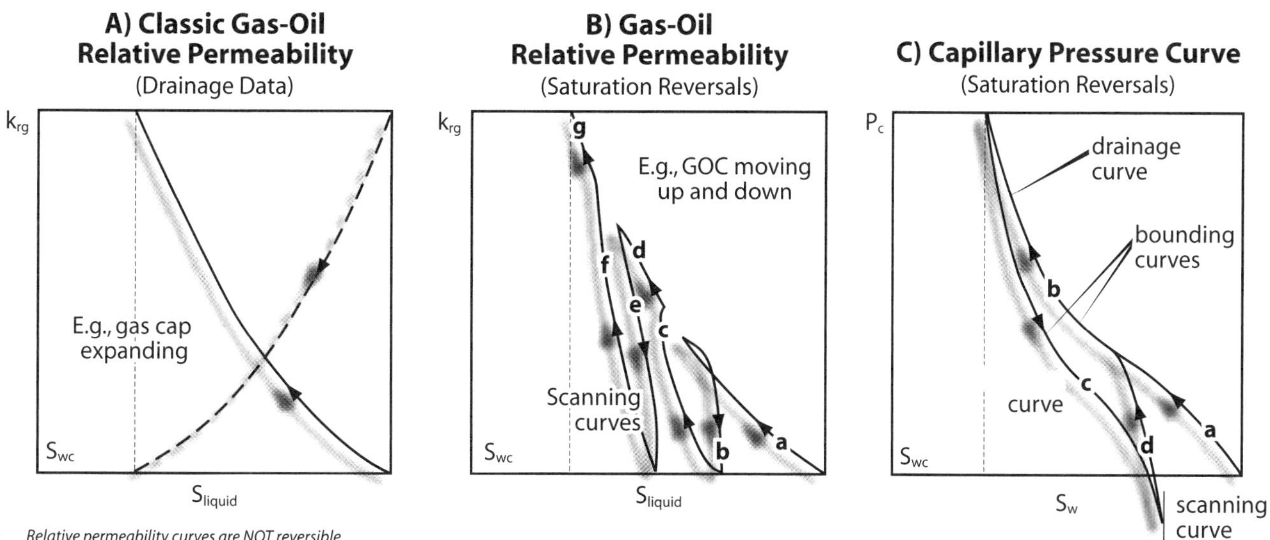

- *Relative permeability curves are NOT reversible*
- *Hysteresis affects nonwetting phase most strongly, wetting phase curves are not adjusted*

A. Saturations change consistently, S_{liquid} decreases.
B. Relative permeability for nonwetting phase takes a new path. These can be estimated using different methods, which produce scanning curves.
C. Capillary pressure is different with saturation changing in different directions. For P_c, scanning curves interpolate between bounding curves.

N.B. Most gas-liquid relative permeability is done with gas-oil drainage tests—with no connate water present. Note both oil and water are considered wetting phases and gas-liquid behavior is considered to dominate capillary pressure. The end-point irreducible oil saturation may not represent $(S_w + S_{or})$ required for three-phase relative permeability. In practice, a manual adjustment will be required to end-point of $(S_w + S_{or})$. The author knows of no procedure in the literature: $(S_w + S_{or})$ can be obtained from water-oil relative permeability. Force k_{rg} end-point to 1.0 at S_w and estimate k_{rg} at $(S_w + S_{or})$.

Methods: Killough, J.E.; Carlson, F.M.; Jargon, J.

Fig. 7–42 Scanning Curves for Relative Permeability

Screening two-phase relative permeability data

Special core analysis, such as relative permeability data, is relatively rare. Some reservoirs, therefore, have only one set of data. There are other times, although rarely, where numerous sets of data will be available. In this case, it is necessary to somehow generate a representative average.

One of the immediate difficulties is most data will have different end-point saturations. It is very helpful to summarize the data available in a table as shown in Table 7–3.

With this kind of data, it is possible to arithmetically average residual oil saturations as well as connate water saturations. It is also possible to group the data by test type.

Connate water saturation data

Another alternative to a table is to take all special core studies from one reservoir and use a histogram. This was done with an electronic spreadsheet as shown in Figure 7–43. The most common water saturation is in the area of 15%. There are a number of points existing from 30–75% and likely represent poorer quality reservoir samples. An end-point connate water saturation was selected from this graph to be 15%.

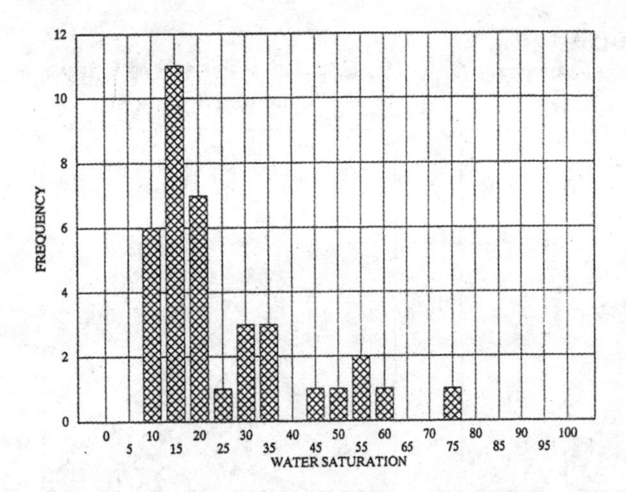

Fig. 7–43 Data Screening Histogram—Water Saturation

Determination of representative samples

It would be also appropriate to make up similar histograms for porosity and permeability, as shown in Figures 7–44 and 7–45. This enables one to determine how representative the data is.

Plot lab data

The first step the author normally takes is to plot all of the data on a linear graph. If there is much data, curves may need to be plotted individually. An example of this is shown in Figure 7–46. This will normally screen out any wiggles and other data indicating a bad or fractured sample.

Next, plot all of the data on a single graph, as shown on Figure 7–47. This is an oil-wet sample, and it is rather hard to make out the range of oil relative permeability curve data.

Plot all of the data on a log scale, as shown on Figure 7–48. With the log scale, all of the oil relative permeability curves become clear. Not all relative permeability data requires the use of a log scale.

Finally, plot k_{rw}/k_{ro} on a log scale, as shown on Figure 7–49. The slopes and shapes of the various curves are very similar, although they are shifted up and down the page.

Again, in general, the end-points of different data sets will be different.

Normalizing relative permeability curves

In order to evaluate and compare the shape of curves, it is possible to adjust the curves to extend only between S_{wc} and S_{or}. This is a simple linear translation. The standard practice is to rescale the data to span the movable hydrocarbon range. This is denoted with S* and is transformed as follows

$$S^*_{n,\,w\text{-}o} = \frac{(S_w - S_{wc})}{(1 - S_{wc} - S_{or})} \tag{7.7}$$

$$S^*_{n,\,g\text{-}o} = \frac{S_g}{(1 - S_{wc} - S_{or})} \tag{7.8}$$

The resulting series of graphs are shown in Figures 7–50, 7–51, 7–52, and 7–53. While this is a common practice, it appears to have no theoretical justification. Still, it is the only alternative available.

Which plot to use

Most people screen on linear normalized data. Others, such as Steven Ko, recommend the data be screened on k_{ro}/k_{rw} ratios. In the author's opinion, the answer is both. Well productivity is controlled by the absolute level of k_{ro} and k_{rw}. This will affect rates directly. Water cuts will be affected by the ratio of k_{ro} to k_{rw}. The consistency of the slopes is a good quality-control indicator.

Table 7–3 Different End-Point Saturations

Lab	Location	Sample	Date	Type	K_{air}	K_{liq}	Phi	S_{or}
Core Labs	6-25-50-10W5M	157	78-10-25	refined oil	16	13	0.132	0.464
Core Labs	6-25-50-10W5M	158	78-10-25	refined oil	53	12	0.147	0.521
Core Labs	6-25-50-10W5M	164	78-10-25	refined oil	20	10	0.139	0.412
Core Labs	6-25-50-10W5M	165	78-10-25	refined oil	53	36	0.128	0.332
Core Labs	6-25-50-10W5M	180	78-10-25	refined oil	81	50	0.118	0.402
Core Labs	6-25-50-10W5M	306	78-10-25	refined oil	2750	2,700.00	0.23	0.212
Core Labs	6-25-50-10W5M	323	78-10-25	refined oil	5.3	0.7	0.148	0.419
Core Labs	6-25-50-10W5M	361	78-10-25	refined oil	21	2.5	0.09	0.361
Core Labs	6-25-50-10W5M	362	78-10-25	refined oil	557	137	0.168	0.461
Core Labs	6-25-50-10W5M	390	78-10-25	refined oil	83	55	0.129	0.506
Core Labs	10-04-51-9W5M	77	78-10-25	refined oil	30	19	0.144	0.497
Core Labs	10-04-51-9W5M	124	78-10-25	refined oil	43	20	0.185	0.143
Core Labs	10-04-51-9W5M	207	78-10-25	refined oil	56	10	0.098	0.411
Core Labs	14-26-51-09W5M	15A	78-10-25	refined oil	25	16	0.107	0.411
Core Labs	14-26-51-09W5M	21A	78-10-25	refined oil	8.8	1.5	0.086	0.537
Core Labs	14-26-51-09W5M	37A	78-10-25	refined oil	4.9	1.4	0.088	0.406
Core Labs	14-26-51-09W5M	52A	78-10-25	refined oil	12	9.8	0.144	0.256
Core Labs	14-26-51-09W5M	64A	78-10-25	refined oil	51	25	0.071	0.479
Core Labs	14-26-51-09W5M	65A	78-10-25	refined oil	2.2	0.5	0.082	0.499
Core Labs	14-26-51-09W5M	73A	78-10-25	refined oil	11	7.3	0.147	0.587
Core Labs	14-26-51-09W5M	96A	78-10-25	refined oil	9.8	3.1	0.066	0.493
Core Labs	14-26-51-09W5M	110A	78-10-25	refined oil	77	59	0.078	0.39
Core Labs	14-26-51-09W5M	161A	78-10-25	refined oil	171	134	0.154	0.646
Core Labs	06-25-50-10W5M	165	79-07-23	ref. to res.	47	30	0.128	0.362
Core Labs	06-25-50-10W5M	180	79-07-23	ref. to res.	97	55	0.114	0.566
Core Labs	14-32-49-11W5M	81-1	79-09-28	ref. to res. _&T	59	45	0.117	0.494
Core Labs	14-32-49-11W5M	279-1	79-09-28	not reported	8.2	0.4	0.077	n/a
Core Labs	14-32-49-11W5M	284-1	79-09-28	ref. to res. _&T	1.2	0.4	0.097	0.392
Core Labs	14-32-49-11W5M	81-2	79-09-28	refined oil	117	105	0.115	0.429
Core Labs	14-32-49-11W5M	279-2	79-09-28	refined oil	4.9	0.1		0.418
Core Labs	14-32-49-11W5M	284-2	79-09-28	refined oil	1.1	0.5	0.094	0.338
Core Labs	06-12-50-11W5M	8A	81-01-15	ref. to res.	3.2	0.3	0.113	0.342
Core Labs	04-35-49-11W5M	16A	81-01-15	ref. to res.	86	40	0.112	0.476
Core Labs	04-35-49-11W5M	24A	81-01-15	ref. to res.	26	1.8	0.08	0.463
Core Labs	04-35-49-11W5M	41A	81-01-15	ref. to res.	21	0.7	0.104	0.448
Core Labs	06-12-50-11W5M	71A	81-01-15	ref. to res.	106	83	0.091	0.166
Core Labs	06-12-50-11W5M	160A	81-01-15	ref. to res.	551	415	0.182	0.446
Core Labs	04-35-49-11W5M	139A	81-01-15	ref. to res.	102	68	0.12	0.545
Core Labs	06-12-50-11W5M	240A	81-01-15	not reported	158	16	0.136	n/a

Influences on relative permeability

A number of factors affect two-phase permeability. Wettability is accepted as having a strong effect. Saturation history (imbibition versus drainage) is also known to have a definite effect on relative permeability. The effect of interfacial tension is less clear; some studies indicate sensitivity while others do not. Viscosity has been shown to have an effect in many studies; however, agreement on this is not universal. Many studies indicated other factors, such as interfacial tension and wettability, which correlate with viscosity, were governing factors. Relative permeability is also affected by the chemical nature of the crude oil. Polar compounds in the crude oil typically affect the shape and type of the relative permeability curve.

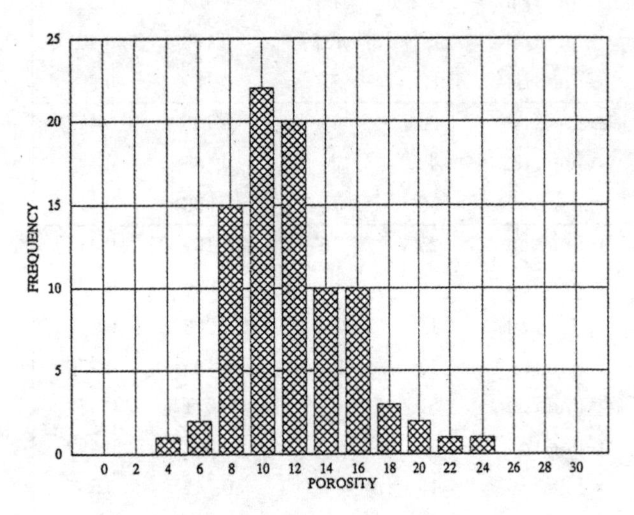

Fig. 7–44 Data Screening Histogram—Porosity

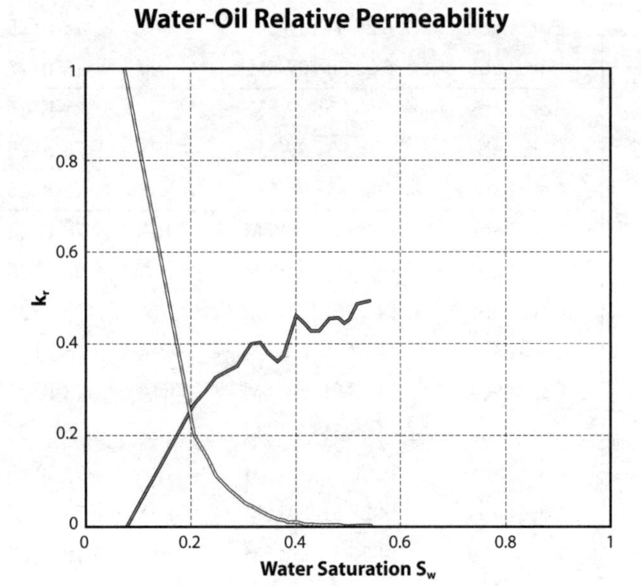

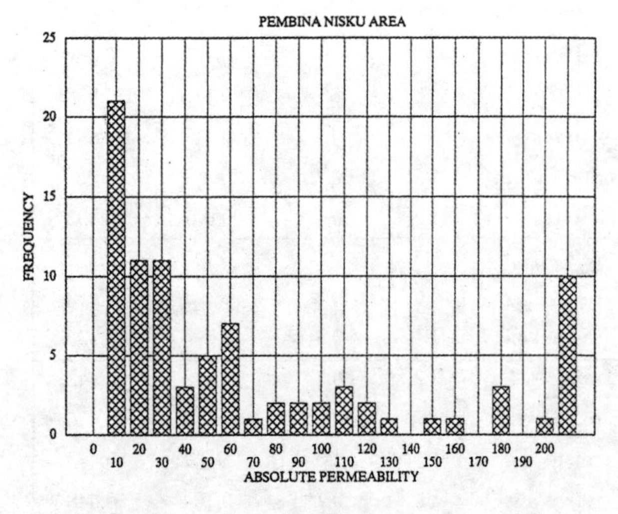

Fig. 7–45 Data Screening Histogram—Permeability

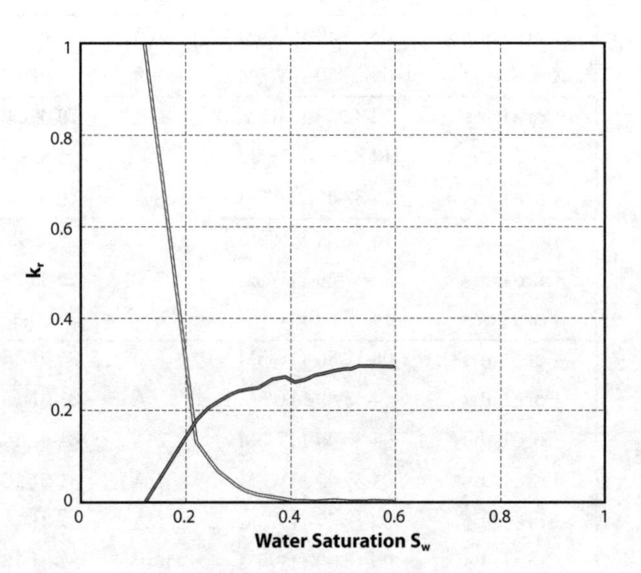

Fig. 7–46 Examples of Individual Relative Permeability Curve Quality

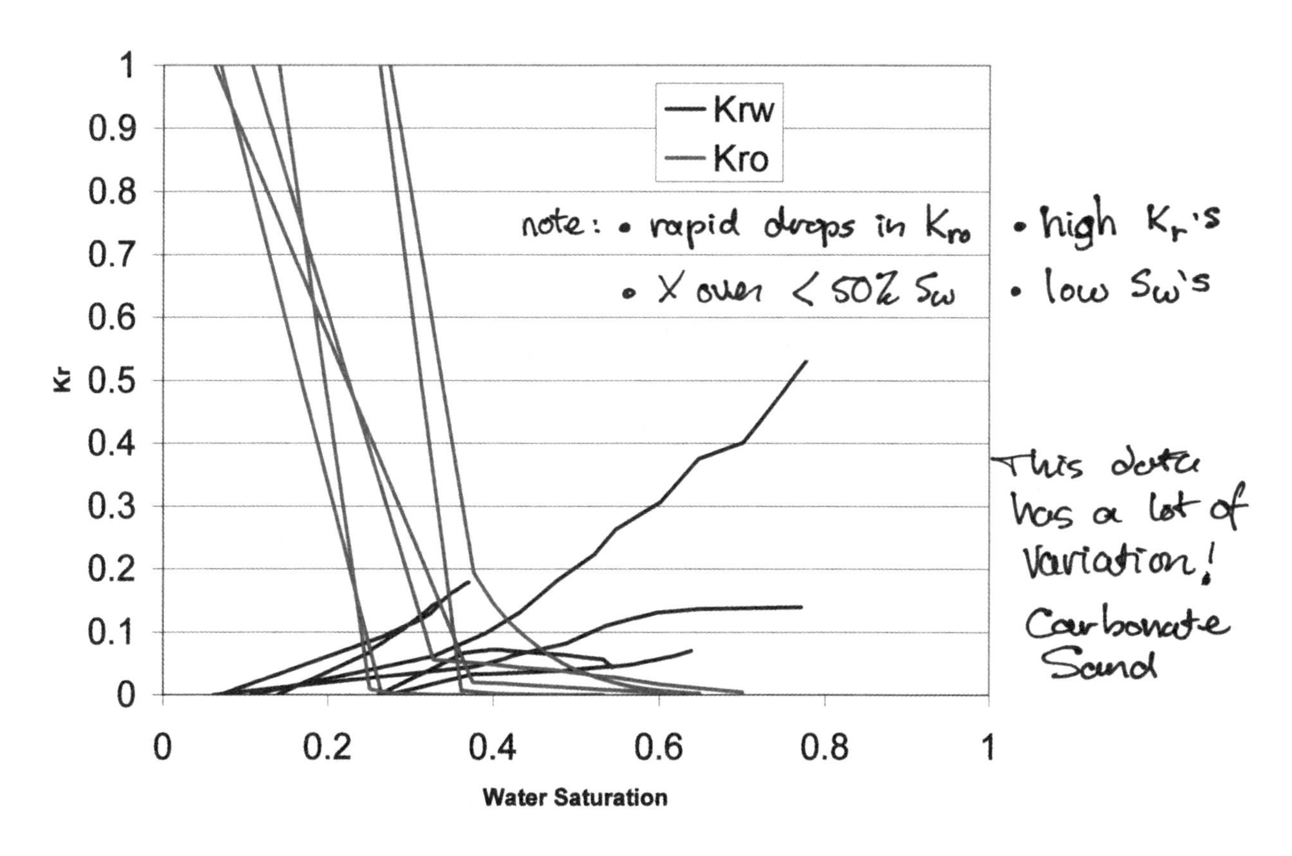

Fig. 7–47 Oil-Water Relative Permeabilities—Field Data Showing Use of Linear Scale

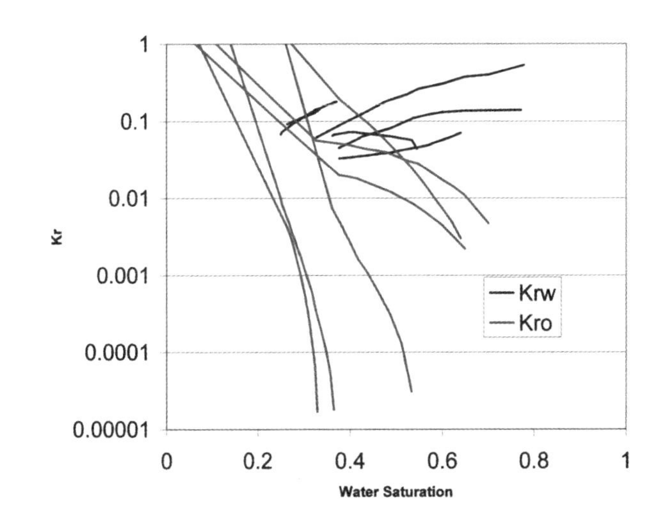

Fig. 7–48 Oil-Water Relative Permeabilities—Field Data Showing Use of Log Scale

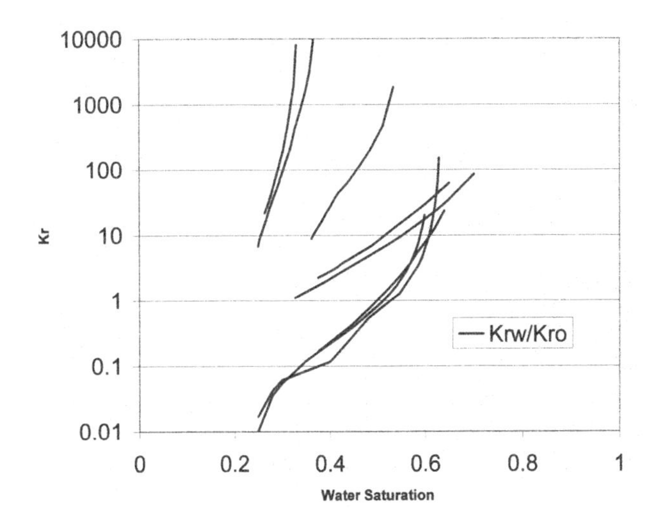

Fig. 7–49 Use of k_{rog}/k_{ro} Ratio—Field Data, Log Scale

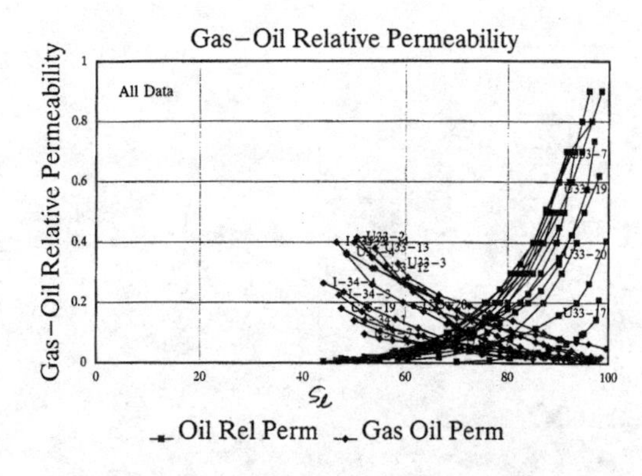

Fig. 7–50 Gas-Oil Relative Permeability vs. S_{liquid} *Showing Raw Saturations*

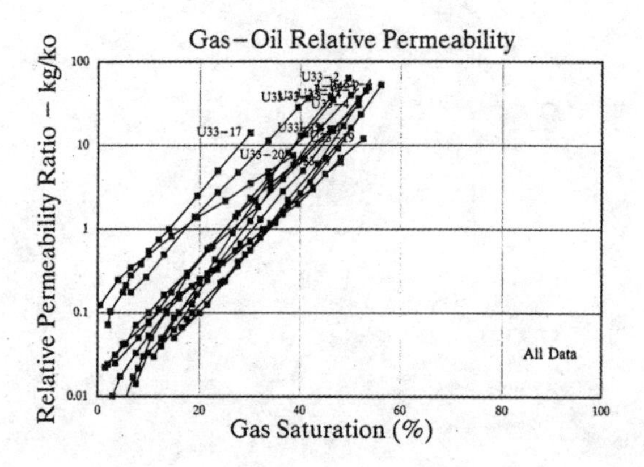

Fig. 7–52 k_{rg} / k_{ro} *vs.* S_l

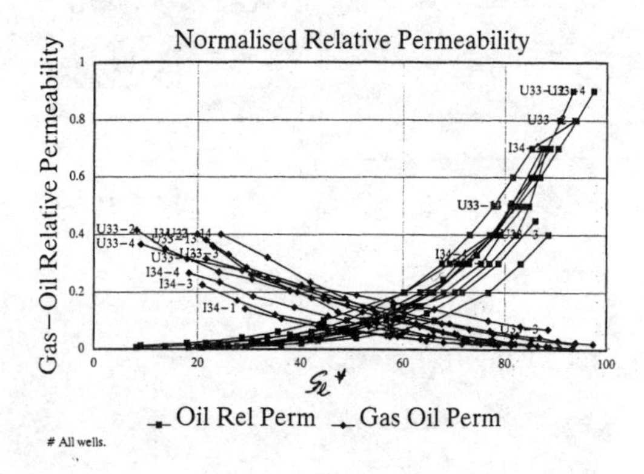

Fig. 7–51 Gas-Oil Relative Permeability vs. S_{liquid} *Showing Normalized Saturations*

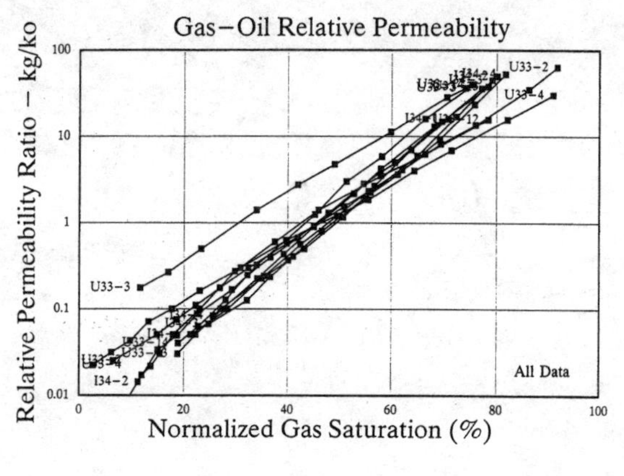

Fig. 7–53 k_{rg} / k_{ro} *vs.* S_l

Waterflood or displacement tests

Earlier in this chapter, relative permeability was introduced using the steady state test. In fact, the majority of data is obtained using displacement or waterflood tests. The primary reason for this is expense. Steady state tests can take several months to perform, whereas waterflood tests can be completed in about a week.

The same kind of holder is used as for the steady state test. However, water is simply injected in one end, and the effluent is measured. These tests require considerable interpretation to generate relative permeability curves. In addition, there are some limitations. The waterflood tests only give relative permeability information after the waterflood front has passed through. Therefore, a full range of relative permeabilities is not provided.

The technique traditionally utilized is graphically based and is usually called the JBN method.[20] The technique is quite sensitive and must be done with great care. Relative permeability is also affected by capillary pressure end effects. This will be discussed in a subsequent section.

One implication of the sensitivity of the graphical technique and the need to include capillary pressure is simulation itself is now being used to interpret relative permeability tests.

Simulation to interpret relative permeability tests

The potential to use a history match and to include capillary pressure effects are potential advantages to using a reservoir simulator. Some reservoir simulators have lab scale units. There is a trade-off here. For the displacement tests, the shape of the relative permeability curve must be assumed beforehand. Normally this means the use of a power law type curve. While this is often representative of real data, there is no requirement this be the case. For this reason, a history-matched relative permeability curve is not necessarily unique. Actually, neither were the graphical interpretations. Steady state tests are the preferred lab data.

The capillary pressure input used in a core flood model should be of the imbibition type, which is rarely obtained. Although an attractive feature, it means more expense and work to measure this data first.

Much of the early history-matched curves did not assume any capillary pressure relation, and so the curves were, and sometimes still are, developed without the effect of capillary pressure. This is rarely documented in the lab report.

One might suspect the quality of the history match might be a good quality indicator of the resultant relative permeability curve. This seems reasonable; however, the relative permeability test reports that utilized this technique have not included history-match plots. Perhaps this will appear in the near future.

Clearly, lab relative permeability curves from displacement tests are subject to more interpretation than the author first realized. Buyer beware!

Capillary pressure end effects

Capillary pressure end effects are not new. The solution to capillary pressure end effects is to operate waterflood tests at higher rates and to use longer samples. Some indicator ratios can be used as rules of thumb for quality control. Unfortunately, it is rare to see these issues actually written up in a relative permeability report.

The use of higher rates raises some questions as to whether the displacement occurring in the test is representative of what is happening in the reservoir. Evidence suggests the increased displacement rates are not sufficiently different to provide significantly different answers. Although there is no reason to doubt this is the case, conclusive and wide-ranging proof appears to be lacking.

Then again, any deviations could be explained by sample variations—a catch-22.

Stacked core samples

The desire to use longer samples and to obtain more samples has led to the use of stacked cores. Historically this practice was avoided due to the potential for capillary pressure end effects at the boundary of each sample.

Mechanical contact at the joints between core samples is not easy to obtain. It is possible to machine samples using a metal lathe. However, this is done at the risk of disturbing the pore structure. Normally, the samples utilize a piece of tissue paper to aid with continuity, as is the practice for capillary pressure tests.

At least one technical article indicates stacking core samples produces satisfactory results, but the author's experience with other stacked samples raises serious questions. Hence, this text does not recommend using stacked samples.

Refined oil versus live reservoir oil

Refined oil was once used on a considerable proportion of relative permeability tests. This practice is no longer common since the industry is much more sensitive to polarity and wettability.

At one time, it was thought most reservoirs were water wet. One of the questions raised by using refined oils was whether this apparent skew in character was due to the widespread use of refined oils.

Refined oil was used primarily to shift the fractional flow curve to the left and, by so doing, increase the range of saturations across which relative permeability was directly measured in waterflood displacement tests. Refined oils were chosen for their high viscosities and were normally designed to be as nonpolar as possible.

In general, live reservoir oils are probably a better choice. Again, so are steady states tests.

Wettability

The contact angle between reservoir fluids and the rock mineral grains can be modified by many drilling mud additives and core cleaning methods. Figure 7–54 shows these effects on capillary pressure data.

The topic can be nicely summarized from the index of *Interfacial Phenomena in Petroleum Recovery*, chapter 11, "Obtaining Samples with Preserved Wettability."[8]

Coring
Wettability alteration due to interactions with drilling fluids
Pressure and temperature changes
Storing
Laboratory Preparation
"No solvent" procedures
"Clean and restore" procedures
"No hysteresis" procedures
Recommendations for cleaning procedures
Experimental conditions

Probably the most interesting section from Morrow et al. concerns the mud system that must be used to achieve a sample with unaltered wetting characteristics. Key points:

1. All oil-based and emulsion muds tested altered water-wet samples so they became oil wet or significantly less water wet. This would have the effect of reducing S_{wirr} on later core tests.
2. Several components, such as CMC, bentonite, and lime in water-based muds, reversed the wettability of oil-wet samples. This analysis did not include obviously strong surfactant agents. Restated, if core had oil-wet tendencies, the laboratory values would be skewed to values of S_{wirr} that were too high.

A perception within the industry is that the best cores are cut with oil-based drilling fluids. Such oil-based cores were given the misnomer *native state* cores, probably because it was believed oil-based fluids maintained the original immobile water saturation in the core. However, as shown in item 1, the wettability of cores is usually altered by oil-based mud additives, giving potentially misleading values of S_{wirr}. This also precludes the designation *native state*.

A paper written by Tom Hamp in the *Journal of Canadian Petroleum Technology (JCPT)* is an excellent example of using oil-based cores.[21]

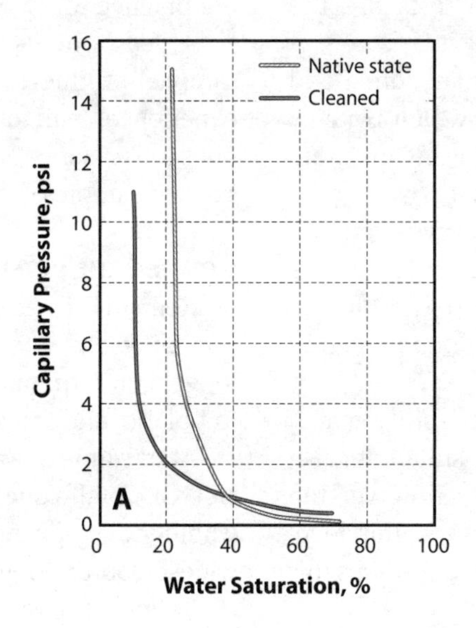

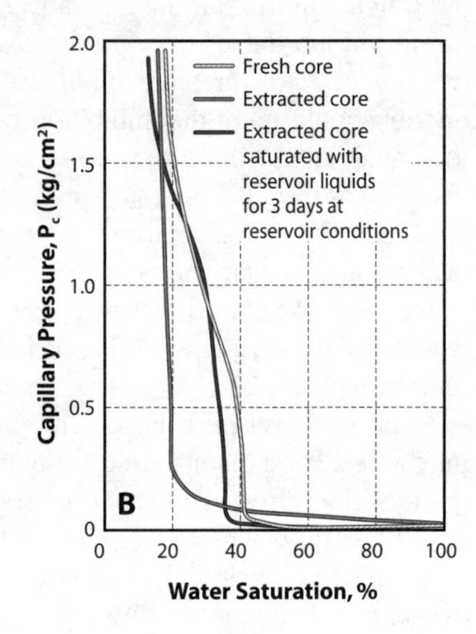

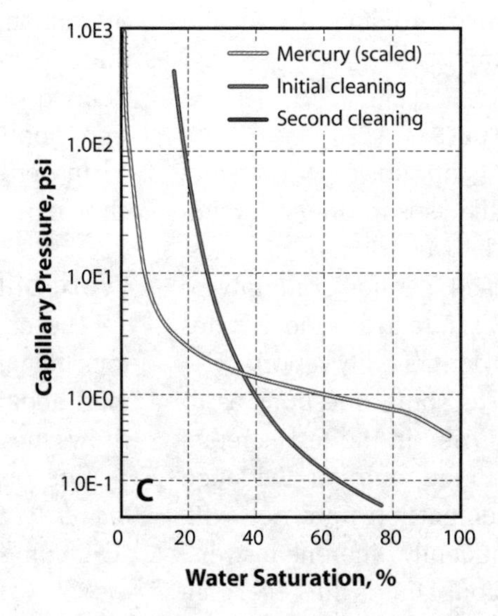

Effect of wettability on form of capillary pressure curves for reservoir core samples: A) native state and cleaned cores; B) fresh, extracted, and restored cores; C) mercury, initial cleaning, second cleaning.

Fig. 7–54 Effect of Wettability on Capillary Pressure

Clay dehydration

Most sandstones contain some clay. After cleaning with toluene and carbon dioxide, the samples are normally dried in an oven. If the oven temperature is set too high, the clays will be thermally altered. This depends on the type of clay present. It might be useful to ask the lab what temperature was used for drying on completed reports, if not specified.

It is also possible to be proactive if you are fortunate enough to be involved in obtaining the data before the study. It is common to *tight hole* service companies, for example, by withholding the XRD or thin section point counts along with the core sample on which you want to do a relative permeability test. Although this may identify problem clays, this oversight is not normally done deliberately. Most oil companies have a policy of not giving out data unless they must. If the service company doesn't ask the right question, then the appropriate precautions may not be taken.

Reasonable confidentiality

Is there more to be gained by being open with the service companies than robotically keeping everything confidential? This is worth asking at the development stage, when most land ownership issues have been resolved.

Will the service company use the knowledge gained from the samples on tests for other companies? Probably, but the funny thing is most companies look for labs with similar experience in selecting their service companies. This works both ways, and the company will gain if this is the practice in the industry. Does an oil company really gain from lower recoveries on their competitors properties? Perhaps the answer is yes; however, the most critical element of competition is in the exploration phase. It would be fair to say oil companies are expected to have some social responsibility for a nation's natural resources. Forgetting this often leads to heavy regulation and, frequently, nationalization.

Quality control

As outlined previously, actual data is obtained in multiple different ways. A high degree of expertise is required to effectively analyze this type of data. A number of years ago Core Labs produced *A Course in Theory, Application and Quality Control of Relative Permeability Measurements*. It was originally developed by Mr. Terry Wong and Mr. Ted Ellefson in Calgary and is strongly recommended.

Table 7–4 is extracted from the course notes. The summary is quite handy since it outlines rock properties affecting relative permeability across the top and variations in testing procedures along the left side. The table does not include the interpretation method, which was discussed in the previous section.

Correlations

If no relative permeability data is available, correlations are obtainable. These include correlations by Wylie, Corey, Land, Knopp, and others. The use of such techniques has a number of potential advantages. It produces smooth curves, which does not necessarily happen with experimental data. It draws on more data by virtue of its development based on a large sampling of many different reservoirs.

Overall, the author does not like the data from correlations. Plotted data, which appeared to have good averages on linear curves, did not fall in the reasonable range on other plots. Therefore, the author's preference is offset data, even if it is from a reservoir that may not be as close as desirable. These issues are demonstrated in the following sections.

Water-oil relative permeability

Returning to the example of representative data discussed with respect to capillary pressure for the Brazeau River and Pembina areas, the available water-oil relative permeability data was summarized in Table 7–3.

For this study, the water-oil relative permeability data was averaged by using an empirical correlation. The technique was extracted from *Relative Permeability of Petroleum Reservoirs*. The equations may also be found in Honarpour et al., "Empirical Equations for Estimating Two-Phase Relative Permeability in Consolidated Rock."[22] Some tuning was applied to the equations to accurately represent the Brazeau River and Pembina area reservoirs. The results are shown in Table 7–5.

The raw data and correlation results were plotted on a single graph as shown in Figures 7–55 and 7–56. The raw data curves with severe wiggles were deleted (two sets). This data was plotted on a k_{ro}/k_{rw} basis as shown in Figure 7–57. Plots were also made using normalized saturations as shown in Figures 7–58 and 7–59. Finally, k_{ro}/k_{rw} ratios were plotted against S_w* (normalized) as shown in Figure 7–60.

Table 7–4 Summary

Experimental Variables	Wet-tability	Rock Pore Structure	Fluid Properties	Saturation History	Flow Process
Field Handling					
• Mud, overbalance, preservation	✔			✔	
Preparation					
• Plug cutting, cleaning, and aging	✔	✔		✔	
• S_{wi}		✔		✔	✔
Test Conditions					
• Temperature	✔	✔	✔		
• Overburden and pore pressure		✔			
Fluids					
• Interfacial Tension			✔		✔
• Viscosity Ratio			✔		✔
• Density			✔		✔
Flow Method					
• Steady State			✔		
• Unsteady State			✔		
• Flow Rate			✔		

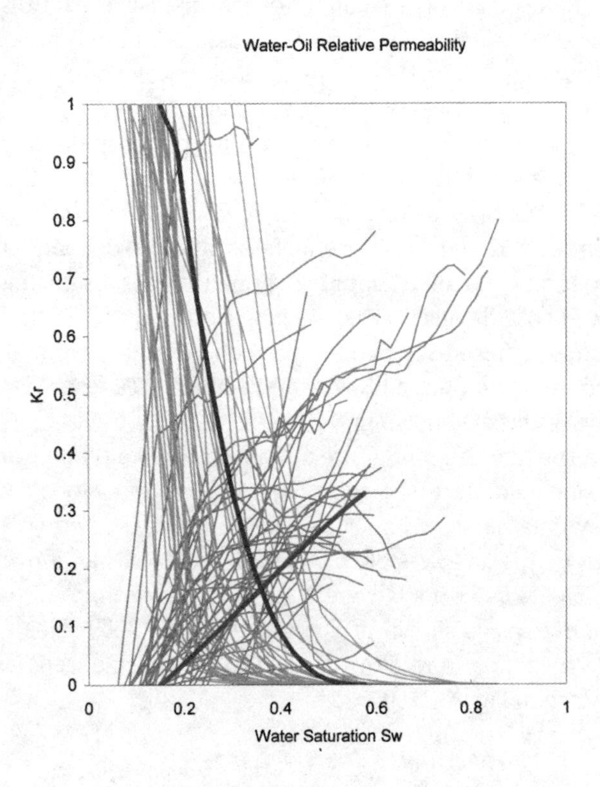

Fig. 7–55 Oil-Water Relative Permeability—Linear Scale

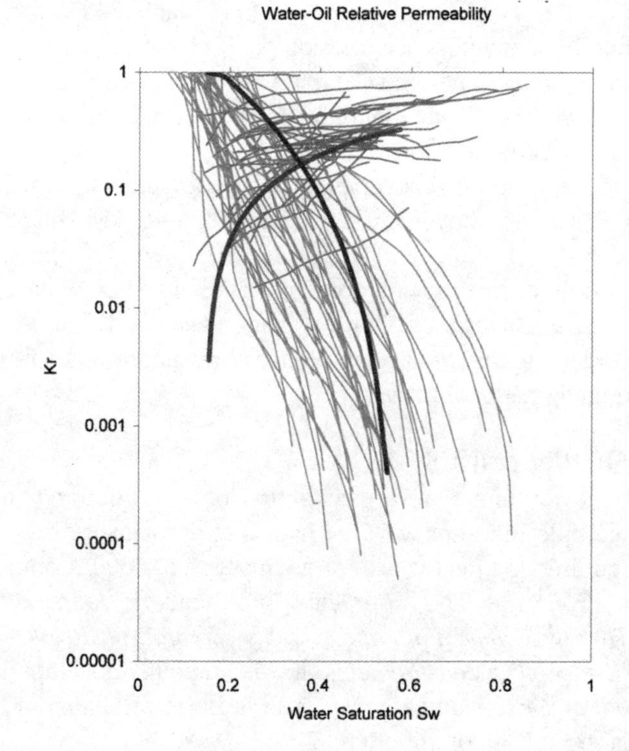

Fig. 7–56 Oil-Water Relative Permeability—Log Scale

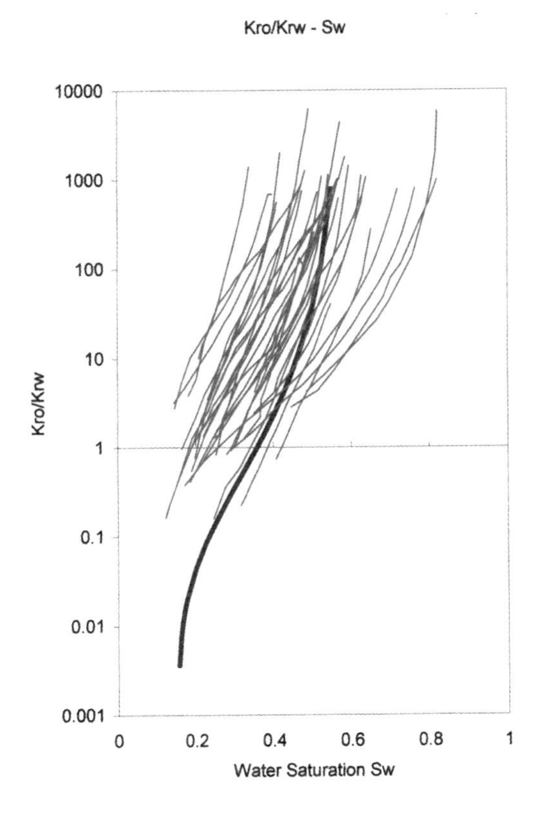

Fig. 7–57 k_{ro}/k_{rw} *Relative Permeability Relation*

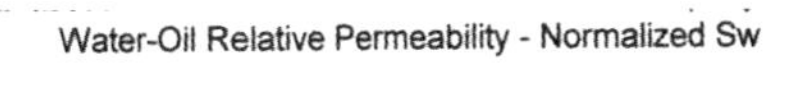

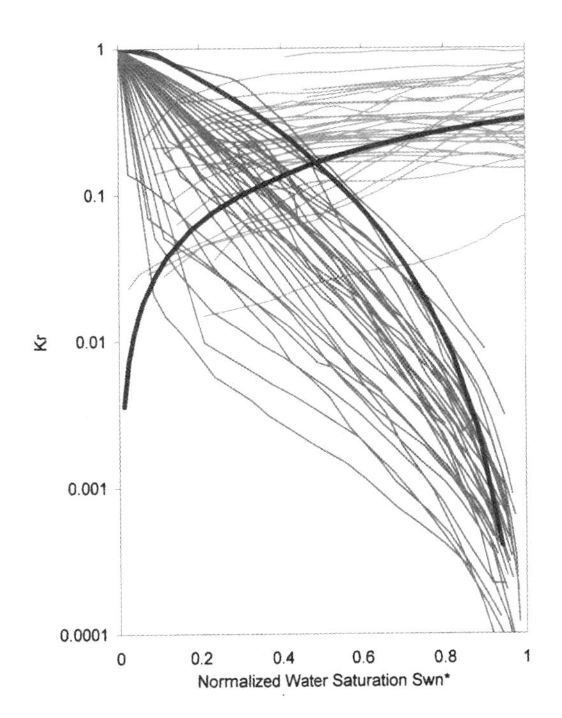

Fig. 7–59 Oil-Water Relative Permeability—
Log Scale vs. Normalized S_w

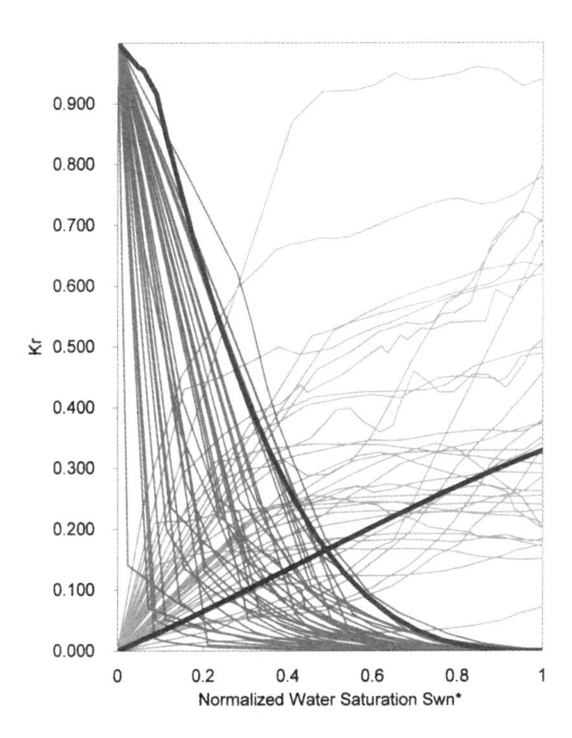

Fig. 7–58 Oil-Water Relative Permeability—
Linear Scale vs. Normalized S_w

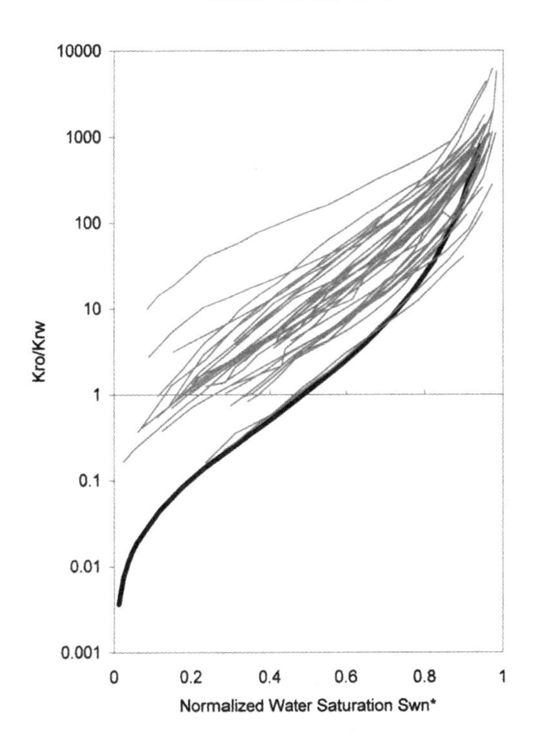

Fig. 7–60 k_{ro}/k_{rw} *Ratio vs. Normalized* S_w

Residual oil saturations were compiled in Table 7–5 from all of the water-oil relative permeability tests. The average residual oil saturation for 37 tests was 42.5%. This value was used as an end-point for the water-oil relative permeability data. This is a high value of residual oil saturation, which typically ranges from 20–35%, with the norm being about 25% or 30%. The relative permeability data indicates an oil-wet system and a high value of S_{or} is consistent with such a system.

Reservoir recovery is quite strongly affected by residual oil saturation, since it is a major control on total displaceable oil ($1 - S_w - S_{or}$) in a waterflood. For instance, for a reservoir with an irreducible or connate water saturation of 15% (typical for an oil-wet system), a change from an S_{or} of 42.5% to 25% increases volumetric displacement by 41.2%.

The empirically derived relationships appear to be fine when plotted against raw saturations. However, they are not as strong when plotted against the entire series of plots, which include normalized saturations and k_{ro}/k_{rw} ratios.

Gas-oil relative permeability

Data obtained from the special core data available for the Brazeau River and Pembina areas has been summarized in Table 7–6.

Gas-oil relative permeability was matched to an empirical correlation taken from Honarpour et al. The results are shown in Table 7–7.

The raw data and correlation results were plotted on a single graph as shown in Figures 7–61 and 7–62. The data has been presented as k_{ro}/k_{rg} ratios as shown in Figure 7–63. These last three graphs were then replotted using normalized saturations as shown in Figures 7–64, 7–65, and 7–66.

Again, the correlations did not match normalized data.

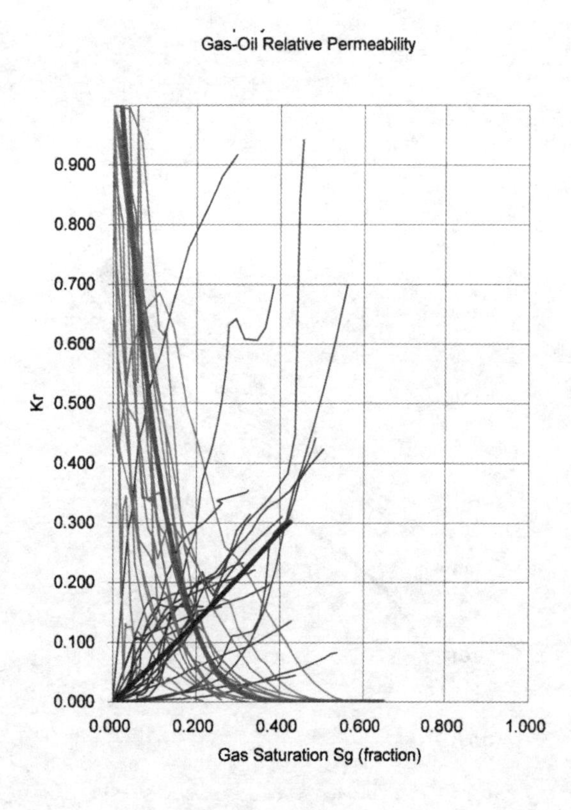

Fig. 7–61 Gas-Oil Relative Permeability—Linear Scale vs. S_l

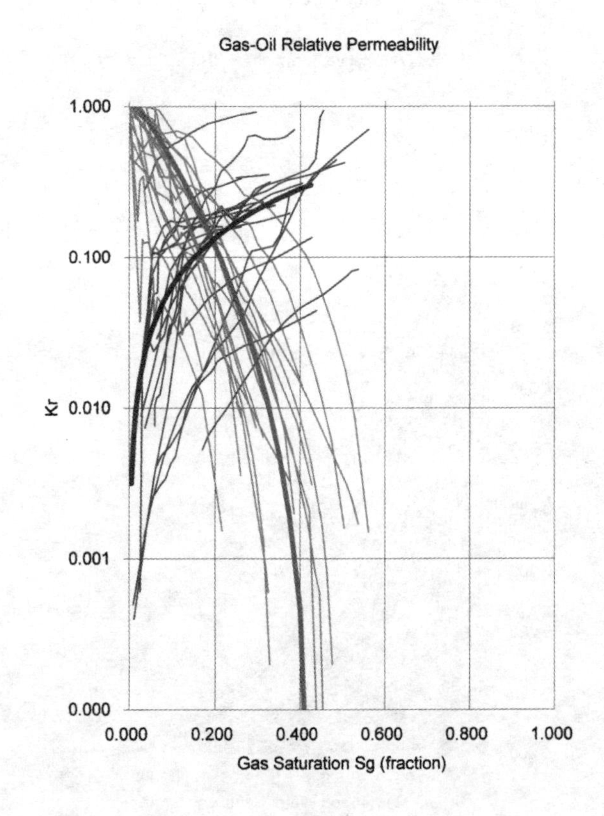

Fig. 7–62 Gas-Oil Relative Permeability—Log Scale vs. S_l

Table 7–5 Results

S_w	S_o	k_{rw}	k_{row}	k_{rw} (modified)
0.15	0.85	0	1	0
0.155	0.845	0.006	0.9999	0.0036
0.16	0.84	0.012	0.9999	0.0073
0.165	0.835	0.018	0.9999	0.0109
0.17	0.83	0.024	0.9999	0.0146
0.175	0.825	0.03	0.9999	0.0184
0.188	0.813	0.046	0.9162	0.0278
0.2	0.8	0.061	0.8315	0.0374
0.225	0.775	0.094	0.6779	0.0569
0.25	0.75	0.126	0.5445	0.0768
0.275	0.725	0.16	0.4299	0.0971
0.3	0.7	0.193	0.3326	0.1176
0.325	0.675	0.227	0.2513	0.1382
0.35	0.65	0.261	0.1844	0.1589
0.375	0.625	0.295	0.1305	0.1794
0.4	0.6	0.328	0.0884	0.1998
0.425	0.575	0.361	0.0564	0.2199
0.45	0.55	0.394	0.0333	0.2396
0.475	0.525	0.425	0.0175	0.2588
0.5	0.5	0.456	0.0078	0.2774
0.525	0.475	0.485	0.0025	0.2954
0.55	0.45	0.514	0.0004	0.3125
0.575	0.425	0.54	0	0.3288
0.6	0.4	0.565	0	0.3441
0.65	0.35	0.61	0	0.3712
0.7	0.3	0.646	0	0.3932
0.75	0.25	0.672	0	0.4092
0.8	0.2	0.688	0	0.4184
0.85	0.15	0.69	0	0.42
0.9	0.1	0.69	0	0.42
0.95	0.005	0.69	0	0.42
1	1	0.69	0	0.42

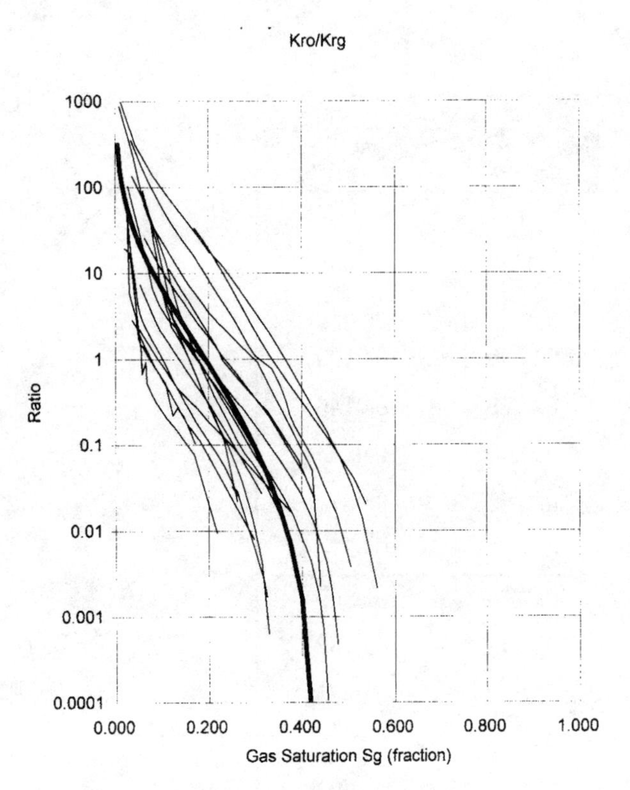

Fig. 7–63 k_{ro}/k_{rw} *Ratio vs.* S_l

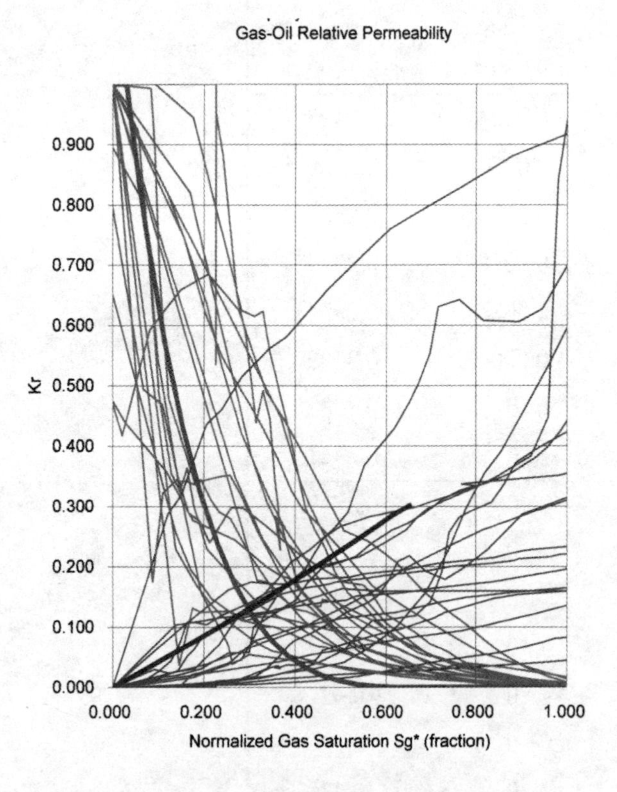

Fig. 7–64 Gas-Oil Relative Permeability—
Linear Scale vs. Normalized S_l

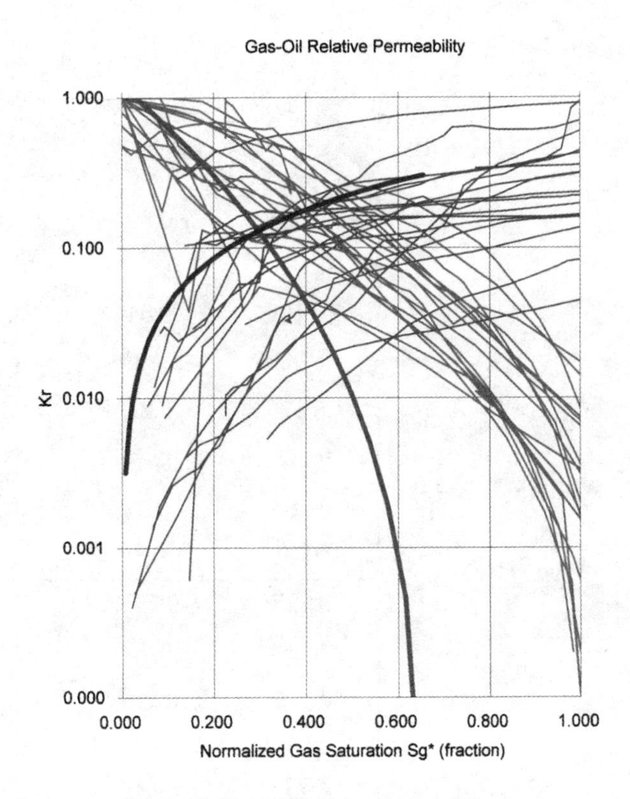

Fig. 7–65 Gas-Oil Relative Permeability—
Log Scale vs. Normalized S_l

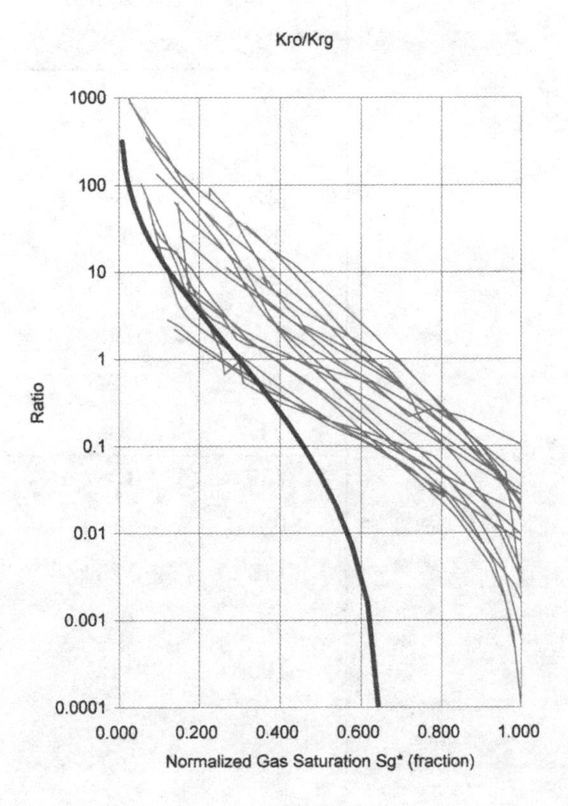

Fig. 7–66 k_{ro}/k_{rw} *Relative Permeability vs. Normalized* S_l

Table 7–6 Brazeau River and Pembina Areas

Lab	Location	Sample	Date	k_{air}	k_o	Phi	S_{wi}	Type
Chevron	01-09-50-12W5M	96(run 2)	79-05-18	213	144	0.154	0.271	k_g/k_o
Chevron	01-09-50-12W5M	110(run 3)	79-05-18	837	805	0.173	0.266	k_g/k_o
Chevron	01-09-50-12W5M	282(run 2)	79-05-18	52	25.8	0.121	0.75	k_g/k_o
Core Labs	14-32-49-11W5M	81	79-09-28	59	54	0.119	0.119	k_g/k_o
Core Labs	14-32-49-11W5M	279	79-09-28	8.2	3.2	0.077	0.166	k_g/k_o
Getty	04-35-49-11W5M	839.1	80-12-22	150	27	0.1	0.476	k_g/k_o
Getty	04-35-49-11W5M	842.67	80-12-22	42.9	9.2	0.092	0.316	k_g/k_o
Getty	04-35-49-11W5M	843.63	80-12-22	38.7	4.8	0.085	0.292	k_g/k_o
Getty	04-35-49-11W5M	849.54	80-12-22	46.9	7	0.088	0.587	k_g/k_o
Getty	04-35-49-11W5M	863.59	80-12-22	600	345	0.127	0.231	k_g/k_o
Getty	04-35-49-11W5M	866.27	80-12-22	900	248	0.104	0.341	k_g/k_o
Getty	04-35-49-11W5M	876.75	80-12-22	22	4.7	0.065	0.324	k_g/k_o
Core Labs	04-35-49-11W5M	16A	81-01-15	86	40	0.112	0.178	k_g/k_o
Core Labs	04-35-49-11W5M	24A	81-01-15	34	1.8	0.08	0.1	k_g/k_o
Core Labs	04-35-49-11W5M	41A	81-01-15	21	2.4	0.104	0.151	k_g/k_o
Core Labs	04-35-49-11W5M	139A	81-01-15	102	35	0.12	0.113	k_g/k_o
Core Labs	15-09-47-14W5M	45A/15-9	81-12-21	6.3	1.72w	0.063	1	k_g/k_w
Core Labs	15-09-47-14W5M	81B/15-9	81-12-21	13.3	7.09w	0.107	1	k_g/k_w
Core Labs	09-01-48-14W5M	37234	81-12-21	173	60.3w	0.168	1	k_g/k_w
Getty	04-35-40-11W5M	839.1	82-11-29	410.5	524.6	0.105	0.535	Res T&O
Getty	04-35-40-11W5M	876.75	82-11-29	14.1	2.73	0.061	0.45	Res T&O

Table 7–7 Empirical Correlation Results

S_g	S_g^*	k_{rg}	k_{rog}	k_{rg} (modified)
0	0	0	1	0
0.005	0.008	0.0105	1	0.0032
0.01	0.015	0.021	1	0.0064
0.015	0.023	0.0316	1	0.0095
0.02	0.031	0.0422	0.9975	0.0127
0.025	0.038	0.0528	0.9501	0.016
0.038	0.058	0.0796	0.8395	0.024
0.05	0.077	0.1065	0.7395	0.0322
0.075	0.115	0.161	0.5684	0.0486
0.1	0.154	0.2163	0.4308	0.0653
0.125	0.192	0.2725	0.3213	0.0823
0.15	0.231	0.3296	0.2354	0.0995
0.175	0.269	0.3875	0.1688	0.117
0.2	0.308	0.4462	0.1181	0.1347
0.225	0.346	0.5057	0.0802	0.1527
0.25	0.385	0.5661	0.0525	0.171
0.275	0.423	0.6247	0.0328	0.1895
0.3	0.462	0.6895	0.0193	0.2082
0.325	0.5	0.7524	0.0105	0.2272
0.35	0.538	0.8162	0.005	0.2465
0.375	0.577	0.8808	0.0019	0.266
0.4	0.615	0.9462	0.0005	0.2858
0.425	0.654	1	0	0.302
0.45	0.692	1	0	0.302
0.475	0.731	1	0	0.302
0.5	0.769	1	0	0.302
0.525	0.808	1	0	0.302
0.55	0.846	1	0	0.302
0.575	0.885	1	0	0.302
0.6	0.923	1	0	0.302
0.625	0.962	1	0	0.302
0.65	1	1	0	0.302

Selecting a curve

After this is split into rock types, some representative average relative permeability curve is selected. As outlined previously, S_{wc} was selected from a histogram. In some cases, if the special core analysis data is not representative of the average, the end-point can be forced to derive a log analysis S_{wc}. Plotting up the data on a variety of techniques gives a more reliable indication of whether a true average is obtained. Generating a curve running through the center of all the data is actually quite difficult. It is simpler and more accurate to choose a best curve. This is, of course, a matter of preference.

In some cases, no data is available, and correlations are necessary. More commonly, a small amount of data is available. The example from the Brazeau River-Pembina area is atypical. There is tremendous variation in this example. The geological mechanism causing this was thermal sulphate reduction, as well as other dolomitization processes. This is probably the worst variation the author has seen. The reservoir fluids in this type of system are extremely sour.

As a rule, carbonates have more variability than sandstones, so the risk is higher on carbonates.

Three-phase relative permeability data

Most classic reservoir engineering textbooks, such as Dake, Craft and Hawkins, and Amyx, Bass, and Whiting, do a fair job of covering two-phase relative permeability.[23,24,25] Strangely, they stop almost completely at three-phase relative permeability. This topic is, however, very relevant to numerical simulation.

Actually, three-phase relative permeability data is extremely rare. There are slightly more than a dozen sets worldwide. Therefore, it is extremely unlikely that three-phase data will be available for most reservoirs. From the data obtained, trends are apparent. First, water relative permeability, or k_{rw} is mostly a function of water saturation. This is shown in Figure 7–67.[26] The lines are shown as relatively straight.

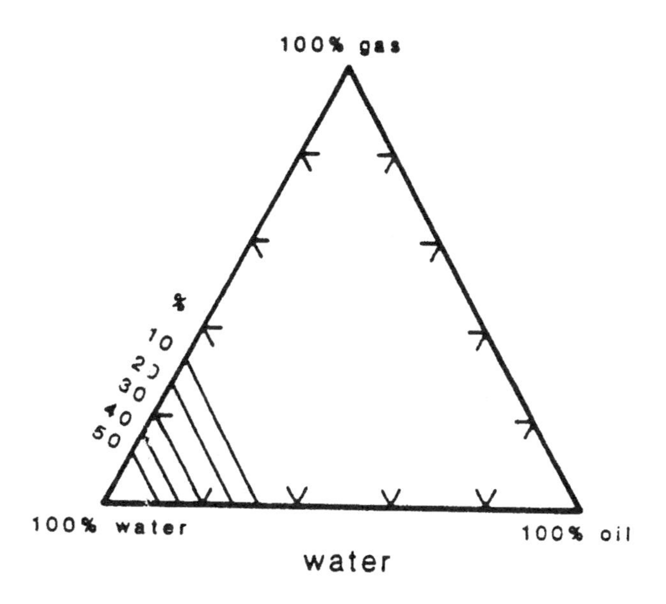

Fig. 7–67 Generalized k_{rw} for Three-Phase Relative Permeability

Similarly, gas relative permeability, or k_{rg}, is most strongly controlled by gas saturation. This is illustrated in Figure 7–68. Some curvature is shown in these lines.

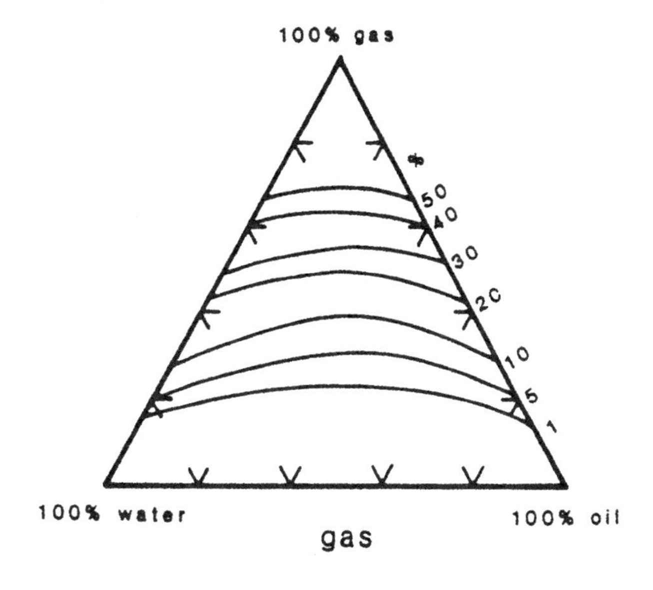

Fig. 7–68 Generalized k_{rg} for Three-Phase Relative Permeability

Oil relative permeability, or k_{ro}, appears to be controlled by both oil and gas saturations and, more particularly, by the product of $S_w \times S_g$. This is shown in Figure 7–69.

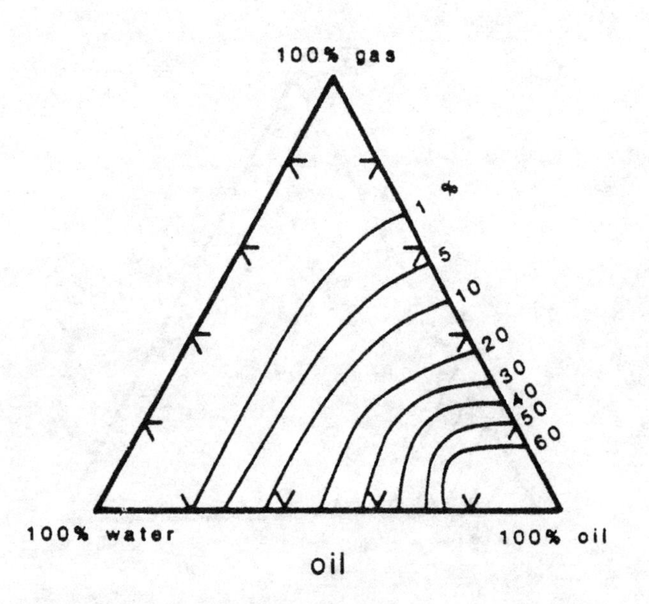

Fig. 7–69 Generalized k_{ro} for Three-Phase Relative Permeability

In summary, some trends in the relative permeability of the most common three phases can be used for a basic understanding.

Relevance of three-phase flow

It is interesting to evaluate where three-phase flow has the greatest effect. The saturations where simultaneous flow of all three phases occurs are actually quite limited. This information is presented in Figure 7–70. The traditional use of two-phase relative permeabilities does have some technical justification.

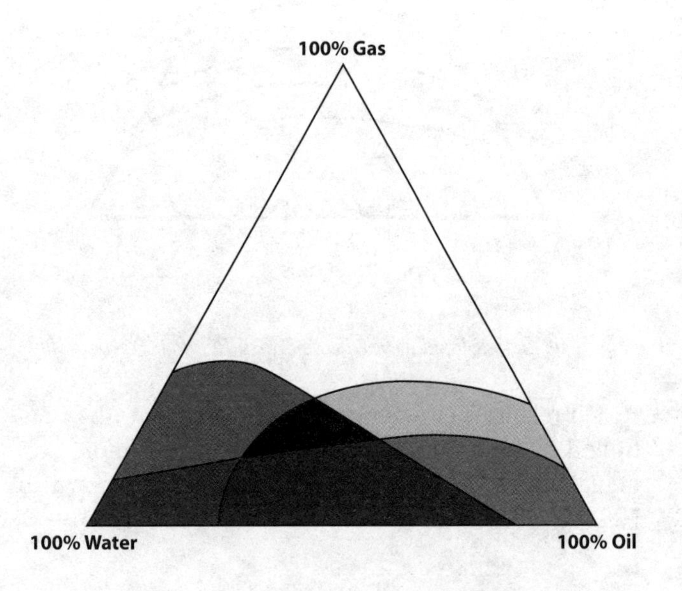

Fig. 7–70 Area Where Three Phases Are Equally Mobile

Three-phase relative permeability correlations

The general trends shown indicate correlations could be of some benefit. The first correlation is due to Stone. The equations have been outlined as follows:[27]

Saturations are calculated as shown in Equations 7.9 through 7.11:

$$S_o^* = \frac{S_o - S_{om}}{(1 - S_{wc} - S_{om})}, \quad S_o \geq S_{om} \tag{7.9}$$

$$S_w^* = \frac{S_w - S_{wc}}{(1 - S_{wc} - S_{om})}, \quad S_w \geq S_{wc} \tag{7.10}$$

$$S_g^* = \frac{S_g}{(1 - S_{wc} - S_{om})} \tag{7.11}$$

Oil relative permeability is then calculated using Equation 7.12.

$$k_{ro} = S_o^* \beta_w \beta_g \tag{7.12}$$

Terms in this equation are defined in Equations 7.13 and 7.14.

$$\beta_w = \frac{k_{row}(S_w)}{(1 - S_w^*)} \tag{7.13}$$

$$\beta_w = \frac{k_{row}(S_g)}{(1 - S_w^*)} \tag{7.14}$$

The mobile oil and definition of S_{om} is shown in Figure 7–71.

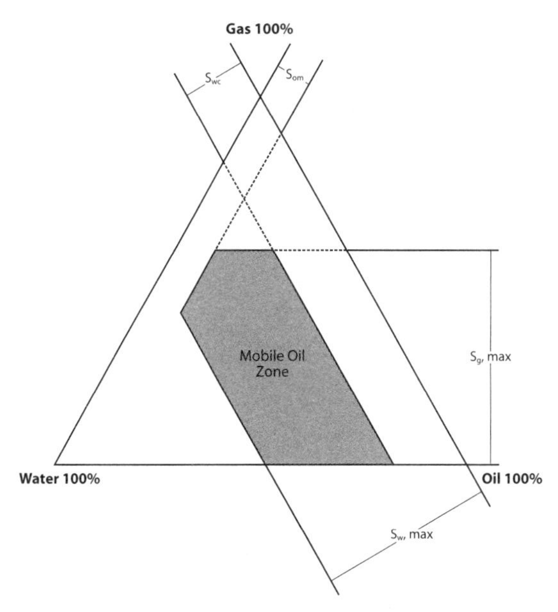

Fig. 7–71 Zone of Mobile Oil for Three-Phase Flow

Stone's second model (Stone II)

Stone also created a second model. This model was in response to a lack of agreement with the dependence of waterflood residual oil saturation on trapped gas saturations.[28]

$$k_{ro} = (k_{row} + k_{rw}) * (k_{rog} + k_{rg}) - (k_{rw} + k_{rg}) \quad (7.15)$$

Dietrich and Bondor adjusted Stone's second model based on a comparison with published data and noticed an end-point problem. This is shown in Equation 7.16.[29]

$$k_{ro} = \frac{1}{k_{rocw}} \left[(k_{row} + k_{rw}) * (k_{rog} + k_{rg}) \right] - (k_{rw} + k_{rg}) \quad (7.16)$$

After this, Nolen, while at Intercomp, further modified Stone's model. This is shown in Equation 7.17.[30]

$$k_{ro} = k_{rocw} \left[\left(\frac{k_{row}}{k_{rocw}} + k_{rw} \right) * \left(\frac{k_{rog}}{k_{rocw}} + k_{rg} \right) \right] - (k_{rw} + k_{rg}) \quad (7.17)$$

Simulators often allow different specifications of these two models. Some linear interpolation options are included as well. The author uses Stone II with the modifications of Nolen where they are available. Some simulators appear to use this without documentation.

There can be a remarkable amount of difference between the two methods. An example is shown in Figure 7–72. Although rarely done, this figure suggests some plotting would be beneficial. Regrettably, there is almost never any three-phase relative permeability data upon which to do quality control checks. Buyer beware!

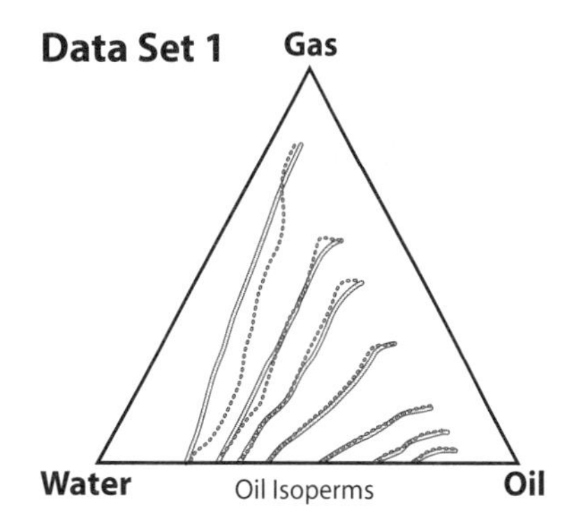

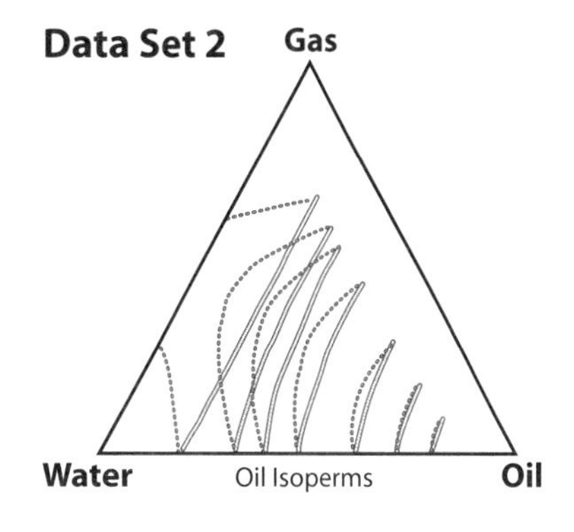

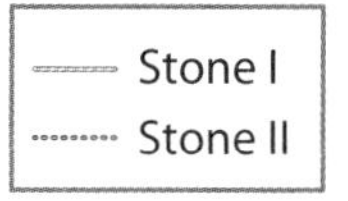

Issues:

- *Shape of oil isoperms depends on relative permeability data and which model is used.*
- *Even if isoperms are the same, is there lab data to show what correct three-phase relative permeability is?*
- *Actual saturation ranges where $k_{ro} \sim k_{rg} \sim k_{ro}$ is normally limited and not extensively traversed in displacement. (Recall that $\mu_{gas} << \mu_{water}$ and μ_{oil}).*
- *Checking for reasonable profiles*

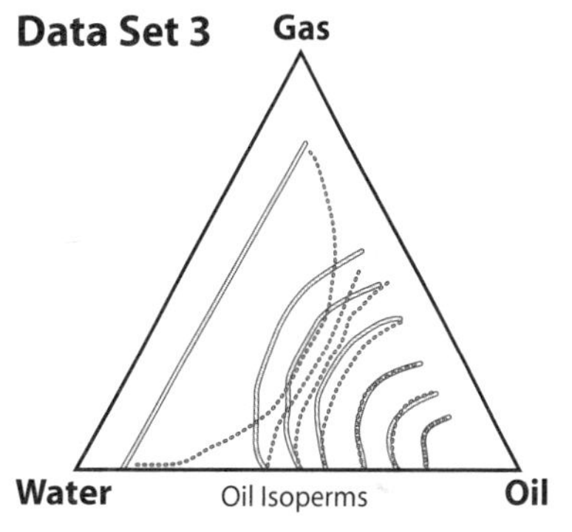

Fig. 7–72 Comparison of Stone I vs. Stone II Model for Various Relative Permeability Data

Input data consistency

Three-phase relative permeability correlations can be very sensitive to end-points. It is important to use the correct convention in entering this data. Recall, as discussed earlier, there are different lab conventions. The requirements for simulator input are shown in Figures 7–73 and 7–74.

Most simulators have included this as part of their data-checking routines. Unfortunately, data-checking routines are rarely documented. It is therefore unwise to rely on this feature. Committing these graphs to memory is the recommended practice.

A rumor suggests one of the proprietary programs, which did not have absolutely complete data checking, resulted in negative relative permeabilities. As the story goes, some mistakes were made in the end-points, most likely related to laboratory conventions. This can create real havoc. Fluids move from lower potentials to higher potentials, leading to rather baffling history-match results.

Some simulators check for negative relative permeabilities through a range of saturations, but others do not. The triangular table Eclipse appears to generate is designed for error checking. Not all simulators put out the three-phase relative permeability calculations, and not all modifications to Stone's methods are clearly documented.

Fortunately, these erroneous results were so baffling and of such a scale that an intense investigation resulted. One wonders what would happen if the results only affected a few wells and were not detected.

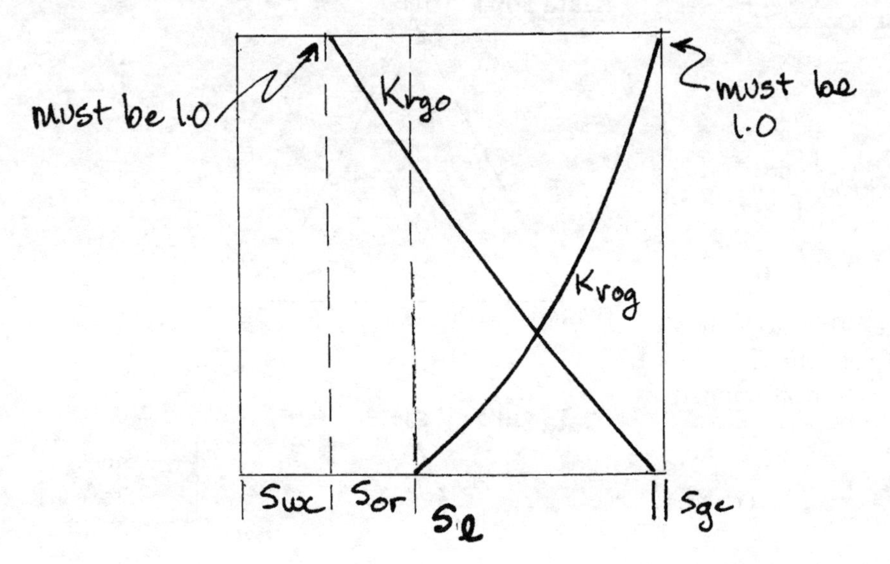

Fig. 7–73 Generalized Gas-Oil Relative Permeability Input

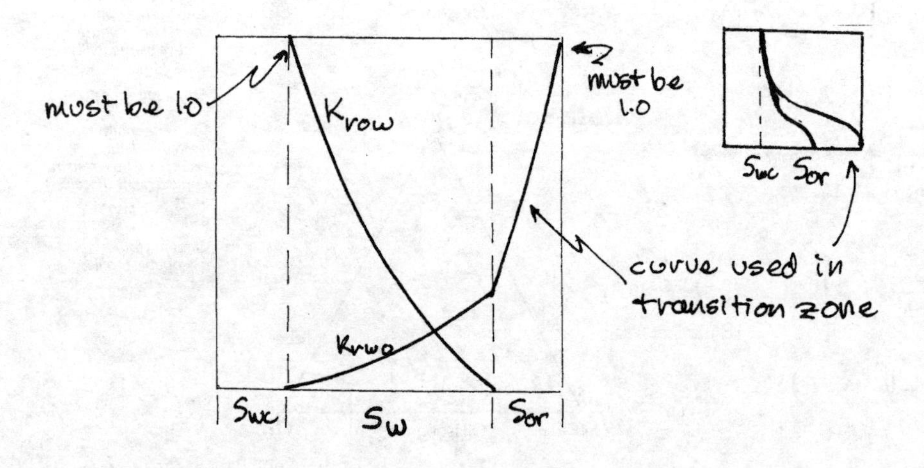

Fig. 7–74 Generalized Water-Oil Relative Permeability Input

Extension of k_{rw} curve between $(1–S_{or})$ and 100% S_w

It is necessary to extrapolate from the end-point of the k_{rw} curve at a saturation of $(1 - S_{or})$ to $k_{rw} = 1$ at $S_w = 1$ on the water-oil relative permeability curve. There is a good reason for this.

Although no core flood would enter this range of saturations, there is a physical explanation of when this range of water saturations will be encountered. The answer is in the transition zone between a completely water-saturated aquifer and the main oil leg, which is at S_{wc}.

Whether this will exist in a reservoir may or may not be true. However, most simulators are generically designed in data checking to spot this potential problem.

No kinks, please

One item that can cause problems is kinks in relative permeability data. The nonlinear or outer iterations use slopes or derivatives to solve for the correct saturations. If there is a kink in the relative permeability data, as shown in Figure 7–75, the solution can be directed to a saturation where the iteration will fail. This type of problem does not manifest itself as a complete failure. It is far more insidious. The problem will shorten your timesteps and require many solver iterations, which causes the simulator to run very slowly. The moral of the story: it is a good idea to plot your data. In addition, you can calculate the slope of the data to ensure the slope does not decrease with increasing saturations, i.e., it increases monotonically.

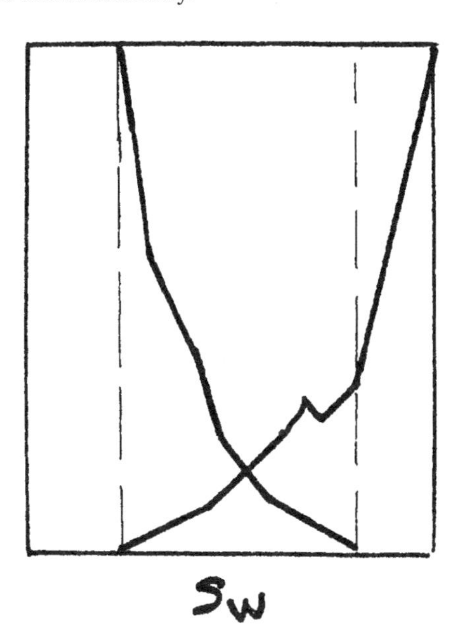

Fig. 7–75 Example of Kinked Data (Bad)

Power law curves

Some people advocate the use of power law curves for relative permeability data to avoid the problem with kinks. With this implementation, it is possible to replace table lookup and use an equation, which may increase speed. Given the effort put into screening, the approach may not be advisable; however, it may make little difference to the results in many cases.

There are circumstances, such as when using pseudo relative permeability curves, where the rules regarding kinks and monotonicity may be broken or bent somewhat. This is not necessarily fatal but may cause longer execution times. It may be a better alternative to a greatly expanded reservoir grid.

Pore volume compressibility

This may seem an unrelated topic; it is included in this chapter since simulation data normally requires this in the same section as relative permeability. When a reservoir is produced, a considerable reduction in reservoir pressure occurs; therefore, some degree of consolidation takes place. This reduction of pore space is referred to as pore volume compressibility. Normal oilfield practice is to consider this process as a change in pore volume per change in pore pressure as shown in Equation 7.18.

$$\phi = \phi^0\left[1 + c_r\left(p - p^0\right)\right] \tag{7.18}$$

Further, it has been assumed this relationship is linear. Three sources of data most commonly used are reproduced in Figure 7–76.[31]

These correlations are not very close because the rocks do not behave at all linearly. Crushing and other nonlinear effects occur. In addition, the behavior of rock and soils is strongly affected by load history. This data is presented since it is generally of the correct order of magnitude and is often better than nothing.

Typically, pores contract with decreasing pore pressure. Mobile water saturation is created, since the compressibility of water is normally smaller than the pore volume compressibility. As a result, a small water cut can increase up to 5% at advanced stages of depletion. Most of these water cuts are below 3% and are not captured in production data, since crude oil has a bottom sediment and water (BS&W) limit of 3%. In some instances, ARE has found the minor water flow has an impact on execution time. In the coning case, this involved a time-intensive compositional simulation. To increase execution speeds, pore volume compressibility was set at the lower end of the correlation ranges.

Pore volume compressibility becomes important in a number of conventional oil and gas situations.

In reservoirs that have undersaturated oil, the compressibility of oil, water, and rock pores are of the same order of magnitude. It is rare to have pore volume compressibility measured. One is therefore forced to use the correlations outlined previously. Since the compressibility could easily vary by a factor of 2 to 3, material balance and simulation results above the bubblepoint are not considered accurate.

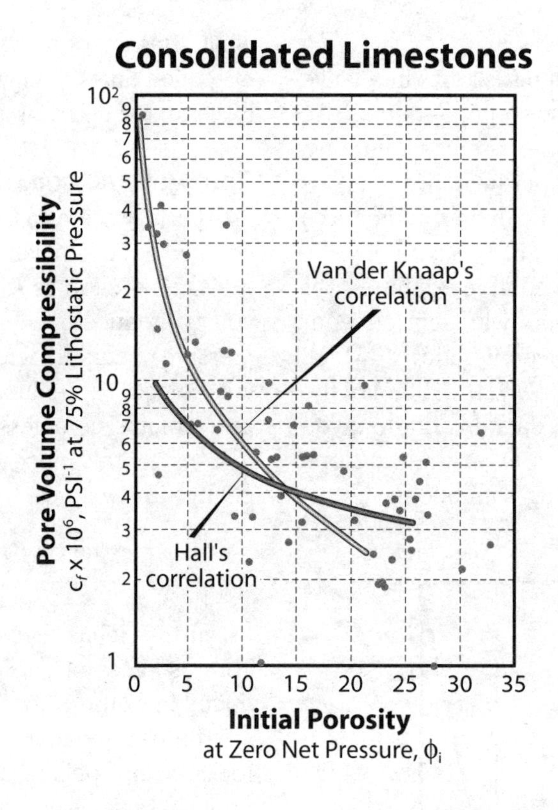

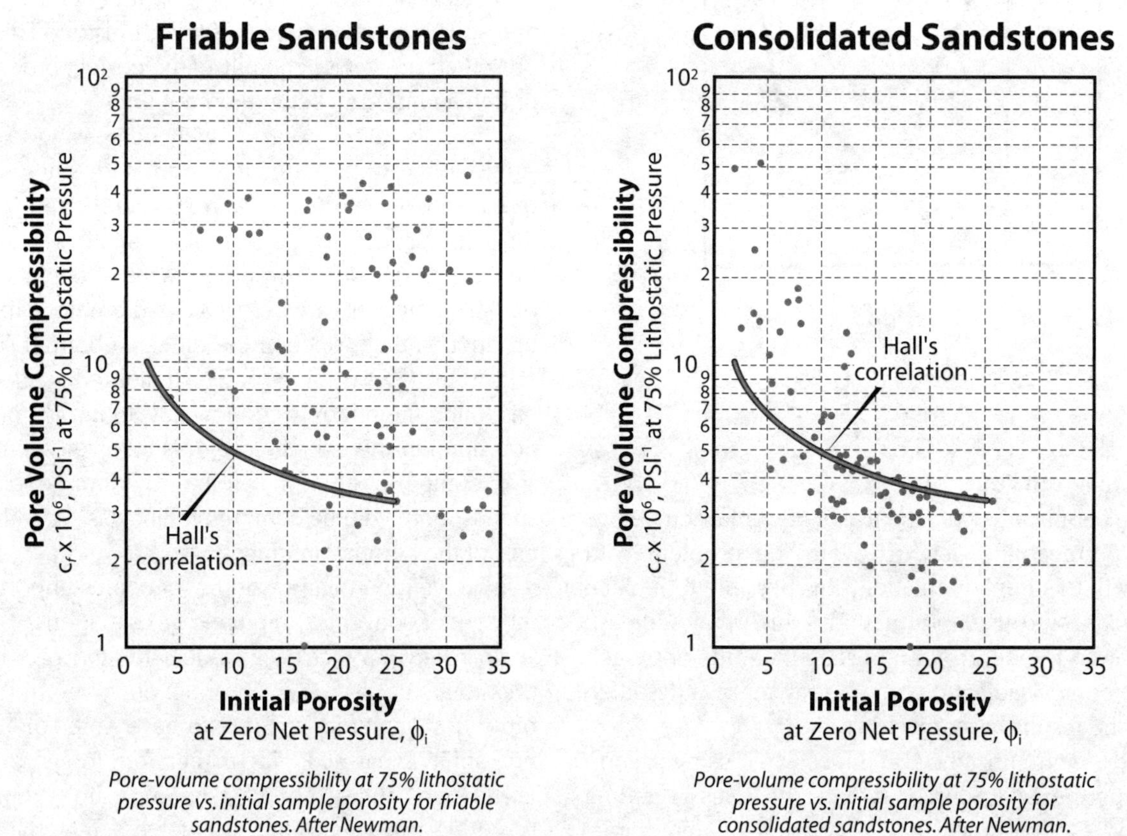

Pore-volume compressibility at 75% lithostatic pressure vs. initial sample porosity for friable sandstones. After Newman.

Pore-volume compressibility at 75% lithostatic pressure vs. initial sample porosity for consolidated sandstones. After Newman.

Fig. 7–76 Correlations for Rock Pore Compressibility

In a number of unconsolidated reservoirs, there is considerable subsidence and compaction. This can have a significant effect in gas reservoirs. Normally, the high compressibility of gas masks formation effects. Formation compressibility has proven to be very important in shallow Venezuelan gas reservoirs. Likely, it is very important in the shallow gas sands in the northeastern section of Alberta in areas such as Leismer.

A number of overpressured reservoirs are in existence. Some of the best examples are in Texas and involve high-pressure gas in carbonates as well as some high-pressure gas in conglomerates. Both types of reservoirs are onshore. At pressures greater than 35,000 kPa or 5,000 psi, the compressibility of gas becomes quite low, on the same order of magnitude as rock compressibility.

Some very famous examples of compaction in chalks come to mind—in particular, the Ekofisk field in the North Sea. The structure of the rock is derived from tiny, hollow plankton shells. Aside from adding reservoir drive, this resulted in ocean floor subsidence with the unintended result that the production platforms started to sink. The Ekofisk chalks are unique since the burial history involved early emplacement of the oil. Most chalks are not as sensitive to these problems.

Significant progress is being made in this area with heavy oil research. Heavy oil wells that have been steamed produce large amounts of sand. Resolving this phenomenon has resulted in improved linking between soil mechanics and reservoir engineering.

Gas condensate relative permeability curves and heavy oil relative permeability

Relative permeability relations for these types of reservoirs are discussed in the second part of this book.

Procedure

- Check to see if or what special core data is available.
- Plot conventional core permeability versus core porosity cross-plot.
- Plot special core data.
- Input into simulator.

Summary

Capillary pressure is difficult to visualize. It is tremendously complex by virtue of the diversity of pore networks and the number of fluids and minerals involved. There are a number of different ways of obtaining this data in the laboratory. Lab measurements have to be adjusted to reservoir conditions; however, the adjustments are crude approximations. In most cases, good estimates of reservoir surface tension and average contact angles are not available. Averaging of mixed lab data types is almost impossible.

Reservoir conditions are not accurately replicated by static capillary pressure tests. This does not appear to be widely recognized. Key concepts are:

- Water moves hydrogeologically through the subsurface. Water movement is pervasive.
- Trapping is by imbibition capillary pressure threshold barriers, not low permeability.
- Emplacement of oil and gas is by countercurrent drainage.
- Connate water saturations are proximal to the end-point on a relative permeability curve and correspond closely to the transition of capillary pressure data from continuous film flow to discontinuous flow.

The best interpretation is obtained from *in situ* measurements, i.e., log analysis. Lithology variations can make this data difficult to interpret. With some judicious averaging, this problem can be overcome.

In the final analysis, lab data is a good starting point. Commonsense adjustments to observed field data or history matching are highly appropriate.

The screening criteria for both capillary pressure data and relative permeability data overlap extensively. Important considerations include the effects of drilling fluids, coring procedures, storage procedures, and sample preparation. The fluids used must be considered as well as the effects of these processes on the surfaces of the rock pore system.

Relative permeability data is screened in a number of different ways. Optimal screening involves looking at raw data in a variety of formats and using normalized curves. This is done on an individual basis as well as a group basis and is a work-intensive process. Checking against offsetting data is very important. Frequently, it is simpler and more accurate to pick one set of data as being most representative. Correlations do not necessarily match experimental data. Quality control on relative permeability is an involved issue. Reviewing a report for all of the relevant data takes time. Often, it is helpful to talk to the individuals who have actually done the work. In the author's experience, lab personnel are extremely helpful for the most part.

Three-phase relative permeability data is very rare. Correlations are used extensively and are not always tested well. For the most part, simultaneous flow of oil, gas, and water is relatively rare. It is very important to get the end-points entered correctly. Data consistency on relative permeability input is a common problem. Check data for smoothness.

Pore volume compressibility was discussed in this chapter since the input is normally included in the same data input section as relative permeability.

References

1 McCordell, W.M., "A Review of the Physical Basis for the Use of the J-function", Eighth Oil Recovery Conference, Texas Petroleum Research Committee, 1955.

2 After Benner, F.C., and F.E. Bartell, "The Effect of Polar Impurities upon Capillary and Surface Phenomena in Petroleum Production," *Drilling and Production Practices*, API, 1944.

3 Hough, E.W., M.J. Rzasa, and B.B. Wood, "Interfacial Tensions of Reservoir Pressures and Temperatures, Apparatus and the Water-Methane System," Transactions of AIME, Vol. 192, pp. 57–60, 1951.

4 Hocott, C.R., "Interfacial Tension Between Water and Oil Under Reservoir Conditions," *Transactions of AIME*, Vol. 132, pp. 184–190, 1938.

5 Livingston, H.K., "Surface and Interfacial Tensions of Oil-water Systems in Texas Oil Sands," Petroleum Technology, 1938.

6 After Swartz, C.A., "Interfacial Tensions?" *Physics*, Vol. 1, p. 245, 1931.

7 Purcell, W.R., "Capillary Pressures—Their Measurement Using Mercury and the Calculation of Permeability Therefrom," *Transactions of AIME*, Vol. 186, 1949.

8 Morrow, N.R., *Interfacial Phenomena in Petroleum Recovery*, Marcel Dekker, Inc., 1990.

9 Leverett, M.C., "Capillary Behavior in Porous Solids," *Transactions of AIME*, Vol. 142, 1941.

10 Dickey, P.A., "Discussion of Hydrodynamic Trapping in the Cretaceous Nahr Umr Lower Sand of the North Area, Offshore Qatar," SPE 18521, *Journal of Petroleum Technology*, 1988, p. 1075.

11 Dullien, F.A., *Porous Media: Fluid Transport and Pore Structure*, 2nd ed., Academic Press, 1992.

12 Lenormand, R., et al., "Mechanism of the Displacement of One Fluid by Another in a Network of Capillary Ducts," *Journal of Fluid Mechanics*, 1983.

13 Bourbiaux, B.J., and F.J. Kalaydiagian, "Experimental Study of Cocurrent and Countercurrent Flows in Natural Porous Media" *SPERE*, August 1990, p. 363.

14 Schafer, P.S., T. Hower, and R.W. Owens, *Managing Water Drive Gas Reservoirs*, Gas Research Institute, 1993.

15 *Alberta Energy and Utilities Board, PVT and Core Studies Index, Guide 14*, 19th ed., Alberta Energy and Utilities Board, 2001.

16 Wright, H.T., and L.D. Woody, "Formation Evaluation of the Borregas and Seeligson Field, Brooks and Jim Wells County, Texas," Symposium on Formation Evaluation, AIME, Oct. 1955.

17 Geffen, T.M., et al., "Experimental Investigation of Factors Affecting Laboratory Relative Permeability Measurements," *Transactions of AIME*, Vol. 192, pp. 99–110, 1951.

18 Chatzis, J., N.R. Morrow, and H.T. Lim, "Magnitude and Detailed Structure of Residual Oil Saturation," *Society of Petroleum Engineers Journal*, Vol. 23, Mar.–Apr., 1983.

19 After Tek, M.R., *Underground Storage of Natural Gas, Complete Design and Operational Procedures, with Significant Case Histories*, Gulf Publishing, 1987.

20 Johnson, E.F., D.P. Bossler, and V.O. Naumann, "Calculation of Relative Permeability from Displacement Experiments," *Transactions of AIME*, Vol. 216, 1959.

21 Hamp, T., et al., "Oil Base Coring for Connate Water Saturation—Utikuma Keg River Sandstone," *Journal of Canadian Petroleum Technology*, 1990.

22 Honarpour, M.M., L. Koederitz, and H. Harvey, *Relative Permeability of Petroleum Reservoirs*, The CRC Press, 1986; and M.M. Honarpour, et al., "Empirical Equations for Estimating Two-Phase Relative Permeability in Consolidated Rock," *Transactions of AIME*, Vol. 273, 1982, pp. 2905–2908.

23 Dake, L.P., *Fundamentals of Reservoir Engineering*, Developments in Petroleum Science 8, Elsevier, 1978.

24. Craft, B.C., and M.F. Hawkins, *Applied Petroleum Reservoir Engineering*, Prentice Hall, 1959.

25 Amyx, J.W., D.M. Bass, and R.L. Whiting, *Petroleum Reservoir Engineering—Physical Properties*, McGraw-Hill, 1960.

26 Leverett, M.S., and W.B. Lewis, "Steady Flow of Gas-Oil-Water Mixtures through Unconsolidated Sands," *Transactions of AIME*, Vol. 142, 1941, pp. 107–116.

27 Stone, H.L., "Estimating Three-Phase Relative Permeability," *Journal of Petroleum Technology*, 1970.

28 Stone, H.L., "Estimation of Three-Phase Relative Permeability and Residual Oil Data," *Journal of Petroleum Technology*, 1973.

29 Dietrich, J.K., and P.L. Bondor, "Three-Phase Oil Relative Permeability Models," SPE 6044, Society of Petroleum Engineers, 1976.

30 Nolen, J.S., "Numerical Simulation of Compositional Phenomena in Petroleum Reservoirs," SPE 4274, Society of Petroleum Engineers, 1973.

31 Newman, G.H., "Pore Volume Compressibility of Consolidated, Friable and Unconsolidated Reservoir Rock Under Hydrostatic Loading."

8

Pseudo Relative Permeability and Upscaling

Introduction

The petroleum literature describes many pseudo properties (pressures, times, permeabilities, etc.) that rarely indicate the true purpose or nature of the concept. *Pseudo relative permeability* is no exception. In the author's opinion, this is the single most misunderstood aspect of reservoir simulation. This technique has a number of distinct purposes. The best way to explain it is from a historical perspective.

Vertical equilibrium

This technique first appeared in "The Use of Vertical Equilibrium in Two-Dimensional Simulation of Three-Dimensional Reservoir Performance," by K.H Coats, J.R. Dempsey, and J.H. Henderson.[1] It can be used in thick reservoirs, where the capillary pressure transition zone is thin—e.g., a sharp gas-oil contact—in relation to the total zone. The implicit assumption is gravity forces are much stronger than lateral forces. Thus, it is always assumed two sharp and distinct phases exist in a grid block. Therefore, this technique can allow the representation of a thick reservoir with a single-layer areal model. Reducing a multilayer problem to a single layer can result in significant economic savings or allow simulation of a large field, which could not be solved any other way. This is also known as vertical equilibrium (VE).

At the time of its development in 1970, computers were much less powerful than today. Hence, the technique offered the potential for both economic savings as well as reducing some problems to grids that could actually be tractably modeled.

The development for displacement of oil by water is developed from the basic situation shown in Figure 8–1.

If we assign the fractional thickness of water as $b = y/h$, the average water saturation can be calculated as shown in Equation 8.1:

$$\overline{S_w} = b(1 - S_{or}) + (1 - b)S_{wc} \qquad (8.1)$$

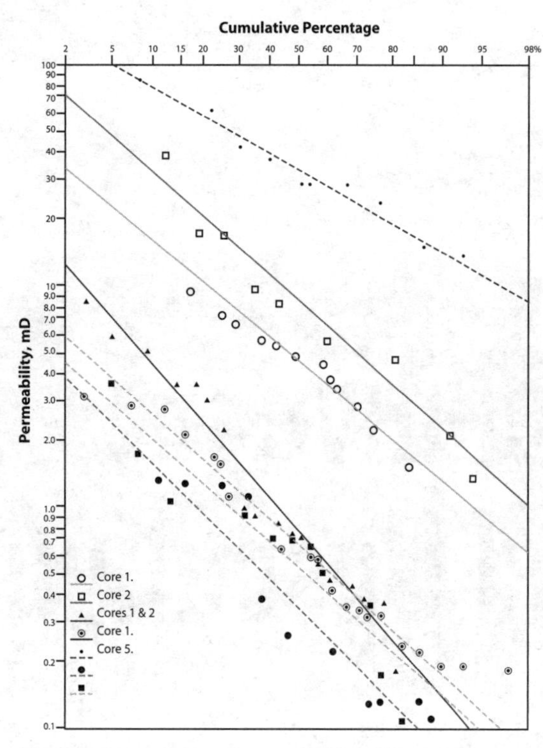

Fig. 8–1 Dipping Reservoir with Thin Water-Oil Contact

This equation can be used to solve for b as shown in Equation 8.2:

$$b = \frac{\overline{S_w} - S_{wc}}{1 - S_{or} - S_{wc}} \qquad (8.2)$$

Since S_{or} and S_{wc} are constants, the thickness-averaged relative permeability to water can be calculated as shown in Equation 8.3.

$$\overline{k_{rw}}(\overline{S_w}) = b\, k_{rw, S_w=(1-S_{or})} + (1-b)\, k_{rw, S_w=S_{wc}} \qquad (8.3)$$

Of course, the relative permeability to water at S_{wc} is zero, and k_{rw}, at residual oil saturations, is the end-point relative permeability to water. Hence, we may derive Equation 8.4:

$$\overline{k_{rw}}(\overline{S_w}) = b\, k_{rw,\, endpoint} \qquad (8.4)$$

Using a similar development, it is relatively easy to show via Equation 8.5:

$$\overline{k_{ro}}(\overline{S_w}) = \frac{1 - S_{or} - \overline{S_w}}{1 - S_{or} - S_{wc}}\, k_{ro,\, endpoint} \qquad (8.5)$$

The resultant *fudged* relative permeabilities look like two straight lines such as depicted in Figure 8–2.

Some degree of vertical equilibrium is common. This can be demonstrated by running sensitivities on a 5-spot waterflood pattern with good reservoir properties, moderate thickness, and relatively low production-injection rates. The author has overheard engineers say, sarcastically, they could use simple straightline relative permeabilities and get decent results from a simulator. Their results were likely an accurate representation of the physics in the reservoir, i.e., gravity segregation effects were dominant over lateral effects.

The technique developed by Coats did not require the transition zone be very thin in relation to the grid block—only that gravity segregation occur more quickly than horizontal flow. Thus, if a transition zone resets instantaneously, one can calculate a pseudo relative permeability set as shown in Figure 8–3.

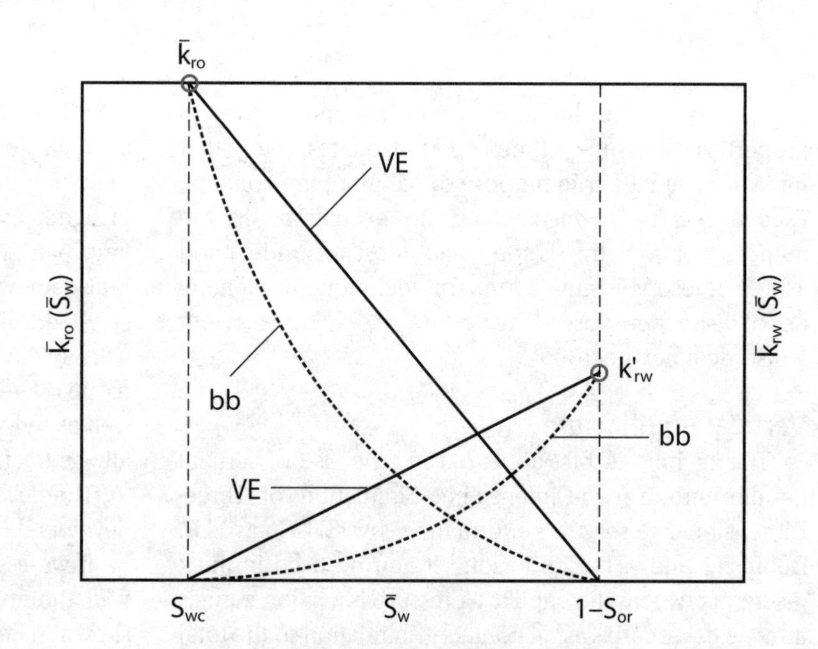

Fig. 8–2 VE Relative Permeability Curves

The calculation becomes somewhat more involved with the inclusion of a transition zone, and the derived pseudo relative permeability will no longer be simple straight lines, as shown in Figure 8–4.

Hearn-type relative permeabilities

It has long been recognized that the effectiveness of waterfloods is strongly affected by layering in the reservoir. In a layered reservoir, the displacing fluid will move more quickly through the most permeable layers. This causes a more rapid and more gradual breakthrough of the displacing fluid. In classical reservoir engineering, this has been accounted for via techniques developed by Stiles and Dykstra-Parsons.[2, 3]

The second pseudo relative permeability technique envisaged was developed by C.L. Hearn, precisely to account for these layering effects. This calculation assumes piston-like displacement in each layer of the reservoir and no vertical communication between layers. This is the opposite of VE. Although there are some gross simplifications in this method, the loss in accuracy can be insignificant when compared to correctly accounting for layering.

Table 8–1 was used in a study the author performed a number of years ago to calculate Hearn-type relative permeabilities. The raw relative permeability data is shown in Figure 8–5. The layering was tied to the Dykstra-Parsons ratio, which was discussed in chapter 3. The resultant curves are shown in Figure 8–6. These curves accelerate water breakthrough. The end-points have not changed. In Figure 8–7, the actual Dykstra-Parsons plots are shown for different wells. The layering is quite severe as reflected by a Dykstra-Parsons ratio slightly below 0.9. This eventually proved to be the correct value in history matching. The various pseudo relative permeability curves were very useful in explaining the effect of layering.

Dynamic pseudo relative permeability

Some time after these two techniques had been developed, it was realized various combinations of capillary pressure transition zones, layering, reservoir thicknesses, and the extremes of horizontal versus vertical velocities could be accounted for dynamically. These dynamic curves can be derived both analytically and via cross-sectional modeling.

Smith, Mattax, and Jacks

Smith, Mattax, and Jacks developed the first dynamic method. It requires a cross-sectional model of the reservoir be constructed. Special computer utilities have been developed to average the relative permeability of each phase for each column of grid-block layers contained within a cross section. The average relative permeability is plotted against the average water saturation for each grid block layer column. The resultant curves become the dynamic pseudo relative permeability curves. These curves need to be used in conjunction with appropriate pseudo well relative permeabilities. A single layer cross section, utilizing both these types of pseudo curves, can duplicate with relative ease the

Table 8–1 Example Calculation for Hearn-Type Pseudo Relative Permeability Curves

V=0.62						
S_w	H	Perm	h*k	k_{rw}	S_o	k_{row}
0.44				0	0.56	1
0.492	0.8	21.167	16.934	0.174	0.508	0.502
0.544	0.8	10.41	8.328	0.26	0.456	0.256
0.596	0.8	5.899	4.719	0.309	0.404	0.118
0.648	0.8	3.331	2.665	0.336	0.352	0.039
0.7	0.8	1.66	1.328	0.35	0.3	0

V=0.70						
S_w	H	Perm	h*k	k_{rw}	S_o	k_{row}
0.44				0	0.56	1
0.492	0.8	15	12	0.174	0.508	0.452
0.544	0.8	6.2	4.96	0.26	0.456	0.225
0.596	0.8	3.5	2.8	0.309	0.404	0.097
0.648	0.8	1.87	1.496	0.336	0.352	0.029
0.7	0.8	0.78	0.624	0.35	0.3	0

V=0.90						
S_w	H	Perm	h*k	k_{rw}	S_o	k_{row}
0.44				0	0.56	1
0.492	0.8	70	56	0.174	0.508	0.18
0.544	0.8	10.7	8.56	0.26	0.456	0.056
0.596	0.8	3.5	2.8	0.309	0.404	0.014
0.648	0.8	1	0.8	0.336	0.352	0.002
0.7	0.8	0.17	0.136	0.35	0.3	0

V=0.80						
S_w	H	Perm	h*k	k_{rw}	S_o	k_{row}
0.44				0	0.56	1
0.492	0.8	27	21.6	0.174	0.508	0.333
0.544	0.8	8.05	6.44	0.26	0.456	0.134
0.596	0.8	3.5	2.8	0.309	0.404	0.047
0.648	0.8	1.48	1.184	0.336	0.352	0.011
0.7	0.8	0.43	0.344	0.35	0.3	0

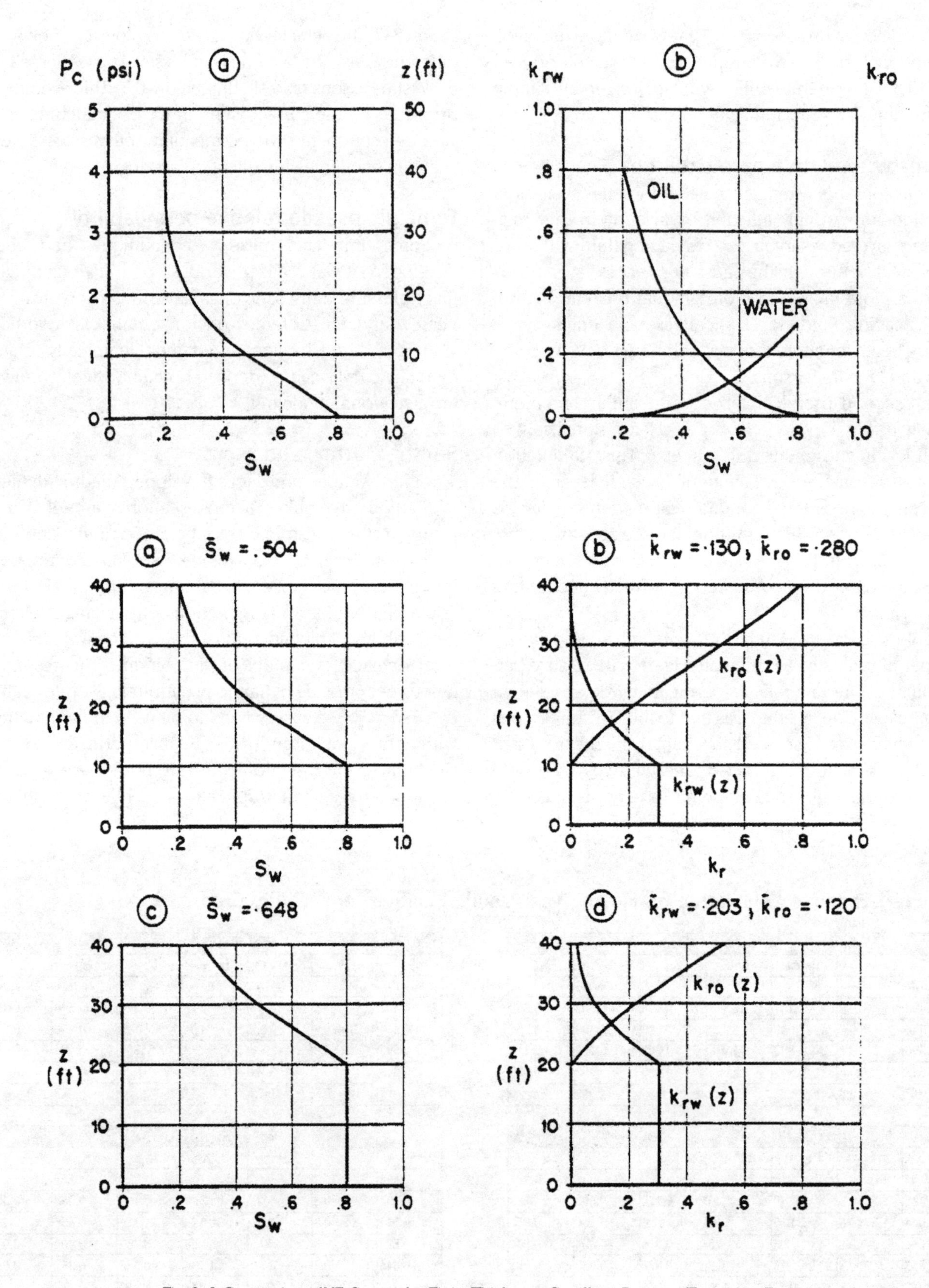

Fig. 8–3 Generation of VE Curves for Finite Thickness Capillary Pressure Transition Zone

z ft	S_w fig. 10.28(a)	k_{rw} fig. 10.28(b)	k_{ro} fig. 10.28(b)
0	.800	.300	0
5	.650	.170	.055
10	.470	.060	.195
15	.350	.020	.370
20	.275	.006	.540
25	.225	.002	.690
30	.200	0	.800
40	.200	0	.800

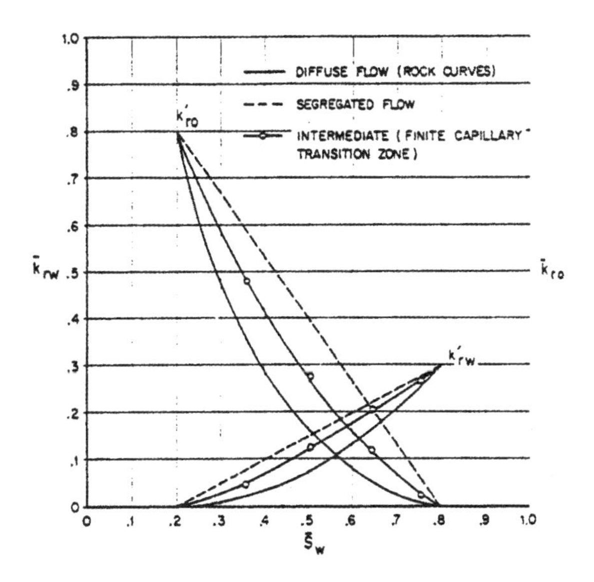

Fig. 8–4 VE Pseudo Relative Permeability Curves for Finite Thickness Capillary Pressure Transition Zone

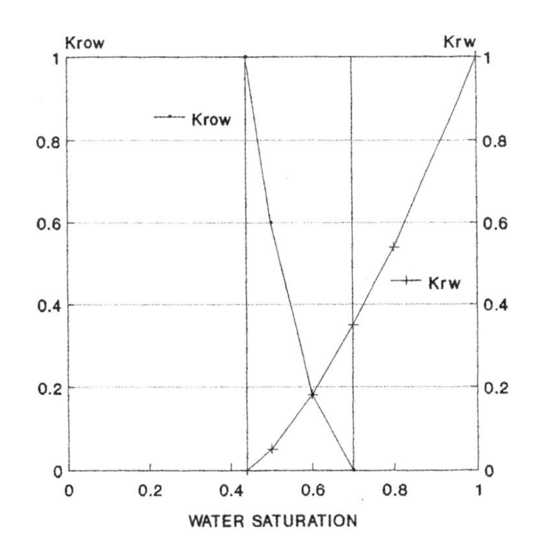

Fig. 8–5 Rock Relative Permeability Curves Used for Hearn-Type Pseudo Relative Permeability Curves

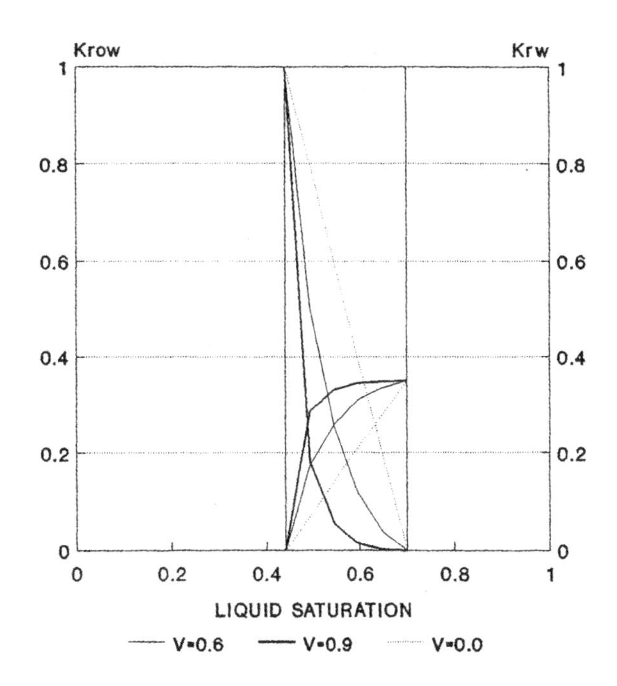

Fig. 8–6 Hearn-Type (Layered) Relative Permeability Curves for Various Dykstra-Parsons Ratios

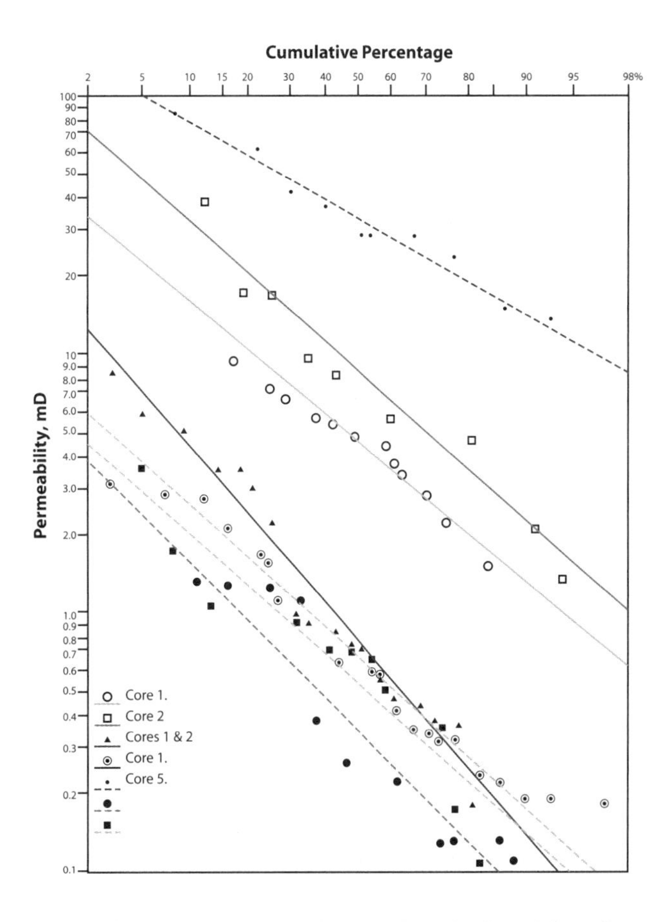

Fig. 8–7 Examples of Dykstra-Parsons Plots for Actual Core Data

results of a multilayer cross-sectional run to within a few percentage points. The cross section Smith, Mattax, and Jacks used is shown in Figure 8–8.[4]

The calculations are outlined in Equations 8.6 through 8.8:

$$\overline{k}_{column} = \frac{\sum_{i=1}^{n} k_i h_i}{\sum_{i=1}^{n} h_i} \tag{8.6}$$

$$\overline{\phi}_{column} = \frac{\sum_{i=1}^{n} \phi_i h_i}{\sum_{i=1}^{n} h_i} \tag{8.7a}$$

$$\overline{S}_{w,column} = \frac{\sum_{i=1}^{n} S_{wi} \phi_i h_i}{\sum_{i=1}^{n} \phi_i h_i} \tag{8.7b}$$

$$\overline{k}_{ro,column} = \frac{\sum_{i=1}^{n} k_{ro} k_{absolute}\, h_i}{\sum_{i=1}^{n} k_{absolute}\, h_i} \tag{8.8a}$$

$$\overline{k}_{rw,column} = \frac{\sum_{i=1}^{n} k_{wo} k_{absolute}\, h_i}{\sum_{i=1}^{n} k_{absolute}\, h_i} \tag{8.8b}$$

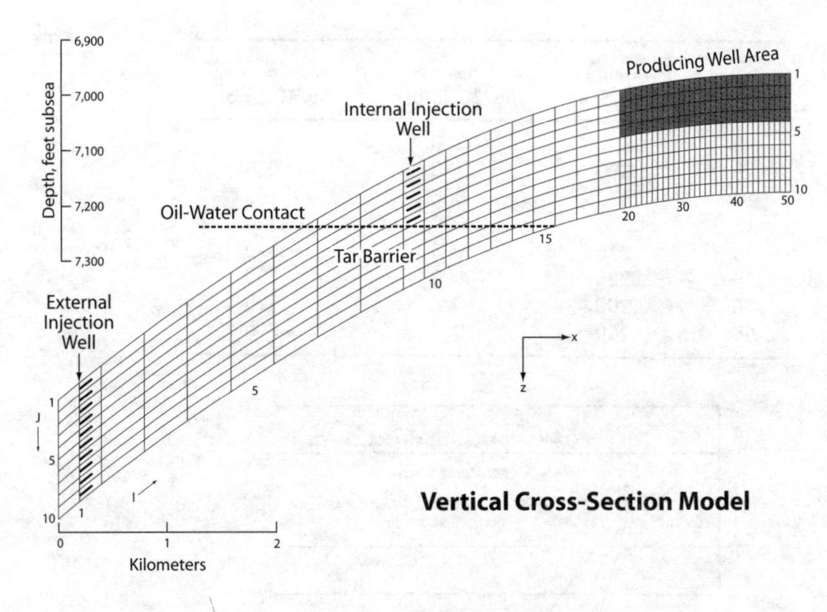

Fig. 8–8 Cross Section Used by Smith, Mattax, and Jacks for Dynamic Pseudo Generation

Kyte and Berry

A second dynamic method has been developed by Kyte and Berry.[5] They averaged saturations and relative permeabilities from laterally adjoining grid blocks to make larger effective grid blocks. This is shown in Figure 8–9 and has the additional advantage of controlling dispersion, provided the original cross-sectional grid is fine enough.

Relative permeability curves used in simulation represent the effects of transition zone thickness, layering, the ratio of vertical to horizontal permeability, and production rates, in addition to the fundamental property of the rock as determined by laboratory experiment.

The details of the calculations are shown in Equations 8.9 through 8.19.

$$\overline{h}_{column} \phi_{column} \Delta x_{column} = \sum_{i=1}^{n} (h_i \phi_i \Delta x_i)_{column} \tag{8.9}$$

$$\phi_{group2} = \frac{\sum_{j=group2\ columns} \overline{h}_j \overline{\phi}_j \Delta x_j}{0.5\left(h_{group1} + h_{group2}\right)\Delta x_{group2}} \tag{8.10}$$

$$\frac{h_{column}k_{column}}{\Delta\bar{x}_{column}} = \sum_{i=1}^{n} \frac{(h_i k_i)_{column}}{0.5(\Delta x_{column} + \Delta x_{column+1})} \quad (8.11)$$

$$k_{group2} = \frac{\Delta\bar{x}_{group2}}{h_{group2}\sum_{i=8}^{12}\frac{\Delta\bar{x}_i}{h_i k_i}} \quad (8.12)$$

$$S_{pw,group2} = \frac{\sum_{group2\,columns} h_i \phi_i \Delta x_i S_{wi}}{0.5(h_{group1} + h_{group2})\Delta x_{group2}\phi_{group2}} \quad (8.13)$$

$$q_{pw,group2} = \sum_{j=1}^{n} q_{wj} \quad (8.14)$$

$$q_{pw,group2} = \sum_{j=1}^{n} q_{oj} \quad (8.15)$$

$$p_{pw,group2} = \frac{\sum_{j=1}^{n} p_{w,datum,8,j}k_{8,j}k_{rw8,j}h_{8,j}}{\sum_{j=1}^{n} k_{8,j}k_{rw8,j}h_{8,j}} \quad (8.16)$$

$$p_{po,group\,2} = \frac{\sum_{i=1}^{n} p_{o,datum,8,i}k_{8,i}k_{ro8,i}h_{8,i}}{\sum_{i=1}^{n} k_{8,i}k_{ro8,i}h_{8,i}} \quad (8.17)$$

$$k_{prw,group2} = \frac{888 q_{pw,group2}\mu_{w,group2}\Delta\bar{x}_{group2}}{k_{group2}h_{group2}p_{w,datum,group2}} \quad (8.18)$$

$$k_{pro,group\,2} = \frac{888 q_{po,group2}\mu_{o,group2}\Delta\bar{x}_{group2}}{k_{group\,2}h_{group2}p_{o,datum,group\,2}} \quad (8.19)$$

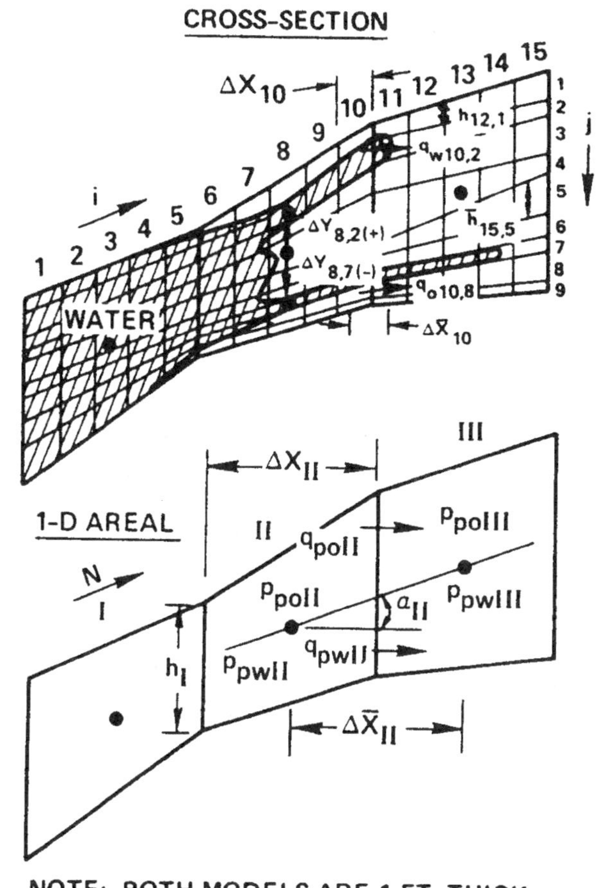

Fig. 8–9 Grid Definition Used by Kyte and Berry for Dynamic Pseudo Generation with Dispersion Control

Curvilinear (streamline) grids

An interesting method to implement and develop pseudo relative permeability curves involves the use of a streamline grid. In fact, this can be borrowed from classic waterflood calculations as shown in Figure 8–10.

The streamlines are actually developed from single-phase incompressible flow, or for incompressible two-phase flow with unit mobility ratio. In truth, the streamlines will move with a displacing fluid, unless the mobilities are identical in front of and behind the waterflood front. This is rarely the case. In practice, the shift in streamlines is relatively subtle compared to the applied pressure gradients in production and waterflooding, and it does not appear to cause a major loss in accuracy. An example of a simple 5-spot adaptation is shown in Figures 8–11 and 8–12.

This technique has been used with considerable success in miscible flooding simulations, where grid orientation effects are a major problem.

Pseudo well relative permeability

For areal studies with multiple wells, one will still encounter situations where multiple fluid contacts exist. The size of the problem will prohibit the use of extremely detailed grids. A single-layer grid model cannot reproduce these effects. The previous example of an element of symmetry from a 5-spot waterflood pattern has been used to demonstrate these effects.

First, the effects of slumping have been quantified with pseudo relative permeability, as shown in Figure 8–13. The use of these rock relative permeability curves is not sufficient to account for the near-well effects.

It is necessary to trick the simulator by assigning a different relative permeability set (gas-liquid and oil-water) to the producing well. This is called a pseudo well relative permeability. This will work provided the saturations in the reservoir change consistently in one direction, which gives a one-to-one mapping for the relative permeability curves. An example of a pseudo well relative permeability is shown in Figure 8–14.

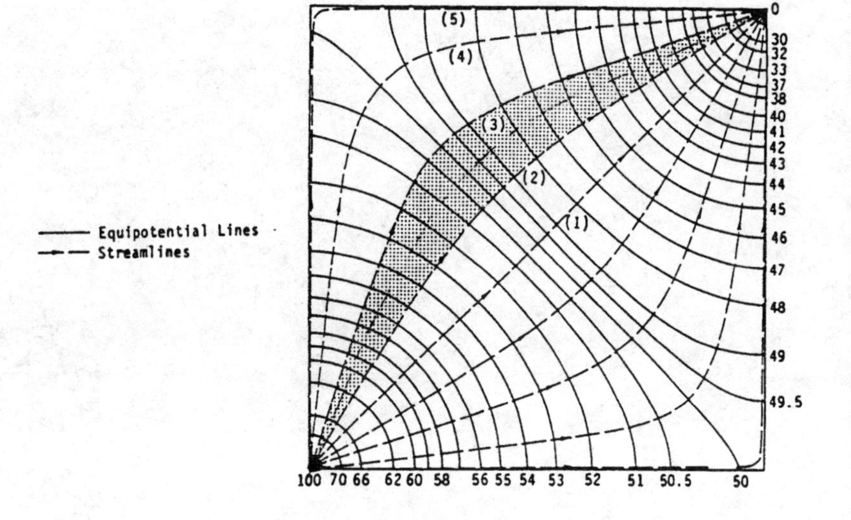

Fig. 8–10 Streamline Grid for Element of Symmetry 5-Spot Waterflood

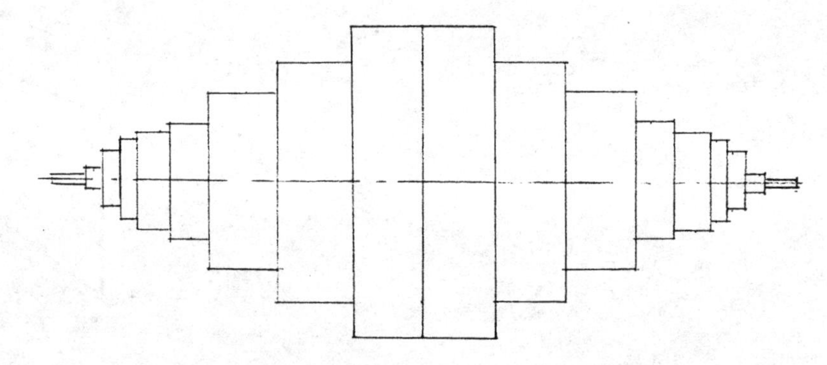

Fig. 8–11 Streamline Grid Model from a 5-Spot Waterflood Pattern

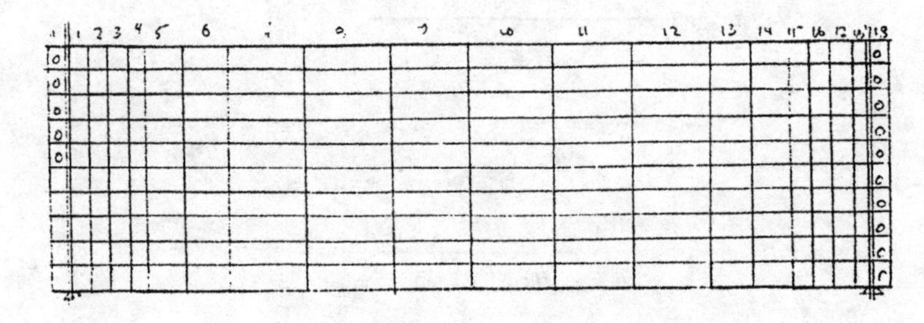

Fig. 8–12 Cross-Sectional View of Streamline Model

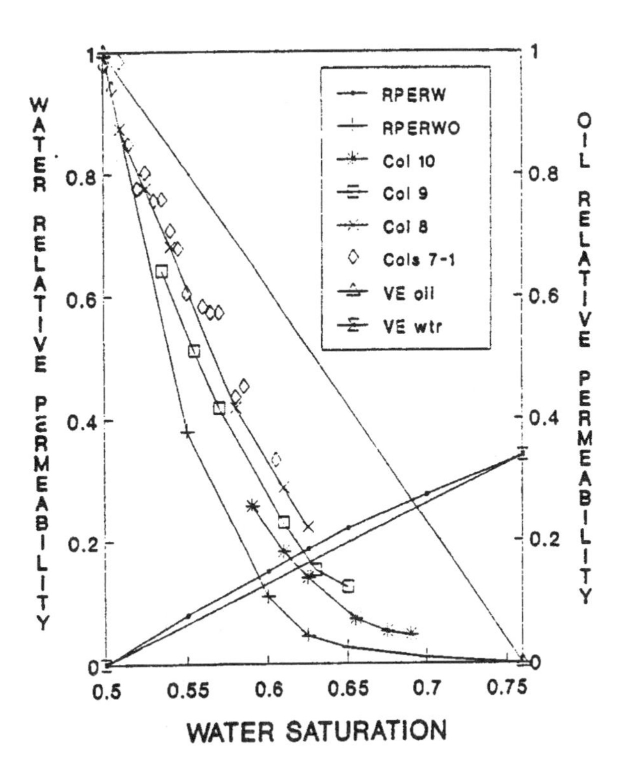

Fig. 8–13 Pseudo Relative Permeability Curves Generated from Streamline Model

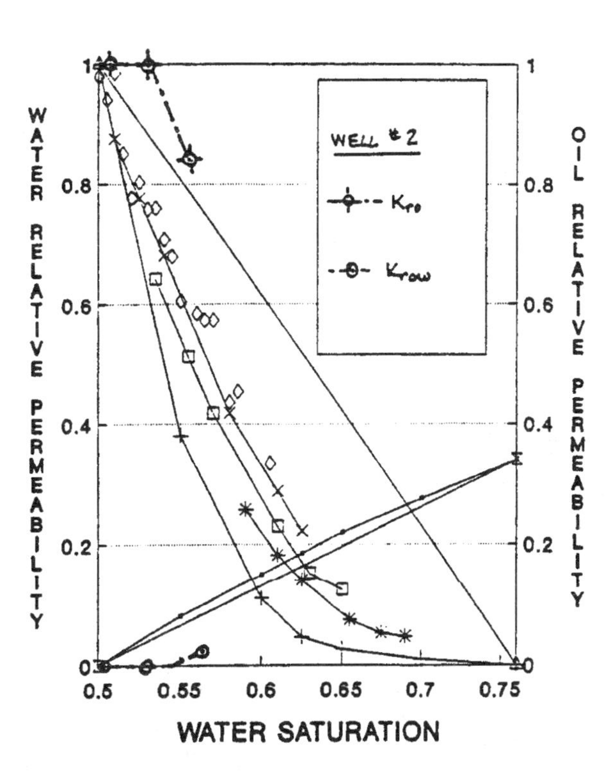

Fig. 8–14 Pseudo Well Relative Permeability Curve Generated from Streamline Model

Note how the character of the well pseudo relative permeability is similar to an end-point saturation shift. This is a typical result. Simulators have an option to input pseudo well relative permeability curves by simply shifting the end-point. This is the way well pseudos are normally input. This implementation is shown in Figures 8–15 and 8–16. The other option is to specify separate relative permeability curves.

The results from coning models can be used as a guide for adjusting the relative permeability for coning effects. This gives a quantitative guide as to the magnitude of the changes.

Analytical calculation

The analytical method of calculating pseudo well relative permeability has been extracted from Emanuel and Cook as shown in Equations 8.20 and 8.21.[6]

$$k_{rp} = \frac{\sum_{i=1}^{n} \left[k_{rp} c_p \left(p_e - p_w \right) \right]}{\sum_{i=1}^{n} c_{pi} \left[\frac{\sum_{i=1}^{n} \left(p_e \phi \right)_i}{\sum_{i=1}^{n} \left(\phi \right)_i} - p_{wa} \right]} \tag{8.20}$$

$$S_r = \frac{\sum_{i=1}^{n} \left(S_p \phi \right)_i}{\sum_{i=1}^{n} \left(\phi \right)_i} \tag{8.21}$$

Dangers of using pseudo relative permeability

The use of such curves has the following dangers:

- It is difficult to transfer the 2-D curves from a cross section to an areal study. Likewise, matching a well pseudo relative permeability can be very difficult.
- The pseudo relative permeability curves can be sensitive to such parameters as the layering and production rates within the reservoir. If these parameters change either areally or in time, the use of pseudos can lead to physically inconsistent predictions.
- Often the range of saturations history matched in a cross-sectional or coning study will not cover the full range of saturations covered in the predictive stages of a study. Some extrapolation will be required.

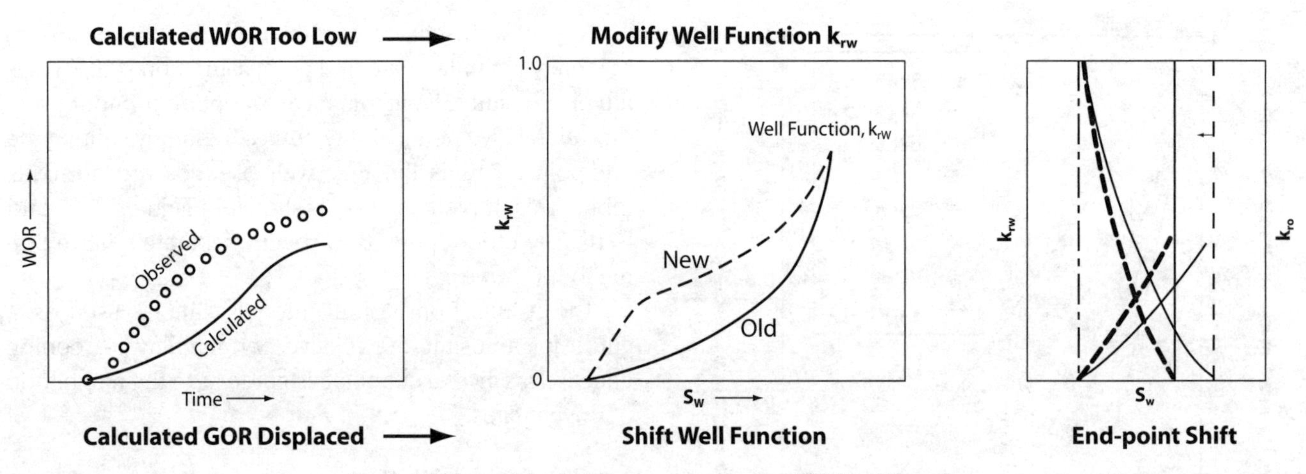

Fig. 8–15 History-Match Adjustment to Correct WOR or Water Cut

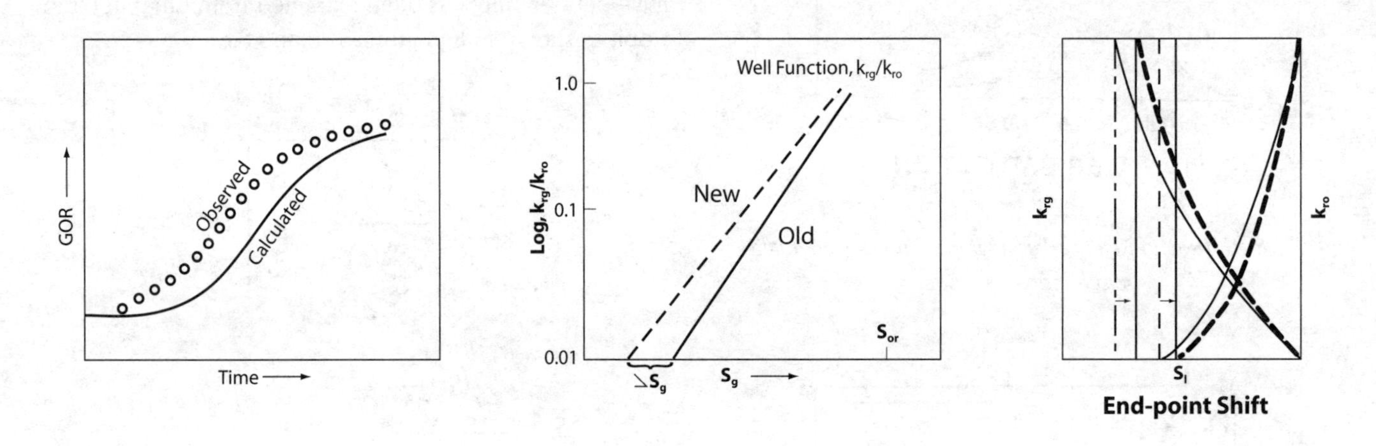

Fig. 8–16 History-Match Adjustment to Correct Well GORs

- It can be difficult to find a good representative cross section through a reservoir. In particular, flow into a cross section from across the reservoir will cause problems. In one cross section prepared by the author, pressures did not drop to the correct levels with actual well production. There was obviously a strong cross-flow in the reservoir. The scope of the study had to be changed.

Evaluation of the pseudo relative permeability concepts is important to determine if they have been correctly implemented. In particular, if dynamic pseudo relative permeabilities or pseudo well permeabilities have been used, the pseudo relative permeability curves will be rate sensitive. Predictions may not be correct if rates change.

Many people are under the impression the quality of a history match is an indication of the quality of the model. *If dynamic pseudo functions have been used, the history match may not be an accurate indication of model quality for predictive purposes.*

Dangers in the use of well pseudo relative permeability

The use of well pseudo relative permeabilities has a number of similar pitfalls.

- A coning or cross-section study should cover the full range of fluid contact movements that will be encountered in the areal study. Frequently, detailed coning or cross-sectional studies only cover the saturation ranges covered during the history-match period and do not cover the range of saturations occurring in the predictive phase.

- If repressuring occurs in the reservoir, the saturations of oil, gas, and water may reverse. This will happen with water or gas injection. Due to hysteresis effects, the rock relative permeability curves and the pseudo well relative permeability curves may not be valid unless hysteresis has been accounted for in the rock relative permeability curves.
- Coning is also a rate-dependent phenomenon. If well rates change significantly, the predicted WORs and GORs will not be correct. In Alberta, significant changes in well rates will probably occur if GOR penalty relief, good production practice (GPP), or increased waterflood allowables have been obtained.

Summary

Most areal simulations involve, to some degree, pseudo relative permeability curves. To the uninitiated, the use of pseudo relative permeability, which results from history matching, appears to be a gross modification of lab data. The previous discussion is intended to provide enough background to understand the basic concepts. A summary of the various techniques and where they should be used is presented in Figure 8–17. The use of these curves is often an economic necessity. In general, it is best to avoid pseudo relative permeabilities if possible.

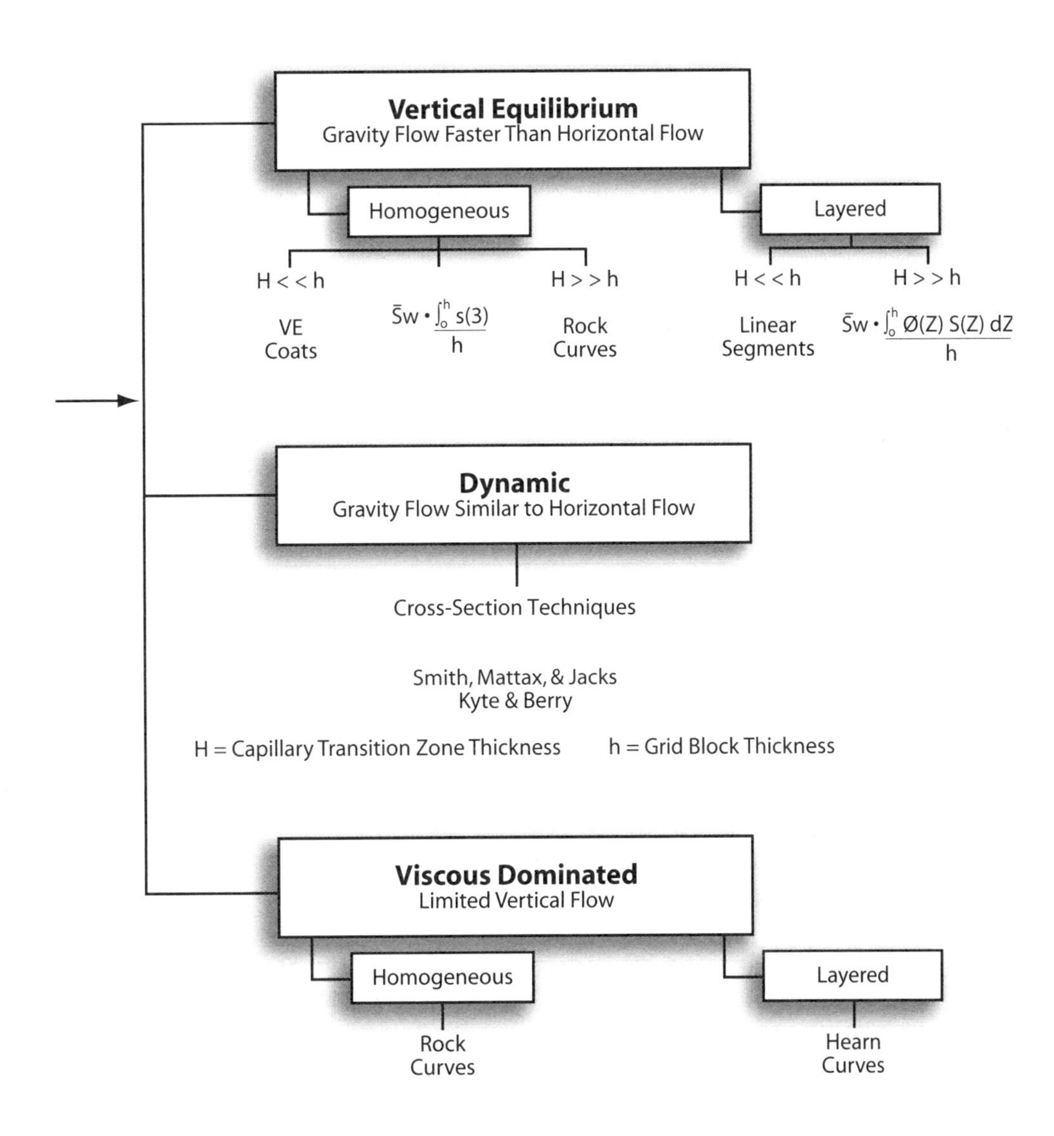

Fig. 8–17 Summary of Pseudo Relative Permeability Curve Techniques

References

1 Coats, K.H, J.R. Dempsey, and J.H. Henderson, "The Use of Vertical Equilibrium in Two-Dimensional Simulation of Three-Dimensional Reservoir Performance," Dallas, Texas: Second Symposium on Numerical Simulation, (February 4–5), 1970.

2 Stiles, L.H., "Optimizing Waterflood Recovery in a Mature Waterflood, The Fullerton Clearfork Unit," SPE 6198, Society of Petroleum Engineers, 1976.

3 Dykstra, H., and R.L Parsons, "The Prediction of Oil Recovery by Waterflood," *Secondary Recovery of Oil in the United States, Principles and Practice*, 2nd ed. American Petroleum Institute, 1950.

4 Jacks, H.H., O.J.E. Smith, and C.C. Mattax, "The Modeling of a Three-Dimensional Reservoir With a Two-Dimensional Reservoir Simulator—The Use of Dynamic Pseudo Functions," *Society of Petroleum Engineers Journal*, 1973.

5 Kyte, J.R., and D.W. Berry, "New Pseudo Functions to Control Numerical Dispersion," *Society of Petroleum Engineers Journal*, 1975.

6 Emanuel, A. S., and G. W. Cook, "Pseudo-Relative Permeability for Well Modeling," *Society of Petroleum Engineers Journal*, 1974.

9

Initialization

Introduction

The primary objective of initialization is to establish the correct amount and distribution of the original reservoir fluids. The majority of simulations are initialized based on static gravity capillary pressure equilibrium. This works well in the majority of cases. However, there is no requirement the fluids in the earth are static and, in truth, this is rarely completely correct.

A number of different methods are used to enter fluid saturations.

Grid-block center capillary pressure initialization

If the grid blocks are smaller than the capillary pressure transition zone, then the saturations of grid blocks can be accurately estimated from the capillary pressure curve at the midpoint of the grid block. This also requires the depth of the relevant contacts be specified.

This was the method used by early reservoir simulators. The distribution of fluids in the grid blocks was assumed to be uniform.

Pseudo capillary pressure

Later, it became possible to use a specialized technique to simplify reservoir problems that involved capillary pressure transition zones smaller than the grid block height. Known as vertical equilibrium (VE), this method required the saturations in the grid block be averaged. Since the transition zone is assumed to be of negligible thickness, the saturation of the block can be calculated using a linear relationship based on the distance from the specified gas-oil contact (GOC) or water-oil contact (WOC). This is shown in Figure 9–1.

Recall from chapter 8 that VE generated straightline relative permeabilities. Capillary pressure was ignored in the displacement calculations. As shown earlier, capillary pressure does affect frontal displacement and introducing this *fudging* strategy to the relative permeability will affect displacement processes in the simulator.

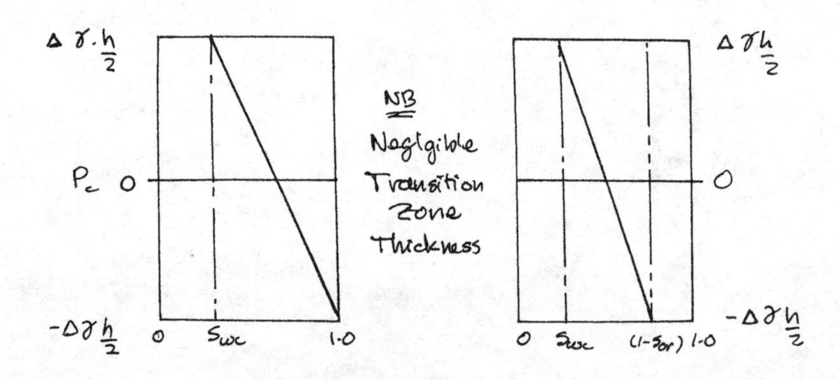

Fig. 9–1 Pseudo Capillary Pressure Curve—VE

Finite transition zone thickness

A problem soon became apparent regarding relatively thick blocks existing with transition zones consisting of a fraction of the grid block. Such a situation is shown in Figure 9–2.

Early efforts involved generating a pseudo capillary pressure graphically. Such a curve has the same appearance as the results shown in Figure 9–3 for the previous data.

This will correctly calculate the original fluids in place. Apparently, it is difficult to program an automatic integration of the P_c curve. This system had two problems. The capillary pressure curve is dependent on grid-block thicknesses. This makes calculation of pseudo capillary pressures difficult where grid-block thicknesses vary; the solution is that one must use uniform grid-block sizes. In addition, the pseudo capillary pressure affects displacement calculations.

Often, getting the correct amount of fluid in place was more important than displacement, such as primary production performance. Modern simulators now use a different approach.

Grid-block subdivision

By arbitrarily subdividing grid blocks during the initialization process, as shown in Figure 9–4, it is possible to pick multiple points on the P_c curve and average them to derive an averaged grid-block saturation.

For most programs, this is user transparent. However, there are programs allowing users to specify how many subdivisions are to be made. It is also necessary to specifically request depth averaging for this type of calculation to be done. This is much easier to implement and should give satisfactory saturation accuracy.

Further thought is needed regarding gridding. With a fluid distribution in the grid block, does the computed pseudo capillary pressure accurately represent the displacement process? The author has not done any tests on this yet, but it may be a serious consideration in constructing a reservoir grid.

Type of capillary pressure data

At present, capillary pressure data for initialization is normally obtained from drainage capillary pressure tests. This is one area where simulation technology may change. As discussed earlier, the real reservoir process is one of countercurrent imbibition. Eventually, with more development, initialization may be based on more realistic laboratory data.

Tilted water contacts

Tilted water contacts have been recognized for a long time. Figures 9–5 and 9–6 show some diagrams for reservoirs where tilted water contacts are well-established.[1] There are many reservoirs with tilted water contacts, which are caused by the subsurface movement of formation fluids. These movements are studied by hydrogeologists.

Hydrogeology

The study of hydrogeology is far from new. Most provincial governments and the federal government are busy modeling surface groundwater flow on several watersheds in the province at any given time. Most of these studies do not extend far past the Pleistocene shallow formations and are intended to manage farm, municipal, and industrial groundwater use. A significant amount of original research on this topic was done by the Dutch to support their system of dykes.

The application to oil and gas was recognized many years ago by M. King Hubbert.[2] He demonstrated the movement of water in the subsurface would alter the location of oil and gas pools and showed how to calculate this effect. He died a few years ago and was widely recognized for his achievements. However, the practical application of this theory has been slow.

Due to the foresight of the Canadian government and industry, all DST data is in the public domain after one year. This source of data is critical to performing basin-scale hydrogeology. Brian Hitchon of the National

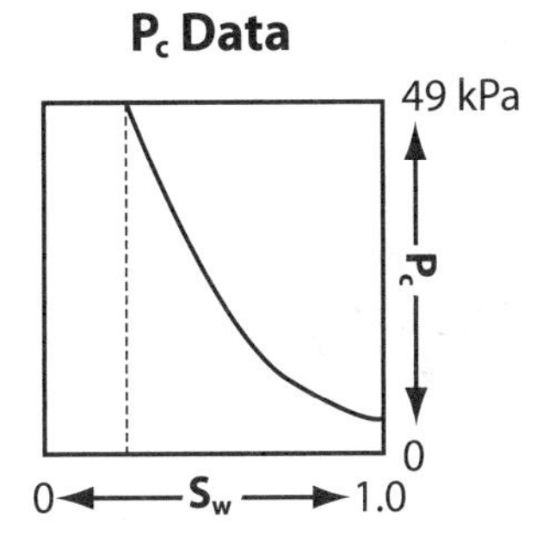

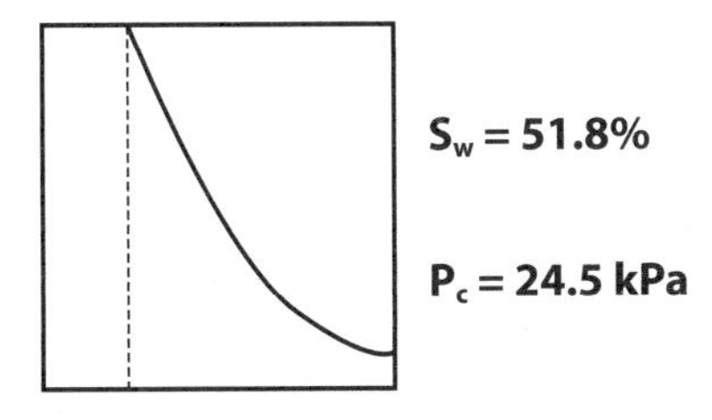

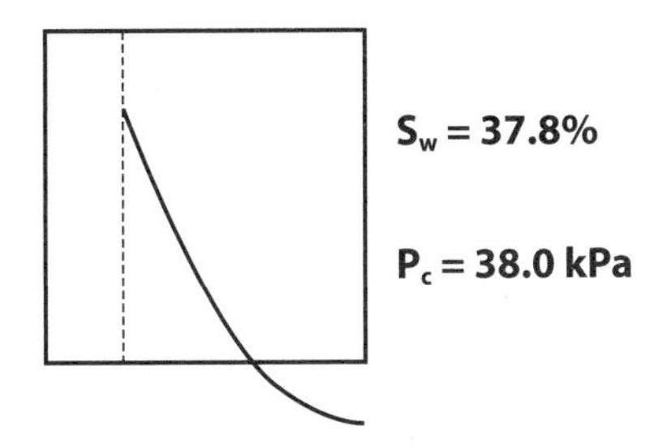

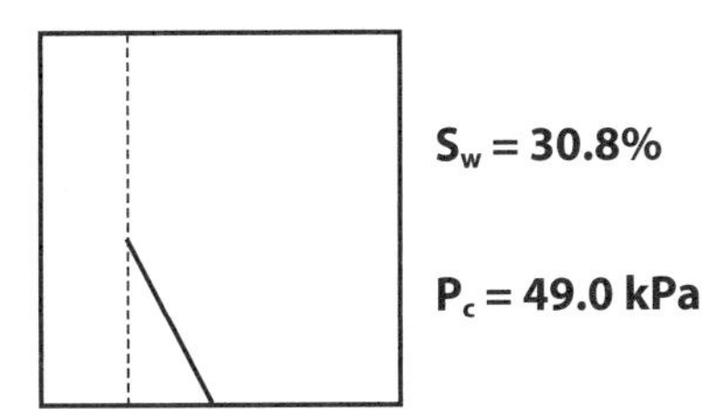

Fig. 9–2 Grid Block with Finite Thickness P_c Zone

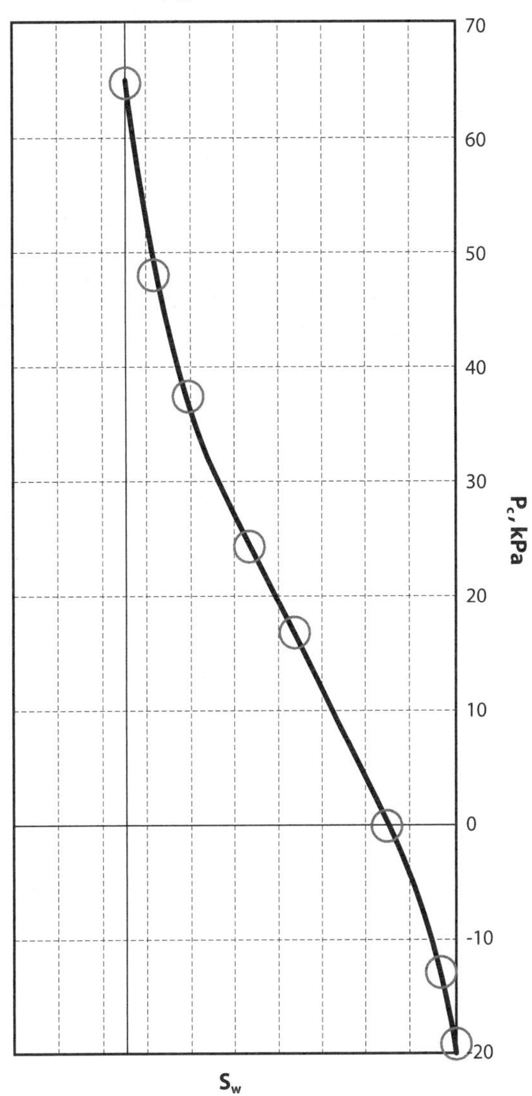

Fig. 9–3 Pseudo Capillary Pressure Curve Calculation—Finite P_c Zone

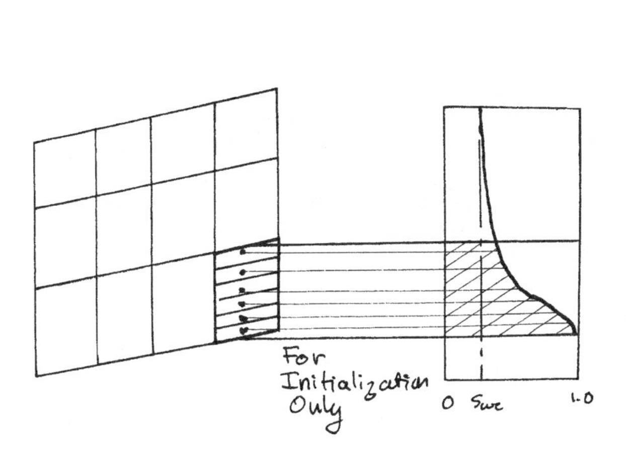

Fig. 9–4 Generation of Average Grid-Block Saturation Using Vertical Grid Refinement

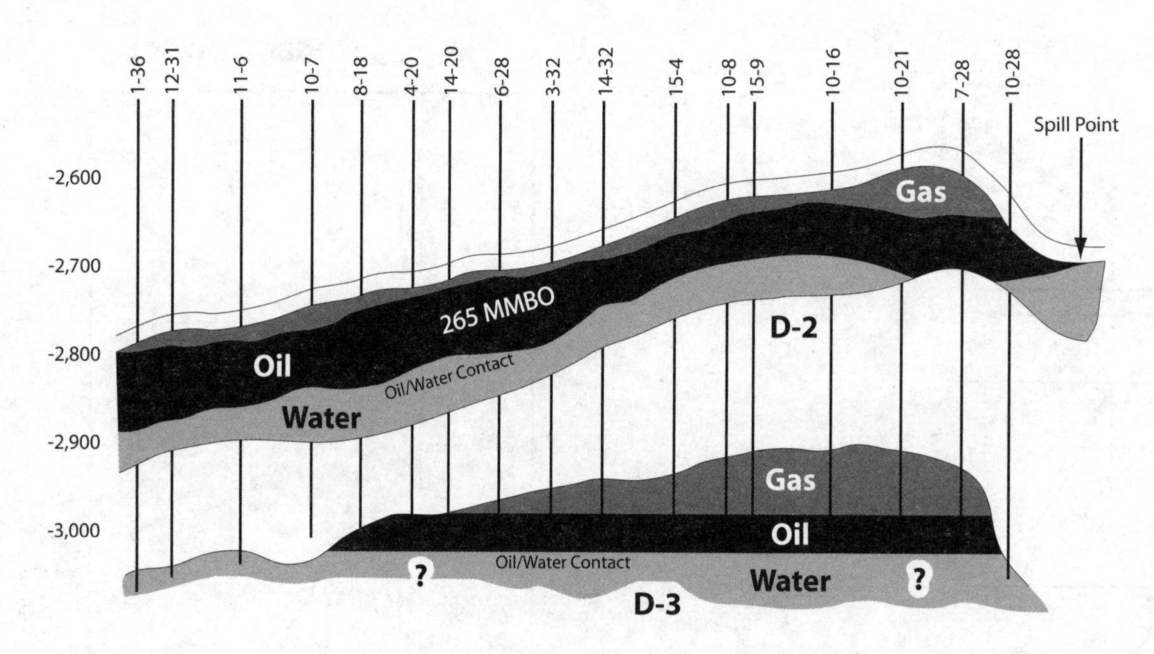

Fig. 9–5 Example of Tilted Water Contact—Leduc Field

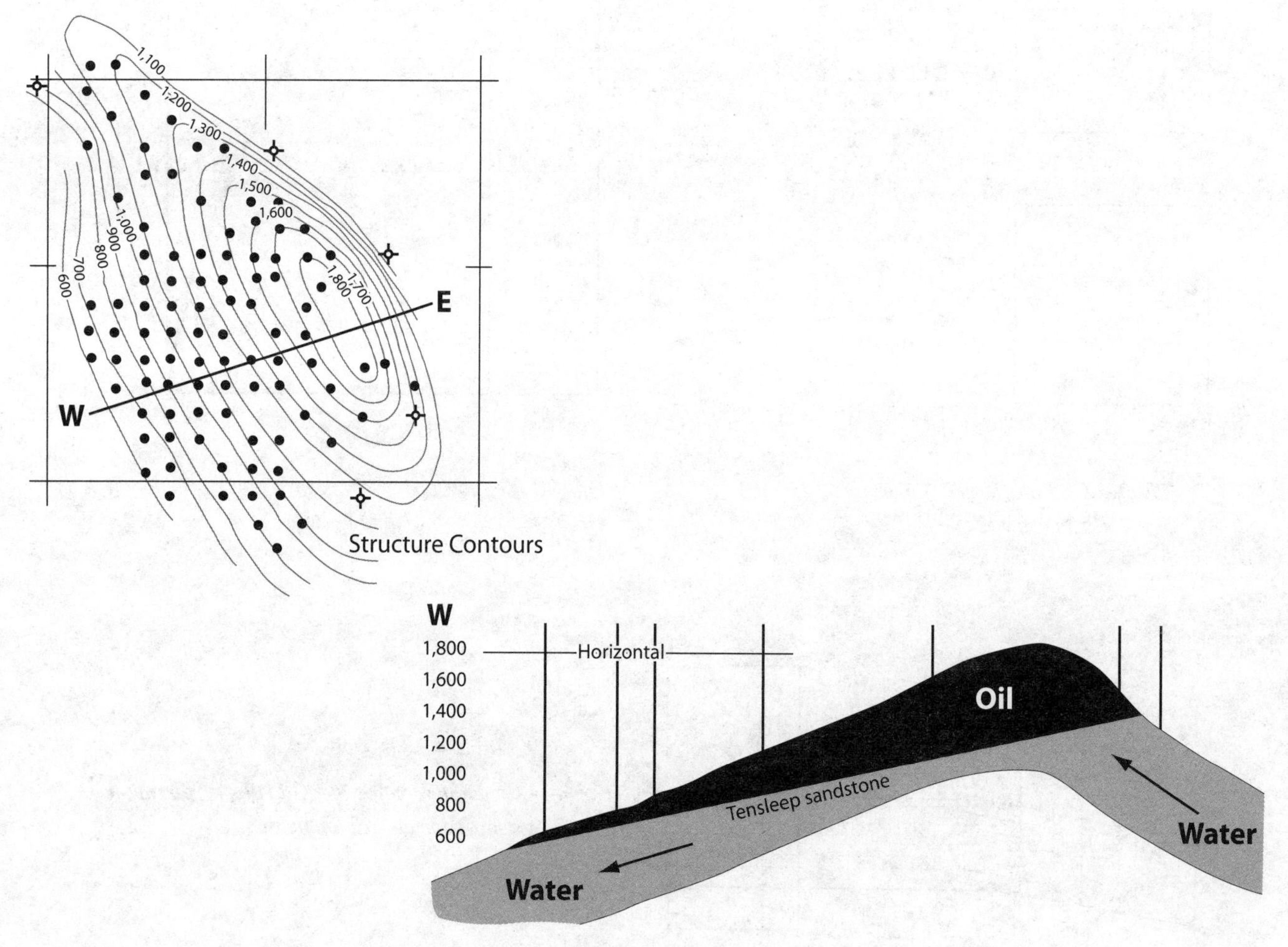

Fig. 9–6 Example of Tilted Water Contact—Montana

Research Council Canada (NRC) has done pioneering work in this area.[3] One company remarketed the DST data with interpretations in a form that had been quality-control checked, but they are no longer actively selling this data. More recently, Hydro Petroleum Canada has prepared complete subsurface potential maps for almost the entire basin, and it includes water chemistry as well. The trapping of most oil and gas pools in the Province of Alberta can be explained based on fluid movements and a number of trapping mechanisms.

Many exploration wells during the years have been termed geological successes but economic failures. Restated, the geology extrapolated was accurate, but no oil or gas accumulation was found. The industry has not, historically, been able to predict the structures, reefs, etc. containing oil or the ones containing oil and gas.

It is much easier to trap gas, and this is why 60% of reserves in Canada on a barrels of oil equivalent (BOE) basis are gas.

Manual input

Water, oil, and gas saturations can be input manually in a simulator using one of many different read options. This would include grid-block by grid-block specifications as well as block or layer reads.

This is normally done either on simple data sets, where it is faster to hardwire a saturation, or when dynamic conditions exist.

In the latter case, it is possible to input saturations based directly from log analysis. At present, simulators do not allow for dynamic initialization. However, a simulator can be run to achieve this objective and the output arrays converted to input arrays. Developing the dynamic initialization may require experimentation, and the reservoir simulator must be run for a sufficient time to emulate steady state conditions.

Reservoir engineering implications of dynamic trapping

Tilted contacts and dynamic conditions are frequently not recognized. Quite bizarre interpretations are often required in order to force (incorrectly) a static interpretation. Some examples are outlined following.

Example 1. If the contact is tilted, the obvious adjustment to the geological interpretation is compartmentalization with separate water contacts. Very often, these end up as the mystery pinchouts and phantom faults, i.e., can't be seen on seismic, such as those depicted in Figure 9–7.

Clearly, when a simulation is done on a reservoir that is actually continuous, it is going to be difficult to history match reservoir performance. In one case, we had a client who was going to stop injection since they felt communication did not exist, despite the fact pressure data proved conclusively that communication was occurring.

Example 2. Very often, most of the oil or gas will be displaced off the top of the structure and reservoir performance will reflect the original distribution of fluids. This is shown in Figure 9–8.

Water influx will occur from updip and downdip, and the best wells will actually be offset from the structural high. This example is based on a reservoir in Saskatchewan and is easily demonstrated based on well cumulative productions. *If a static gravity equilibrium is used here, it will be difficult (impossible?) to match the cumulative production, since the oil will not be located in the correct location.*

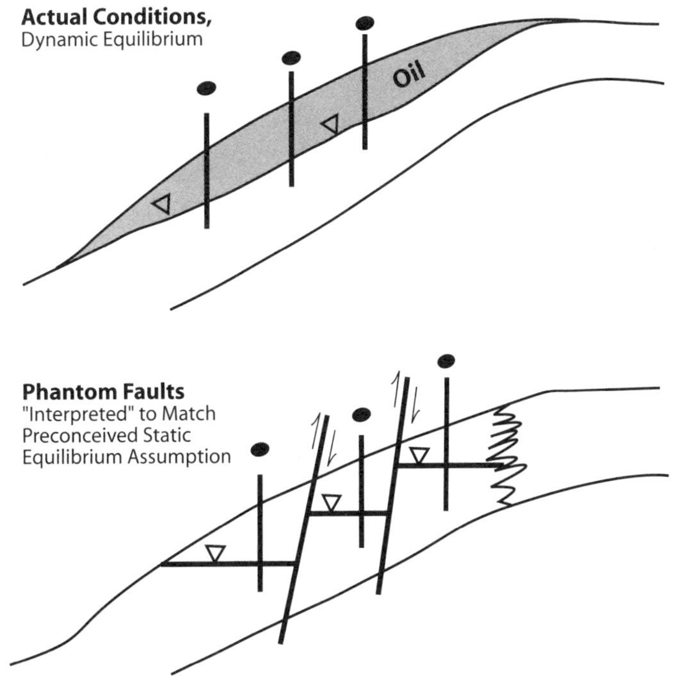

Fig. 9–7 Effect of Misinterpreting a Tilted Water Contact as a Series of Barriers

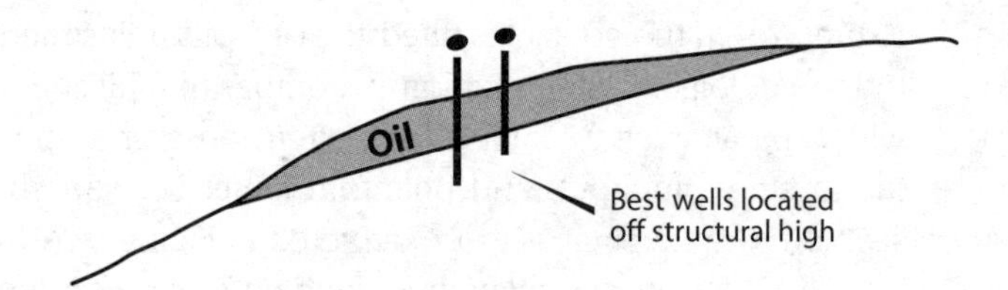

Fig. 9–8 Effect on Well Placement When a Tilted Water Contact Has Been Misinterpreted

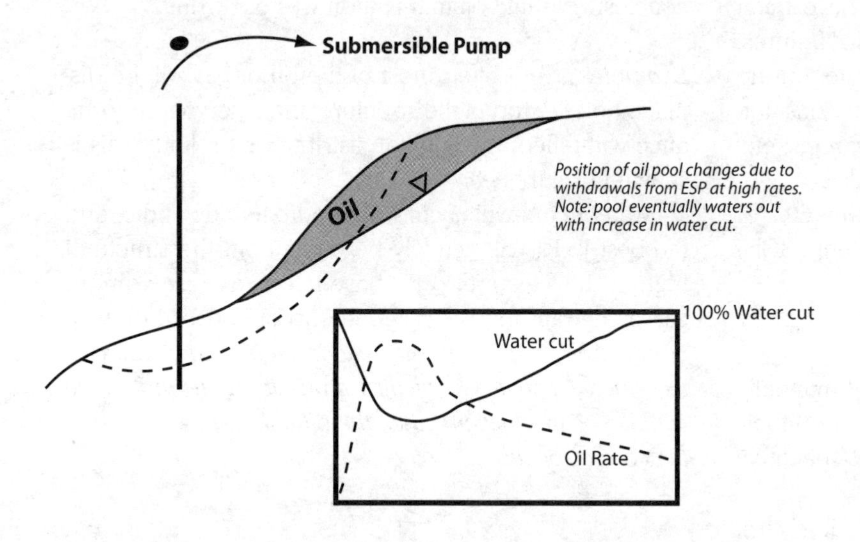

Fig. 9–9 Explanation of Wells that Originally Produce 100% Water Followed by Increasing Oil Cuts

Conventional waterflood approaches have shown placing an injector in the center of high-permeability in an isolated pool will give the highest flood efficiency. Based on the assumption used, this is true. However, if one places injection in the center of a hydrodynamic pool, then the oil is ejected out of the trap. This is depicted in Figure 9–11.

This has happened in a few pools. *Waterflooding with incorrectly placed injectors in a hydrodynamically trapped pool will reduce recovery.*

Simulation failures

These are examples of so-called simulation failures. Of course, the simulator is an inanimate object and is not responsible for anything. The failures that have occurred did so because the underlying physics were not understood. Sadly, the poor understanding often dates back to the 1950s. However, in Canada, the wealth of public data has made available detailed mapping of the sedimentary basin, and the full implications of hydrogeology can be applied.

Example 3. In some reservoirs, the water cut will go down when the well is put on production at high rates. The situation is shown in Figure 9–9.

The gradients induced by reservoir production are much stronger than the potential gradients associated with trapping. The performance described previously is common in the Sawtooth in the southern part of Alberta. Often, these wells will mysteriously produce many times the OOIP for their drilling spacing unit. *A gravity static equilibrium is unlikely to calculate the correct cumulative oil, since the updip pinchout location is unknown and the strong aquifer precludes accurate material balance calculations.*

Example 4. From an areal perspective, oil trapped hydrodynamically in high-permeability formations will naturally shrink to a central point. The influx occurs from all directions. Water displacement will progress as shown in Figure 9–10.

Modeling dynamic equilibria

Geologists are quite interested in hydrogeology, and the author has prepared a talk on reservoir engineering implications for the Canadian Society of Petroleum Geologists (CSPG). The positions of oil-water and gas-oil contacts were generated using a reservoir simulator. Most hydrogeology calculations utilize steady state flow assumptions and could solve this kind of problem faster than a reservoir simulator. However, a model specifically for initialization has yet to be built.

An anticlinal-like structure with a production well located on the structure is shown in Figure 9–12. At present, the fluids show a distribution of fluids that would be obtained via static gravity equilibrium. Capillary pressure is included in this model. Note the gradual change in the water-oil transition.

The situation has been made a little more realistic by including layering in the problem. The permeability distribution could change to something similar to Figure 9–13.

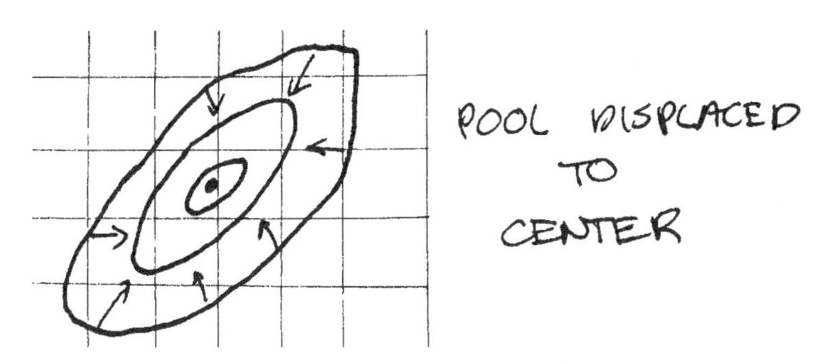

Fig. 9–10 Natural Effect of Water Influx with an Active Aquifer

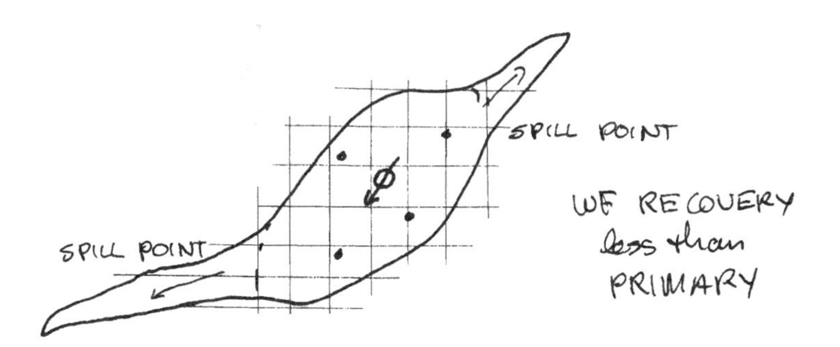

Fig. 9–11 Effect of Misplacing an Injector in Dynamically Trapped Pool

To obtain dynamic equilibrium, the simulator was deceived by using an injection and production well combination in the aquifer. The author has used a high horizontal potential gradient to make the point more clearly. The resultant potential distribution is shown in Figure 9–14.

The potential field is not identical to the pressure distribution, which is shown in Figure 9–15.

The locations of the fluids are shown in Figure 9–16.

The direction of water flow can also be reversed. The effect of this is shown in Figure 9–17.

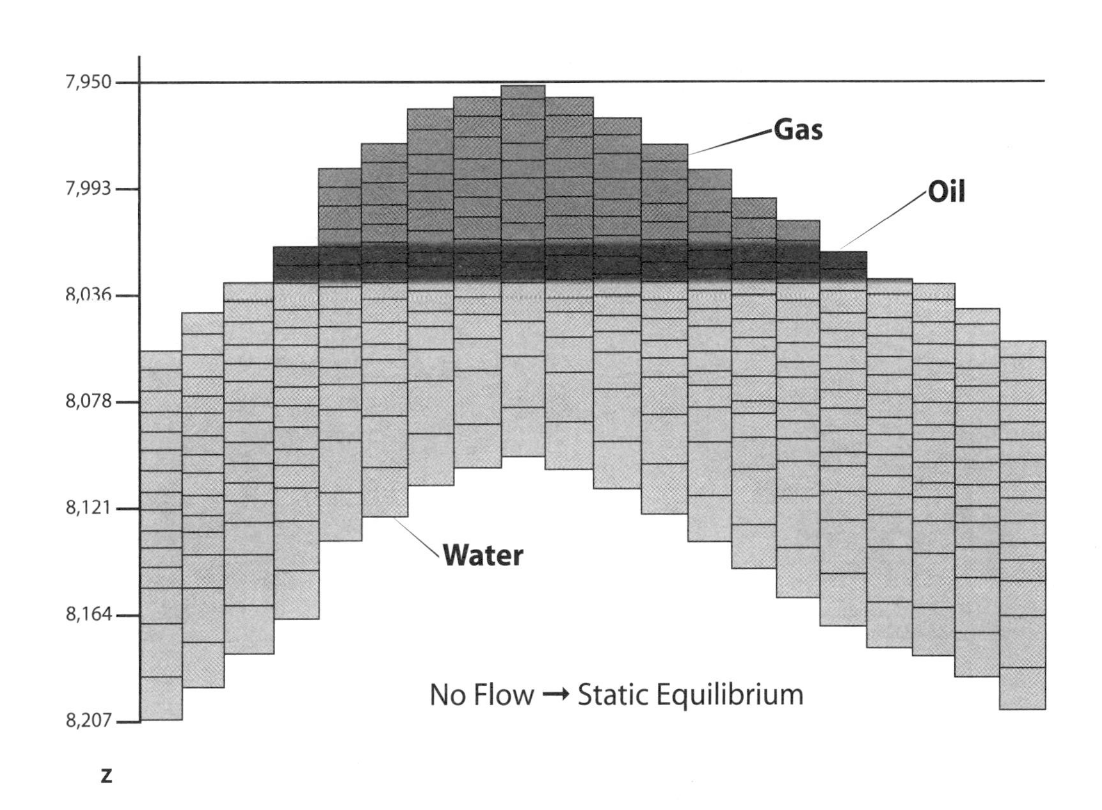

Fig. 9–12 Reservoir Simulation Tests for Dynamic Equilibrium—Static Initialization

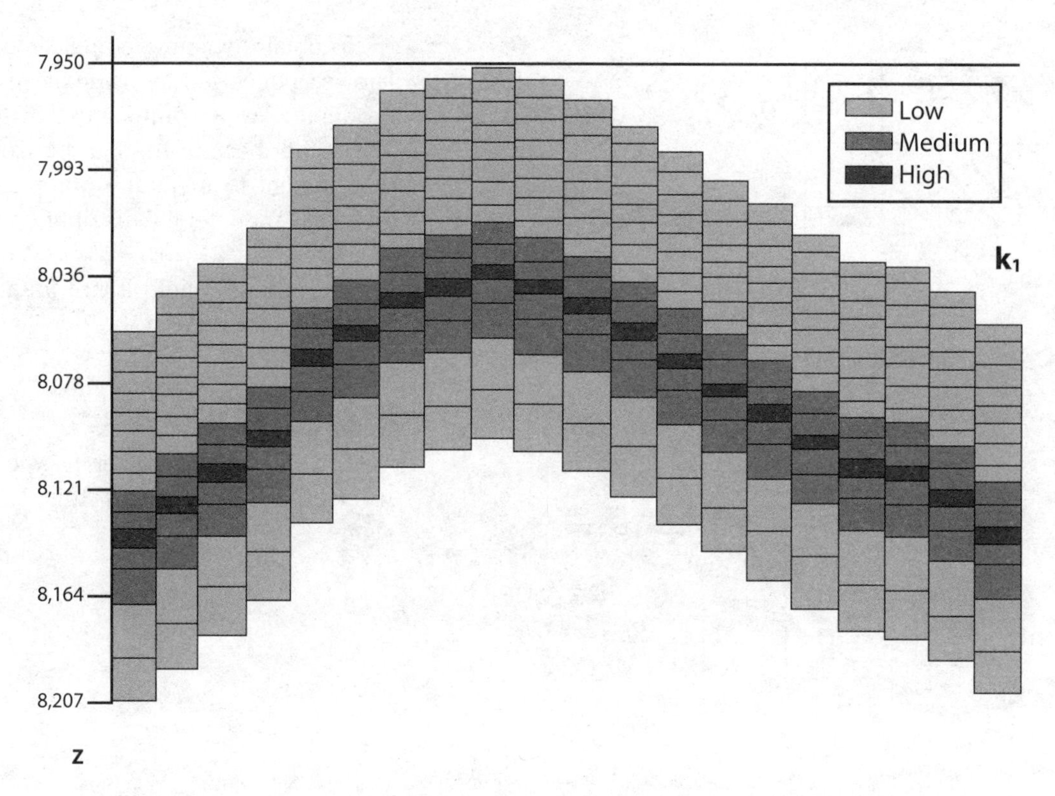

Fig. 9–13 Reservoir Simulation Tests for Dynamic Equilibrium—Permeability Distribution

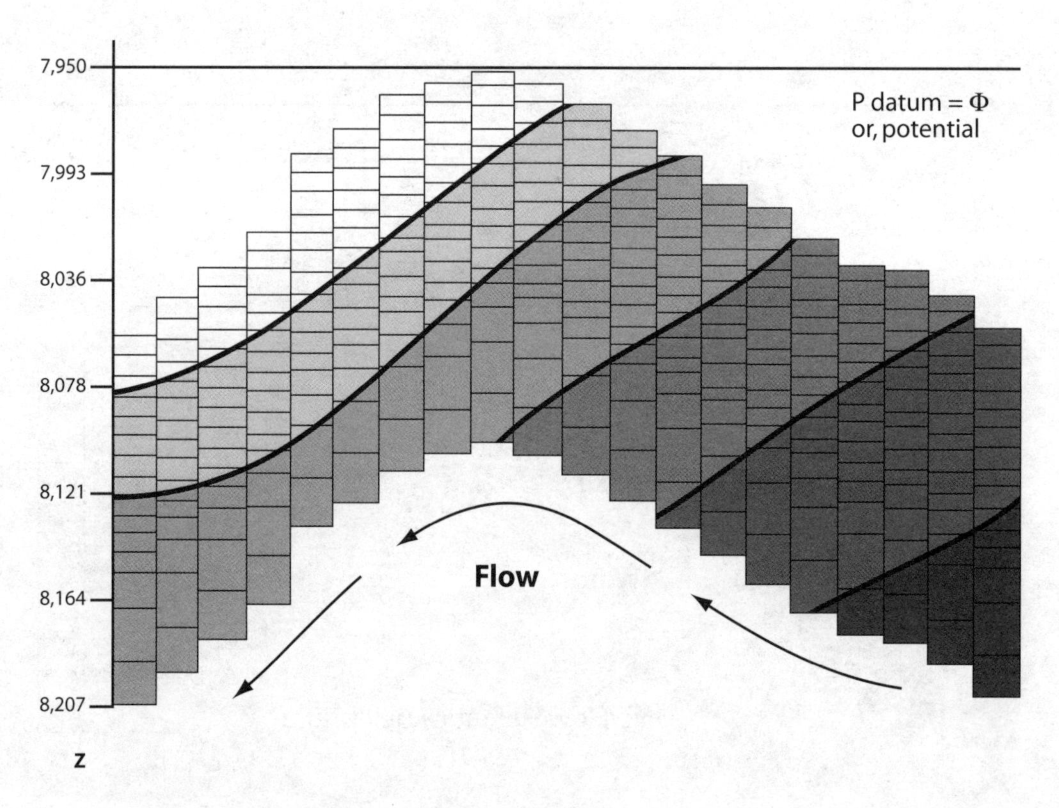

Fig. 9–14 Reservoir Simulation Tests for Dynamic Equilibrium—Flow Potential (Datum Pressure)

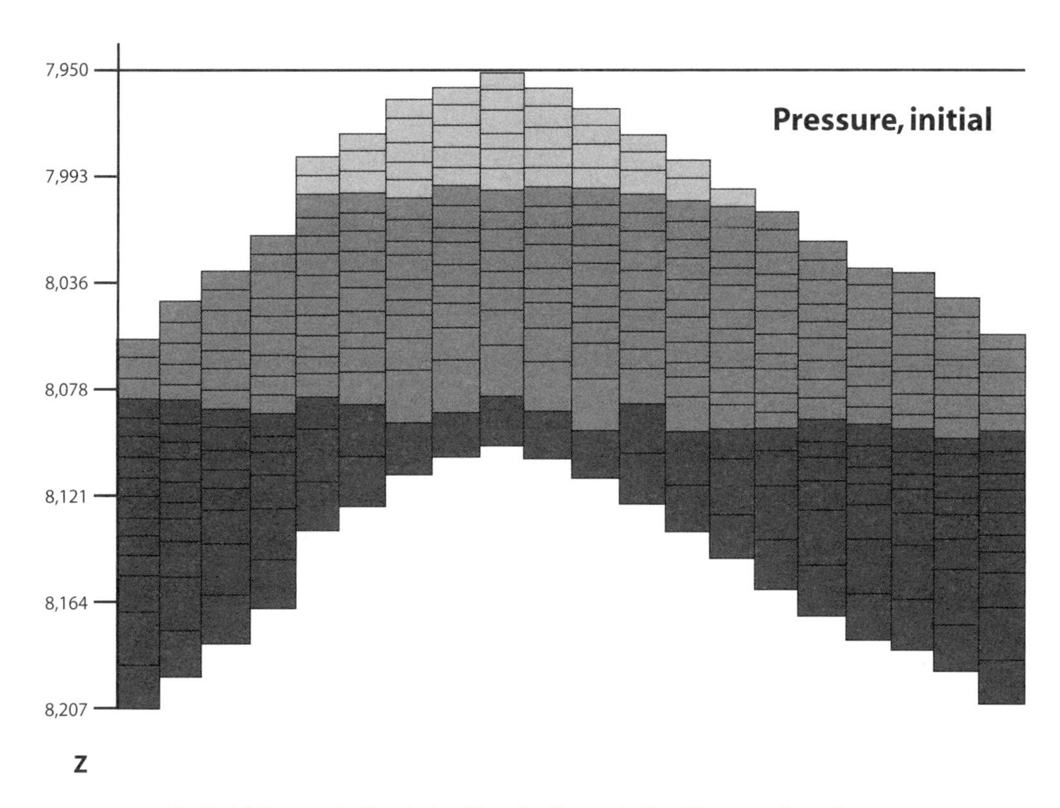

Fig. 9–15 Reservoir Simulation Tests for Dynamic Equilibrium—Raw Pressure

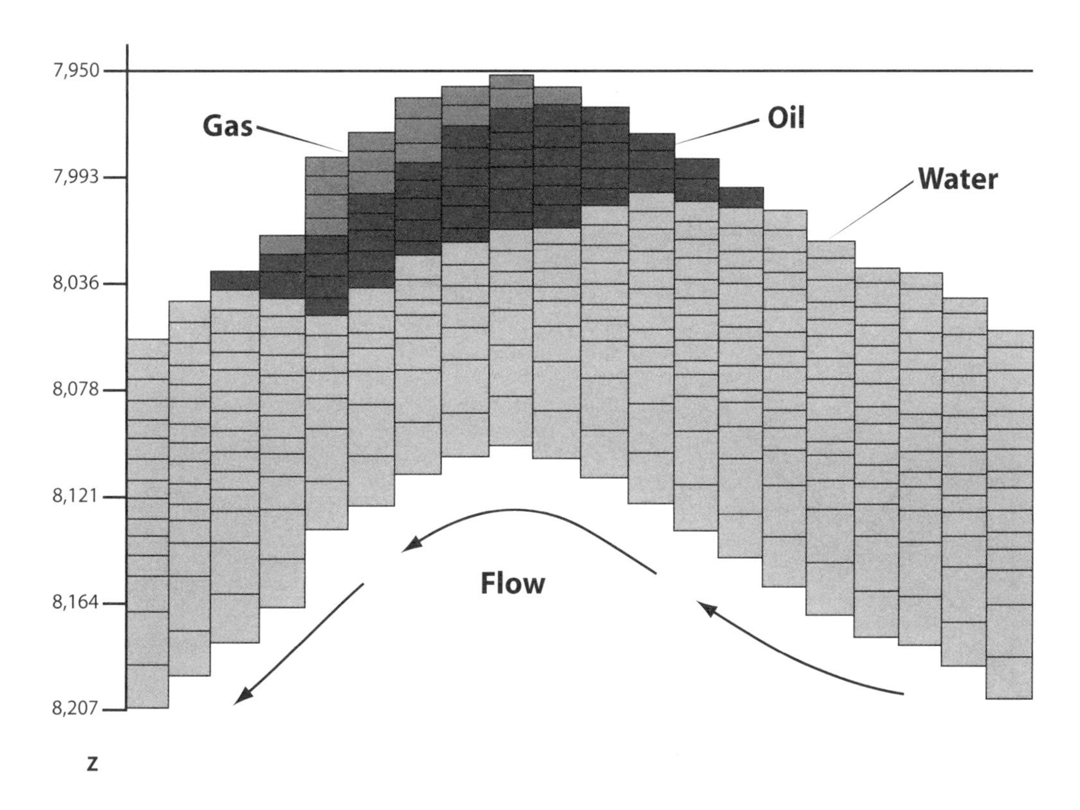

Fig. 9–16 Reservoir Simulation Tests for Dynamic Equilibrium—Location of Fluids

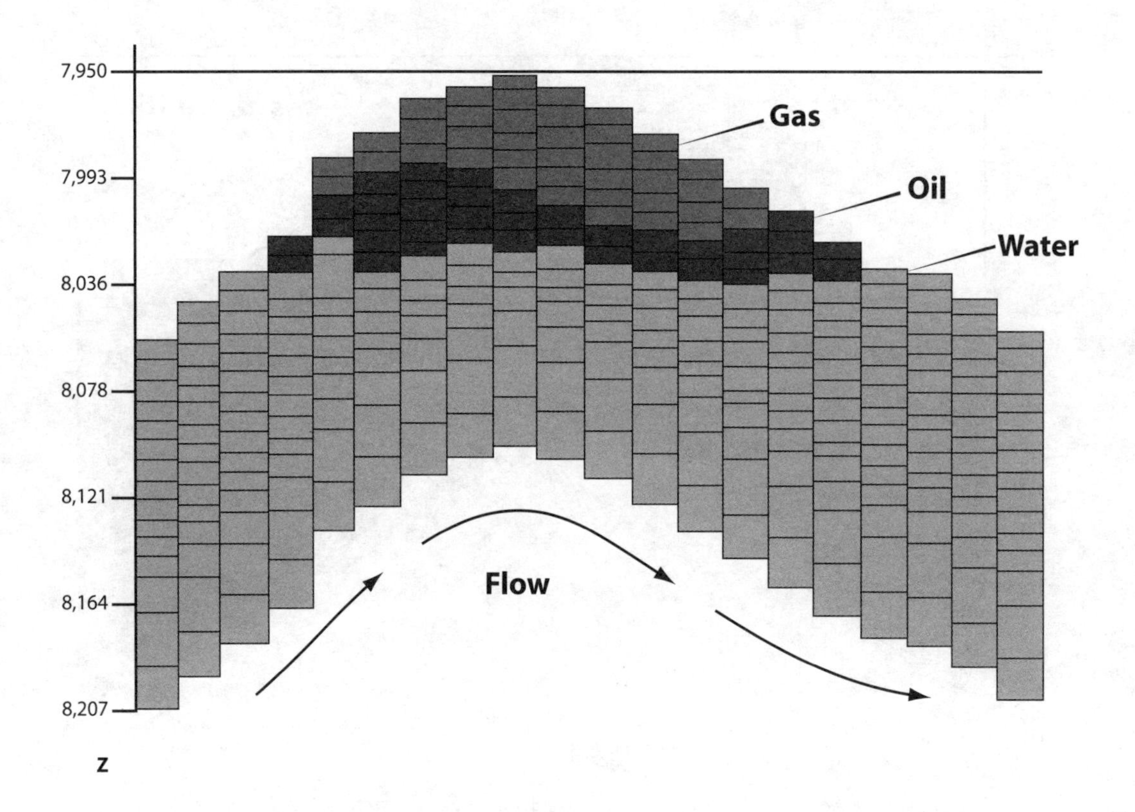

Fig. 9–17 Simulation Tests for Dynamic Equilibrium—Reversed Flow Location of Fluids

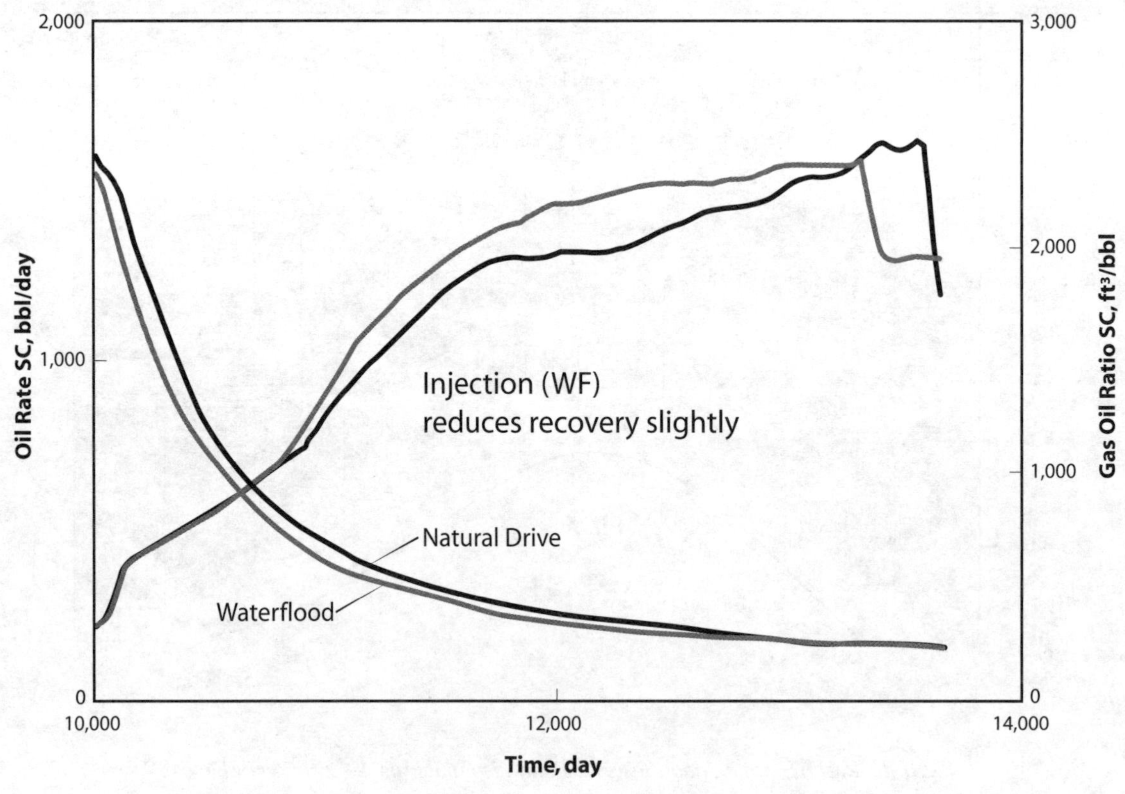

Fig. 9–18 Reservoir Simulation Tests for Dynamic Equilibrium—Waterflood Sensitivity

The location of fluids is determined by the potential field and, when steady state is achieved, is independent of the absolute permeability. Layers will affect location of the hydrocarbons. You can increase permeability to speed up the development of the saturation distribution. The output from such a run can be used to manually specify the saturation array in the simulator.

Water injection

The effects of water disposal were modeled using a restart file. Cases were run for both initial conditions in which water flow starts from both directions. In both cases, recovery from the production well went down with waterflooding instead of increasing. The production forecasts are shown in Figure 9–18.

The effect of the high-permeability layer is shown in Figure 9–19. Note the drop in oil productivity and the change in the GOR profile. The waterflood starts at 10,000 days, which was the time required to develop the steady state flow conditions for initialization.

Hydrogeology

Recently this problem has been resolved by Henning Lies of Hydro Petroleum Canada (HPC). At the time of this writing, most of his material remains unpublished, which is unfortunate. HPC has created regional hydrogeological maps of most of the Western Canadian Sedimentary Basin. They have examined what causes the trapping of probably 80% of the pools in Alberta, British Columbia, Saskatchewan, and Manitoba.

Another key area where HPC has done practical research is in modeling the fluid movement in the subsurface. At present, only one of these projects has been published: "Hydrogeological Trapping Model of the Milk River Gas Field" by H. Lies.[4]

No production runs

During the early days of simulation, it was common to run the simulation with no well production. The objective of this test was to see if gravity equilibrium had, in fact, been obtained. The output log would indicate no material balance changes and no saturation changes in any grid blocks. With depth averaging, this may not always be the case and will definitely not be

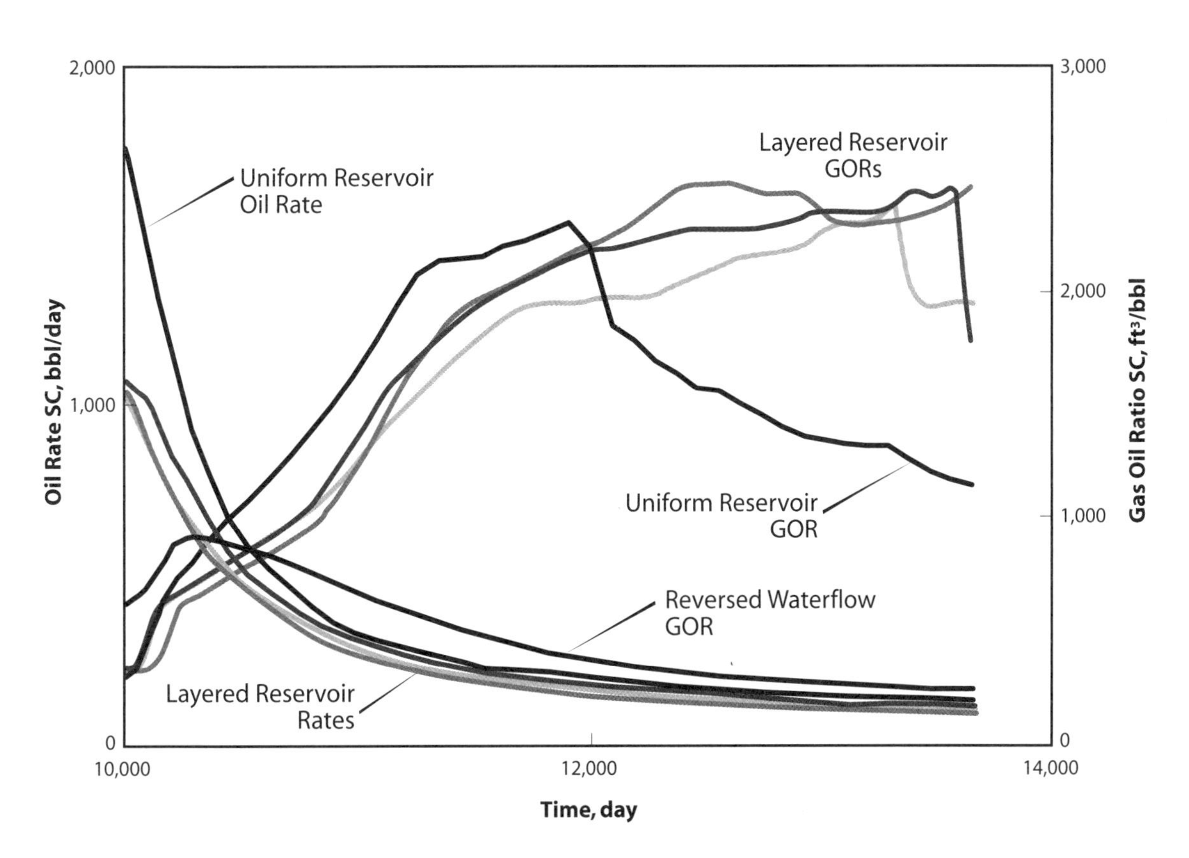

Fig. 9–19 Reservoir Simulation Tests for Dynamic Equilibrium—Effect of Layering on Waterflood

true if a dynamic system is modeled. It is still useful. It will indicate how much effect the depth-averaging calculations have on the reservoir as fluid distributions equalize. The reservoir simulation calculation is still based on grid-block average properties that mathematically occur exactly at the grid-block center.

Summary

Initialization appears to be a straightforward matter on the surface or initially (two bad puns). Gravity initialization is relatively simple to implement with modern features. An engineer must be aware of what features are available. There is no law, however, stating that static gravity equilibrium must exist. In fact, the opposite is true. If one has trouble getting a history match, this may be a good place to reexamine basic assumptions. Dynamic conditions are common in the Sawtooth of Southern Alberta, all of Saskatchewan, Indonesia, and in the Middle East. Qatar has some strong examples of this.

References

1. Dahlberg, E.C., *Applied Hydrodynamics in Petroleum Exploration*, 2nd ed., Spring-Verlag, 1995.

2. Hubbert, M.K., *The Theory of Ground-Water Motion and Related Papers*, Hafner Publishing Company, 1969.

3. Hitchon, B., et al., "Culling Criteria for 'Standard' Formation Water Analyses," *Applied Geochemistry*, Vol. 9, 1994.

4. Lies, H., "Hydrogeological Trapping Model of the Milk River Gas Field," *Journal of Canadian Petroleum Technology*, (December): 25–30, 1995.

10

Integration and Gridding

Introduction

Putting the simulation together was reserved deliberately for this later point in Part I. There are a number of steps at this stage.

- Until all of the input has been evaluated, the model cannot be finalized. The key point to emphasize here is to keep an open mind at this stage. The study scope can change significantly depending on the data available.
- Frequently, material balance or other ancillary studies are conducted as well. These are discussed briefly. Sometimes they provide useful insight, and, at other times, the results can be misleading or ambiguous.
- It is also important to review the original objectives in terms of the data reviewed and the latest understanding of the reservoir.
- The final step in this chapter concerns creating a grid.

Integration

At this stage, you should have a good idea of where the study is heading. Normally, the most critical aspects are the geological and production performance review. All of the data should have been screened. If some of the data is weak or unavailable, part of the study may be to determine the influence of the assumptions. This is a pivotal point, which the author calls the integration phase. The result of this phase will be a complete data set. Gridding is normally determined at this point. Each of the preceding four points will be discussed.

Data

Missing data

In a number of cases, individuals felt studies should not be conducted because some element of data was missing. In particular, this seems to apply to relative permeability data. Although this is undoubtedly important data, the truth is that limited data is not much better than no data, since there can be a lot of variation.

In one study for retrograde condensate systems, there was no relative permeability data available at all. Using sensitivities, the author demonstrated the relative performance between fraced and unfraced wells was similar regardless of the relative permeability data assumed. This is a serious consideration since, even for low levels of condensate dropout, a frac can significantly reduce productivity impairment caused by retrograde liquid condensation. The simulation could not predict the absolute level of production (test data did), but it could indicate the direction and trends. Restated, an engineering solution and a scientific solution are not the same. It may be sufficient to put an upper and lower bound on an answer rather than provide a precise solution.

Scientific results versus pragmatic results

I wrote a paper on the study described previously. The paper was severely criticized due to the "flaky" nature of the data, which relied on general gas-oil relative permeability correlations developed by Arps.[1] As a practicing reservoir engineer, I was horrified by this response. The client had spent considerable money on exploration and needed to make an expensive nomination for a sour gas plant on short notice. A nonanswer was untenable. They needed a reasoned, educated guess. To put the simulation in perspective, the land and wells were drilled based on substantially higher levels of risk.

An engineer's job will sometimes be to give the bosses or clients the best answer possible with the data available. At such a point, they need a risk assessment with the answer, i.e., a qualified response.

Ancillary Studies

Material balance calculations

This calculation only works well in reservoirs with good communication, i.e., in those reservoirs having high permeabilities. This is not just a matter of calculating an average pressure, although this is a significant issue. There are also other considerations. Material balance studies normally have considerable uncertainty when a gas cap and/or a water leg is involved.[2]

In this area, one often finds contradictory advice in the literature. The author has worked on a number of reservoirs where the material balance study provided inaccurate results and actually provided results very different from simulation results. The material balance does not take into account pressure variations existing in moderate- to low-permeability reservoirs. Having said this, ARE recently did a study, and an individual who worked on the project matched the OOIP and gas cap volume using a material balance program. The reservoir in question had braided channels, and the material balance was of no help in determining these locations. Based on previous experience, the author had been extremely doubtful that accurate results would be obtained.

Material balance studies have obscured or misdirected solutions as often as they have provided useful insight. In general, the author no longer does material balance studies as part of presimulation work, unless the reservoir has very high permeability and is of limited size. However, this is not an absolute rule. Reservoir engineers' views on this matter will vary according to the proportion of success they have observed.

Objectives

Importance of objectives

Depending on its design, a numerical simulation will take into account various factors to varying degrees. Tradeoffs are usually required. Hence, most studies are normally aimed to address specific issues.

At first glance, determining the objective might seem a trite comment, but, in many cases, problems are defined in terms of the solution. For example, if the production performance from a reservoir is proving to be enigmatic, one will be requested to do an areal simulation. However, the underlying problem is that reservoir mechanisms are not understood. In this case, the real objective is to propose possible reservoir configurations, fluid properties, or geological models to explain the observed behavior. A smaller model can be used to determine physically plausible proposals.

Common situations

Here are a number of common situations:

- The most common situation concerns a reservoir on primary production, which an operating company has now decided to waterflood. They are looking for a waterflood reserves and production forecast based on the results of a simulation.
- Another situation involves a recently drilled well in a new field, which has a gas-oil, gas-water, or water-oil contact in close proximity to the completed interval. A simulation has been contemplated to support a recovery factor as well as to predict GOR or watercut trends and, possibly, to evaluate mitigation strategies.
- An operating company has a mature field they have decided to model to optimize production. Small increases in overall recovery factors may represent major increases in remaining reserves. Often, it is hard to distinguish incremental reserves from production acceleration.
- A large developed gas field exhibits retrograde (liquid condensation) behavior. A study has been commissioned to determine natural gas liquids (NGL) and gas recovery.
- A large miscible flood has been installed. The operating company has decided to perform a simulation. Since limited production history exists, a simulation is often the best method of engineering analysis available to estimate recovery.

Each of these situations involves somewhat different issues that must be resolved. In truth, one cannot get all of the details of the real reservoir into a simulator. It is necessary to get the important things right so the mechanistic behaviors—the physics—are captured correctly.

Critical issues

The following is a list of critical issues likely to arise with the situations listed previously.

- For the waterflood implemented after a period of primary production, one of the most critical aspects of model design is characterizing layering or heterogeneity. This can be done either with detailed layering built into a model or via pseudo relative permeability curves.
- For a well drilled into a pool with a gas-oil, gas-water, or water-oil contact, the critical modeling issue will likely be accounting properly for near-wellbore effects. This usually involves a detailed coning study, defining layers, covering the correct range of production rates, and including the effects of hysteresis.
- For a mature field optimization, study the most critical aspect for the most likely way to build a balanced model. This could require preliminary coning and cross-sectional studies that must be integrated later into an areal model. Integration is the major issue.
- For a retrograde gas condensate reservoir, a compositional simulator is likely required. Proper PVT characterization will comprise a major portion of the study.
- For a miscible flood, a number of critical issues are possible. The same simulator may not address all of the issues effectively. If miscibility is the critical issue, a compositional simulator is in order. If maximizing sweep efficiency is the critical concern, then a pseudo miscible simulator is probably the most effective solution.

Other factors, such as well spacing, laboratory relative permeability, PVT properties, injectivity, and timing of wells are still important.

Multiphase studies

Historically, a single model could not address all issues. Therefore, it has been common to assemble many different models to deal with specific issues. A large-scale simulation may be split up as follows:

- Coning Study. A detailed radial grid study would be performed to evaluate coning. For instance, a reservoir appears to have no gas coning despite a 7,000 kPa or 1,000 psi pressure drop. Experience levels vary, and not everyone may find the solution obvious. A relatively cheap coning study can determine the issue and cause subsequent elements of the study to be changed. If a gas cap really does exist, it is possible to use this information to evaluate well pseudo relative permeability curves.
- Cross-Sectional Study. The objective of a cross-sectional study is similar to the coning study, that is, to simplify the problem to gain general understanding. Often, these studies are used to generate pseudo relative permeability curves, as shown in the problem set in chapter 20 of this book. The use of these curves can simplify a multilayer problem into a single-layer model, which may be used to develop an areal model.
- Areal Study. This is probably the most common type of study. It allows complex waterflood patterns, which seem to occur as an accident of history, to be evaluated. However, with the larger computers available today, it is getting easier to make multilayer, full-scale 3-D studies.

Each of the stages will build on the earlier stages to develop a full field model.

Gridding

Realistic grids

Ideally, a very fine grid could always be used. However, as the dimensions of a grid increase, the amount of calculation time increases rapidly. Determining the amount of work varies depending on the mathematical implementation used in the simulator. In general terms, the amount of computer time increases to the fourth power of the grid dimension (n^4) for a common direct matrix solver implementation and a 2-D square grid, i.e., n cells in each dimension or a total of $n \times n$ cells in the grid. With layers, the amount of time increases by the cube of the number of layers (l^3). A large grid can, therefore, exceed the capacities of a supercomputer for a large reservoir; hence, constructing a model usually involves some compromises.

General guidelines

Some engineers use the number of acres in a grid block as a guideline, but this is not recommended. Others use two grid blocks between wells, but this is neither adequate nor recommended.

A reservoir simulation is often equated with a material balance; this is true, but only in a limited sense. In the numerical model, the blocks are linked via a mass transfer equation (Darcy's Law).

By virtue of this interblock transfer, a reservoir simulation is designed to calculate the transient behavior across the grid. Therefore, it is entirely different from a material balance calculation, which presumes tanklike conditions of equilibrium at each timestep.

Flow visualization

The key difference between the reservoir simulator and a material balance equation is interblock transfer. The best guideline is a flow net. Muskat frequently used flow nets in his work, *Flow of Homogeneous Fluids through Porous Media*.[3] They are relatively easy to draw by hand. Two examples are shown in Figure 10–1 and 10–2.[4]

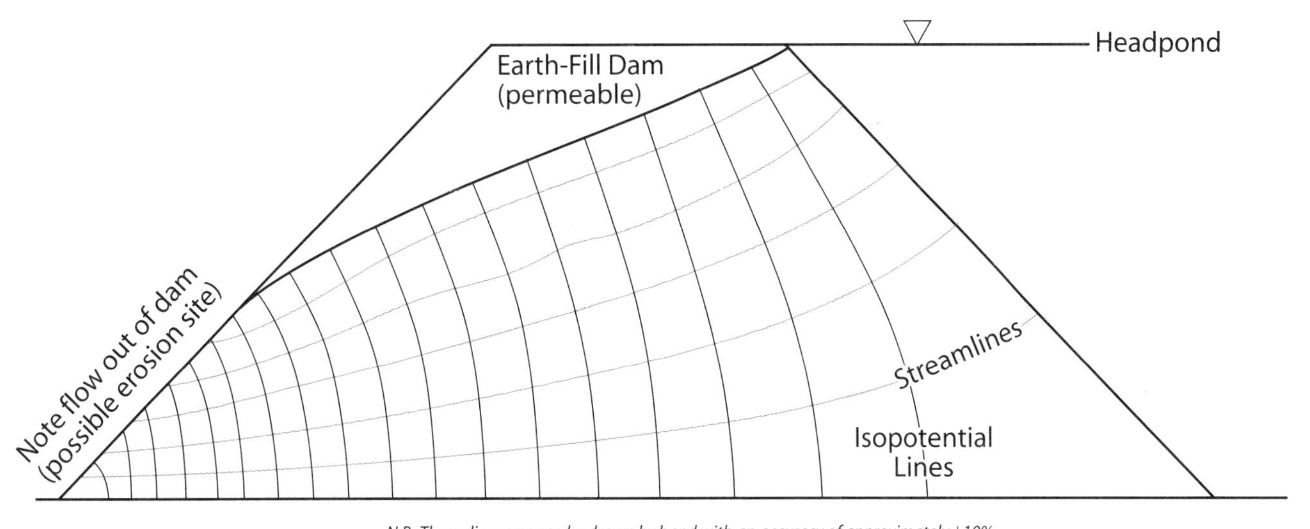

Fig. 10–1 Flow Net for Earth-Fill Dam

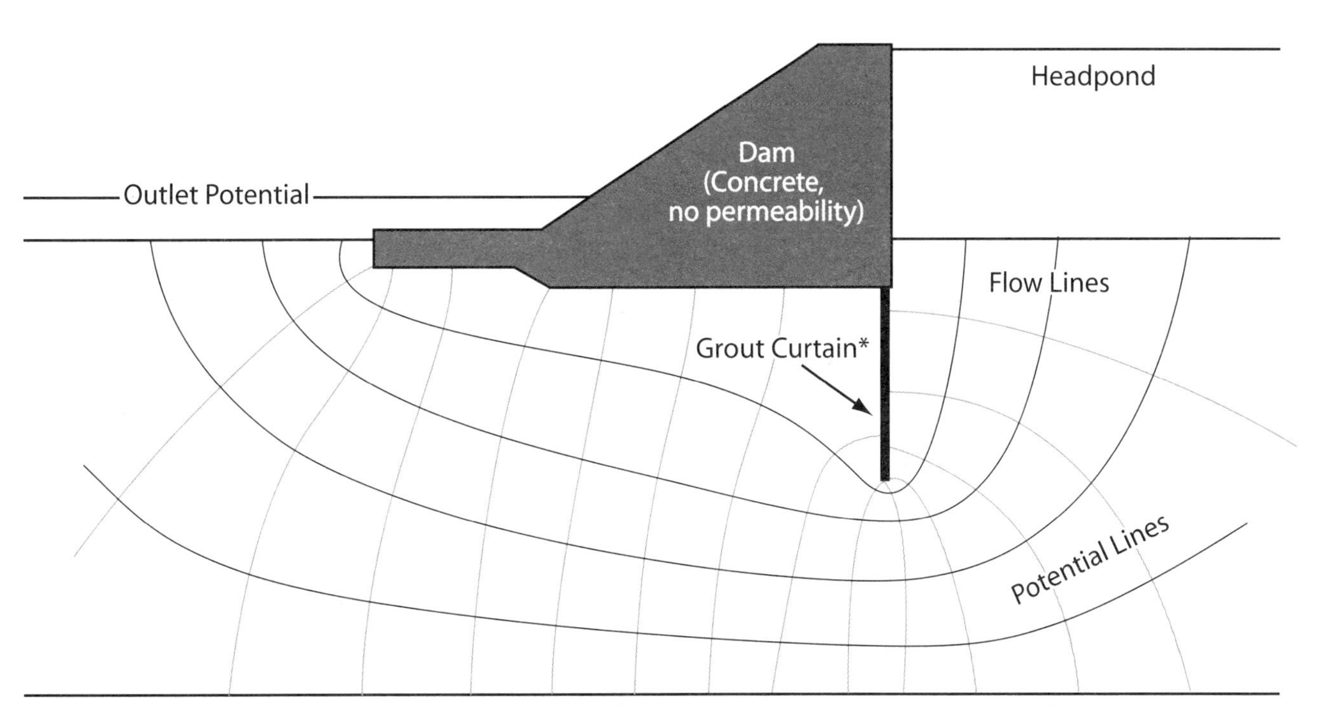

Fig. 10–2 Flow Net for Earth-Fill Dam with Grout Curtain

Irregular grids

The use of flow nets suggests the use of irregular grids. Where the flow net is fine, more grid blocks are required. In fact, it is possible to use the grid from a flow net to make a curvilinear grid system. This can produce accurate results. For example, a grid for the leakage under a dam is shown in Figure 10–3. From this information, a finer grid could be constructed where flow concentrations exist.

Grid sensitivities

One cannot analytically calculate the amount of error from grid discretization. However, it is possible to determine how the error varies. In practice, a grid sensitivity is required. By changing the grid dimensions and making successive runs, it is possible to show the model is accurate to within an approximate percentage. Figures 10–4 and 10–5 show a gas-oil and water-oil coning grid sensitivity done by ARE during a reservoir simulation.

Sadly, grid-sensitivity checks are rarely run in field studies. The example in Figures 10–5 and 10–6 shows the areal simulation is only accurate to within ±10%. The mathematicians are horrified with this level of accuracy; however, the brutal truth is most reservoir simulation input is not known with this degree of accuracy. With the capability of computers continuing to increase, this may be less problematic in the future.

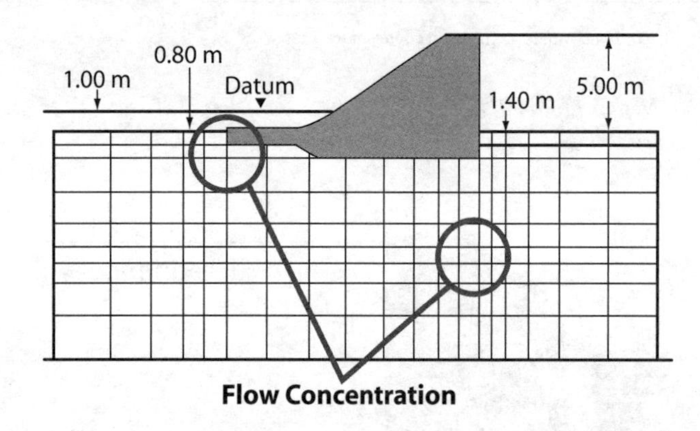

Fig. 10–3 Flow Concentration for Earth-Fill Dam with Grout Curtain

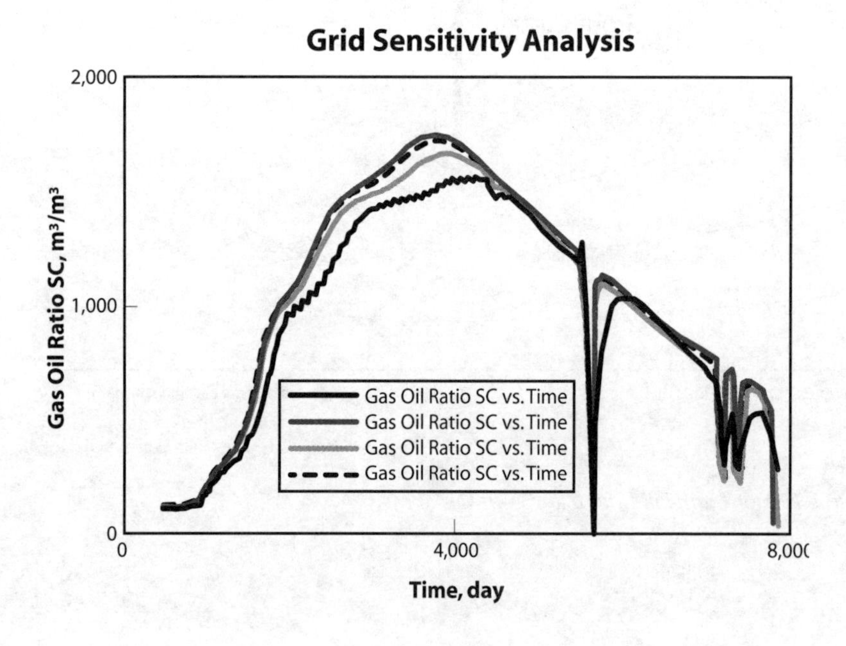

Fig. 10–4 Grid Sensitivity GOR from Field Study

Insufficient grid blocks

Reducing the number of grid blocks can be taken too far. The following example occurred in a study proposal submitted to the author. The objective was to determine horizontal well potential in a tight sandstone reservoir. Production in vertical wells was uneconomic due to low rates and GOR penalties. The study objective was to determine if a horizontal well would increase productivity sufficiently to enable economic production and if GORs would be low enough to avoid penalties. As discussed previously, a flow net is a useful tool for visualizing flow in the reservoir. The proposed grid and the expected flow pattern are shown in Figure 10–6.

Since the reservoir is tight, less than 1 mD, localized gas saturations would be expected to arise. A fine grid made with small grid blocks and thin layers would be required around the well. An element of symmetry could be utilized through the center of the well to reduce computation.

The proposal for the situation described previously recommended three layers with the wellbore centered in blocks with 100 ft. horizontal x and y dimensions. Although these large grid blocks could not correctly represent the physics in the reservoir, the model would still produce results. The proposal was not considered favorably by the author.

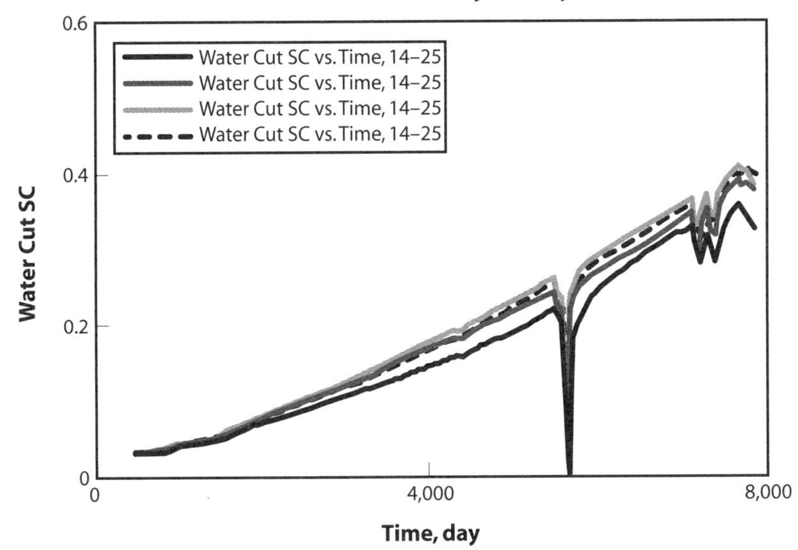

Fig. 10–5 Grid Sensitivity Water Cut from Field Study

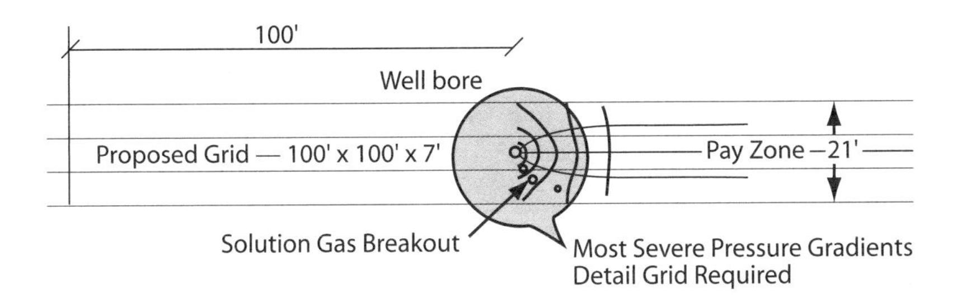

Fig. 10–6 Flow Concentrations around a Horizontal Well

Actually, this problem is quite common in the author's opinion. Simulation output should have a warning at the top of the output file:

Warning:
Simulation may give nonphysical results if too coarse a grid has been used

In reality, there are no warnings at all, although Eclipse documentation is reported to have some warnings. The fact is the simulator runs much faster and gives fewer problems if too coarse a grid is used. Buyer beware!

A good example of these differences can be found in the SPE comparative solution on horizontal wells. One of the simulators used a high-permeability grid block to represent the horizontal well. Figure 10–7 shows the predicted GORs for a horizontal well.[5] The author does not recommend the approach in the example given.

Two basic approaches to gridding

Based on experience with reservoir engineers new to simulation, gridding is not intuitively obvious. There are two basic approaches for setting up a 3-D grid. The first is based on setting up a spatial grid and then assigning rock properties based on geometrical position. The second is to assign grid blocks based on geology and then alter the dimensions of the grid block. These two methods are shown in Figures 10–8 and 10–9.

In the author's experience, it is rare to assign grids based on fluid type, i.e., oil, gas, and water. Nevertheless, fluids can be a consideration.

Capillary pressure transition

Normally capillary pressure curves are relied upon to give the fluid saturations in a grid block. However, if flow visualization is applied to wells in a transition zone, it may have an important effect on reservoir performance. For instance, would the two situations shown in Figure 10–10 be equivalent?

Grid orientation effect

Grid orientation can be summarized quickly with Figure 10–11. Grid orientation becomes significant at high adverse mobility ratios. This commonly occurs whenever there is gas injection. Since most hydrocarbon miscible floods (HCMF) use enriched gas, grid orientation is important in EOR design. It is also a major factor in thermal simulation; however, these models typically incorporate a nine spot finite difference molecule to control this problem. This effect will not show up as a material balance error or anywhere else for that matter.

The best method of handling this type of problem is to develop a streamline model and then make detailed cross sections. For miscible floods, detailed heterogeneity can be incorporated using geostatistical realizations. This technique has been used by Chevron in the Mitsue Gilwood Unit No. 1.[6]

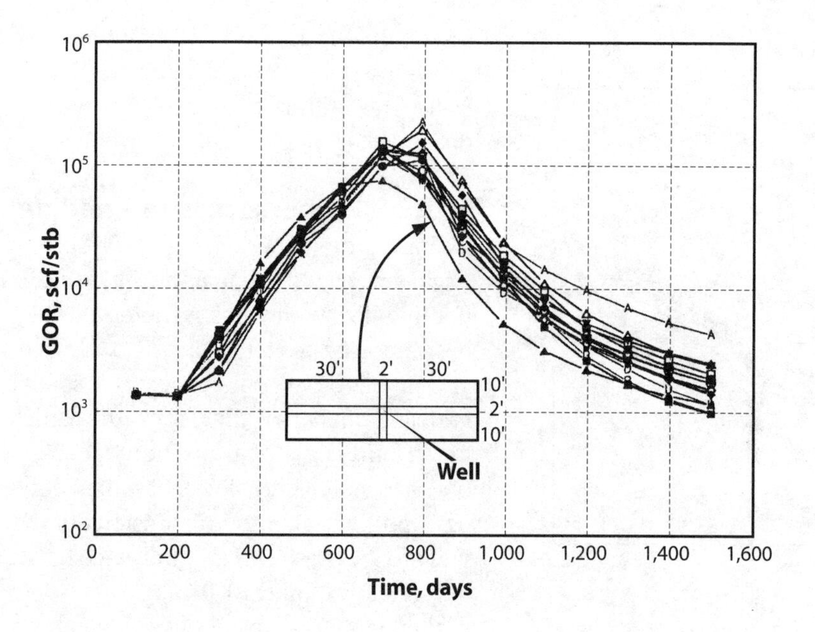

Fig. 10–7 SPE Comparative Solution—Varying Grid Refinement

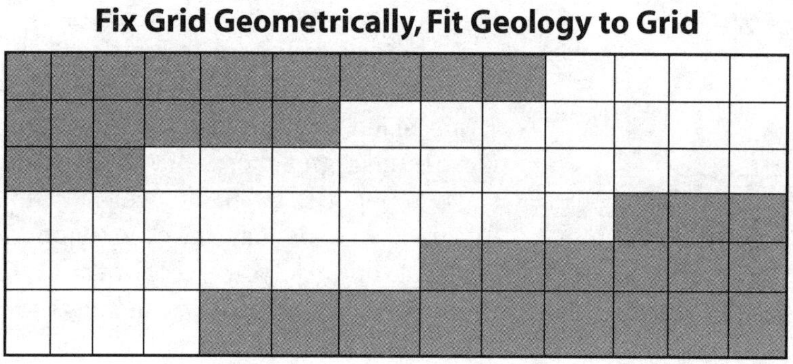

Fix Grid Geometrically, Fit Geology to Grid

Fig. 10–8 Basis for Grid Construction—Fixed Grid in Space

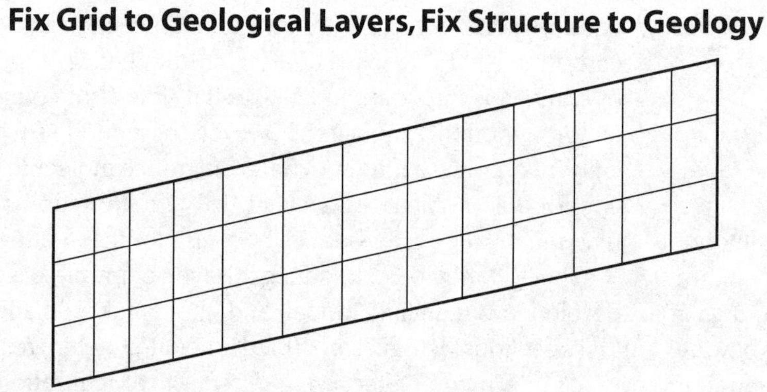

Fix Grid to Geological Layers, Fix Structure to Geology

Fig. 10–9 Basis for Grid Construction—Fit Blocks to Geology

Reality check

The perception of where good engineering is required can be somewhat distorted. For example, consider the case of a massive high-permeability carbonate reservoir above the bubblepoint, which has strong bottom water influx, and a production history water free, since the wells are completed away from the water influx. No gas coning is involved because the reservoir is undersaturated. Large grid blocks with lab relative permeability curves will work very well. The pressure differences between grid blocks will not be that large due to the high permeability. Water influx is nearly vertical because of the high permeability. Furthermore, geological layers stabilize vertical displacements rather than accentuating breakthrough. Costs are not a real concern. Frequently, the biggest problem is getting the tubing pressure drop curves matched. Simulation, in some situations, is actually quite easy.

If the same simulation engineer who has been working on the reservoir described previously later works on a large carbonate reservoir with low permeability and with water and gas contacts, he (or she) will find life gets very hard. The gas will cone. GORs will increase rapidly due to the low permeability. The water will cone in downdip wells. There will be significant differences in pressures between grid blocks due to the steep pressure gradients in the reservoir. A waterflood will be implemented in a pattern to provide for voidage replacement. Large changes on an individual well basis are required for well pseudos on gas cap wells and downdip water producers. The rock relative permeability curves will be changed to a pseudo relative permeability. This is required, since layering controls the breakthrough of the displacing fluids for horizontal floods. It is the reservoir from hell.

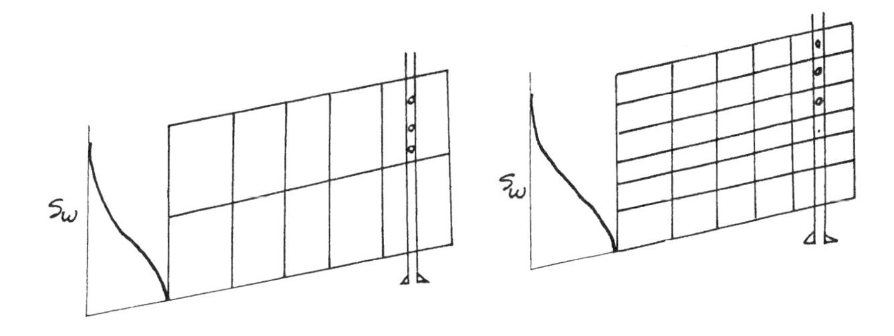

Fig. 10–10 Basis for Grid Construction—Distribution of Fluid in Reservoir

Not only that, but everyone is constantly concerned about costs and when the project will be finished. The engineer may ask, "Why am I working here?"

Simulation and reservoir engineering requirements are often more severe in low-quality reservoirs (read mature basins) than in massive high-quality (high-permeability) reservoirs. This is often opposed to the economic margins. This is frequently true of the more classical reservoir engineering techniques such as material balance.

Limitations of a simulator

A simulator cannot direct an engineer to the right solution. It will answer yes or no, but is incapable of a creative process. On the positive side, the creativity is really the fun part, so perhaps it is best the computer can't do this. Management is unlikely to replace us computationally challenged biological units with a computer for a long time.

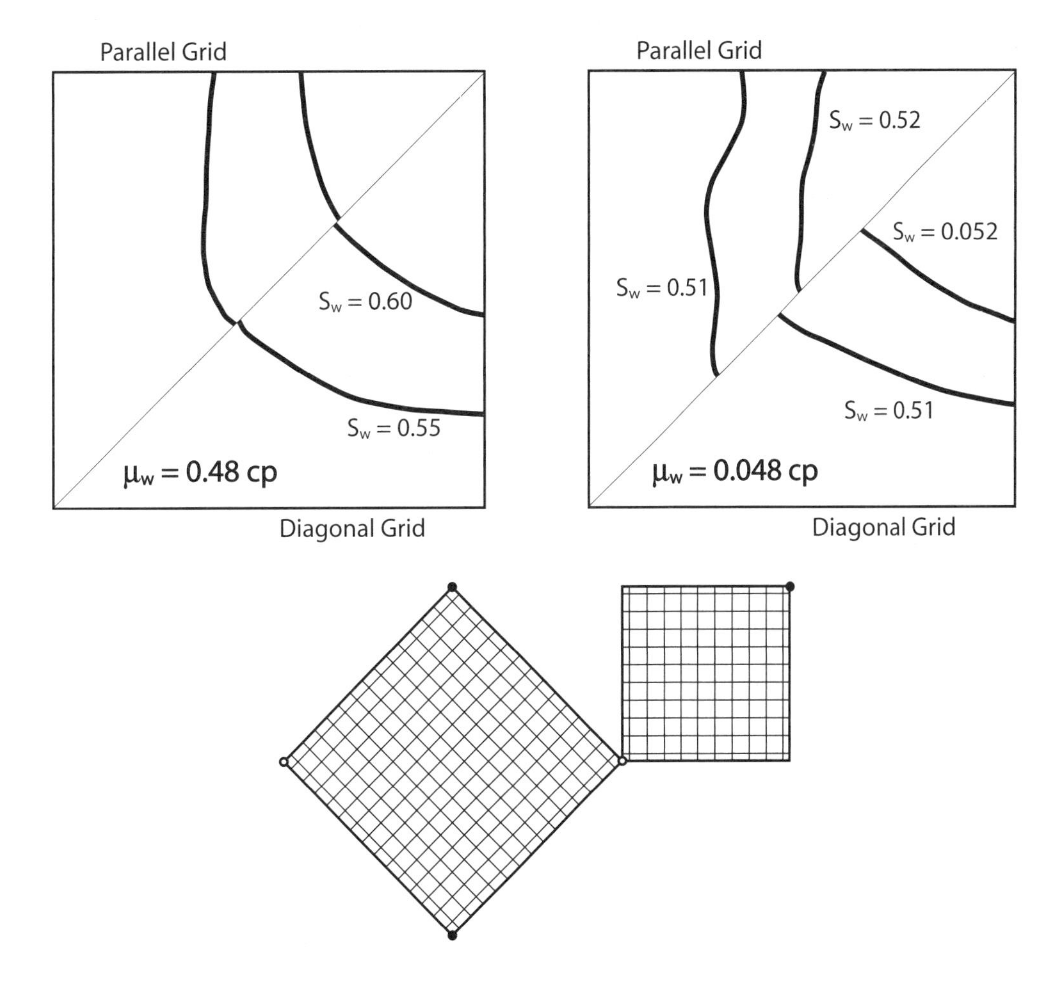

Fig. 10–11 Grid Orientation Effects

Summary

In order to provide effective answers, review all of the data, production performance, and geology to determine all of the issues that must be resolved. This will require some flexibility. It is easy to fall into traps if the problem is described in terms of a potential solution. Lack of data need not be a deterrent. Either a series of sensitivities may show a range of solutions, or it may conclusively show the missing data does not affect overall results significantly.

Experience has shown there are a number of well-known critical issues for certain situations. Be aware of these problems and address them in developing the approach. Breaking the study into different components allows one to evaluate various aspects in detail. Certain situations allow specific creative solutions. Radial sector studies often work well on salt dome structures; a Cartesian grid can be used for fractured wells and non-symmetric reservoir discontinuities.

Developing a grid requires a number of tradeoffs. The capacity of your computer will have a major impact on how fine a grid you can use. Be careful of some of the rules of thumb. It is impossible to base grid requirements on blocks of *x* acres or two grid blocks between wells. Grids depend on many factors. Flow nets are probably the best tools for indicating grid requirements. The only way to tell if it is "right" is to perform a grid sensitivity.

Setting up a grid can be difficult, particularly for novice users. Again, new engineers can cheat, i.e., look up grids used in existing papers and textbooks. Grid orientation is important in a number of advanced applications. Often, miscible flooding is best handled with streamline models.

Finally, fundamental grid errors are user transparent. The simulator can and will calculate physically meaningless results with incredible precision. Moreover, it will run faster and give you fewer problems as it merrily leads you to perdition. Buyer beware!

References

1 Arps, J.J., "Analysis of Decline Curves," *Transactions of the AIME*, Vol. 160, 1945.

2 Carlson, M.R., "Tips, Tricks and Traps of Material Balance Calculations," *Journal of Canadian Petroleum Technology*, 1997.

3 Muskat, M., *Flow of Homogeneous Fluids Through Porous Media*, McGraw-Hill, 1937.

4 After Wyckoff and Reed, *Physics*, date and issue unknown.

5 Nghiem, et al., "Seventh SPE Comparative Solution Project: Modelling of Horizontal Wells in Reservoir Simulation," SPE 21221, 11th SPE Symposium on Reservoir.

6 Behrens, R.A. et al., "Implementation of a Streamline Method for Flow Simulation of Large Fields," *Journal of Canadian Petroleum Technology*, 1996.

11

Basic Data Set Building

Introduction

Building a data deck (input file) for the first time is daunting, since a great deal of information must be specified. The difficult part is making sure the data is realistic. This depends more on a general knowledge of reservoir engineering than the peculiarities of a simulator. The objective of this chapter is to identify the practical tricks that make life easier.

Data handling

One of the most important skills for a reservoir engineer is the ability to handle large amounts of data. While conventional reservoir analysis requires a simple file, the simulation process becomes more demanding. Simulation requires detailed well information in addition to overall reservoir information, such as relative permeability, capillary pressure, permeability, and porosity.

Try to use pictures. Summarizing data such as well log responses on a cross section or making a color-coded wall map can reduce pages of data to a single page. The author finds the process of putting this information together helps to develop a mental image and to observe trends. Getting projects approved depends on management understanding the concepts. The use of pictures or graphics is, in turn, an essential step for one to become confident with these concepts.

For simulation projects, the author makes three sets of three-ring binders. The first set contains field-wide data such as PVT analyses, relative permeability lab tests, pool production plots, pressure surveys, and material balance calculations. The second set contains all of the previous reports on reserves, technical papers on the pool and/or nearby offsets, as well as geological and petrophysical reports. The third set contains well information. This normally consists of a well summary card, a completion diagram, a photocopy of relevant sections of the logs, pressure survey reports, and analyses on the wells.

For example, the largest simulation the author has done to date had more than 100 wells. It required a box of three-ring binders to file all the wells. When the history-match plots were reviewed, one could flip open the tab on the well. It was then easy to correlate which part of the zone was completed, verify which pressure points had good buildups and to look at production performance. Frequently, the base production of oil and gas is not put on the plot. To review a history match on this project took a day and a half. A colleague had previously done an even larger simulation with close to 300 wells. In this case, he ordered a set of filing cabinets to house all the information.

Reservoir engineering supervisors usually are more impressed when staff has well-organized data. This normally means it can be retrieved in the course of a conversation.

Data deck construction

Putting comments in the data deck is important. Additionally, choosing a logical layout of data fields can be important. There are a number of reasons for this:

- It is much easier to make changes to the simulator with comment lines. This prevents having to look up flags in the documentation as one goes along.
- It is easy to make clerical errors when handling large amounts of data. By laying out the data in a logical fashion, it is easier to spot incorrect values.
- Clients (and remember, the boss is a client too) will normally see some part of the data deck. An organized approach is much more likely to produce a positive impression.

Clerical errors

One consulting engineer used to track his clerical errors. He was extremely thorough at checking his work. He could estimate the number of data points that would have an error. His power of prediction was impressive. The engineer observed this author double-checking a data set. After looking at the size of the input file, he correctly estimated the number of bugs the author was *about* to find.

While this is probably extreme for most people, checking data is important. It is not much fun to have to redo several months of work when a simple mistake is found in a well completion, relative permeability set, or some grid values of porosity, permeability, or net pay. Professionals continually check their input.

Maintaining backups and data sets

Inevitably, one of several things will happen:

- The computer crashes and destroys the data.
- After a number of runs into the history match, one needs to go back to an earlier version of the data deck.
- Space on the computer disk is limited, and it can only accommodate a limited number of data sets.
- A backup will be corrupt or not made.

Most system administrators are good at methodically maintaining data sets; however, it is best to maintain a set of independent backups at critical times. Examples include the final initialization run, the final history-match data set and output, and a permanent copy of all predictions. When the project is finished, put these files(usually stored on tapes) in a box for archiving. This makes starting up a project again much easier.

Logic in data deck scanning

The earliest simulators required fixed format data entry. This required every piece of data, including the decimal place, be on the exact card and in the exact column in order to work. This was an extremely painful and tedious process.

Formatting requirements subsequently became less stringent. User demand also caused a change in the logic. The next generation of models, with free format lines, would scan the entire input file (in the required input order) for each type of data. Users could literally put any data in any place. Most modern simulators are not quite this flexible. Now, within each module, such as Input-Output, Grid, PVT, Numerical Controls, and Run, the data can be input in any order. If a model won't read the data, consider how the code is scanning the input file.

Some simulators will scan key words in their thirst for data. Therefore, if a row of array numbers is missing, it might start reading permeability as grid dimensions. Then it will start in the correct place for the permeability, since this is a completely new scan. At other times one can have an array read problem on the last set of data in the section. Looking for the missing data in the vertical permeability array when the actual problem is in the grid dimension or structure top array can be extremely frustrating. If one is aware of this logic, a quick review of the output file will often give the requisite clue as to what is missing. Think like the programmer who put the code together.

Dimensioning

Until recently, most simulators were written in Fortran 77. This program, while very efficient at numerical calculations, requires memory allocations be done in the source code—the executable. Therefore, when running large problems, it is sometimes necessary to regenerate a custom executable. At the time of this writing, most suppliers have converted to the Fortran 90 standard, which allows for dynamic memory management. This feature has been available in the C programming language for quite some time. If dealing with an older model or one that has not been upgraded during the last few years, a user may need to arrange for a custom executable to be compiled.

Some software companies provide models that are matched roughly to memory requirements. Use the smallest memory model possible, since it will be the most efficient use of computer resources. Conversely, if the model doesn't run, try swapping the executable file for one prepared for larger problems.

Simulation input

Experience on the simulator hotline indicates a high percentage of problems are actually reservoir engineering input problems. Part of this is the lack of familiarity with data. For instance, relative permeability is traditionally used in Tarner-Tracy type material balance calculations and waterflood calculations, such as those developed by Craig, Geffen, and Morse. The author has yet to use one of these for a real problem. Reservoir simulation, which allows for areal variation in reservoir properties, has replaced these older methodologies.

Three-phase relative permeability had almost no place in classical reservoir engineering calculations. Its main application is in reservoir simulation. PVT data is used primarily in material balance calculations and in reservoir simulation. It is rare for the average exploitation engineer in Western Canada to have experience with more than one or two oil material balances.

Today, the brutal truth is the vast majority of advanced reservoir engineering calculations are done almost exclusively with a reservoir simulator. For those doing a simulation for the first time, it is likely their first time to use special core analysis results and PVT data.

Computer prowess

One of the most common barriers for people new to simulation is handling the computer. Perhaps this will change in the near future. It appears more work will be done on the personal computer (PC) in the future than was done previously on Unix RISC (reduced instruction set computer) workstations. General PC expertise is steadily increasing. This is one area where there will be a correction. There is a strong trend towards making the graphical user interfaces very complex. In comparison to DOS (disk operating system), Windows 95 was an unfriendly system. Although DOS clearly had limitations, it was very simple. Although later versions of Windows have some automated features, such as plug-and-play, these programs are so complex there is a good chance individuals will find it too much to administer by themselves, which, in the author's opinion, defeats the purpose of a PC.

Inside the IBM PC

Except in the larger companies, there will be few people in your company who are familiar with reservoir simulation. Because there are so few people with this skill, it is likely the systems department will not be able or want to support the simulator. This means, in all probability, the engineer will have to load and support it. Reservoir simulators are complex programs compared to most applications. In addition, due to the smaller size of the user base, they are never as well debugged as a standard worksheet or word processor.

As a rule, above-average computer skills are required for reservoir simulation. As an example, look at the job advertisements for simulation specialists for the Middle East. At the least, acquire some knowledge of how the hardware works, and read Peter Norton's book, *Inside the PC*.[1] This is an old book, but it has been rewritten and updated a number of times. You can find current copies in computer bookstores. A number of books are available on how to upgrade and repair PCs. If you work for a company, the systems department will object—vigorously—if an employee takes a screwdriver to one of their machines; however, it helps to know where and why problems are occurring. Computer users can get help more quickly if they can explain the logic behind their diagnosis. For those who own their own PC, these books will pay dividends in reduced maintenance costs and faster repair times.

Unix systems

Many major simulations are run on workstations. Although the speed of the PC on a MIPS (million instructions per second) basis appears to be close, there is a big difference between a PC and a Unix workstation. The latter has much more sophisticated peripherals, more caching, normally more memory, and better and larger buses. Unix was designed as a true multitasking operating system. It is also widely debugged and tested.

Buy a book on the basic commands. It shouldn't take too long to get up to speed. The common commands such as *ls* (dir), *mv* (ren), *cp* (copy) and *vi* (edit) are all two letter combinations. Superior typing skills are not required. Earlier versions assumed a high level of knowledge by all users. Regrettably, the more recent versions seem to be saddled with command controls for system security that make efficient work more difficult. Getting back on track again, once past basic Unix, read O'Reilly's *Unix Power Tools*.[2] This book provides sufficient information to join the advanced users group quickly.

Finally, buy a book on Unix systems administration. It's a bit of a dry read no doubt. It reveals how a Unix system works and what a user can, or should, request. These books are available in computer bookstores. The author recommends three books:

- Tim O'Reilly, *Complete Windows NT & Unix System Administration Pack*[3]
- Nemeth, Snyder, Seebass, and Hein, *Unix System Administration Handbook*[4]
- Thomas and Farrow, *Unix Administration Guide For System V, Version IV*[5]

Have the systems administrator grant rights to make and read tapes at will. This is necessary when archiving at critical points in the study and allows one to exchange information with the software supplier for debugging if necessary. Normally, floppy disks are too small to hold these files.

Get as much authority from the systems administrator as possible. Many of the author's clients have unwittingly wasted days of time, on a cumulative basis, while waiting on a systems administrator to reallocate space or make minor changes. Another current change is hardware manufacturers recognize administration is being done by individuals who are the primary users of a computer. Silicon Graphics, Inc. (SGI) in particular has simplified the administration by using a clever screen menu system.

Working with simulation data sets

The author does not use preprocessors for manipulating the data set. Rather, preprocessors are used for digitizing and grid property definition. There are so many potential bugs in a preprocessor that it is not worth risking a data set to uncontrollable factors. This comes from grievous personal experience, albeit on some early versions of these programs. It means the engineer has to know the input syntax and, especially, become proficient with a text editor.

Virtually every company and computer vendor comes with an editor that is, in some way, significantly better than the Unix *vi* editor. On the other hand, *vi* is available on each machine, and it is in the exact same form. It is part of the standard ATT Unix code. It is not a bad editor and was designed for the basic VT100 terminal. However, it is powerful and, in view of its ubiquitous presence, it is hard to recommend any other editor. Buy O'Reilly's book on the *vi* editor and keep it nearby at all times.

When working on the PC, the author found he often exceeded the capacity of the DOS edit program and used KEDIT for editing on the PC. It is capable of handling large data sets and is somewhat like *vi* in capabilities, although it is not quite as easy to learn.

Utilities

When working with a hundred wells in the data set, manually specifying production input is not something one should seriously consider. Yet, data bureaus are generally quite thoughtless because they do not supply production data in the myriad of formats that a limited number of users wish to use for direct input into a reservoir simulator. A utility program is required to make these adjustments. Nightmares of the first Fortran course taken 15 or more years ago will return.

May useful utilities are available. The author has some that round array values to lower levels of precision. In some cases, this doesn't do more than keep data sets to reasonable sizes when the preprocessor has been overdesigned for the project, but this is not a trivial issue. Data checking is important. Some utilities check net to gross pays. Minor differences often occur due to the nature of the interpolations routine. If the difference is small, the utility ensures the two values are consistent. This reduces the number of warning messages. While these messages can indicate trivial details, important messages can get lost in the multitude. It helps to eliminate the trivial messages.

At least one simulation software development company has many of these utilities prebuilt and supplies some of them as part of their general release.

System space

Simulations produce abundant output. In the past, systems administrators objected if users took more than 50 megabytes (MB) of space. Large gigabyte (GB) drives are common now, and people tend to keep many data sets these days. The largest files are binary output files for the postprocessor. It is not necessary to archive each run; one to three runs back in a history match is all that is necessary. Store them to tape. One person uses 4GB of system space by storing more than a dozen previous history-match runs. The person(s) responsible actually convinced the systems administrator they needed this much space. The author normally keeps or archives to tape each data set and the simulator executable to reproduce bugs if necessary. With these files, a previous history match can be reproduced by running the data set again.

Development environment

For the most part, simulator development is likely to stay on Unix machines for some time to come. The debugging tools are much more powerful on these platforms. These environments are impressive. On the Silicon Graphics machines used by the author, this is called IDO. It can make flow charts from source code.

Display some initiative

Quite often, a new software release cannot find a file it requires. The error message is usually kind enough to tell you the name of the file. A simple *Find File* command, available on both Unix machines and PCs, usually finds the file in some other directory. Copying the file into both directories eliminates the problem. A little thought often solves such problems.

Run data

Run data is one of the least glamorous parts of simulation. Wells are defined, completions are entered, and well controls are included. Output control is normally specified in this section of the simulator data file. It is common to generate a history-match file simultaneously, which is used in graphics to check if the actual reservoir performance matches the output from the simulator.

Still, it is possible to go astray in this area.

Base product

The essence of history matching is found in the run section. The base product is input, and the model predicts the amount of secondary products as well as pressures. For a typical oil pool simulation, the oil production is input directly. The simulator calculates the amount of gas production and any water production. It also determines reservoir pressure. The output format is shown graphically in Figure 11–1.

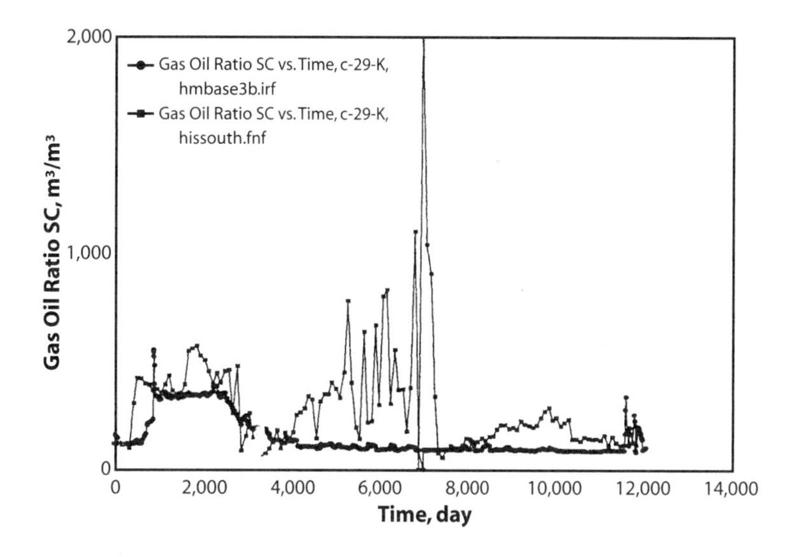

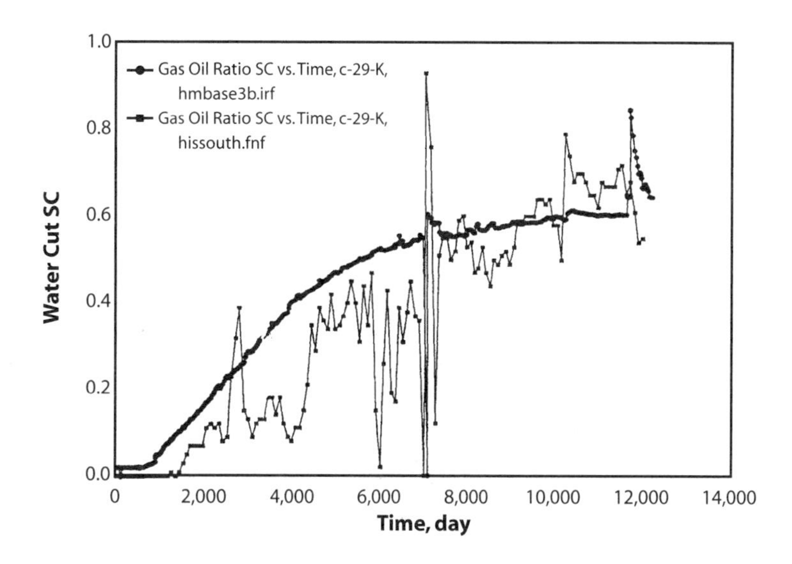

Fig. 11–1 Example History-Match Plot—Linear Scales—GOR and Water Cut

For gas reservoirs, the amount of gas production is input and the amount of condensate or water produced is calculated by the simulator. If there is an oil pool with a gas cap, a mixture of the two controls may be used.

Include all wells

Though it sounds trite, it is important to have all the wells in the pool entered properly. Invariably, the client or the boss asks, "How does the history match on well xxx look?" Even if the well has negligible production, it doesn't look or sound impressive if the well is missing. Clients normally expect *all* wells be included. This is something one catches with experience; however, it is surprisingly easy to miss a well. *Use a well list.* The best simulation engineers perform numerous cross checks while working on data sets.

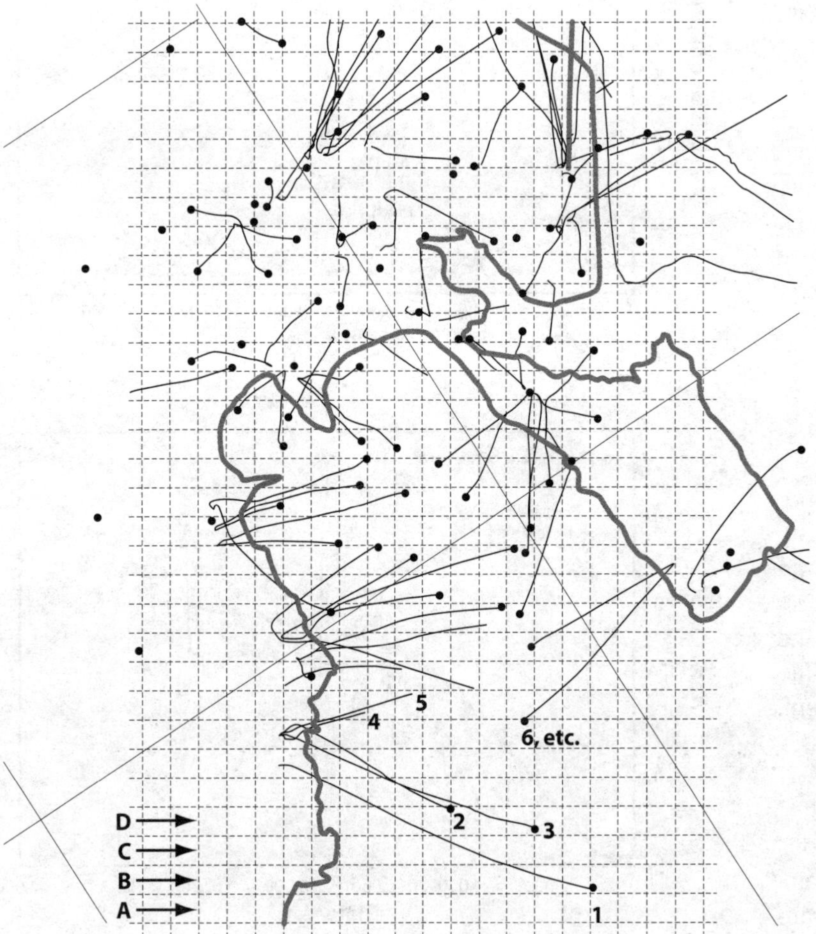

Ordering wells

Another little trick to consider is ordering the well list. This is not easy. The graphics output can be controlled in a number of ways. It may go by well number, which is entered in the data file, or alphabetically. This is where foresight can be useful. For large simulations, it is helpful to proceed in an orderly fashion. This allows one to mentally correlate what is happening in certain sections of the reservoir as one goes though the data deck. Figure 11–2 shows the well locations from a large (more than 150 wells) simulation. The well names are outlined in order of drilling and are not sequential. To make analysis easier, it helped to start at the south end downdip and then proceed along the reservoir to the north. This required a conscious decision at the beginning of data input to utilize the strategy.

Associated with this strategy is a base map. Use these and a series of coloring crayons later in the history-match process. For instance, three dark red check marks may denote too high a GOR and three light red check marks may denote too low a GOR.

Fig. 11–2 Example of Numbering System

Spreadsheets

Another alternative to the utilities described earlier for assembling well data is to use spreadsheets. The author prefers using utilities, but this may reflect his experience in programming, which predates the development of spreadsheets. Learning to use utilities is worth the effort.

Newer engineers tend to use spreadsheets and, in particular, macros. This is understandable since they have used computers and spreadsheets throughout school and their professional life. The author is familiar with three engineers who used large spreadsheets. The spreadsheets included color shading, multilevels, and large sets of data. On the surface, the spreadsheets should have been able to handle the data input. Unfortunately, they ultimately led to some alarming data input mistakes using macros. In one case, many weeks of history matching were wasted after it was discovered the production data had serious errors. Production from different wells had been juxtaposed.

Although spreadsheets can definitely save time, they can become overdeveloped. Macros do not execute efficiently in many cases. A utility can update production data input or well completions in seconds, while a spreadsheet may take hours to do the same thing. At this point, spreadsheets clearly consume enormous amounts of time and drastically reduce efficiency. Because of experiences like these, the author does not trust large spreadsheets or macros in particular. However, spreadsheets can be reused for pre-prepared material balances.

It is better to use a larger number of smaller and simpler spreadsheets to screen data. In addition, it is easier to check data and determine what someone else has done if the spreadsheet is smaller and simpler. If a spreadsheet takes more than three days to create, it's probable the designer has become carried away.

Production averaging

For long simulations, such as a 20 plus year history match, it is common to average production across three-month periods. During the last two or three years of history-match time, more detailed production schedules are used, typically on a monthly basis. The author is not aware of whether there is any mathematical justification for production averaging. One cannot use these time-averaged rates for pressure transient analysis; however, experience indicates it does work for long history matches. Specifying production on a quarterly basis reduces the size of data files considerably. Regrettably, it doesn't make checking any easier. Most production in Canada is tabulated monthly.

With regard to checking, neatness counts. Organize data in orderly arrays and in a fixed format.

Completions

Completions input is one of the most time-consuming areas of data deck preparation. It requires the well files be reviewed on each well and a history created. For a large simulation, this means large amounts of work. Simulation often gives some insight into the success of cement squeezes and reperforations. History matching indicates some perforations are not effective. This may be either a wellbore problem or a $k{\times}h$ problem. Recall that many $k{\times}h$ predictions are based on core permeability versus core porosity relations. This correlation normally varies by an order of magnitude from the best permeability to the worst permeability for a given porosity. Good comments in the completions section help recall what has been done when it is time to write a report.

This is one area where a good set of data binders really pays off.

Bottomhole pressures

In Western Canada, most oil wells are pumped. The most common practice is to specify the bottomhole pressure. Although it is easy to measure, it is difficult to know what the bottomhole pressure was during a given time period. Ideally, one would have a series of sonologs to indicate the fluid level and density, allowing a realistic specification of the bottomhole pressure. Therefore, with many bottomhole pressure data points, it is possible to tabulate this data and enter it in the simulator. However, the author has never worked on a reservoir with this level of data.

Most of the time, an engineer does not have the bottomhole pressure, or the data is so sparse as to be of limited value. Under these circumstances, one solution would be to estimate the bottomhole pressure as being 10% of the original reservoir pressure. In practice, it is not possible to completely pump off wells. If the well is over-pumped, the pump may become gas locked, and the well will stop producing. Therefore, most oil wells will have some column of fluid in the annulus. Figure 11–3 shows these concepts.

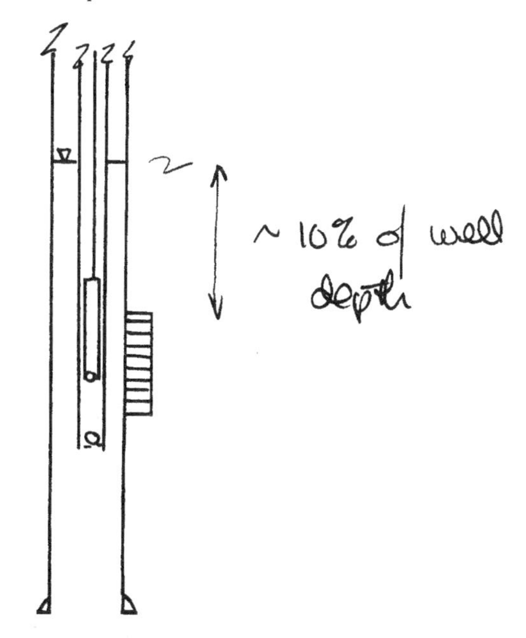

Fig. 11–3 Estimating Fluid Level

Tubular performance

In some instances, wells are flowed and the bottom-hole pressure cannot be directly estimated completely. This is normally the case offshore. The surface pressures and the rates are well-known. A tubing performance curve is used, such as that found in the *Technology of Artificial Lift Methods* by Kermit E. Brown.[6] These tables are input on a well-by-well basis. An example of a tubing pressure drop graph is shown in Figure 11–4.

More recently, tubing pressure drop curves can be generated with PC programs. This allows the specific completion for a well to be analyzed. These programs are easy to use and include multiphase flow and inclined wellbores.

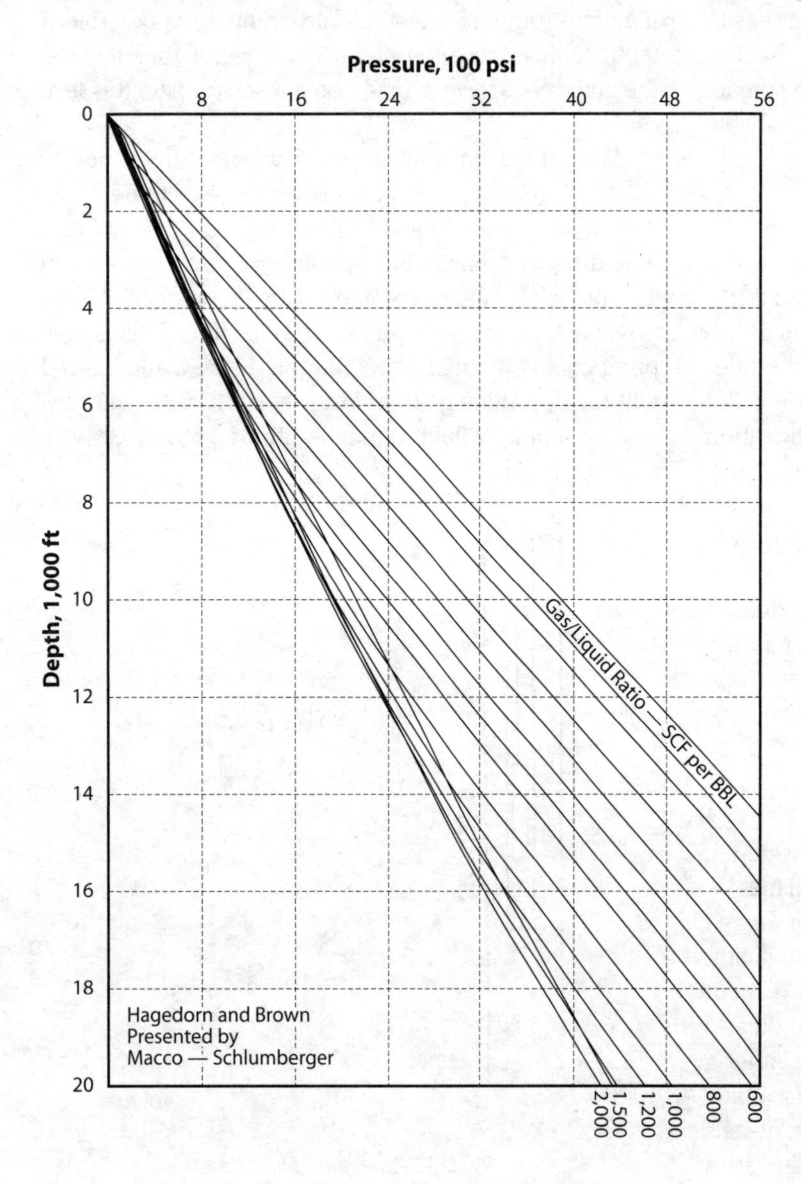

Fig. 11–4 Flowing Pressure Gradient Table

Surface facilities models

More sophisticated ways of analyzing this type of problem exist. Some programs offer a combination of surface facilities models and a reservoir model. The gas surface facilities models have been around for a long time. Originally, they were based on p/z versus cumulative production plots and periodic AOF tests. Such a process worked well on high-permeability gas reservoirs. Reservoir simulators were developed separately—exclusive of surface facilities.

At least one model from the early 1980s included both gas surface facilities and a reservoir simulator. It did not address condensate and had limited capability in handling water. Subsequently, it was found to be more efficient to couple detailed surface facility and reservoir simulation models. The reservoir simulator calls for a solution from the surface facilities model after a set period or number of timesteps.

When surface facilities are included, the engineer must know flowline sizes (IDs), lengths, and elevations. Data will also be required for line heaters, dehydrators, compressors, etc.

It is also necessary to tune surface facility models. Surface facility models are steady state models, whereas reservoir simulators are transient models. The normal practice is to identify a day when all of the wells are on production and the system is relatively stable. Portions of the surface facility can be matched to mimic pressures recorded throughout the system.

It is possible to model transients in pipelines, although this is rare in practice. Some problems, such as looping, are difficult or impossible to solve.

Intermittent production

An unusual problem arose on one project. The project involved modeling a gas gathering system, including wellbores, along with the reservoir. The operating company wanted to upgrade its surface facilities to meet a higher gas contract rate that had been negotiated. The production plots had been reviewed as part of the proposal stage. However, the wells were producing intermittently. In fact, they were not capable of sustained production.

During the early life of the pool, it did not matter that all of the wells could not all go on—there was excess capacity. Production data is normally reviewed from government sources, and the data is grouped by month. Under these circumstances, and given the known overcapacity, there was no indicator during the proposal process that intermittent production was occurring.

This became known about midway through the study. The wells continued to be shut in during the history match. An example of this is shown in Figure 11–5. It took a day or so to determine the problem. (After all, we knew the wells were producing.) It was necessary to take all of the government data and manually adjust the input to deliver the same volume of gas in partial months. Because this was a time-consuming process, the project went over budget correcting this problem. Problems like this are difficult to spot, so be forewarned.

Include files

Most simulators now permit the use of *include* files. Therefore, the data deck can be quite sparse with little more than headings and include files for grid properties, PVT data, relative permeability functions, numerical controls, and run data. However, different people seemed to have different preferences.

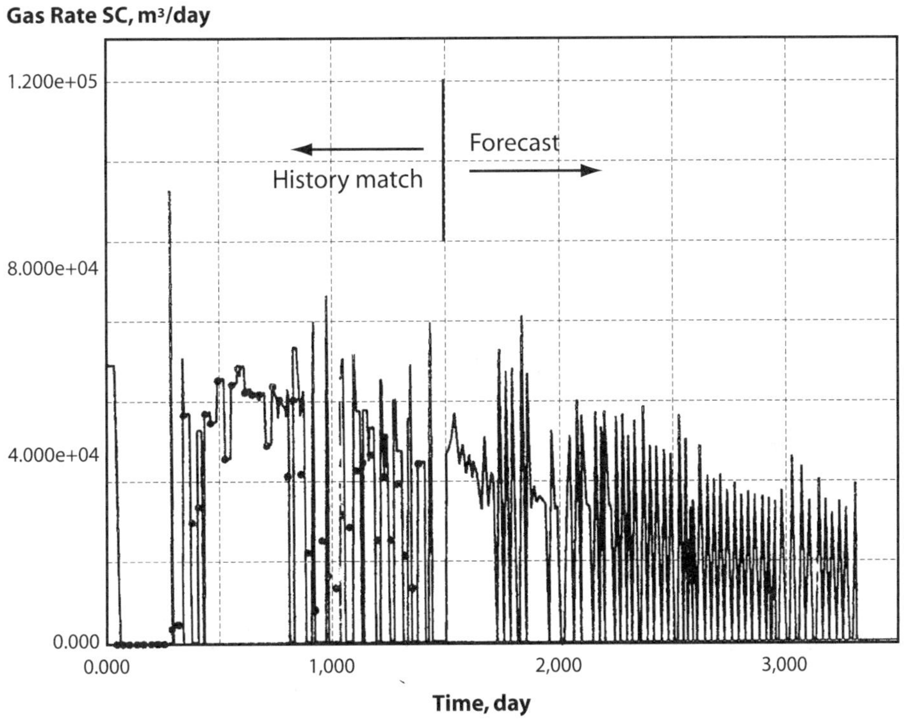

Fig. 11–5 Well Cycling—Combined Surface System and Reservoir Model

The author's preference is not to use *include* files—not even on large data decks. The proliferation of file names in a directory is distracting. However, not everyone has the same preferences. In the final analysis, both work well. Eventually, ARE developed a consistent naming convention, so the relation between the various files became obvious and different people could work on a project at various times.

Error checking and formatting

The run section contains an enormous amount of data if there are many wells and a long production history. Again, checking is very important. Check that the cumulative totals are accurate. This will ensure no mistakes were made on any individual wells.

Well specifications

A number of peculiarities are associated with run sections. In some simulators, the engineer has to specify all of the wells in the beginning, and in others, wells can be specified as the engineer decides to use them.

Output specification

On small simulations, it is easy to use the default output, but large simulations are a different matter. The largest files, where most of the storage space is consumed, are in output files. A number of types and configurations are used, depending on the software used.

One software company uses a single large *.mrf file in a binary output form. Another splits their files into *.sxxxx and *.nxxxx files with xxxx representing timestep numbers; the former one contains summary (production rate) data and the latter contains graphical (array) data. With the smaller files, the engineer can cull out uninteresting timesteps and reduce space. The larger files are more prone to corruption. If the graphics postprocessor crashes with a file open, not all of the data will be destroyed.

With a large simulation, use space more effectively. This may mean tracking only some of the array data and not all production data. In some cases, this is specified with flags and in some cases with keywords—or a mixture of both.

History-match output

It would be more logical to include this topic in the chapter on history matching. However, it is necessary to set up history matching in advance—in the run section. Different people use different formats for history matching. Some plot combinations work better than others. This is a matter of style and personal preference. Therefore, if the reader is new to simulation, start with what has worked for the author, and develop your own style later.

It is easy to make plots too busy. The author prefers a layout similar to typical production history plots. This is a semilog graph of oil rate, GOR, and water cut. Plot pressure separately. This type of plot can be misleading in that, although they look close, the rates can be in error on total volume by a great deal. Since the reservoir simulator is largely a material balance calculation, these errors can be severe. Therefore, it is prudent to run a second set of plots featuring cumulative oil, cumulative gas, and cumulative water production. Normally, these two sets of plots can be generated from a single graphics batch file. This combination yields good total accuracy as well as a feel for reservoir performance in a format typically used for analysis. These plots are done on a field basis as well as an individual well basis.

Keep these plots in a sequential series in binders with a separate binder, or section of a binder, for each history match run. Sometimes this order is set in the simulator. In this case, ordering the well name lines will save a considerable amount of sorting time. A set of example graphs is shown in Figures 11–6 and 11–7.

Datum pressure

In a problem similar to the coning problem in chapter 20, a mistake was made. Pressures were analyzed in order to determine where the fluid was moving. This caused a problem. Logic would indicate the reservoir fluids would move from areas of higher pressures to lower pressures. Normally, everyone agrees with this. On this basis, however, the reservoir fluids would logically

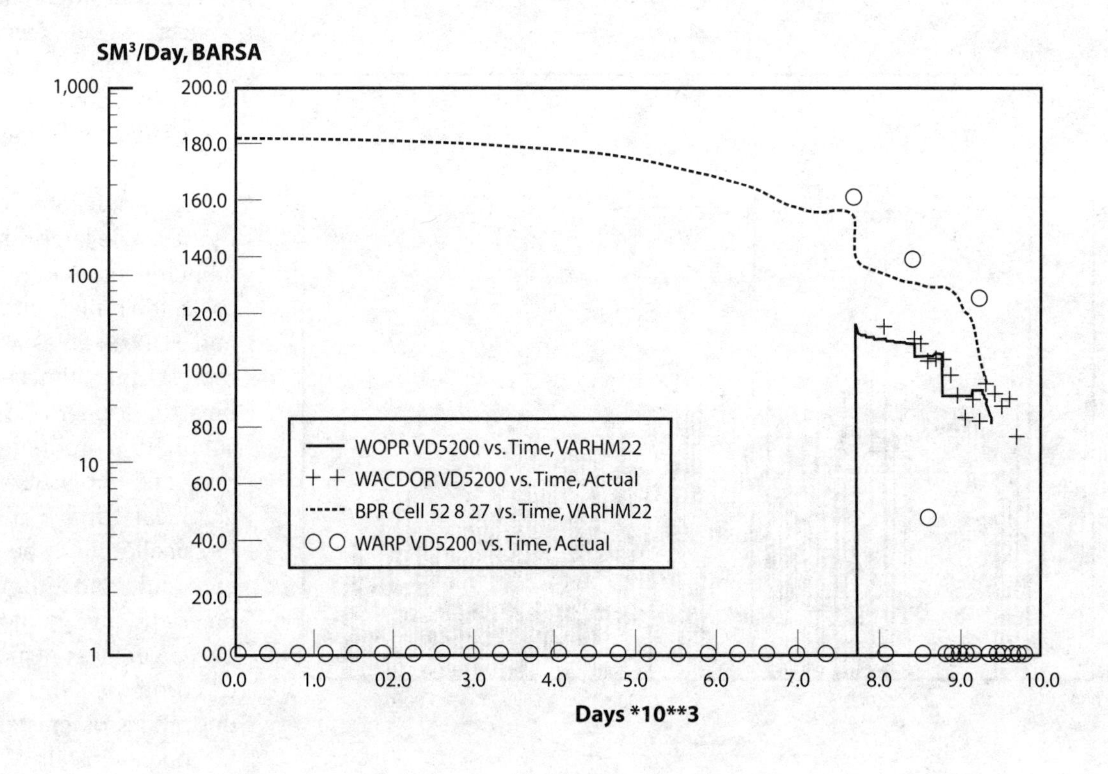

Fig. 11–6 History-Match Plot—Input Well Rate and Reservoir Pressure

be flowing upwards prior to production—in the reservoir initialization. The initial pressures increase with depth due to hydrostatic gradient.

Of course, the latter is usually a static situation, and the fluids are not moving. In fact, the fluid moves based on potential, which is pressure plus the distance from a datum times the fluid density. The output arrays are not called potentials but rather datum pressures. Grid-block pressures are useless in determining flow direction, whereas datum pressures work quite nicely. Therefore, in array outputs, use datum pressures to determine flow patterns.

If using pressures from well test analysis, which are frequently referenced to the midpoint of perforations (MPP), the grid-block pressures are more useful. Therefore, use grid-block pressures for tracking reservoir pressure. Use caution, because some data is adjusted to a pool datum.

Numerical controls

Most simulators allow one to change numerical controls on the fly. This can be useful in some situations. In one case, increased solver tolerances were required during water breakthrough when modeling pattern waterfloods with large hydraulic fractures that were directly modeled. Once breakthrough had occurred, larger timesteps and normal solver tolerances could be used again. The use of short timesteps and tight tolerances was not required during the entire model, and this increased the speed of solving the whole simulation accurately.

Decreased timestep sizes can also be required when wells are initially put on production. This is particularly true when doing single well model studies for drawdowns, buildups, and coning.

Most software support lines are alarmed at changing solver tolerances and tell users they never need to do this. It is possible to cause problems by unwittingly setting the outer tolerances higher than the inner iterations. This is not a difficult concept to understand. Changing tolerances is necessary at times. That is why the adjustments are user controllable.

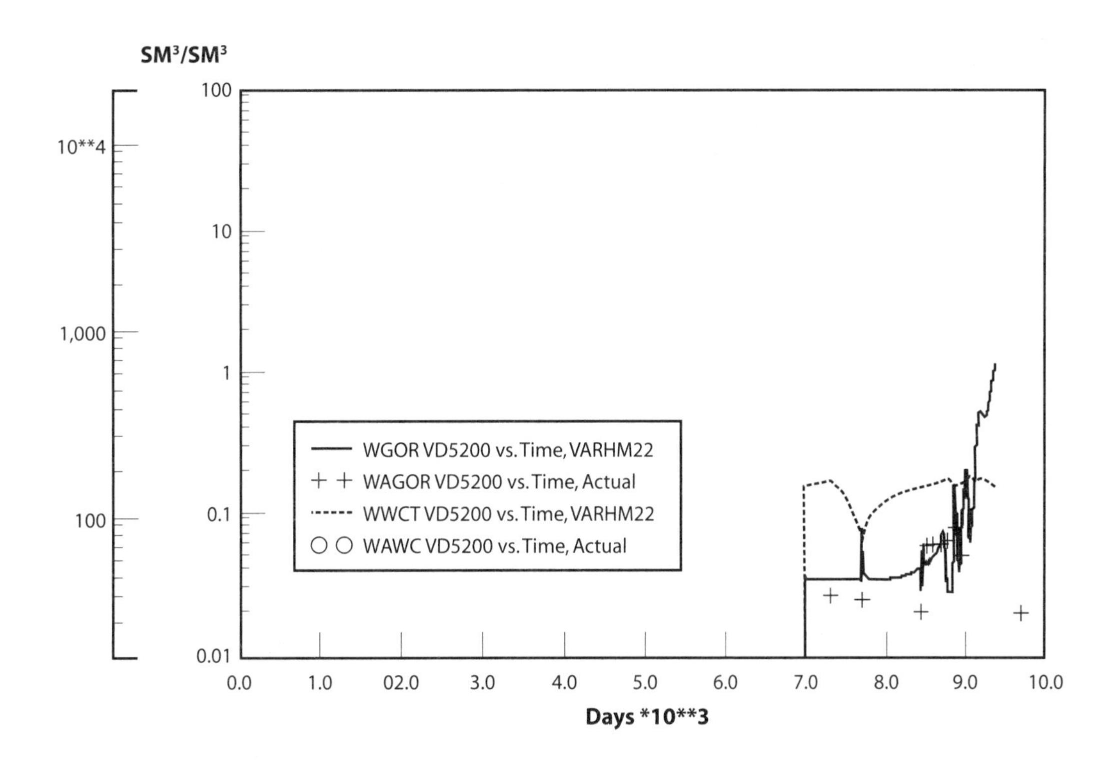

Fig. 11–7 History-Match Plot—GOR and Water Cut—Log Scales

Summary

This chapter outlines some of the practical aspects of assembling an input file. Organization is a prerequisite. Effort must be made to control data input errors. A number of practical tips were presented. Finally, data input is likely an unfamiliar process, which is not supported well in textbooks or the literature.

If nothing else, the author hopes this chapter can be used to convince reluctant managers that engineers really do deserve to take computer courses. This would include basic and advanced operating systems, Fortran, C, C++, and graphical user interface code. Engineers with these skills may be perceived, both inside and outside the company, as power users. This is not always a matter of hours of experience, although this helps; it is a reflection of how one approaches problems. It might also prevent cursing at others and vice versa.

Although the run data section is droll from a reservoir engineering perspective, it requires a substantial amount of work. This is a bad combination, and discipline is necessary to keep this part of the data file both neat and accurate. One of the best simulation engineers kept track of his error rates. He could predict, for a given number of lines, the approximate number of mistakes he would likely make. Part of the reason he was so good was he was meticulous in eliminating data input errors.

Some forethought can make the mechanics of running multiple history-match runs and analyses more effective. The author does not always use the exact same combination, and the reader's requirements may be different based on the problem or preferences.

Sometimes it is necessary to change numerical controls for limited periods during a simulation run. This can be a trial-and-error process.

References

1 Norton, P., et al., *Inside the PC,* 8th ed., Sams Publishing, 1999.

2 Peek, J., et al., *Unix Power Tools,* 2nd ed., O'Reilly Publishing, 1997.

3 O'Reilly, T. and Associates, *Complete Windows NT & Unix System Administration Pack*, O'Reilly Publishing, 1998.

4 Nemeth, E., et al., *Unix System Administration Handbook,* 3rd ed., Prentice Hall, 2000.

5 Thomas, R., and R. Farrow, *Unix Administration Guide for System V, Version IV*, Prentice Hall, 1988.

6 Brown, K.E., *Technology of Artificial Lift Methods*, PennWell Publishing Co., 1980.

12

History Matching

Introduction

History matching is the phase of a study that will take up the largest portion of study time. It is one area where mystical prowess is sometimes attributed to individuals. There have been many attempts to use statistical techniques and try to reduce this to a systematic method. For the most part, these efforts have not been successful. This author has never used commercial statistical history-matching software packages. However, statistical minimization is dangerous for even the simplest material balance calculations.

The majority of simulations still involve considerable manual input. Most of the existing guidelines on history matching have three basic cautions:

- Use the simplest model possible. Since so much is unknown, the simplest is likely the best.
- Keep the changes to the factors with the most influence on the answer. This is in keeping with the attempt to find the simplest change.
- Keep the changes to the factors with the highest uncertainty. For example, if permeability data from the core is available, but there is no relative permeability data, alter the latter.

Reservoir characterization

The first caution has led to some serious problems. Many of these problems became clear in the North Sea developments. The capital commitments are sufficiently large on North Sea projects that errors in production forecasts, which have occurred and been ignored for onshore projects for a long time, became embarrassingly obvious to both company shareholders and the government. This led to the development of a large volume of geostatistical data and spawned a new discipline—reservoir characterization. In essence, simple models misled management, and, invariably, they overestimated production performance. Ironically, a recent geostatistics course the author attended suggested the simplest geostatistical model and the simplest semivariogram be used when faced with limited data.

The principles used by ARE state reservoir models must be made as realistic as possible, as early as possible. This is not easy. Because there is relatively little data in development stages, the data must be invented or created based on analogy from other reservoirs. Naturally, this makes everyone uncomfortable. This requires experience.

On the surface, this would appear to be a classic paradox or catch-22. However, this is not the case, and it does not mean outrageous guesswork. A considerable amount of experience exists in the literature, government data, and company files. This approach requires the use of a geological model. Moral: if you don't have direct experience then cheat—use someone else's experience.

Contradictory advice

As suggested previously, much of the advice on history matching is contradictory. For example, the two extracts following are from *Modern Reservoir Engineering—a Simulation Approach* by Crichlow and *Reservoir Simulation* by Mattax and Dalton.

Crichlow

There are several parameters, which can be varied either singly or collectively, to minimize the differences between the observed data and those calculated by the simulator. Modifications are usually made on the following:

1. *Rock Data Modifications*
   - a. *Permeability*
   - b. *Porosity*
   - c. *Thickness*
   - d. *Saturations*
2. *Fluid Data Modifications*
   - a. *Compressibilities*
   - b. *PVT data*
   - c. *Viscosity*
3. *Relative Permeability Data*
   - a. *Shift in relative permeability curve*
   - b. *Shift in critical saturation data*
4. *Individual Well Completion Data*
   - a. *Skin effect*
   - b. *Bottomhole flowing pressure*[1]

Mattax and Dalton

Those reservoir and aquifer properties appropriate for alteration, in approximate order of decreasing uncertainty, are (1) aquifer transmissibility, kh, (2) aquifer storage, φhc_t, (3) reservoir kh (including vertical restrictions and directional variations), and (4) relative permeability and capillary pressure functions.

The following additional properties must sometimes be altered, but they are usually known with acceptable accuracy: (5) reservoir porosity and thickness, (6) structural definition, (7) rock compressibility, (8) reservoir oil and gas properties (and their geographic distribution within the reservoir, if properties are not uniform), (9) water-oil contacts (WOCs) and gas-oil (GOCs), and (10) water properties.[2]

The difference between these two is Mattax and Dalton have outlined a prioritization and suggest some changes should be rare. This is better in some ways because it discusses a systematic approach to field history matching, which is not included in the Crichlow excerpt. Crichlow includes a discussion of how rock properties, relative permeability end-points, k_{rg}/k_{ro} ratios, and communicating areas affect simulation results. For instance, if the rock properties are changed this way, this is how the flood front advance changes in the reservoir. Mattax and Dalton even state:

Making changes by guessing or by following one's intuition can be expensive and usually will prolong the history matching phases. The decision to use such an unstructured approach may result in the impression that experienced reservoir engineers develop a "feel" for the "art" of history matching. Experience is valuable of course, but only because it increases one's understanding of reservoir mechanics. Using that understanding to improve the accuracy of a reservoir's description should not be viewed as an art.

Author's views

In this author's view, both of the excerpts provide guidance that can be viewed in various contexts of being right *and* wrong. First, neither considers which factors apply in particular reservoir situations. Second, some weighting must be given to the accuracy of the data—neither addresses the issue. Third, some understanding is required of how reservoir simulators are constructed.

Changes in the history-match model must be viewed in the context of the assumptions and construction of the simulation program. Fourth, the changes have to be considered in a global context, more specifically, considering the geological setting. To elaborate:

- In an undersaturated reservoir, formation and fluid compressibilities will have a major influence on pressure behavior. In a reservoir below the bubblepoint, these factors are of almost no consequence. Formation compressibility will contribute to minor amounts of water production as pressure depletion occurs. Under these circumstances, changing formation compressibility will have no effect on reservoir pressure, but it *will* cause minor water production.

  Experience will show simulators are sensitive to changes in pore volume—i.e., porosity and thickness—and they are surprisingly insensitive to reservoir permeability. This is not to be confused with well $k \times h$, which is often manually specified. On the surface, this would appear to represent familiarity with simulators. **What it really amounts to is a fundamental grasp of reservoir engineering.** There is a corollary to this: learning simulation may make one a better engineer, but it cannot make up for lack of basic understanding of reservoir engineering.

- Ideally, core data is immutable, however, it changes with the stress applied and can be altered with improper handling. More frequently, porosity is obtained from log analysis correlated to limited core data. There is a significant error in individual wells and zones because of these assumptions. It would also be fair to point out that a 4-in.-diameter core samples an infinitesimally small volume of the reservoir and, by virtue of removal, is no longer relevant.

  There can be a considerable amount of variation in relative permeability data due to measurement and sampling errors as well as a failure to correctly establish a restored native state. Yet, some engineers stubbornly stick to a single sample of data because it happened to be a "real measurement." **A sound knowledge of the data source and the accuracy of the data is very important.**

- Using a pseudo relative permeability curve in a detailed coning model is inappropriate. This technique was developed to model near-well core effects or to account for layering in areal studies. This is not to say relative permeability data in a coning study should not be changed; however, any changes should be consistent with known ranges of data.

  Perhaps more commonly, pseudo relative permeabilities are implemented without including rate changes that result from waterflood allowables (government rate restrictions) or good production practice. In some cases, the development of relative permeability curves is not extrapolated into the predictive phase at all. **A sound knowledge of simulation technique is required; history matching must be consistent with modeling limitations.**

- In a number of studies, bizarre combinations of faults were used to achieve a history match. Actually, in one of them, combinations were brilliant. From an overall geological perspective, however, these changes did not make sense. On the other hand, the result was a surprisingly good match of pressure, GORs, and water cuts. It's fair to point out the operating company's geologists did not have any other solution to offer. **A sound knowledge of geological input is required to assess and modify, directly or indirectly, geological descriptions.**

Finally, contrary to the Mattax and Dalton view, reservoir engineering is definitely an art. So is mathematics. Very often, the correct solution to a differential equation, for example, is recognizing it fits a certain class of problems that can be solved in a certain way. Collecting variables can be aided by the use of little symbols, and this will reduce the chance of your making an error. An engineer's ability to solve these problems becomes faster and more consistent with experience. In the case of reservoir engineering, the solution to the problem may not be directly related to a knowledge of simulation technique, but it is quite likely to be related to a general knowledge of geology, reservoir engineering, completions, geostatistics, and log analysis, etc.

Conceptual models

The first "law" of history matching is if one has trouble obtaining a history match or finding help, then most likely the problem is with basic conceptual implementation. Go back and determine what is missing. This may entail returning to the beginning and completely reconstructing the model.

Off the wall—the "Far Side" of history matching

ARE did a study involving two reservoirs along a north-south trend. The reservoirs were thought to be similar, although the quality—permeability and net pay—of the more northern pool was thought to be lower. The two pools were thought to be separated by a fault. We did the south pool first and were successful in getting a history match and understanding the production mechanisms. Armed with this experience, we marched on to the north pool confident of a fast history match. Nothing we learned on the south pool applied to the north pool.

After being stumped for about two weeks, we photocopied all of the history-match plots—carefully archived in our history-match binders—using a 50% reduction and placed a number of our sensitivities on a large wall chart, which looked like Figure 12–1.

We mounted a series of these on the wall. The history-match plots were placed on a background that included an outline of the reservoir and structural contours. After wall gazing for a time, it became obvious some of the history matches worked well in the south part of the north pool and others worked well in the north part of the north pool. It is doubtful we would have spotted these trends without being able to look at the big picture.

Eventually we were able to settle on an explanation as to why the two reservoirs behaved differently, although there was never any conclusive proof our suggestions were correct.

Use of graphics

History matching is greatly accelerated by the use of modern graphics. This is both a blessing and a curse. Sometimes it lets one get to the heart of a problem quickly. The graphics are as important as how well the simulator works.

At other times, it blinds you. On large models, the picture on the screen is very small—even with a 26-in.-monitor, and it's hard to see the detail. This happened on the East Swan Hills Unit No. 1 study.

Coloring crayons

I was stuck on the performance of a waterflood immediately north and south of a channel. No reasonable change of pore volume or permeability would fix the problems. At this point, I got out the coloring crayons and spent most of the afternoon coloring about 2,000 tiny grid blocks. More than one person asked what exactly I was doing. Halfway through the afternoon I was coloring the saturations in the tidal channel that separated the north pool from the south pool. Actually, the geologists in 1960 knew the channel was there. There was one D&A in the

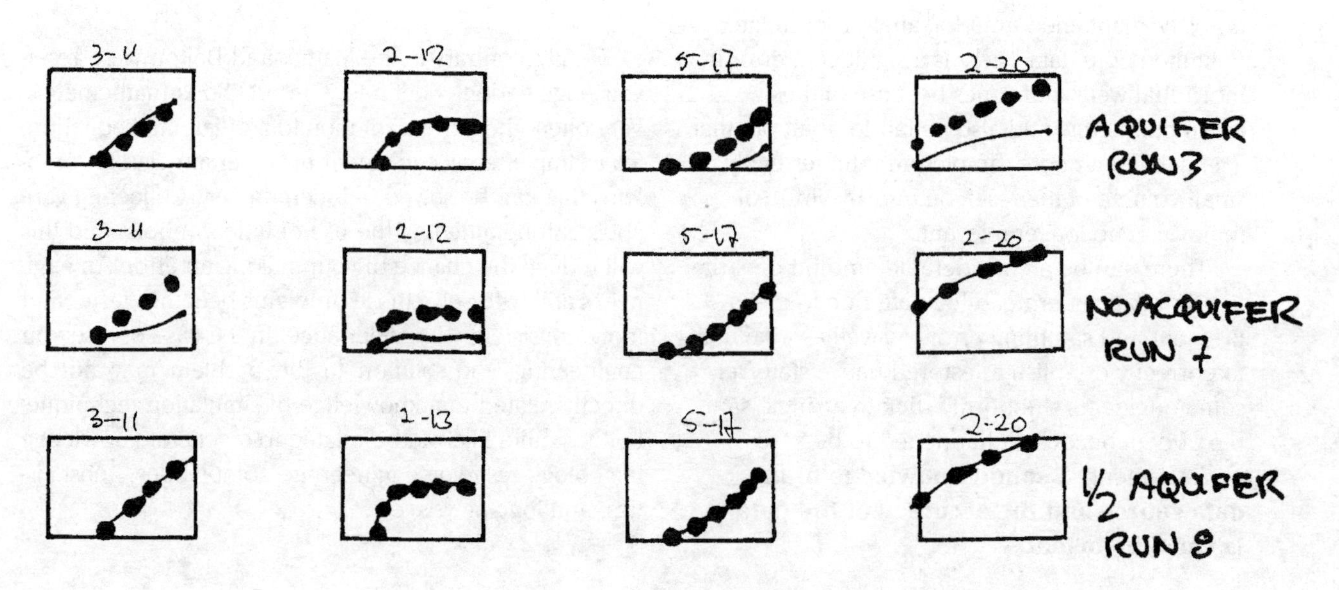

Fig. 12–1 Use of Well History-Match Diagrams to Spot Trends

northeastern extreme of the unit. The geologist had assigned low net pay and porosity but had contoured low values from the north and south sides.

By now, I was getting a little cranky. Then it struck me. The contours were reasonable extrapolations, but what if there were *no* permeability in the channel? The channel was filled with lime mud. I reran the simulator with the entire channel reset with null grid blocks. This gave dramatically different results. After reversing most of the changes in the rest of the grid I had made to permeability and pore volume, the match was nearly perfect.

I never did complete the coloring. Although coloring has a mindless quality, it does force one to slow down and see things that would be missed by racing through simulation results with a postprocessor.

More importantly, the correct solution could not be derived with any of the systems described earlier. Recall Crichlow's list of rock data, fluid data, relative permeability, or individual well completion data; and Mattax & Dalton's aquifer transmissibility, aquifer storage, reservoir $k \times h$, relative permeability, and capillary pressure functions.

Trial-and-error process

History matching is a trial-and-error process. Despite this, there are still admonishments one will hear, such as, "Use simple models and hand calculations when they are cost-effective, but balance the manpower and computer costs." This is entirely unrealistic. People are going to make mistakes. Accept it.

Go for the extreme

Try extremes early. One can spend computer time and waiting time trying progressive changes. Going to the extreme of the change, an engineer can determine quickly if such a change will achieve the desired results. If it works, focus on the ideal value with a simple linear extrapolation. If, however, the changes don't work, proceed quickly to a new approach.

Organized approach

The use of an organized approach will help greatly, particularly on areal simulations. The flow chart used by the author is shown in Figure 12–2.

The general strategy is to start with the overall picture and work down to progressively more detailed matching. No hard-and-fast rules exist to set up a guideline. Probably the best thing to do is read about field cases in the literature. Relatively few of these are published. Historically, the SPE has discouraged many field cases, but a large number are available in conference proceedings.

Faster run times

Many people who do large-scale areal simulations have observed execution times seem to decrease the closer they get to a good history match. This is consistent with the author's observations—with some exceptions. If a grid refinement around a well is needed, the results will improve along with the execution time. For detailed pattern models, the author has found that cutting timestep sizes when injectors go on improves results and increases simulation efficiency.

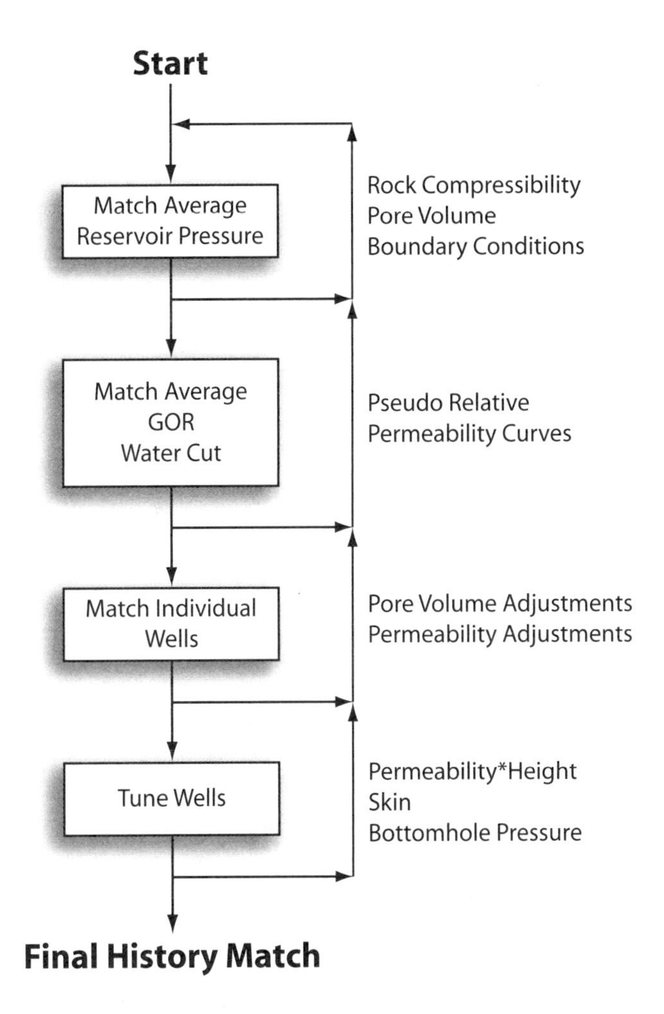

Fig. 12–2 A Systematic Approach to History Matching

Detailed comments

I taught a course overseas, and a number of students asked me to be more specific in a discussion on history matching. We had spent nearly a week discussing various input parameters, and I thought either something was missing or my teaching skills were lacking. I asked for the specific input parameters that produced their question, and in the course of reviewing the material, I realized this is not a simple process. In response to their request, I have identified the most likely problems. Here are my detailed comments with thanks to the students.

Permeability

Permeability was at the top of the list. Strangely, simulators are sensitive to $k \times h$ only in the near-wellbore areas. In most studies, this will show up as wells that have their production clipped by an inadequate $k \times h$. Since the rate of the primary product is fixed during the history match, it will not identify *all* wells that have an incorrect $k \times h$—only those that are *too low*. Otherwise, the simulator will be quite insensitive to small changes in permeability. This struck the students as counterintuitive. I agree.

Larger changes in permeability will affect the simulation results, but the author suggests changes in multiples of 2, 5, 10, and 100. Permeability is normally important, in an overall sense, to the order of magnitude. The biggest effect is normally on GOR for oil systems, condensate-gas ratio (CGR) for gas condensate systems, and water production, if coning is present.

If we go back to the source of our permeability data, it is usually derived from a core porosity versus core permeability transform. In most cases, this is the classic linear relationship occurring empirically. Some will argue the fine points on gamma distributions. Typically, the range of data spans one order of magnitude, as shown in Figure 12–3. This data is for air permeabilities, and we apply normal fudge factors to correct this to field *in situ* permeabilities. Overall, this is far from a precise process.

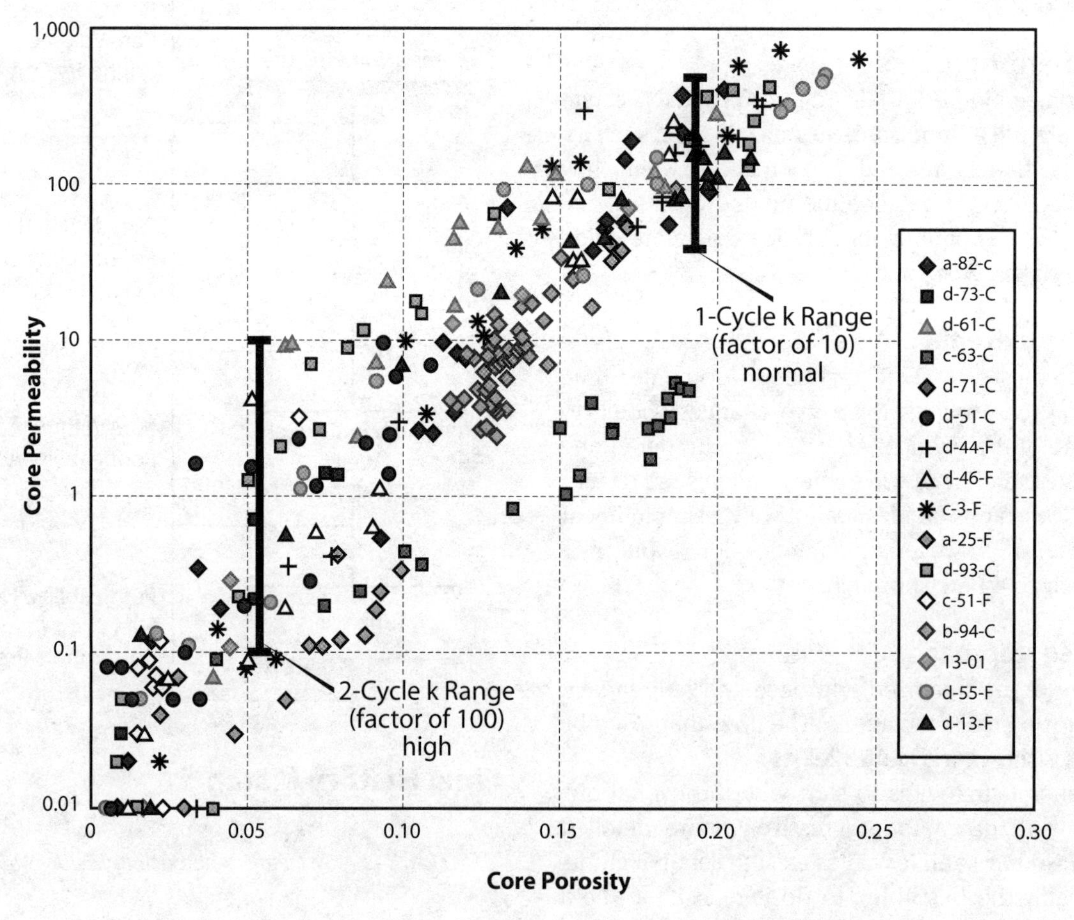

Fig. 12–3 Core Permeability vs. Core Porosity—Typical Permeability Data Range

A case can be made, if the general adjustments are made correctly, that exactly one-half of the well productivities should be clipped during the history-match stages. The range of corrections should be less than ± one order of magnitude since there should be data point averaging for an entire well versus individual data points.

One engineer known to the author thinks a slight bit of clipping on well plots indicates the permeability in the model is nearly correct. This twist of logic actually makes sense.

Having said this, many people use air permeabilities directly and get great history matches, since the simulator is not particularly sensitive to this parameter. Sometimes the errors are too large, and, for example, the GORs will be too low since the reservoir permeability is too high. If other incorrect, compensating changes are made to adjust the history match, such as adjusting pore volume, then the model is inaccurate. Sometimes one gets away with murder, sometimes one gets murdered.

Overstated permeabilities should be caught in the tuning stage. Practical experience has shown this doesn't always happen, and incredibly optimistic predictions occur. On the positive side, lives are rarely at stake. Financially, the picture is not so pretty.

Directional permeability

Again, assuming that well $k \times h$ values have been manually specified, simulation performance predictions are not strongly affected by directional permeability. In the author's experience, during history matching (where the production is input directly), the ratio of Kx to Ky must be on the order of about 30:1 or greater to show significant changes in reservoir performance.

Porosity

Porosity is normally determined from logs. The quality of the prediction then depends on how accurately the logs work and how well the results are tuned to core data. In general, log interpretations are accurate to within 10% or better. Simulators are quite sensitive to pore volume, and history-match changes will be done to correct this. This will usually show up as a pressure mismatch or as an incorrect water breakthrough time in waterflooding. Unfortunately, in almost all circumstances, the simulator cannot tell the difference between porosity problems and thickness problems. Most simulators allow one to alter $\phi \times h$ and do not force a specification to thickness.

Decline rates are also sensitive to total pore volume.

Thickness

Thickness is normally interpreted from logs and interpolated between wells via geological interpretation. This is not nearly as accurate as porosity, and most of the pore volume unknowns are probably related to thickness.

As suggested previously, simulators are normally very sensitive to $\phi \times h$, and one is unlikely to get a history match without some $\phi \times h$ changes. As a rule, changes of 10–20% in selected areas should not cause alarm and can be regarded as normal. In some limited areas, 30% changes may be required. This is not usually a cause for concern.

On an overall level, total pore volume changes with reasonable mapping should be accurate to within 10–15%. Larger changes indicate a geological interpretation problem. Interpretation problems of more than 15% can and do occur. However, this should not be a surprise. This frequently occurs near the exploration phase and may be grounds to investigate possible water drives or other factors.

Saturations

Generally, saturations are determined from well logs—similar to porosity—and are reasonably well-known. Some exceptions exist for water saturations.

- The largest exception, in terms of saturation error, relates to unconsolidated sands and damaged or dilated core samples. This really applies to a limited number of reservoirs, the most significant of which are the tar sands in northeastern Alberta.
- One of the more common problem areas is shaly sands, where saturations are difficult to determine or where the resistivity of water is low, in combination with a shaly sand.
- In chalky or fine-grained carbonates, it is difficult to determine water saturations. Chalky reservoirs have nonstandard resistivity relationships.

There can be nonstandard a, m, and n values (which are the Archie equation saturation exponents used to calculate water saturations) for coarse lithologies and fractured reservoirs.

Residual oil saturations can be a problem if there is relatively little relative permeability or laboratory waterflood testing data. Sometimes sample sets are skewed and contain either high- or low-quality samples. Sample sets can also be limited in size. Waterflood recovery will be quite sensitive to the location of end-points, and this is important. Unfortunately, water-oil S_{or} has no real effect on primary history matches, and consequently, history matching provides no guidance to this parameter at all. Predictions can be made with a range of residual oil saturations. The real solution to this problem is to perform laboratory waterflood tests, which are usually quite inexpensive.

Capillary pressure data

As outlined earlier, there are many potential problems with capillary pressure data. Generally, log calculations will control the irreducible water saturation assumed, if for no other reasons than shear volume of sampling.

The capillary pressure transition zone thickness can, and should, be a history-matching parameter. Figure 12–4 shows a situation that may require P_c adjustment. If the water contact level cannot be adjusted to fit the downdip and updip rows of wells, it may be necessary to adjust the transition zone thickness.

This change will also affect the original oil or gas in place. Therefore, use caution.

PVT data

Errors in PVT data are quite common; however, these problems are normally spotted in the data-screening stage. The most common problem is a mismatch with field GORs and the PVT laboratory data GORs. This was discussed extensively in chapter 6.

It is possible to have a material balance bug in the PVT data. This also has been discussed. The problem is a procedural one and usually is restricted to certain labs. When the problem does occur, it causes great mischief; however, it should be spotted in screening.

If there is no lab data and a correlation is used, there is room for adjustment. Again, this would normally be caught in the screening stage. In low-permeability formations, the initial GOR can be difficult to spot, and PVT properties may have to be adjusted to match initial GORs. Although the simulator will show the R_{si} for a day or two, real production data is record-

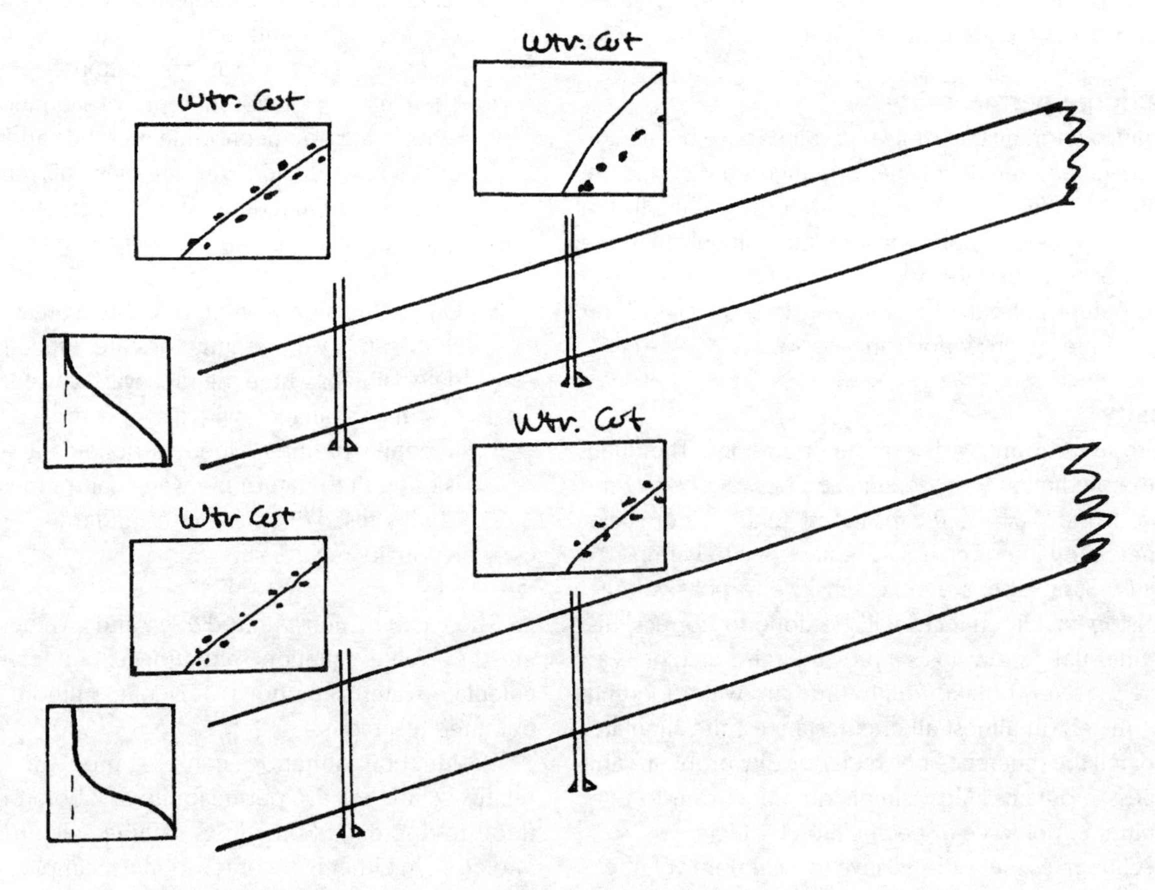

Fig. 12–4 Adjusting Capillary Pressure Transition Zone Thickness to Correct Water Cuts

ed during a period of a month. GORs will rise within days in low-permeability formations. PVT adjustment may be required in these situations.

Compressibilities

Above the bubblepoint, liquid and formation compressibility are important. Generally, fluid compressibilities are well-known. In rare circumstances, fluid compressibility will need to be estimated from correlations.

Even in the latter circumstance, the uncertainty will not be a major issue. Unfortunately, rock compressibility is of roughly equal importance as fluid compressibility in most cases. Most of the formation compressibilities (also known as pore volume compressibilties) are estimated from standard correlations, which were outlined in chapter 7. These are unreliable, and there is a considerable range of data. Restated, fluid compressibility is not normally the obstacle.

In practice, rock compressibility can represent a real dilemma. Although rock compressibility tests are not particularly expensive, they are rarely done. This is compounded by a fair amount of variability among sample specimens. The latter has made this type of test unpopular with management, despite its importance in certain circumstances.

The author recommends a series of tests, due to the variability of rock samples.

Once below the bubblepoint, gas exsolves (comes out of solution), and fluid and rock compressibilities are then rarely important in comparison to gas compressibility (gas compressibility is approximately proportional to inverse pressure, so the farther below bubblepoint the system is, the less important fluid and rock compressibilities become). Many oil companies prefer waiting and uncertainty to obtaining this type of data. Pressure decline above the bubblepoint is *very* rapid, and, in truth, waiting may not take much more time than performing the tests. If the bubblepoint is very low in relation to the initial reservoir pressure, this is can be a vital issue (because, in such a case, a substantial amount of oil recovery will occur prior to reaching the bubblepoint). If this is the situation, make the case to management early.

Compressibilities can be difficult to match with an EOS and, in particular, for near-critical fluids.

Although in general, compressibility is not a major problem, the author has encountered situations where this is the case. One reservoir with a light fluid was miscible flooded. The EOR scheme utilized injection substantially above the bubblepoint. It was also implemented early in the pressure history. Although some high-quality rock pore volume compressibility data was obtained, an insufficient number of samples were obtained to overcome data variability.

Viscosity

The importance of viscosity is twofold. Oil viscosity will determine productivity via Darcy's Law, and it is important for displacement—really the mobility ratio—in waterflooding. Generally, well testing should provide adequate information for oil viscosity. Dead oil, which is normally tested, requires adjustment for solution gas. Most PVT analyses include this information. If a PVT sample is unavailable, there are correlations to estimate live oil viscosity.

Barring this relatively rare condition, viscosity is not an issue. Water and gas viscosities are quite adequately estimated by correlations.

Relative permeability curves

Changes to relative permeability curves are common. Having said this, the author strongly discourages these changes. With modern computers, most reservoirs can be matched without resorting to pseudo relative permeabilities.

Most relative permeability changes involve increases in the water relative permeability curve. If one is making these changes, it is likely because the layering has not been correctly handled or the grid is too coarse. This means a fundamental problem in the way the problem was set up, which is where the changes should be made. The author recommends changes to layering rather than to relative permeability.

In general, I have found that reservoir simulators are less sensitive to gas-oil relative permeability than most engineers, including the author, would anticipate. There are some comments on whether to screen on k_{rg}/k_{ro} ratios or linear curves (use both) in chapter 7. The ratio will control the initial GOR. After this initial production period, reservoir simulation performance predictions are surprisingly insensitive to this input for a solution gas drive. The amount of gas is controlled as much by the amount of solution gas available, i.e. R and how much solution gas has been released. The latter is a function of pressure and which, in turn, is most strongly affected by *pore volume*. It is recommended that every simulation engineer should do a radial grid sensitivity to develop a feel for these responses.

If the reservoir is extremely large and these situations do exist, then pseudo relative permeability curves are acceptable. Beware of the dangers outlined in other sections of this book.

If one is faced with a limited number of samples, some changes may be quite reasonable. Carbonate reservoirs inherently have a lot more variation in relative permeability data than sandstones. For this reason, the author is more amenable to altering data in carbonate reservoirs.

Individual well completion data

The two areas of most concern are skin data and bottomhole flowing pressures. Most all wells will have skin. Skins will change in time, normally increasing (i.e., reflecting increased flow restrictions due to fines migration, scale buildup, wax deposition, etc.). Pressure transient analyses can be used to estimate the skin.

Bottomhole pressures are not well-known for most producing properties in western Canada where most oil wells are pumped. Therefore, data on skins is rarely available. Most drawdowns are localized and do not affect the reservoir simulation or pressure as a whole. The pressures of greatest interest are those that are fully built-up. This also tends to be the most accurate data.

Simulators are quite sensitive to the amount and pattern of withdrawal. Restated, from a reservoir engineering perspective, the most important aspect is that of material balance. *Has the right amount of material been removed?*

Thus, skin and bottomhole pressure are of secondary importance during history matching. Consequently, the author ignores this data unless it affects cumulative production. However, skin and bottomhole pressure will affect cumulative production in some circumstances—for instance, offshore wells that are normally controlled by surface pressures and tubing performance curves. Restrict changes on skin and bottomhole pressure to those wells where cumulative productions are in error.

This may appear counterintuitive to some. If it causes consternation, employ some simple sensitivities. For instance, if the $k \times h$'s are kept constant in the grid exercise outlined in chapter 20, compare the effect of average reservoir pressure on the pressure distribution in the reservoir. Similarly, it is possible to change different combinations of bottomhole pressure and skin. Provided the withdrawal rates are the same, one will find the simulator quite insensitive to these changes.

Although skin and bottomhole pressure are not of great importance in history matching, they are of critical importance during predictions. This will be discussed in the next chapter.

Geostatistical modeling

It is worthwhile to consider the implications of the preceding sections on the many conditional simulations that are now commonly run for some types of projects. For the most part, these concentrate heavily on permeability distribution, which is usually developed from a porosity array. Where should one concentrate one's efforts? Pragmatic history-matching experience would concentrate on pore volume and determining well $k \times h$ values. Most performance is not really strongly affected by permeability distribution. This suggests that performance variation could be simplified to two factors.

Layering, or more generally heterogeneity, strongly affects displacement processes, i.e. waterflooding. In this regard, it is the relative flow path lengths that are critical. This will control breakthrough times and the rate at which watercuts increase. Therefore, after primary production is established, simulations involving displacements should concentrate almost entirely on the distribution of flow path lengths.

Chapter 3 concentrated on the physical layout of the reservoir, and this definitely exerts a strong influence on the pattern of flow in the reservoir. This suggests that where objects (such as channels) are located is of critical importance. Object modeling requires some quantitative ranges and will allow multiple possible positions to be evaluated.

History-match output

History-match output was discussed in chapter 11, since it is normally set up in advance.

Program bugs

To effectively develop large-scale software requires an archiving system. All changes are carefully dated and tracked. A special software system is used that effectively date stamps each line of code. It also allows the version of executable code to be recreated on the computer if the date or release number of the program is known. In this way, it can track whether the bug has been fixed. It is not possible to reissue software for every minor bug found, and this can be annoying to those who have wasted time working with a buggy option in the software. In some cases, a bug fix can introduce unforeseen bugs elsewhere in the program. It is also possible that it is a completely new bug and somebody will be assigned to trace it, test it, and add the correction to the archive system.

Problem-solving skills

Having handled a number of user problems for software, I am astounded to find weak problem-solving skills. Perhaps one develops this skill by handling these calls. First, the most important objective is, like the three laws of real estate, *location, location, location*. Therefore, do everything possible to identify exactly where the problem is occurring. This is a systematic process. It is appropriate to point out the software hotline will likely go through the same steps. Here are some examples:

- Does the problem start when a well is turned off or on? This will lead you back to well control statements or perhaps timestep size changes.
- Have you made a series of changes? If so find out which change causes the simulator to crash. You might have to go back and do your changes stepwise to find out what statements or combination of statements is causing the problem.
- Does the problem occur when a certain pressure is reached? Maybe the pressure range in the PVT table has been exceeded at some stage during a waterflood.

To expand on this last point, when history matching the East Swan Hills Unit No. 1, the lab relative permeability curve did not take into account the water breakthrough taking place. The model did not include detailed layering. Consequently, the model overpressured compared to the actual reservoir. A pseudo relative permeability curve solved this problem, even though the error identified by the simulator appeared to indicate the pressure range needed to be expanded. This is an example of indirect indications.

This same message also occurred in injection well grid blocks. In this case, the PVT data needed to be expanded because high pressure did occur in the well grid blocks with injection. This was a far more direct indication of where a minor problem existed.

As a rule, the error messages in a simulator provide hints at the location of the problem. It may not be direct, but it definitely narrows the range of possibilities. This is what the *simulation hotline* calls a data set problem. Requests for help with these types of problems are really asking for a solution to reservoir engineering problems. This is the engineer's job and not a job for the hotline. It will likely annoy them. In fact, they may charge for consulting time. This is not always the case.

Difficult bugs

Probably the worst bug the author ever encountered concerned a grid refinement for a horizontal well. The problem started with an ambiguous statement in the documentation, which he interpreted to mean 2, 3, 4, or up to 10 refined grid rings could be used around the well, i.e., a series. The statement was intended to mean a total number of 10 horizontal well refined grids could be used with up to four rings in each refinement. Cryptic documentation is not a new a problem.

There was no error-checking routine to identify if more rings than four were used. No one anticipated someone would misunderstand the documentation, but users are incredibly creative. There is no limit to the preconceived notions users have that cause different interpretations of the documentation, and this includes the author. The simulator actually ran, although very slowly. This was noticed, but, since reasonable-looking results were obtained, it didn't raise any immediate flags. More importantly, the results were urgently needed for a presentation, which was produced on time. The project was then at the end of the budget.

Later, the data set was used again on a PC. The PC executable was generated using a different compiler, which was evidently not as forgiving or, depending on your perspective, was better. The data set refused to run at all on the PC. Nor were there any sensible error messages from the simulator. The PC compiler, being much less developed than its Unix counterparts, gave one of those cryptic numbered error codes. The end user consequently had no idea of why the data set was failing.

The input-checking routine should have detected this. It is surprising an *out of range* error did not occur on one of the DO loops in the simulator code in all of the compilers.

Bugs lacking an error message to provide a few clues are rare, but they are difficult to find when they do occur. Of two simulators with identical speed and capability, the one with the better debugging statements is preferable. In fact, most simulations engineers are happy to give up some speed in favor of better debugging. Simulator input checking will not find every error. This is an example of how timing and business requirements contributed to missing a simulation bug. The project was to be used as a data set example, and it eventually failed. The presentation and the example solution went very well. It was not used as the basis for field development. When 10 refined grid rings was changed to 4, in a few lines of the data input, the problem ran very well.

The last word on bugs

Program inconsistencies and bugs are a fact of life on large-scale programs. The quality of software has improved during the last 20 years of the author's experience. Superior debugging tools and programming have become more structured during this time.

Engineers will likely encounter a bug with a simulator at some time. Sadly, software support people spend a significant amount of time answering "bonehead" questions. Following are some hints in dealing with them.

First, if possible, ask someone in an adjoining office if he or she can spot what actually may be a user error in the input data. Sometimes it is as subtle as a spelling error of an input parameter name, for which computers are notoriously intolerant. This assistance saves embarrassment and deserves a reciprocal favor.

Second, think through the reservoir engineering implications of what is happening, as outlined earlier. This will likely save not only embarrassment but also some consulting fees.

Third, users are likely angry—at least frustrated—when they encounter a problem. In some cases, it is justifiable; however, it is a fact of human nature that venting frustration will not endear one to the support person who can assist with a problem. Spend 30 seconds thinking pure thoughts before dialing the number for help.

Fourth, demonstrate knowledge of the situation. Outline the problem and steps taken to solve the problem. If possible, indicate the documentation is at hand. Even for relatively simple user errors, support personnel will likely forgive an occasional problem if the user has made a sincere effort to solve the problem.

Fifth, the software support person will try to determine if the problem is a real bug or a simple user mistake. They will normally ask a series of questions. If the problem is sufficiently complex, they will ask for the version of the program and a copy of the data set.

Many problems are caused by completely cryptic documentation. One software application contained an error message that said, in essence, *Read the manual, dummy*. The author found a bug demonstrating a mismatch between the documentation and the program. The error message was removed.

Summary

Regardless of the source, including this book, don't believe all the advice on history matching.

Getting the model right is important, and it is not simple. The integration between geology and the reservoir model is the area most likely to cause problems. Engineers are blamed for bad reservoir simulations, not the geologists—even if geology is the most critical input. Reservoir engineering skill is the most important element in history matching. An organized approach will greatly increase the chance of success and will make the engineer look much better. Automated history matching, whether via statistical minimization or expert defined process, will not systematically identify all problems. Much of history matching is a sophisticated *what if?* process. Intuition will guide the engineer on the appropriate *what ifs*—not cookbook procedures. Knowledge of the issues the geologist interpreted based on the limited information available is of genuine value.

Although advanced graphics packages are a great help, or even a necessity, intense thought and detailed analysis generally will resolve history-match problems. One cannot zoom through history-match information.

Finally, software problems will occur. This chapter provided some guidelines on how to make the problems as painless as possible.

After evaluating many simulations as part of economic evaluations, the proportion of truly good history matches is disturbingly low. Getting an exceptional history match is an admirable achievement; however, most reservoirs can be accurately history matched. When engineers cannot get a match, more than likely there is a formulation problem; for example, a major mechanism has been overlooked, the grid is not fine enough, the geology is not correct, or the model is too simple. Persistence is important.

Finally, a little luck never hurt anyone—just don't count on it.

References

1 Crichlow, H.B., *Modern Reservoir Engineering—A Simulation Approach*, Prentice Hall, 1977.

2 Mattax, C.C., and R. L. Dalton, *Reservoir Simulation*, SPE Monograph, Vol. 13, Henry L. Doherty Series, 1990.

13

Predictions

Introduction

Predictions are the final objective of a simulation study, and this portion of the study causes many problems. Following are a number of detailed comments on this subject.

Case 1, Case 2, Case 3, etc.

One thing that causes frustration and annoyance is the common usage of titles such as Case 1, Case 2, Case 3, etc. to denote various sensitivities. These designations are meaningless. Use a descriptive title such as Pilot 5-Spot Pattern, Run Down Case, Base Case, Maximum Infill, Inverted 9-Spot, 5-Spot WF with Increased S_{or}, etc. Abbreviated names still are used. The use of a numerical series means nothing to anyone except, perhaps, the person doing the simulation.

When reviewing a report, the author suggests going to the graphs and writing the English translation of what the sensitivity represents on various case sensitivities. Use this technique for the diagrams as well.

Tuning

The first step to tuning is to adjust the *permeability* × *thickness* products of the computational cells ($k \times h$ values) so the initial forecast production continues on the existing decline trends from the history match. This is not the only way to go about this; some prefer to adjust bottomhole pressures to achieve a similar objective.

Note that $k \times h$ values do not necessarily remain constant during the life of a well. Wax deposition, scale buildup, and fines movement can all degrade well $k \times h$ in time. Similarly, subsequent treatments will improve apparent $k \times h$. In injectors, dissolution of carbonates can lead to increased $k \times h$ in time—not to mention communication behind pipe. Similarly, bottomhole pressures can and will change if there are allowable restrictions, artificial lift changes, and reservoir pressure depletions. In practice, it is hard to capture all of this data.

Bottomhole pressure

The adjustment of $k \times h$ values still requires a bottomhole pressure be used. In reality, this information is frequently unavailable. This information is more likely to be found on offshore wells, which are more commonly flowing wells. Tubulars are a major constriction on high-capacity wells and optimization is usual under these circumstances.

For most onshore Canadian oil wells, a beam pump is used. In some cases, sonologs will be available, but this is generally not the case. The question then becomes, "What bottomhole pressures should one use?" Assume most pumpers are conscientious and their wells are running at good efficiencies. If the pump is run too fast, it will pump off and shut down. The author views a reasonable fluid column as roughly 10% of the well depth. Most reservoirs in Western Canada are slightly underpressured, and the author generalizes this to about 10% of the original reservoir pressure.

For gas wells, it is easy to use gradient curves that may be obtained either from the CIM well gradient monograph, *Gradient Curves for Well Analysis and Design*, Aziz et al., or from *Technology of Artificial Lift Methods* by Brown.[1,2] The 10% rule works fairly well here, so, when one is short on time or information, this is a good starting point.

Production optimization

Although these assumptions are reasonable in a global context, they are not necessarily true for individual properties. Some pumpers, operating companies, or field offices are run more efficiently than others. Perhaps as the result of an acquisition, production engineering and field operation can show substantial improvements in production. Simulation engineers are far removed, for the most part, from firsthand knowledge of this type of operation. As mentioned in other chapters, operational experience is a valuable background in performing simulations, and this is an example of where this knowledge can be applied.

Tuning strategy

It is normal to use a restart to do the tuning. This is an option allowing the reservoir simulation to be continued—without rerunning the problem through the history match. The author does not run the predictions for long periods, normally about three years of simulation time. Tuning can proceed quite quickly with this methodology.

The author prefers an iterative process. First, a run is made with $k \times h$ used for the history match and then a correction factor. This normally takes three to four runs. In some cases, this overcorrects. Dampening the changes by approximately 30% can help.

$$k\,h_{new} = k\,h_{previous} \frac{Q_{avg,\ actual}}{Q_{avg,\ previous}} \tag{13.1}$$

As a precaution, use production graphs on semilog scales to make these changes. There are minor variations in rates, which occur as a natural part of field operations. Tuning to the last month or even the last three months may not provide good extrapolations. At best, it may have to be redone; at worst, it can cause some serious biases in the prediction.

Decline rates

Decline rates should be monitored as well. Decline rates are controlled primarily by effective pore volume and permeability. If a change in decline rate occurs, then something is either wrong in the model, or it represents some change in process in the reservoir. Justify that this change in decline corresponds to an actual reservoir process. Additional infills will accelerate decline, as can water breakthroughs and other reservoir changes.

Record tuning changes

The author uses a systematic method of recording these changes in the data, which is shown in Figure 13–1. Many modifications are cumulative.

In addition, keep track of these changes on an overall basis. Generally, the permeability array is determined from log analysis. In many cases, it will be further filtered to some degree in the mapping process. There are systematic biases in this data, which is inevitable. For prediction wells, the author averages the $k \times h$ correction factors and uses this average correction factor on all of the prediction wells. In general, these corrections tend to be less than 1, an issue discussed later in this chapter. Sometimes the average tuning factor will be as low as 0.3, even if bulk average adjustments have been made to correct air $k \times h$ values to liquid $k \times h$ values. For that matter, there is no such thing as an average correction factor. Most individual core samples vary considerably, and core choices frequently have a bias.

```
*alter    5
               200
*perfv    5
**   first tune multiply by 0.1
**   second tune multiply by 0.393
**   third tune multiply by 0.8
    4           3.236      **    102.92
    5           0.506      **     16.1
    6           0.616      **     19.6
    7          28.610      **    910.0
    8           9.002      **    286.3
    9           0.594      **     18.9
*alter    6
               200
*perfv    6
**   first tune multiply by 0.048
**   second tune multiply by 2.00
**   third tune multiply by 2.1
    1           3.986      **     19.92
    2           6.644      **     33.22
    3           6.644      **     33.22
    4           2.659      **     13.29
*alter    10
               200
*perfv    10
**   first tune multiply by 0.045
**   second tune multiply by 6.429
    1           8.711      **     30.10
    3           2.025      **      7.00
```

Fig. 13–1 Tuning Technique—Data Set Documentation

Sometimes these changes will reflect stimulations as well. Well completion experience has shown each well has its own peculiarities that are often accidents of operational history. Different service companies are used, and treatments of different sizes are utilized and performed during different weather conditions.

Table 13-1 shows an example from a study the author performed. In this case, the wells were stimulated. All of the well cards had a skin of –2. Anything larger (i.e., more negative) would cause the numerator in the well equation to become negative.

Systematic bias

The use of averages can be dangerous. First, averaging dampens natural variation. This is similar in concept to the difference between a conditional simulation and kriging, which was discussed in chapter 4. The productivity of well predictions will represent an average, at the expense of masking the production variations that wells will actually exhibit, generally reflecting real differences in local reservoir properties. In addition, averaging may represent

Table 13–1 Tracking of Well Log/Mapping *k x h* and Well Test *k x h**

00/d-59-I

Zone/ Company	Top/ Year	Bottom	Gross	Net	Phi*h	Perm*h	k_{avg}	k_{geo}	k_{liq}/k_{air}	Skin
D_unit	1069	1071	0.457	0.457	0.041	0.213	0.467	0.464		
C_unit	1072	1074.4	2.286	2.286	0.251	4.55	1.99	1.781		
B_unit	1075	1076	0.152	0.152	0.018	0.498	3.266	3.266		
A_unit	1076	1081.2	5.334	5.334	0.702	62.234	11.667	7.993		
Cocina	1083	1083	-	-	0	0	-	-		
Basal	1083	1087.5	0.457	0.457	0.04	0.332	0.727	0.702		
				8.686	1.052	67.827	7.809			
Haidey	1991			7		18.24	2.606		**0.249**	-5.84
Haidey	1996			7		17.4	2.486		**0.257**	-6.84

00/a-21-J

Zone/ Company	Top/ Year	Bottom	Gross	Net	Phi*h	Perm*h	k_{avg}	k_{geo}	k_{liq}/k_{air}	Skin
D_unit	1040.5	1046.3	5.75	5.75	0.193	17.3	1.275	2.81		
C_unit										
B_unit	1046.6	1050	3.25	3.25	0.484	12.63	3.075	3.08		
A_unit	1050	1054	7.25	7.25	0.856	20.4	3.806	2.02		
Cocina	1058	1058								
Basal	1058	1058								
						50.33	3.097			
Fekete	1990			14.123		27.6	1.954		**0.548**	-5.1
Opsco	1980			14.242		28.2	1.98		**0.56**	-4.91
Haidey	1996			14.2		44.9	3.162		**0.892**	-5.18

**With Adjustment Factors for Air to Liquid Permeabilities*

a bias carried through from the history-match process. Wells that are $k \times h$ deficient will show up in the history match with a shortfall in cumulative production, which should be caught. For $k \times h$-inflated wells, no shortfall will occur, and this will not cause a change in the specified $k \times h$. Thus, the commonly used history-matching process introduces a liberal bias in well $k \times h$ values.

There should be a final history-match stage in which well $k \times h$ values are minimized. This will not be a guaranteed solution. Wells that are allowable (government market rate) limited may be reduced too much, since they are not producing at low bottomhole pressures (which is what the author normally assumes). A comprehensive bottomhole pressure history signifies a great deal of work for the simulation engineer. Recall earlier comments regarding $k \times h$ changes with time: tuning will always be required. Strangely, as a consultant I have found that the details of tuning do not seem to be widely known. This process is detailed and unfortunately occurs at a time that budgets are much more obvious and widely understood.

Predicted well rates

The critical implications of this, which directly affects predicted well rates, are:

- History matching and tuning can impart a systematic and optimistic bias.
- The variations in core to air permeability correction do not allow precise well rate predictions. Nonetheless, reasonable estimates can be made.

Base case

Most simulation forecasts are run for a maximum of 20 additional years of production. This represents a reasonable economic horizon. After this period, cash flow discounting renders later production irrelevant. For most base cases, only the existing wells will be put on. This represents continued operation of the field as it is currently developed (i.e., a "do nothing" scenario).

Evaluating infill locations

Experience has shown the most effective way of evaluating potential infills is to enter all possible drilling locations in the simulator. Optimal locations require one to consider the following:

- Saturations of gas, oil, and water
- Reservoir pressure
- Porosity
- Permeability
- Potential drainage effects
- Coning (sometimes from above and below)

Although some locations are obvious, evaluating all of the above simultaneously is difficult and requires a great deal of time and evaluation. Of course, the simulator has been designed to take all of these factors into account. Therefore, it is simpler to input all possibilities and examine forecasted production performance.

Wells can be ranked based on productivity and projected recovery. Wells that are clearly not economical can be commented out in the data set in future—second—development sensitivities. With this methodology, it is possible to obtain an optimized case within two runs of completing the base case forecast.

Maximum density case

All drilling spacing units (DSUs) permitted are completed in the first forecast run. As shown in the discussion on the base case, some adjustment is required to $k \times h$ to accurately estimate productivity and the effects of hydraulic fractures. One of the implications of this technique is it dampens the real productivities that can be expected, since a range of corrections was required for the base case.

Results of maximum density case

An example of the total production from a unit and total production with maximum infills is shown in Figure 13–2. This shows the effects of increased infill drilling on existing producers. It is possible to generate a true incremental case using tabular data, i.e., the gains on infill drilling minus reduced performance on existing wells.

Screening criteria

As a rule, the economics of individual wells are most affected by the production forecast; incremental cases normally assume the same price forecast. Therefore, the author prefers to screen on rate rather than recovery. Table 13–2 shows the rates of forecasted wells from a study.

In the reservoir forecast report from this study, qualitative analysis indicated forecast productivities might be increased to adjust for downtime. The rate could be increased by as much as 45%. However, this off-production time was somewhat related to cyclic flow limitations, so it was questionable whether upgrading the productivity was appropriate.

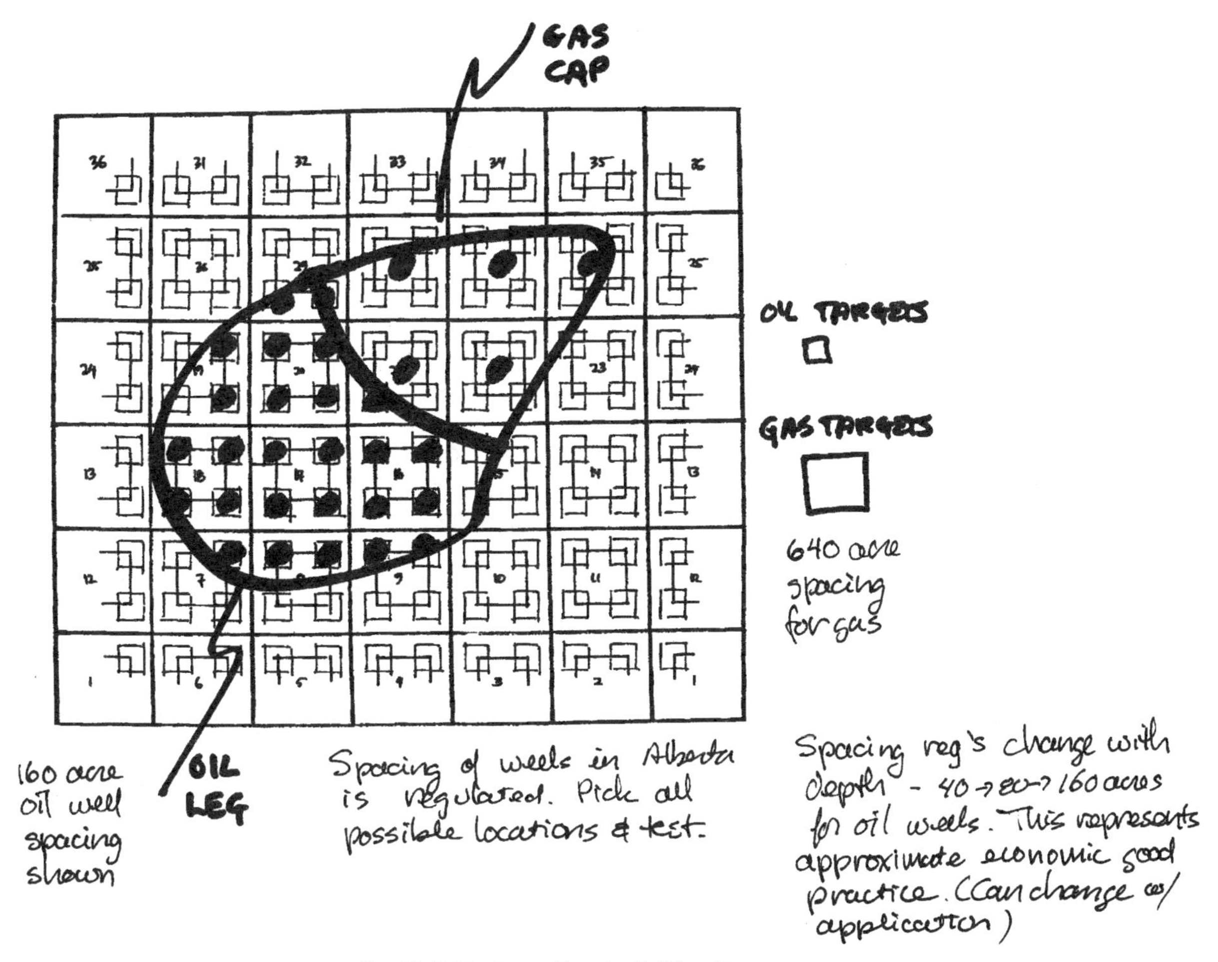

Fig. 13–2 Maximum Density Drilling Case

Table 13–2 Primary Depletion with Infill Wells

Year	Prod. Wells	Inj. Wells	Infill Wells	Oil Rate M^3/d	Water Rate M^3/d	Gas Rate EM^3/d	Water Rate M^3/d	P_{avg} kPa
1997	23	0	0	104.93	7.95	109.29		6257
1998	23	0	5	248.22	27.4	246.2		5241
1999	23	0	5	159.73	16.44	161.69		4585
2000	23	0	5	120.55	12.6	124.48		4086
2001	23	0	5	99.18	10.96	102.88		3676
2002	23	0	5	82.74	9.59	86.23		3332
2003	23	0	5	70.41	8.49	72.94		3041
2004	23	0	5	60.27	7.4	62.38		2790
2005	23	0	5	52.05	6.58	53.48		2576
2006	23	0	5	45.75	6.03	46.29		2391
2007	23	0	5	40.27	5.48	40.31		2229
Cumulative HM Production (E3M3)				14.5				
Cumulative F/C Production (E3M3)				58.4				
Total Recovery (E3M3)				72.9				

Presentation to management

Initially, clients protest when the approach of evaluating the maximum infills case described above is introduced. Often, the whole simulation is commissioned to justify an infill location someone has proposed. Invariably during the presentation for well approval, a manager will ask if the chosen location is "the best" location. It is easier to justify a choice when one can say every conceivable location was reviewed and the proposed location is "the best." By evaluating each possible location in a piecemeal fashion, a run is required for each possible infill. This requires large amounts of time, and the engineer must compare how the other infills affect the other wells chosen in a potential multiwell or multiyear program as well.

In perhaps 50% of the cases, the targeted locations that eventually prove to be the best are those identified by common sense. Simulation is not necessary in every case.

The other 50% of cases are subject to a major shift in the perceived producing mechanisms. These are not minor differences. The saturation pressure may be altered, the presence of gas caps proved or disproved, or sources of water production correctly identified—e.g., bottom water versus edge drive. Most studies the author has worked on ultimately represent less than 1% of capital expenditures.

Waterflood implementation

Frequently, predictions will include waterflooding. In this case, you will have to envisage a number of processes. The power of a numerical model is hypnotic because it can evaluate a production scenario rigorously; however, it is incapable of identifying which scenario must be input.

In many cases, the existing wells will force one into a situation requiring the implementation of nonstandard patterns. This can be difficult to visualize. As a first-order approach, the author suggests the gross swept volume be evaluated, as suggested by Slider in his book *Worldwide Practical Petroleum Reservoir Engineering Methods*.[3] The reservoir simulator will provide a more comprehensive answer. This is an excellent starting point.

Check for other possible schemes

Consider other likely schemes. For instance, one study reviewed assumed a 5-spot injection pattern. The analysis indicated, given pool geometry and offset performance, fewer injectors were actually required. Since no predictive run was made with this scenario, it was not evaluated. The economics of the waterflood were not favorable due to lost production. A nine spot implementation would most likely have allowed for an economical project. On the other hand, I have observed the opposite on a number of occasions as well (i.e., more injectors required than "conventional wisdom" might suggest).

Injection pressures

One area that can cause many problems is water injection rates. As a rule, rates used in simulations have a tendency to be too high. Bottomhole pressures need to be below fracture pressures. An assumption of 0.8 psi/ft. is not universally applicable. The Alberta Energy Utility Board's Guide 51outlines gradients for a number of zones and fields.[4] This is probably a good starting point. One is likely to face these limitations—barring well testing to demonstrate otherwise to the board.

It is also true that much injection occurs above parting pressures. It is not altogether clear whether this is good or bad. Practical examples exist showing injection above the parting pressure has no discernible effect on production performance.

Settari has developed simulators propagating fractures with injection. To date, the author is aware of one or two applications of this model. One is outlined in "Waterflood-Induced Fracturing; Water Injection above Parting Pressure at Valhall," by Eltvik, Skoglunn, and Settari.[5] It has been applied to the Pembina Cardium Formation also.

Well control schemes

The need for well controls occurs quickly in predictions. Perhaps the most common example is when one decides to implement a waterflood involving more than a single injection well. Normally there is an indication of the maximum water supply rate, but how is the injection water divided between the wells?

Typically, one might decide to allocate the water equally between the two wells. The problem is the size of the patterns is different and more water is needed in one of the injection wells. The first pattern has more cumulative production and a lower reservoir pressure. The simulator should allocate more production to one well than the other, based on reservoir pressure; therefore, a ratio of distribution can be set.

Voidage replacement ratio (VRR) controls

In another case, one waterflood pattern is known. Although currently at the same reservoir pressure as the second pattern, it is limited by production facilities. Hence, one expects to have more production from the pattern and wants to match water injection with the production that will occur. This time the objective is to match reservoir withdrawals with water injection, i.e., voidage replacement. This is simple enough when the reservoir is above the bubblepoint, but if it is below the bubblepoint, then gas production needs to be considered. This requires a slightly more sophisticated form of voidage replacement calculation. Simple oil production will not suffice as a guide. This is when a VRR constraint is used.

The sophistication of VRR constraints can be a serious issue, because not all calculations in different simulators with different key words are created equally. Read the documentation, and use caution with VRR constraints. These issues can be a concern in compositional simulation as well, where determining different volumes with different compositions can be difficult.

Constraints

In some cases, the government restricts the maximum GOR that can be produced. This is normally designed to prevent venting a gas cap and losing the beneficial effects of gas cap drive. In cases like this, a GOR limitation is required. Oil production must be reduced until the required maximum GOR limitation is met. This is generally termed a constraint.

Recycle control

Another type of control has to do with what is available. For instance, with gas injection schemes, it is generally desired to input a fixed amount of gas that has been contracted for injection. For an oilfield, any gas produced will usually be stripped of liquids and re-injected. This represents a recycle volume. The volume available for injection depends on the production from the previous or same timestep. Since we are relying on the simulator to calculate how much gas production there will be, this needs to be calculated as the prediction proceeds.

For actual areal or field studies, many possible combinations may be required.

Hierarchical structure

Well controls are not as easy as one might think. For a single well study, well controls are usually easy to input, since there are defaults in the simulator transparent to the user. Once past this, the user will discover well control schemes are very complex.

The complexity starts with the need to take into account different production groupings. This can be related to physical limitations, such as satellite facilities, an offshore platform, working interest changes due to differing leases, or different jurisdictions. Resources are provincially controlled in Canada, and in one case, a single reservoir is actually in three different provinces. In other cases, a reservoir can straddle national boundaries. Such examples occur in the North Sea.

Most reservoir simulators have settled for three levels of control, which are similar to Figure 13–3. At the top level is *field*, which normally covers the entire simulation grid. Following this is a *sector* or *region*, and the last level is normally called *group*.

This entire formal structure must be input to the simulator to make pool simulations work. This is true even though not all parts of the hierarchy are used. Most simulators cannot invoke the entire control structure on the fly. Normally, changes to the structure are not a problem. However, the basic structure and assignments have to be made at the start of well definitions in the history-match run. During the history match, the author does not consider the predictions that will be made and often has to rerun the history match with the hierarchy added in at the beginning.

The author has found documentation on these features to be rather cryptic. Making the controls work is often difficult and frustrating. The software support personnel for more than one commercial package have admitted some inconsistencies are inexplicable, and they have not been able to resolve them. This invokes the need to use "witchcraft." Once a "spell" (data set) works, try not to mess with it. Witchcraft is invoked when true understanding has not been obtained or when faulty, incomplete, and unnecessarily complicated documentation exists.

Since there may actually be some inconsistencies in the simulator, checking is a very good idea. In this area, software development changes and significant improvements continue to occur. Therefore, the foregoing advice may become dated.

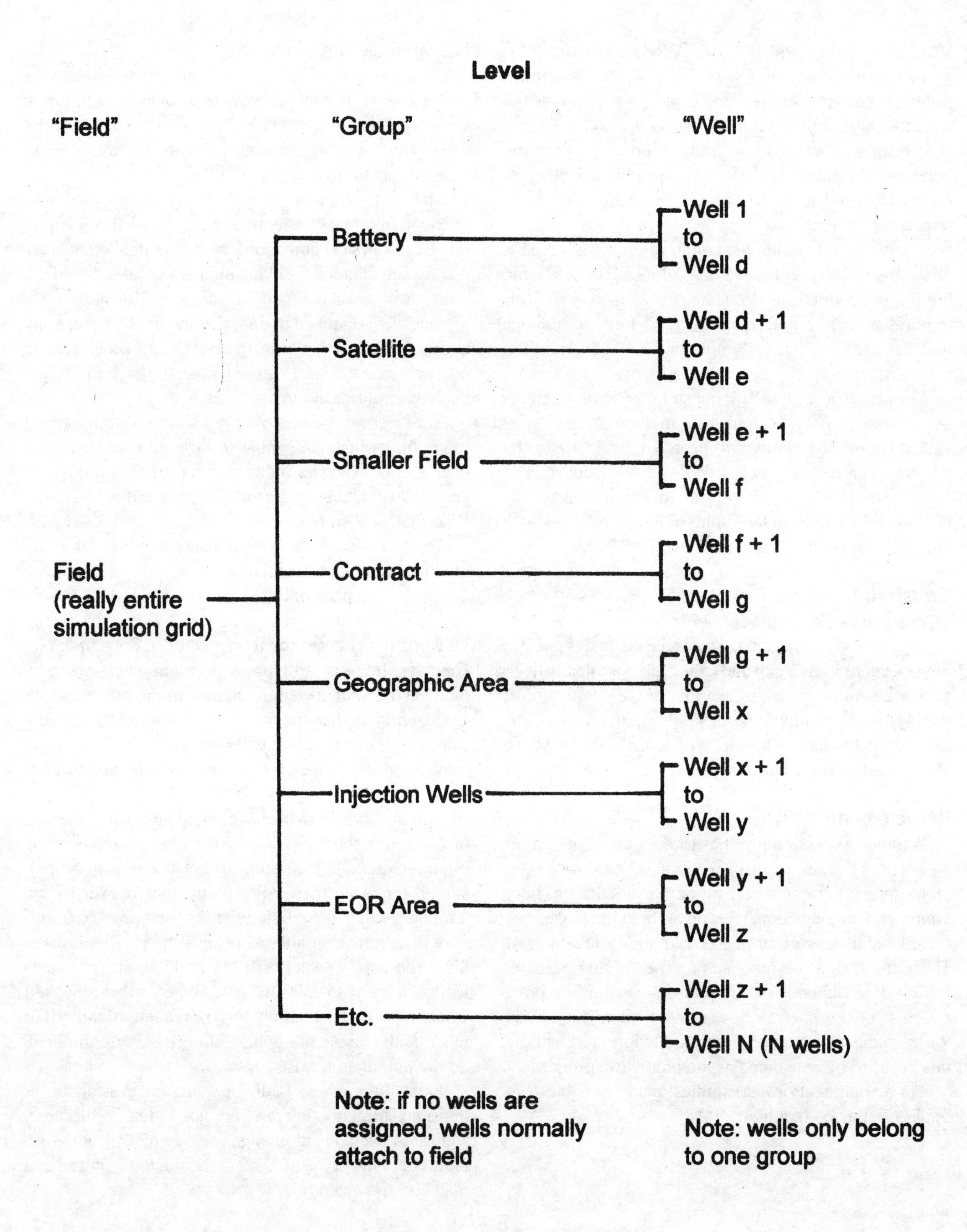

Fig. 13–3 Well Control Hierarchies

The simulator graphics will normally print, plot, or display the *field* level automatically. The subsequent, more detailed levels or subdivisions generally must be requested separately.

There are then two types of controls. Ones applying to groups—i.e., levels one and two—and ones applying to individual wells. Most software uses separate commands for these two classes.

Relative permeability for injectors

This may seem to be a strange feature. Some injection well rates are controlled by relative permeability while others are not. This is an area where there are differing opinions. Relative permeability tests are usually conducted at rates intended to be representative of displacement rates. This varies with position in the reservoir. Near the well, rates are high and, yes, relative permeability is dependent to some extent on rate. If you really use high rates, and inject enough fluid, residual saturations may be stripped. This may occur not necessarily by rate alone, but by turbulence, dissolution of the rock in fresh water—i.e., with carbonates—or simply because the residual phase is dissolved away. And, yes, the solubility of oil in water is very low, but it is finite.

Since the injectivity is controlled by the relative permeability in the rock adjacent to the reservoir, there are two schools of thought. Some people believe relative permeability should not be included, and others believe it should.

The author believes the high rates and turbulence do affect the near-wellbore regions, carbonates do involve some dissolution, and residual saturations will disappear if injected with enough fluid. Nevertheless, the author uses the mobility-weighted model because it is conservative. Waterfloods are sensitive to how much water is injected, and it is easy to overestimate injection performance. This doesn't mean the other view is wrong.

One final point needs to be made. At connate or irreducible water saturation, the relative permeability is zero. Ergo, according to the relative permeability relation, the simulator should not accept any water in the wellbore grid block. This issue is similar to the upstream weighting requirement in waterflood displacement discussed earlier in this book. Most simulators use an upstream weighting on the injection well—and throughout the grid—to overcome this problem. This is true of all the larger commercial simulator software applications.

Some simulators don't do this, and one has to fool the simulator by putting in a water saturation above the zero water relative permeability. There is nothing wrong with this software, but the developer assumes the users are experts. The irreducible water saturation is determined by flooding water-filled core with oil until no more water is recovered. Although we call this the irreducible water saturation and assign k_{row} to be zero, we have no difficulties injecting 100% water in a displacement process. This is the commonly performed waterflood test, which is an imbibition process. When we displace the core to oil, it is a drainage process. These two processes are fundamentally different.

Mathematical implementation

In simplified terms, a maximum well rate is usually determined with a slack variable. In other words, an equation is added to the matrix. This equation adds the well rate to the slack or dummy variable and is set equal to the maximum rate. The slack variable must be positive. Therefore, each well control statement adds another equation to the solution matrix.

Difficulties caused by well controls

Well controls are often difficult to handle, as discussed below. Well controls are not a trivial problem numerically. These controls can produce what mathematicians call *stiff* matrices. A slight shift in productivity will trigger additional wells being put on production or other operating changes. It is easy for the simulator to cycle across one of these set points and become unstable. Dampening may be required in some cases for the matrix solver to become stable again. In addition, well equations can disrupt the orderly structure of the matrix and make it more difficult to solve.

Even simple well and group controls make the solution of the predictions much slower.

Surface elevations

On one study, the model involved a combined surface facilities network as well as a reservoir simulation model. In the process of making the predictions, it was necessary to put in the elevation changes for the surface flowlines. Generally, we use the KB elevations from well tickets for this purpose.

However, for pipelines, the intermediate points are needed, and we generally use topographic maps. These are obtained from the Canadian government for nominal fees. The contour plots are based on air photographs that have been calibrated to ground surveys. Likely, the newer ones are based on satellite—at least in part. Although freely available in some countries, other countries have significant restrictions on the distribution of maps.

We found the elevations on the National Topographical Service (NTS) map associated with some of the KB elevations did not seem to match wellhead elevations. Part of our data was apparently sourced from a map provided by surveyors, which had some generalized contours. Tabulating the values derived from the surveyors' map and NTS maps showed the NTS maps seemed more consistent. Alarmingly, the contours between the two maps were also quite different. The NTS data was used in preference to the survey map data. Later, serious bugs were found in the surveyor's maps and elevations.

Once, when mapping a gas-water contact on a large gas field, the author found holes in the gas-water contact. When we rechecked the surveys, the wells were located 3 to 4 meters below the level of an adjacent lake. Since the area was relatively flat (muskeg), it became obvious the surveys were wrong.

Although surveyor errors are rare, they occur with sufficient frequency to merit checking.

Summary

There are some easy pitfalls in presenting simulation results. Please use names with some meaning. Tuning is an important process that is not understood well. The author cannot explain this. It seems to fall in the shadows of the history-matching limelight. This area lends itself to systematic techniques—more so than history matching.

Prediction technique utilizes some averaging that removes some of the natural variation expected from infill predictions. Make sure this is understood and explained in your report. The author uses an overkill and back-off methodology, which is more useful than a straight intuitive *what if* approach. This is in contrast to history matching.

Well control schemes present some challenges and are more difficult to implement than one would think initially. The well controls also require considerable solver resources and can cause nasty instabilities. This means users have to be consistent in what they have requested.

Predictions are normally the final product in simulation. Here the limitations in what the simulator can and cannot do should be fully explained. Management's job is, in large part, to account for risk from various inputs from different departments. This is the time to be honest about how accurate things really are. In most cases, the simulation engineer is best able to provide the assessment of the limitations. To do this well, one needs to be a reservoir engineer first and a simulation engineer second.

References

1 Aziz, K., et al., *Gradient Curves for Well Analysis and Design*, Canadian Institute of Mining, Metallurgy and Petroleum Special Series, Vol. 20, 1979.

2 Brown, K.E., *Technology of Artificial Lift Methods*, PennWell Publishing Co., 1980.

3 Slider, H.C., *Worldwide Practical Petroleum Reservoir Engineering Methods*, 2nd ed., PennWell Publishing Co., 1983.

4 Alberta Energy and Utilities Board, *Injection and Disposal Wells—Well Classifications, Completions, Logging, and Testing Requirements, Guide 51*, Alberta Energy and Utilities Board, 1994.

5 Eltvik, P., T. Skoglunn, and A. Settari, "Waterflood-Induced Fracturing: Water Injection above Parting Pressure at Valhall," SPE 24912, Society of Petroleum Engineers, 1992.

14

Study Management

Introduction

A number of issues related to simulations are not directly technical. These are grouped in this last chapter of Part I. The issues are significant. The areas covered are as follows:

- Simulation involves much input, and, frequently, the quality of a simulation is only as good as the confidence the people who use it have in the study. More importantly, success in corporations often has to do with communication skills. That issue is addressed in this chapter. Good communication skills help improve the quality of an engineer's work.
- A short section is included on how to select software.
- Many reservoir engineers will not actually perform a reservoir simulation study. Due to the specialized nature of this type of work, such studies are often contracted out to a third party or performed by a different department. Therefore, many reservoir engineers will be interested in the issues associated with the management of a study. How does one choose a consultant? What results are expected? The author has included a checklist in Appendix B that may seem like a review of previous material in this text. It is; however, reversing this process requires some thought. Having a separate procedure to check off a study can be helpful.
- Reservoir simulation is technically complex, and it is easy to get lost in the details. One section is dedicated to identifying problem areas and, in particular, where simulators are weak. The material in this chapter is well-known to reserves evaluation engineers, where the emphasis is heavily oriented to historical data and analogy. This provides important balance in the reporting of simulation results.

Reports

Communication

The true value of work performed must include both a communication factor and a knowledge retention factor. To maximize these two factors requires:

- formal presentation of results at the end of the project.
- a clearly documented and readable report. Sufficient detail must be present to enable one (or one's successor) to continue the project at a future date.

In the job ads of the local paper, Mr. Employer invariably asks for "superior communication skills." This applies to both oral presentations and written study results.

A common philosophy has been to take for granted that the work is done correctly and a minimalist approach to documentation is appropriate, i.e., only the important results need to be presented. Some people have suggested the details would bore expert readers and it would detract from the report.

The author's experience is exactly the opposite. If the reader really knows the subject, a thorough, well-written report provides reassurance the report was prepared using a high level of detailed information and a thorough background. In most cases, silence on issues will raise suspicions in the expert's mind rather than reassure him or her. This author suspects a minimalist report covers gaps in knowledge, and all the correct material will be ignored. Restated, experts normally feel compelled to check everything has been done correctly.

This is particularly clear in government hearings, where undocumented work is automatically seen as a chink in the armor and usually results in re-creating the results. Frequently, this leads to detailed questions in the form of *Information Requests*, which can be very detailed.

It is easy to say, if there are questions, people will simply ask. Again, the author's experience is not consistent with this. In most cases, if the reader of a report does not know, he or she is more likely to remain silent than admit ignorance. The less people understand your report, the more silent and suspicious they are likely to be.

Complete documentation is not a luxury—it is essential.

Quality

Reservoir simulation studies vary in quality. Some simulation studies give correct results, and some are totally misleading. A wrongly posed study or incorrectly run simulation can actually result in the wrong implementation of a reservoir depletion strategy. Lost opportunities in these cases can cost millions of dollars and greatly exceed the cost of the study.

The most important differences between high- and low-quality studies are: accurate reservoir description, proper layering, and comprehensive checking of input and output data. Most predictions should be checked against offset analogues. It is quite possible studies are not implemented correctly; i.e., an inappropriate simulation technique is used.

Evaluations report

During the past 15 years, improved quality of reports is the trend in the evaluations sector. More than a decade ago, it was sufficient to provide a property-by-property computer output of economics plus a summary table. A modern evaluation report now includes a descriptive write-up and a comprehensive listing of assumptions. This provides two main benefits:

- It gives a technical overview of the property. Not everyone who uses evaluation reports has a technical background, and the overview provides useful information to many readers.
- It provides the expert reader an accurate appreciation of the assumptions used. This allows more accurate risk assessment—without having to repeat the evaluation.

To date, this does not appear to be true south of the 39th parallel (the border between Western Canada and the United States). Most information there is proprietary, so it is harder to obtain. Success in the U.S. market will also strongly correspond to having a wide client base. The reports are normally based on knowledge gained from multiple operators. However, each individual company does not intend to share its data with other offset operators.

A corporate evaluation represents years of exploration and development for a company. In this sector, it is often taken as a given that the manner in which the material is presented to financial institutions affects the perception of the company. For this reason, evaluation reports seem to have a better finish than most technical reports.

Without doubt, the finish on a report affects others' perception of a simulation study.

Approach

The general elements that should be in a report are as follows:

- Comprehensively prepared reservoir descriptions. The first step is a good understanding of the depositional environment. Normally, detailed layering calculations are required. This involves specifying either the correct layers or the correct development of pseudo relative permeability curves.
- Checks against offset information. This aids in identifying critical modeling parameters. Frequently, unrepresentative models can be weeded out by comparison to offsetting production.
- A reasonable review of reservoir performance. This includes a description of decline trends, water rates, and GORs, including other information, such as DSTs, pressure buildup analyses, pressure surveys, and material balance studies.
- A copy of the basic data input deck. There is no other method of reproducing the results in the future without this information. It should also describe how the data was screened and why particular data was either kept or rejected.
- A run-by-run history match log. This is a positive sign of a well-thought-out strategy. A description of the logic used in history matching is even more important. The changes should be described so the reader will know what decisions and assumptions were made.
- Results displayed graphically. Graphic display allows rapid assimilation of the results and provides a vital quality-control check.

It is easy to spot the cost on consultants' reports, since, unlike salaries, they are usually approved separately. However, if a significant amount of time is spent trying to determine exactly what the consultant did, it is an expense in time and real money that could have been saved with a better report.

The author includes a basic description of the techniques used. It is easy to assume the reader is familiar with simulation techniques; however, not everyone has studied simulation or done the research. At ARE, we cut and paste these segments from previous reports and do not have to re-create them from scratch for every report.

Self-improvement

One supervisor in the author's experience taught good report writing. The supervisor had worked as an evaluations consultant for an industry-leading firm, and he earned a reputation for being demanding. Writing reports is difficult; one has to tie everything together. Inconsistencies become much more obvious in a written report. Writing is as much a process as an end in itself, and the writing process makes the quality of the work better.

If people say writing the report will not add value, they are wrong. It is easy to make up excuses to avoid writing reports.

The writing process

One of the biggest problems is the expectation one can sit down and write a letter or a report from the beginning to the end correctly the first time. This is the same as the admonishments that history matching can be done automatically as efficiently as possible. In reality:

- Many writers find it helpful to plan and organize their report by writing an outline. In some cases, the outline in point form can be directly expanded to full text. This is the classical approach, which is normally taught. Others know the general outline, and do not spend a lot of time on this stage. Consequently, they don't find this methodology helpful.
- In the first draft, put down all the ideas in any way possible. Some people write at this stage as if explaining to a friend. Don't worry about the order, just write down as much as possible.
- Realistically, if one is doing a significant amount of simulation then the time spent using a computer justifies learning typing by touch. Word processors make it easy to rearrange report text.
- Make at least one serious edit and rewrite. More likely two or perhaps three revisions will be necessary.

- Give the edited report to someone to proofread for you. Having a professional technical editor do this is worth the money in many cases. Some large companies have technical writers who can do this.
- Make final corrections, and then read it again.

Many people are turned away from the writing process because they have unreasonable expectations of what is required.

Outline

ARE uses a standard outline for reports that follows this general theme:

1. Title Page
2. Table of Contents
3. Cover Letter
4. Introduction
5. Summary
6. Conclusions
7. Recommendations
8. Geological Review
9. Production Performance Review
10. Pressure Transient Review
11. EOS Tuning
12. Model Construction
13. Grid Refinement or Single-Well Modeling
14. Cross-Sectional Modeling
15. History Matching
16. Predictions
17. Report Preparation
18. References
19. Appendices

Most reports follow a similar theme, with varying degrees of detail.

Graphical output

Graphs are important. They convey information quickly. The majority of simulation engineers seem to prefer linear rate versus time plots to semilog rate versus time plots. Evaluation engineers usually prefer the latter, since they commonly use decline analysis. In order for one's experience base to have its full impact, the results of the study should be plotted in the format to which one is accustomed. The author strongly recommends the use of semilog production plots.

The predictive phase and the history-match phase of a study are usually performed at different times using a separate computer run. Often, the history-match plot and the prediction are not plotted together. A combined plot can be very informative. Plotting the prediction and history match on consistent plots for individual wells is recommended. The plotting routines available in the post processor will not always allow one to do everything. Do not hesitate to use a spreadsheet to make custom graphs.

Large output

The author produces many large maps and outputs. Many people do not realize it is easy to paste up a large diagram on a sheet of paper from a flip pad. Paste on text from the word processor using a label sheet. Send this to your local oil patch printer, such as Petro-Tech or Riley's, and they will photocopy a vellum. It costs less than $10.00, and they can accommodate large sheets. Blue line or black line prints can be made in large quantities for limited costs.

What parts of simulator output are believable?

It is partly in the nature of people who are attracted to simulation to be strong technically and analytically. It is difficult to state in a report, "Oh, by the way, there are some big parts of this simulation output that are really not accurate." A pet peeve in this regard is gas injection, which is discussed later in this chapter. In many cases, reports have had absolutely no qualifiers or caveats in this regard.

It is the responsibility of the simulation engineer to point out limitations. For people who are strong analytically, it is often contrary to their value system to denigrate what is really, in other applications, a useful tool. It is hard to write something critical of one's own work. Here is a suggested solution—stick to the facts. For example, "The cases presented involving gas injection represent an optimistic prediction of performance. Factors not included in the simulator include viscous fingering, large mobility ratios..."

How does the simulation engineer correct for the things the simulator cannot do? One has to put aside the simulation documentation and look at practical experience. One can be definite in saying, "The average recovery in all horizontal gas floods implemented in the Province of Alberta is X.X%, which is 10% of the Y.Y% predicted in this simulation model."

Management is paying, in part, for the engineer's opinion. This is probably an ethical requirement if one is a licensed engineer. There is nothing wrong with stating an opinion, e.g., *In my judgment the recovery will be Z.Z%, which is above/below the average projected recovery for the Province of Alberta because this reservoir is better/worse than the average reservoir in the following respects:...*

This is an example where knowledge restricted to simulation alone is extremely dangerous. It is a concern to the author when individuals call themselves simulation engineers as opposed to reservoir engineers who happen to do simulation.

Summary

Producing output requires a great deal of creativity. Many people will give the job to a secretary to muddle through. The author looks at the success of a study as a two-part score. The score is calculated by technical quality multiplied by presentation quality. If one does a mediocre job technically and an excellent job of presenting the results; the score equals 0.55×0.9 or 0.495. If one is brilliant technically but does an abysmal presentation, then the score is 0.9×0.28 or 0.252. The mediocre technician with a quality presentation appears to be twice as knowledgeable as someone who is technically brilliant. Remember, the objective is for people reading the report (the boss and, in particular, his boss) to assimilate results quickly and in a manner that will give them confidence. Those reading your report are not always experts in simulation. It is appropriate to explain exactly how and why the study was conducted. Other readers may not be simulation experts, and a clear explanation of the limitations of simulation belongs in the simulation report.

Choosing Simulation Software and Hardware

After one reads a text on reservoir simulation, someone will likely decide the reader should be involved in choosing software. An engineer may be asked to analyze simulation software for acquisition by their company. The software will be used by both company and outside personnel (contractors). Likely, one will be hiring people who must have the technical skills and resources to accomplish the company's needs.

The main issues in acquiring software are as follows:

- technical capability
- ease of use
- hardware requirements
- market penetration
- client/corporate/partner programs
- service record (help desk and customer support)
- proximity of service

In some of these issues, there will be fundamental tradeoffs. For instance, technical capability always conflicts with ease of use. Other issues, such as market penetration and service, often go hand in hand—but not necessarily. With larger market penetration, service should be better and more widely available. There is rarely a unique answer. The objective is to find the right balance of these factors for the individuals and the business.

Some main sources of software are listed as follows:

- software companies
- research organizations (government and industry)
- integrated consulting companies
- self-developed

The following sections expand on the advantages and disadvantages of these sources.

Software companies

This is normally a good source of software. The software tends to be well-written, since this is their primary business. One of the reasons PCs are so powerful is much of the software is written by specialists for specific purposes. Either companies in this category produce good quality software or they go out of business.

One problem with software companies, particularly those in the technical area and reservoir engineering software in particular, is their pricing is not based on the size of the user base. This reflects the domination of their perceived market—large oil companies. As a result, many of these companies set their purchase price at the same level for a one-man consulting company as they do for the world's largest integrated producer. There is no price differentiation for hundreds of users versus a single user. From a commonsense point of view, this situation seems ludicrous. However, it is true in many situations.

PCs have had the advantage that the *shrink-wrap marketing* approach automatically produces a product designed to be priced on a per user/machine basis. This is fair enough for operating systems, word processors, and spreadsheets.

The extensive use of networks means many companies often buy a few copies to cover active users rather than the total number of machines. This sometimes means loading times for software are slow because they come across the network. It also means some of the more expensive data programs are not really single-use licenses but rather distributed licenses. Many software companies have difficulty understanding why a two or three person consulting firm, where the software is installed on one machine, is reluctant to pay the same amount as a company with 50 technical people using a network.

One approach to providing software to smaller customers is to strip off capability, but this approach does not address the issue of usage level completely. In running a smaller consulting company, ARE works on and solves problems as difficult, and often more difficult, as almost any oil company or large consulting company. It appears oil companies often elect to have someone else work on long, involved technical projects.

Some software companies have developed strategic relations with consulting companies. In one situation for economic evaluations, a consulting company and the software company teamed up. This produced efficient and user-oriented software that dominated the market for more than a decade.

In addition to product capabilities and support, look for three things in a software company:

- What is their level of customer sensitivity?
- Do they have a realistic pricing schedule?
- Have they capitalized on strategic relations?

Research organizations

There are a number of government and private research organizations. This can be an excellent source of software. In general, they are happy to have as many users as possible. In areas where development is important, this can be a particularly good source of software.

The biggest concern with research organizations is the lack of ease of use of their software or "user friendliness." Historically, some of this software was difficult to use. For this reason, test the software before making a commitment. If one is a relatively advanced user, this may not present a problem. If time is limited or one is not inclined to work with an unfriendly program, another source of software may be more suitable. In general, the level of technical capabilities—if not user friendliness—of research organization software is pleasantly surprising.

In a number of cases, the government has supported the development of programs that were not freely available to all sizes of companies due to proprietary development. For example, the United States government paid for the development of BOAST, a black oil simulator. Despite some reputed bugs in the original government issue, it did promote competition in the market. The base program, which came as raw source code, has been used to develop subsequent PC versions with many improvements, including graphics.

Recently, a full suite of geostatistical software was made available at moderate cost (free) from Stanford. The software, GSLIB, does not have a fancy graphical user interface, but it has a reputation for being technically sound. This software was written by Dr. Clayton Deutsch and several books have been published supporting its use.[1, 2]

Check the Internet. There are a number of free programs available on the World Wide Web.

Integrated consulting companies

Many companies making simulation software are also in the consulting business, or perhaps, they make software for their consulting business. Somewhere along the way, they decide to market their software independently of their consulting services.

A company using the software they develop normally builds more user-friendly software that more closely meets technical requirements. On the other hand, the level of code development can suffer if the program was developed by users rather than IT professionals. Engineers have a reputation for code that works, although rarely elegant in structure or efficiency. In truth, the best software is a compromise between programming efficiency and logical usage. Software from these companies will cover the entire spectrum, from the very best to rough-around-the-edges. Thus, the software from one of these companies must be tested and analyzed.

The goals of an integrated consulting company may compete between providing consulting services and providing software. Some do not want to provide software to small consultants whom they perceive will take away consulting business. In some cases, particularly when selling to large operating companies, this does not seem to be an issue.

In reality, a large portion of consulting transpires based on personal contacts. In this case, an individual has an advantage simply because the quality of the work he or she does is known to the producing company. There may be individuals with better qualifications who are not even considered. (This is one of the major reasons many operating companies hire ex-employees or staff who were previously laid off. This is a risk-management issue in which the producing company follows the adage, *Better the devil you know than the one you don't.*) To return to the point, obtaining revenue on the software is normally better for the integrated consulting company than no revenue at all.

It is rumored a major reservoir simulation software supplier quickly revised its policy on this issue when its customers hinted the company would lose a substantial part of its user base if it attempted to shut out small consultants and contractors. The author's interpretation is the operating companies wanted a more competitive market for sourcing simulation expertise. The author has worked for a large integrated software/ consulting company, large and intermediate producers, on contract, and as an independent consultant. The largest determinant of success remains the skill of the person who is doing the work. Good people work in all of the categories listed.

An integrated consulting company is potentially a good source of software. First, approach the company and determine its attitude toward consulting competition. This is a function of management attitudes. If the approach is mature, chances are consultants and small businesses can do business with the company.

In-house software

Engineers who work for a major company may have no choice. For the most part, the author's experience with in-house software is that it rarely has the polish of commercial products. However, it is not all bad. For example, one can talk directly to the person in research who is developing and maintaining the program. If they have time, they will add custom features for your project.

The author worked on one such suite of programs. The software was developed as a research tool and is highly touted in technical papers from the company that developed it. It is entirely different from any software in the commercial sector. The data input requires knowledge of C programming. The documentation for these programs was significantly out-of-date. A large investment in time was required to learn this program. The final results were mixed. Some of the output was quite good and other parts were not. Some of the features on the subsidiary programs were useful and impressive. The latter were not available, to the author's knowledge, in the commercial sector. This program cannot solve some of the difficult compositional problems, but a number of commercial programs can solve them. It is difficult to find consultants familiar with this software; the company's market is effectively restricted to former employees. On balance, the technical pros and cons struck the author as being equal, a potential advantage in some situations and a disadvantage in others

Where to find different sources of software

There are many software sources. Weeding them out will take time. Following are sources the author has used to identify suppliers:

- The directory of professional services at the back of technical journals.
- The yellow pages. This is one of the most overlooked sources of information. The cost of putting an ad in the yellow pages is almost trivial. Most companies have a yellow pages listing.
- Computing journals. Magazines such as *Geobyte*, *PC Magazine*, *Byte*, and *Unix World* often have reviews as well as a full slate of advertising.
- Many industry journals have an annual software review. *Oilweek*, for instance, has an annual review. PennWell puts out a worldwide review. *The Journal of Canadian Petroleum Technology* now provides a software summary in their "Computer Applications" issue, which is usually the April issue.
- The Internet is very useful. Some oil industry specific sites list suppliers by business category.
- Normally, word of mouth will provide a great deal of information. Ask friends and contacts what software is available and what they think of it.

Local sources

The search for software does not have to be worldwide. First-class software can be developed locally. Canadian software from different suppliers has set the pace, particularly in the area of economic evaluation, for a considerable period. A number of companies in Calgary lead the pack internationally in certain areas of simulation.

Hardware, software, and peripherals

Software must always be considered in conjunction with hardware. This is particularly true whenever there is graphical output of any sort. Getting the program to communicate with various screen standards, plotters, and printers is an enormous headache. This is surprisingly easy to overlook. The problem does not always become apparent until after the machine is hooked up, the software loaded, and one is partway through the first analysis.

Actually, this is becoming less of problem. Most of the Unix machines have adopted PC standards for printers. Postscript printers are now much cheaper in relative terms, and most digitizers now do reasonable emulations of all of their competitors.

Output

When the author started using computerized log analysis, he worked for a company that had the software, a suitable computing platform, and a suitable output device. However, the output device was not dedicated to log analysis use, which made it impossible to make significant progress until a strip plotter, dedicated to log analysis, was obtained. Results are of little use when one cannot look at nor communicate the information to others.

It is easy to spend too much on the base platform and spend too little on output. The client never sees whether one has a 166 MHz Pentium I or a 2.4 GHz Pentium IV. However, they will always be able to tell if the output comes from a ninepin dot matrix printer or a 1,600 dpi laser printer. Subconsciously, the quality of the work will be measured to some degree by the appearance of the output.

First, determine what the output will look like, and then work backwards. When the compromises are made, give up a bit of speed in favor of good quality output.

Personal computers versus workstations

The greatest thing about the PC is that the degree of standardization is extremely high. For this reason alone, the author favors software that runs on a PC.

Until recently, the exception to this was for computationally intensive programs. Reservoir simulation fell into this category. Other examples of this included geophysical applications, geostatistical applications, geological mapping, and log analysis applications. In these cases, the RISC- (reduced instruction set code) based architectures offered a serious advantage. Most of these ran on a Unix operating system, but recently, some of the software tool packages are available on PCs.

During the last few years, the sophistication of PCs has increased, and they are now faster than the Unix RISC machines. Still, Unix RISC machines are very expensive, so there is virtually no contest. PCs have become the platform of choice.

The development of C++ and other programming tools seems to have drastically improved the speed and quality with which graphical user interfaces can be developed. The author expects to see rapid improvement in GUIs in the near future.

Floating point performance and integer performance

A number of years back a Unix workstation was produced with phenomenal floating-point performance. Most people felt it would be the best machine on the market, but the author was concerned it might not be as fast as hoped. It appeared there was too much contrast in integer and floating-point performance. Although floating-point calculations take longer than integer calculations, each floating-point calculation requires several index manipulations to accomplish the correct multiplication. On this basis, the two types of calculations are both important. In truth, the author's computer, which had good floating point and integer performance, achieved better performance than the Unix workstation.

RISC versus CISC

A recent article on this issue was unclear on some of the differences between RISC and CISC (complex instruction set computer) systems. The CPU of a computer comes with certain instructions hard-wired into the circuit board. A compiler, for a high-level language such as Fortran, uses combinations of these instructions to perform operations and built-in functions.

Historically, most large computers were expected to perform numerous database manipulations. The complex instructions were designed to ease data manipulation. This increased the performance of computers and was a very good thing.

Subsequently, particularly for calculation-intensive programs, it was found that the compiler could be programmed to use more steps that were simpler in the CPU. However, this optimization relied on sophisticated compilers to devise better and more direct machine instruction sets. The implication of this is that a hardware vendor's compilers have as much to do with achievable computer performance as the actual speed of the internal clock. Therefore, when evaluating software and hardware, investigate the efficiency of compilers.

The improved performance of recent PC CPUs probably has a lot to do with the RISC expertise that Hewlett Packard and Intel shared. Incidentally, different software companies can use different compilers in generating code for their PCs.

Multiprocessors

Multiprocessing is developing along a number of different paths. Some of the RISC workstation manufacturers have been working on massively scalar machines.

This same process is now taking place with PCs. It is possible to order a two-processor machine off the shelf, with reasonably high clock speeds. Two processors do not double simulation times. An increase of 30% is about normal. Many of the larger simulation software packages come with code designed to take advantage of this feature. There are also some attempts to use a large number of processors for large models. There are some SPE papers on this. For the most part, the increased power of the newer machines is quite adequate for all but thermal simulations and some of the large compositional simulations.

Cost-effectiveness

The cost-effectiveness of hardware varies considerably. The truth is that the majority of us are unlikely to fully appreciate the factors that affect performance. Whenever we do not understand the inner workings of a device, we are vulnerable as consumers. In these circumstances, we end up buying perceptions rather than underlying value or performance.

The author is not an expert in computer architecture, but he understands some of these differences. Most people who use the IBM PC or IBM clone are using an Intel chip for the central processing unit. Variations in performance occur depending on how information is passed between various components in the computer, and the configuration of the hard disk. This means that how memory is accessed (memory bus) and the data bus strongly affect performance. The new Rambus memory increases performance considerably. This memory probably decreases actual execution times by 30% compared to a standard PCI bus.

Cost effectiveness varies considerably. One can easily pay 30% more for a machine that performs 30% slower than a competing product. Overall, the PC market has become much more competitive, and these differences will diminish in time. Many local companies are now getting out of the hardware side of the business.

One item that also varies considerably is the cost of add-on peripherals. The first PC the author owned was an IBM model 60, which featured the micro channel architecture (MCA). To add a 5.25-in. floppy disk would cost $700.00 since the computer had the newer (and supposedly better) MCA. At the time, a standard AT type external drive cost $350.

What is the moral of the story?

- Learn as much as possible about how computers really work. It can be hard to find this information in a readable form. The situation is improving radically. The books by Peter Norton and PC magazines were good at explaining the basics.
- Look through PC magazines and check out the ratings. Reading these requires some subtle interpretation. Since magazines depend on advertising from hardware manufacturers for revenue, they never give a machine an outright bad review. However, if one reads between the lines of various shades of acclaim, one can determine the better choices.
- A number of websites post ratings. The top PC magazines also have informative Internet sites. Some specialty web sites also offer good sources of information.
- Many of the cheaper, less well-known computer brands with smaller reputations may offer superior cost performance. The CPU chips are all made by the same company, probably Intel. The differences in physical reliability are, in the author's opinion, not that great.
- Many of the suppliers who sell or market directly on the Internet seem to have high-quality equipment and good ratings in the PC magazines.

Maintenance

The author does not recommend buying maintenance programs. Normally this costs about 12% of the computer cost for large Unix RISC machines. There has to be a major failure to recover that investment. Since the electronics are quite reliable, the author recommends getting a quote from an insurance company on electromechanical failure. The cost for this is normally about 2–3% of the computer price. This is radically more cost-effective.

The hardware vendor will say that, without maintenance, the user will not get operating system upgrades. As a consultant, who is a user of software, the author has found that this is actually an advantage. In the past, new releases of operating systems have corrected bugs that had been found, but a work around or patch had already been developed for these old bugs. However, the introduction of the new and improved system often introduced a number of new bugs. This involved more expense to identify and fix. Colloquially, the *leading edge* of technology has often become known as the *bleeding edge* of technology.

The exception to this rule is when a company's main business is software development. In this case, the company has to be aware of the new releases and the bugs that go with them. The company will not fix the bugs that have patches, because some of their customers will still be using older versions of the operating system; however, the company does need to identify and contain the new bugs.

Both disk drives failed in the author's first Unix computer. The computer manufacturer refused to replace drives that failed in less than a year even though the manufacturer normally warrants the drives for a year. A local source replaced the drives for substantially lower prices than directly from the manufacturer. The author had not purchased the maintenance but was justified—even in these circumstances. The cost of both disk drives was about half the annual maintenance fee.

There is an exception to the "don't buy maintenance" rule. If the entire business depends on the continued function of one computer, then maintenance may be worthwhile. There are a limited number of companies in this position, but it is not critical for the average oil industry simulation engineer or consultant.

Variations in performance

At a technical conference, one of the engineers from a large Canadian company proudly announced that the company he worked for was using the software that is the most widely used worldwide. The devil made me do it, and I had to ask, "How do you think the performance compares to other simulators that you've used?" There was a long and somewhat dark pause, at which point the individual was forced to admit that he had never used any other simulation software.

For black oil reservoir simulation, the author has tried a number of software applications. He used similar projects and thoroughly tested the capabilities of the programs. All of the models used produced similar results with similar levels of performance. Differences in speed were relatively minor.

However, everything said in the preceding text about black oil models is probably not true for compositional models. There are big differences in performance for compositional simulators. The implementation mathematically is far less standard. Probably, there are big differences in the performance of thermal simulators as well.

Commitments to platforms

Simulation software companies tend to make strong commitments to one type of platform or another. They were unwilling to attempt installations on other platforms and expressed serious doubt about being able to compile even a standard Fortran 77 program. In the author's experience, the latter concerns were unfounded. The biggest problem encountered was the use of SGI (or sgi) as a variable name. To the reservoir engineer, this is obviously initial gas saturation. On Silicon Graphics, Inc. machines this is a reserved variable name and is used for system type calls. The initial gas saturation variable was easily renamed S_{gasi}. Changes were required in two or three modules with a find-and-replace command. It was hardly a serious problem.

As for the graphics on pre- and postprocessing, the software companies simply laughed. The software for these features apparently had radically different structures.

Tough problems

Technical capability commonly seems to be a major concern. Surprisingly, the author has found people unaware of what constitutes a difficult problem. The toughest simulations the author has run are for hydraulically fractured wells, with the frac included in the grid.

The contrast in transmissibilities makes the problem difficult for the matrix solver to handle. One cannot put in the frac with actual dimensions. If the frac width is no less than one foot (0.3 meters), the program works very well. Low compressibility fluids are harder to work with; hence, oil problems are more difficult than gas problems. Layering makes the problem even more difficult. There are multiphase effects for oils below the bubblepoint. In the past, some simulators could handle this and some could not.

Traditionally coning was the most difficult problem. A high degree of implicitness is required to maintain stability. This requires implicitness not only for the reservoir formulation, but also for the well equations. Here one has high grid-block throughput due to converging flow and multiple phase effects, which are nonlinear.

Recently, many of these difficulties have been resolved by local grid refinement. This definitely slows down a simulator. The regular structure of the matrix is disrupted, requiring more time to solve.

Combined coning and hydraulically fractured or horizontal wells can result in problems that are more difficult. It is only recently that this issue has been addressed for compositional simulations. Some interesting results were obtained. In fact, much of this work is still being done with black oil (condensate) simulators such as the work by Settari, Bachman, et al.

Well controls are often difficult to handle, as discussed in chapter 13.

Features

Strange to say, features may dominate the choice of software. This is not a trivial issue. Most of it is concerned with well control schemes. Some useful features include voidage replacement calculations and group controls. These will make an engineer's life much easier.

Another area, particularly relevant to offshore work, is related to production engineering options. Most offshore wells have high rates and, under these conditions, tubulars and completion efficiency will have a major influence on cash flow. Problems must be solved considering both of these issues.

Most simulators now have local grid refinement. This was not always the case even a few years ago, but it is a useful feature. Similarly, a horizontal well grid refinement, which normally uses ring-type refinements, is also useful. On high-rate horizontal wells, in high-permeability formations, wellbore hydraulics (horizontal flow through the wellbore) may become significant.

Graphics

Graphics are actually an important consideration. Evaluating simulation results can take time. A good graphics postprocessor will have a significant impact on how much can be done. Given that the simulators seem to have similar levels of performance, input modules and output processing will be the deciding criteria. There is no substitute for sitting down for an afternoon or two and playing with it.

Graphics actually put more demands on hardware than the simulator. This relates to the graphics toolkit, which makes huge demands on memory for 3-D realizations. Therefore, consider that the graphics might be more demanding on computer resources, such as memory, than the simulator.

One area where the author has noted variation in performance relates to digitizing geological maps. Some programs are easier to use than others. This area consumes much time. The interpolation schemes used to generate grid cell values from contours vary in quality. Manual correction is very time-consuming and should be avoided if possible.

The author does not recommend automatic data set builders. If an engineer knows reservoir engineering, they are more of a deterrent than a help. If unsure and tempted to rely on the data set builder, slow down and learn the input. An engineer risks serious errors by not understanding the process.

Geostatistics

Geostatistics is becoming increasingly important. Investigate this system in conjunction when choosing a reservoir simulator.

Testing

The SPE has commissioned a number of interesting comparative simulation projects. All of them were developed to test various components of reservoir simulators. In truth, some of the issues, such as vertical movement of gas, are no longer major technical issues. On the other hand, a number of comparative simulation projects are now used both for debugging and as benchmarks. In the latter category, the older tests are somewhat small to be much of a challenge for today's computers. However, these tests are still used as benchmarking standards, since no one has come up with problems to replace them.

A variety of tests is optimal. Convert some real problems to a new data set. To date, the author is not aware of any SPE comparative solutions that concentrate on well controls nor is there a review of graphical user interfaces.

Business issues

Some software companies are quite flexible and will rent their programs by the month. Normally there is a premium for doing this compared to leasing for the entire year. The code can be purchased outright, normally in executable format. In this case, there will be an annual maintenance fee.

One software supplier wanted a deposit for the consulting help he perceived a user would need to run their simulator. Engineers who have worked on a large number of different models require little, if any, support, but help is definitely required in some instances, for example, when there is a bug or the documentation is wrong or cryptic. In these cases, the better choice is to find other software.

The oil industry often involves joint interests on producing properties. One issue that arises is the ability to trade simulation results and input. For example, in a unitization proceeding, there is usually a technical committee. It is rare that there is unanimity among owners on technical, economic, and operational issues. Simulation is often used to determine an operating strategy and to determine optimal development. Many of these technical issues have tradeoffs, which, when fully analyzed, allow companies to agree on participation, if not on the actual technical input. Therefore, if the majority of the partners are committed to a particular software package, there are advantages to using this software.

It sometimes appears that operating companies are concerned about getting different results from different simulators. While there are differences in features, this is a matter of efficiency. Different simulators should not, and in the author's experience do not, produce different results for black oil studies. If differences exist, most likely the input to the simulator has changed. More often, this represents differences in geological input, grid-block sizes, and other significant input.

Incidentally, the differences in input between companies can often be dramatically different. This can be a real learning experience. Integrating the different perspectives from several operating companies, both large and small, generally results in superior engineering. This can be a difficult process politically. In some situations, different companies have different objectives and consensus cannot be achieved.

Summary

With judicious shopping, it is possible to find a combination of technical excellence, hardware compatibility, and realistic business terms. Technical capability is normally not the dominant factor in choosing software. In the author's experience, sometimes the order is reversed.

Managing Studies

Realistically, a significant number of reservoir engineers will never do a simulation study. More commonly, a consultant, either internal or external, is retained. In these circumstances, an engineer will be faced with a number of difficulties that are not entirely technical. These include bidding out a job, selecting somebody who will do the job correctly, and meeting a deadline. Having worked on both sides of the fence, the author has some suggestions, which may be helpful.

There is a book on this general topic: *Choosing and Using a Consultant, A Manager's Guide to Consulting Services* by Herman Holtz.[3] This book is unique in that it is written from a client's perspective.

Choosing a consultant

In one situation, ARE submitted a bid for a complex coning study for $15,000. The operating company obtained a bid for almost half the cost at $8,000. The company personnel knew me and asked if I could do the job for less. In our opinion, the job could not be done for less, and we declined to reduce the price. Of course, the other consultant was awarded the job. Several months later, we learned that the budget had been spent, and the consultant had announced that the problem required a dual porosity simulation, which he did not have. All work was halted. Since the consultant had done all he outlined in the bid, he was paid. Nevertheless, this is an indication of where not spending enough can result in no benefit whatsoever.

ARE was awarded the next job and many more follow-up projects with virtually no questions asked.

Bid preparation

Bid preparation can take a long time. A moderately sized simulation can take several days of billable time to prepare comprehensively. Being realistic doesn't always pay either; the low-cost bidder normally gets the job.

If the job is underbid, the operator is in a catch-22. It is unlikely that a consultant will continue if he knows that continued work will lead to a larger loss. It is preferable not to be paid for one month of work and obtain payment for the next two months than spend three months completing the project and being paid for one month. This leaves the operator with a choice of hiring somebody new from scratch, and paying for him or her to get up to speed on what the existing consultant has done, or continuing with the existing consultant.

This does not mean that consultants do this deliberately. It is human nature to underestimate engineering completion time. This systematic bias has been known for a long time.

In many cases, it is impossible to anticipate what results will develop during the course of a study. There are simply too many variables. The issue that arises is whether consultants should suffer financially for something that neither the operator nor consultant could foresee. Legally this would be viewed as an "unjust enrichment." The operating company normally agrees to an outline of the approach before allowing the study to proceed.

In many cases, it is impossible to determine how many of the possibilities really contributed to a study going over budget. However, the system of choosing low bids almost guarantees that overruns will occur—either by accident or by design.

You may not win by being on budget

In one case, the author reviewed a previous study and wondered why certain prediction schemes had not been run. In this case, a five spot waterflood had been evaluated. The project wasn't economical because too much production was lost before waterflood production response was achieved. The permeability was sufficiently good that a nine spot would have been sufficient, if placed correctly. It is likely that further investigation was stopped by not seeing this possibility or the matter was wrapped up without further thought simply to stay on budget and look good. The project was never implemented, so the company would not have had an obvious engineering failure. On the other hand, they may have walked away from a waterflood that would likely have added significant value, since a limited number of injectors had worked well on an offsetting pool.

Lack of information

In a number of circumstances, bids go over budget because the operating company failed to provide pertinent data. One such example involved the simultaneous modeling of a surface facility system and the reservoir. The wells were loading up with fluids, whereupon they were shut in until the pressure built up sufficiently. Overall, the property was contract constrained so the reduced hours on the wells did not raise any flags.

The problem did not materialize in the reservoir model either. There was no liquid dropout in the reservoir, and the low rates did not cause any problems. When surface facility modeling was implemented, a substantial portion of the wells was shut in by minimum rate limitations. At this point, the reservoir engineering staff realized that the field staff were concerned about this issue and had been requesting a surface/reservoir model. A considerable amount of time was required to adjust rates to a weekly or half-monthly basis.

Was the client treated unfairly? No. There was no other choice but to adjust the production rates manually. It is not certain that they were leaving out information deliberately, but a number of people within the operating company knew that this was a problem. It is quite easy to assume that someone else had filled in all of the details. In the end, this doesn't work well for anybody. The consultant ends up stressed, which does not contribute to efficiency, and the company personnel end up with a cost overrun that does not look good to their boss.

Most company personnel are unlikely to rush into their supervisor's office and proclaim that they made the mistake that led to a cost overrun.

Joint venture bids

One situation that develops is working interest partners in a unit decide a simulation is needed. The usual procedure is to have each company choose a consultant. All of the consultants are asked to bid. All of the bids are evaluated, and the objective is to pick the best submission. In one case, the operator evaluated all of the bids and determined that ARE was the best prepared. Despite this, the job was awarded to another consultant. The operator agreed to use a different consultant as a concession to another reluctant partner in order to get a simulation done at all. The engineer at the operating company was very honest. Although ARE was disappointed, the political situation was understandable. He made a decision in the best interests of his company.

These forces are completely beyond a consultant's control, and the presence of a large number of bidders automatically reduces the chance of success. A strong case can be made for spending less time on a bid in which many companies are bidding.

Contracts

Contracts are a source of real problems. The most common problem is the use of an inappropriate "off-the-shelf" contract. Invariably, somebody hands over one of these contracts and announces that this *must* be signed. Here are some of the problems ARE has experienced:

- The purchasing department's newly standardized contract was designed for companies that install field facilities. One of the main requirements was that fireproof clothing was to be worn at all times and that the contractor and all his equipment be insured. Yes, this contract was deemed appropriate for contract engineers and was given to ARE. None of the company's personnel was wearing fireproof business suits. Nor were any of the contractors. If the matter ever ended up in court, the million dollar question would be, "How would a contractor know which clauses were to be ignored and which were to be followed?" Under these circumstances, the courts may set aside the agreement on the basis that it was common knowledge that signing the contract was an absurd formality, implying that no one expected the contractor to follow the entire contract.
- ARE was a subcontractor on a large simulation project for a company that also wrote software. The contract we received was written for people who were preparing software code and working in-house as contractors. In this case, the contract was one-sided, with no protection afforded to the consultant. This contract was close to making ARE personnel direct employees. Under these circumstances, labor law may automatically supercede any direct terms. This is a complex area of the law, and it is easy to get blindsided, unless you're a lawyer working in this specialty. In this particular contract, some of the clauses were also contradictory.
- Another project we did was an independent evaluation for the purchase of a property swap. In this case, we were provided with a contract that was intended for another operating company that was reviewing information in a data room. The main objectives of the contract proffered were to get all information back (including all copies) and to absolve the operator of the properties from responsibility for the data provided. As an independent consultant, we hold the operating company responsible for providing all of the relevant data (see the disclaimers in the report). Moreover, we had every intention of retaining a copy of the data in a permanent record. This was our due diligence record in the event that the deal turned sour and one of the two parties decided to sue somebody. One of the companies was based in the United States, where litigation is a common part of doing business.

In all of these cases, the contracts were appropriate for the original circumstances for which they were designed. However, ARE would not sign the contracts since they were clearly inappropriate. Sadly, in all of the cases, there was high initial resistance because these contracts had become company policy. In the end, ARE convinced the individuals involved that the inappropriate contracts did both sides more harm than good. In fact, they were better off with no contract at all than with a "wrong" contract. There are many remedies through common law and existing precedents. In all but one of these cases, the client company decided against rewriting their contracts. Contracts are fine, but make sure the contract is appropriate for the type of work being done.

Value of contracts

It is hoped that the process of making up a contract leads to a common understanding, but this is not always the case. In the final analysis, contracts may be of great help when in litigation. However, at this stage, one is in damage-control mode, i.e., having a contract may not avoid damage. Getting to a good understanding, with the right consultant, will likely avoid the problems up front. To this end, letters are much faster to prepare, get to the point more quickly, and are strong backup if one ends up in court.

Realistic costs

One area that seems to cause consternation is consulting fees. Many employees compare their wages with consulting fees and feel that consultants *must* be gouging them or their companies. From a consultant's point of view, things have been competitive on the engineering side since 1986. Most of this problem stems from employees' lack of understanding as to how much they cost their company. They see only their direct salary and easily forget the benefits, the cost of the office space they occupy, the cost of equipment, such as computers, the money they are paid for vacation, and holiday time. Most oil industry consultants are well-educated and quickly find full-time jobs in other industries, in the event that their fees don't cover overhead costs and time between jobs. Thus, there is a billing efficiency adjustment to bring a consultant's rates in line with full-time work. Consulting is inherently riskier than being an employee, and the rate of return should be correspondingly higher.

An article extracted from the SPE's *Journal of Petroleum Technology* covers this topic.[4] The consulting fee you are paying does not directly correlate to the money that the individual consultant is putting in his pocket.

Realistic expectations

As a rule, operating company personnel expect a great deal from consultants. Having worked in both environments, the author recognizes that the expectations of some operating companies are not realistic. Nevertheless, knowing that people will rigorously try to find fault with your work has a tendency to make one sharper.

Emphasis on G&A costs

During the past 10 years, there has been tremendous pressure to reduce general and administrative (G&A) costs. Controlling these costs has always been a good idea; however, one wonders if this has been taken too far.

In general, many companies are quite willing to spend millions of dollars on horizontal wells and in deep drilling areas. Yet, the same companies resist spending more than $10,000 to $15,000 on a reservoir study. Part of this may be due to a percentage of studies that, for various reasons, do not produce usable results. However, it seems that a good general objective has been taken to an extreme of late. In our experience, it is not uncommon for production to double based (sometimes partly) on the results of a technical study. The key is always to link the potential upside versus a realistic cost for a study.

In this regard, there is always an element of salesmanship in dealing with management. When working for a major company, the author was not always able to find the right sales points to justify an expense. These points are not easily quantified and making them can take a lot of creativity. Sometimes no amount of salesmanship will work if management has entrenched views. The key is to realistically and honestly demonstrate that the incremental benefits justify the cost of a study.

Mistakes

A professor brought a guest lecturer from one of the largest engineering and procurement (E&P) firms in Canada to a university engineering class, and his lecture left a lasting impression. In his presentation, he outlined a story in which a subcontracted engineering firm that was famous in its specialty had made a serious engineering error. Six months after submitting its report and recommendations, the subcontractor found the error. They approached the E&P firm and admitted they had made a grievous error. Further, they admitted that they were responsible and most likely, if the E&P firm elected to litigate, that the subcontractor would be forced out of business. It was also suggested that they could recover a significant amount of money by dissolving the company. The E&P firm realized that this error would cause them to lose a significant sum of money, since a considerable amount of engineering time had already progressed based on the faulty report supplied. On the other hand, they would have lost a great deal more had the mistake not been brought to their attention and allowed to propagate to completion of the project. At this extreme, a noticeable negative effect would occur to the bottom line for the E&P firm.

In the end, the E&P firm decided to let the subcontractor redo the work and paid the company for it. The subcontractor's promptness and honesty had saved them an enormous amount of money and the embarrassment of a failed project. The E&P firm could have put the subcontractor out of business and recovered a good portion of the cost. Other consultants in the field likely could have done the work. On the other hand, this would have resulted in more delay, since the existing subcontractor was already familiar with the project. The E&P firm also stood to lose in the end by removing a well-qualified firm in a market with relatively few competitors. The E&P firm also realized that it had a good relation with the subcontractor. They could trust that problems would be promptly brought into the open in the future, where damage could be contained at minimum cost.

Consulting is a relationship

This is significantly different from buying a hamburger at MacDonald's. The best results are obtained when the consultant and the client have a good understanding of what is expected and required. There are a number of good books on this subject such as *Flawless Consulting* by Peter Block and *The Consultant's Calling* by Geoffrey M. Bellman.[5, 6] Obviously, the author read these from the perspective of running a consulting company. However, if you can understand the other person's position, it is often easier to communicate.

Similarly, consultants can make staff feel awkward and threatened, particularly when the staff is placed in the position of relying on an expert. This is often unconscious. To the consultant, it can be baffling. One sometimes gets negative reactions although one's actions are no different from other similar engagements.

With experience, it is possible to tell when this is occurring, and one does get better at avoiding this problem. At other times, a contractor can feel as if there is nothing to do or say to relieve the tension.

Finally, there are some instances where clients are simply rude to consultants as a matter of course. Eventually, people who fall in this category will find themselves dealing exclusively with second-rate consultants. The good ones will find better work and will not come back. Moreover, consultants do talk to each other, and companies develop good and bad reputations.

Meetings

As a rule, have regular meetings with clients when a major simulation is underway. The requirements vary somewhat depending on the stage of the study. Normally, there will be a meeting at the beginning of the bid process when the problem is outlined. A second meeting falls when the bid is presented.

During the early stages of data gathering, relatively few meetings are required while PVT and relative permeability data are being processed. A review after the first run is normally sufficient to cover this issue.

History matching on a large project requires meetings at least weekly or biweekly. Similarly, the prediction phase normally involves regular contact. Shorter simulations require less time.

Summary

Picking a consultant is not easy. The issues involved are only partly technical. In general, people are less comfortable discussing these issues than technical content. In the final analysis, getting the right answer is the most important objective. Under most circumstances, the implication of the work that is being done vastly exceeds the cost of study. If this is not the case, then it is likely that one should be directing attention to other projects.

Problem Areas for Simulation

A number of problems seem to recur from year to year despite practical experience to the contrary. These include:

- Gas injection
- Gravity override
- Layering

Several reasons can cause these same problems. Invariably, the problems relate to the people doing the work and management.

Gas injection

Gas injection scenarios are commonly run in black oil simulations. Gas injection is handled poorly by a simulator. Simulators are poor at predicting horizontal gas displacements for a long list of reasons:

- At high mobility ratios, which occur with gas injection, grid orientation effects become very significant.
- Viscous fingering is not accounted for in the formulation of a black oil model.

- Gridding requirements are more demanding. Smaller scale heterogeneities must be accounted for and more layers are required than for water-oil displacements.

These limitations are known by simulation engineers. The problem is more probably that numerically oriented simulation engineers have difficulty adjusting their results with something that cannot be quantitatively determined.

Gravity override

Another problem area is gravity override, which can occur quickly even in thin reservoirs. This problem is widely recognized in steam injection. It was more slowly recognized with miscible flooding. Some intense debates have taken place regarding how much gravity override would actually occur. In the latter case, the amount of gravity override was seriously underestimated.

One major company, as part of a government application, submitted hydrocarbon miscible flood predictions for a major carbonate reef reservoir that were one-third to one-half less than predictions prepared by other companies. Significantly, royalty holidays and other economic incentives were determined based on estimated recovery. Thus, the other operators were unhappy with this break from their more optimistic recovery predictions.

The source of some of these problems relates to limitations of computing, which again means that comprehensive quantitative predictions could not be made.

Actual reservoir performance has proven the lower recovery factor to be closer to the mark. Employees of government agencies have noted, in private, that applications seem to be systematically optimistic. They have also noted, publicly, when industry participants were on opposing sides of technical issues, an extreme spread in expected recoveries for the same reservoirs.

Heterogeneity/layering

Experience has shown that geological effects are consistently underestimated. The displacement of oil and gas is reduced by streaks of high permeability. While reservoir engineers are familiar with this concept, their training does not include sufficient geology to deal with the solution easily. Only a limited number of people have both geological and engineering training and can deal directly with these issues. This is one of the major reasons for multidisciplinary teams. However, with multiple people involved in attempting to resolve complex issues, it is difficult to achieve seamless integration. Furthermore, it is inherently difficult to extrapolate between reservoirs and create a realistic model. Models tend to be simpler than they should, and this systematically overestimates production performance and recovery.

Consequently, this issue arises repeatedly.

Solutions to recurring problems

Simulation results need to be tempered with practical experience. Ideally, most reservoir engineers would only be allowed to do simulation when they have accumulated a certain amount of practical reservoir experience along with some geological background.

This is not consistent with the way people actually approach work on simulations. Often, the more technically oriented engineers gravitate to this field during the latter part of their undergraduate education, or they pursue detailed training at the masters level. They are less familiar with how to find this information or are less likely to have direct experience with the production performance of a large number of reservoirs. Conversely, reserve evaluation engineers, who normally gain massive amounts of experience analyzing production performance, often are not interested in detailed technical study. Reserve evaluation engineers can develop a jaundiced view of simulation, often with good cause.

The solution to these problems requires understanding in both the reserve evaluation and simulation disciplines. Since this is a simulation book, we have identified areas where simulation engineers can be more conversant with practical experience. The current discussion is limited to gas flooding, since it is by far the worst culprit. The following sources of data can be used for both primary recovery and waterflooding.

Interstate Oil Compact Commission

There are two excellent sources. The first is the Interstate Oil Compact Commission (IOCC):

> *In those reservoirs having high permeability and high vertical span, gas injection may result in high recovery factors because of gravity segregation.*
>
> *If the reservoir lacks sufficient vertical permeability or relief for gravity segregation to be effective, a frontal drive similar to that used for water injection can be used. This is often referred to as "Dispersed Gas Injection." In general, we may*

expect dispersed gas injection to be more successfully applied in reservoirs that are relatively thin and with little dip.

Injection into the top of the formation (or into the gas cap) will be more applicable to reservoirs having high vertical permeabilities (200 mD or more) and enough vertical relief to allow the gas cap to displace the oil downwards.

In most proposed secondary projects gas will do neither of these things (conformance, displacement and sweep efficiency) as well as water.[7]

Gas injection does have its application. However, the conditions are restrictive. You must have high vertical permeability and high vertical span. Note that waterflooding is usually more effective than gas flooding.

AEUB

The IOCC's comments are largely qualitative. Another source on this subject is the Alberta Energy and Utilities Board (AEUB). In this regard, the AEUB has *the* bottom-line answer—years of experience.

Some interpretation of the figures in Table 14–1 is required. First, the gas flood reservoirs comprise about 12 pools, compared to more than 4,500 pools on primary, 2,700 waterflood pools, and 53 solvent floods. The primary recovery factor in these pools is very high at 32%. These recovery factors include reservoirs with favorable gravity segregation conditions of high vertical span and high vertical permeability. Yet, the incremental recovery for these gas floods is only 7%, compared to 14% for waterfloods. Gas flooding increases primary recovery by 0.07/0.32 = 0.22 and waterflooding increases primary recovery by 0.14/0.15 = 0.93. Since gas flooding can give very high recoveries in the right application, it may be deduced that horizontal gas floods give very low recoveries.

This information is extracted on a pool-by-pool basis as shown in Table 14–2.

Five reservoirs have 10% or higher gas flood recoveries. If the two high vertical relief Rainbow 'F' and 'N' pools are eliminated, only three reservoirs remain. Two are Cardium pools, which by virtue of the porosities are likely high-quality conglomerates. The Aerial Mannville pool, with a porosity of 22.3%, is also likely of a very high quality.

Golden Spike is a well-documented failure of vertical gas injection. There is a *JCPT* paper on this failure. The main problem concerned an areally pervasive tight horizontal layer across the main section of the reservoir. Vertical communication only existed around the reef rims, which meant a vertical displacement was not possible. This is really an example of where inadequate geological description resulted in the expenditure of large sums of capital inappropriately. The gas flood area is listed as having a pay of only 6.1 meters; the entire reef has a height of closer to 136 meters.

The AEUB data doesn't include partial pressure maintenance schemes. For instance, the author worked on the Wood River D-2 'C' pool, which had crestal partial pressure maintenance. This was not particularly effective and is not included in the previous figures.

Many gas floods don't work very well.

Incidentally, the most common form of EOR in Canada is hydrocarbon miscible flooding—i.e., enriched gas. In the United States, the most common form of EOR is CO_2 gas flooding. Many gas flood problems apply to advanced processes.

Other sources of practical experience

The Petroleum Society of CIM Monograph Number 1 titled *Determination of Oil and Gas Reserves* is an excellent source of information.[8] This monograph was prepared several years ago and has sold well throughout North America. Recently, the SPE has produced a handbook on reserves, which is also comprehensive.[9] The styles of the two documents are quite different and both are recommended.

Summary

These recurring problems are possible to solve, but they are not easy to rectify since resolution involves an overlap of different disciplines. Engineers are encouraged to get as broad-based experience as possible. Working in reserves and geology are constructive experiences. In any event, these problems are well-known, and constructive steps can be taken to insure they do not recur.

Table 14–1 Summary of Reserves of Conventional Crude Oil

	OOIP (E6M3)	*Recovery Factor (fraction)*			
		Primary	Secondary	Tertiary	Total
Light-Medium					
Primary	3,560.9	0.22	0.00	0.00	0.22
Waterflood	2,626.7	0.15	0.14	0.00	0.29
Solvent Flood	880.8	0.28	0.18	0.13	0.59
Gas Flood	149.5	0.32	0.07	0.00	0.39

Summary of Reserves of Conventional Crude Oil Attributable to Various Recovery Mechanisms as at December 31, 1995.

Table 14–2 Recovery Factors for Gas Flood Pools in the Province of Alberta, Canada

Reservoir	OOIP (E3M3)	Primary R_f	Gas Flood R_f	Height, meters	Porosity Fraction
Aerial Mannville	1.188	0.12	0.14	2.62	0.223
Cyn-Pemb Cardium D	0.900	0.20	0.10	4.86	0.102
Golden Spike D-3	49.600	0.53	0.05	6.10	0.068
Grande Prairie Halfway A	3.682	0.14	0.05	7.00	0.102
Kakwa Cardium A	5.163	0.15	0.19	2.44	0.124
Pembina Ostracod G	0.400	0.18	0.04	1.10	0.100
Rainbow Keg R F	37.000	0.41	0.10	61.60	0.080
Rainbow Keg R N	2.940	0.30	0.13	28.31	0.037
Rainbow Keg R H2H	0.200	0.35	0.04	9.70	0.060
Ricinus Cardium A	7.137	<0.12	0.03	8.98	0.140
Ricinus Cardium O	4.850	0.13	0.02	8.81	0.120
Willesden Green Cardium A	35.600	<0.07	0.07	4.83	0.111
Windfall D-3B	0.810	0.10	0.04	13.31	0.050

References

1 Deutsch, C.V., and A.G. Journal, GSLIB: *Geostatistical Software Library and Users Guide*, 2nd ed., Oxford University Press, 1997.

2 Deutsch, C.V., *Geostatistical Reservoir Modeling*, Oxford University Press, 2002.

3 Holtz, H., *Choosing and Using a Consultant, A Manager's Guide to Consulting Services*, John Wiley and Sons, Inc., 1989.

4 Truman, R.B. II, "Outsourcing Engineering Services: the Business Side of the Business," *Journal of Petroleum Technology*, (October): 915–918, 1996.

5 Block, P., *Flawless Consulting: A Guide to Getting Your Experience Used*, United Associates, Inc., Jossey-Bass Publishers, 1999.

6 Bellman, G.M., *The Consultant's Calling: Bringing Who You Are to What You Do*, Jossey-Bass Publishers, 1990.

7 Interstate Oil Compact Commission, *A Study of Conservation of Oil and Gas in the United States*, Interstate Oil Compact Commission, 1964.

8 Aguilera, R., et al., *Determination of Oil and Gas Reserves*, Petroleum Society Monograph No. 1, 1994.

9 Chronquist, C., *Estimation and Classification of Reserves of Crude Oil, Natural Gas, and Condensate*, Society of Petroleum Engineers, 2001.

Part 2

Selected Advanced Topics

15
Compositional Modeling: Gas Condensate and Volatile Oil Reservoirs

Introduction

The course from which this text was drawn has been in preparation for a long time. In the original versions, the author would not have broached this topic; however, in recent experience, the use of compositional simulators is increasing. A number of reasons explain this; probably the most important is the increased power of Unix-based RISC workstations and PCs. It is fair to point out that compositional simulators have undergone some important improvements, both in compositional solutions algorithms and in matrix solvers in general. A number of gas condensate reservoirs are under development at present, and this has increased usage. Sour gas systems will be discussed in this chapter, since they are frequent candidates for compositional analysis in Western Canada.

Objectives

The first objective in this section is to create a reservoir characterization for use in an equation of state (EOS) simulator. A number of issues arise in achieving this objective:

1. Equations of state are not sufficiently accurate to be predictive. Therefore, the input data must be tuned. The quality of tuning depends on the quality of lab data, the type and detail of results, and the range of conditions that the tests cover.
2. Compositional simulators require a great deal of computer resources and typically run much slower than black oil simulators. However, the amount of time that is required decreases significantly if the number of components in the EOS decreases. This is a simulator design tradeoff. The objective is to develop a description that uses the least number of components yet reproduces PVT behavior within reasonable tolerances.

This area is relatively new and is still developing. Although there have been numerous papers on systematizing this process, there is no consensus on the best way. There is, therefore, a significant element of trial and error.

Since much of this work depends on compositional calculations, a brief outline of an equation of state and how it works is included in this chapter. Only an outline is presented; there are no thermodynamic derivations. Pragmatic tests, using an EOS package, have been prepared by ARE.

Cubic equations of state (EOS)

The basis of the equation starts with the ideal gas law:

$$pV = nRT \tag{15.1}$$

$$p = \frac{nRT}{V} \tag{15.2}$$

The ideal gas law is based on the assumptions that gas molecules collide like perfectly elastic spherical molecules in space, that the volume of the molecules is negligible, and that there are no chemical attractions between the molecules.

Although these assumptions are adequate at low or room temperatures and pressures, the assumptions are not true at reservoir conditions. The ideal gas law is frequently corrected using a Z factor. This is sometimes called the gas deviation factor or supercompressibility factor as shown in Equation 15.3.

$$pV = znRT \tag{15.3}$$

The ideal gas law can be modified to account for the finite volume of the molecules and for chemical interaction as shown in Equations 15.4 and 15.5.

$$p = p_{repulsion} - p_{attraction} \tag{15.4}$$

$$p = \frac{RT}{(v-b)} - \frac{a}{v} \tag{15.5}$$

This is Van der Waal's equation. Mathematically, it is normal to combine the effects of repulsion and attraction into a single Z value. This is calculated by solving for roots of a cubic equation (in Z). This solution dates from 1873.

If pressure is plotted as a function of volume for a fixed molar quantity, then a graph is obtained as shown in Figure 15–1.[1] Below the critical point, there is an S-shaped wiggle. In reality, the system does not behave this way. Between the points, there is a mixture of liquid and gas. It has been determined that, by balancing the areas that are shaded in the figure, an isotherm can be drawn from the dew point line to the bubblepoint line.

The previous equation was improved upon by Redlich and Kwong in 1948. They noted that, as system pressure becomes very high, the molal volume approaches 26% of the critical volume. The resulting equation is shown in Equation 15.6:

$$p = \frac{RT}{(v-b)} - \frac{a}{v(v+b)T^{0.5}} \tag{15.6}$$

This equation was further improved by Soave, who changed the temperature dependent term $a/T^{1/2}$ to $a \times \alpha$. Alpha (α) was defined as a function of temperature and an acentric factor as shown in Equation 15.7.

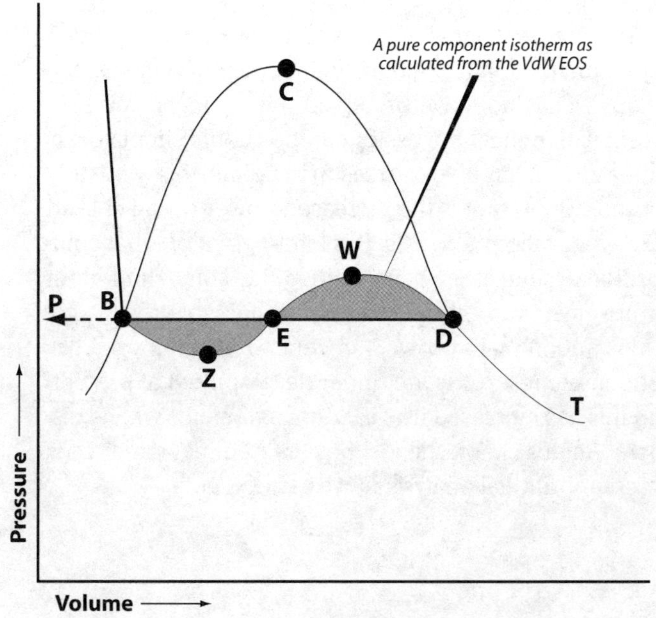

Fig. 15–1 Volumetric Behavior of a Pure Component by EOS

$$\alpha = \left(1 + m\left(1 - T^{\frac{1}{2}}\right)\right)^2 \quad (15.7)$$

The acentric factor corrects for nonsphericity of actual molecules. This introduces additional temperature dependence. The equation is once again solved for Z by finding roots of a cubic equation. This modification was introduced in 1972.

Not long thereafter, in 1975, Peng and Robinson showed that an improvement was required in the Soave-Redlich-Kwong (SRK) equation. Their equation is similar in form as shown in Equation 15.8.

$$p = \frac{RT}{(v-b)} - \frac{a\alpha}{v(v+b)+b(v-b)} \quad (15.8)$$

It is similar to the SRK equation. However, the errors from this equation are more balanced. This is shown in Figure 15–2 in which the error in predicted molal volume—e.g., for one mole of normal butane—is plotted against pseudo reduced pressure.[2] It is important to note that the data shown is for a single or pure component. The Peng-Robinson errors do not reach as high a deviation, although the overall pattern of error is essentially identical. The errors are higher for the liquid phase than for the vapor phase. In the liquid state, the intermolecular forces are larger than accounted for. This is not surprising, since the EOS was designed from the ideal gas law.

Multicomponent systems

In order to accommodate multicomponent systems, the properties of all of the individual components are averaged by using a mixing rule. It is logical to conclude that, if the EOS is imperfect for pure substances, it will be even less perfect for mixtures.

In long-standing engineering tradition, adjustment factors are available to include molecular interaction effects. These adjustment factors increase when there is a large contrast in molecular weight. Similar compounds of similar weight should have a zero binary interaction coefficient. This fudge factor increases between nonhydrocarbons, such as CO_2, H_2S, and N_2.

Other equations of state

There are other equations of state than those discussed previously, for example, the Benedict-Webb-Rubin equation. These methods are more calculation intensive. Normally, it is assumed that a reservoir is in thermodynamic and chemical equilibrium, i.e., that composition is uniform throughout the reservoir. In practice, this is not necessarily correct.

Equations of state package

An EOS general phase behavior package has been utilized by ARE for compositional studies. It uses either the SRK or Peng-Robinson cubic EOS for two- or three-phase reservoir fluid calculations. Up to 11 components may be specified in this package.

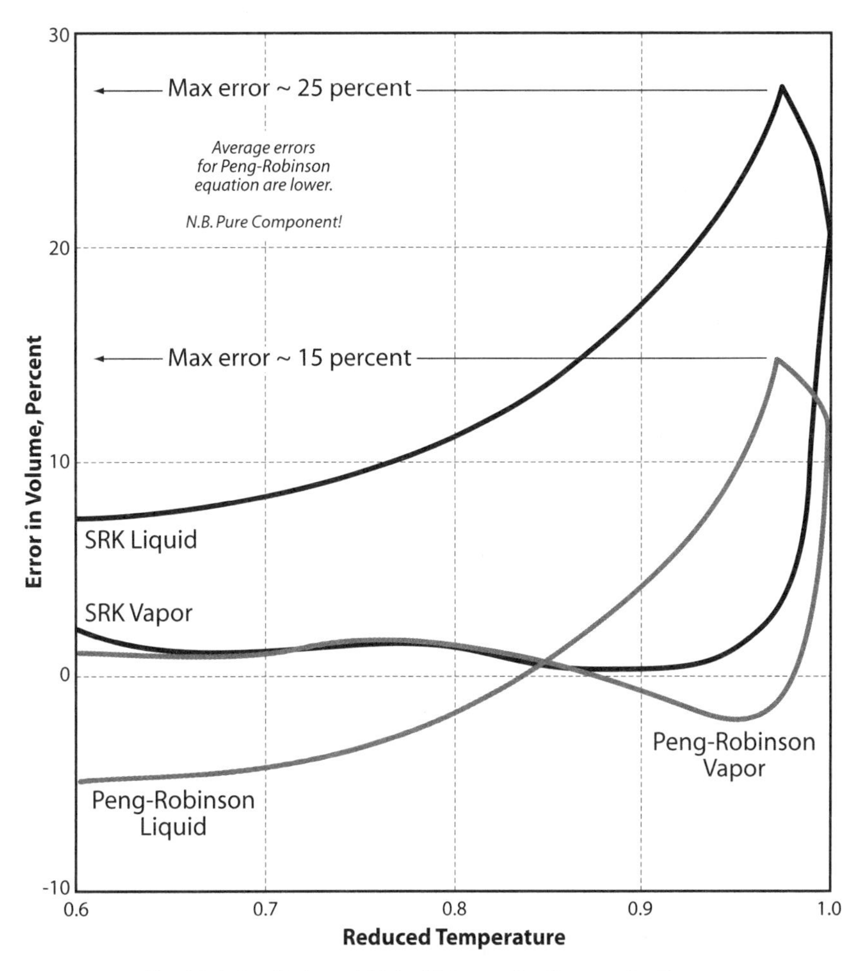

Fig. 15–2 Prediction of Molar Volumes for Saturated n-Butane

Accuracy of the equation of state

The accuracy of phase behavior predictive techniques should be considered in conjunction with the accuracy of the technique. For instance, the GOR ratios on Standing-type calculations have a standard error of around 15%, which is relatively large. Most Z factor calculations are accurate to ±3%. Yet, it is common to see Z factors reported to three decimal places. For example, a reported Z factor of 0.82 is really $0.79 < Z < 0.85$. This has a noticeable effect on reserve calculations. Examination of Figure 15–2 shows the liquid volumes for a pure component can be in error from -5 to +15%.

Binary mixture checks

Some simple checks were performed against phase diagrams in some reservoir engineering texts. These are shown in Figure 15–3.[3] For these simple systems, the equations of state that utilize binary interaction factors appear to work quite well. Unfortunately, the empirical binary interaction factors do not necessarily apply to more complex multicomponent mixtures.

Tuning equations of state

It is also interesting to compare how results might vary between different EOS programs and variations in input. An example was taken from *Properties of Oils and Natural Gases*, by K. S. Pedersen, A. Fredenslund, and P. Thomassen. This example is for a gas condensate reservoir from the North Sea, the composition of which is shown in Table 15–1.[4]

In their textbook, they tuned on liquid drop out. The tuning was done by altering the critical temperatures and pressures for the heavy ends as shown below. The results are shown in Figure 15–4 and tabulated in Table 15–2.

Following this Pederson et al. used the tuned parameters to compare against other physical properties. Figure 15–5 shows that the concentration of methane (C1) is less accurate after tuning. Figure 15–6 shows that the Z factor is accurate and nearly unaffected by the tuning. Figure 15–7 shows that the shape of the phase envelope is altered at lower temperatures. The upward curve is attributed to binary interaction factors and is not thought to represent the real phase behavior at low temperatures. Their conclusion is that tuning physical properties will give good matches to experimental data. However, the changes in physical properties cause subsequent errors when other properties are analyzed. In their final section, they recommend tuning the molecular weights of plus fractions (heavy ends) rather than tuning EOS parameters.

Extrapolating tuning based on limited data

An alternate heavy-end grouping was created by ARE based on the same data as was described previously as shown in Table 15–3.

Note that the "FC" in the components are library components from the software used to perform the tests. Although this could be taken to mean "Fictitious Component," it more likely refers to the data of Firoozabadi and Katz, who published a series of typical component properties and which are used in the EOS software library.

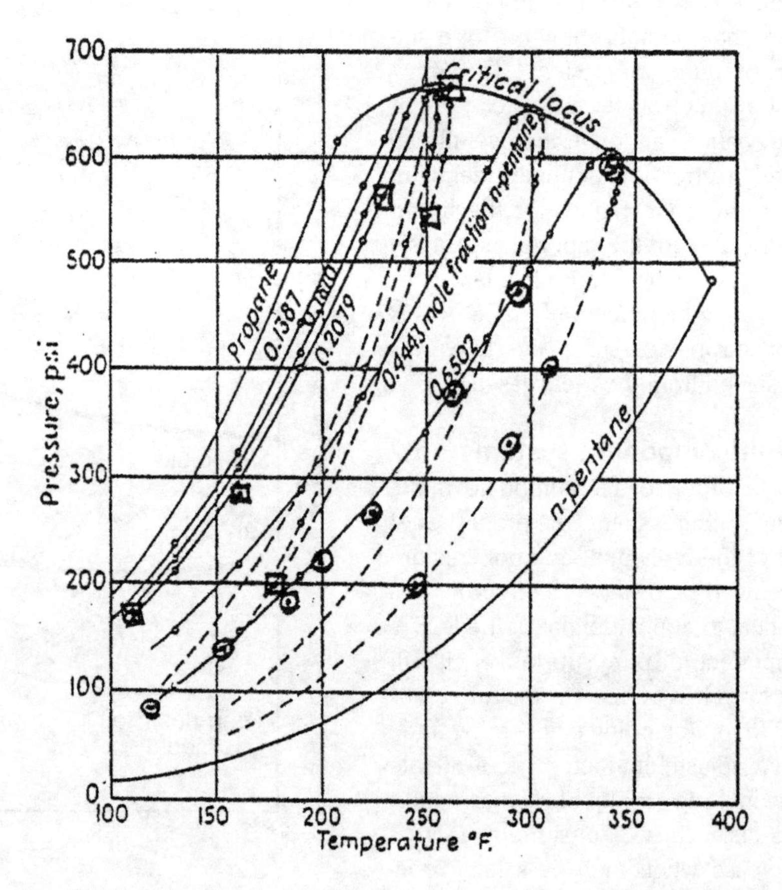

Fig. 15–3 Two-Component Systems—Lab Data for Binary Interaction Coefficients

Table 15–1 Components Used for Example Tuning of Equation of State

Component	Mole Percent	Component	Mole Percent
N_2	0.42	C9	1.43
CO_2	2.98	C10	0.6
C1	66.36	C11	0.48
C2	8.44	C12	0.4
C3	5.12	C13	0.39
iC4	1.04	C14	0.34
nC4	2.35	C15	0.3
iC5	0.84	C16	0.21
nC5	1.12	C17	0.23
C6	1.36	C18	0.16
C7	2.14	C19	0.14
C8	2.2	C20+	0.95

Table 15–2 Parameters Utilized To Achieve Tuned EOS

Component	Mole Percent	T_c	T_c'	P_c	P_c'	ω	ω'
C7-C9	5.77	560.7	571	3,007	2,700	0.433	0.30
C10-C15	2.72	658.3	629	2,100	2,100	0.748	0.60
C17-C60	1.48	802.3	782	1,216	1,450	1.300	1.35

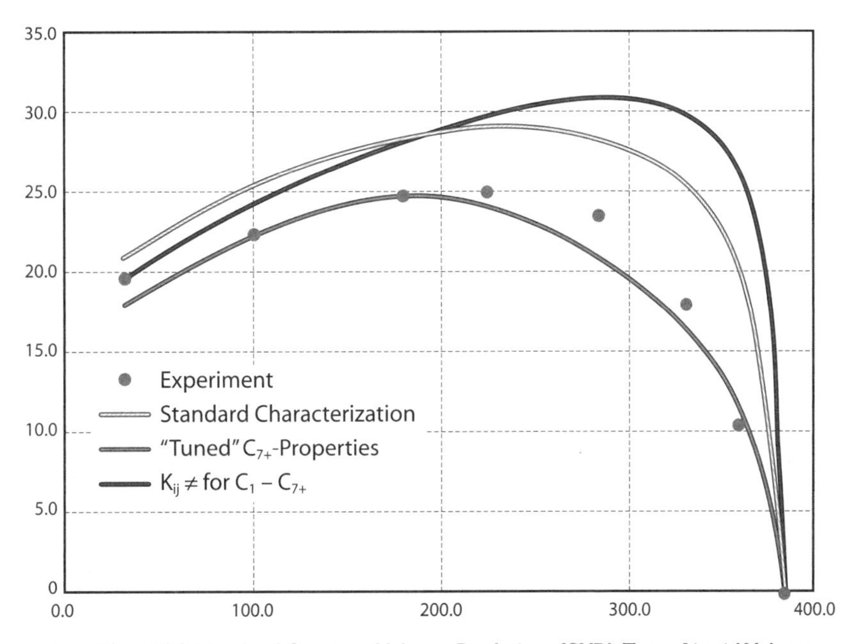

Fig. 15–4 EOS Match of Constant Volume Depletion (CVD) Test—Liquid Volume

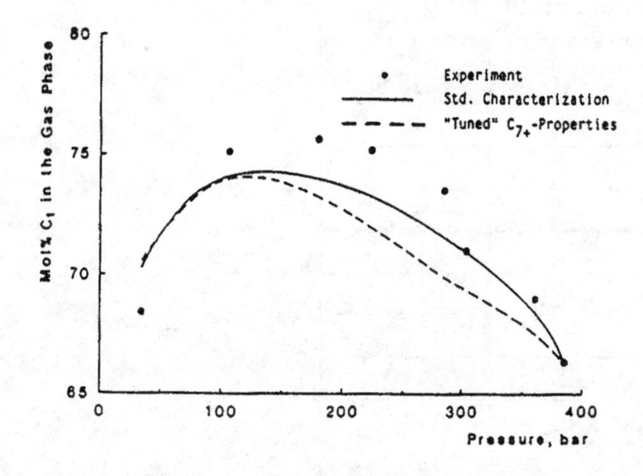

Fig. 15–5 EOS Match of CVD Test—Mole Percent C1

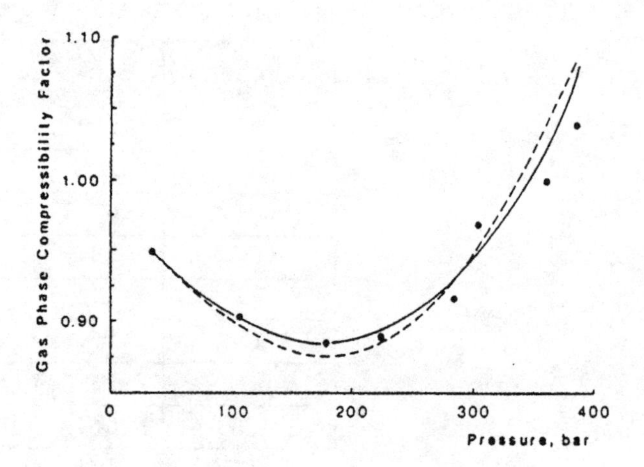

Fig. 15–6 EOS Match of CVD Test—Z Factor

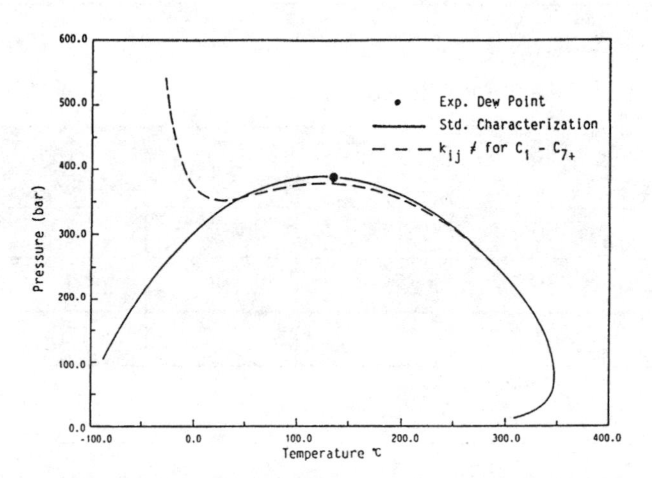

Fig. 15–7 EOS Match of CVD Test—Phase Envelope

Table 15–3 Components Used by ARE to Replicate Example Tuning

Component	Mole Percent	Component	Mole Percent
N_2	0.42	C5-C6 (NC5)	3.32
CO_2	2.98	C7-C9 (FC8)	5.77
C1	66.36	C11-C16 (FC13)	2.72
C2	8.44	C17-C19 (FC17)	0.53
C3	5.12	C20+ (FC20)	30.95
C4	3.39		

It is common to designate groupings of components as pCx-y, where the "p" represents pseudo and the "x" and "y" represent the number of carbon atoms. [5]

The alternate tuning results are shown in Figures 15–8, 15–9, 15–10, and 15–11. Compared with the previous tuning, the matches are clearly lacking. It can be seen that the same trends result. However, the levels are different. The difference represents different default properties for the heavier components and their individual components. It is also common that the library components can be custom designed for various areas of the world where the local components vary from a world wide average. Some EOS software is promoted based on the internal library of components.

Recall that obtaining a match is normally a trial-and-error process. In addition, the reservoir temperature and pressure are close to the critical point: large errors can be expected for this example.

This example also shows that *good EOS tuning depends on having lab data.* In addition, all relevant parameters must be looked at to ensure that the characterization is correct. It may also be further interpreted that regression should not be done on a single parameter. The best regressions will use Z factor, molar concentrations, and liquid volume calculations.

Effects of low saturates/aromatics

Although the physical structure of the molecules varies considerably, there are roughly similar numbers of hydrogen atoms for each carbon atom. This reasoning has been used to justify average properties (e.g. Firoozabadi and Katz). Pederson et al., in their text, concentrate on tuning the heavy-end components based on the percentage of paraffins, olefins, naphthenes, and aromatics (PNA). Differences in properties are shown in Figures 15–12 and 15–13.

Differences in the properties of these different classes are readily observable on a convergence diagram. The Firoozabadi and Katz "critical properties" can be plotted. Note that the GPSA convergence diagram has the critical points of pure alkanes (i.e., paraffins) already plotted. It is also possible to add the aromatics in a trend that includes benzene, styrene, naphthalene and terphenyl. All of these sets of properties show a linear trend on the diagram. Note that this trend is well above the trend of the alkanes and Firoozabadi and Katz properties. It is reasonable to deduce that the Firoozabadi and Katz properties really represent a typical PNA distribution. This is shown in Figure 15–14 in the bottom right-hand corner.

Thermal sulphate reduction

Thermal sulphate reduction (TSR) is one of the geological processes that cause sour gases. The TSR process also has a propensity to retain or create aromatics. This has been noted by others (personal communication) as key in characterizing similar reservoirs. Therefore, tuning sour fluids requires some changes to the hydrocarbon components as well. A high aromatic content should increase the critical pressure required for the heavy-end characterization.

Standard gas chromatographic analyses do identify some of the lighter cyclic and aromatic compounds, but do not provide detail all the way to C30. A special test (PONAU) can be made to about C12. A detailed procedure, which allows for estimating the proportion of PNAs, is outlined in the text of Pederson et al. Part of EOS tuning accounts for variations in PNAs, as well as overcoming the limitations of the technique.

Limitations of gas chromatography

It is easy to forget the complexity of the GC techniques, particularly since the vast majority of lab reports do not discuss this. First, multiple chromatographic columns, are required to cover the range of compositions that are commonly supplied (to C30). There is overlap on the components recovered by the different columns, and integrating these results is an important issue. At higher carbon levels, there is overlap on structural isomers that cannot be distinguished and which are sometimes estimated. The technique is not yet capable of giving complete PNA breakdowns.

Material obtained from column manufacturers also indicates that the standard detector used for hydrocarbons does not accurately detect sulphur compounds. Therefore, many of the reaction products described previously will not be detected in an industry standard suite of tests. Hydrogen sulfide contents are determined by Tutweiller titrations, since an accurate peak (sufficient for quantitative calculation) is not obtained on the standard hydrocarbon GC. The balance of hydrocarbon compositions is adjusted to match this H_2S concentration.

Another alternative is the use of gas chromatography-mass spectrometer (GC-MS) analysis.

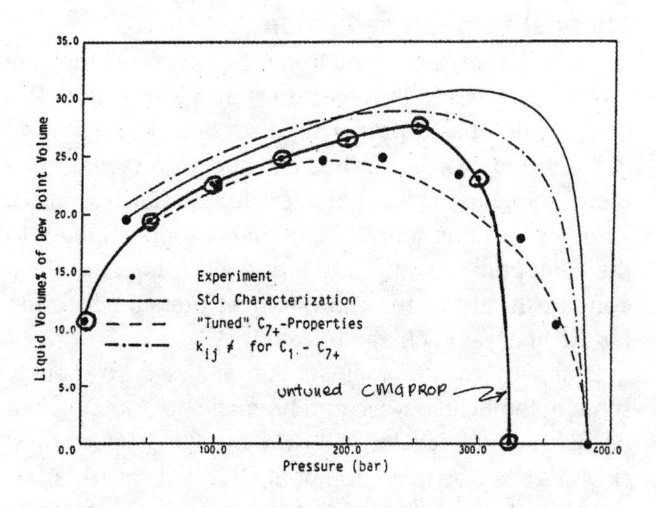

Fig. 15–8 EOS Match CVD Test—
Liquid Volume, with Different Library and EOS

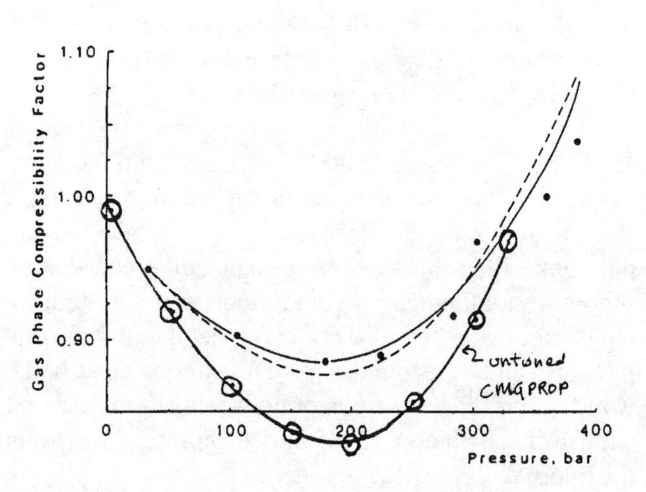

Fig. 15–10 EOS Match CVD Test—
Z Factor, with Different Library and EOS

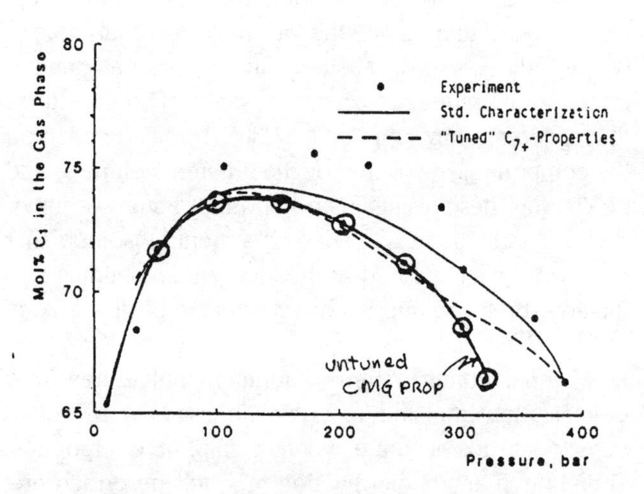

Fig. 15–9 EOS Match CVD Test—
Mole Percent C1, with Different Library and EOS

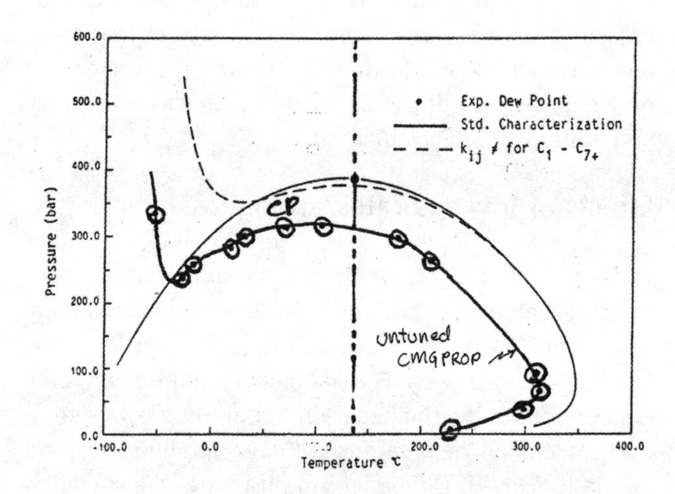

Fig. 15–11 EOS Match CVD Test—
Phase Envelope, with Different Library and EOS

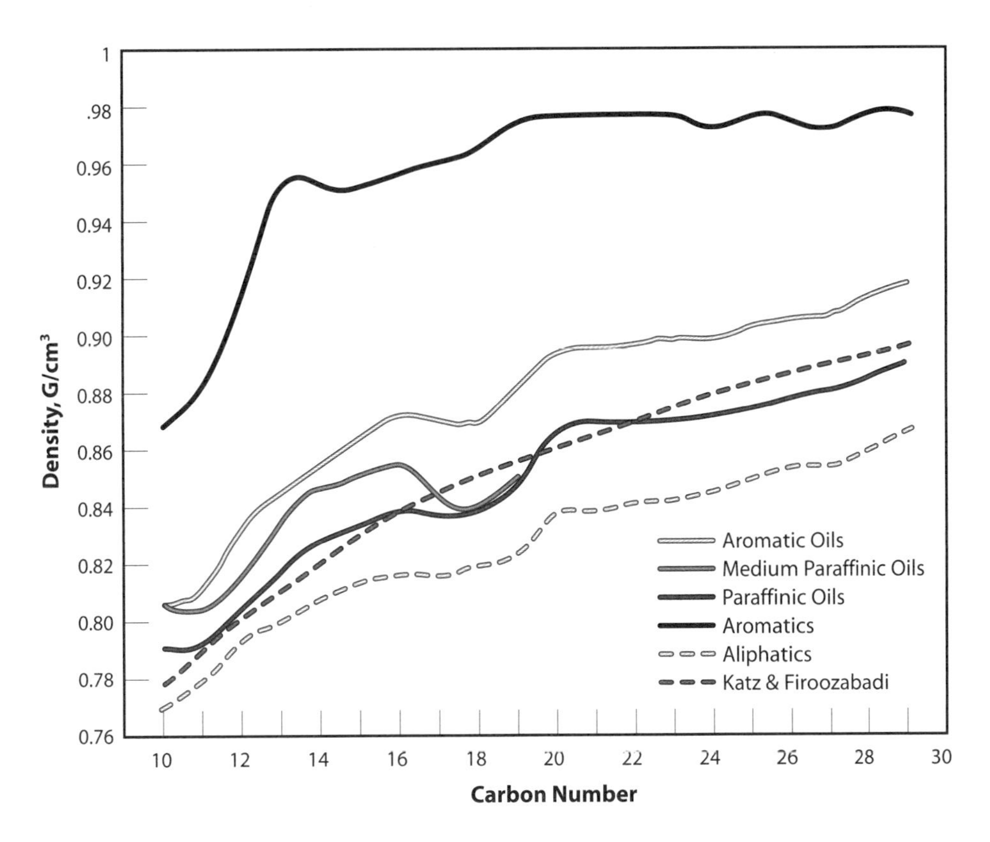

Fig. 15–12 Density Profiles of North Sea Oils—Showing Varying Compositions

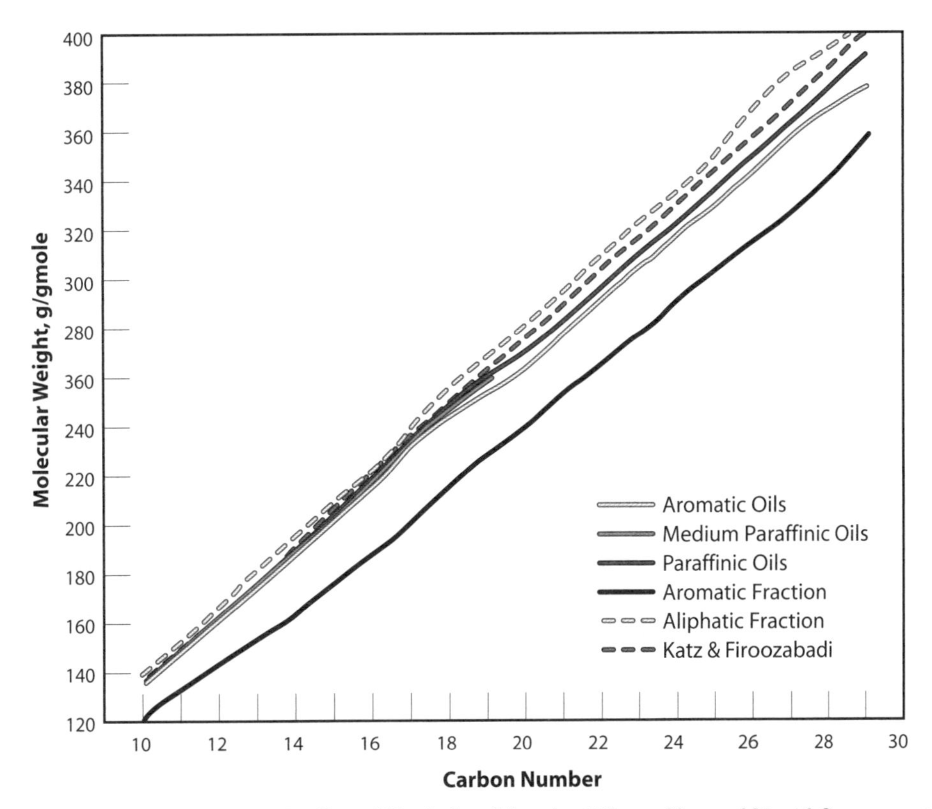

Fig. 15–13 Molecular Weight Profiles of North Sea Oils—for Different Types of Liquid Components

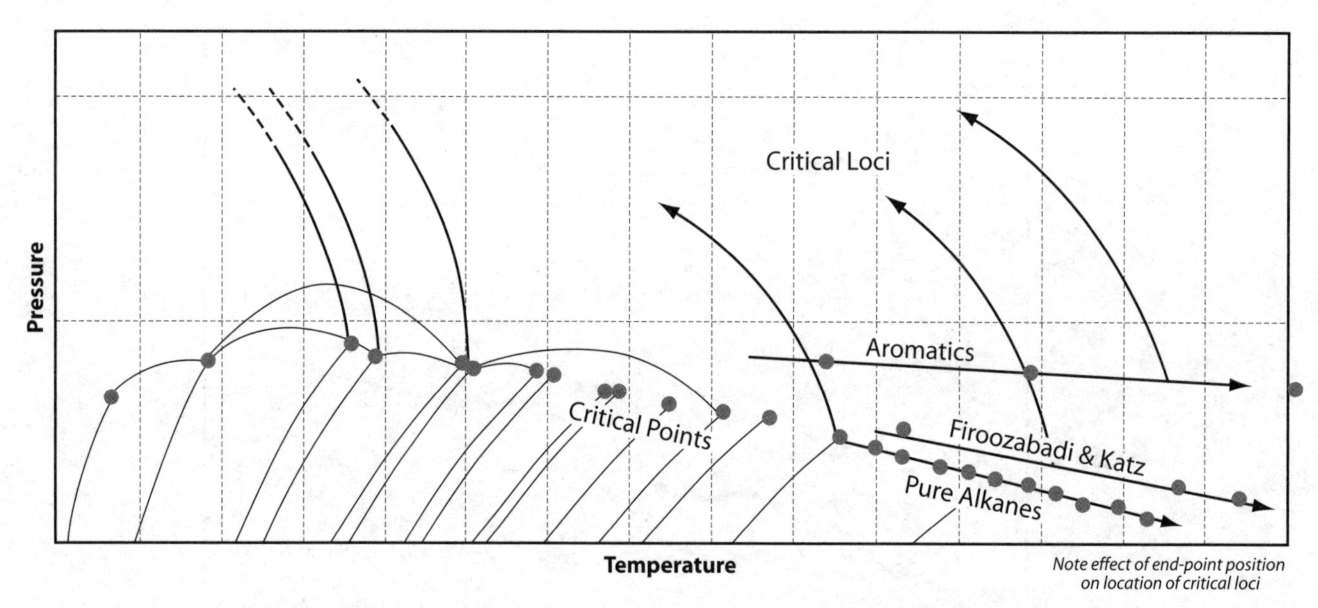

Fig. 15–14 Convergence Diagram—Modified from GPSA Data Book

Influence of aromatic content

The effect of changing the aromatic content was investigated by changing the cyclohexane pseudo component in one of ARE's studies. There is a pronounced sensitivity to this concentration as shown in Figure 15–15. The relative changes from 7.1 mole % are outlined in Table 15–4, as well as the absolute changes:

If the concentration of cyclohexane is increased, the system behaves more like a liquid. In view of its critical temperature and pressure (refer to convergence chart), this is not a surprise. In summary, the system studied previously was sensitive to the aromatic content.

Compositional variations during sampling

For sampling fluids, the surface production facility conditions control how much of the well effluent ends up in which sample container (liquid or gas). Obtaining an accurate recombined composition is the real objective. It is common to evaluate GORs to see if well production has stabilized. However, the variation in produced GOR only tells part of the story. GORs are an accurate indication of well effluent composition only if surface facility conditions remain constant (rarely the case).

In Figure 15–16, the variation in well effluent H_2S (recombined liquid and gas) is shown for a highly sour dew point reservoir for which a four point AOF test had been run. The concentration varies in the range of 20.9% +/- 0.2% (plus or minus 1% on a relative basis). The results from the GC analysis or Tutweiller titrations are probably not sufficiently precise to spot these differences. A similar analysis was also performed for an oil well. Since the results parallel the retrograde gas condensate system, a separate discussion has not been included. The variations in the FC11 pseudo component are plus or minus 0.15% on a molar composition 2.2% (a relative change of 6.8%), which is still a relatively small difference. Note that only the variation with rates (not separator condition) is shown. The variation of the FC17 pseudo components concentration is about 0.1% out of 0.73% molar composition (a relative change of 13.7%), as shown in Figure 15–17, which is probably just noticeable. The concentration of methane in the well effluent ranges from about 56–58%, a relative range of about 3.5%.

For the most part, these differences in composition would be quite difficult to discern on a GC analysis. The following conclusions were derived based on well test modeling:

- There will be large changes in GOR due to changes in surface operating conditions in addition to reservoir effects.
- There will be compositional changes that are significant immediately after an abrupt change in production rate.
- At the time the reservoir fluid samples were taken, no significant changes in well effluent composition should be expected. Note that the sampling process is not instantaneous.

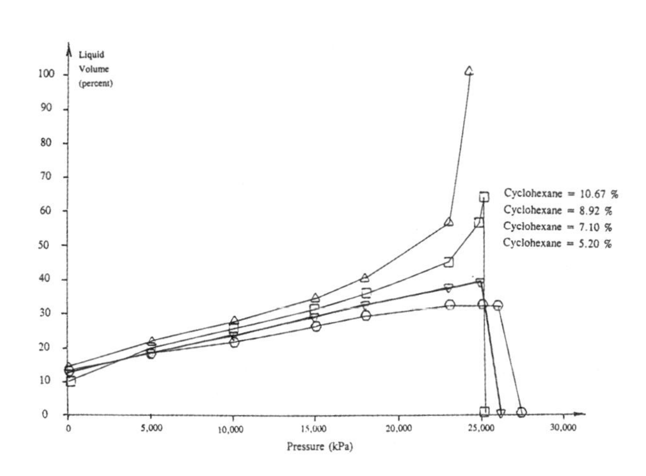

Fig. 15–15 EOS Sensitivity to Aromatic Composition

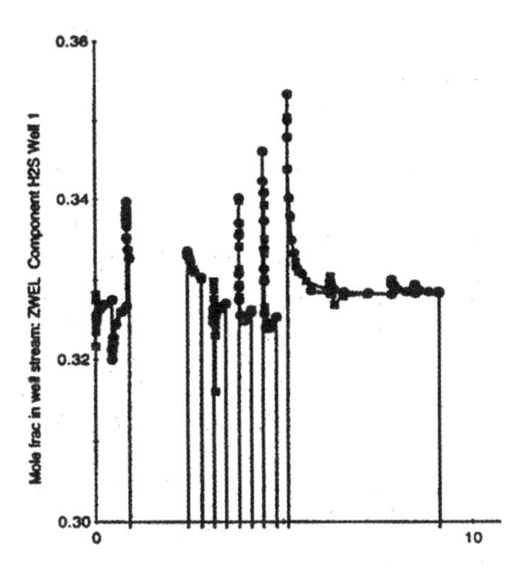

Fig. 15–16 Variation of H_2S during Well Test for Volatile Oil-Gas Condensate

Table 15–4 Table of Input Compositions for EOS Sensitivity to Aromatic Composition

C6H6 Mole Percent	Absolute Change	Relative Change
10.67	3.57%	50%
8.92	1.82%	26%
7.1	0.00%	0%
5.2	-0.73%	-27%

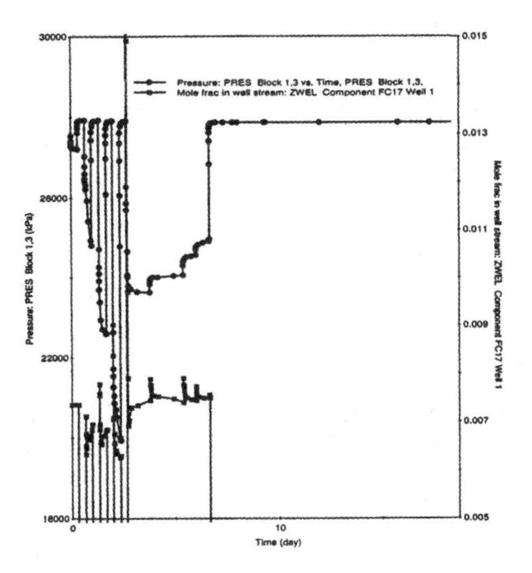

Fig. 15–17 Variation in Heavy-Weight Component during Well Test for Volatile Oil

Overall, compositional modeling indicates that the four point AOF test used should not have caused the changes in PVT results that were observed. More sophisticated analysis, in which adjustments can be made for changing surface conditions, to determine composition in the reservoir, was recommended based on this sensitivity.

Gas condensate reservoirs

Although there is a considerable amount of progress being made in this area, gas condensate reservoir analysis has always been difficult. The author has taken the liberty of concocting a brief history of reservoir engineering trends:

- The classic analysis of gas condensate reservoirs was based on constant volume depletion (CVD) PVT analysis. By using this data and ignoring the pressure gradients in the reservoir, it is possible to calculate the condensate gas ratio and composition of reservoir effluent.
- Practical experience has indicated that this calculation has not proved to have been particularly accurate. The author can think of one field in particular, in the Brazeau River area, where predicted performance for reservoir bookings proved quite unreliable. In the past, these problems have been explained by sampling problems and variation in

PVT properties. The matter would probably have been addressed earlier, except that gas condensate reservoirs are relatively rare.

- It was also found that predicting the productivity of gas condensate reservoirs was difficult. This has particularly been a problem in lower permeability reservoirs, where natural gas liquids condensing in the reservoir obscured the already small and tortuous pore throats around the well, reducing the effective permeability to gas. This can reduce deliverability by up to 40–60%.
- Gas reinjection was quite popular for a number of years. In this process, the reservoir pressure was maintained by injecting dry gas. It was intended that this would prevent natural gas liquids from condensing. Most of the wet gas was intended to be recovered at high pressure, allowing the NGLs to be recovered in a gas plant and then sold. Actual results were, in many cases, far short of predictions. The chief cause of these problems was the early breakthrough of dry gas. Continued injection served only to promote recycling. An example of this was in the Kaybob area of Alberta, Canada.

The development of condensate reservoirs has been fraught with problems.

PVT representation

As outlined earlier, the main technique of predicting gas condensate reservoir PVT behavior was CVD experimental data. There are methods of handling PVT data to enable simulation. For nonretrograde reservoirs, it is possible to use a "reversed" black oil PVT description. In this method, the condensate contained in the gas is a function of pressure. This system provides adequate results for systems that contain relatively low amounts of condensate, say 30 bbl/mmscf or less. This will normally correlate to systems with approximately a 5% or less by volume liquid dropout.

For richer systems, it is possible to predict liquid and gas volumes using equilibrium ratios or *k* values. This has provided adequate results in a number of cases.

Finally, it is possible to use an equation of state (EOS) simulator. This methodology can be used to describe richer systems yet, and is reasonably accurate across a wider range of compositions. However, even this methodology has problems with near-critical reservoir fluids. This is probably the most difficult system to model, even more so than hydrocarbon EOR processes.

Gas condensate relative permeability

The most significant issue in the new understanding of gas condensate systems, in the author's opinion, was experimentally derived gas condensate relative permeability. This work was done by both French and British researchers.

The fundamental mechanisms in a gas condensate reservoir as compared to an oil reservoir are quite different:

- For a typical oil reservoir, which is initially at the bubblepoint and being depleted by solution gas drive, a gas saturation begins to form as the reservoir pressure drops due to withdrawals. As gas bubbles form, the amount of pore space open to flow decreases, making it harder for liquid to move in the reservoir. When the gas bubbles expand to the point where the gas phase becomes continuous between the pores, both oil and gas flow simultaneously. Eventually, the oil phase will become discontinuous and/or the much more mobile gas phase bypasses it. At this point, in essence, the oil stops flowing.
- The fundamental mechanism in the reservoir is completely different for a gas condensate system. Heavy ends will condense on the surface of the rock as the pressure is reduced, as shown in Figure 15–18. Dropout may occur as droplets or as a thin film. This condensate decreases the amount of pore space open to flow, making it harder for gas to move in the reservoir. Condensate may not move in the reservoir until sufficient dropout occurs such that a continuous condensate phase becomes present. This is shown on a gas condensate relative permeability relationship, as shown in Figure 15–19.

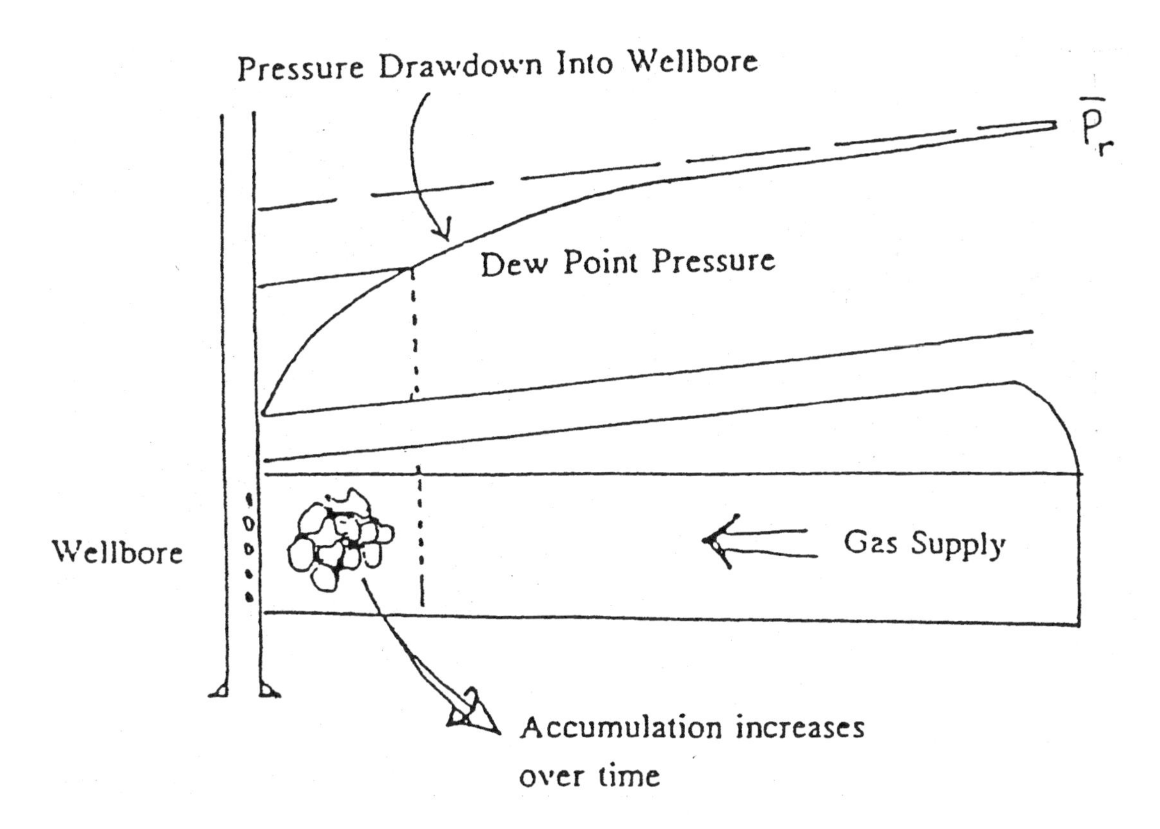

Fig. 15–18 Reservoir Pie Diagram for Gas Condensate System

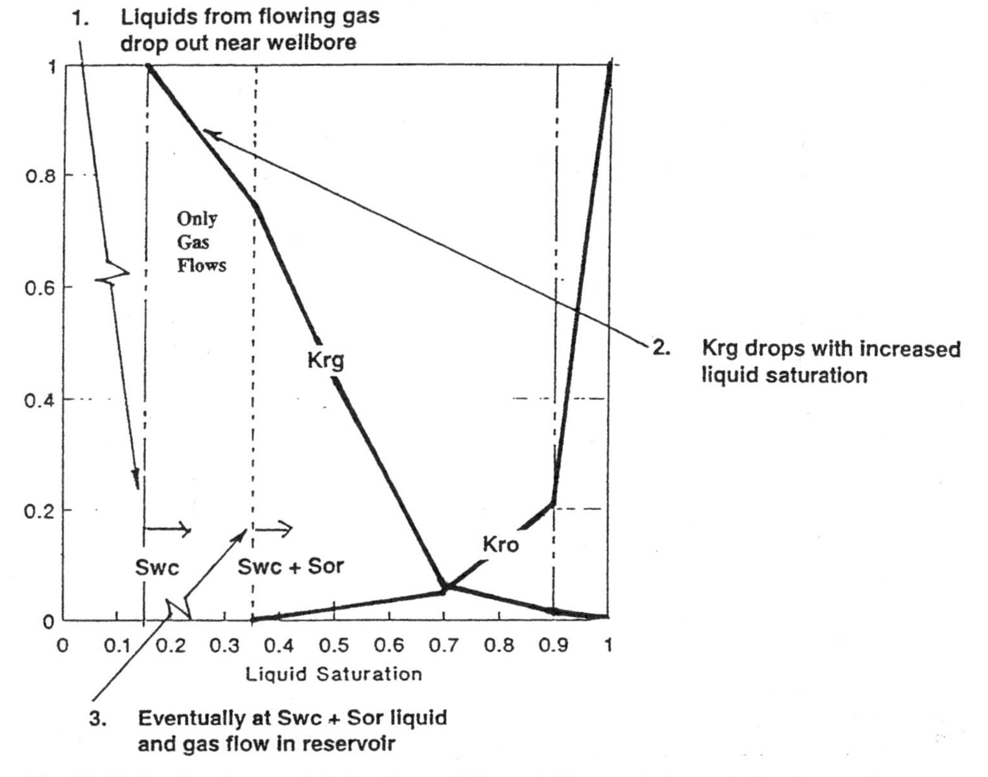

Fig. 15–19 Gas-Condensate Relative Permeability and Changes in k_{rg} with Condensate Dropout

Near-wellbore condensate dropout

By far the most important effect of the relative permeability effects is the change in well productivity. This is shown for two well geometries in Figure 15–20. The example shown is for a reservoir fluid that has a very low CVD liquid dropout. Therefore, impairment occurs even for very lean retrograde condensate systems. The example shown is for a homogeneous reservoir. Note that the condensate impairment is quite severe for the radial flow case. The hydraulically fractured well is not significantly affected.

There are two caveats that the author should point out. First, this was originally thought to only happen with low-permeability formations. Second, that "conventional wisdom" indicates that if the fluid is less than 0.5 percent CVD dropout or less, then the effects are not noticeable. Based on the author's experience and the comments in the recent literature, neither of these two "rules" are probably directly true; in fact, well productivity in gas condensate systems involves subtle details that can have large effects. This will be outlined in the following.

The development of a condensate ring occurs over time, as shown in Figure 15–21. This gives a material balance/buildup aspect to well performance. It also explains why low levels of CVD dropout can still result in significant productivity impairment.[6]

One possible method of relieving this impairment was suggested historically, which was to shut in the well and allow the condensate to revaporize. Detailed study, such as that shown in Figure 15–22, shows that while revaporization does occur, it is not sufficient to restore productivity. This suggests that phase interactions are important.

Phase interactions

Novosad, of Shell Canada, has done detailed evaluation of the changes in composition in the condensate bank that surrounds a well.[7] Her work addressed moderate- to high-permeability (50–250 mD) systems in addition to low-permeability systems. Figure 15–23 shows reservoir liquid saturations at various distances from the wellbore, expressed as a function of reservoir depletion.

Figure 15–24 shows the overall fluid characterization in a grid block (i.e., grid-block global composition) a short distance from the wellbore (5 to 9 m or 15 to 30 ft.). This very nicely demonstrates that the system changes from a condensing to vaporizing system. Note that the gas phase is moving all the time so there is, in fact, no static "overall average fluid" in a grid block.

Note that Novosad has used an unusual terminology in her paper. Although the terms she used were carefully defined, they mirror exiting terminology that may cause some confusion. Specifically, she refers to some systems as near critical. In fact, the path shown on this figure does not traverse pressures, and the location of the path is constrained by the flowing gas supply and the liquids that have dropped out and which are relatively immobile. Note that, if the axes are reversed, the diagram will look somewhat similar to CVD test results.

The previously described results demonstrate that phase effects are continually occurring and that the condensate ring is very dynamic in composition. The only part of the reservoir that actually follows a true CVD path is the outer ring of the reservoir.

Traditional gas condensate calculations

Historically, the CVD test was viewed as an accurate method of predicting well effluent composition. Figure 15–25 shows the classic prediction of behavior and the change in GOR. In view of this, this methodology is clearly flawed. Figure 15–26 shows a reservoir simulation prediction of GOR, which does not show the GORs decreasing with time after peaking. Such production performance is consistent with the author's previous field experience. This is typically a very important issue economically. The revenue from a gas condensate pool is normally heavily dependent on the liquids that are recovered.

Rate sensitivity

Earlier in this chapter, variations in composition were discussed for tests that had been supplied to the author. In the following, some active steps to try and control compositional variation are discussed, specifically for gas condensate systems.

The first sensitivity concentrates on production rates. Various rates, expressed as a fraction of initial productive capability, are shown in Figure 15–27. The key to characterizing systems for an EOS is typically the heavy ends; hence this is the criterion that will be discussed first. Referring again to Figure 15–27, the concentration of C7 plus components does vary with rate.

Note also that the system is skewed to understated heavy-end component concentrations. For this reason, most samples will underestimate the dew point pressure. Note that the effect of heavy ends is to widen the phase envelope, which normally (though not necessarily

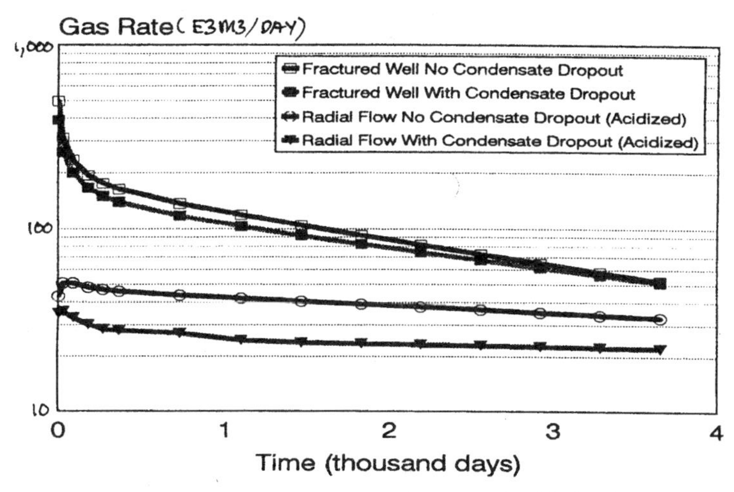

Fig. 15–20 Production Performance Comparison—Hydraulically Fractured Gas Well vs. Nonstimulated Well (Radial Flow)—with and without Reservoir Retrograde Condensation

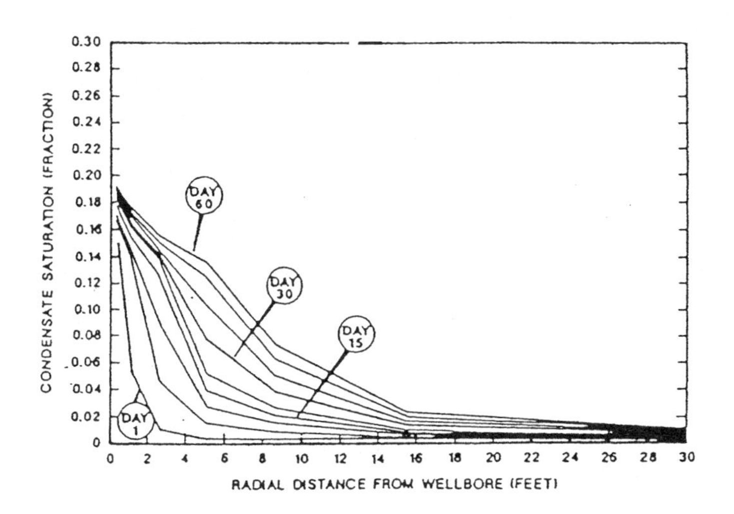

Fig. 15–21 Condensate Saturation with Distance from Well for Gas Condensate Profile with Radial Flow

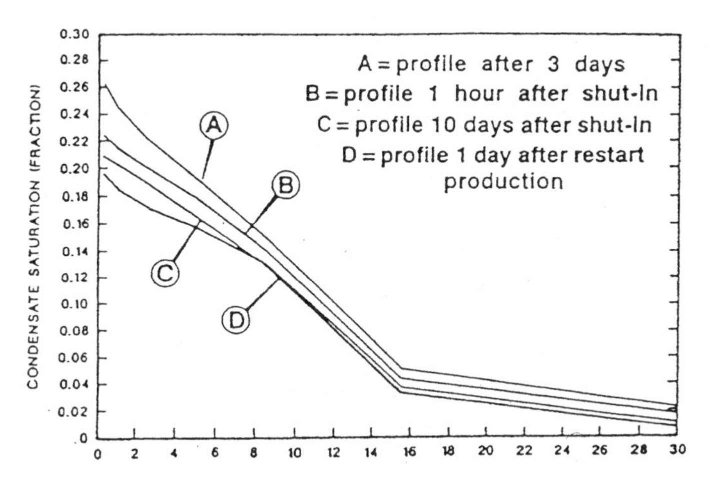

Fig. 15–22 Effect of Well Shut-In Time on Condensate Saturation Surrounding Well with Radial Flow

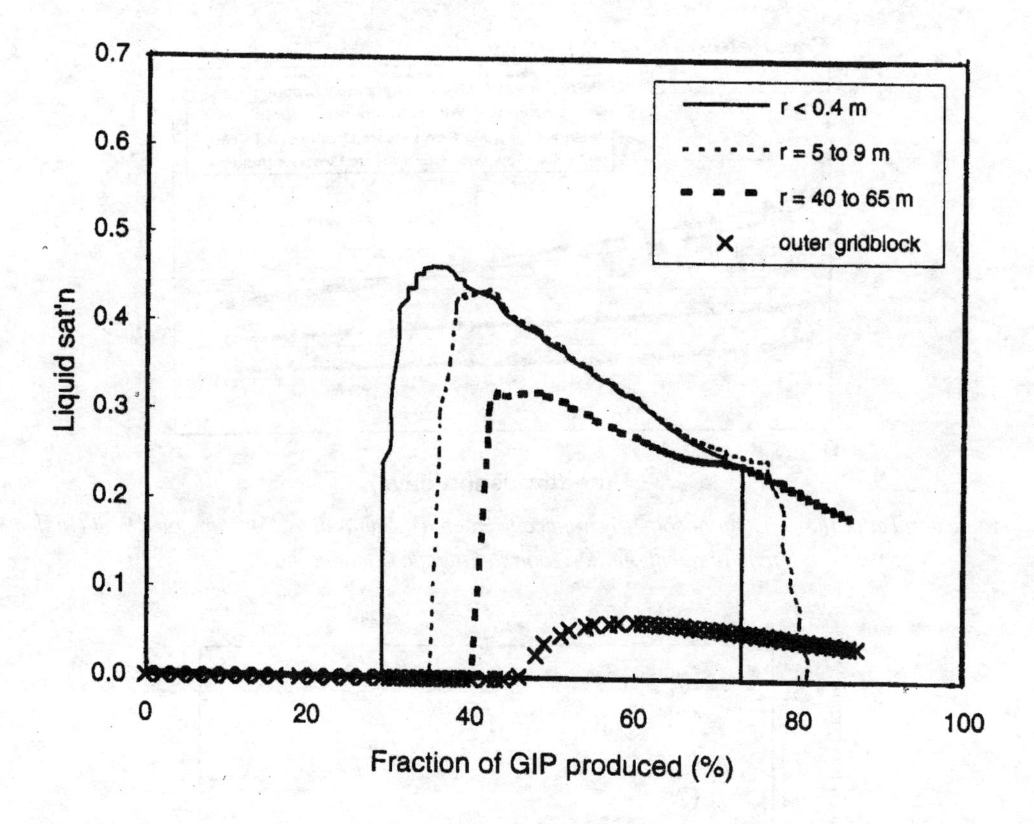

Fig. 15–23 Changes in Condensate (Liquid) Saturation with Depletion

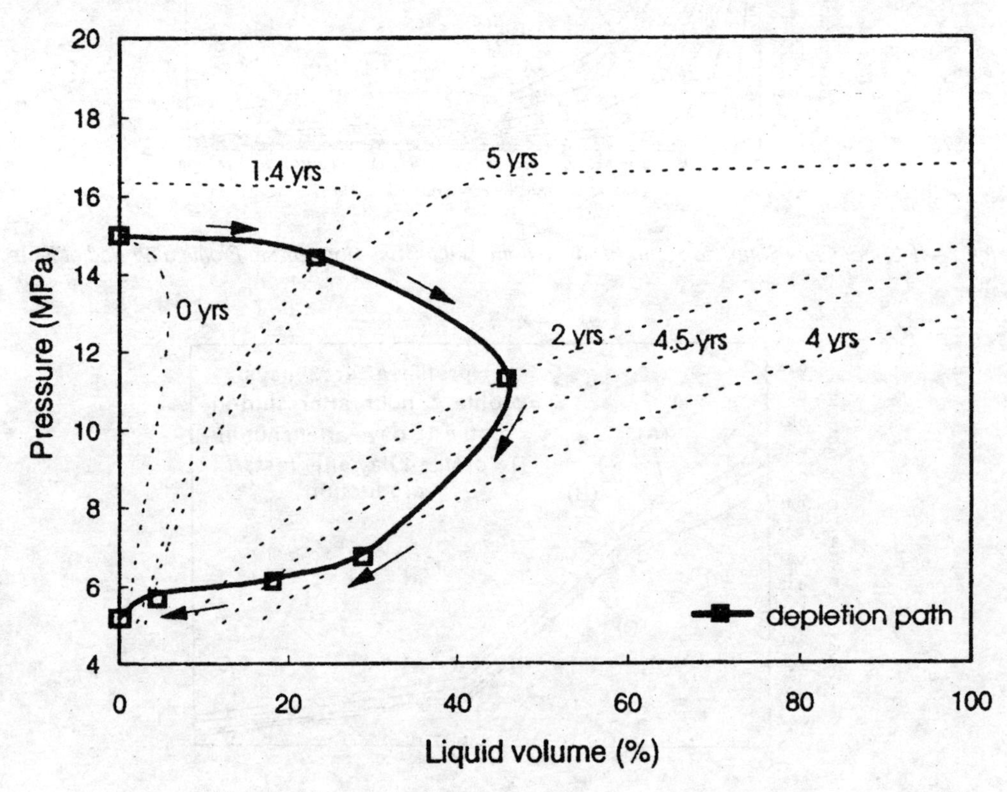

Fig. 15–24 Evolution of Phase Behavior with Depletion in Gas Condensate Reservoir

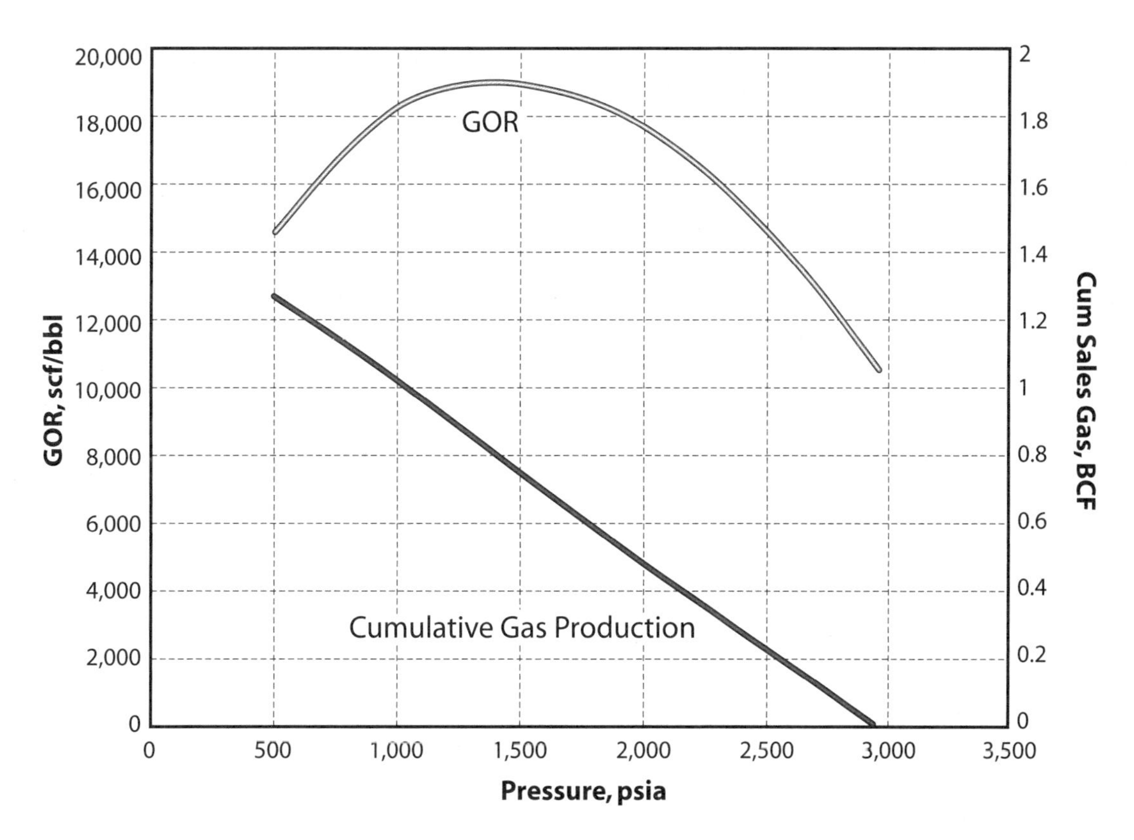

Fig. 15–25 Classic Gas Condensate Reservoir Behavior—Assuming Limited Pressure Depression Surrounding Producing Well

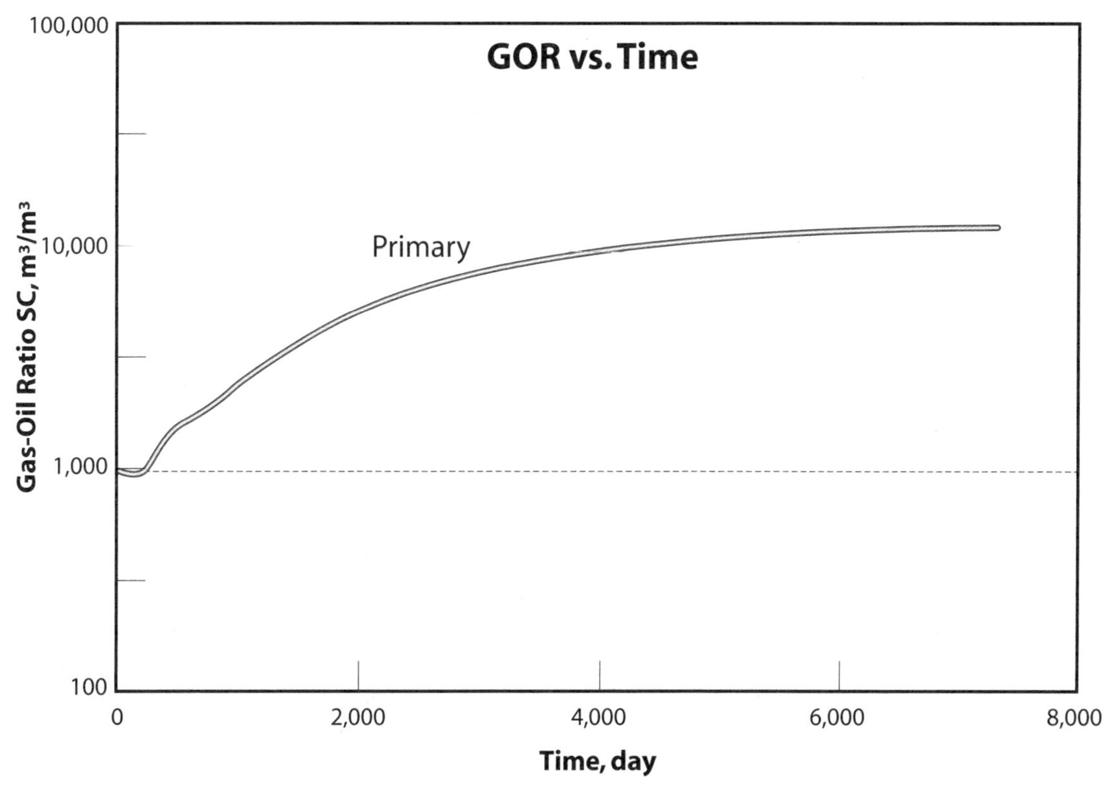

Fig. 15–26 Simulator Prediction of Gas Condensate Reservoir Performance—Pressure Depression Surrounding Well Included

always) increases the height of the phase envelope. This effect can be predicted by using a convergence chart, an example of which was shown in Figure 15–14.

The GOR is also affected as shown in Figure 15–27. This does not necessarily lead to a bad sample *per se.*

Referring to a second set of three graphs in Figure 15–28, the effects of a single-step change in rate are evaluated. There is a time lag in the system to adjust for significant changes in rate. Next, in Figure 15–29, a "noisy" production profile results in "noisy" heavy-end concentrations in the well effluent stream and on the GOR.

Reservoir heterogeneity

Continuing with a discussion of the work described previously, layers can concentrate condensate buildup as shown in Figure 15–30. McCain and Alexander concluded that sampling should be done sooner and with lower drawdowns.[6] As will be discussed in the next two chapters, hydraulic fractures, horizontal well geometry, and high-permeability zones of limited areal extent can have quite dramatic effects on well effluent composition.

General comments on sampling

In general, the issues that have been discussed previously have a parallel in black oil systems and are known as well conditioning. This was discussed in chapter 6. The issues have been explicitly outlined in API standard RP-45. Black oils are not that sensitive compositionally (ergo the simplification), and the problems that are well-known for these systems are exacerbated for gas condensate situations.

First, it is not possible to obtain the exact original reservoir composition using a stabilized well rate. Early sampling is important. To some degree, back calculation using a compositional simulator can be done. Dew points are systematically underestimated.

Although an EOS is not predictive of performance, the author recommends that testing and sampling for gas condensate systems should be modeled with a compositional simulator before the test. Plan for some sample redundancy. PVT lab work and sample containers are trivial in expense within the overall cost of well testing.

Recent gas condensate study

A study that is more recent has been done specifically for gas condensate reservoirs. The results can be summarized briefly:

- Where low interfacial tensions exist, gravity effects become important.
- At low levels of interfacial tension, the relative permeability relationship does not follow the classic gas-oil curves. Lower interfacial tensions occur at temperatures and pressures near the reservoir fluid critical points. Under these circumstances, the liquid and vapor phases become near miscible. The relative permeability curves become intermediate between miscible (straightline) and conventional gas-oil relative permeability curves.
- At more normal reservoir conditions, in which there is a moderate to high interfacial tension, the relative permeability relationship is similar to a normal gas-oil system.
- The latest data indicates rate sensitivity at low levels of interfacial tension (under 0.3 dynes/cm). Whether this is related to turbulent flow in the pores is not clear.

There are relatively few experimental sets of data currently available (about two sets). It is common practice to use gas-oil data or to use Corey correlations.

Low interfacial tension (IFT) effects

There are three effects that occur at low interfacial tensions. Note that IFT drops as the pressure approaches the critical point. This should resemble near-miscible conditions. Further, as IFT drops, the surface spreading tendencies between the rock and the liquid condensate will change. This is known as critical point wetting. Finally, at low interfacial tensions and at low viscosities, there were suggestions that the condensate would be mobile along the sides of the rock grains and that relative permeability would be rate sensitive.

Richardson and Dawe, of Imperial College, developed a technique for altering the relative permeability functions from fully miscible (straightline sticks) to the more regular gas-oil relative permeability curves. This permits modeling when these conditions change during depletion.

Shell did some tests with alcohol-water systems that featured such low IFTs in the lab. They showed some rate sensitivity. Following this, Henderson et al. showed steady state test results that were rate sensitive. An example of this is shown in Figure 15–31.[8] It has been proposed that the pores scale ratio of IFT to viscous forces will govern how much rate dependency is likely to exist.

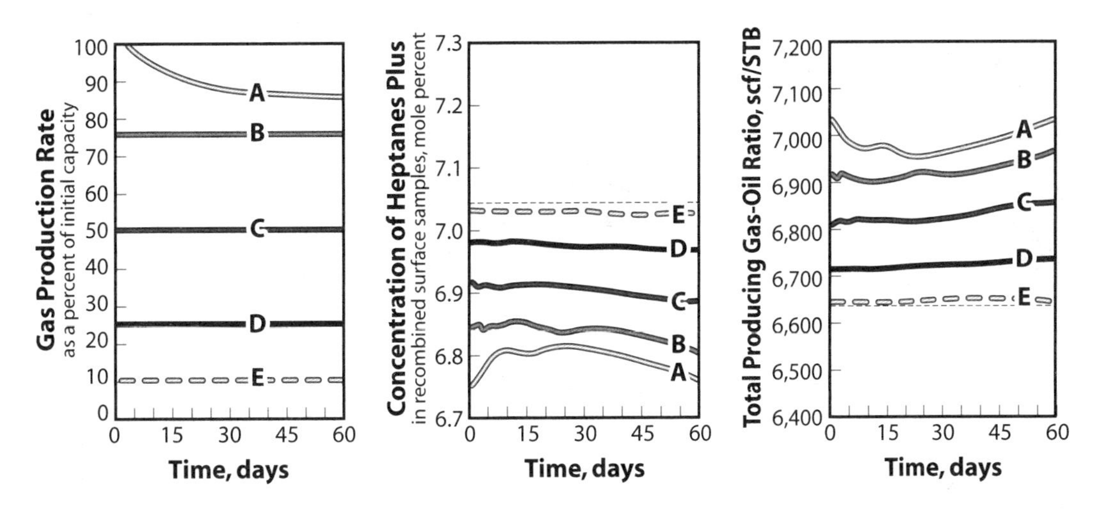

Fig. 15–27 Effect on Well Effluent Composition Due to Varying Rates for Gas Condensate Wells

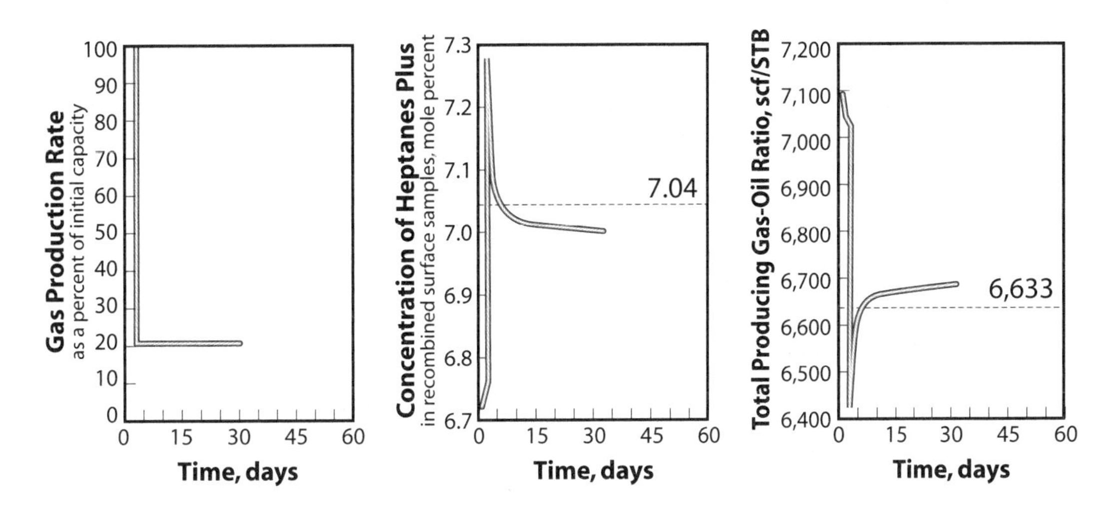

Fig. 15–28 Effect of Step Rate Change on Well Effluent Composition, Gas Condensate Well

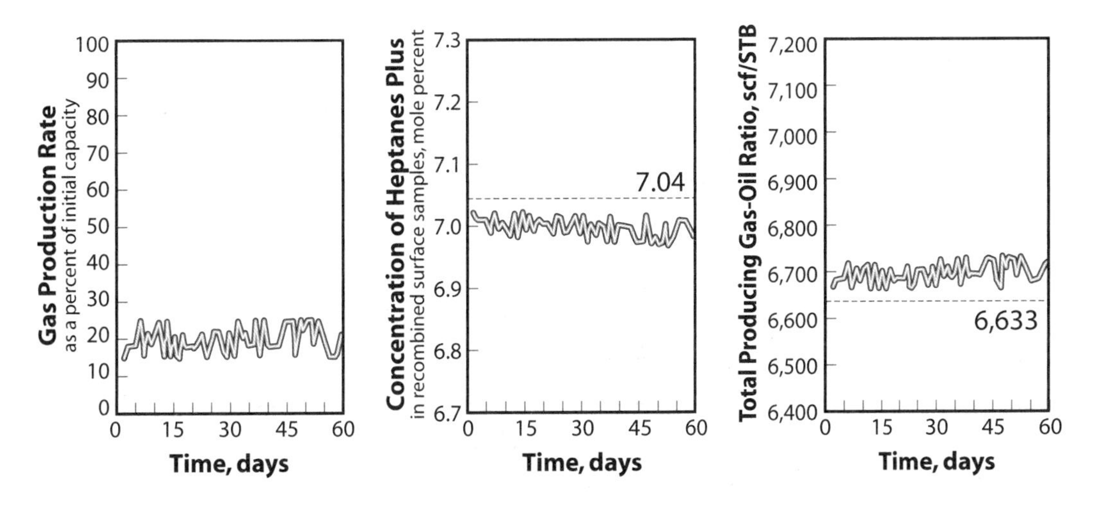

Fig. 15–29 Effect of Randomly Varied Rates on Well Effluent Composition, Gas Condensate Well

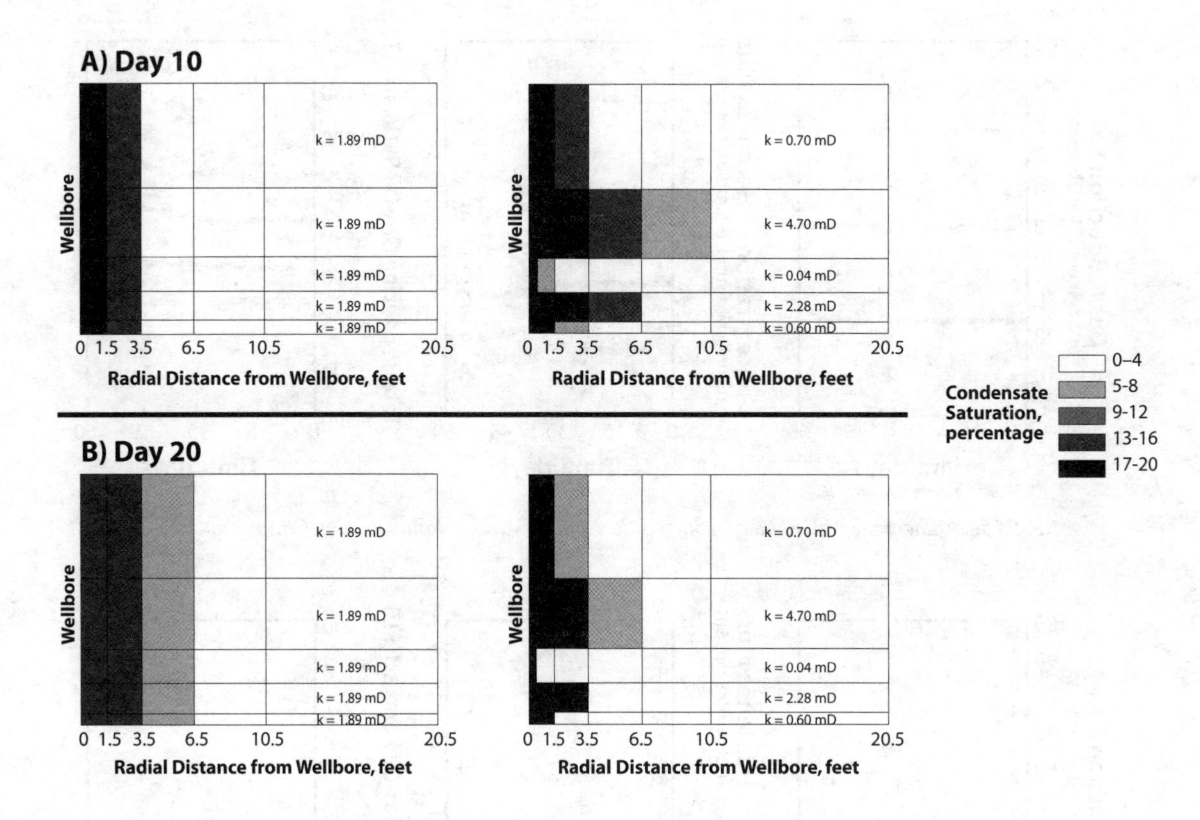

Fig. 15–30 Effects of Layering on Condensate Saturation Surrounding Producing Well—Radial Flow

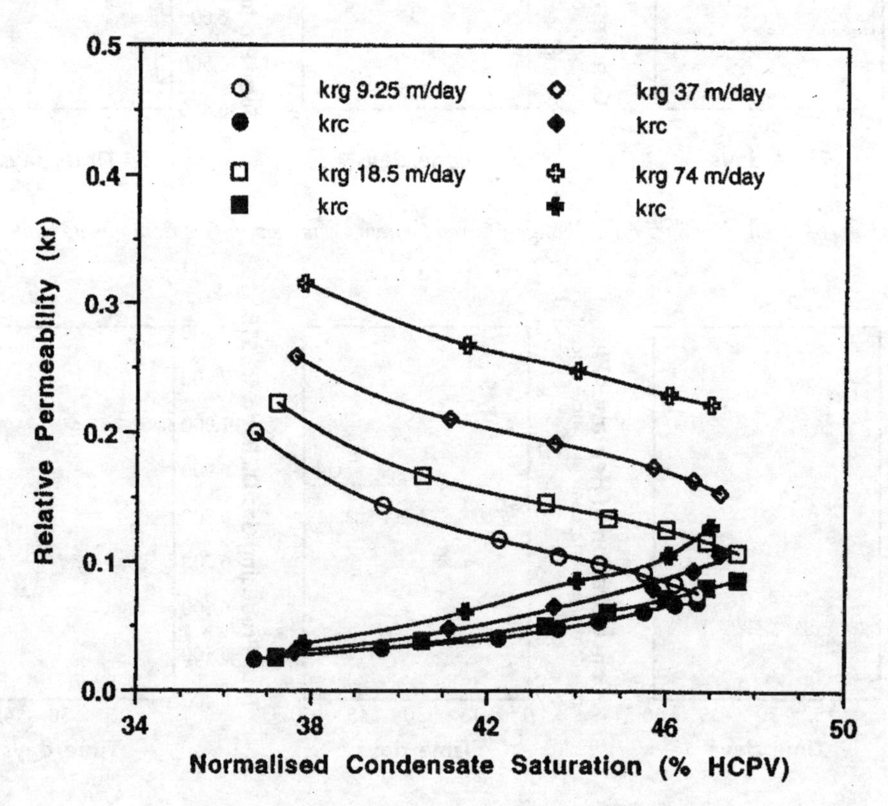

Fig. 15–31 Effect of Varying Rate on Steady State Relative Permeability Curves

Relative permeability sensitivities

Carlson and Myers, Settari et al., and Whitson indicate that the rate performance of individual wells is not strongly affected by the shape of the relative permeability curves.[9, 10, 11] The first two studies make this conclusion based on sensitivities. Whitson contends that the conditions in which lab tests result in rate sensitivity are not representative of true near-wellbore reservoir conditions. He also performed a detailed analysis showing how each part of the reservoir where condensation occurs affects productivity. Novosad, in one of her papers, shows that only a portion of the relative permeability curves are traversed, as shown in Figure 15–32.

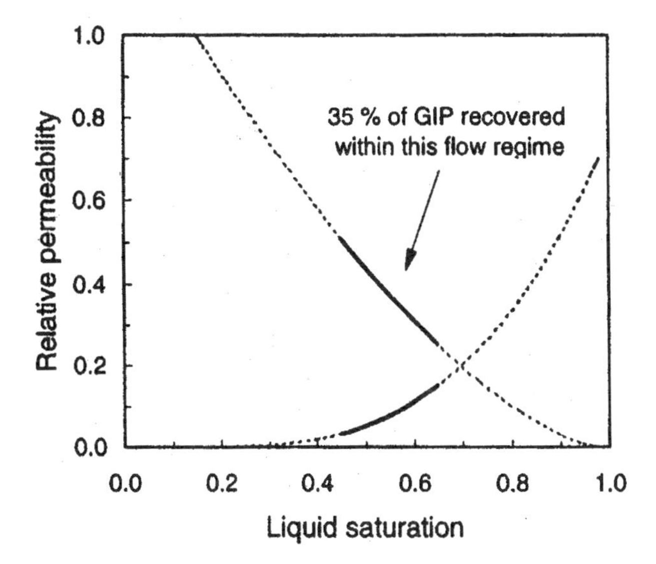

Fig. 15–32 Actual Range of Saturations Traversed in Gas Condensate Well

Turbulent flow

It was envisaged that the reduction in pore throat cross section will make gas condensate systems more prone to turbulent flow conditions. This, in turn, will also affect the amount of condensate that builds up in the pores. Turbulent flow would also be more capable of stripping liquids off the sides of the pores.

There is some debate as to whether turbulent flow, i.e., not laminar, really occurs in pore spaces. Some people are of the opinion that this is an inertial effect rather than a change in flow regime within the pore (Scheidegger; Happel and Brenner).[12, 13] This flow reduction may have a "blasting" effect. Darcy's Law is normally thought to be the effect of fluid molecules slipping past each other or a viscous effect. Others are more of the opinion that true turbulence does occur, depending on the geometry. Darcy's Law does break down at higher rates, even if it is not known exactly why.

It is possible to include non-Darcy flow in a simulator. This is still a fairly rare implementation. It has been done in Western Canada for single-phase productivity systems (see Flores).[14] It has also been included in some pressure transient analysis software (Duke Engineering). Eventually it will be included in compositional simulators. Some thermal simulators feature non-Darcy flow for steam.

The issue of condensate flow concurrently with gas will not be resolved until some fundamental physics and detailed lab experiments of flow in porous media are resolved.

Analytical analysis

Blom and Hagoort have done a very elegant analysis of near-critical fluids with non-Darcy flow.[15] The rate dependence of relative permeability was correlated against N_c or capillary number. The solution was generated via a semianalytical procedure and was solved using a variation of Newton's method. They looked at:

- dry gas, assuming Darcy's Law, i.e., "no dropout, no inertia"
- dry gas, assuming the Forcheimer equation, i.e., "no dropout, with inertia"
- wet gas, with gas-oil relative permeability "dropout, k_{rg} independent of N_c"
- wet gas, rate-dependent permeability, and Forcheimer's equation "combined effect"

The results are interesting. Figure 15–33 shows the pressure profiles in the reservoir for these cases. Figure 15–34 shows the condensate saturation profile in the reservoir. Note that the fluid studies had a CVD maximum dropout of close to 40%, which is a very rich condensate system. Figure 15–35 shows the capillary number in the reservoir. Finally, Figure 15–36 shows the bottomhole pressure for all of the cases.

The last graph is most interesting. At 5,000 kPa (50 bars or 725 psia) bottomhole flowing pressure, the well will produce about 3,000 E^3M^3/day (106 MMscfd), ignoring condensate dropout and turbulent flow. The inclusion of turbulent flow has almost no effect on rate without condensate dropout in the reservoir. The presence of condensate impairment will reduce the rate to about 1,800 E^3M^3/day (63.6 MMscfd) or by 56.7%. The inclusion

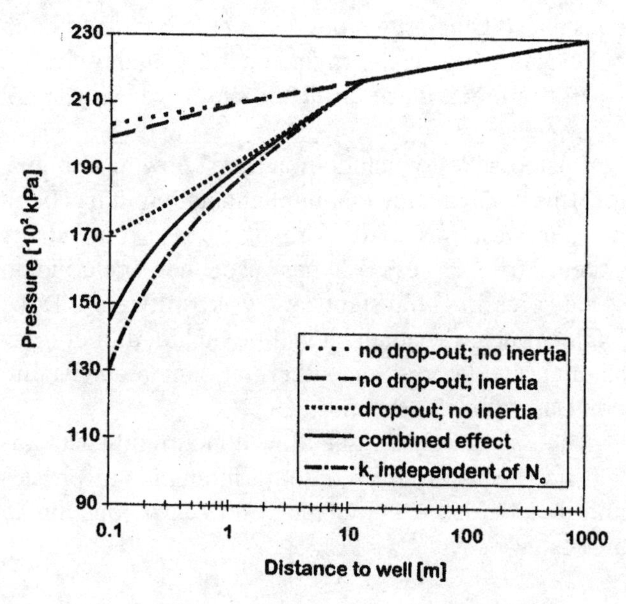

Fig. 15–33 Theoretical Calculation Showing Effects of Inertia and Condensate Dropout Effects

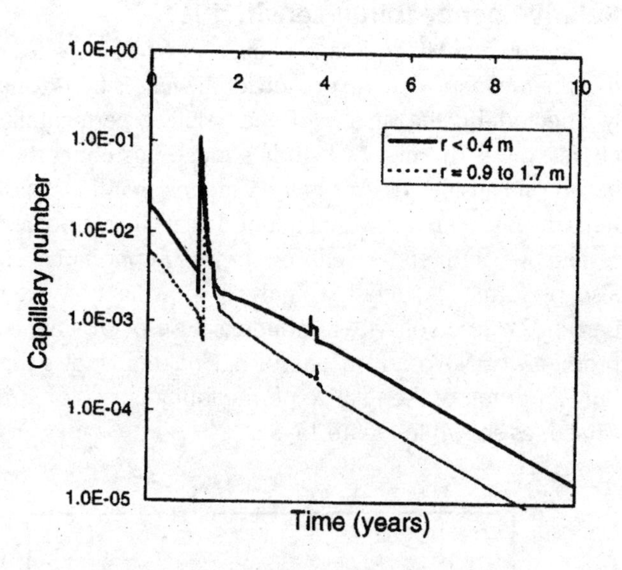

Fig. 15–35 Variation of Capillary Number with Distance from Well

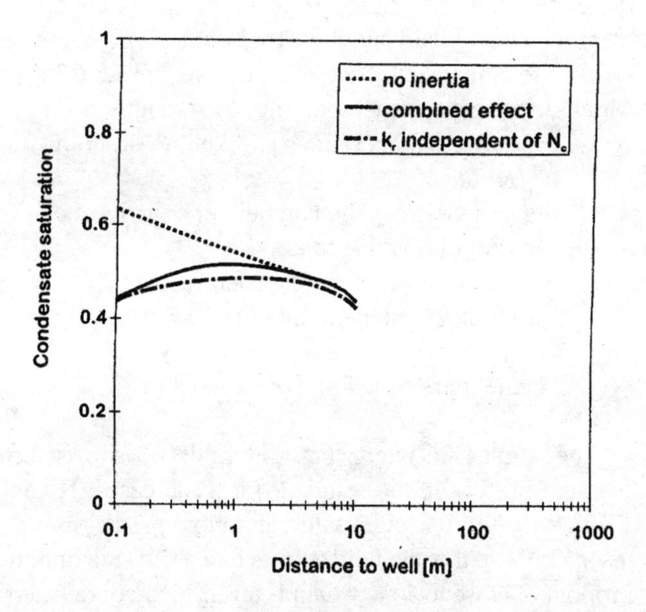

Fig. 15–34 Saturation Profiles in Reservoir from Theoretical Calculation

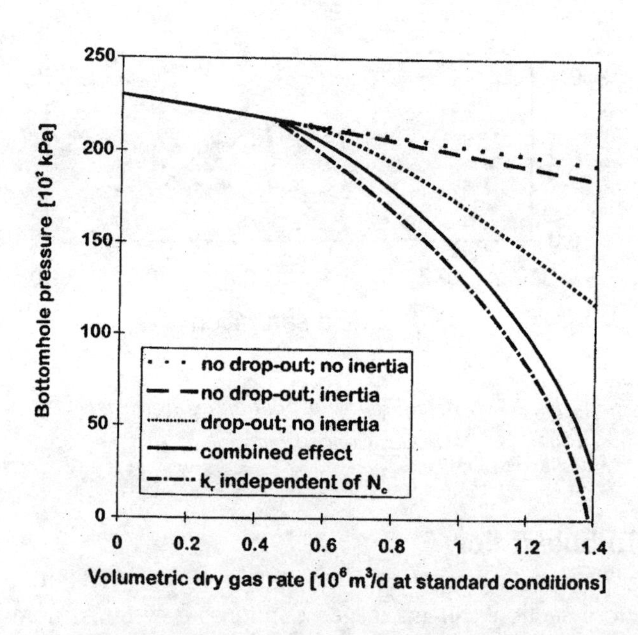

Fig. 15–36 Bottomhole Pressure vs. Dry Gas Rate, with and without Inertia, Dropout

of turbulent flow with condensate impairment present knocks the productivity down to 1,300 E^3M^3/day (45.9 MMscfd), another 27.7%. Rate-dependent relative permeability relieves the impairment, increasing the rate to 1,350 E^3M^3/day (47.7 MMscfd), an uplift of 3.9%.

From the previous discussion it would seem that condensate dropout is the most important factor, followed by turbulent flow. The example involves fairly high rates. The level of CVD dropout indicates a near-critical fluid. In practical terms, the effect of rate-dependent relative permeability is unnoticeable. One's chances of having laboratory, rate-dependent, relative-permeability is very low for most projects. This is not an insurmountable barrier; the previous case indicates that sensitivities might well show this to be a low-order effect.

Condensate viscosity

Liquid viscosities are very important in gas condensate systems. There are newer correlations available for calculating this property, which give significantly lower results. Figure 15–37 is one of the more modern correlations of Pederson et al.[16] Note the sensitivity to the method of estimating density. This is an area of weakness for cubic EOS equations.

Pressure transient analysis

The condensate ring, which has a lower relative permeability to gas, does affect pressure transient analysis. Not surprisingly, the system can be modeled with a concentric ring model, with the inner ring having lower permeability. An example of a type curve is shown in Figure 15–38. This was prepared with a compositional simulator and a well test package.

An analytical procedure was developed by Raghaven and Jones.[17] Although clever, it is actually quite a bit of work to utilize this technique. Relative permeability data is also required. In the author's opinion, the use of a simulator is more straightforward and can be used subsequently for more general sensitivities.

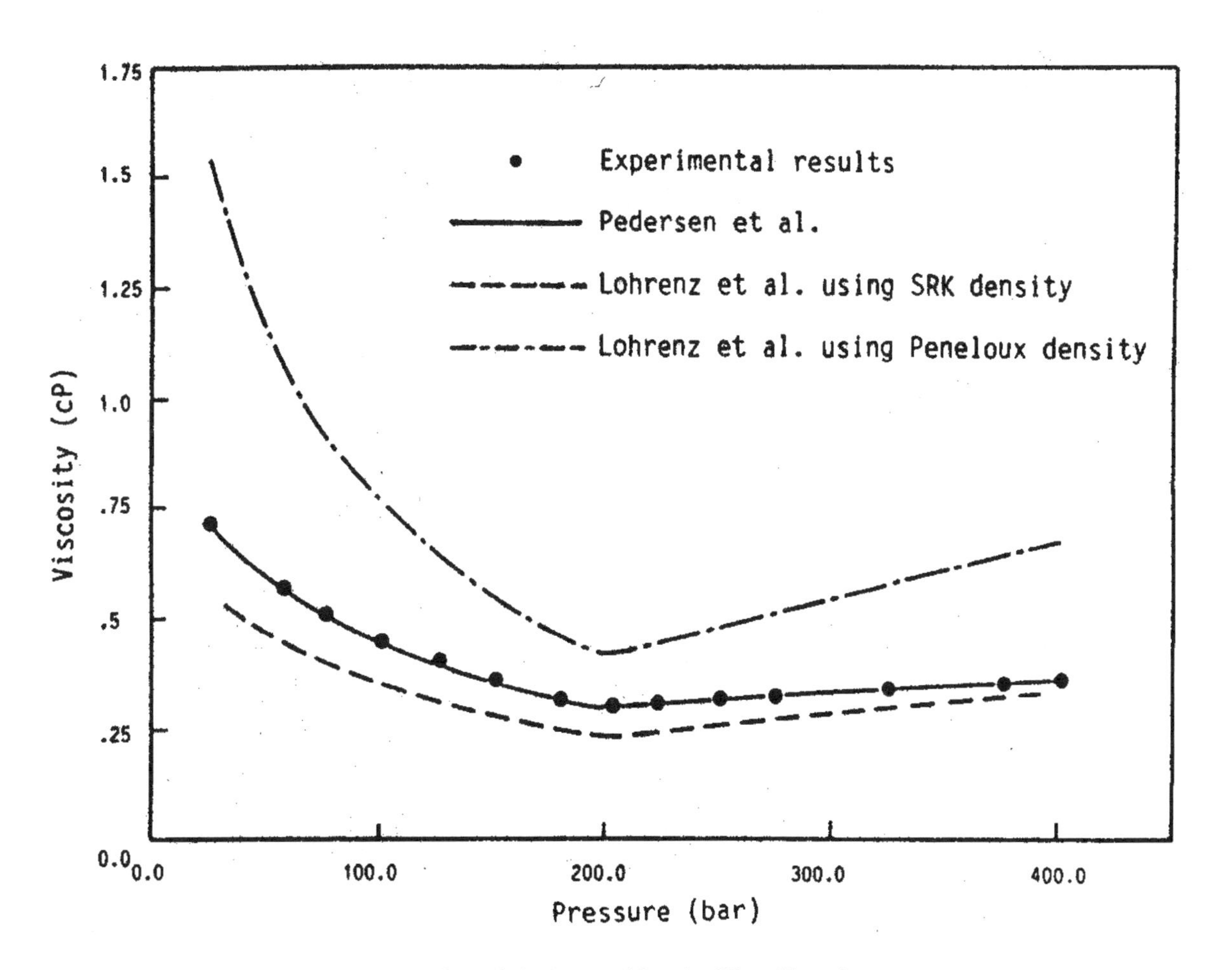

Fig. 15–37 Calculation of Density Effects Viscosity

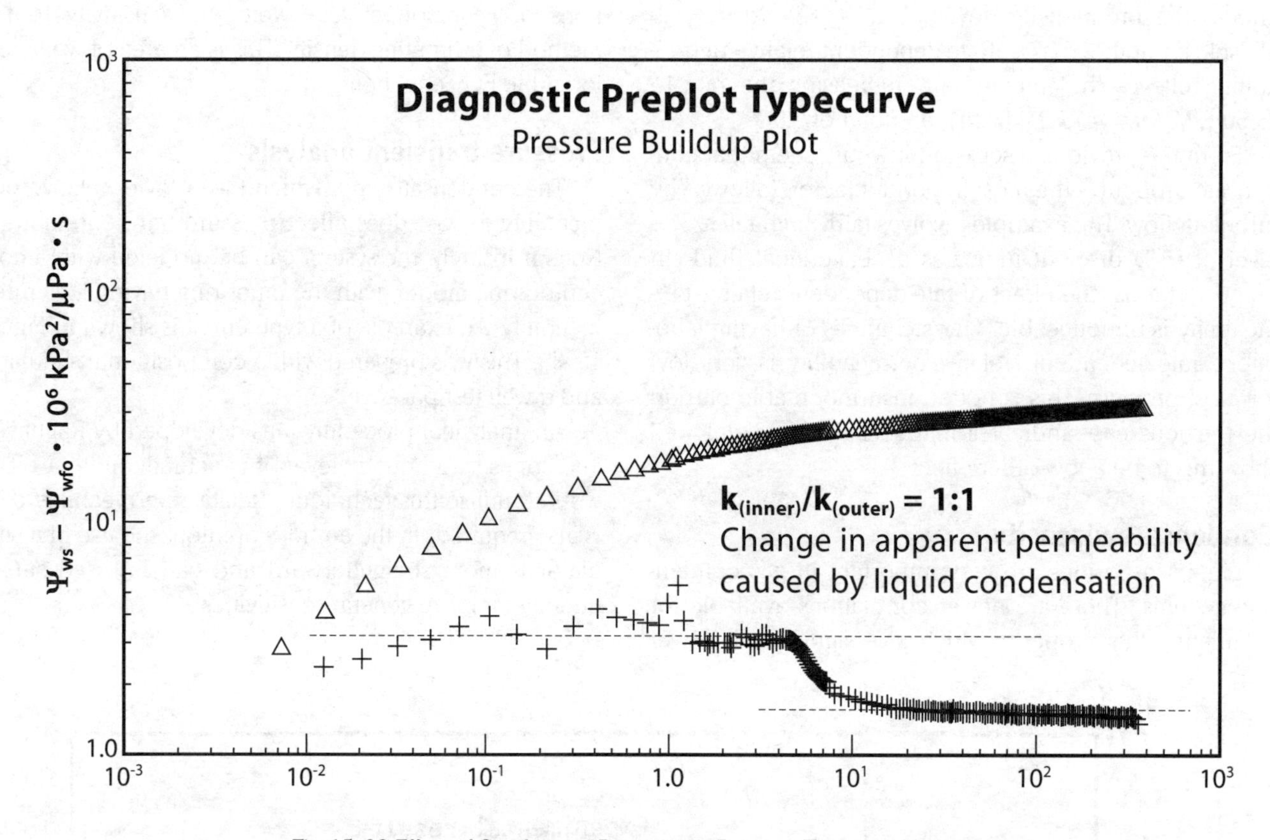

Fig. 15–38 Effect of Condensate Dropout on Pressure Transient Analysis

Near-critical (oil and gas) reservoirs

As suggested previously, near-critical reservoirs also present a number of technical difficulties. Is the reservoir an oil reservoir or a gas reservoir? Due to our training, we are inclined to regard a reservoir that is slightly below the critical point to be an oil reservoir and one that is slightly above the critical point to be a gas reservoir. For the most part, reservoir engineering calculations for oil and gas reservoirs utilize fundamentally different approaches.

However, the production performance of near-critical fluids is not really that far different. Shown below is a phase diagram, Figure 15–39, which shows near-critical reservoirs. Note that there is also the possibility that a reservoir can exist exactly at the critical point:

If we follow a reservoir path from A, which is a highly volatile oil system and compare it with B, which is a high dropout retrograde gas condensate, you will find that the dropout profile is really not that different. Therefore, from the point of view of designing a depletion strategy, determining whether the reservoir is slightly above or slightly below the critical point is moot. Such near-critical reservoirs do not really behave like an oil reservoir or as a gas reservoir. The performance is intermediate. In truth, it is probably closest to a gas condensate reservoir.

Whether the reservoir is an oil or a gas reservoir can affect ownership, since separate oil and gas leases are sometimes sold by governments. In some cases, companies split production rights, via farm outs or other legal agreements into separate oil and gas leases. This can, and has, lead to some interesting legal disputes. These reservoirs also fall on the dividing line of government legislation, which also complicates getting regulatory approvals for this type of project. In this situation, determining whether the reservoir is a gas or an oil can be of critical importance (pun intended). Near-critical reservoirs can therefore be associated with a host of economical, technical, ownership, and political issues.

The best tool available to handle this kind of problem is an EOS simulator. The cubic equations of state used perform weakly near the critical point. Therefore, from an engineering perspective, our ability to differentiate a reservoir that performs along path A or path B is also weak. Furthermore, as outlined earlier, sampling of near-critical reservoirs is difficult. This introduces further uncertainty to our analyses.

Limited background material

In the author's experience, there is not a lot of information about volatile oil systems and near-critical reservoirs. In fact, they are relatively rare. The author was only able to find a small number of papers on this subject in the literature. There is some older material from 1951, 1955, and 1957 that covers Elk City, Oklahoma "Reservoir A" (i.e., confidential) and a Smackover, Arkansas reservoir. There are also references to some North Sea developments in Brent and Dolphin.

Relatively recently there have been attempts to modify the oil material balance equation to include liquids entrained in gas for volatile reservoirs (see Walsh).[18]

One of the most comprehensive papers on this subject is "Full-Field Compositional Simulation of Reservoirs with Complex Phase Behavior" by M.S. El-Mandouh et al.[19] This relates to two Mobil-operated reservoirs in Nigeria. One of the nice features of this paper is that it addresses different possible production scenarios. A number of different scenarios are possible:

- Primary Depletion
- Dry Gas Injection
- Miscible Gas Injection
- Gas Recycling
- Waterflooding

In addition, there are all of the normal issues in simulation, which include faults, heterogeneity, water production (coning), and surface facilities.

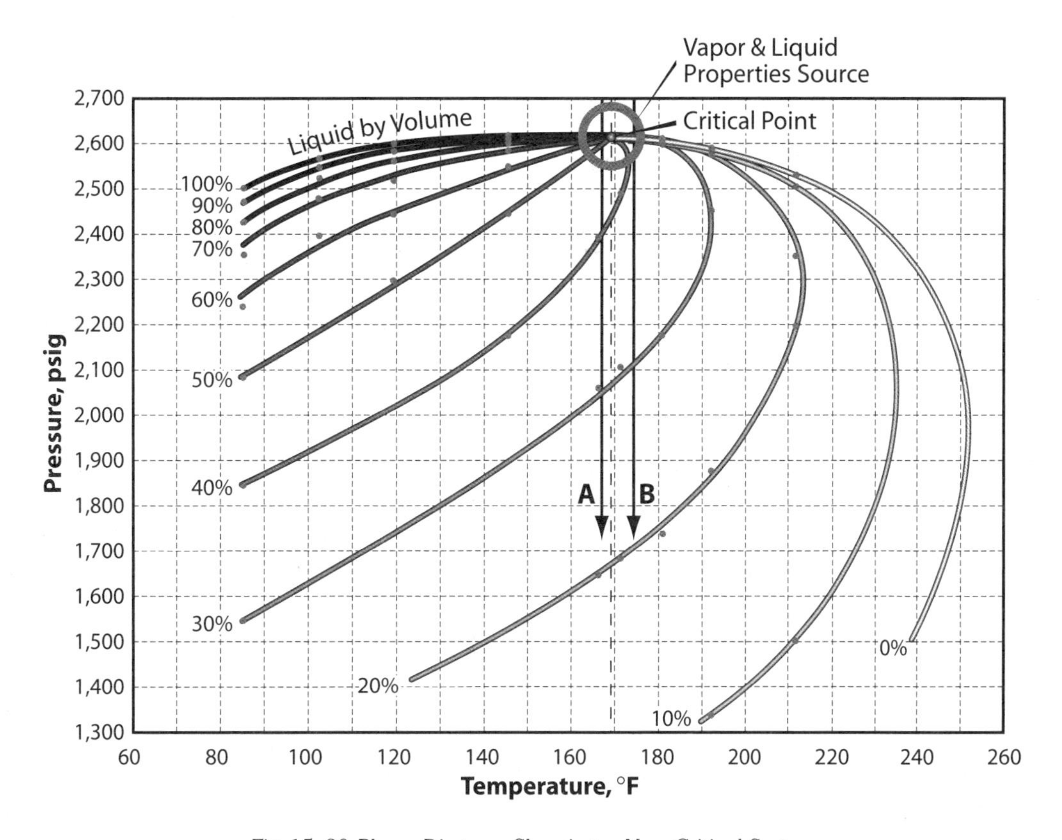

Fig. 15–39 Phase Diagram Showing a Near-Critical System

Production facilities

Volatile fluids are quite sensitive to the manner in which oil and gas are separated. Although typical black oil can be adequately separated in a single stage, cascading separators will result in significant additional liquid recovery for near-critical systems. For instance, a three-stage separation process can be used as outlined in Table 15–5.

Table 15–5 Separator Stages for Calculating OOIP

	Pressure (kPa)	Temperature (°C)
First Stage	7,000	40
Second Stage	400	25
Third Stage	101	15

Currently, the author would assume that a vapor recovery unit would be installed and that all gas, even from the third (atmospheric) stage, could be recovered.

Case history—volatile oil reservoir

With few examples to go around, a case study of a reservoir that ARE modeled has been described in the following. This was a difficult project due to problems with the PVT data. Many of these problems were related to the fact that the reservoir was very sour. A paper was written on these problems and it is indeed the author's opinion that there are considerable opportunities for service companies who can improve sour gas handling.[20] The study was used as the basis for a government application and is therefore in the public domain.

Despite the sampling problems, some "typical" calculations were made for both dew point and volatile oil systems. Some interesting results were obtained. From the experience described, it would appear that there are three key factors to consider with these types of reservoirs:

- It is important not to lose sight of the geological fit and "bread and butter" (mundane?) aspects of reservoir engineering. The additional complexity of EOS calculations, particularly with difficult fluids, can distract one from other fundamental and important aspects.
- The prediction run cases are considerably more complex. They require more "fiddling" to ensure that realistic scenarios are developed.
- More qualitative analysis is required to compare results.

Water coning

The water zone indicated on the logs indicated the potential for water production. Preliminary modeling using a 13-layer radial grid indicated that the water would indeed cone. A log section is shown in Figure 15–40. This modeling indicated that water production would not be indicated on a short-term production test.

Subsequent areal modeling was based on different layering and a higher completion. The results are shown in Figure 15–41. The latter model was designed to concentrate on compositional effects above the water contact.

Reservoir map

A reservoir map was developed from 3-D seismic, as shown in Figure 15–42. This model was digitized and input into the compositional simulator. The reservoir is not that large; the grid used was 16 × 16 × 7 layers, for a total of 1,792 grid blocks. Nonetheless, this model did not run quickly. There are eight components. The minimum grid size was 44.2 meters.

Depletion schemes

A number of different schemes were evaluated to determine the optimal depletion plan for the reservoir:

Primary depletion. The base case for development was primary production. Note that the formation volume factor is considerably lower than the differential liberation formation volume factor. In Alberta, there is a basic well rate (BWR) and a preliminary rate limitation (PRL). The wells in this reservoir would be limited by the PRL. The well eventually goes off the rate control and then declines exponentially on a bottomhole pressure control.

Recycle gas case. Reinjection was investigated based on 100% recycling of gas recovered from the three stages of separation. Normally reinjection is not carried on indefinitely. There is an economic trade-off where the additional cost of continued injection does not justify incremental benefits. Ordinarily, ARE would optimize this timing using discounted cash flow streams. Time did not permit this.

Oil production was estimated based on the initial well test. Note that the rate for this case was increased compared to primary depletion. The recycling would increase

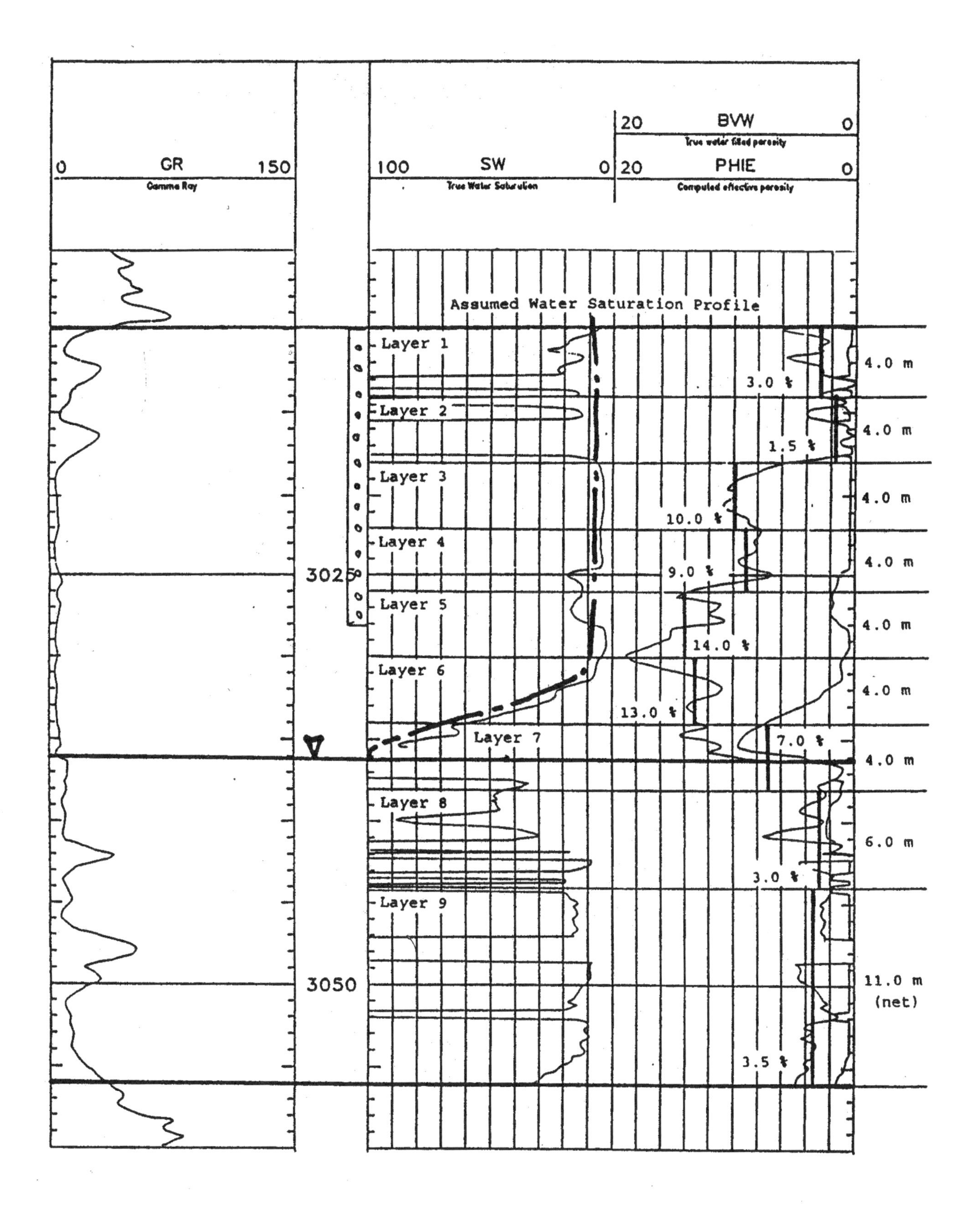

Fig. 15–40 Log Section Showing Water Saturations

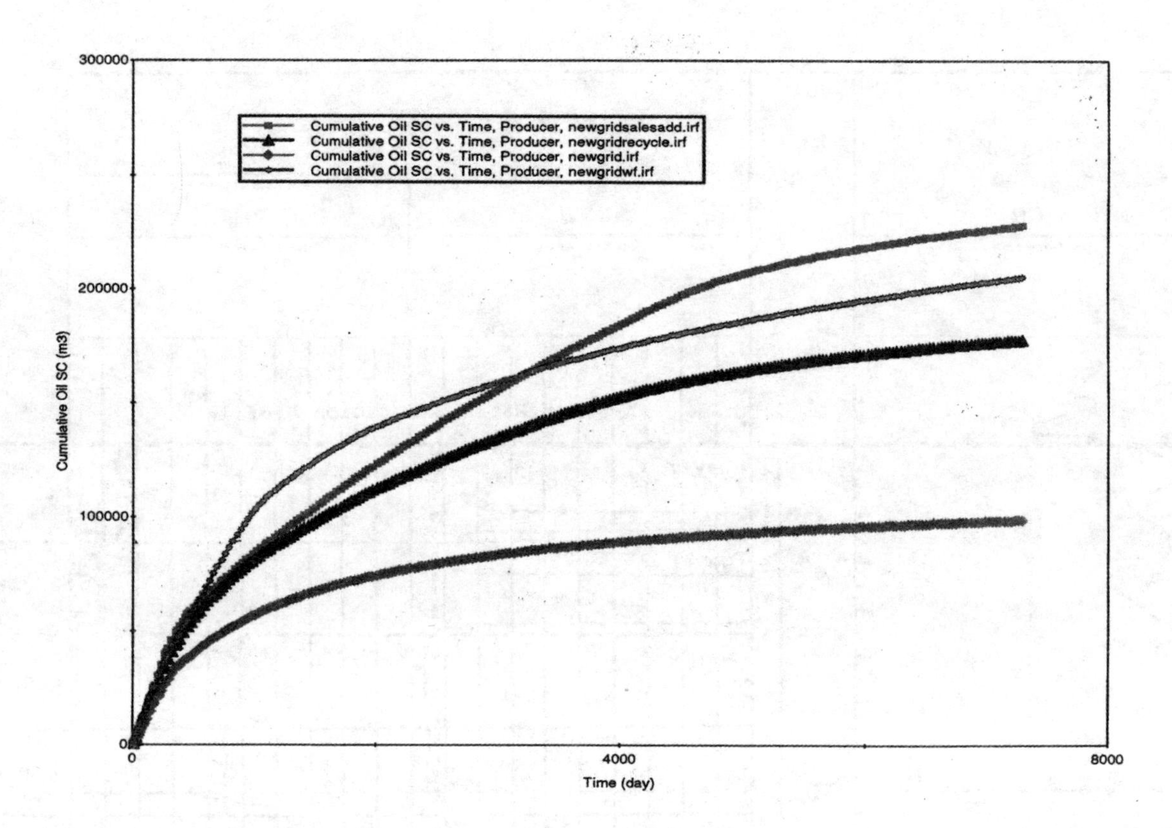

Fig. 15–41 Modeling Results for Compositional Sensitivities

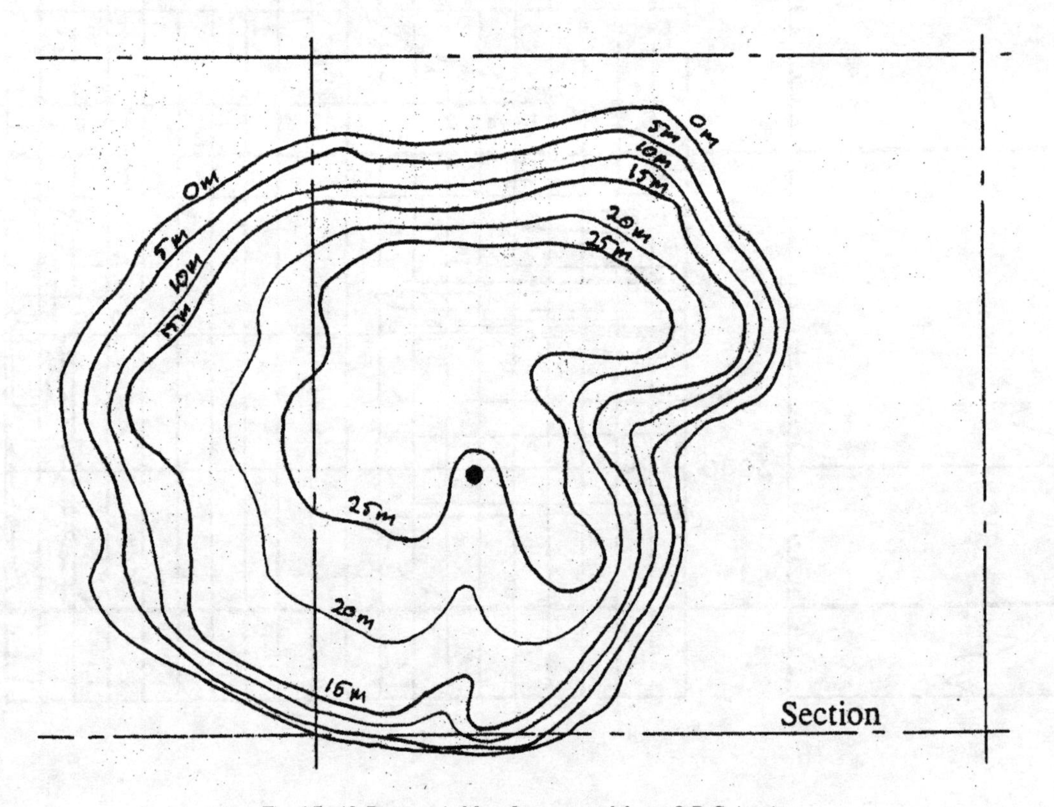

Fig. 15–42 Reservoir Map Interpreted from 3-D Seismic

the allowable maximum production rate. Production was scheduled to be constant for just under half a year before decline begins. Decline for this case is less steep than for the primary production case. The production rate drops at 10 years (3,650 days), when recycling is stopped. The GOR performance shows a slower increase than with the primary case, but follows the same general trend. Changes in produced gas composition and the start of GOR increases begin at roughly the same time at about 460 days or 1.25 years

The reservoir pressure is maintained at a reasonably high level during injection. The H_2S content steadily increases in the gas produced. This occurs slightly after the increase in methane content at 695 days or 1.9 years, compared to 460 days or 1.25 years. Since H_2S is probably the main component with the potential to increase recovery, this indicates that some phase behavior benefit is probably being derived.

Ideally, it might be possible to strip off some of the methane and inject more H_2S to act as a true enriched gas solvent. The proposed facilities did not have this capability. Alternatively, it may be possible to obtain some additional high-H_2S gas from the existing plants in the area. This could allow for improved pressure maintenance, as well as a true miscible displacement process. In view of weak sulphur prices, this may be an attractive alternative to pouring sulphur to blocks. From an operational/safety perspective, running a high-H_2S content line may not be attractive.

Sales gas makeup case. The intent of the sales gas makeup case was to increase the reservoir pressure. From a phase behavior perspective, it is unlikely that additional methane will help recovery. However, additional injection of sales gas will maintain reservoir pressures at higher levels and therefore boost total productivity. Oil production drops below the recycle gas oil rate for the first 306 days or approximately the first 0.85 years. This drop in oil productivity is because the lean sales gas diminishes the benefits of miscibility. Productivity then increases due to higher reservoir pressure. When gas cycling is stopped, the oil rate also declines in a similar fashion to the recycle case.

Sales gas is added for 12 years or approximately 4,380 days. The sales gas makeup case then tracks the recycle case, although at higher pressures. This has an effect on the final recovery. The gas recovery on the sales gas makeup case is understated in comparison to the other cases, where the reservoir pressure has dropped to lower levels.

From a GOR perspective, the sales gas makeup case tracks primary production for the first 1,067 days or 2.9 years. After this, the sales gas makeup case has a lower GOR. The GOR increases again after gas recycling is stopped. The methane content from the production well increases from about 45 mole % – 70 mole % after the 460 days or 1.25 years, indicating that gas breakthrough occurs early. Similarly, the H_2S content of the produced gas drops from about 33% to a low of 20% because of sales gas injection.

If this reservoir warrants reexamination, it might be possible to add sales gas during later stages of recycling. This might give a better combination of phase behavior advantages during the early stages of recycling plus the advantages of increased reservoir pressure when near miscibility is lost.

Waterflood case. The waterflood case was set up to replace reservoir voidage. This was done through a control card. An allowable production was assumed during the first year. Oil production holds up well during the early stages of the waterflood. This is a critical time economically, and this may make it more favorable than other cases that give higher ultimate recoveries.

Waterflood technology is much better developed than is gas injection. It is also significantly easier to implement than gas injection. Pumps are not only cheaper to make, but the increased hydrostatic head available in injection wells requires less surface pressure.

Water breakthrough occurs at 968 days or 2.65 years. This is slower than the breakthroughs for the other schemes, as summarized in the Analysis section below.

With these secondary schemes, the well control options available with a simulator are important. This saves a considerable amount of time.

Rate sensitivity

Production rate sensitivities were also run on the primary depletion case. These runs indicated that there would be minimal effect on ultimate recovery with varying production rates.

Analysis

Gas injection typically suffers from viscous fingering effects, which, as discussed in earlier chapters, is not accounted for in the reservoir simulators used. The waterflood prediction is therefore more realistic than the gas injection cases. The primary production prediction will also be more accurate than the gas injection cases. Lean gas will actually breakthrough earlier than the model predicts. There are also errors associated with the mathematics of the water displacement. These are considerably less serious, however.

Also, recall that layering is important to model directly whenever there is displacement, i.e., the rule applies equally to gas injection as it does to water injection.

The breakthroughs for all of the different cases are outlined in Table 15–6. With the layering that exists in this model, the simulator does predict earlier breakthroughs for gas.

Table 15–6 Breakthrough Times for Various Prediction Cases

Case	Breakthrough	
	Days	Years
Recycle	460	1.3
Sales & Recycle	695	1.9
Waterflood	968	2.7

This should not be surprising. There is greater viscosity contrast between the dry gas and the near-critical reservoir fluid than between the water and the near-critical reservoir fluid. Therefore, more adverse mobility ratios lead to earlier breakthroughs.

A recovery comparison between the different cases has been tabulated in Table 15–7 by product and by component in Table 15–8.

Note that the previous results do not adjust for the amount of gas injected.From an overall recovery point of view, these are intriguing results:

- The primary depletion case results are perhaps not too surprising; the oil recovery is the lowest of all the cases evaluated, and the reservoir performance suffers from a high ultimate GOR, which averages more than 20,000 scf/bbl. The traditional "rule of thumb" for differentiating a gas well is based on an initial GOR of about 1,800 m^3/m^3 (10,000 scf/bbl), so volumetric expansion is not expected to be the dominant oil depletion mechanism. However, given the high oil formation volume factor of more than 4.0, and the high solution GOR, it is not surprising that a significant portion of recovery should occur as volumetric expansion of the evolved gas phase. In fact, on a BOE basis the recoveries indicate gas expansion recovery to be twice that of the oil phase.
- Recycling the gas increases oil recovery compared to the primary depletion case by 79%, while net gas recovery drops 23%. Overall, BOE recovery increases 11%. Some benefit is clearly received from displacement and PVT effects by recycling the gas. However, it is doubtful that the extra 11% in recovery will enhance the rate of return on capital, although if the additional volume were large enough, it would increase net present value (NPV). If a full $20.00/bbl could be recovered (NGLs sell at a discount relative to WTI)—i.e., assuming no incremental operating cost, which is unlikely—the maximum additional capital that could be tolerated can be estimated. Since only a fraction of that amount could be recovered, after royalties, it was deemed highly unlikely that reinjection equipment and operating expense could be justified on this small reservoir.
- Sales gas makeup injection did not appear to increase recovery compared to the primary recovery case. In fact, on a total BOE basis, it drops by 5%. This might indicate that adding additional light ends does not help from a phase-behavior and displacement-efficiency perspective. Methane is not a solvent (based on preliminary ternary diagrams), and immiscible gas displacement is normally not efficient (except in some gravity-stabilized vertical flood cases, which does not apply here). The recovery is lower, in part, since the high reservoir pressure means there is more material to be recovered from the reservoir (see the final reservoir pressure). From an economic perspective, however, production after 20 years is of negligible value on a discounted cash flow basis. If recovery is analyzed by component, the sales gas case is the best at recovering the heavier, and generally more valuable, components. Further optimization of this case could make it more attractive. Shortening the gas injection period would likely allow the reservoir pressure to be depleted to levels similar to that utilized in the other runs.

Table 15–7 Comparison of Recovery by Product for Various Prediction Cases

Case	Oil & NGL Rf%	Gas Rf%	Total BOE	Relative BOE
Primary	16.1	59.8	0.67	1.00
Recycle	28.8	46.2	0.52	1.11
Recycle & Sales	37.0	16.5	0.13	0.95
Waterflood	33.3	33.2	0.37	1.06

Table 15–8 Comparison of Recovery by Component for Various Prediction Cases

Pseudo Component	Primary %	Recycle %	Sales Add %	Waterflood %
C1	67.57	50.39	29.76	33.25
C3	50.71	42.14	33.93	33.25
H_2S	50.51	41.58	33.3	33.25
FC6	29.63	35.38	39.34	33.25
C6H6	25.69	33.83	39.24	33.25
FC13	13.00	27.35	37.19	33.25
FC20	10.18	24.45	34.61	33.25
FC38	9.24	19.63	25.37	33.25
Mole Percent Hydrocarbons	54.35	43.35	39.64	33.25

- Waterflooding achieves slightly more than twice the oil recovery of the primary depletion case. From a practical experience point of view this seems to be a reasonable multiple. Although a recovery factor of 33.3% is relatively high for the Western Canadian Sedimentary Basin, the oil is relatively light and, with a viscosity of 0.12 cp at the bubblepoint, the mobility ratio is unusually favorable. Note that complete pressure maintenance was not maintained: the cumulative total GOR is well above the original solution gas GOR. The incremental recovery compared to the primary depletion case, on a BOE basis, is 6%, a small increment.

If additional makeup H_2S could be obtained, then it seems likely that miscible-type behavior could be maintained, which could increase the recovery of the recycle and sales makeup cases.

Using a BOE conversion factor of 10:1 (as used here) is not necessarily the most accurate indicator of actual values. Historically, on a value basis, the ratio is often higher, in the range of 12 to 14 mcf per barrel. The government often puts more weighting on energy content than industry, which is a ratio of 6 mcf per barrel of oil.

Results of predictions

The results of the predictive runs indicate that reservoir recovery is not that sensitive to the process implemented on a mole percent of hydrocarbons basis. Since there does not seem to be a major advantage to implementing any of the schemes outlined, the most logical alternative is to minimize capital expenditure and carry on with the existing primary depletion scheme. In view of the problems with the PVT sample, this is also the most prudent option from a risk point of view.

The wells in question were put on production on primary depletion. The first well watered out after 11 months of continuous production, albeit at somewhat higher rates than used in the simulator. The area is relatively deep, and the sour gas content makes wells expensive to drill and operate. The wells evaluated probably paid out.

Summary

A brief outline of compositional modeling using EOS characterization has been outlined. As with other numerical simulators, knowledge of the mathematics involved is generally not required to run a computer program. However, it is the author's opinion that some background is necessary to avoid drastic mistakes. The emphasis on the material presented is to use the PVT package and to check how it works. Start with single component systems, proceed to binary, and then attempt some multicomponent systems that have already been characterized.

Considerable detailed material on various characterization schemes is available in the literature. To date, no foolproof nor rapid implementation system seems to have been developed. The author has found the use of convergence diagrams to be of considerable help. Interestingly, the EOS is often thought to replace the older techniques, which relied on the use of convergence charts.

EOS characterization is quite demanding. Once again, it is easy to become so absorbed in the process of doing the mathematics or running the program that it is easy to forget about sample reliability, the effects of testing, and that there are some real limitations to this fashionable technique.

Many concrete steps can be taken to check the accuracy of PVT samples. Whether the separator was in equilibrium can also be tested. Material balance calculations will identify problems. It is also possible to test trends against an EOS; however, the technique really isn't accurate enough to predict performance, as demonstrated earlier in the chapter. One can check for internal consistency by plotting CCE and CVD dropouts on the same graph. This also implies that the quality check can be anticipated and built into your lab program. Finally, you can check against offset PVT behavior. For sour systems, you will see a lot of variation.

Reservoir conditioning has a big influence on gas-condensate and volatile oil systems. Sampling and hence EOS characterization is linked to reservoir performance. BUYER BEWARE!

If management is reluctant to spend money on lab tests or include some quality-control duplication, you need to sell the importance of these steps. A number of problems are listed previously that can be used. As mentioned in earlier chapters, it is rare that companies voluntarily come forward and publish the mistakes they've made. In this regard, the petroleum literature is not much help. A few (true) rumors might help sway management.

There are a number of points with respect to gas condensate reservoirs:

- The old style of reservoir analysis using a CVD experiment is not really sufficiently accurate to be used any more. EOS simulation is now probably the best way to predict future condensate composition.
- Dry gas breakthrough continues to be a problem for gas cycling schemes. This reflects variance in gas properties (unstable displacement) and heterogeneity. This is no different from waterflooding. If these effects are not accounted for, simulation results will be substantially optimistic.
- Waterflooding is being investigated for North Sea rich gas condensate reservoirs. It is probably too early to tell how well this will work. It certainly has potential for high dropout fluids.
- The compositional process in the reservoir is more complicated than was previously assumed. This will place more emphasis on EOS modeling. Important compositional effects occur in the wellbore area.
- Local wellbore phenomena affect productivity. This therefore has a big influence on project economics. The biggest problem occurs in homogeneous formations. Local heterogeneities can mask this problem. Fracturing can be used to alleviate these problems.
- Sampling is a major problem for gas condensate systems. Get samples early! Get lots of samples! Even with ideal sampling, some "back calculating" will be necessary. The proper sampling is even completion dependent. This is a big difference from oil systems, which are much less sensitive.
- Gas condensation affects pressure transient analysis. It is only during the past decade that these effects were identified. Analysis for hydraulically fractured reservoirs has been around for only a few years.

- The understanding of gas-condensate relative permeability has now increased immensely.
- In many cases, gas condensate reservoirs will be affected by non-Darcy flow. Until recently this was a rare simulator implementation. It has been done in the past in rare instances in Western Canada for productivity prediction.
- Gas condensate systems are still an "open book" area for further research, particularly for fractured reservoir systems.

The analysis of near-critical reservoirs is more difficult due to the compositional effects. There are a number of dangers:

- EOS and compositional issues may override traditional and important reservoir issues such as layering, coning, and heterogeneity. This can be exacerbated when modeling is done by pure EOS specialists.
- The processes that are often associated with gas injection into near-critical reservoirs fall right in the areas where a simulator has many weaknesses. Unstable displacement occurs with dry gas injected into a gas condensate reservoir resulting in viscous fingering. Dry gas breakthrough has historically been a problem in gas condensate cycling schemes.
- There is a propensity for management to sometimes say, "Just give me the final results." Although this can be difficult, the limitations of the techniques need to be clearly explained or documented to management. The author has no problem with adjusting the simulation results, using engineering judgment, to provide management a simple set of data upon which to make a decision. This is the difference between being a simulator jockey and a reservoir engineer.
- Waterflooding is a realistic alternative for near-critical fluids. Although there are also simplifications in the mathematics to predict water displacement, they are not as severe (i.e., not subject to as much error) as with gas displacements.
- Gravity-stabilized injection has worked very well in a large number of reservoirs, provided there are no horizontal barriers, such as encountered in the Golden Spike Leduc reservoir in Alberta as discussed in chapter 14. These reservoirs still need considerable reservoir surveillance/management to optimize their depletion. It is usually necessary to control and balance injection to maintain a level movement of the gas-oil contact.

Gas condensate projects are one of the author's favorite interests. Almost all of the work that is currently being done is based on EOS simulation. Finally, since the author has only found a limited number of studies that are documented in the literature, he hopes that approaches outlined previously may be of some guidance to others in the future.

References

1 Ahmed, T., *Hydrocarbon Phase Behaviour*, Gulf Publishing, 1989.

2 After Peng, D-Y, and D.B. Robinson, "A New Two-Constant Equation of State," I&EC Fundamentals, Vol. 15, no. 1, pages 59–64, 1976.

3 Sage, B.H., and W.N. Lacey, "Pressure-Temperature Phase Diagram for the Propane- n-Pentane System." *Ind. and Eng. Chemistry*, 32, p. 992, 1940.

4 Pedersen, K.S., A. Fredenslund, and P. Thomassen, *Properties of Oils and Natural Gases*, Gulf Publishing Co., 1989.

5 Firoozabadi, A., and D.L. Katz, "Predicting Phase Behavior of Condensate/Crude-Oil Systems Using Methane Interaction Coefficients," *Journal of Petroleum Technology*, 1978.

6 McCain, W.D., and R.A. Alexander, "Sampling Gas Condensate Wells," *SPE Reservoir Evaluation and Engineering*, August 1992.

7 Novosad, Z., "Composition and Phase Changes in Testing and Producing Retrograde Gas Wells," SPE 35645, Calgary, Alberta, Canada: Gas Technology Symposium (April 28 – May 1), 1996.

8 Henderson, G.D., et al., "Measurement and Correlation of Gas Condensate Relative Permeability by the Steady State Method," SPE 30770, ATCE, 22–25 Oct., 1995.

9 Carlson, M.R., and J.G. Myer, "Reduced Productivity Impairment for Fracture Stimulated Gas Condensate Wells," *Journal of Canadian Petroleum Technology*, Special Edition Vol. 38, no. 13, 1999.

10 Settari, A., et al., "Productivity of Fractured Gas Condensate Wells: A Case Study of the Smorbukk Field," SPE 35604, Calgary, Alberta, Canada: Gas Technology Symposium, April 28–May 1, 1996.

11 Fevang, O., and C.H. Whitson, "Modeling Gas Condensate Well Deliverability," SPE 30714, SPE ATCE, October 22–25, 1995.

12 Scheidegger, A.E., *The Physics of Flow Through Porous Media*, 3rd ed., University of Toronto Press, 1974.

13 Happel, J., and H. Brenner, *Low Reynolds Number Hydrodynamics, with Special Application to Particulate Media*, Englewood Cliffs, New Jersey: Prentice Hall, 1965.

14 Flores, J., and P.M. Dranchuk, "Non-Darcy Transient Radial Gas Flow through Porous Media." *PetSocCIM 80-31-41*, Calgary, Alberta, Canada: Petroleum Society of CIM, 31st ATM, May 25–26, 1980.

15 Blom, S.M. P., and J. Hagoort, "The Combined Effect of Near-Critical Relative Permeability and Non-Darcy Flow on Well Impairment by Condensate Drop Out," SPE 51367, Calgary: SPE Gas Technology Symposium, March 15–18, 1998.

16 Pederson, Fredenslund, and Thomassen, 1989.

17 Jones, J.R., D.T. Vo, and R . Raghavan, "Interpretation of Pressure-Buildup Responses in Gas Condensate Wells," *SPE Reservoir Evaluation and Engineering* (March): 93–104, 1989.

18 Walsh, M.P., et al., "The New, Generalized Material Balance as an Equation of a Straight Line: Applications to Undersaturated Volumetric (Part 1)/Saturated Non-Volumetric Reservoirs (Part 2)," SPE 27684/27728, Society of Petroleum Engineers, 1994.

19 El-Mandouh, M.S., et al., "Full-Field Compositional Simulation of Reservoirs with Complex Phase Behavior," SPE 25249, Society of Petroleum Engineers, 1993.

20 Carlson, M.R., and W.B. Cawston, "Obtaining PVT Data for Very Sour Retrograde Gas Condensate and Volatile Oil Reservoirs: A Multidisciplinary Approach," SPE 35653, Society of Petroleum Engineers, 1996.

16

Hydraulically Fractured and Horizontal Wells

Introduction

In a number of circumstances, where the effects of hydraulic fractures must be considered, the cases are somewhat akin to water and gas coning, and detailed near-wellbore calculations are necessary.

The physics of horizontal wells is similar to fractured wells. The former has undergone a revolution in the last 15 years and is now maturing. Most people are surprised to learn that a large percentage of the horizontal wells in the world are located in Western Canada.

When hydraulic fractures must be modeled directly

In the past, McCracken determined that if the fractures were less than a quarter of the well spacing distance then hydraulically fractured wells could be represented in field simulations by radial wells with negative skins (i.e., stimulated wells with effective wellbore radii greater than actual). Whether this one-quarter ratio is exactly correct for using such an approach, the author has not determined. However, in general, fractured wells can be represented in many field situations without difficulties.

The examples that follow are instances where direct modeling of the fracture is required. This is a difficult task for a numerical simulator and must be done with care. The examples given relate to a number of situations:

- a case where the fracture is a significant proportion of the interwell distance
- a case where the effects of layering and fractures are combined
- a case that involves condensate dropout around gas wells. Condensate can greatly impair the productivity of gas wells, even with low levels of CVD dropouts, if the formation is of low overall permeability.

Horizontal wells

Horizontal wells have a number of advantages, but horizontal wells should always be viewed in competition with hydraulic fracturing. Fracing a well is usually the better choice from both a reservoir flow and a cost point of view. Restated, horizontal wells are advantageous where fracing a well will create problems. Good candidates for horizontal wells include:

- situations where fracturing will intersect a gas cap or water leg
- situations where a fracture will go out of zone or may cause unwanted communication

These situations occur with sufficient frequency to make horizontal wells a useful technology.

Expected flow patterns for horizontal and hydraulically fractured wells

The expected flows around a horizontal and fractured well are shown in Figure 16–1.[1] From this figure, it can be seen that the principal difference between a fractured well and a horizontal well is the amount of vertical contact with the producing formation. Therefore, fracturing is generally preferable to drilling a horizontal well. The flow into a horizontal well converges vertically, which does not happen with a fractured well.

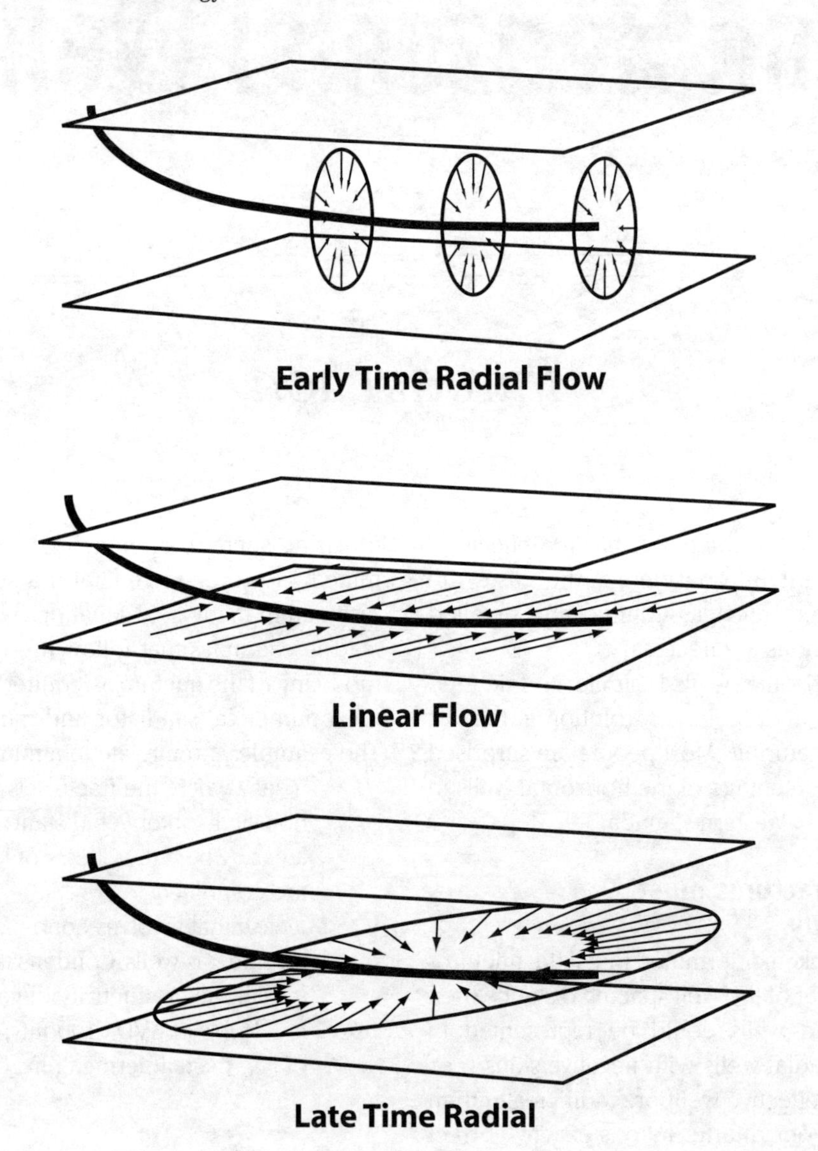

Fig. 16–1 Expected Flow Surrounding a Horizontal Well and a Hydraulically Fractured Well

Hydraulically Fractured Wells

Options for modeling hydraulic fractures

Historically, a number of modeling approaches have been tried. One is to enhance the permeability in the expected orientation of the fracture. Such an approach will help to a degree. It is also possible to represent a fracture by a group of wells in a row and apply group controls. All of these techniques were used in the past and, for a long time, they were the only options that actually worked.

The capabilities of simulators have increased during the past 20 years to the point where the grids can be input more directly. The mathematical formulation of the differential equations has not changed. The increased capability represents improvements in the matrix solvers.

Typically, a type model approach is required, because a fine grid must be used to include the effects of fractures directly. To implement a field study using such fine grids would result in a massive number of grid blocks that, even with a modern computer, would require excessive computing time.

Issues unique to modeling hydraulic fractures

Modeling fractures directly involves a number of unique issues. The first issue is stability. It can be difficult to get the models to function or to converge quickly to a solution. Second, input data is problematic because so many of the parameters are not well-known. Some way to input the fracture information is required. Here are the issues:

- grid construction and sensitivities
- fracture length
- fracture width
- fracture conductivity
- fracture volume
- turbulent flow
- solver problems
- fracture propagation
- tectonic stresses

Local grid refinements

In general, hydraulic fractures do not lend themselves to using local grid refinements. Direct modeling of a fracture causes numerical difficulties, and these would be exacerbated using a local grid refinement. Another significant issue pertains to the transition from the refined grid blocks to the outer larger grid block. Ordinarily, the transfer is based on weighting the different subelements and the adjoining grid block based on an assumed flow distribution. The flow around a high-permeability hydraulic fracture would disrupt this assumed weighting, unless one specifically designed for it. The latter would require research and coding changes to the simulator.

Injection at pressures in excess of fracture parting pressure

An issue of debate for a long time has been, "Is injecting over fracture pressure a good or a bad thing?" The short answer is, "Yes."This can be good, and it can be bad. The conventional wisdom on this is simple—some fields require injection over fracture pressure to maintain pressure. This is a practical necessity and is, without question, the only way to proceed in many cases. However, traditional waterflood calculations indicate that deviance in flow pattern from design will occur if fracture pressures are exceeded. This too, is also true without doubt.

Considerable debate is ongoing about whether this deviation can be used to some advantage. The author has worked on a number of pools where well performance indicates line drives would be appropriate. Later analysis has not supported this stance after conversion, and in some circumstances, neither the north-south nor east-west orientation could be shown to be favorable. This depends on determining the actual orientation of the fractures *in situ*.

Fracture properties

Determining fracture properties can be difficult for a variety of reasons. First, formations that benefit from fracturing and where the technology is applied are normally tight. Thus, pressure transient analysis requires long buildup times to obtain relevant formation permeabilities. Historically, it was difficult to get recorders that could be left downhole for the entire buildup period. The old clocks would stop or become nonlinear at extended times. The latter leaves a tail on some extended buildup interpretations.

More generally, many assumptions are included in pressure transient theory. Infinite conductivity fracture solutions have been replaced by finite conductivity solutions; however, these still assume uniform fracture properties along the entire length of the fracture from well to tip.

Since real fractures are tapered, the true length is always underestimated by solutions from pressure transient analysis. Probably, a numerical simulator can give better results than analytical solutions.

Fracture treatment designs

During the early part of the 1980s, a number of computer programs were built to predict fracture treatments. At the time, the author was a lease production engineer and, shortly thereafter, a completions engineer. These simulations modeled the fracture treatment processes and predicted the amount of sand that could be injected, the sand concentrations in the frac fluid, the degree of vertical growth of the fracture, and how much sand would settle out.

Since this is a reservoir simulation text, this is not the best place to discuss the intricacies of treatment design. Better references are available for providing information about fracture dimensions and conductivities.

The author designed a number of programs using one of these models. Suffice to say that there are many input assumptions. One of the most important assumptions concerns the shape of the fracture. This was a major design issue in the earlier developments, and the equations that were used presupposed a fracture shape. More recently, some work has been done to find the shape of the fracture from rock properties using finite element or other techniques.

Unless extensive testing programs are carried out, the exact stress state in the earth and the physical properties of the rocks in the formation and directly above it are not known with certainty. These inputs are normally estimated. Therefore, some experience in the area and with this type of modeling helps considerably.

Fracture treatment reports

Out of this process, a number of techniques were developed to monitor fracture treatments as they progress. This is done on a real-time basis while the treatment is being performed. The actual technique uses a Nolte plot, named after the gentleman who developed the technique. Engineers can investigate if the fracture treatments followed the intended design. Most service companies now include this analysis as part of their treatment report, and it can be found in the well files.

The author finds this data to be helpful.

Actual dimensions of fractures

Most engineers' conception of fracture widths tends to be somewhat overoptimistic. The author continues to think of them as between 0.25–0.5 in. wide. The treatment reports are normally on the order of a few (~3) millimeters. The frac sand has high permeabilities, on the order of 50,000 to 70,000 mD. This results in high conductivities ($k_f \times w_f$).

Grid construction

Simulators generally will not work if one uses actual dimensions and permeabilities directly. The minimum practical dimension for modeling is about 0.3 meters (1 foot). The conductivity (fracture width × permeability) must be adjusted accordingly.

It is immediately obvious that the volume of the fracture will not be correct. In practical terms, this volume is insignificant. However, there will be a volume of flush production from the fracture, which is overstated when compared to reality. On the other hand, the simulator does not normally consider wellbore volume, which exists in reality. It is not surprising that these minor differences are not discernible from production data.

Relative permeabilities

Some people feel compelled to change the relative permeability functions to straightline relations. On the studies made by the author, this has not resulted in a significant difference. An engineer can choose one of these or run a sensitivity.

Solver modifications

Often, introducing pivoting in the matrix solution is necessary. This is a feature available on most, but not all, simulators (see chapter 2). It may be necessary to tighten solver tolerances. Strangely, tighter tolerances usually speed up the runs on these problems. This probably prevents the solution from drifting due to round-off error accumulation. In addition, it may be necessary to increase the bandwidth on preconditioning.

Normally, user manuals advise dire consequences if one dares to adjust solver parameters. Modeling of hydraulic fractures is one of the reasons that adjustments are included as keywords. If it doesn't work the first time, boldly ignore these warnings, and experiment with these changes.

Timesteps may have to be shortened in order to achieve convergence. This is normally worst during the startup phase of the well. In some cases, the author has had to bring the wells on in stages at progressively higher rates or drawdowns. This can make modeling well buildups somewhat difficult. If one is performing waterflood calculations, short timesteps may be required as the water breaks through at the tip of the fracture. This instability usually does not persist, and the run usually will speed up when this reservoir phase has passed.

Horizontal Wells

Issues unique to modeling horizontal wells

A parallel may also be made to the factors described with fractured wells. They have been recast as follows:

- Pressure variations may occur in the wellbore.
- Fine grids are required around the well.
- The orientation of the well needs to be considered.
- Variations occur in well level—"waviness" or "undulations"—within the producing zone.
- Extremely high well indices often cause numerical instabilities. This is described more fully in chapter 19.

One other point should be mentioned: horizontal wells will accelerate depletion of the reservoir. In primary recovery situations, the additional capital cost of the horizontal well may offset the net present value gain from more rapid recovery, making a horizontal well a less economic alternative. One situation where this is not the case—i.e., where it is economically prudent to use a horizontal well—occurs when there is pressure maintenance, whether natural or induced. One of the reasons horizontal wells work so well in the Province of Saskatchewan is a combination of natural water drives providing pressure maintenance and the necessity to minimize water coning.

Interestingly, these same conditions are common to Indonesia.

Analytical horizontal well equations

Some fundamental issues relate to the analytical solutions that have been generated. To solve the differential equations for a horizontal well, basic assumptions have been used to specify boundary conditions that allow constants of integration to be determined. The first was an assumed constant pressure in the wellbore, and the second was constant flux along the well.

Since the well is not considered directly, a quarter element of symmetry is used, and the fluids in pressure transient solutions move to the middle of the well. Restated, horizontal well analytical (pressure transient) solutions are based on a butterfly well.

The question of which solution was correct led to some acrimonious debate in the literature. In this regard, simulation provides some opportunities to test these two assumptions. The use of a constant pressure assumption clearly shows a nonuniform flux along the well. If one refers to the finite conductivity solutions contained in the literature with respect to hydraulic fractures (Cinco-Ley), a constant flux is unlikely.[2]

A comparative solution for numerical simulation relating to horizontal wells was done by the SPE. Various equations are used in the wellbore to predict the pressure drop. The prediction of pressure drops is only as good as the assumptions in the wellbore equation. These assumptions include volume fraction, density, viscosities, velocities, wellbore height, diameter, completion equipment, and roughness.

Finite pressure drops in wells

One of the most significant results from this comparison was that wells produced at high rates in high-permeability formations have finite pressure drops in the wellbore. As one may expect, various simulators use different wellbore equations, and this means that results from different companies are not the same. The more sophisticated wellbore equations yield better predictions of performance. As time goes on, simulator technology will evolve to provide a choice of which equation the user wants to employ. This process seems well underway.

Based on this, the constant pressure assumption along the well is the closest assumption upon which to derive a solution. Clearly, this assumption is not universally correct, and there is more to the story.

Due to finite wellbore pressure drops, the flow patterns in the horizontal well casing will not be symmetric along its length. This precludes quarter elements of symmetry.

As a practical matter, skin damage seems to be much more severe near the heel of the well than at the tip. This appears to be related to the amount of time that the drilling fluid has contacted the reservoir and is difficult to measure quantitatively. Reservoirs are rarely close to homogeneous. More specifically, permeability varies significantly across short distances and, as a result, the chance of having a uniform flux is even more remote. The implication of these factors is that horizontal wells almost never have anything that approaches an analytical solution—using either assumption.

Pressure transient analysis of horizontal wells normally indicates well lengths considerably shorter than indicated. This issue occurs with hydraulically fractured wells, which was discussed in detail in this chapter. Most of the latter reflects problems in defining the physical properties of the fracture. The author has not spent significant time using the simulator to develop pressure transient responses. The issue in this case is slightly different because the deviations of analytical solutions from practice probably are affected by variations in reservoir properties. This is an area where research might be helpful.

Radial solutions, i.e., vertical wells, are affected by the wellbore area, and the physics of flow tend to dampen reservoir variations. Fortunately, it is also the area with the strongest effect on well productivity. Pressure transient analysis of horizontal wells may not be as inherently diagnostic.

Effect of depletion

Many of the analytical equations used by the industry predict well indices or steady state rate predictions. As mentioned previously, the more rapid depletion associated with a horizontal well has a major effect on economics. Therefore, basing economic decisions on well productivity equations is potentially dangerous. The analytical equations have been cast for long-term production, such as the work of Poon.[3] This is an area where a simulated solution may be more accurate than an analytical solution. A simulator can be set up to include more complex boundary conditions and accounts for depletion, or material balance, effects.

Horizontal well grid refinement

The biggest difference between horizontal wells and hydraulically fractured wells is that grid refinement is required vertically for the horizontal well. This is due to gravity and the ratio of k_v/k_h. Horizontal wells are demanding from a gridding point of view. One interesting observation is that the k_v/k_h ratio gives a clue as to the vertical grid dimensions that are required. The lower k_v causes steeper pressure gradients, which, in turn, requires more grid blocks. This is a bit oversimplified in that it ignores gravity and capillary pressure transition zone thickness. Nonetheless, it seems to work well in practice and is a useful starting point for grid refinements.

Orientation

Horizontal wells can be most sensitive to orientation. The example that comes to mind is the Weyburn area of Saskatchewan and involves a fractured reservoir. Normally, one would expect that it would be best if wells intersected fracture systems at right angles to maximize productivity. This does not work in Weyburn, where lower water cuts result in better economics if the wells are parallel to the fractures. The point is that orientation can be a subtle issue, and trial and error in the field can produce unexpected results.[4]

Waviness and undulations

Most horizontal wells are rarely true horizontals. The undulations in the well can be included by placing the well in different grid block layers, and this affects productivity. Deciding on the correct grid block lengths along the well can be difficult. This is best evaluated with a grid sensitivity study. A flow net will show there are flow concentrations at the tips. Overall, productivity is not affected by slight changes in a homogeneous reservoir. It is a particularly sensitive issue with steam-assisted gravity drainage (SAGD) because the distance between injectors and producers varies. Detailed comments are made on this topic in chapter 19.

Case Histories

Approach

In a departure from other chapters in this book, a number of these points will be demonstrated using case studies. The three examples included in the following have a number of points in common.[5, 6, 7]

Case history—Dodsland Saskatchewan

The Dodsland field, located in southwest Saskatchewan, was discovered in 1953. Light oil (36° API) production is obtained from the Cretaceous Viking sandstone. Delineation of the field occurred rapidly during the 1950s. Unitization and waterflooding followed in the mid-1960s. A resurgence in development occurred in the late 1970s and early 1980s due to high oil prices. Development in the mid-1980s was also fueled by attractive government incentives.

Lithology

The Viking consists of an interbedded shale and sandstone. The sandstone is a poor quality reservoir rock; however, its shallow drilling depth and large area have permitted extensive development.

The more productive areas of the reservoir are comprised of 4.6 to 7.6 m thick sandstones with discrete shaly partings. In the less productive areas, the shale partings are dense and the pay consists of thin lenses of sandstone. This layering does not show directly on well logs, since it is too fine. The sandstone appears to be dirty or has a high gamma ray reading in the cleaner sands. Identifying this layering depends on visual examination.

The clay minerals have a strong effect on reservoir performance:

- The amount and type of porosity is controlled by shale content.
- Permeability is controlled by the amount and type of clay.
- The clays are water sensitive.
- Vertical permeability is restricted by the shaly partings.

Core analysis

Data from a number of wells was combined, as shown in Figure 16–2. This figure shows core porosity versus core permeability. The slope of the graph is extremely steep— there are large variations in permeability for small changes in porosity. It has been found empirically that if log core permeability is plotted on normal probability paper then a straight line is obtained. The slope of this line is described by the Dykstra-Parsons coefficient. Most reservoirs exhibit a Dykstra-Parsons ratio between values of 0.5 to 0.9. The larger number represents more severe layering. Dodsland core data was plotted as shown

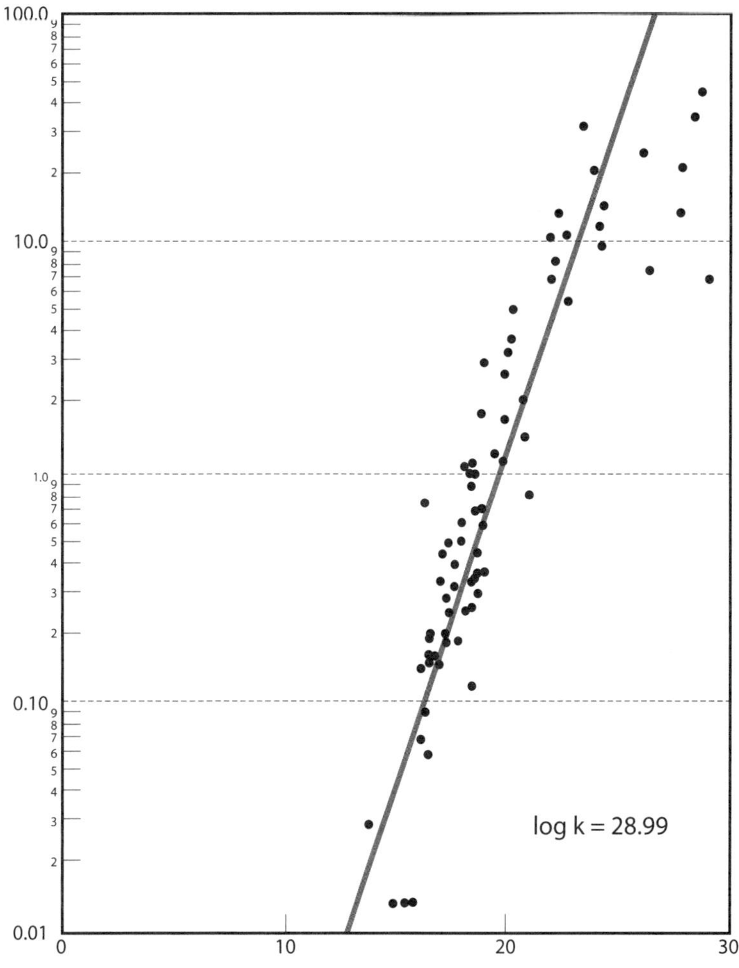

Fig. 16–2 Core Permeability vs. Core Porosity—Dodsland Area

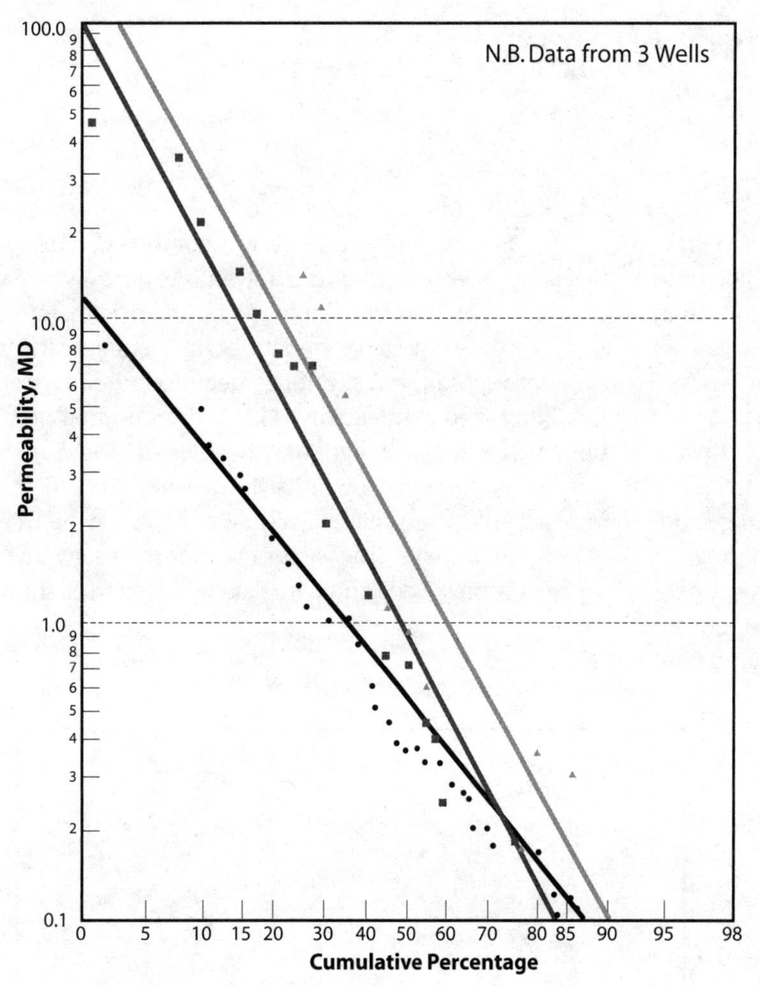

Fig. 16–3 Dykstra-Parsons Plot—Dodsland Area

in Figure 16–3. This data shows an extremely high permeability variation with a Dykstra-Parsons coefficient of 0.92.

Production profiles

Based on economic evaluation experience in the area, a distinct period of flush production is seen, which is typical of low-permeability formations. Production decreases rapidly during the first year, exhibiting a hyperbolic decline shape. After two years, the wells decline exponentially. A characteristic production profile is shown in Figure 16–4.

Extended primary production

The best examples of extended primary production are in the eastern portion of the Dodsland field, since the western portion was placed on waterflood. The eastern area, which is of lower reservoir quality, was originally drilled during the 1950s on 64-ha (160 acres) spacing. Production from this area has sometimes been considered representative of primary production performance, regardless

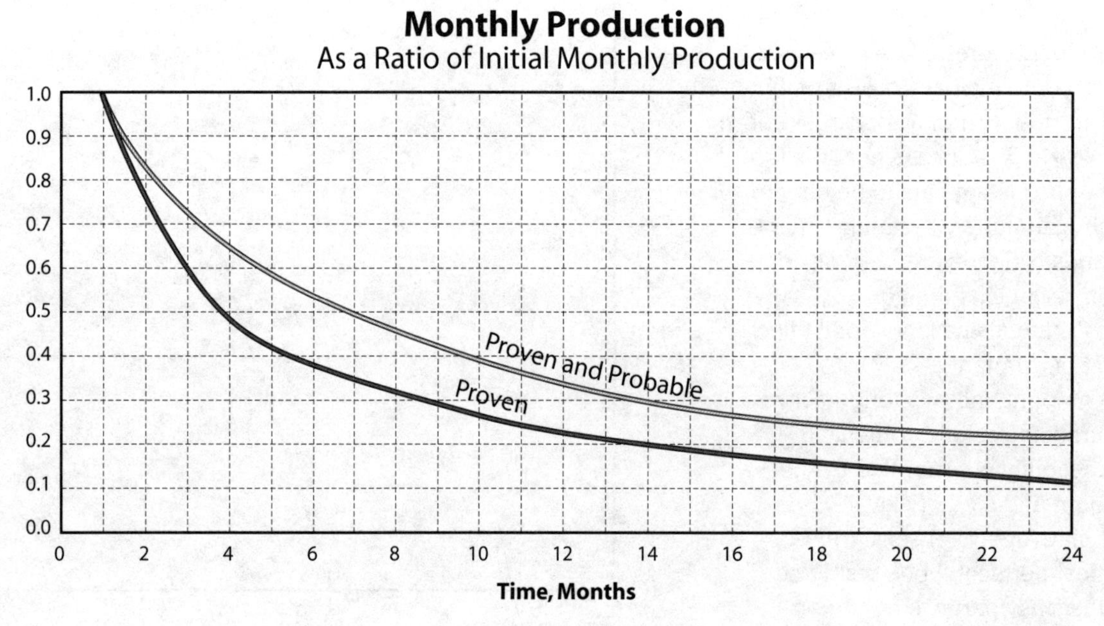

Fig. 16–4 Characteristic Production Profile—Dodsland Area

of spacing, due to the lower overall permeability of the Viking. This assumption had been extrapolated to the western waterflood area.

Hydraulic fracture treatments

Since the reservoir is of poor quality, successful development required that all wells be hydraulically fractured. Treatments of 23,000 to 32,000 kg (50,000 to 70,000 pounds) of sand were typically used. Due to the importance of the fracture treatments, they were simulated numerically. Fracture half-lengths of approximately 76 to 91 meters (250–300 feet) from well to wingtip were generated.

Ideally, pressure transient analysis could be used to verify these dimensions. However, buildups are difficult to obtain on pumping wells. The reservoir is tight, and extensive shut-in times are required to get a full analysis. With low production rates, extended shut-ins are hard to justify. No suitable well test analysis has been found by the author. Therefore, the data from the fracture simulator was used as input for the reservoir simulator.

Therefore, it was necessary to review production on a well-by-well basis. Few of the existing wells show a discernible production increase after the implementation of the waterflood. Performance can only be described as poor. The only positive indication of waterflood response has been a reduction in produced GORs, to near-solution gas oil ratios.

Grid construction

Grid design is a compromise between greater accuracy achieved by using smaller grid block sizes and minimizing the amount of time required to run the simulation. It is not possible to analytically determine the amount of error produced by a grid. Therefore, the "appropriate" degree of grid refinement must be determined parametrically by grid sensitivities.

Initially, a very fine grid was used. This grid was used as a base case for developing the justification for applying a coarser grid.

Paleo stress

The orientation of the hydraulic fractures is controlled by the residual stresses in the earth. Adams and Bell have documented fracture trends, as shown in Figure 16–5.[8] Fractures in the Dodsland area will trend east-west.

Existing waterfloods

Most of the Dodsland field was put on waterflood during the later part of the 1960s. There is considerable production data to analyze. These waterfloods were implemented simultaneously with extensive infill drilling programs to 16-ha (40 acres) spacing. This has made it difficult to distinguish, on a unit-wide basis, how much production resulted from infill drilling and how much production increase resulted from waterflood response.

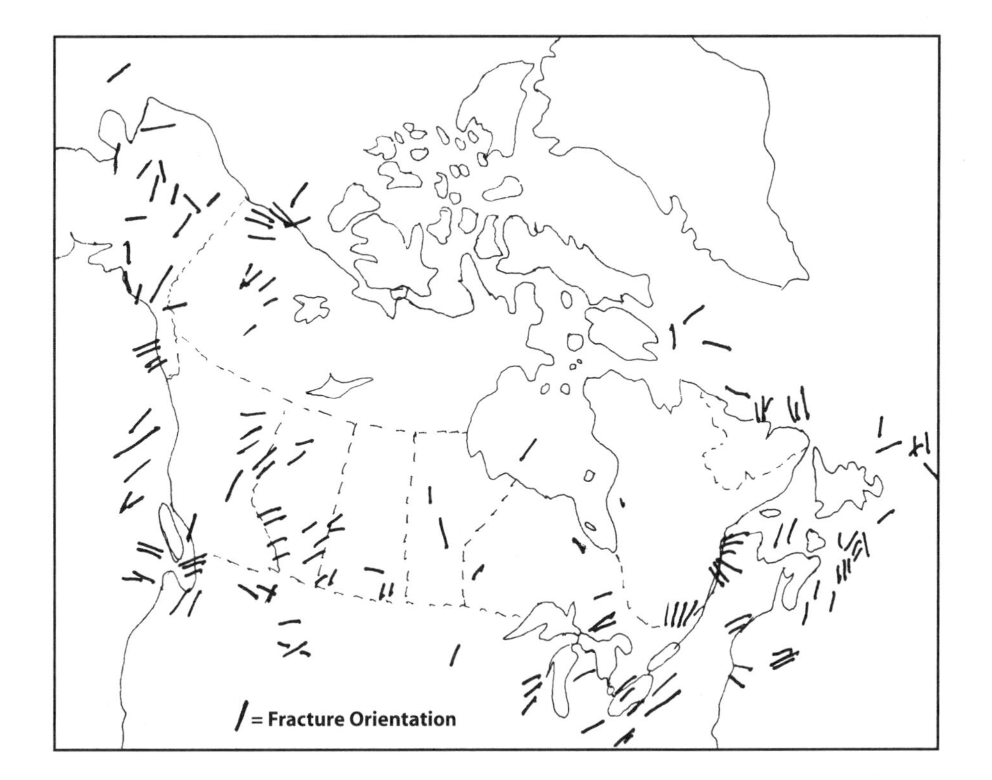

Fig. 16–5 Fracture Trends in the Western Canadian Sedimentary Basin

Grid properties

A net pay of 1.2 meters (4.0 feet) was used. The zone thickness from logs averages 2.75 to 4 meters (9 to 13 feet). It is customary in the area to reduce the pay from logs by one-half since the sandstone is layered with alternating shale and sandstone.

Porosity was input using an average of 19%.

Permeability was estimated from core data, offset experience, and pressure buildup analysis. Core data from the unit wells indicated an average air permeability of 8.5 mD. The liquid permeability was adjusted to be 4.25 mD. Pressure buildups indicated permeabilities from 1 to 4 mD. These tests were not the best, due to the low permeability, but once again, were the most accurate data available. A final value of 3.75 mD was used, representing a minor reduction in core values of overburden effects, to match pressure buildup analysis.

No capillary pressure data was available. The data used was from correlations and offsetting reservoirs.

Multiple layers

An attempt was made to run a multilayer (3-D) simulation. The layer properties were input as described in the preceding section (*see* Table 8–1, V = 0.9). The time requirements for this case were enormous. The variation in layer properties from hundreds of millidarcies to fractions of a millidarcy was severe. This variation in transmissibility was too high for a problem that was already difficult by virtue of the presence of a frac. For this reason, the use of Hearn-type relative permeabilities was necessary. (Note that program capability has improved since this work was done.)

Hearn-type relative permeability curves

The curves were calculated by arbitrarily assigning five layers. The permeability of the layers was then assigned by positioning a line on a log core permeability versus log normal probability paper with Dykstra-Parsons slopes of 0.6, 0.8, and 0.9. Permeability was taken from the graph (at the 0.10, 0.30, 0.50, 0.70, and 0.90 cumulative percent points) and input in a spreadsheet.

The effects are shown in Figure 8–6. The relative permeability to water is strongly increased, and the relative permeability to oil is decreased. The utilization of Hearn-type relative permeability curves is the most important difference between this and previous modeling.

Grid refinement

The grid was coarsened, as shown in Figure 16–6, and test runs were made for comparison. The results of the two runs are presented in Table 16–1. The new comparisons are made between waterflood cases. These figures indicate that the numerical errors are of the order of 2.45% (4.80–2.35%) on primary production and up to 11.47% (4.80–6.67%) on waterflood. Originally, these comparisons were made only on primary production. However, this gave a somewhat optimistic indication of grid errors.

The results from these two runs are also compared graphically in Figure 16–7. The primary portions of these runs overlay; however, the water breakthrough on the coarser grid occurs much earlier. This represents numerical dispersion. Most of the differences occur at low water cuts and are not as important as the graph might suggest initially. The timing of water breakthrough is normally important in establishing a history match, and some loss of accuracy has been accepted. The saving in run time from 159.65 to 11.12 minutes is significant. Considering the acceptable loss in accuracy, the coarser grid was used for the balance of the study.

Primary production

Primary production was modeled prior to proceeding with waterflood predictions as shown in Figure 16–8. The initial rate for the second month of production was 1.9 m^3/day (12 bbl/day). The classical Dodsland area flush production is seen before stabilizing to an exponential decline. The exponential decline is 21.7% per annum, which was slightly higher than the actual field values for the unit studied. Final recovery, at a final rate of 0.8 m^3/day (5.0 bbl/day) was 2,509 m^3 (15,780 bbl).

Effects of spacing

As discussed earlier, a number of wells in the Dodsland area were drilled during the 1960s on large spacings. The production decline from these wells was used to a greater or lesser extent to predict the performance of wells drilled in recent years. Figure 16–9 shows the results for the same run on 64-ha (160 acres) spacing. The decline slows to an exponential decline rate of 8.5% per annum. The cumulative recovery after 16 years was 5,078 m^3 (31,809 bbl). This was compared to actual well performance (on wells from the 1960s) of 4,647 m^3

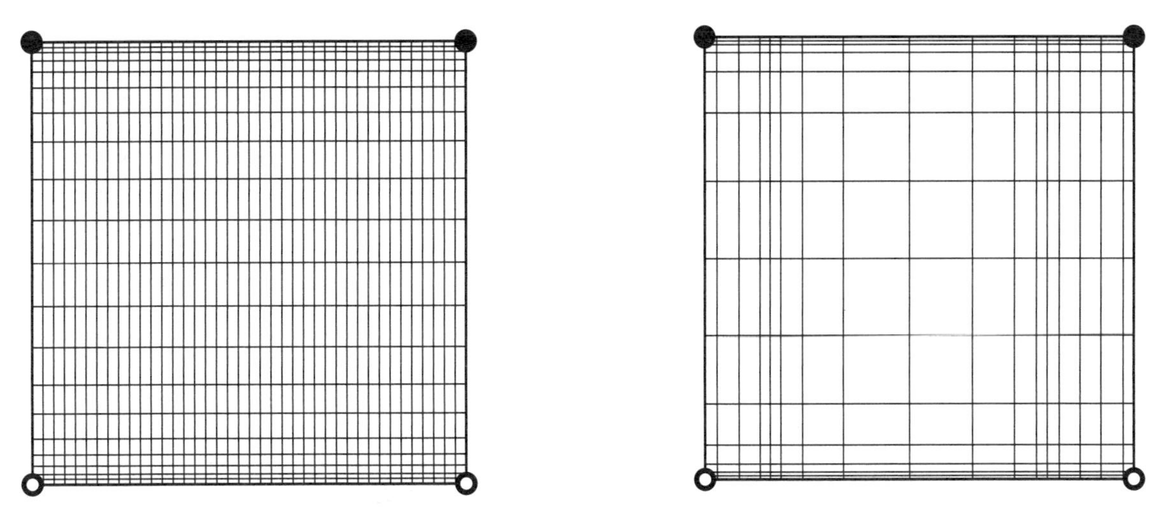

Fig. 16–6 Simulation Grids Used in Grid Sensitivity

Table 16–1 Grid Refinement Results—Waterflooding Hydraulically Fractured Well

		Fine Grid	Coarse Grid	Difference %
Pore Volume	rb	223,041.00	277,781.00	4.78
OOIP	stb	120,830.00	126,919.00	4.80
Primary Recovery	stb	12,591.00	13,038.00	3.40
Recovery Factor	percent	10.42	10.27	
Pressure	psia	489.00	501.00	2.35
WF Recovery	stb	38,895.00	37,421.00	-3.93
Pressure	mstb	907.00	901.00	-0.67
Water Produced	stb	43,758.00	42,242.00	-3.59
Water Injected	mstb	48.00	45.00	-6.67
Gas Produced	mscf	12,583.00	12,457.00	-1.01
Run Time	minutes	159.65	11.12	-93.00

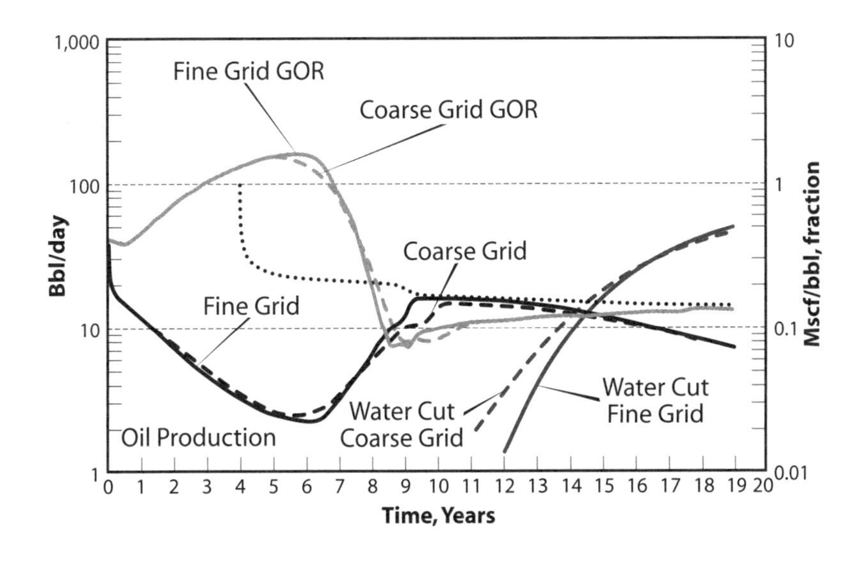

Fig. 16–7 Grid Sensitivity Results

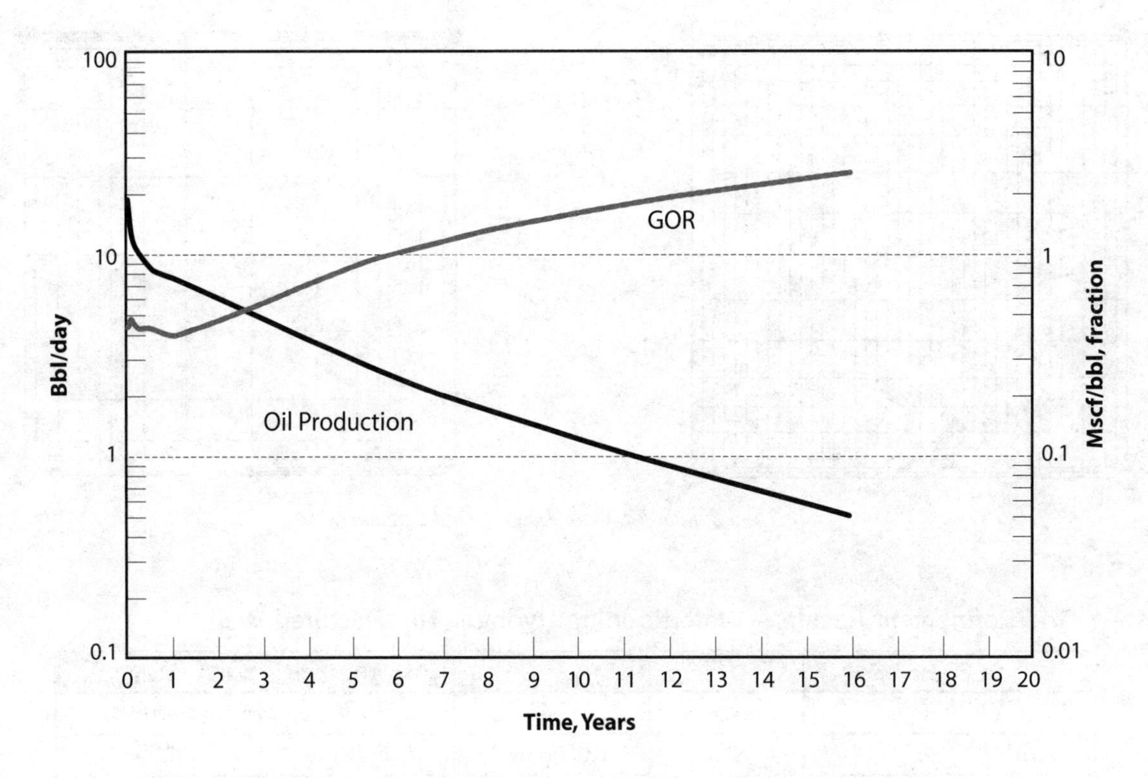

Fig. 16–8 Primary Production Model Results—40-Acre Spacing

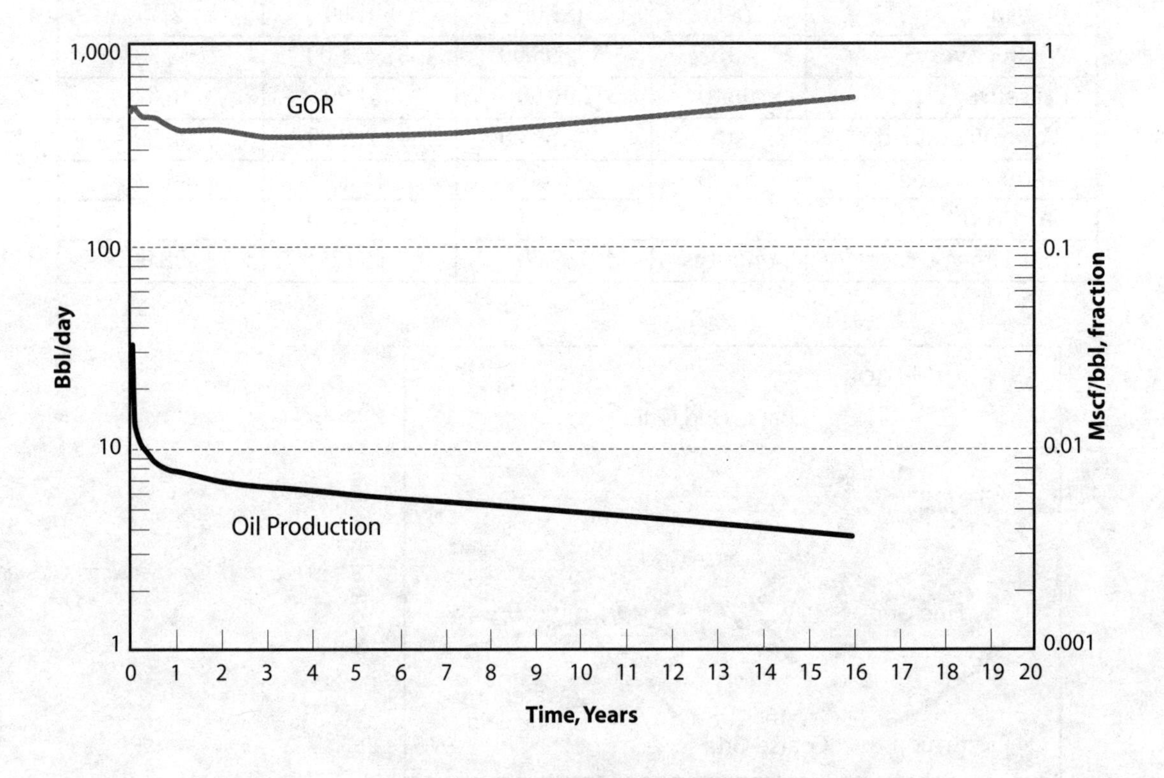

Fig. 16–9 Primary Production Model Results—160-Acre Spacing

(29,228 bbl). This is 92% of the simulator prediction. Since infill wells have been drilled during the 1980s, this model is likely accurate.

Note that the GOR shows an initial rise, then stabilizes at slightly above the solution gas GOR. The GORs increase less quickly than on 16-ha (40 acres) spacing.

The ultimate recovery has been compared as follows:

- 16-ha (40 acres) 2,137 m^3/well (13,440 bbl/well)
- 4 wells 8,547 m^3 (53,760 bbl)
- 64-ha (160 acres) 6,310 m^3/well (39,691 bbl/well)

The difference for 64-ha (160 acres) is 2,237 m^3 (14,069 bbl) or 35.4%. Restated, infill drilling down to 16-ha (40 acres) clearly increases recovery by about 559 m^3 for 16-ha (3,517 bbl for 40 acres).

Waterflood predictions

A five spot water injection pattern was modeled for a uniform reservoir. The "top right" well was placed on injection after four years of primary production. Note that the oil rates respond slowly. Oil production levels for 3 to 4 years and does not peak until 11 years after the start of injection. Water breakthrough does not begin until after 14 years. The GOR does not respond for two years, after which it decreases rapidly to near the solution gas-oil ratio.

Layering sensitivities

Sensitivities were run based on Dykstra-Parsons ratios of $V = 0.6$, 0.8, and 0.9. In each run, the peak oil response becomes progressively lower as summarized in Table 16–2 and as shown in Figures 16–10, 16–11, and 16–12.

The peak oil rate declines rapidly when V is greater than 0.8. Water breakthrough is greatly accelerated by moderate layering.

In the last case, with the most severe layering (V=0.9), waterflood performance consisted of a reduction in the decline rate to low levels. Later, the production resumed decline at about one-half the initial decline rate, approximately 12%. Note that the GOR in all cases decreases rapidly.

GOR response

If the layering in the reservoir is severe, no increase in oil production may occur. The only concrete indication that communication occurs and there is a waterflood response is the decline in GOR, which the majority of Dodsland wells and projects exhibit. For many waterfloods in layered reservoirs, this may be the only definitive indicator of response.

Start of waterflooding

The effects of starting the waterflood later were also investigated. Due to the advanced primary decline, a much higher gas saturation exists in the reservoir. The gas that has come out of solution must be compressed and redissolved or be displaced out of the reservoir, which delays waterflood response. This has a profound negative effect on waterflood economics, which depend on an increased oil rate to generate an economical rate of return.

Effects of injecting over fracture pressure

It is not possible to model fracture extension using a conventional reservoir simulator. However, an upper limit can be calculated by doubling the fracture length of the injector and increasing the downhole injection pressure from 11,035 kPa to 17,930 kPa (1,600 psi to 2,600 psi).

The effect is shown in Figure 16–13. There is a short and rapid production increase followed by a production decline similar to primary production decline.

Table 16–2 Comparison of Peak Rates with Different Levels of Heterogeneity

Dykstra-Parson Ratio V (Fraction)	Peak Rate (M^3/day — quarter element) (bbl/day —full well)	Breakthrough Time 20% Water Cut (years)
0.6	0.87 (21.9)	10.0
0.8	0.65 (16.4)	6.5
0.9	0.40 (10.1)	4.5

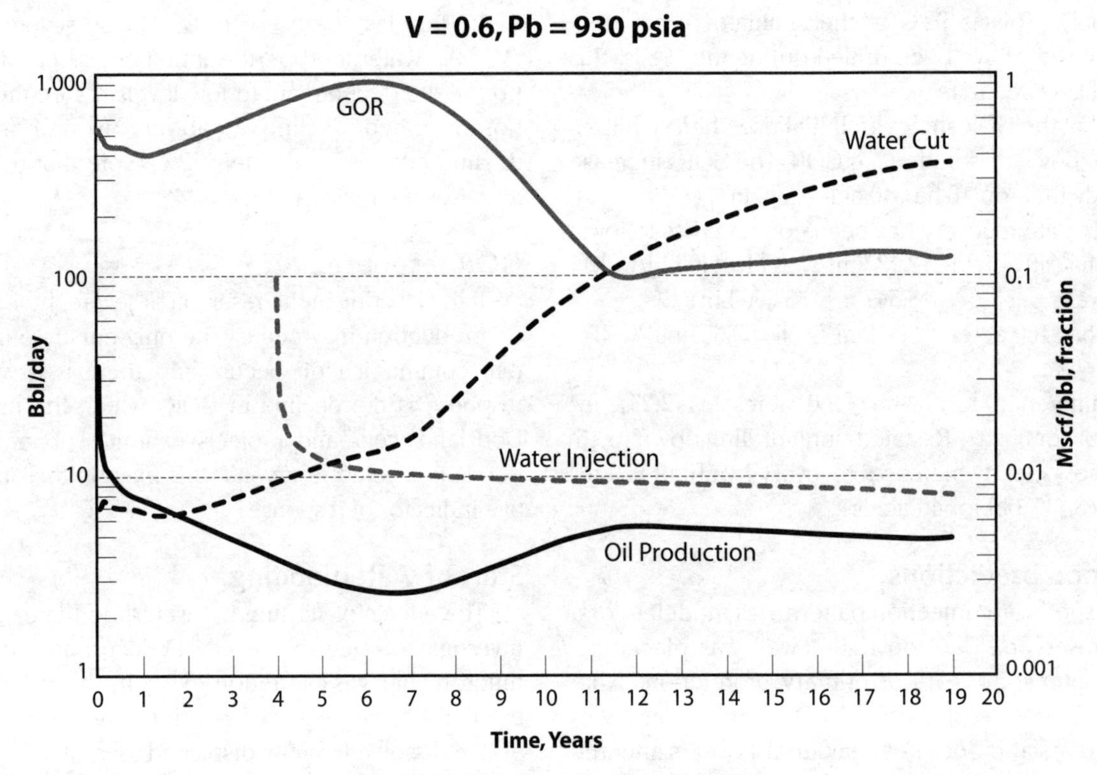

Fig. 16–10 Waterflood Recovery V = *0.6*

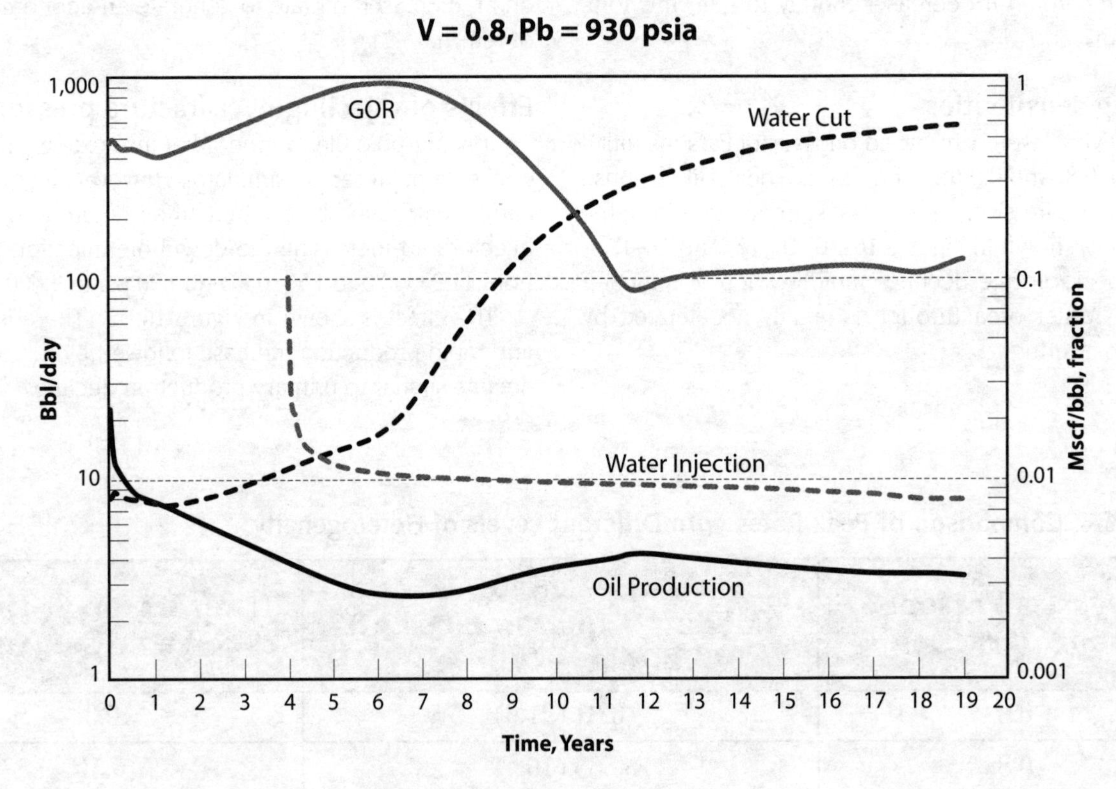

Fig. 16–11 Waterflood Recovery V = *0.8*

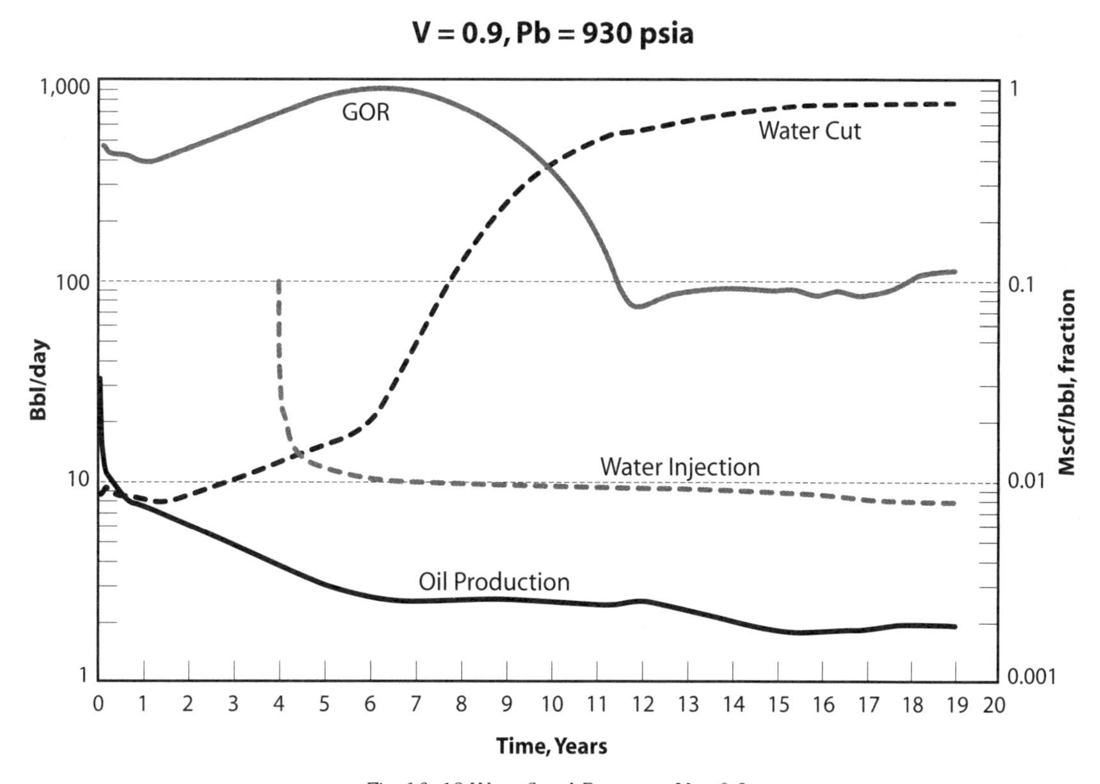

Fig. 16–12 Waterflood Recovery V = *0.9*

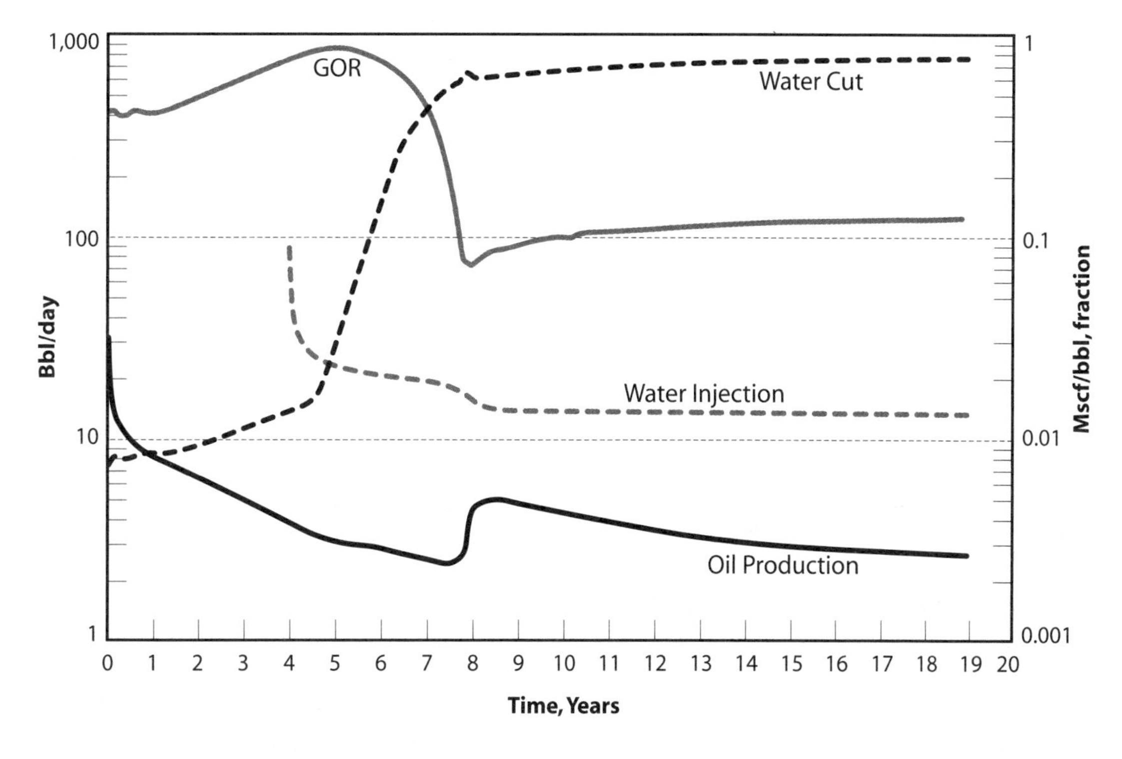

Fig. 16–13 Effect of Injecting at Over-Fracture Pressure

Model accuracy

The production response at high levels of heterogeneity shows some "wiggles" in the output. This likely represents numerical errors. Numerical stability is a function of the relative permeability curves. The use of pseudo relative permeability has increased the level of error since the grid refinement discussed earlier. The previous grid refinement sensitivity indicated the accuracy of waterflooding modeling was about 11.5%. The last model was estimated to have been accurate to roughly ±20%.

Grid sensitivities were not rerun. The model achieved its basic purpose in showing the effects of layering. In reality, the various input data—i.e., relative permeability curves, permeability, porosity, fracture length, etc.—are known approximately or taken from analogous reservoirs. In view of these uncertainties, which most likely exceed ±20%, further grid sensitivities were not performed.

Modeling summary

- Grid sensitivities were used to generate a more efficient grid. The approximate accuracy is ±11.5%.
- A reasonable model of primary production is obtained.
- The early wells on 64-ha (160 acres) spacing give an optimistic indication of recovery on primary production.
- Layering sensitivities indicate that, at high levels of permeability variation, waterflooding will result in only a leveling of production.
- A decrease in GOR indicates that the reservoir is responding to waterflood, even if there is no increase in production rate.
- Delaying waterflooding has a strong effect on how rapidly production response is seen. This is most significant from an economic perspective.
- The effect of fracture extension will be a short, rapid increase in oil production, followed by oil production decline rates that are similar to those seen on primary production.
- The use of pseudo relative permeabilities decreases the accuracy of the simulator. However, the loss in accuracy is compensated by accounting for layering, which is a critical parameter. Adequate accuracy is achieved for an engineering solution.
- The current technique is not a completely accurate representation of the mechanics in the reservoir. However, it is the best available solution, without resorting to substantially more sophisticated techniques, such as geostatistical modeling.

Points demonstrated from Dodsland studies

The presence of hydraulic fractures results in a nonradial flow pattern in the reservoir.

The severe layering in the reservoir diminishes waterflood response.

This study is essentially a post analysis of waterfloods that have been in place for a long time. For the unit studied, it was possible to identify that there would not be a future increase in oil production due to waterflooding.

Hearn-type pseudo relative permeability curves can be used to efficiently solve this type of problem. Accounting for layering or heterogeneity is critical.

Case history—Deep Basin of Alberta, Canada

This work was originally completed to forecast production from wells in a new field, which was being developed in the Deep Basin area of Alberta, Canada. The work concentrated on the most important technical point: the productivity of wells that are hydraulically fractured may not be as adversely affected by condensate precipitation as was previously reported. However, in any actual study, a significant number of factors are derived from different disciplines within petroleum engineering that must be integrated. The main results are placed within an abbreviated outline of the process used in solving an actual field problem.

PVT characterization

A PVT characterization was required that could be used in a compositional simulator. A general phase behavior package was utilized for this study. The Peng-Robinson equation of state was used throughout this work.

The original reservoir temperature and pressure were approximately 81° C (178° F), or 354° Kelvin (637° Rankine), and 18,875 kPa (2,738 psi), respectively. The critical point of the reservoir fluid was calculated to be 231° Kelvin (415° Rankine) and 5,070 kPa (735 psi), which was considerably below the original reservoir conditions. Hence, on first look, it appeared that this reservoir was at highly supercritical conditions and therefore unlikely to exhibit retrograde condensation. There was a relatively low concentration of intermediate hydrocarbon components (C2s are below 5%); however, there is more than 10% H_2S. Sour gases are more prone to exhibit retrograde condensation, which prompted a partial gas-condensate lab study.

Initially, a PVT package was used to get a general feel for the problem. For this, a number of simplifications were made to the reservoir fluid composition. First, the isopentanes and normal pentanes plus all other structural isomers were input as the normal component. Second, a small correction was applied to the recombination so that the sum of the mole fractions of the reservoir fluid was exactly 1.0000. Third, the heavy ends (C7+) were characterized as: nC7, nC8, nC9, and nC10. The compositions used are outlined in Table 16–3.

The results of these sensitivities were plotted on a convergence chart, as shown in Figure 16–14. This chart is normally used for estimating parameters for traditional flash calculations using equilibrium ratios; however, it is useful in tuning equation of state parameters as well. The critical points of individual components are shown, including the approximate locus of critical points for binary mixtures. Note that the critical pressures and temperatures of C7 (heptane) through C10 (decane) are on the right side of the diagram—far to the right. The critical point of the mixture remains fairly close to the left side near methane.

The liquid dropout curve from the partial gas-condensate lab analysis is shown in Figure 16–15. A prediction utilizing the PVT package is also shown. The calculated amount of liquid dropout is different by almost a factor of two. Although this is a large error in relative terms, the overall amount of liquid dropout is relatively small.

Table 16–4 Final Components Used for Reservoir Fluid Characterization

Component	Mole Fraction
C2	78.02
H_2S	10.9
C2	5.78
C3	3.26
pC8	0.83
pC10	1.24

Table 16–3 Scoping Sensitivity Reservoir Fluid Compositions

Case	1	2	3	4
Component	Mole Percent			
H_2S	10.9	10.9	10.9	10.9
C1	78.04	78.04	78.04	78.04
C2	5.73	5.73	5.73	5.73
C3	2.11	2.11	2.11	2.11
nC4	1.15	1.15	1.15	1.15
nC5	0.54	0.54	0.54	0.54
nC6	0.29	0.29	0.29	0.29
nC7	1.24			
nC8		1.24		
nC9			1.24	
nC10				1.24

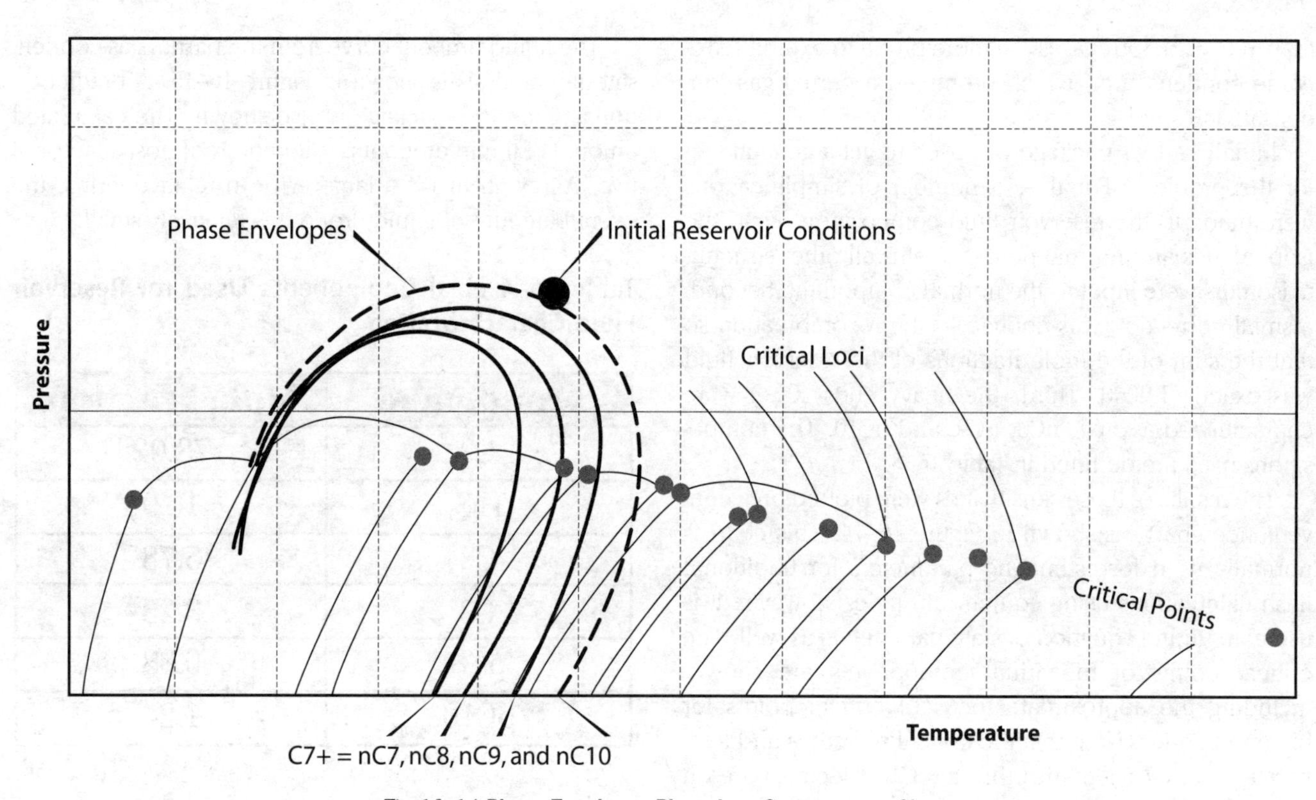

Fig. 16–14 Phase Envelopes Plotted on Convergence Chart

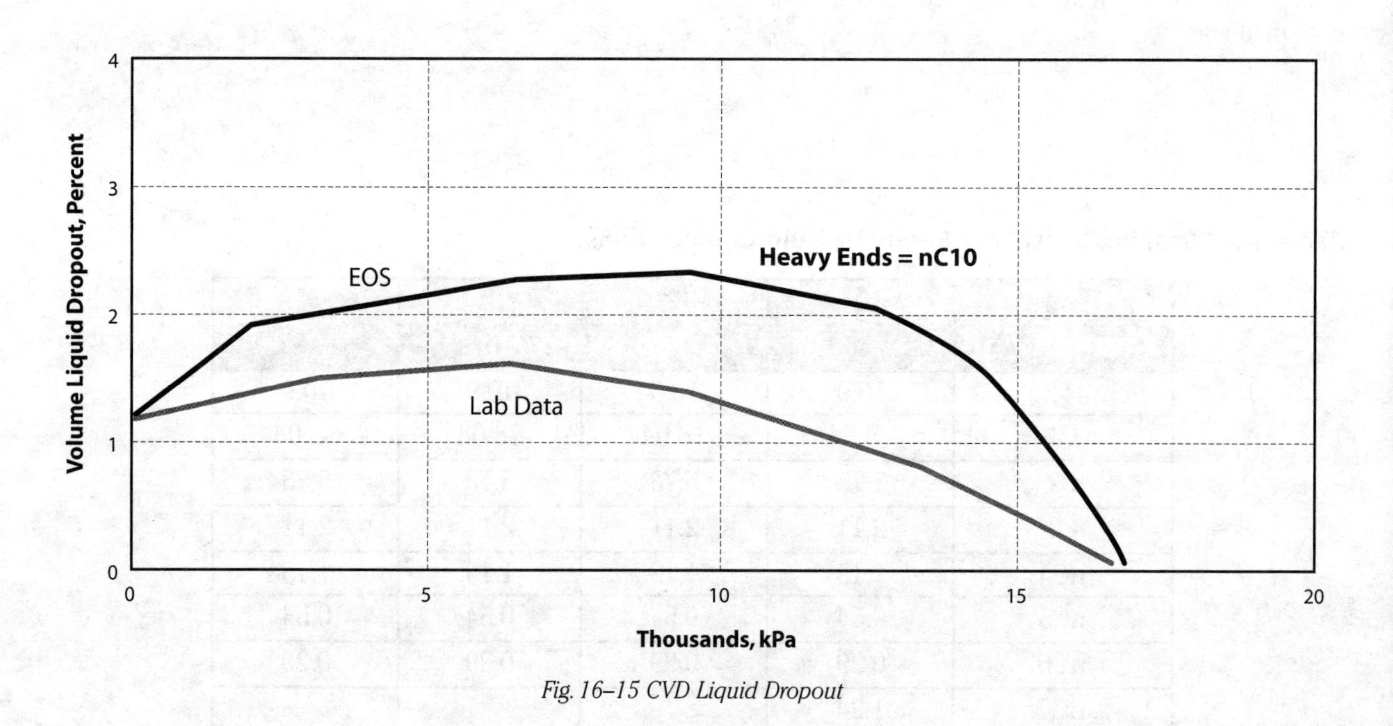

Fig. 16–15 CVD Liquid Dropout

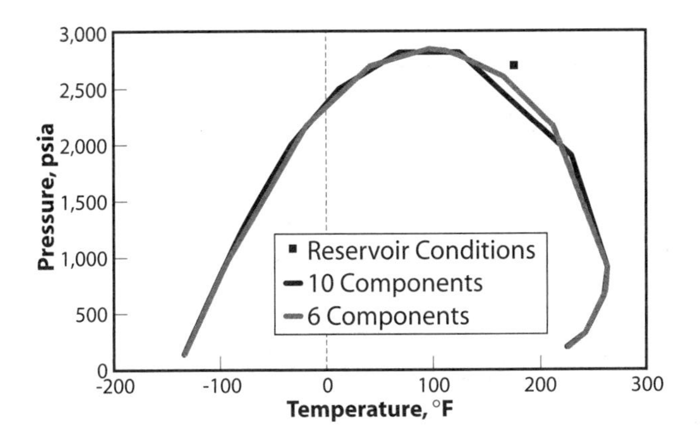

Fig. 16–16 Comparison of 10-Component vs. 6-Component System

In order to reduce compositional model simulation execution times, the original 10-component system was simplified to 6 components. These are shown in Table 16–4. Note that some of the components were also tuned. The phase diagram for this system matches the original 10-component system closely as depicted in Figure 16–16. This characterization was used initially for the compositional modeling.

Further tuning was done on the equation of state parameters. One way to split the characterization is to group the molecules by general chemistry and geometry. Paraffins (P), napthenes (N), and the aromatics (A) can be split to improve characterization. This distribution was applied to the C7+ fraction. Tuning was also done using a regression option on the critical temperature, pressure, and volume as well as the acentric factor of the C7+ components. Following this, even more sophisticated tuning was done by allowing regression on the binary interaction coefficients between H_2S and the C7+. The improvement was not as great. Most of the liquid volume predicted for constant volume depletion is accurate to within ± 20%, an acceptable result.

Only the initial 6-component system was used. Although the final PNA interaction tuned model was considerably better, particularly in relative rather than absolute terms, the results indicated that liquid precipitation would not be a problem at 2.5% of pore volume, which was higher than the lab-measured liquid dropout volume. Therefore, rerunning the model was not justified.

Model construction

The quality of any simulation is only as good as the input data. Much of the input was common to the two models used. In the following summary, the order used in the data file was retained as much as possible.

For modeling a fracture-stimulated well, a detailed grid was required. A fine grid and quarter element of symmetry were used as depicted in Figure 16–17. The grid was set up to represent single-section (259 ha or 640 acres) spacing with an overall grid dimension of 805 m by 805 m (2,641 feet by 2,641 feet). Variations of this grid have been used and tested extensively in the past.

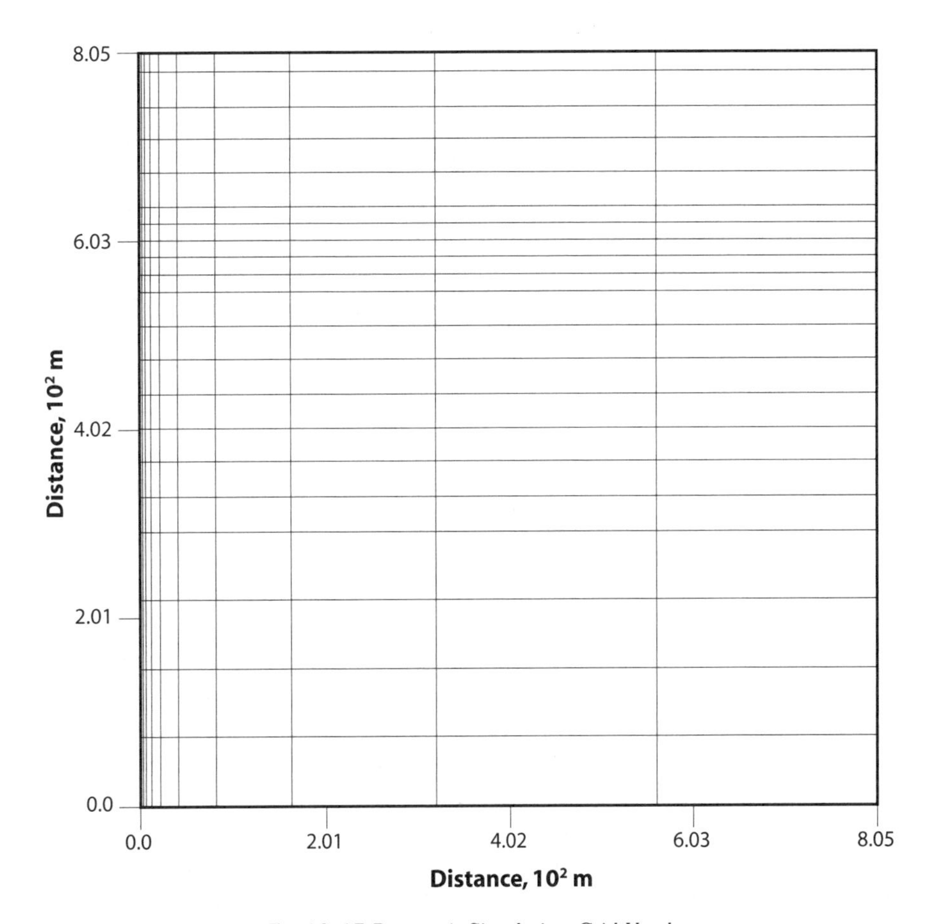

Fig. 16–17 Reservoir Simulation Grid Used

Table 16–5 Hydraulic Fracture Properties from Stimulation Design

Propped Half Length	210 m	689 ft.
Average Conductivity	197 mD*m	646 mD*ft
Propped Width at Well	3.4 mm	0.134 in.
Average Fcd	1.2	1.2
Average Propped Width	2.8 mm	0.11 in.

Distance Along Frac		**Propped Dimensions**					
		Width		**Height**		**Conductivity**	
m	ft.	mm	in.	m	ft.	mD*m	mD*ft
53	174	2.8	0.11	30	66	451	1479
105	344	3.5	0.138	24	79	138	453
158	518	2.9	0.114	21	69	123	401
211	692	1.9	0.075	18	59	76	249

Net pay and porosity were input based on log analysis. Since the model was largely conceptual, these values were rounded off. In addition, since a type model was being used, parameters were input as constants for the entire simulation grid.

Fracture dimensions for the well were determined from a treatment design prepared by a well service company. Pertinent data is outlined in Table 16–5. The treatment design does not, of course, necessarily represent the fracture that was placed in the formation. This could be due to lack of cement bond integrity or due to large height growth. What actually happened was uncertain. Detailed analysis was performed on the treatment log. No indication of vertical height growth was seen; however, bottomhole pressure measurements were not available. Therefore, the interpretation was not conclusive. On this basis, the design parameters were accepted as being at least reasonable indications of fracture dimensions.

Well test modeling

Well test analysis can provide an estimate of formation permeability and hydraulic fracture properties. The majority of pressure data that was obtained exhibited the character from a bilinear flow regime. Various combinations of fracture conductivity, fracture length, and formation permeability yield similar results. Therefore, no unique result was obtained. A complete analysis cannot be obtained without pressure data that comes from the pseudo radial or transition to pseudo radial flow regime. A least squares type error minimization was used to determine the "best" interpretation. An investigation was made to see if the minimization procedure was sufficiently sensitive to converge to correct formation permeability and fracture properties, even if complete conventional analysis (straightline) methods could not be used.

One way to test this was to generate a buildup via simulation with known input, and then analyze the test. For this segment, a black oil model was used assuming a dry gas. Hence, condensate dropout and gravity segregation were not accounted for.

Several slight deviations were made in the final rates between the actual field test and the simulation model. These changes were not significant for testing the minimization process; however, the test interpretations shown do not represent the actual well test.

The fracture treatment prediction program indicates a tapered fracture. This was included in the numerical simulation model and would be expected to yield a significant effect, since the flux into a finite conductivity fracture is concentrated at the ends. The analytical solution used in the pressure transient analysis assumes a constant fracture conductivity.

Input parameters included a formation permeability of 0.4 mD, a fracture conductivity of 308 mD × m (1,011 mD × ft), and a fracture length of 201 m (669 feet).

Buildup-type curves were generated through an option in the pressure transient analysis program and were allowed to regress on x_f, $k_f \times w_f$, and k. The results are shown in Table 16–6. The results of the previous minimization changed depending on the initial guesses of k, x_f, and $k_f \times w_f$.

A buildup was then generated in the pressure transient analysis package using the simulator input parameters and compared against simulator output. Results are shown in Figures 16–18, 16–19, and 16–20. The least squares error was 52.5 kPa (7.6 psi). Although not as good as the earlier minimization, this is still a relatively low level of error.

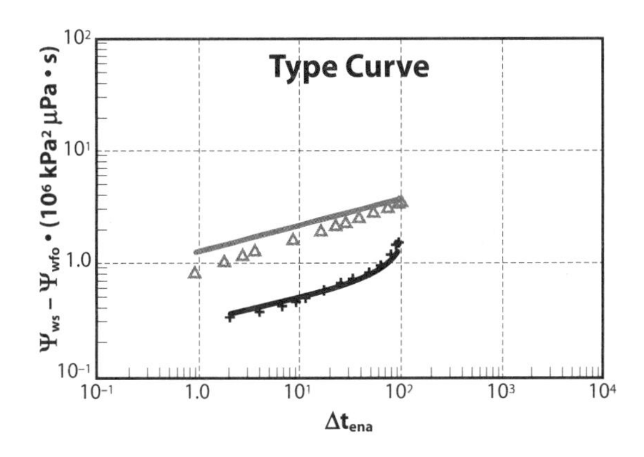

Fig. 16–18 Pressure Buildup Results—Type Curve

The well test did not extend into pseudo radial flow nor into the transition to pseudo radial flow; hence, reliable formation and fracture properties were not expected. In the author's experience, such tests should nevertheless be analyzed, to get as much information as possible from them. The simulation and well test interpretation were not directly comparable, since a tapered fracture was included in the simulation. No rigorous error analysis was made. Although simulator tolerances were tightened, analytical pressure transient solutions are inverted from Laplace space to real time via numerical inversions and, therefore, have a numerical error.

From a pragmatic point of view, the tools were compared as they would normally be used.

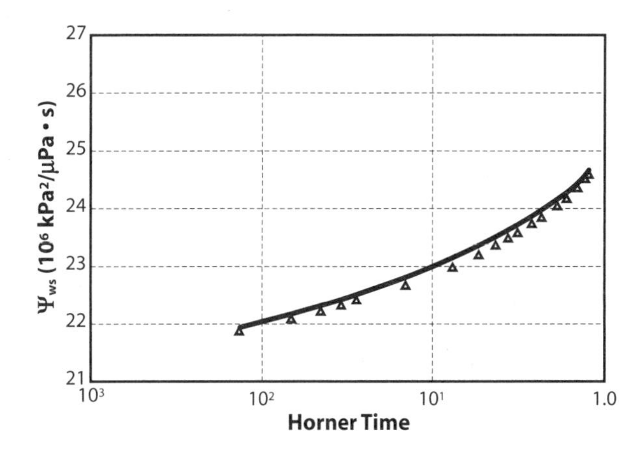

Fig. 16–19 Pressure Buildup Results—Hortner Plot

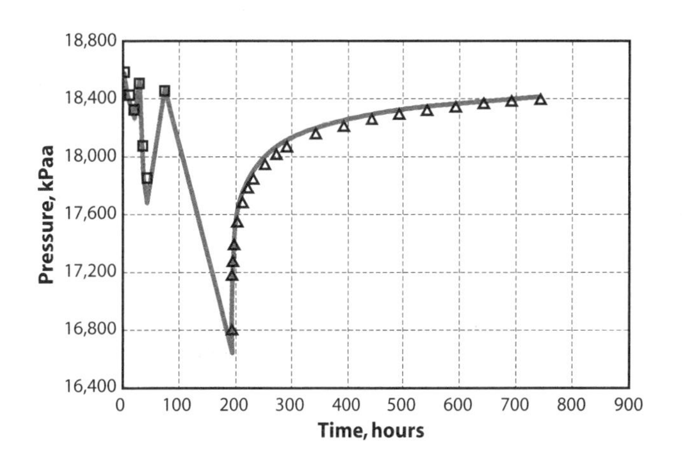

Fig. 16–20 Pressure Buildup Results—Wellbore Pressure Strip Chart

Table 16–6 Reservoir and Fracture Properties Derived from Pressure Transient Analysis (Finite Conductivity)

Permeability k	Fracture Half Length xf		Fracture Conductivity kf*w		Error	
mD	m	ft.	mD*m	mD*ft	kPa	psi
0.44	149	489	536	1759	13.1	1.9

Modeling the pressure transient analysis gave an indication of a systematic bias in the estimation of formation permeability. With this, it was possible to combine the different sources of data to make a better estimate of formation permeability. It should be pointed out that a reasonable order of magnitude estimate of formation permeability was derived.

Although a reliable interpretation of fracture properties could not be made from well test analysis, there was no indication that the information from the fracture treatment prediction was incorrect.

Results of predictions

The condensate saturations of the two cases were contoured and compared as shown in Figure 16–21. The buildup of condensate is concentrated near the wellbore; consequently, this figure is difficult to read. To make as direct a comparison as possible, the dropout profiles shown were obtained at approximately the same average reservoir pressure. For the radial case, this was at a time of 7,300 days, at which point the average reservoir pressure was 15,340 kPa (2,224 psi). For the fractured case, the average reservoir pressure was 15,310 kPa (2,220 psi).

The dropout profile on the fractured case appears, at first glance, to be quite surprising. However, it is known from pressure transient theory that flux into a finite conductivity frac is not uniform. Condensate saturations reflect the flux pattern into the fracture. The contours on the radial case are more concentrated (only every second contour was shown on this plot to make it more legible). In the radial case, the maximum dropout was 36.6% immediately adjacent to the well. For the well with a hydraulic fracture, the maximum dropout was 25.4% in and adjacent to the end of the frac. In the radial flow case, the liquid saturation adjacent to the well was higher, but did not involve as large a volume of total fluid. The flow is more concentrated, and the dropout has more effect. With the hydraulically fractured case, the drawdown is not as concentrated. Liquid saturation levels adjacent to the fracture are not as high, but involve a much larger volume of total condensate.

Points from Deep Basin case study

A hydraulic fracture treatment reduces the amount of drawdown in the well and results in a less-concentrated condensate precipitation. Significant impairment does not occur during the first 10 years of pro-

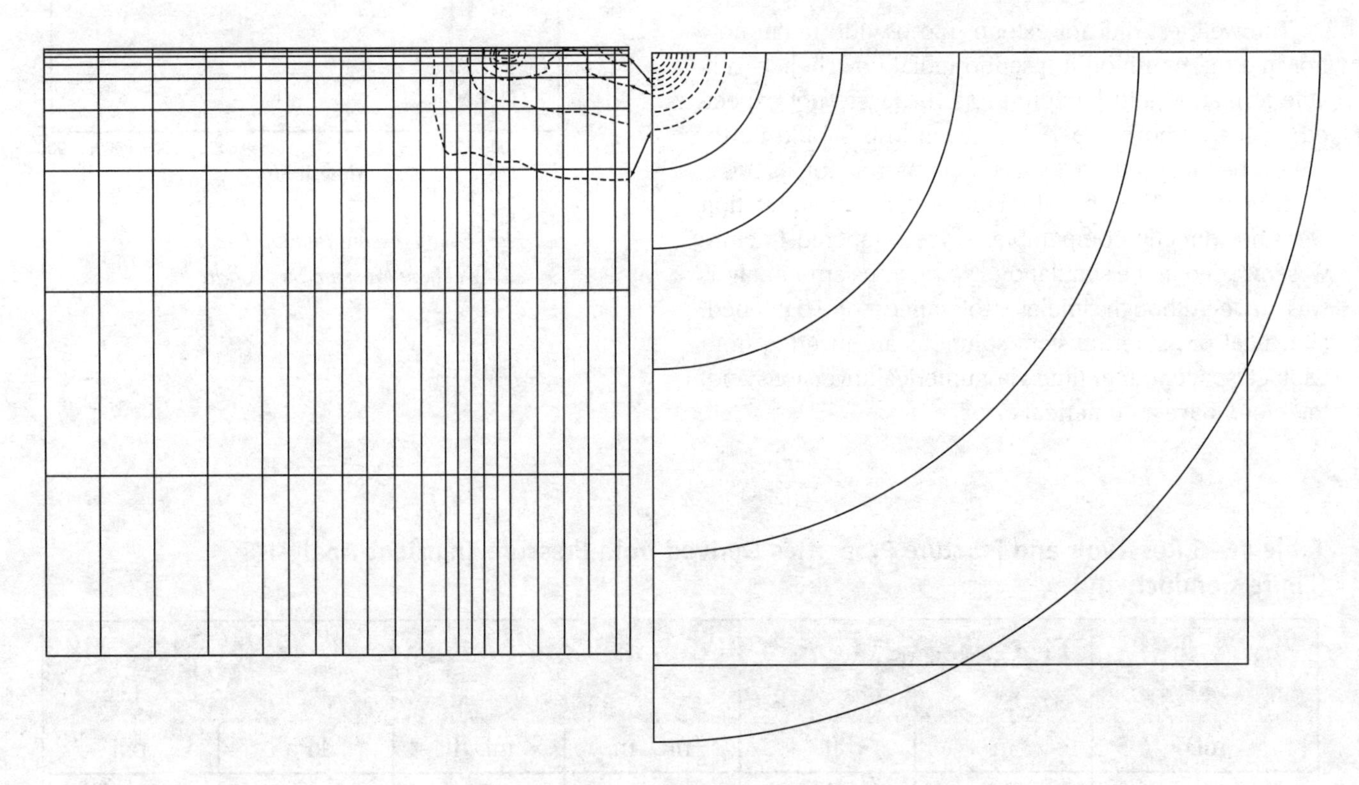

Fig. 16–21 Comparison of Condensate Saturation Contours—
Flow Into a Wellbore vs. Flow Into a Hydraulic Fracture at the Same Level of Depletion

duction for the subject well. Modeling the effects of a hydraulic fracture requires that the fracture be included in the grid. This study demonstrated and relied upon the increased power available with more modern compositional simulation techniques.

Recently, the author has rerun these early cases with a more current version of the compositional simulator, and it now runs much faster and with fewer numerical difficulties. The results have changed slightly with the newer and better model.

Case history—Glauconite channel

In another project, direct modeling of a hydraulic fracture produced some interesting results. Glauconite formation is often found as fluvial channels that feature high-permeability. In this particular case, an oil pool had been discovered, and the well was hydraulically fractured. This was probably due to damage—most high-permeability formations do not require fracturing. The well then produced at high GORs, the source of which was not apparent on logs. Material balance calculations and reservoir simulation coupled with volumetric mapping indicated a gas cap located just above the interval in the discovery well. Seismic indicated that the well would be relatively close to the edge of the channel, and a history match could only be obtained by having the fracture adjacent to the boundary. The GOR prediction, based on about a month and a half of daily production data, was extremely accurate. The GOR reached a critical level and then leveled off. The history match somewhat understated the OOIP. This is unusual, since simulation typically overestimates OOIP.

Water and gas coning

Gas coning is normally much more severe than water coning due to the low viscosity or high mobility of the gas. This is only partly offset by the increased density contrast between gas and oil versus water and oil. In some gas-over-oil-over-water situations, it may be advantageous to place horizontal wells in the water leg and suck the oil down into the well. This reverse coning strategy increases the standoff distance between the gas zone and the well and uses the water to run interference with the gas since it reduces the relative permeability. This strategy may increase recovery because it avoids or delays production of the gas, thereby maintaining the gas cap expansion drive. Venting this gas should be avoided if possible.

In one project that ARE worked on, this was the case. A few other cases are known where simulation demonstrated that this would be successful. Often clients and management are taken aback by this concept. It is an "out of the box" solution. Even though it is a little outlandish, from an engineering perspective this is a sound conclusion.

Summary

This chapter was intended to identify where detailed analysis of hydraulically fractured wells is necessary. Detailed information has been presented on how to develop reasonable input data. Horizontal wells and fractured wells were lumped together because the author views the two technologies as direct competitors and because they involve similar issues.

Quite a few "tricks of the trade" are involved in taking advantage of these technologies.

Hydraulic fracturing is a common stimulation technique and is, therefore, an important topic in evaluating reservoirs.

References

1 Horne, R.N., *Modern Well Test Analysis,* 2nd ed., Petroway Inc., 1997.

2 Cinco-Ley, H., et al., "Transient Pressure Behavior for a Well with a Finite Conductivity Vertical Fracture," *Society of Petroleum Engineers Journal* (August): 253, 1978.

3 Poon, D.C., "Decline Curves for Predicting Production Performance From Horizontal Wells," *Journal of Canadian Petroleum Technology*, 1991.

4 Mullane, T.J., et al., "Actual Versus Predicted Horizontal Well Performance, Weyburn Unit, S.E. Saskatchewan," *Journal of Canadian Petroleum Technology*, 1996.

5 Carlson, M.R., and R.J. Andrews, "A Dodsland-Hoosier Viking Waterflood Prediction," *Journal of Canadian Petroleum Technology*, 1992.

6 Carlson, M.R., "The Effect of Reservoir Heterogeneity on Predicted Waterflood Performance in the Dodsland Field," *Journal of Canadian Petroleum Technology,* (December): 31–38, 1995.

7 Carlson, M.R., and J.G. Myers, "Reduced Productivity Impairment for Fracture Stimulated Gas Condensate Wells," *Journal of Canadian Petroleum Technology 50th Anniversary Special Edition*, Vol. 38, no. 13, 1999.

8 Adams, J., and J.S. Bell, "Crustal Stresses in Canada," *Neotectonics of North America*, ch. 20, D. B. Engdahl, et al., eds., Geological Society of America Decade Map Vol. 1, 1991.

17

Advanced Processes

Introduction

The PennWell *International Petroleum Encyclopedia* contains a good summary of enhanced oil recovery (EOR) floods in the world, with a summary of the reservoir characteristics as well as the status of the floods. Of the total number of reservoirs in the world, the EOR reservoirs are listed on approximately 14 pages of 343 total pages. More flood projects are in the United States than anywhere else. This covers 170 active projects. Note that EOR includes thermal, chemical, and miscible gas (other than immiscible solution/dry gas reinjection) floods, as well as *in situ* combustion.

In the Western Canadian Sedimentary Basin, there are 26 nonthermal EOR projects. Most are hydrocarbon miscible floods (HCMF); the most important are projects in the large Beaverhill Lake carbonate reservoirs. There have been successful miscible floods on the Holmgren-Rimbey trend, notably Wizard Lake. Smaller projects exist in pinnacle reefs in the Rainbow area and the Brazeau River-West Pembina areas. Most of these are enriched gas drives. Other projects include Ante Creek and Brassey. EOR projects are relatively rare. Collectively there is a lot of experience in Canada; however, few people have designed a number of EOR projects.

Lately, a trend in HCMF is the use of horizontal wells to improve sweep efficiency, and this has been done in the high-quality reef rims of the Beaverhill Lake reservoirs. These horizontal wells are set up almost in an SAGD. The major difference is that the HCMF upper solvent injector and the lower producer are separated vertically by a larger distance. This avoids coning and viscous fingering through these lighter oils.

Carbon dioxide flooding has been the method of choice in the United States, due mainly to the extensive supplies of solvent. Two carbon dioxide floods are in Canada, and both are in southern Saskatchewan.

Overall exposure

As indicated, only a small percentage of reservoirs have EOR projects. The author has worked directly on the Swan Hills Unit No. 1 flood, a West Pembina flood, Goose River, and the Brassey field—all HCMF projects. This included injection pressure forecasting, water supply issues, economics, and detailed compositional simulation for one of the projects. From a reserves perspective, the author has looked at a significant number of different EOR projects.

Based on today's technology, HCMF, CO_2, and SAGD will constitute the majority of installed EOR processes in the future. Since this text is practical in intent, these processes are described from that point of view. SAGD is discussed in chapter 19.

Other esoteric topics on the subject include *in situ* combustion (fire floods), microbial EOR, polymer floods, alkaline floods, and other chemical floods as well as research and development on these processes. However, one should know that the essential aim of most of these processes is the recovery of additional oil by reducing the residual oil saturation, often by reducing the interfacial tension. In some respects, knowing the fundamentals of miscible processes lays the groundwork for understanding these more esoteric processes.

Classic miscible processes material

For the most part, a large volume of material is available on all EOR topics, particularly in the area of miscible flooding. The SPE published a monograph by Stalkup on *Miscible Displacement* and a textbook on *Enhanced Oil Recovery* by Green and Willhite.[1, 2] One of the better works theoretically is Larry Lake's *Enhanced Oil Recovery*.[3] Other references include *Enhanced Oil Recovery I and II* by Donaldson, Chilingarian, and Yen as well as Van Poollen's book *Fundamentals of Enhanced Oil Recovery*.[4, 5, 6] The fundamental concepts are relatively straightforward and will be discussed briefly in the next few sections.

Initial conditions

Three classic miscible processes are miscible slug, enriched slug, and lean gas injection. Note that the diagrams that accompany these processes assume that flooding is started at tertiary conditions. Although this is often the case, this situation is not necessarily the starting point for miscible flooding. The three processes are described in the following.

Miscible slug (first contact miscible) process. Miscible slug displacement refers to the injection of some liquid solvent that is miscible upon first contact with the resident crude oil. Typically, the process involves injecting a 2–5% of hydrocarbon pore volume slug of propane, or other solvent, tailed by natural gas, inert gas, and/or water. A diagram of this method is shown in Figure 17–1.[7] The diagram assumes tertiary conditions; i.e., the primary depletion and waterflooding stages have already been completed.

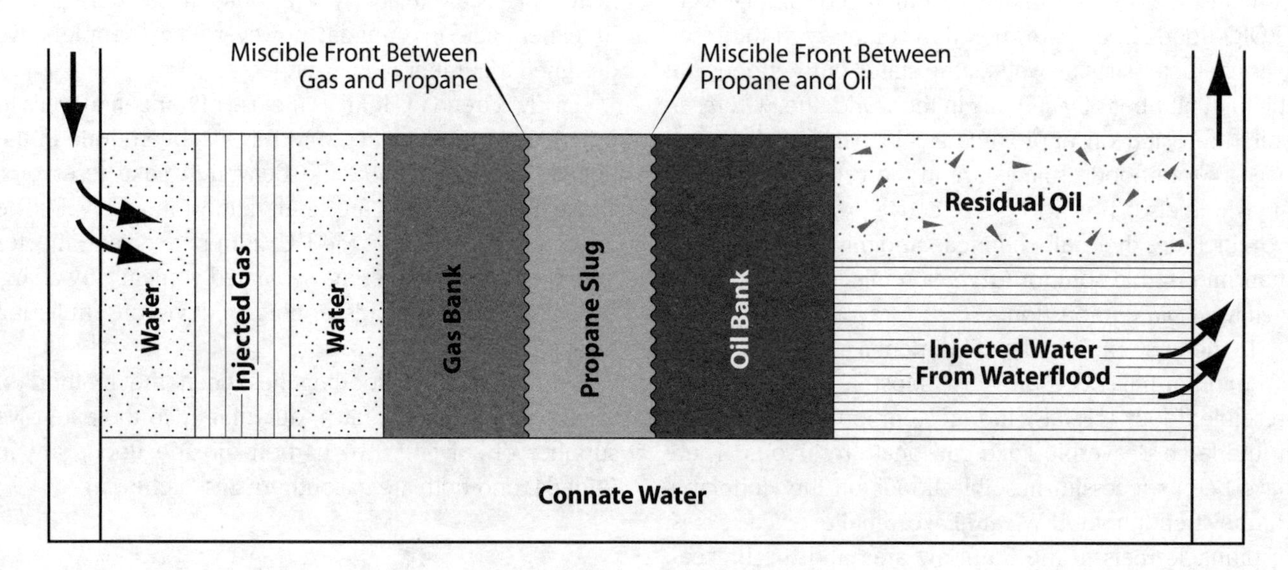

Fig. 17–1 Liquid LNG Slug Process

Enriched slug (condensing) process. In the enriched gas process, a slug of methane enriched with ethane, propane, or butane comprising 10–20% of hydrocarbon pore volume is injected followed by lean gas and/or water. As the injected enriched gas contacts virgin oil, the enriching compounds are stripped from the injection stream and absorbed into the oil. Continued injection of fresh enriched gas and stripping of the light ends around the wellbore form a zone rich in C2 to C4. If a sufficient quantity of enriched gas is injected, the enriched oil band around the wellbore will become miscible with the injected enriched gas. This process is shown in Figure 17–2.

Unlike propane, a typical miscible slug injectant that is miscible on first contact with the reservoir oil, the enriched gas drive process relies on multiple contacts between the oil and enriched gas to develop a miscible slug *in situ*. Although this process uses less-expensive material (ethane is cheaper than propane), more is needed and higher pressures are required to attain miscibility. There is more viscosity and density contrast in the system, which results in more gravity override effects and a less-efficient displacement process. As discussed later in this chapter, gravity override is a phenomenon that is easy to underestimate, and anything that increases its effect must be carefully weighed in the overall economic process assessment.

Lean gas (vaporizing) process. The lean gas process involves the continuous injection of high-pressure methane, ethane, nitrogen, or flue gas into the reservoir. Intermediate hydrocarbon fractions (C2 to C6) are stripped out of the oil and absorbed in the lean gas phase.

If the reservoir liquid is rich in intermediate fractions (C2 to C6), the leading edge of the gas front will become saturated with the reservoir oil's light ends and become miscible with it. Miscibility is not attained at the well face but at some distance away from the injection point. This distance, depending on injection pressure and oil composition, may vary from a few to hundreds of feet before enough intermediates are absorbed into the gas front to become miscible with the oil. Thus, there is a ring of residual saturation of stripped oil around the well. This process is shown diagrammatically in Figure 17–3.

Objectives

EOR is an area that generates many questions in courses taught by the author. The two most common question areas relate to ternary diagrams and miscible flood simulation. The third most common topic is EOR screening. Even though there is a considerable volume of material on this topic, it seems difficult to summarize. Following is the author's simplified version of design issues:

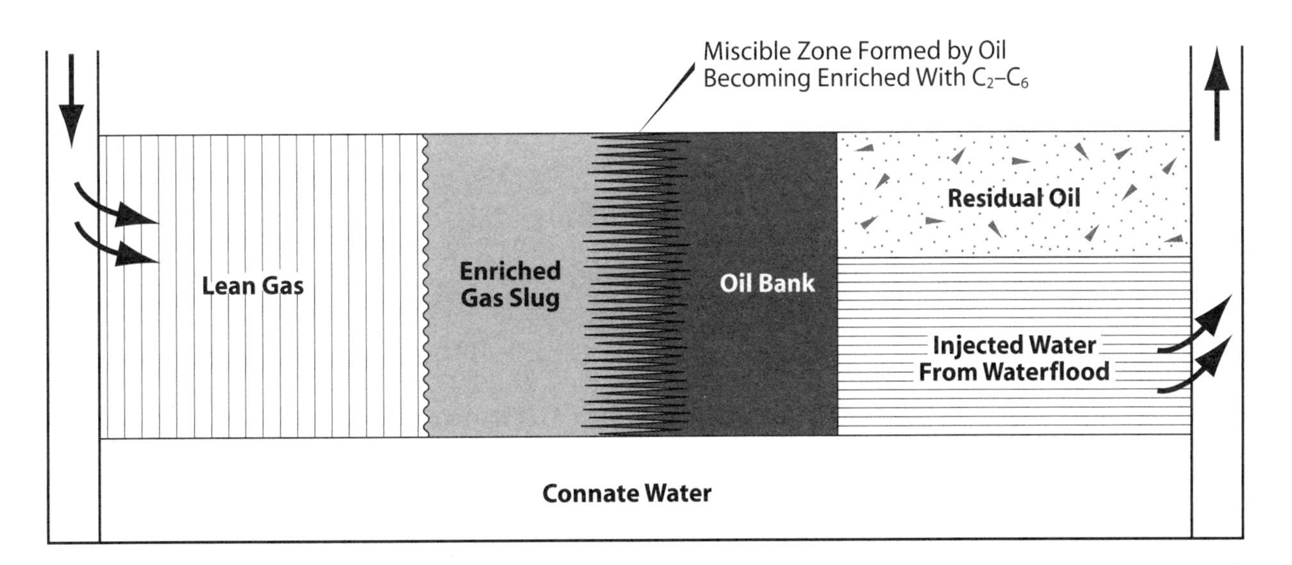

Fig. 17–2 Enriched Gas Drive Process

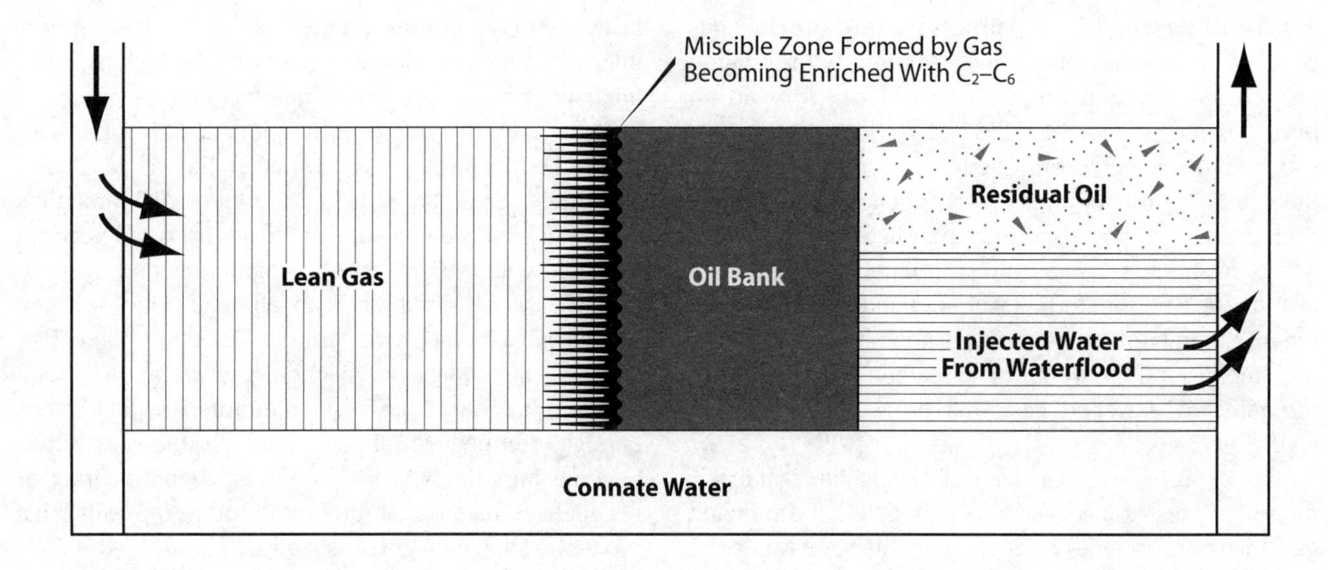

Fig. 17–3 Lean Gas Process

- We inject solvent into the ground and must determine if the solvent will miscibly mix with the oil. If this is the case, then residual oil saturations will be reduced by solvent displacement. It is also necessary to determine over what pressures and temperatures miscibility will take place.
- We must determine the amount of solvent required. Ideally, a small amount of the hydrocarbon pore volume is injected. However, practice indicates that more is required to overcome sweep irregularities. Ideally, it would be possible to use about a 5% hydrocarbon pore volume slug. Practice indicates that somewhere in the range of 15–20% may be required. Restated, we need to know when to stop injecting.
- Determining the additional amount of oil that will be recovered is required to determine economics. Just as importantly, knowing when the oil production increase will arrive is important. The oil production increase is also known as the oil bank. Restated, a production forecast is required. This is for use in discounted cash flow analyses.
- Experience has shown that that gas is not a good displacement medium, therefore it is necessary to take active steps to control this—water alternating gas injection is normally used.
- It is necessary to displace or propagate the solvent slug through the reservoir's oil zone in the direction the design specifies. This can be horizontally, vertically, or somewhere in between. Either water or gas can be used to achieve this.
- All of the previous technical considerations interplay with the economic viability of the process, with the result that very detailed economic models are built.

One of the most readable versions of this overall process can be found in *Enhanced Oil Recovery, Part II* by Donaldson, Chilingarian, and Yen. The methodology uses some generalized correlations, but it is constructive in that it links together the key issues in a readable fashion and in bite-sized chunks.

After this simplified view, miscible flood design becomes very ugly. A discussion of this topic leads immediately into numerous sophisticated techniques, which don't work very accurately. The following expands on making the best use of the design objectives listed previously.

Step 1. When mixing hydrocarbons, the option of using an equation of state (EOS) is a natural starting point. It is involved technically, and this topic alone consumes entire books. In the end, an EOS is not accurate enough to be predictive, which leaves one reliant on lab data. It helps to have an idea of where to start, and this leads back to the

EOS to design the lab program. Alternatively, one can use the rough design graphs outlined in Donaldson, et al. Realistically, the design charts, information from offset HCMFs, and the EOS are used.

Lab tests are expensive, and, to optimize this process, it is best to have a quick review or screening test before committing to major expense. The rising bubble apparatus combined with the use of a video camera yields a significant amount of information about the process that is occurring. This sets the stage for slim tube tests and swelling tests, which are in addition to the basic PVT experiments for the oil.

Step 2. The next stage is to determine the solvent slug size. This is a natural topic for a reservoir simulation discussion. Unfortunately, simulators are not capable of handling viscous fingering directly or from first principles. This is one issue that must be addressed—in addition to reservoir heterogeneity. These issues have a bearing on sweep efficiency, productivity, overall recovery, and the economic viability of the process. In addition, during this stage, it is necessary to evaluate how miscibility will be developed and maintained if a multicontact process is chosen.

In addition, there are two basic types of processes. Originally, it was thought that the intermediate components in enriched gas drives were absorbed *into the oil*, which is termed a *condensing gas drive*. Subsequently, it has been suggested that the heavier components of the oil will be absorbed *into the solvent*, enriching the solvent near the oil. This is termed a *vaporizing gas drive*. The latter set of problems are compositional issues and do not lend themselves to being solved with a single simulation run. The sweep issues are best identified with a modified black oil simulator, while the latter are best addressed using a compositional simulator. Intermediate between these two issues is the issue of gravity override. The densities of the fluid are PVT controlled, and the tendency for segregation is controlled by reservoir properties or sweep issues.

Step 3. The amount of oil remaining in the reservoir is not easy to determine. In large part, this depends on relative permeability end-points, as well as vertical and areal sweep efficiencies. Since the investment in a miscible flood is capital intensive, significant effort is expended on these points. There are a number of ways to go about this.

As with the issues described previously, it would be unusual to rely on a single method. Some techniques include well logging, recovering core from infill wells, log inject log techniques, a review of existing core relative permeability and capillary pressure data, detailed core flooding tests, and revised mapping to minimize uncertainty on the OOIP.

Intrinsic with the issue of available residual oil is the issue of how water saturations from the waterflood affect solvent access to the residual oil saturation. Theoretically, a miscible flood should leave zero residual oil saturation; in practice, a minimum amount cannot be recovered. Reservoir simulation can be used to estimate unswept areas of the reservoir. For the majority of EOR reservoirs, this amounts to classic waterflood engineering before getting to the nuts of bolts of EOR design. Therefore, as much conventional reservoir simulation as specialty reservoir simulation is used in one of these projects.

Step 4. In order to minimize viscous fingering and the effects of heterogeneity, solvent and gas injection are normally alternated. This has traditionally been designed with a reservoir simulator, since the objective is to take advantage of a relative permeability effect. Sufficient practical experience exists to validate water alternating gas (WAG), and an argument can be made that this design step is not necessary. Nevertheless, most designs will revisit the issue with detailed simulation.

Step 5. For most HCMF drives, chase gas was initially envisaged. This can be done with water also. Chase gas is usually WAGed as well.

Step 6. Due to the capital-intensive nature of the technique, economics are extremely important. Suffice it to say, an engineer who embarks on a major HCMF project will have plenty of opportunities to make presentations to management and possibly to shareholders, the oilfield press, and regulatory authorities.

In summary, a number of different but sophisticated techniques are used, most of which have fundamental limitations. This is no different from the rest of reservoir engineering, but the problem is more severe with this technology. Step 1 alone requires a good knowledge of EOS, which is discussed in chapter 15. Steps 2 and 3 involve two kinds of modeling in addition to conventional reservoir simulation—compositional and pseudo miscible. Both address different issues. Historically, since much of the work in this area was driven by miscible flood requirements, developments were taking place in this area as designs were proceeding.

The pseudo miscible options are unique to EOR design.

When teaching classes on this subject, the author has found that the last day seems to have an "ah ha" moment. Although the topics are outlined at the beginning, the way various elements fit together is not necessarily clear until after all the work components have been covered. However, as indicated, miscible flood predictive technologies are sophisticated and still evolving. Therefore, the coming together of all six steps of a miscible flood design described earlier may not result in as much practical enlightenment as expected—the soft answers may not lead to an "ah ha" moment. That is sometimes the nature of such leading edge technologies.

The reservoir management issue comprises knowing the issues that are important as well as recognizing that different sources of data must be weighed and a conclusion (right or wrong) must be made. Making correct decisions respecting miscible flooding is fundamentally a highly technical process, which requires balanced experience in a number of disciplines.

The balance of this chapter follows the previous outline. Since there is overlap in steps 1 through 6, some topics have been arbitrarily placed in one category instead of other. One other area that is design related is monitoring and surveillance. Experience has proven that this is an important part of achieving maximum performance.

Miscibility and Minimum Miscibility Pressure

Screening

Screening is normally done during the early stages of evaluation—prior to detailed design. As with waterflooding, the best responses come in high-quality reservoirs. This means high permeability and porosity. HCMF is sensitive to gravity override. Generally, CO_2 is less sensitive because it is denser than the corresponding hydrocarbon injectant that would be used at the given reservoir temperature and pressure. Therefore, the best miscible flood candidates have high vertical relief and take advantage of gravity segregation—i.e., the miscible flood is conducted in a top-down vertical configuration. It is best if the reservoir is relatively homogeneous. In support of this point, compare the performance of two reservoirs that are virtually side by side—the Golden Spike and Wizard Lake pools. The former had an extensive barrier in the middle of the pool not previously discovered until it broke up the solvent bank as it descended through the oil column. The latter is relatively free of barriers, with the result that excellent sweep efficiency and recovery were obtained with the vertical HCMF process.

Although it may be tempting to think that sweep efficiencies will improve with miscible flooding as opposed to waterflooding, this is not the case. Although the sweep from miscibility appears to be better, it will not be realized due to heterogeneities, viscous fingering, and less-favorable mobility ratios. Instead, improvements typically come from the recovery of a greater amount of the residual oil in the previously swept portions of the reservoir.

Similar problems with gas flooding

Most HCMF processes use enriched gas. Restated, it is a version of gas flooding. Historically, the performance of gas floods has been disappointing. In order to overcome the negative connotations of gas flooding, some people in the industry referred to HCMFs as being fundamentally different and said that LNGs were injected.

While miscible flooding is different from immiscible flooding, it is not correct to characterize solvents as a liquid. The injected fluids are a gas and are not injected in the liquid phase. The pumps used for injection are made by Ingersoll Rand, a compressor company, and not by National Pumps, a water injection pump manufacturer. Some tubular pressure drop calculations have been based on liquid propane properties, but the truth is that HCMF projects suffer from the same risks as dry gas injection. The author strongly discourages this type of terminology.

Design curves

The design curves of Benham are handy because they give an idea of the trends and likely recovery for a miscible flood. This is a good place to start and a good overview of the process.[8]

Equation of state

The use of the cubic EOS has been discussed extensively in previous chapters. As always, optimization for use in simulation revolves around minimizing the number of pseudo components and comparing data to lab tests. For most miscible flood systems, four components should be sufficient unless the fluid is near critical. In the latter case, up to nine components may be required. Note that a three-component characterization is used in ternary diagram descriptions—a useful simplification.

Ternary diagrams

The most confusing thing about ternary diagrams is that the choice of components is arbitrary. The use of a ternary diagram is a chemical engineering simplification of a more complex system. If one's choice of pseudo components is not realistic, the ternary diagram will not help. Perhaps what is less obvious is that, for the majority of oils, the actual intermediate component is ethane. If one is analyzing a CO_2 flood, the positions of critical properties of CO_2 and C2 on a convergence diagram are similar. CO_2 has a higher critical pressure than C2. However, CO_2 is the intermediate component of choice for ternary diagrams.

In terms of defining first contact and multicontact miscibility, Lake's explanation is the most succinct. His diagrams are reproduced as shown in Figures 17–4, 17–5, 17–6, 17–7, and 17–8.

Some ternary diagrams are developed from untuned parameters and some are created from lab data. The more realistic data will be supported by lab tests. The ternary diagrams from Lake may be of use in evaluating the series of contacts.

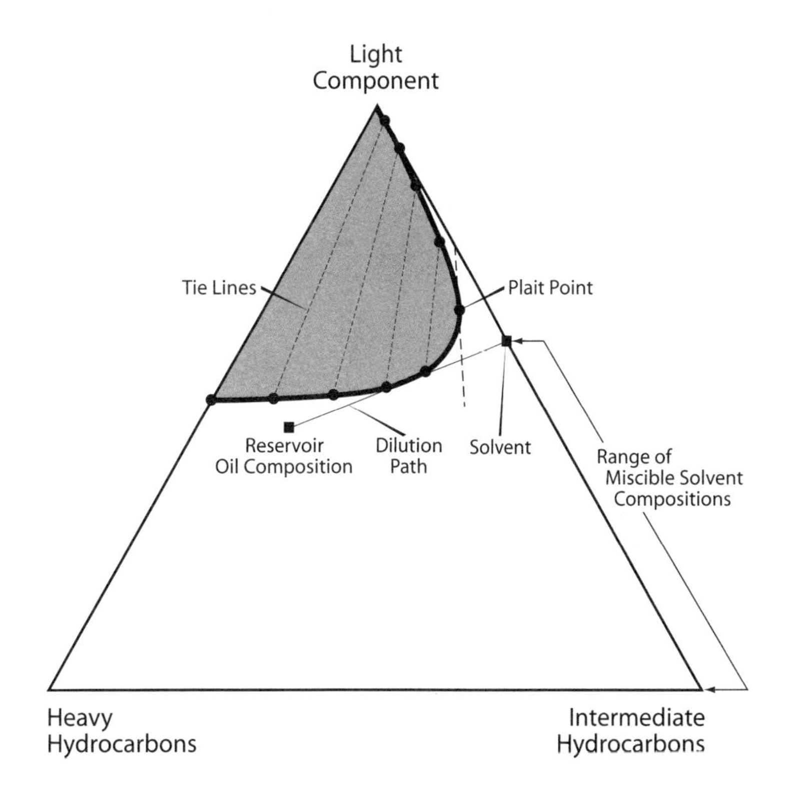

Fig. 17–4 Ternary Diagram A

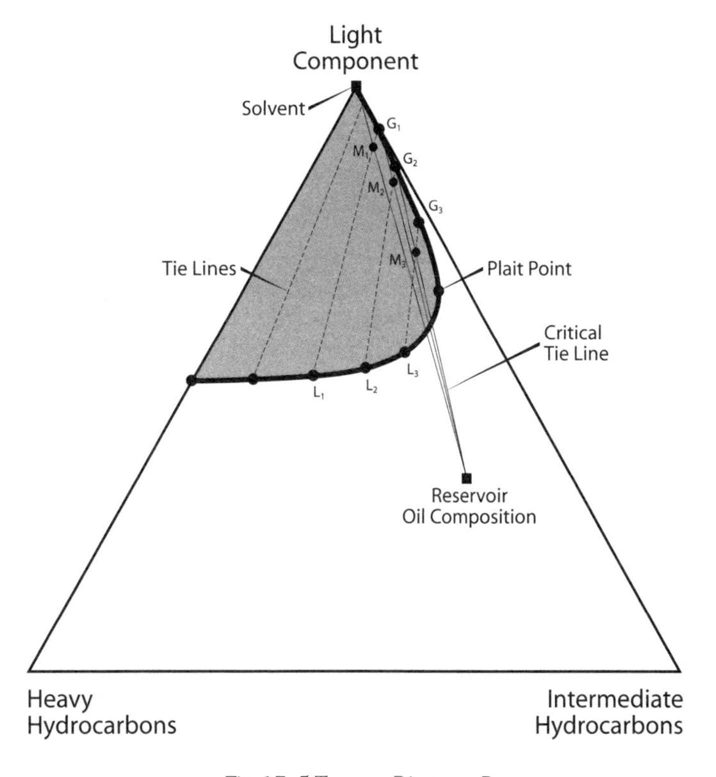

Fig. 17–5 Ternary Diagram B

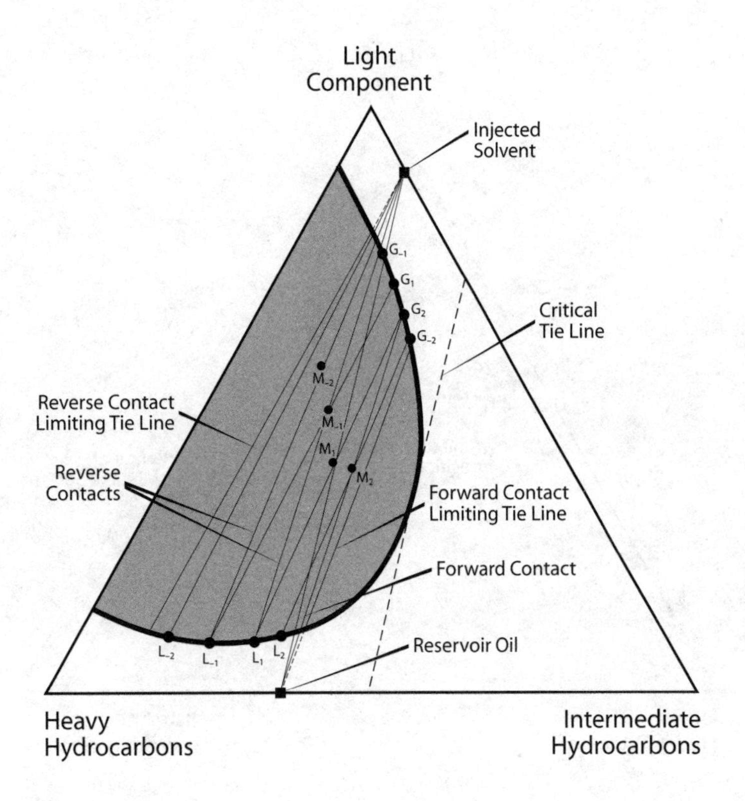

Fig. 17–6 Ternary Diagram C

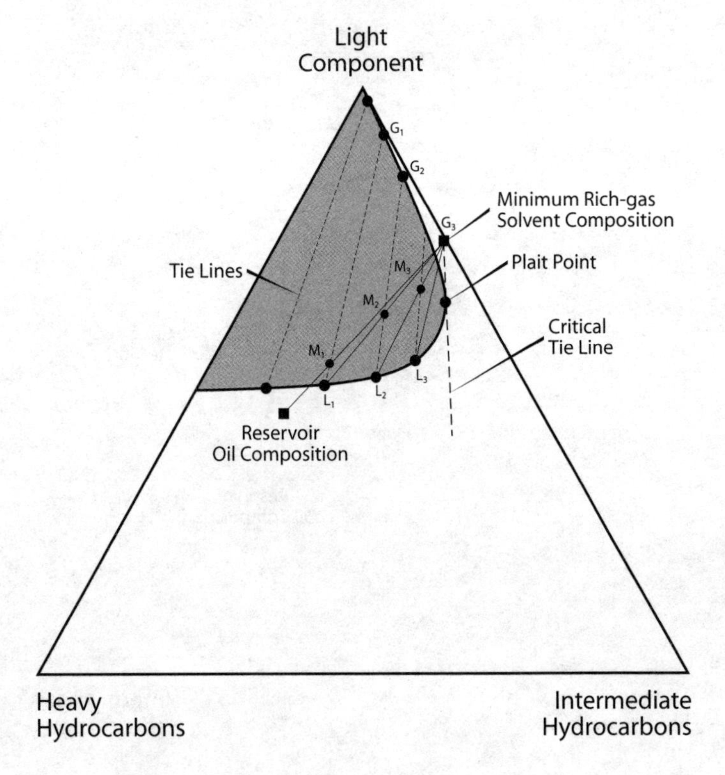

Fig. 17–7 Ternary Diagram D

Condensing versus vaporizing processes

The use of ternary diagrams may not result in unique solutions. Novosad and Costain of Shell Canada did some interesting work on how the phases would interact. This work was related to the miscible flood for the Virginia Hills Units No. 1 and a Beaverhill Lake reservoir. The methodology uses pseudo ternary diagrams that were generated using an EOS.[9]

Swelling tests

ARE evaluated swelling tests for one reservoir, which was subsequently used in a full field model. The number of components was successfully reduced from 10 components from an original study to 5.

Initially, the EOS characterization used the original oil PVT data. The work proceeded to include the results of swelling tests and slim tube modeling. Figure 17–9 shows the before and after picture of the characterization. This is another example where having data that spans the entire expected range of composition space, mathematically speaking, is necessary for good tuning. The obvious implication is that more lab data is better than no lab data.

Reducing the number of components to five was difficult. This took three days of working with the data. More than one person attempted this characterization. The other characterization involved eight components. What is the moral of this story? Keep trying, and a bit of luck helps.

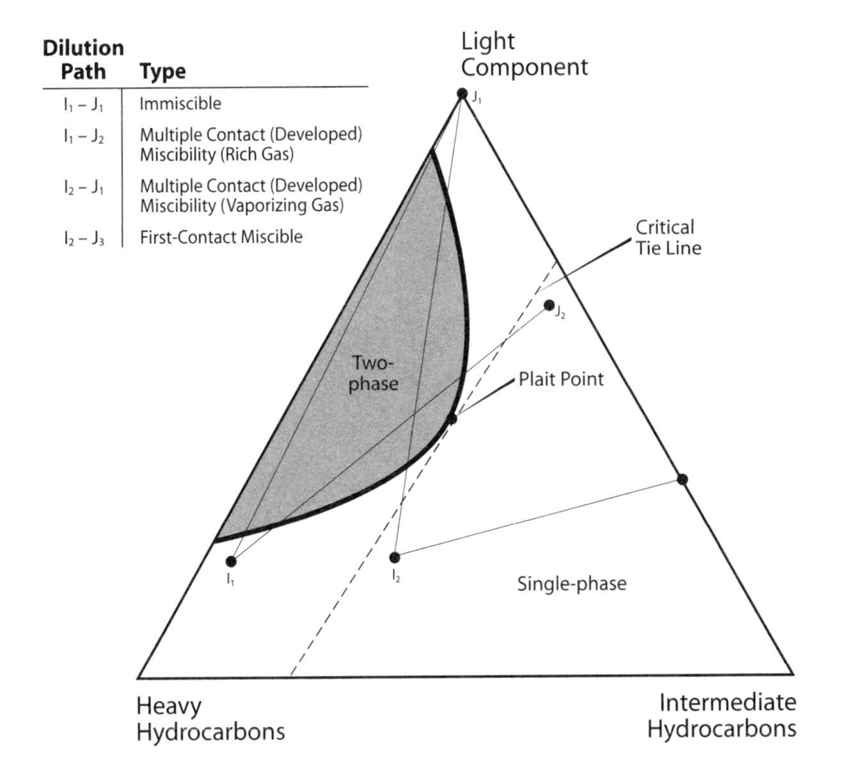

Fig. 17–8 Ternary Diagram E

Nitrogen

Nitrogen is one component that is often found in hydrocarbon reservoirs. Nitrogen has a higher critical pressure and critical temperature than methane, which will raise the minimum miscibility pressure (MMP). Therefore, nitrogen significantly raises the MMP.

Hydrogen sulphide

Hydrogen sulphide is known to depress minimum miscibility pressures as shown in Figure 17–10.[10] Traditionally, hydrogen sulphide has not been popular as a solvent. It is corrosive, constitutes a considerable safety hazard, and can be converted to sulphur—a saleable product. Lately, however, the supply of hydrogen sulphide has exceeded the demand for sulphur, and waste disposal is being investigated. Some of these projects have suggested that the useful properties could be utilized in miscible fashion.

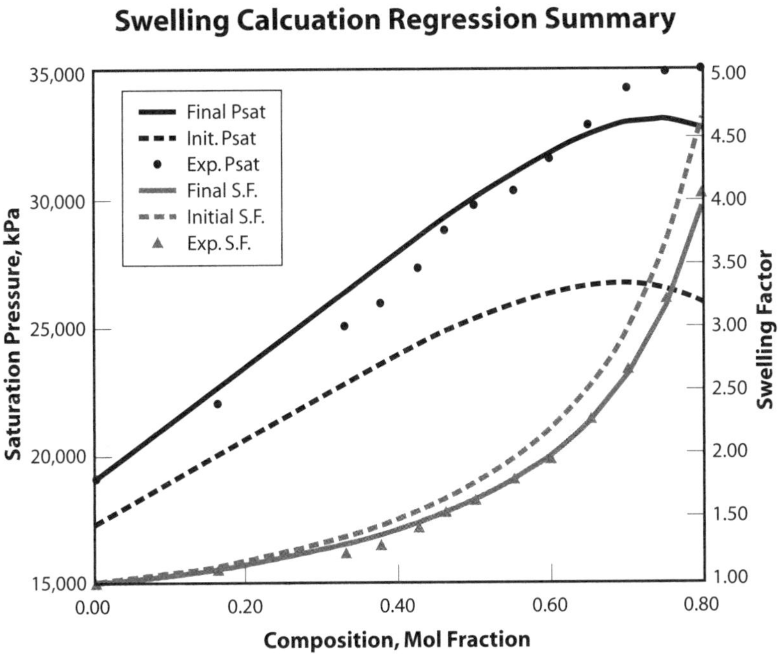

Fig. 17–9 PVT Tuning with and without Swelling Data

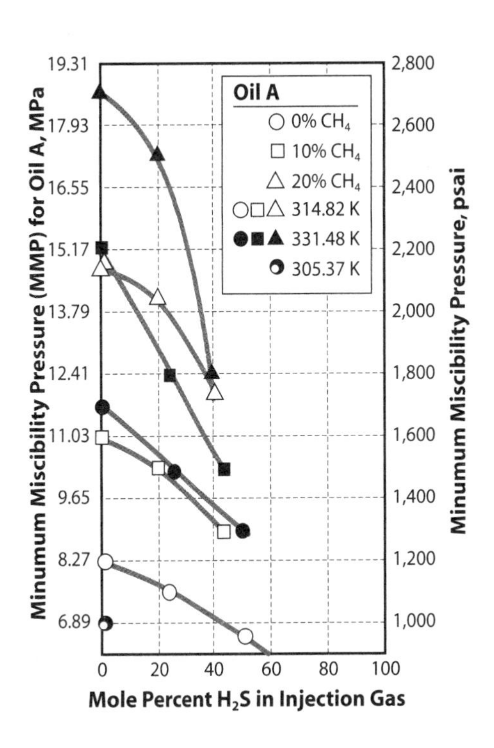

Fig. 17–10 Effect of Acid Gas on MMP

Slug Sizing

Diffusion

The mass transfer process is controlled by the difference in concentration between the various components, such as C1, C2, C3, etc. Most reservoir calculations assume that thermodynamic equilibrium has been reached and that there are no time-dependent effects. However, the injection process is rapid, and this assumption of equilibrium may not be true in miscible-flood-type processes.

On a larger scale, current thinking is that reservoir diffusion effects are much smaller than those caused by reservoir heterogeneities. This is based on work by M. P. Walsh and E. M. Withjack and presented at the 44th Annual Technical Meeting of the Petroleum Society of CIM. The standard Berea core is much more uniform than reservoir rocks with a Dykstra-Parsons ratio of as low as 0.3. Traditionally, most core floods have been done on shorter lengths of core. Walsh and Withjack's work involved core samples that were significantly longer. Analysis of the core showed that more dispersion was caused by heterogeneity than by molecular diffusion.[11]

Viscous fingering

Tests in a Hele-Shaw cell, which consists of two parallel plates of glass, show that the displacing fluid fingers through the formation as shown in Figure 17–11. Originally, this behavior was attributed to minor heterogeneities in the porous media (actually glass plates) that acted as a perturbation and started an instability. More recent work indicates that the fluids *themselves* may be responsible for the fingering and do not require any heterogeneity to initiate fingering (Brownian motion?). Currently, glass plates can be made with flat surfaces, and the distance between the plates can be controlled to within 1/10,000 of an inch (4 microns). Restated, the possibility of heterogeneity between two plates of a significant scale is remote. Lean and enriched gases have substantially lower viscosities than the fluid that they are displacing, an adverse mobility condition that makes viscous fingering inevitable.

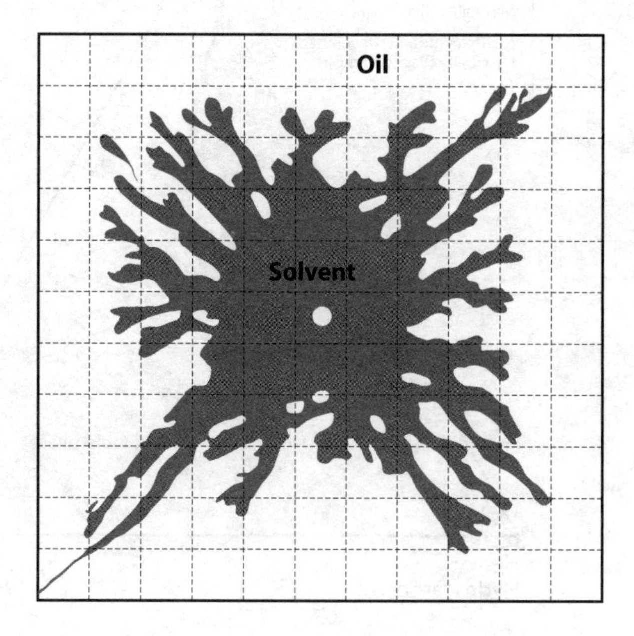

Fig. 17–11 Hele-Shaw Cell Results

Black oil models

Black oil models assume multiphase flow (usually oil, gas, and water), three components, and negligible dispersion. It is further assumed that the oil component can exist only in the oil phase; the gas component can exist in the oil and the gas phase, but not in the water phase; and that the water can exist only in the water phase.

Black oil models can be used in cases where limited miscibility exists or where there is a limited amount of natural gas entrained in the oil. In general, they are not appropriate for miscible flood analysis.

Pseudo miscible displacement models

Pseudo miscible displacement models usually assume single-phase flow, two components called oil and solvent, and significant dispersion. Dispersion is normally included via Fick's Law of Diffusion, in which the mass flow rate for the oil component is proportional to the concentration gradient multiplied by a dispersion coefficient tensor. Convective flow takes place via Darcy's empirical law.

Pseudo miscible displacement simulators suffer from numerical dispersion, which often exceeds the actual dispersion via concentration gradients (Fick's Law). As a result, detailed grids may be required to minimize these effects. Even with refined gridding, some numerical dispersion is inevitable.

In addition, it is possible to adjust capillary pressures and relative permeability to reproduce first contact processes, which are always miscible. However, this method does not allow viscous fingering to be accounted for, which is usually a significant effect.

Techniques of Koval and Todd/Longstaff

In 1963, Koval developed an analytical solution to predict the performance of unstable miscible displacements. The ultimate effect of this technique is to alter the relative permeability relations. The effective viscosity relationships can be altered by adjustment of a single empirical mixing parameter.[12]

Todd and Longstaff also developed a technique for use in a black oil simulator for first contact miscible displacement with dispersive effects (viscous fingering). This involves a modification of the gas-oil relative permeability curves and capillary pressure as well as a method to calculate effective viscosities between the oil and gas phases when mixing occurs.[13]

Both pseudo miscible techniques assume first contact miscibility. In practical terms, this would also include systems that are rapidly multicontact miscible. Phase behavior effects are greatly simplified—no volume change occurs on mixing, and density and viscosity depend on solvent concentration. One other major problem is selecting the appropriate mixing factor, k or ω, depending on whether the Koval or Todd and Longstaff technique is used, respectively.

Determining mixing parameters

Although the methods of Koval and Todd and Longstaff do the same thing, it seems that the Koval technique and the use of ω is more pervasive. Values of ω can be estimated from history matching but cannot be directly calculated. As yet, there is no prediction theory for viscous fingering. The exact value of ω depends on gridblock sizes, the mobility of the fluids, and the mobility ratio. Some papers have tabulated values of ω; however, the surveys are not complete. Nevertheless, unless a company has direct experience on the field one is working on, some source of data is needed.

The papers on this modeling approach emphatically state that CO_2 and HCMF performance can be predicted with simulation. That is, provided one has the right fudge factor (ω). Verbal reports indicate that a value of 0.6 is a ω factor that works fairly well. Now, there are many different opinions on this, and the author cannot argue from direct experience. Even three or four similar design projects do not qualify as a worldwide average. Stalkup, in his SPE monograph, suggests a range of 0.5 to 0.7. However, values as high as 0.8 have been used. Therefore, to do this type of model, it will be necessary to do some research on values of ω used on offsets and simulation sensitivities to different values of ω. The petroleum literature expands every day, so it is necessary to collect all the data available at the time. In general, some operating companies may be reluctant to reveal their experience. However, there is data in the literature.

Compositional model

Compositional models assume multiphase flow (oil, gas, and water), two or more components, and negligible dispersion. The distribution of components between the vapor (gas) and liquid (oil) phases is determined via k values—equilibrium ratios of vapor versus liquid phase component mole fractions. The latter can be input as functions of pressure and composition or derived from equations of state (EOS).

Compositional simulation is the natural choice for flooding applications where complex phase behavior and compositionally dependent properties, such as phase density, interfacial tension, and viscosity, are more critical to predicting displacement efficiency than viscous fingering and areal sweep effects. Dispersive behavior—i.e., viscous fingering and molecular diffusion—is ignored.

Hybrid compositional

Other simulators have been created that mesh the best features of black oil, pseudo miscible, and compositional models. They do not partition components in as exacting a manner as a conventional compositional simulator. Component transfer can be neglected when the block pressure exceeds a preset miscibility pressure. Segregation of fluids is controlled through a mixing parameter approach. It is possible to include an immobile residual oil saturation, solubility of CO_2 in the water phase, and asphaltene dropout.

Stream tube

Another method that has been used successfully is the generation of a series of streamlines with a simplified reservoir simulator. Each streamline is modeled as a separate cross-sectional simulation. This method allows detailed reservoir descriptions to be incorporated. This is work intensive, and all of the streamlines must be summed to reach a final field level forecast. The method is not 100% accurate. The streamline locations are controlled by the mobility ratio of the solvent and by the reservoir heterogeneities that exist along the path. The compromise is successful because the streamline locations do not move appreciably and the reservoir heterogeneities have a significant effect on total performance.

Chevron successfully used this technique in the Kaybob field. A paper was written on this by Behrens that outlined the general process. As mentioned earlier, reservoir performance is strongly affected by reservoir heterogeneity. The use of these stream tube models allowed geostatistical properties to be included in the simplified vertical cross sections. All of the stream tube models are added to get total reservoir performance. This approach requires significant logistics to store, generate, and run all of the different models.[14]

This paper did not address many of the issues in selecting the stream tubes nor the way a pattern of stream tubes may vary with time or varying production rates. However, stream tubes are now being incorporated in commercial software. The technique appears to be gaining popularity as software becomes more widely available.

Choice of method

No model exists that covers all aspects of miscible displacements simultaneously. Pseudo miscible models ignore phase behavior effects but take into account viscous fingering. Compositional models take into account phase transfer processes but ignore viscous effects.

Klins has prepared a diagram for CO_2 floods that summarizes the previous concepts into different pressure and temperature regions. The diagram he presents, which is included as Figure 17–12, is a function of the oil. Hence, each oil must be examined on its own merits.[15]

Heterogeneities

The high contrast in viscosity or high mobility ratio M make miscible processes sensitive to reservoir heterogeneities. This is not new; the inclusion of layering in waterflood design is well-established. The increased sensitivity of EOR processes to finer scale heterogeneity has promoted recent interest in the detailed descriptions available through geostatistics.

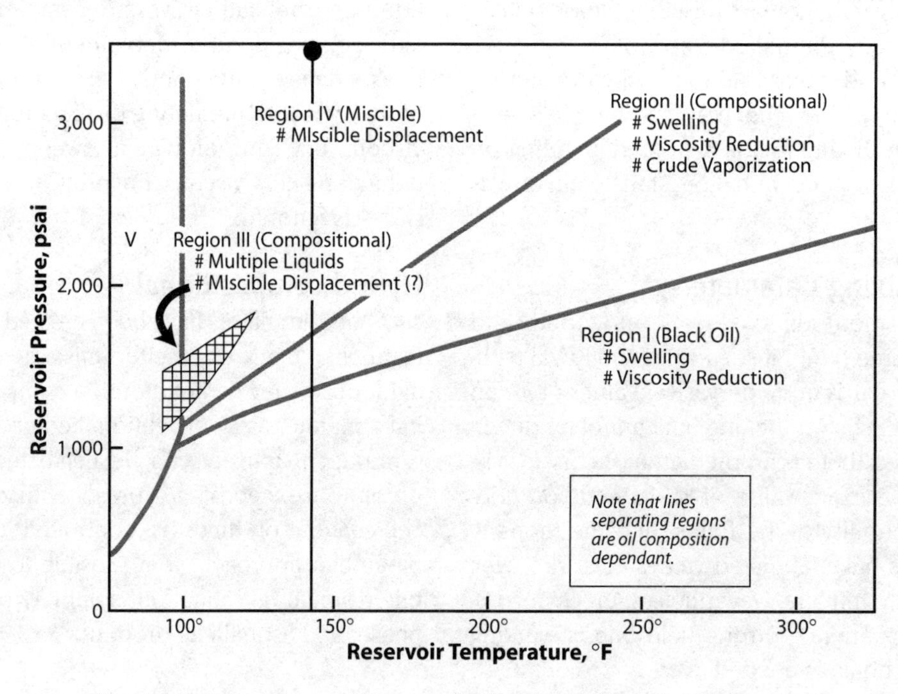

Fig. 17–12 Temperature and Pressure Regions for Carbon Dioxide Flooding

Sweep Efficiencies

Frontal advance

As an immiscible fluid is injected, a displacement takes place that is controlled by the classic gas-oil relative permeability curves. Mass transfer then takes place between the two phases, which coexist spatially in the reservoir. As time progresses, the fluids become progressively more similar. Relative permeability can be affected by the interfacial tension between the oil (liquid) and vapor (solvent) phases. These changes are shown in Figure 17–13.[16] Hence, mixing is governed partially by relative permeability, which changes with distance and time along the displacement path.

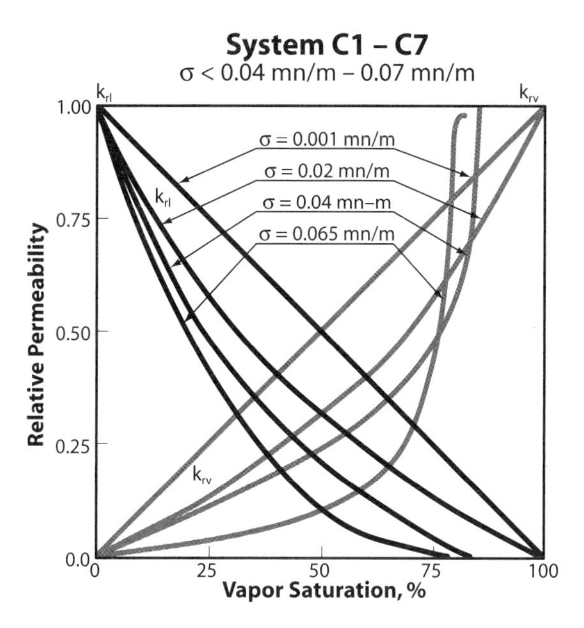

Fig. 17–13 Effect of IFT on Relative Permeability

Slim tube testing

The miscible flood process is quite complex; therefore, it is normal that advanced lab testing is conducted. Normally, this takes the form of a slim tube test. A slim tube apparatus consists of a long tube in which 100 to 200 mesh glass beads or crushed quartz has been packed and which is 100% oil filled. The absence of connate water is one of the general limitations of this test. Water is not thought to have a large effect on PVT properties, and this is the primary purpose of the test. However, connate water saturations can be an important consideration in displacement processes. Generally, the tube diameters are relatively small, typically 0.250–0.375 inches. The thin cross section is intended to minimize gravity segregation and later dispersion. The tube is normally 6–12 meters (20–40 feet) long. Generally, the tube is coiled, so that it can fit into an oven. There are two ways to do this. One is to use a coil that is similar to a regular spring, and the other is to use a spiral coil. The latter is perfectly level, whereas the first method results in a slight slope.

Oil recovery is then plotted as a function of pore volumes of solvent injected; an example is shown in Figure 17–14. Performance is gauged based on breakthrough occurrence (for a miscible displacement, this will typically be later than 75–80% pore volume injected) as well as the amount of recovery at 1.2 pore volumes injected (if miscible, this should be greater than 95% of the OOIP). However, oil recovery from a single slim tube test is not sufficient for determining either MMP or solvent composition. Stalkup suggested that setting a minimum recovery level as a criterion for miscibility is not correct; instead, he suggested that the breakover point from a series of displacements be used. For example, if MMP is sought, the recovery at 1.2 pore volumes versus pressure of the respective slim tube experiment may be plotted. The sharpness of this breakover point, which depends on solvent composition and temperature, and recoveries above the breakover point can help to identify where MMP has been attained. This is shown in Figure 17–15.[17]

Slim tube recoveries are *known* to be extremely high, but they are not used to predict reservoir recovery levels *per se*. The shape/character/parametric response to pressure and composition are what are used in reservoir process design.

Other parameters can be examined, such as lean gas mole fraction and produced fluid density during the course of the experiment, which help to identify that a process is miscible and that it behaves as first contact versus multicontact miscible. The slim tube process can be simulated numerically. ARE has done slim tube modeling. The results are shown in Figure 17–16. In order to do this, the simulator must have conventional relative permeability curves.

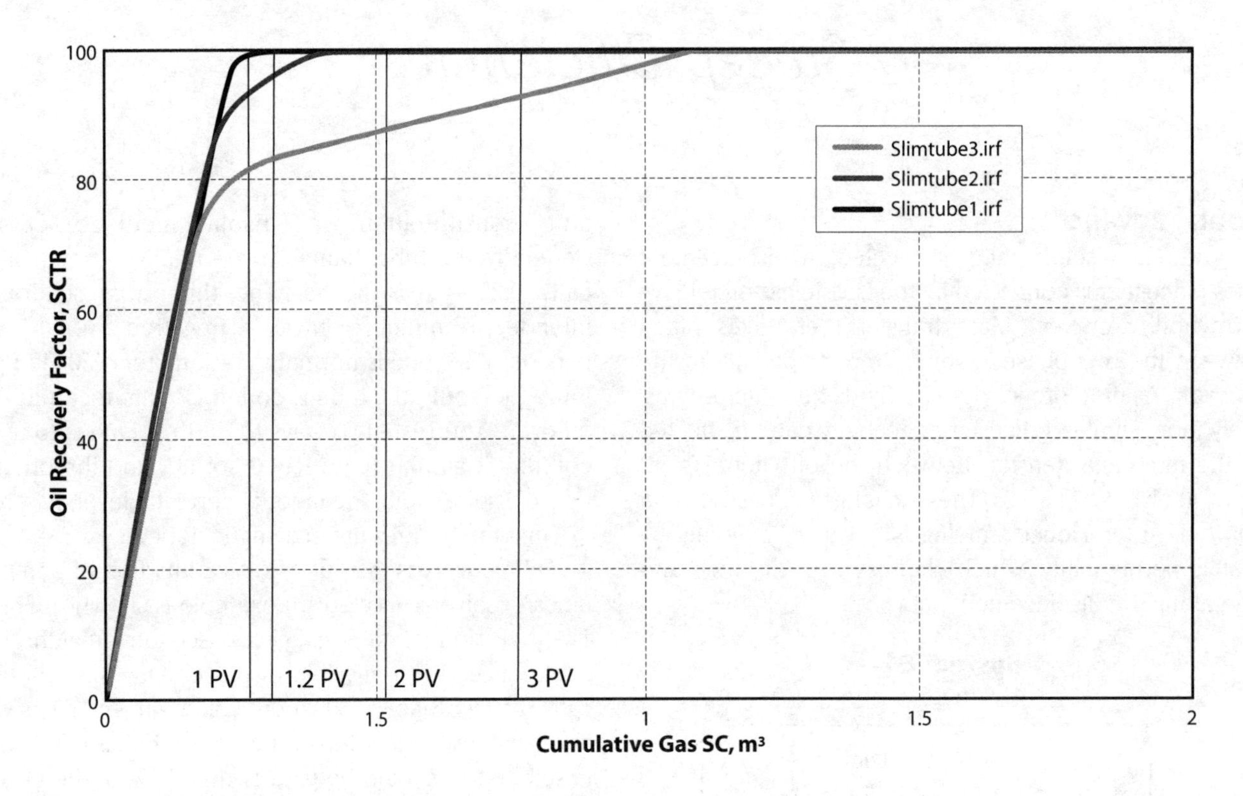

Fig. 17–14 Slim Tube Recovery as a Function of Oil Recovery

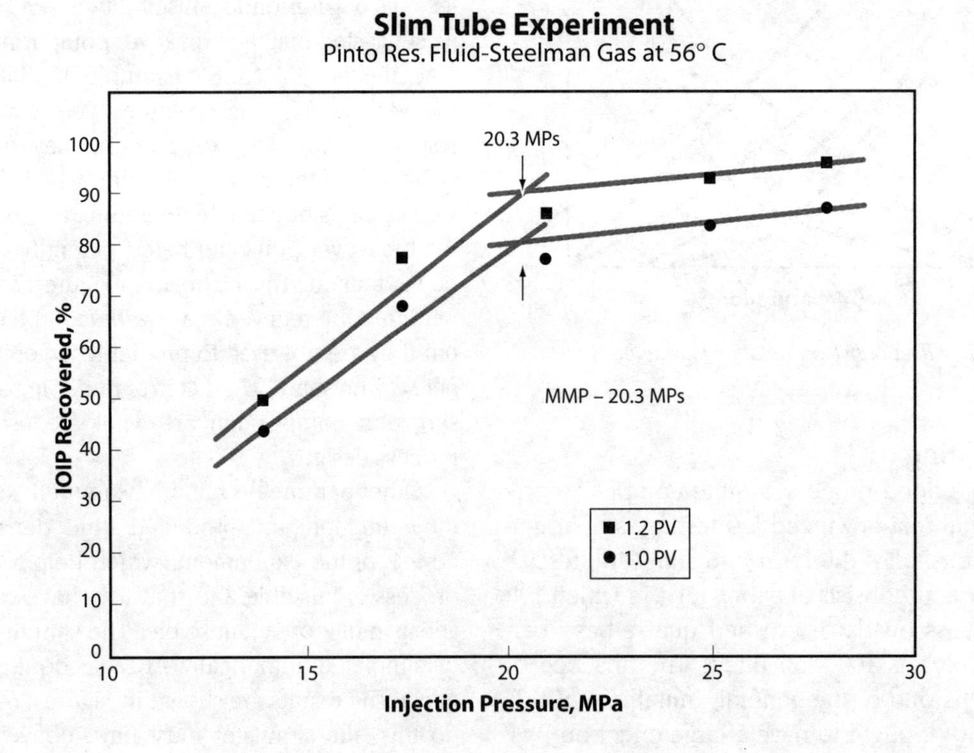

Fig. 17–15 Estimation of MMP from Multiple Slim Tube Tests

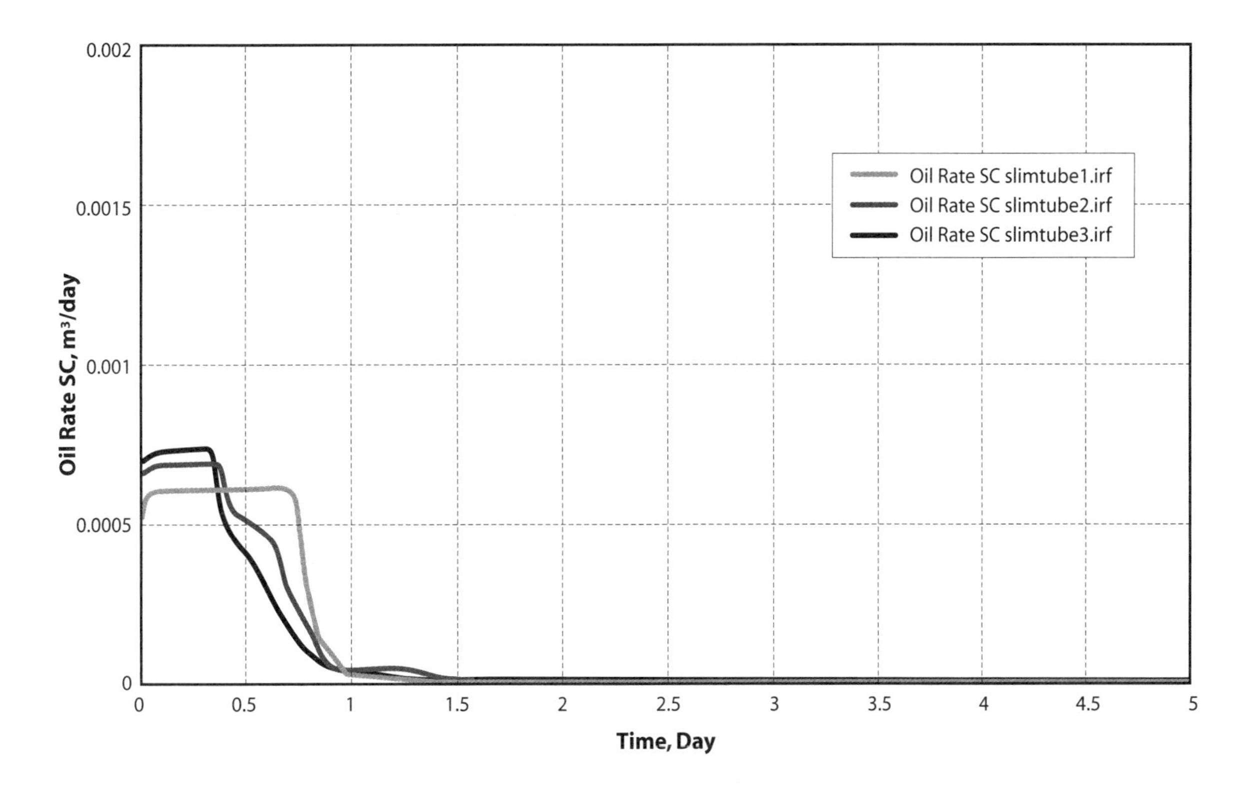

Fig. 17–16 Slim Tube Modeling Results

Compositional modeling and relative permeability

Consider the implications of relative permeability in compositional modeling. When the solvent and oil come into contact in a simulation grid block, the two are assumed to mix thermodynamically and instantly. If the two are miscible, then there will only be one resultant phase, either a gas or a liquid. From a relative permeability standpoint, the grid block will be at one of the end-points of the relative permeability curves (ignoring water for the moment).

If the solvent and oil are not miscible, two phases will result, and the movement into the next simulation grid block is governed by the relative proportion of gas and oil, i.e., relative permeability of each phase is involved. In fact, the exact compositions and properties of the fluids in the adjoining grid blocks are precisely what are required to predict whether multicontact miscibility occurs or to predict what the next contact will produce. The mixing will be affected by grid-block size.

Some of the author's students were eager to learn about using Koval-type mixing factors in a compositional simulator. This is restricted to pseudo miscible modeling. In the pseudo miscible model, the solvent is a separate species, and it is a gas. In fact, the Koval technique was based on the assumption of first-contact miscibility. Furthermore, there is no volume change. The solvent and oil coexist in the same grid block, even though they are not mixed. The Koval technique alters the velocity of the solvent by changing the relative permeability curves based on solvent saturation.

A number of questions arise. Since the Koval technique requires a solvent saturation, how does one apply the label "solvent" when the two phases are exchanging mass? Restated, how does one label solvent when the grid-block compositions are changing from one grid block to the next? Similarly, how does one label solvent in a grid block where miscibility has occurred? There is now only one fluid.

In essence, the use of the Koval technique interferes with the calculations of composition, which is the intent of using a compositional simulator. The net result is that the compositional simulator deals with issues that are different from a pseudo miscible model. Most field models of miscible floods will be of the pseudo miscible type.

Effect of saturation history

One of the major evaluation firms analyzed a major Beaverhill Lake reservoir HCMF flood on behalf of a non-operating working interest owner. They were not prepared to assign miscible flood reserves since they felt that the high water saturations in the reservoir from waterflooding would block access to the residual oil. A number of people were sent from the operating company to discuss this matter. The external consultant never did change opinions, despite core evidence to the contrary.

The presence of the external evaluator indicated that someone was interested in either buying or selling one of the working interests. From their position, it appears that they were representing a buyer. Of course, at that time, most major operators with Beaverhill Lake reservoirs were in the process of implementing HCMF projects and regardless of the evaluator's reserves assignment, general industry's perception was that miscible flooding would enhance the value of the property.

The operator's reserves evaluator was less concerned about this issue and gave higher reserve estimates. In the end, the HCMF worked well, so the concerns expressed did not prove to be correct.

A number of points can be derived from this story. First, the concern of contacting the oil was valid. The solution that was used to evaluate the concern at that time was core flood tests; although one of the major tenants of this text is that lab data requires significant quality control. The reader may conclude that the author specifically recommends this type of test as part of a miscible flood design. Second, there can be differing opinions amongst experts, and the riskier areas, such as miscible flooding, will result in a higher divergence of opinion.

Third, the SEC has specific rules about booking EOR reserves, which apply to waterfloods as well. Many people in the industry feel these rules are unreasonable. From a technical perspective, the author agrees. However, the main purpose behind the SEC definitions is to protect investors and not to promote oil companies. Yes, somebody made some really big mistakes in the past, and that incident caused the rules to be put in place. It was not as big a problem as Enron, but it was big enough. These rules are a fact of life, and the SEC rules serve a useful purpose. EOR has to be proven within the subject reservoir before *proven reserves* can be assigned. Of course, management would like to book additional reserves as soon as the facilities are in the ground or the capital is spent. The brutal truth is that the rules don't allow for this. The implications of this favor the installation of pilot floods and support phased development of a reservoir. When production response is proven, the company can book the reserves additions.

Gravity override

One of the major issues that surfaced at that time was gravity override. A number of optimistic HCMF submissions were made to the government. However, a major company deviated from the rest of the industry, and this caused a heated debate and some bad feelings as well. This was because its submission was much more conservative—an incremental HCMF oil recovery factor less one-third of the prevailing wisdom of the time. Of course, the government was basing royalty holidays on the amount of additional recovery. Eventually, the incremental miscible flood reserves were found to be closer to the lower value than the higher value.

The difference between the two positions broke down to gravity override. It is easy to underestimate this phenomenon. Detailed modeling should address this issue.

Gravity segregation

As suggested earlier, the best HCMF candidates involve gravity segregation. Indeed, gravity override is a problem to be avoided, so it is advantageous to conduct the flood in a top-down vertical configuration. Ironically, maintaining a consistent downward horizontal contact in a vertical miscible flood can be difficult. In the Rainbow area, uneven gas injection does not equalize sufficiently in the gas cap. The areas that are overinjected have lower gas-oil contacts, and this results in coned solvent and chase gas production.

In order to maximize recovery, it is necessary to match voidage in the different parts of the reservoir. This can be done by local computations of voidage replacement ratio (VRR). Detailed simulation can also aid in this process. The latter option is more difficult to process on a day-to-day basis, and the former is generally preferred. In particular, tight streaks can cause shelving of the oil and blockage of the solvent. The location of such layering can be determined by detailed geological modeling and should be considered in designing an injection program.

Economics

Custom-detailed models

For purposes of reserves evaluation, it is common practice to use one of the commercially available economic programs. The results from the models that the author has used have proven to be quite accurate. However, for the purpose of design, a more detailed model is useful. This allows more flexibility in conducting sensitivities for different solvent compositions, varying slug sizes, and different production forecasts.

Historically, the model used most by the author was an IFPS model. This program used a Fortran-like syntax that developed what is, in essence, a large spreadsheet. These days, this type of analysis would probably be done with a common spreadsheet package.

Solvent optimization

At first, it might seem that solvent optimization is required only during the initial stages for design purposes. However, with the volatility in the prices of different products, such as ethane, propane, butane, and pentanes plus, solvent compositions do not follow the original design closely. Instead, they change in time because of cost/contract considerations in acquiring solvent components.

Thus, the economics program may be required for a considerably longer period than may have been anticipated originally. Along these same lines, the tuned EOS model may be needed to evaluate variations in solvent composition.

Government incentives

In Canada, the provincial governments introduced incentive programs in the form of royalty holidays in order to induce oil companies to use large-scale miscible floods. This took the form of a complex royalty reduction largely determined by the additional recovery of the HCMF, most commonly known in Alberta as "Section 4.2" relief. In effect, the government essentially assumed the risk of the large miscible floods, and the gamble paid off. Eventually, the total royalties were larger, even if at a lower rate. There is one notable exception to this generality of receiving government incentives—Amoco's HCMF of the South Swan Hills Unit No. 1. Amoco and its staff at the time were the true pioneers of the Western Canadian HCMF, implementing their flood much earlier and without the initial benefit of the royalty relief. Now that the technology has been proven, these holidays are much more difficult to negotiate.

Risks

Whether government incentives are a good or a bad thing will not be debated in this book. However, it is clear that, with the capital expenditures involved and the risks associated in a complex process, risk management is a significant issue in project design. This needs to be an integral part of the design process.

To put this into other concrete requirements, almost all miscible flood programs are phased. This limits both risk and capital exposure. It may take a little longer to implement on a total field; however, this is prudent practice. Miscible floods on large projects should be planned initially with a progressive approach.

A definite learning curve is required when applying this technology, and a phased approach has the added benefit of giving sufficient time to evaluate early results and incorporate experience gained in subsequent phases.

Monitoring/Surveillance

Compositions

The author worked on a project where the operator had gone to a great deal of trouble to monitor well effluent compositions. A system was introduced that, at the time, was quite modern, and that included storing all of the data on a computer. The use of this data was not a major issue until it was time to look at blowdown. The operator had used advanced technology for the time, but PCs go out of date quickly and so does database software. It was difficult to retrieve the data from an older PC. In the end, the available data came from transcriptions of paper copies of analyses.

In hindsight, a more organized system of paper copies would have helped. Intense monitoring had occurred up to the time of breakthrough, and this was reasonable. After that time, the changes in composition slowed and frequent readings were not required.

Today, the changes in computer technology seem to be slowing somewhat, but no one uses 5.25-in. disks anymore. Consider what will happen to the CD/DVD medium, which is probably the most popular portable storage format at the moment.

Injection

As outlined previously, solvents tend to change with market conditions, and keeping track of injection compositions is important later. This may mean designing software to keep track of the compositions on a well-by-well basis. A program of injection monitoring is also worthwhile.

Since WAG is used and the solvent composition varies with market conditions, not all wells will have the same solvent history. Furthermore, in order to maintain voidage, considerable scheduling is required. This will vary with each pattern since well productivities do vary. Maintaining a consistent pressure above MMP is also important for the miscible flood to continue working.

Therefore, sophisticated monitoring is very important. T. Okazawa et al. of Esso Canada produced a succinct paper on this issue based on their miscible flood experience.[18] This type of programming can take considerable lead time and should be included in miscible flood design. The author is not aware of any commercial products that do this type of work.

Tracers

On one project, the operator also included a tracer program. A variety of tracers can be used and, in order to monitor them, a large number were used on the project. This worked well until tracers started to show up in unexpected places—at the wrong end of the field. The solvent was being recycled, and the tracers were being carried into the overall injection gas. Ergo, all of the tracers eventually showed up in all of the wells. This was not necessarily a bad thing; however, it came as a real surprise. More information could have been obtained if the injection gas had been monitored as well as the produced gas. Thus, two pulses could have been captured to provide more information. Therefore, one should monitor the injection gas continuously to determine when wide distribution of the tracers occurs.[19]

The tracers turned out to be critical in analyzing reservoir performance. This did not prove to be quantitative. Viscous fingering means that the assumptions used in calculating tracer movements are not accurate. Nevertheless, the tracers were pivotal in demonstrating communication throughout the reservoir.

Pattern-by-pattern monitoring

The larger Beaverhill Lake reservoirs are of sufficient size that it is difficult to simulate them due to the size of the grid required. The interiors of the reefs also have many layers, and this leads to nonunique history match solutions. It was found by experience that it was faster to simply make a pattern-by-pattern conventional reservoir engineering analysis. This process involved using production and injection plots in conjunction with well-to-well log correlations showing perforations. It is not elegant, and it is time-consuming; however, it produces the same results in less time than simulation.

Predicted performance could not be reached without this detailed kind of monitoring. As with many other endeavors, it is easy to underestimate the actual amount of engineering and geological effort required.

Ironically, it seems that one can spend years optimizing a miscible flood and never use a reservoir simulation. A number of the author's acquaintances are in this position.

Summary

Many reservoir engineers think they would like to work on miscible flooding. Some of the people who have worked in this area call it miserable flooding. These projects are large, take incredible amounts of time and effort to get going, and involve substantial risk. Most engineers working on an HCMF don't get to do exotic calculations using simulation nor is this the majority of the work required. Most of it is common-sense reservoir engineering done on a pattern-by-pattern-basis. Usually, an extensive number of water-flood type simulations are made.

If the engineer is lucky, he or she may work on the early design work, which is more or less outlined previously. Although these calculations are sophisticated, they really provide soft answers. The substantive point of this chapter is to discuss the tradeoffs in designing the simulation process used. At present, no single technique is capable of answering all of the relevant issues simultaneously. The results of the simulation must be interpreted, risked, and adjusted to derive a realistic projection of reservoir performance.

Many people are disappointed that the work turns out to be similar to conventional reservoir engineering. Since these projects are rare, working on this type of field would otherwise provide one with skills that have limited applicability. The results being sought are optimizations, and there can be a lot of satisfaction from optimizing production. These results will become evident in reasonably short time frames. On an overall basis, however, the size of these projects means that longer time scales are required to see ultimate recovery results, and one should be prepared for this.

Miscible floods require a great deal of management. This is a systematic process and suitable for people who enjoy setting up and working through these procedures. In the postanalysis of a number of large miscible floods, ARE concluded that as many incremental reserves came from infill drilling and conventional detailed reservoir engineering as from the miscible flood. Installation of the miscible flood was constructive in galvanizing detailed action on a large scale.

References

1 Stalkup, F.I., Jr., *Miscible Displacement.* SPE Monograph, Vol. 8, Henry L. Doherty Series, Society of Petroleum Engineers, 1983.

2 Green, D.W., and G.P. Willhite, *Enhanced Oil Recovery,* SPE Textbook Series, Vol. 6, Society of Petroleum Engineers, 1998.

3 Lake, L.W., *Enhanced Oil Recovery*, Prentice Hall, 1989.

4 Donaldson, E.C., G.V. Chilingarian, and T.F. Yen, *Enhanced Oil Recovery, Part I: Fundamentals and Analysis*, Developments in Petroleum Science 17A, Elsevier Science Ltd., 1985.

5 Donaldson, E.C., G.V. Chilingarian, and T.F. Yen, *Enhanced Oil Recovery, Part II: Processes and Operations, Developments in Petroleum Science 17B*, Elsevier Science Ltd., 1989.

6 van Poollen, H.K., and Associates, *Fundamentals of Enhanced Oil Recovery*, PennWell Publishing Co., 1981.

7 Herbeck, E.F., R.C. Heintz, and J.R. Hastings, "Fundamentals of Tertiary Oil Recovery," *Petroleum Engineer International*, nine-part series, January 1976 – February 1977.

8 Benham, A.L., D.E. Dowden, and W.J. Kunzman, "Miscible Fluid Displacement—Prediction of Miscibility," *Transactions of AIME,* Vol. 219: 229–237, 1960.

9 Novosad, Z., and T. Costain, "New Interpretation of Recovery Mechanisms in Enriched Gas Drives," *Journal of Canadian Petroleum Technology* 27, No. 2 (March–April): 54–60, 1988.

10 Yellig, W.F. and R.S. Metcalfe, "Determination and Prediction of CO_2 Minimum Miscibility Pressures," SPE 7477, *Journal of Petroleum Technology*, January 1980.

11 Walsh, M.P., and E.M. Withjack, "On Some Remarkable Observations of Laboratory Dispersion Using Computed Tomography (CT)," *Journal of Petroleum Technology*, 1994.

12 Koval, E.J., "A Method for Predicting the Performance of Unstable Miscible Displacements in Porous Media," *Society of Petroleum Engineers Journal*, 1963.

13 Todd, M.R., and W.J. Longstaff, "The Development, Testing, and Application of a Numerical Simulator for Predicting Miscible Flood Performance," *Journal of Petroleum Technology*, 1972.

14 Behrens, R.A., "Tertiary Recovery Modelling by Hybrid Simulation for Kaybob Beaverhill Lake 'A' Pool Hydrocarbon Miscible Flood," *Journal of Canadian Petroleum Technology,* Vol. 33, no.10, (December), 1994.

15 Klins, M.A., "Oil Production in Shallow Reservoirs by Carbon Dioxide Injection," SPE 10374, Society of Petroleum Engineers, 1981.

16 Bardon, C., and D.G. Longeron, "Influence of Very Low Interfacial Tension on Relative Permeability," *Society of Petroleum Engineers Journal*, pp. 391–401, October 1980.

17 Huang, S.S., P. Dewit, R.K. Srivistava, and K.N. Jha, "A Laboratory Miscible Displacement Study for the Recovery of Saskatchewan's Crude Oil," *Journal of Canadian Petroleum Technology*, Vol. 33, no. 4, April 1994.

18 Okazawa, T. and F.S.Y. Lai, "Volumetric Balance Method to Monitor Field Performance of Gas-Miscible Floods," *Journal of Canadian Petroleum Technology*, 1991.

19 Zemel, B., *Tracers in the Oil Field, Developments in Petroleum Science 43*, Elsevier, 1995.

18

Naturally Fractured Reservoirs

Introduction

Fractured reservoir engineering has become quite popular during the last two decades. This is mostly due to some rather "sexy" analysis that has become available in pressure transient analysis. This has generated a fantastic amount of sophisticated mathematics. This mathematical approach has been extended to reservoir simulation.

However, there are those who have serious doubts about the description developed for pressure transient analysis and simulators. Some of this has spawned intense debate in the literature. Like many new areas, the final determination likely will take many of the different positions into account.

This chapter does three things:

- Outlines the classical background on this issue. Some elements of the material are generally agreed upon.
- Presents dual porosity issues. There are consistency problems with the use of dual porosity descriptions contained in simulators. In the author's opinion, the coding is correct and customers demand that this be included in simulation packages; however, some fundamental issues have not been resolved.
- Presents the author's experience. This view is supported with modeling. The author acknowledges that this view is only one explanation; however, it seems to fit a number of cases and, in particular, conditions found in thrust-fault environments.

This chapter is divided into these three areas.

Prior to getting into the details, the issue of terminology must be addressed. This seems to arise in many different areas within petroleum engineering because a number of terms are similar or have different meanings in different contexts.

Dual porosity

In the most general sense, dual porosity indicates that the reservoir is comprised of a rock that has a complex porosity system that can be divided into fracture porosity (literally the void in the cracks) and a smaller scale porosity that exists in the pore spaces within the rocks. Vugs can fall in either category, depending on their relative scale. In some cases, this is referred to as a double porosity system.

Reservoir behavior will vary, depending on which of the two porosity types dominates. This has been categorized as type A, B, and C class reservoirs. In the A category, all of the storage is in the matrix, with the effective storage of the fractures being relatively small. The B category represents reservoirs where the storage is approximately equal between the matrix and fractures. In the final category, all of the storage capacity is in the fractures.

In a mathematical sense, the use of dual porosity implies that a continuum exists. This has some physical implications, in that in the development of calculus we add up representative elementary volumes (REMs) and, by the use of limits, it is possible to derive an exact answer. In math class, these elements vanish in size. However, if we were to have infinitesimal REMs, we would have to track individual molecules. In engineering practice, REMs are a question of scale and not absolutes. Clearly they need to be small; however, we use bulk properties like viscosity and compressibility that don't have a direct meaning at infinitesimally small scales. This is what the author calls an engineering continuum. This may not be entirely consistent with the literature in general; however, this is intended to indicate that the topic under discussion is the practical mathematical concept approach, rather than whether there are intrinsically different fracture and matrix properties.

Having said this, it is generally assumed that the fracture system is connected and that it can be described with a bulk property, like permeability and porosity (ergo, the fracture system must be of a scale smaller than the REMs). In the dual porosity concept, the matrix blocks are assumed to contact each other only through the fracture system, which is the *only* effective large-scale permeability.

The transfer of fluids from the matrix blocks to the fractures is governed by an analytical equation. This is generally based on a steady state approximation and is discussed in more detail later.

Dual permeability

In the pressure transient business, the mathematical solution to two layers, one with higher permeability and the other with lower permeability, gives the same appearance on interpretation plots as the mathematical solution for fractures and matrix blocks. Pressure transient analysts call this solution a dual permeability solution in order to differentiate the two physical meanings.

Recently, the mathematics in simulators has been altered so that communication can take place between the grid blocks that are adjacent, as well as within the fractures. Here is where the mischief arises: this also has been named a dual permeability system. This could have, and perhaps should have, been dubbed *dual communication elements.*

Subdomains

One other element of this terminology should be added. As indicated previously, transfer from the matrix to the fractures is governed in simulation by an analytical equation based on steady state assumptions. Pressure transient solutions are sometimes based on transient solutions, which involve a different analytical transfer equation. In order to approximate this, nested grid blocks have been developed that are designed to capture the pressure gradient towards the outside of the grid block. These are called subdomains.

Classical Fractured Reservoir Engineering

A number of books are dedicated to this topic and include *Fundamentals of Fractured Reservoir Engineering* by Van Golf-Racht of Norway, *Naturally Fractured Reservoirs* by Roberto Aguilera, *Reservoir Engineering of Fractured Reservoirs (Fundamental and Practical Aspects)* by Saidi, and Nelson's book, *Geologic Analysis of Naturally Fractured Reservoirs*. There is a surprising variety of approaches suggested in these books.[1, 2, 3, 4]

The author divides these books into two different subcategories. A number of them place a heavy emphasis on reservoir description. The two that stand out in this regard are Aguilera's book and Nelson's book.

The second category deals with the quantitative aspects of reservoir descriptions. The author includes Van Golf-Racht's book and Saidi's book in these categories. Note that all of the books deal with both issues to greater or lesser extents.

Fractured reservoir analysis

Starting with the first category, evaluating a fractured reservoir (after Nelson) involves four main steps:

1. *Interpreting the origin of the fracture system. This information allows one to predict geometry and the extent of communication.*
2. *Determining petrophysical properties of the fractures and matrix. This allows for prediction of the variation in reservoir response. The relative storage (i.e., porosity) must be determined as well as effective permeabilities. Another important property is compressibility.*
3. *The flow interaction between the matrix and fracture system is evaluated to determine ultimate reserves from the reservoir.*
4. *Classification of the reservoir. Depending on the type of flow interaction the reservoir will fit one of several depletion strategies. Note that most of the variations in strategy apply to waterflooding oil reservoirs.*

In this area, it would seem that both authors, Aguilera and Nelson, despite somewhat different approaches, are getting to the same point with their own distinctive style. The author has found these approaches work well and concurs with the importance of the geological approach.

Mechanisms in fractured reservoirs

Proceeding to the second category, Saidi outlines a number of factors that should be considered in analyzing fractured reservoirs:

- *Convection (in fractures)*
- *Diffusion*
- *Solution gas drive*
- *Gravity drainage*
- *Block-to-block flow*
- *Turbulent flow (in fractures)*

Covering these topics in detail requires considerable discussion.

In summary, Saidi, of the National Iranian Oil Company (NIOC), has championed the view that the most important aspect is matrix-fracture interaction and often uses "type columns" for his analysis. The other view has been championed by researchers from the United States and utilizes the dual continuum approach.

A number of people have serious doubts about the latter description. One of the earlier ones was Odeh who has had a profound influence on reservoir engineering.[5] There were also some intense debates in the literature regarding the fundamental physics. The debate took place in the *Journal of Petroleum Technology*. The author has not found a more intense debate elsewhere in the petroleum literature.[6, 7, 8, 9, 10, 11]

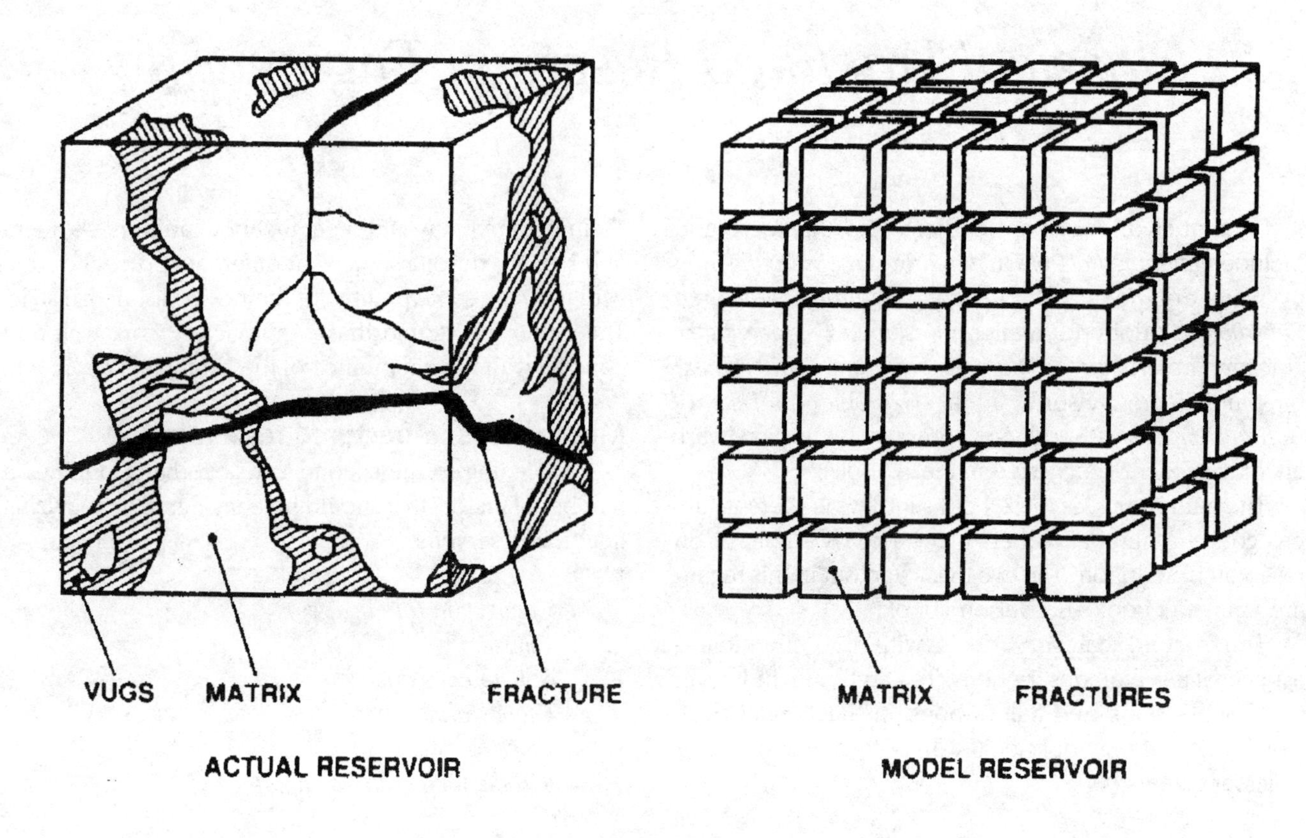

Fig. 18–1 Basic Model Developed by Warren and Root

Fundamental development: the dual porosity model

Warren and Root developed the basic model. The reservoir is conceptualized as a series of sugar cubes. This is shown in Figure 18–1. It is assumed that the fractures extend infinitely. In addition, it is assumed that the fractures are evenly spaced. All flow to the well takes place in the fractures. The fractures are, in turn, supported by a matrix of lower permeability. The original work concentrated on Horner analysis and has been extended in a number of stages by various authors, such as Bourdet and Gringarten, to include the more modern use of derivative analysis.[12, 13, 14]

Pressure transient analysis transfer functions

Two other types of assumptions are important. The first is that the transfer from the matrix to the fractures can be represented by a pseudo steady state calculation, and the second is transient interporosity flow. The pressure response in the well for these two cases is different in the transition from where flow is dominated by the continua next to the well (which governs during early time) and matrix support (which governs in late time). In fact, for a large number of practical well tests, the first data is obscured by wellbore storage and can be difficult to see.

The pseudo steady state solution is determined by a constant times the difference in pressure between the matrix block and fracture. The transient response is based on heat flow equations (from Carslaw and Jaegar). Analogy with heat flow equations is based on the assumptions of fluids having low compressibility and single-phase flow, as are most pressure transient analysis equations.

Pressure transient proof of Warren and Root behavior

Few examples of dual porosity interpretations have been published. This lack of practical application has not been widely discussed. Two authors have braved the issue. The first is Sabet, in his book *Well Test Analysis*. The other is Streltsova, in her book *Well Testing in Heterogeneous Formations*. Sabet's comments are as follows:

Based on the author's experience, it is highly unlikely that the semi-log plot of the test data obtained from wells producing from naturally fractured reservoirs would exhibit the characteristic curve predicted by Warren and Root, de Swaan, Najurieta, or Streltsova. The same conclusion was reached earlier by Odeh (1965) who studied several fractured reservoirs. Rather, the given examples were all related to layered reservoirs, and Streltsova's examples were drill stem tests in which the flow rates cannot be stabilized.

The pressure derivative method is theoretically very powerful, but in reality it cannot be applied in the case of field data except in ideal situations where the semi-log analysis would be more than adequate. Furthermore, if there is a linear boundary in a homogeneous reservoir, the derivative method will give the impression of a naturally fractured reservoir.[15]

Streltsova concurs with Sabet's observations. Her book offers some definite reasons for this being the case.

Discontinuous network of fractures (after Streltsova)

Implicit in these developments is the assumption that the fracture network is continuous and of uniform permeability throughout a well's drainage radius. Present day exploitation of hydrocarbon deposits has shown that fractures within a reservoir are not necessarily interconnected over a well's drainage area. Reservoirs with fractures of limited lateral connectivity are, in fact, not uncommon.

The pressure behavior of a formation with a discontinuous network of natural fractures is entirely different from that of a formation with uniform fracture permeability. If a producing well intersects a localized fracture system, the initial well behavior is indicative of the fracture system response. Generally, a fracture network imparts an anisotropy to the formation's permeability: Parallel to fractures the permeability is at its highest value, whereas normal to fractures the permeability is at its lowest value. If the fracture network is continuous, a single-well test yields an effective transmissibility equal to the square root of the product of the transmissibility values along the major and minor axes of the anisotropy. (Streltsova references the solution of anisotropic reservoirs). *If, however, the fracture network intersecting a well is disjoined, then the preferentially oriented high permeability extends only a certain distance away from the wellbore. Beyond this distance, flow takes place through the lower permeability matrix that connects the fracture network to its neighboring networks. Given sufficient flow time, the pressure response of a formation with disjoined fracture networks assumes the character associated with radial flow. Consequently, the pressure curve has a shape similar to that of either a* hydraulically fractured well, *or a well in a* layered system with laterally terminating layers. *The former type of behavior occurs when only a single fractured zone is open to the well; the latter is usually observed whenever multiple zones are open to the well.*[16]

It would appear that, based on these authors' interpretations, the Warren and Root type model has a number of severe limitations. Part of this problem is that a multilayered system produces the same mathematical solution as the dual porosity system.

Parametric analysis of fracture spacing

Because of these problems, the author made some parametric evaluations of the dual porosity model using a simulator. Starting with a single-phase gas system and using a radial grid, he looked to see how much pressure difference would exist between the matrix and the fractures using the classic original Warren and Root type model. This work was for a fractured carbonate in the foothills of Alberta.

The starting permeability contrast was about 100:1 for the fracture system permeability versus the matrix permeability. Indeed, this is a permeability contrast level predicted or consistent with pressure transient expectations, which are succinctly outlined by DaPrat but may be found in other technical papers.[17]

For the transfer function, which governs flow between the matrix and fractures, the default values in the simulator were used.

There was virtually no pressure difference between the fractures and matrix until grid-block dimensions were as large or larger than 10 meters (33 feet). This is a very large grid spacing and was much larger than what the author could determine accurately from core examination. It is also considerably larger than the measured fracture spacings in the technical papers reviewed by the author.

Similar model results are shown in Figures 18–2 and 18–3.

A search through the literature at that time did not reveal any material on spacing sensitivities.

Parametric analysis of permeability contrast

The next issue was the contrast in permeability that gave the highest differences between the fracture system and grid-block pressures. In this regard, the predictions from pressure transient theory and the simulator were consistent. The results are shown in Figures 18–4 and 18–5. These sensitivities confirmed that a permeability contrast of approximately 100:1 maximized the differences in production performance.

With permeability contrasts less than 100:1, the system behaves as a single porosity system and above 100:1 the system is dominated completely by the fractures.

Discussion of parametric testing

This work has some important implications. First, typical grid-block dimensions will have virtually no effect on history matching. In fact, the transfer seems to be so rapid that the author questions whether steady state was a reasonable assumption; i.e., the transfer is essentially instantaneous, so there is time lag during which steady state matrix-fracture flow occurs. If the transfer is this quick, then the inclusion of the dual porosity option does not seem to be productive. One gets the same results with a single porosity system, i.e., a conventional simulator.

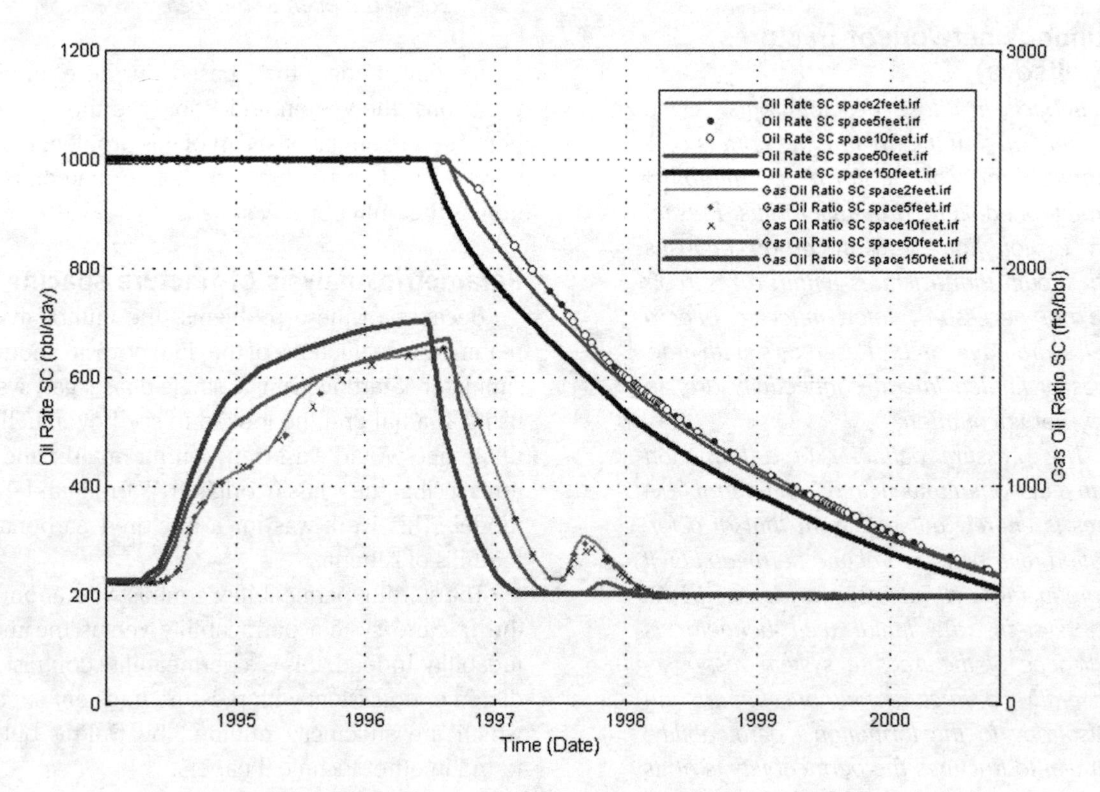

Fig. 18–2 Effect of Matrix Block Dimensions on Oil Rate and GOR

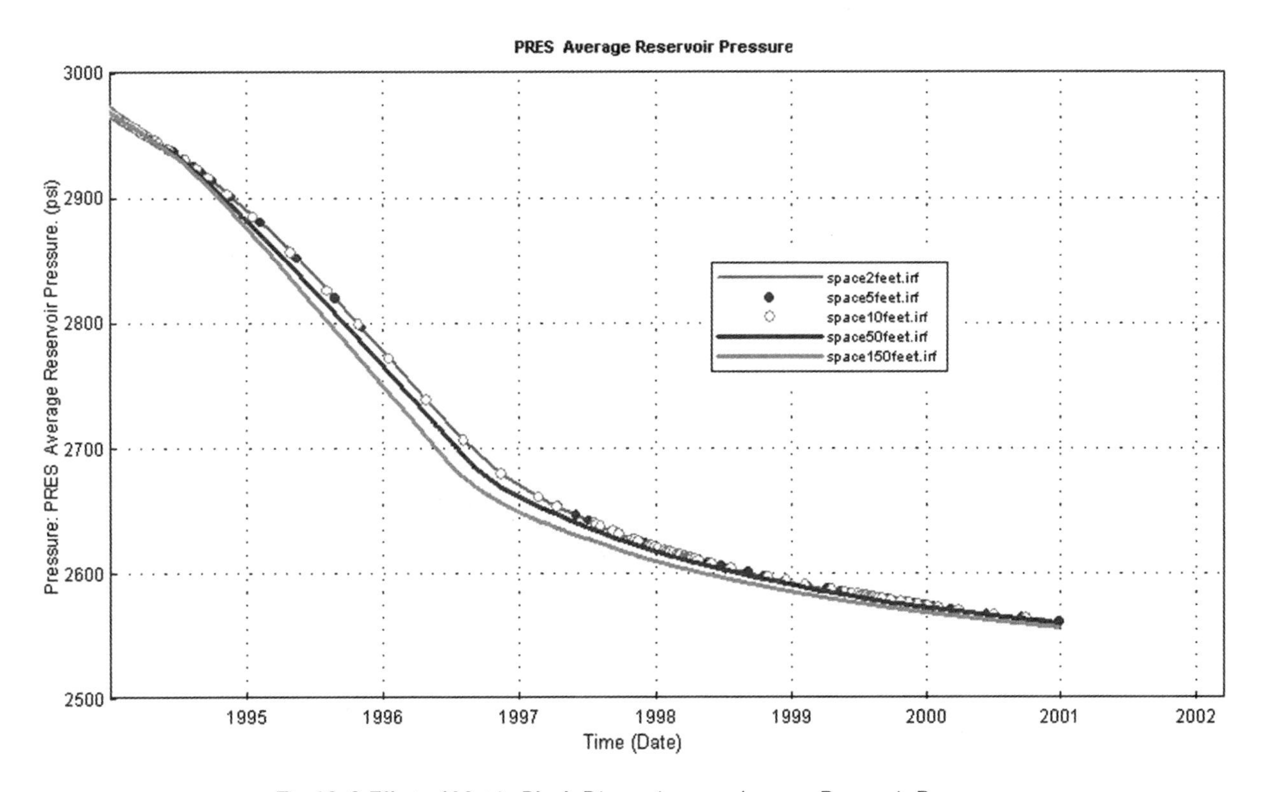

Fig. 18–3 Effect of Matrix Block Dimensions on Average Reservoir Pressure

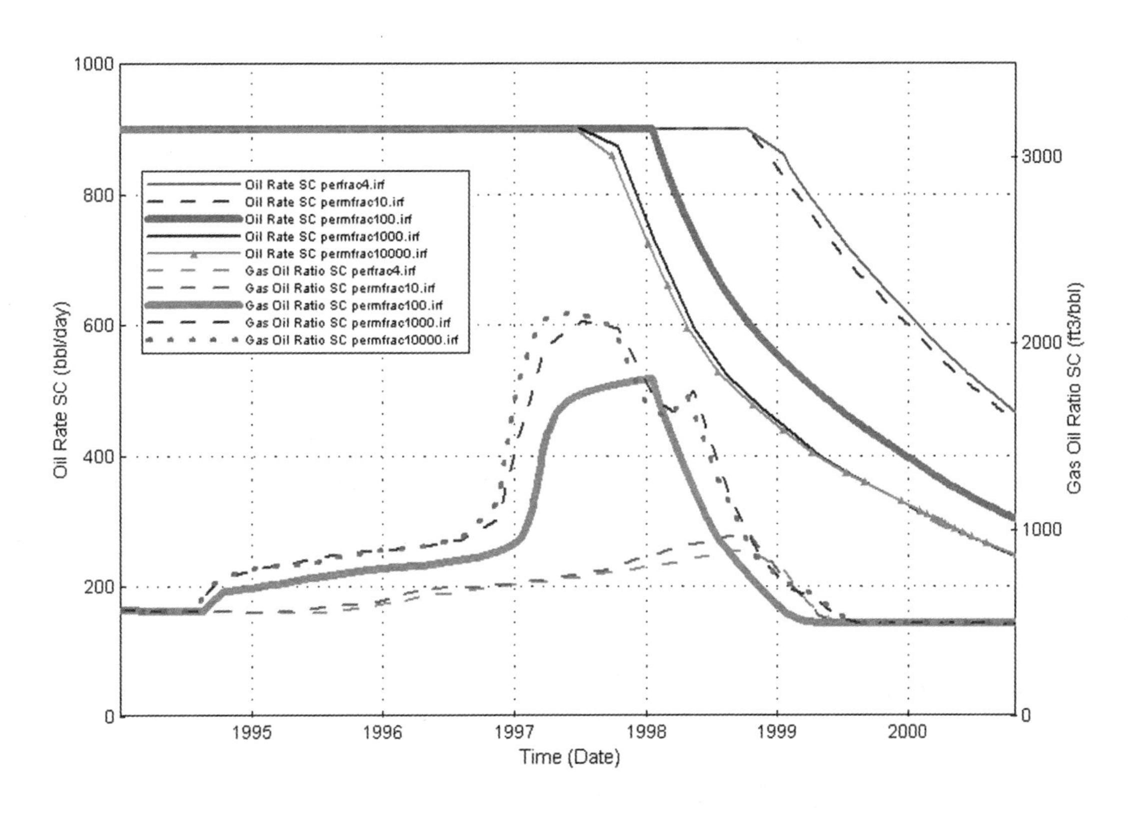

Fig. 18–4 Effect of Permeability Contrast on Oil Rate

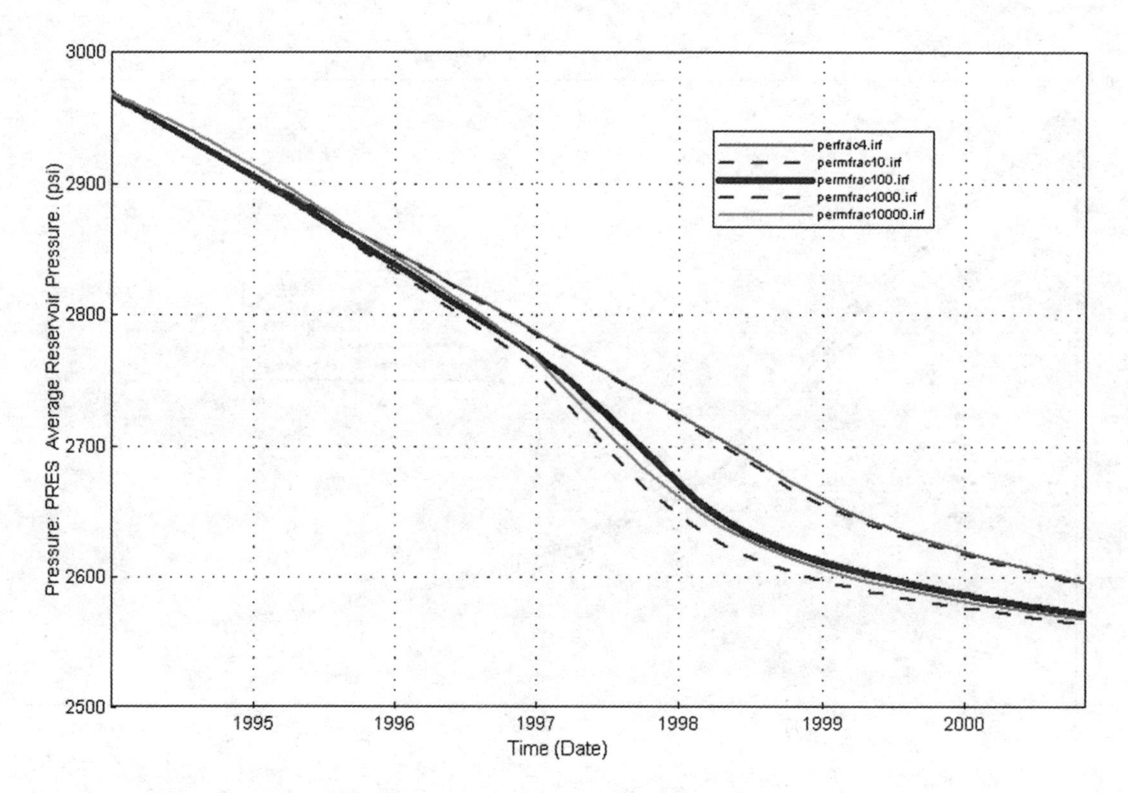

Fig. 18–5 Effect of Permeability Contrast on Average Reservoir Pressure

Some clients noticed that they saw similar effects when history matching. The geologists would revise their average fracture spacing from 3 to 3.5 meters (9.8 to 11.5 feet). The resultant change did not alter the history-match results. In view of the previous sensitivities, this is hardly surprising. The author believes they abandoned the dual porosity simulator option not long after taking his course.

More detailed parametric analysis of performance

Next the author made up a problem that would have gas-oil and water-oil contacts and invoked a major change in strategy. In the previous parametric studies, the objective was to spot the differences between pressures in the fracture system and the matrix blocks. The analysis was restricted to single-phase systems. Although this is inherently interesting, it does not address the bottom line—reservoir performance prediction. A small element of symmetry in a more realistic system was selected to test if detailed modeling of the grid blocks gave the same solution as the dual porosity model, in this case, a five-spot waterflood pattern.

The modeling was rather crude. For the dual porosity modeling to be complete, the creation of multiphase transfer functions was needed.

Multiphase transfer effects

The issue becomes more complex for multiphase situations. This is shown in Figure 18–6. This is an example of an oil reservoir producing under primary production. There is rapid gravity segregation of liquids and gas within the fractures. Fractured oil reservoirs under primary production are associated with rapid increases in producing GOR.

From a primary production point of view, oil would be ejected with pressure decline from within a matrix block, similar to the primary production model assumed for homogeneous reservoirs. How does the reservoir simulator account for this? The basic matrix-fracture transfer is based on the Warren and Root assumption; i.e., Equation 18.1.

$$\tau = \frac{\sigma V k_{rj} \rho_j \left(p_{jm} - p_{jf}\right)}{\mu_j} \tag{18.1}$$

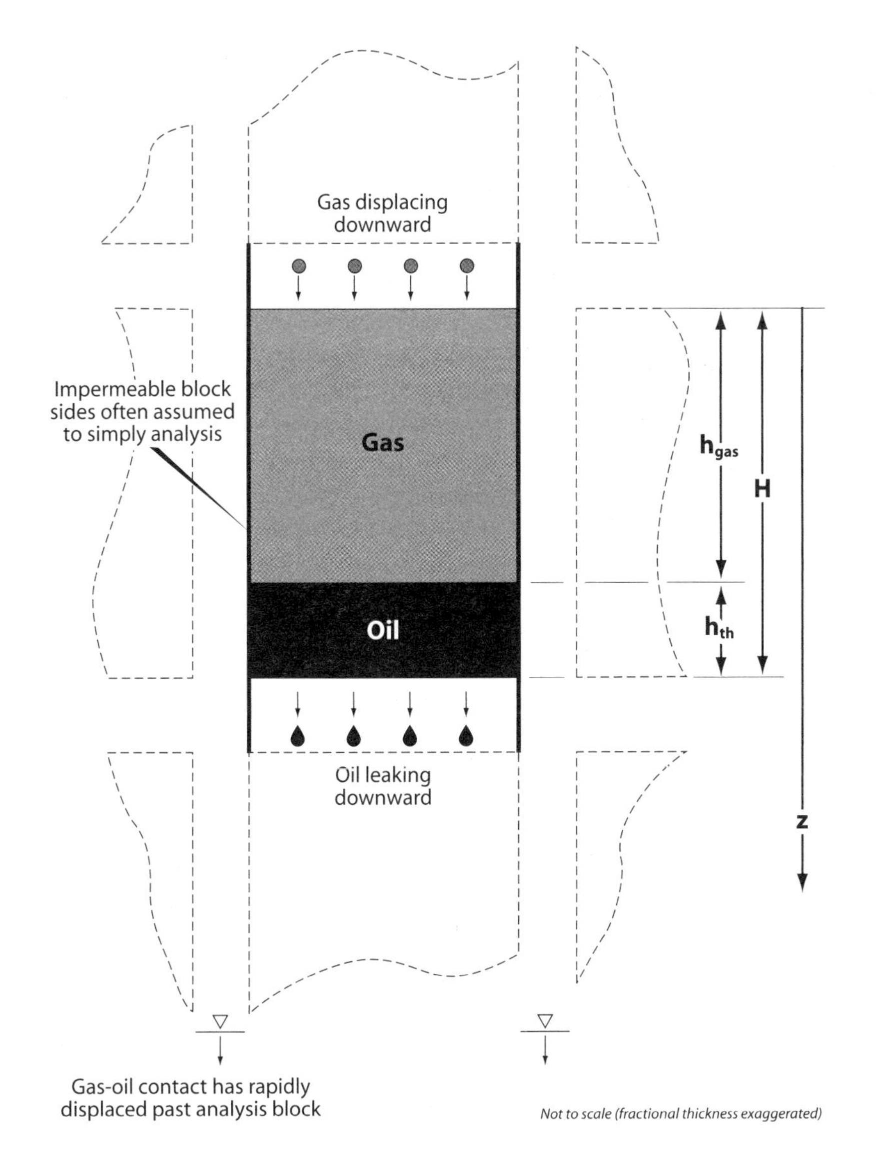

Fig. 18–6 Oil Reservoir Producing under Primary Production

Waterflooding fractured oil reservoirs

The more subtle transfer of different fluid saturations via imbibition/drainage effects is controlled with pseudo relative permeability or capillary pressure curves. An example of the processes occurring in a grid block is shown in Figure 18–7. This figure depicts the upward movement of fluids under gravity effects. These processes are determined based on detailed modeling of grid blocks.

Note that blocks can be surrounded by gas, oil, or water. In some cases, a contact between two phases will be traversed. Different block transfer calculations must be used for each of these combinations. This seems to indicate that the actual mathematical relationship on how pressure is transferred to the matrix would change. Indeed, this could be done; however, the simulators with which the author is familiar do not allow such a function to be specified. Instead, the standard steady state assumption is used and a pseudo relative permeability or a pseudo capillary pressure is used to fudge the desired outcome.

Actual results

The plan was relatively straightforward: generate transfer functions (pseudo relative permeabilities) for use in the dual porosity simulator based on the results of the simplified model.

After running the detailed model, it was obvious to the author that there was a problem. The waterflood front was not following the fractures at all, contradictory to a fundamental assumption in the formal development of the dual porosity model. Water was moving across the grid blocks. If this were the case, then the dual porosity model would break down completely and would not apply.

Permeability contrast sensitivities with detailed model

At this point, additional sensitivities were performed on the detailed model of grid blocks and fractures in a five-spot program. The author found that, with such detailed modeling, the contrast between matrix and fractures must be on the order of 1,000:1 or greater in order for the flow in the reservoir to be dominated by fracture flow. There are two obvious problems:

- The basic assumption of fracture-dominated flow is not correct.
- The transfer functions do not account for across grid block flow.

This immediately derailed any attempts to generate transfer functions. It seemed to the author that there were fundamental problems with the development of the mathematics.

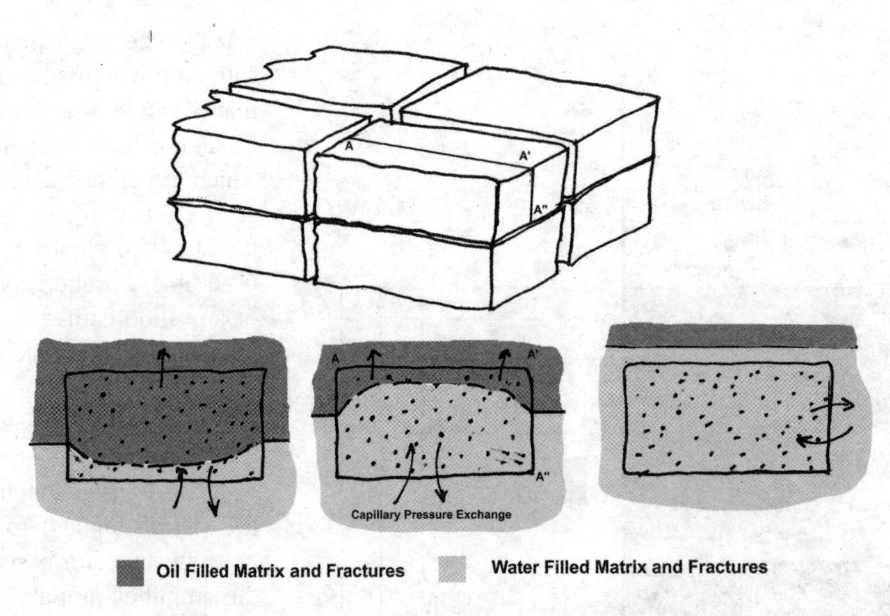

Fig. 18–7 Processes Occurring in a Matrix Block—Waterflooding

History of inconsistencies, or why hasn't this been noticed?

It is natural to ask whether these inconsistencies have been noticed before. In the author's opinion, there are a number of answers to this.

- There are oblique references to problems with the use of representative elementary volumes and dual porosity theory. Some people clearly disagree with this approach.
- The original work on dual porosity theory was completed at a time when detailed simulation was not possible computationally. In fact, Gilman and Kazemi did compare their simulation results with the Warren and Root pressure transient analysis and extremes to the solution (no effective matrix). However, if the same (possibly erroneous) assumptions are used in the different formulations, similar results should have been and were obtained. In this respect, pressure transient analysis and simulator theory are intricately linked. The fact that there are serious problems with the pressure transients observed and those predicted by theory is a significant inconsistency.
- There has been some direct modeling of fractures and matrix blocks. In at least some of these cases, the matrix blocks of 30 meters (98 feet) are a single grid block and the fractures were modeled as 4.5 cm (2 in.) wide. The coarse simulation grid in the matrix blocks is therefore not sufficiently refined to actually calculate the correct pressure and saturation profile in the grid block.
- Gilman and Kazemi did include some effects of the gradients between fractures in one paper.
- There are indications in the literature that this problem has been partly known. Although not explicitly stated, the very fact that MINC and other nested matrix block solutions exist indicates that there was some awareness of limitations with the existing assumptions.[18]
- Recently, simulator formulation has allowed block-to-block transfer, apparently due to physical contact. The sensitivities described previously would indicate that cross-fracture flow occurs between the blocks, regardless of whether there is physical contact.
- The author has talked to one engineer recently who worked in one of the major reservoir research departments, and that company has instituted a policy of including block-to-block transfer on all their simulations.
- Finally, the mathematics has been very popular and it is possible that authors are reluctant to go against established trends.

This strongly suggests that we look at basic fractured reservoir engineering theory with a jaundiced view.

Original work of Aronofsky et al.

The basis of this early work was relatively straightforward. An experiment was conducted in which a rock in a cylinder, open on one end, was placed in water. The amount of oil that came out was monitored. It was found that the amount of oil that came out varied exponentially with time, and so a simple transfer rate could be calculated. This transfer mechanism was then scaled up on a gross scale. Clearly, this method does not address capillary pressure effects, gridblock size, nor does it include gravity effects in detail; however, it does provide a practical indication of how imbibition would manifest itself. Aguilera has extended this basic design calculation to a more sophisticated approach based on exponential decline.[19, 20]

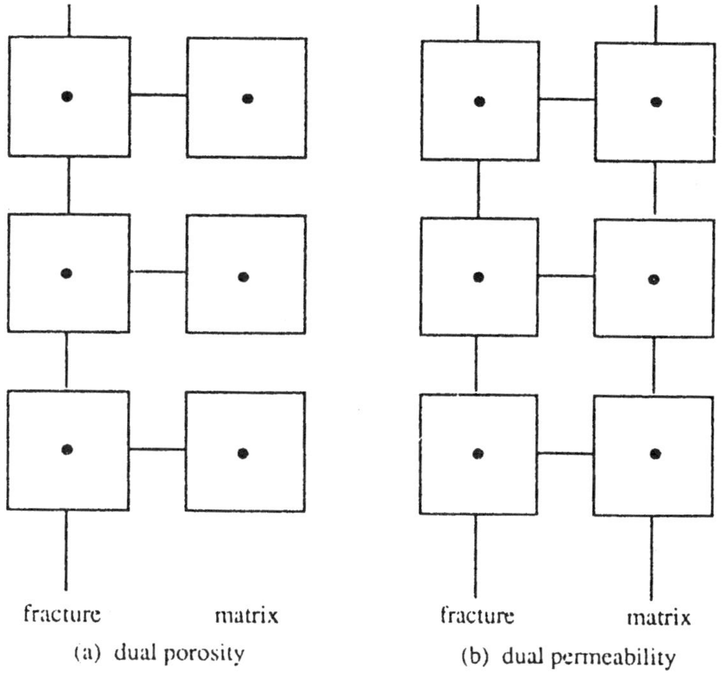

Fig. 18–8 Dual Permeability and Dual Porosity Conceptual Diagram

Production decline

Much of the subsequent work was based on empirical observation of the production decline of fractured reservoirs. The exponential transfer has one unsettling characteristic. It is likely that this relation will show on a rate plot as a semilog straight line. Therefore, it is not really distinguishable from a regular exponential production decline. Therefore, the effects of the matrix-fracture transfer will be difficult, if not impossible, to observe.

Dual permeability (Fung and Collins)

In many instances, transfer within large matrix blocks was thought to be significant, which has led to variances on the more basic Warren and Root model. These include matrix subdomain methods as well as dual permeability methods (again, not to be confused with the normal context in pressure transient modeling that means layered). These two methods are shown in Figure 18–8. The use of such methods increases computer storage (memory) requirements dramatically.

Wellbore effects and large fracture spacings

In Part I of this text, a methodology was proposed that included the use of flow nets and diagrammatic representations of flow. This would be a logical place to start with an independent analysis of dual porosity theory. The author has concentrated on those situations where dual porosity solutions vary from single porosity solutions.

We start with a reservoir with fracture spacings of more than 10 meters and evaluate near-wellbore effects. The reservoir immediately adjacent to the wellbore has a strong impact on reservoir performance and is consequently a major consideration in implementing reservoir simulations correctly. Further, the "sugar cube" model is utilized. This analysis takes into account the relative dimensions of the wellbore and examines a number of possible solutions.

In Figure 18–9, we have a well located in the middle of a grid block at well location A. In this case, the reservoir is assumed to be less than 10 meters in thickness, and the top and bottom of the formation are terminated with an impermeable shale. Note that the well may not intersect a horizontal fracture unless the zone is more than 10 meters in thickness. Thin zones are relatively common in Western Canada. In this case, the dual continuum would have greatly overpredicted productivity; in fact, the productivity of such a well would not be high, since a fracture is not intersected. Extremely variable productivity is often attributed to fractured reservoirs.

From a pressure transient analysis perspective, the radius of investigation is inversely related to formation permeability. With low permeability in the matrix block, it is likely that the fractures will not even be seen on a well test.

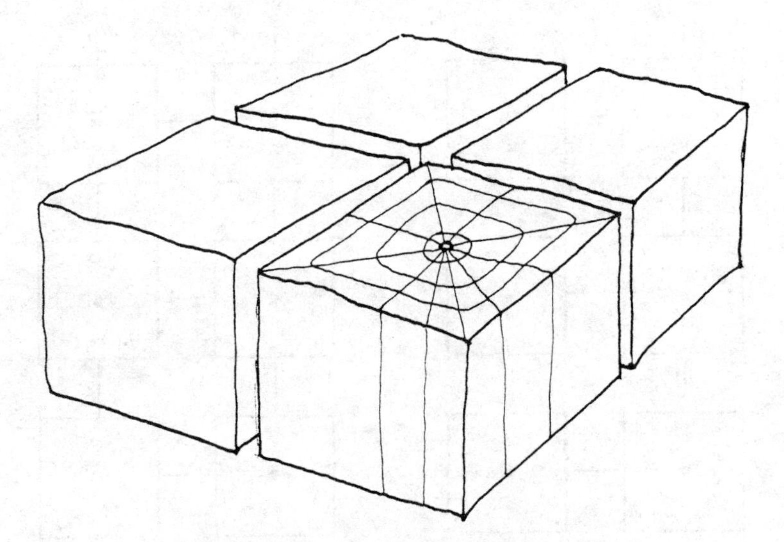

Fig. 18–9 Well in the Middle of a Matrix Block

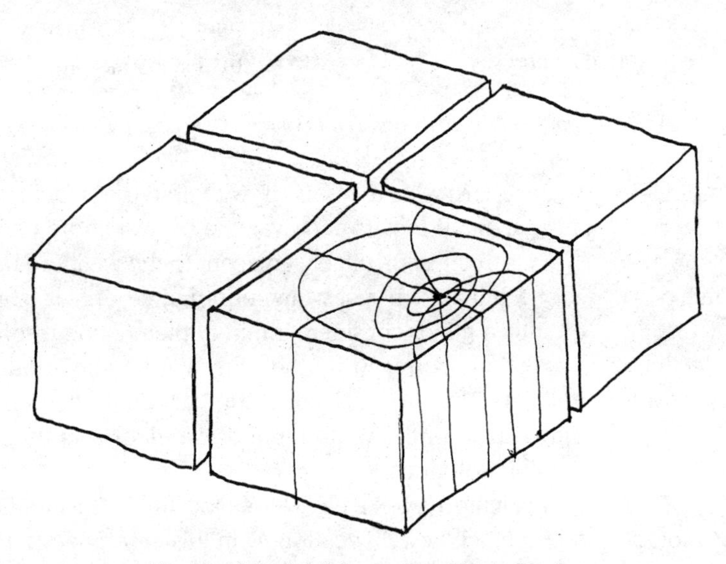

Fig. 18–10 Well Off Center in a Matrix Block

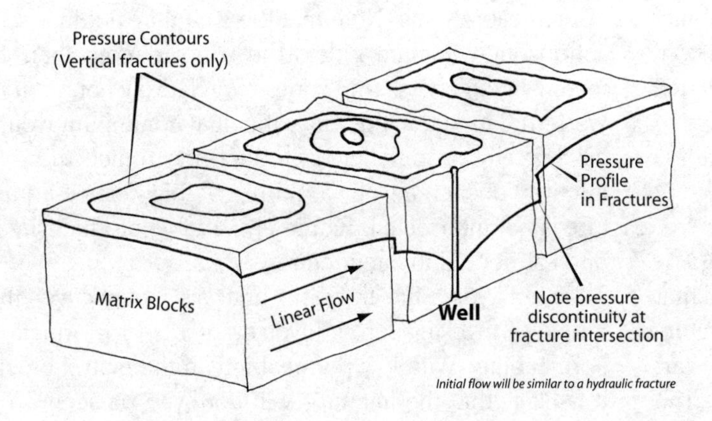

Fig. 18–11 Well Intersecting Natural Fractures

If we locate the well off to the side, then a different flow net will appear. This is shown in Figure 18–10. The well response here will be very different. If located sufficiently close to the fracture, some increase in productivity will occur. However, there will be a large pressure drop between the well and the fracture system. This would appear, at the very least, as a large skin factor. Note, however, that there must be more than two orders of magnitude difference in permeability between the fracture and the matrix block in order for the dual continuum model to differ from single porosity behavior.

Wells intersecting fractures (with large fracture spacing)

If we position the well in a fracture, there will be a different response, which is depicted in Figure 18–11. It would be reasonable to expect a pressure gradient, similar to that seen in a hydraulic fracture (in this case a fracture that branches or bifurcates in several places).

It is worthwhile to consider the relative probability of intersecting a fracture. Figure 18–12 shows a 10-meter block. Assuming a wellbore radius of 0.25 feet (0.0762 meters), there is approximately a 3.3% chance of encountering sugar cube fractures oriented vertically. On this basis, high-productivity wells in fractured formations should be relatively rare.

Continuum versus discrete modeling

At this point, it is necessary to go back and reexamine the use of a dual continuum. We assume that the fractures can be modeled as a field property. In some massive Iranian reservoirs, there may be 200 meters (656 feet) or more of pay. With a fracture spacing of 10 meters (33 feet), there will be 20 fracture intersections vertically. Most Western Canadian reservoirs have net pays of 30 meters (98 feet) or less. Intersections with fractures, then, are the same scale as the wellbore, and a continuum obviously does not exist. It is clear that, at the wellbore, fracture intersections are discrete events.

The author suggests that at least five intersections would be required in a grid block with a well to approximate a continuum or field property. This means a minimum grid-block size of roughly 100 meters (328 feet). When a simple grid sensitivity is run, results show that such thick grid blocks have significant errors when gas-oil and oil-water contacts are considered.

The dual continuum model (Warren and Root) is not, in the author's opinion, an internally consistent model. The question then remains, *How do these reservoirs really behave?*

The answer to this question leads to a number of issues, which are outlined in the following.

Most fractured reservoirs can be represented with single porosity

As we have seen, large contrasts in permeability, approximately two orders of magnitude, and large fracture spacings, approximately 10 meters (33 feet), are required for there to be an appreciable difference between fracture and matrix pressures, assuming the current dual porosity models are used. Restated, with small contrasts in permeability or at close fracture spacing, the pressures in the matrix and fracture will be identical. Since there is no time lag for matrix-fracture flow to occur, small fracture spacings are equivalent to a single—i.e., traditional—porosity system.

At this point, the latter issue of multiphase effects should be discussed.

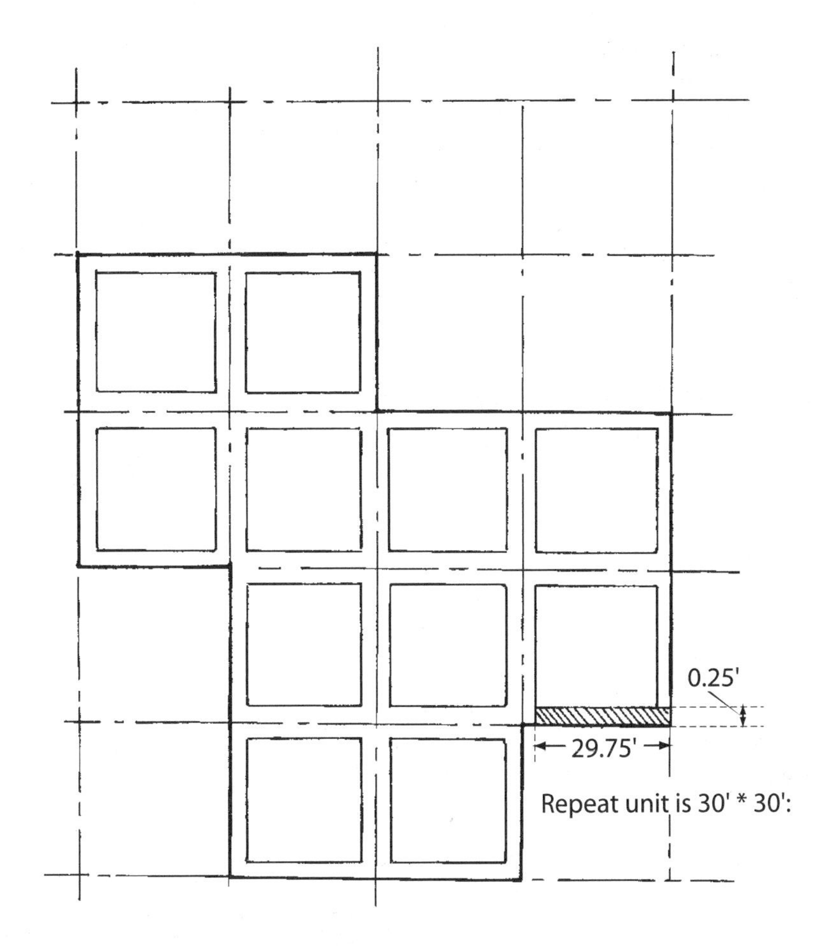

Assumptions:
Well diameter = 0.25'
Fracture spacing = 30'
Fractures are vertical
Diameter of well is 0.25'

*Repeat unit is 30' * 30':*

Area = (30')(30') = 900 square feet

Area around edge (4 strips of 29.75' long by 0.25' wide):

(4) (30 – 0.25) (0.25") = 29.75 sq. feet

Chance of intersection = 29.75/900
= 3.3%

Fig. 18–12 Relative Chance of Encountering a Fracture

Example multiphase gas condensate mechanisms

Consider what happens in a gas condensate reservoir. It is known that the primary production of a homogeneous reservoir follows the CVD depletion path only at the outermost extreme of the reservoir. As pressure declines, a buildup of condensate around the well occurs. The gridblock primary production decline is envisaged as shown in Figure 18–13. The situation is actually turned inside out. The CVD process path will occur in the innermost section (center) of the block. Condensate will build up around the rim of the block until it becomes mobile; i.e., at saturations above S_{oc}.

Gravity segregation should also be considered if a large continuous fracture system is present. On an overall reservoir, the system scale is depicted as shown in Figure 18–14. Grid blocks could be surrounded by condensate, by gas, or by gas-condensate or water-condensate contact.

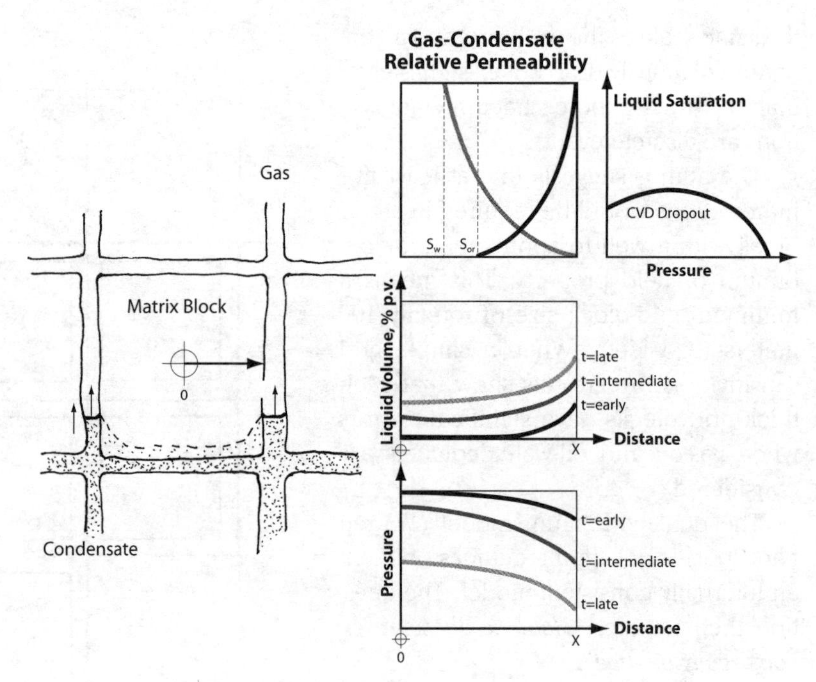

Fig. 18–13 Condensate Dropout in a Matrix Block

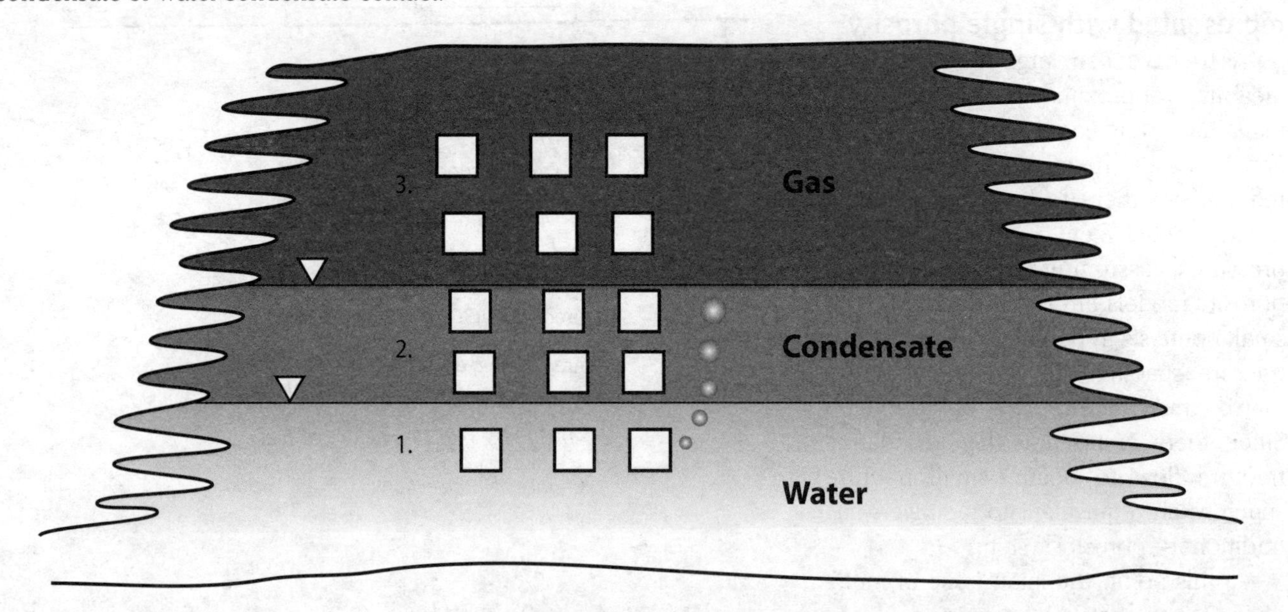

1. *Water-immersed block*
   - *water influx displaces gas and condensate upwards*
   - *gas bubbles out and condensate separates*
2. *Condensate immersed block*
3. *Gas-immersed block*

Fig. 18–14 Overall Scale of Reservoir

Further, if gas injection is introduced, is the condensate displaced back into the matrix block? If the gas injection is dry, how much of the matrix block condensate will be revaporized? Does gas injection rapidly break through the fractures and recycle?

Analytical derivation of multiphase transfer

Most analytical developments involve some simplifications in order to make the mathematics more tractable, and these equations are quite complex. Significantly, the equations used in the multiphase transfer model developed are not included as part of simulator input. Because of this, the input of these equations into a reservoir model is not a straightforward process. The author has never seen a study report that detailed exactly how these curves were generated.

Pseudo relative permeability and pseudo capillary pressure curves

Alternatively, it is possible to generate pseudo relative permeability curves and pseudo capillary pressure curves based on fine grid-block modeling. The author has not done this; however, as outlined in other sections of this chapter, the situations the author has encountered did not lend themselves to dual continuum modeling. The point is that extensive detailed matrix block calculations are required. The modeling described previously suggests that, since flow gradients likely exist across grid blocks, this should be included in the modeling. Most of the analytical work that the author has reviewed does not include this directly.

Realistic geometries

The previous calculations are all based on highly regular geometries and assume uniformity in the fracture geometry/aperture and that matrix blocks are isotropic and homogeneous. Of course, this is never the case in a natural system. Porosity is a basic rock property that varies spatially in natural reservoir rocks, and this means that matrix blocks will not be uniform.

Kazemi's model

Geometry that appears realistic from the point of view of the well is Kazemi's model, which is shown in Figure 18–15. Kazemi tested and developed his model in conjunction with a reservoir simulator, and this accounts more realistically for actual flow into the well. This solution is identical to a layered system. From a physical geometry point of view, the question must be asked, *How does the fracture stay open?* The answer is fracture roughness. The fractures rest on asperities, and there is enough space in the grooves between for fluid flow. Based on this proposition, it is likely that horizontal fractures are more highly stressed than vertical fractures and are less likely to be conductive. Nevertheless, Kazemi's model is a mathematically and physically sound solution.[21]

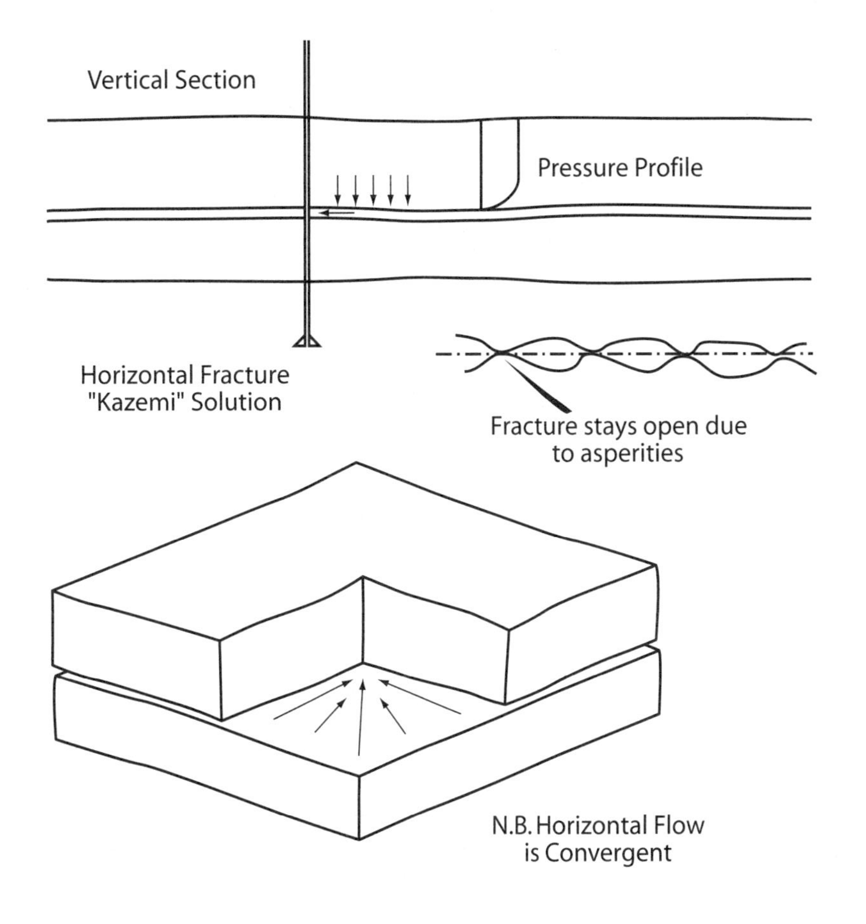

Fig. 18–15 Kazemi's Model

As a side note, the system described previously is easy to model incorrectly with a reservoir simulator. Note that the pressure gradient into the fracture is transient and nonlinear. Such a system cannot be accurately modeled with a two-layer reservoir model. Multiple layers must be used in the low-permeability layer. Logarithmic spacing is the most efficient, with the thin layers adjacent to the high-permeability conduit.

Discrete blocks

This leads to some interesting work regarding completion treatments by Harper and Last. Their work describes a new type (to the petroleum industry) of mechanical model—discrete element analysis. Geotechnical engineers developed this type of analysis. The design of tunnels, mines, and other excavations is frequently governed by rock joints (discontinuities).[22]

This type of model has been coupled with a fluid flow model. There are a number of key points. First, it is not possible to hydraulically fracture a jointed rock mass. The fracture will follow the existing discontinuities.

Figures 18–16 and 18–17 show a number of different situations that have different joint patterns, principal stresses, and fluid injection. Bizarre results are obtained, which do occur. High-rate injection gives a high breakdown pressure (*see* Fig. 18–18) and low injection pressure. A different orientation of principal stresses does not require a significant breakdown pressure but requires higher pressures on a long-term basis.

Even more interesting is that the blocks shift position, and these deformations or changes are permanent after high-pressure injection. This has implications for the design of fracture treatments in naturally fractured reservoirs.

It is said that this model is capable of modeling a lottery-type rotation of objects and accounts for friction between the blocks as well as the rotations and associated stresses.

Fracture modeling

Some models deal with fractures only. Many of these models utilize statistics to describe the fracture pattern. One such model was developed by Golder and Associates, and the author has seen some nice matches to a pressure transient analysis. Most of this work was oriented to North Sea problems. This model was shown at an SPE forum on fractured reservoirs. The author encourages engineers interested in the subject to attend these forums. The presentations are high quality, and there are many opportunities to meet interesting people and to socialize. Do not be intimidated in applying to attend. Most of the big names are pleased to talk to anyone on their favorite topics. Furthermore, the "experts" learn a great deal from attendees who have practical reservoir experience.

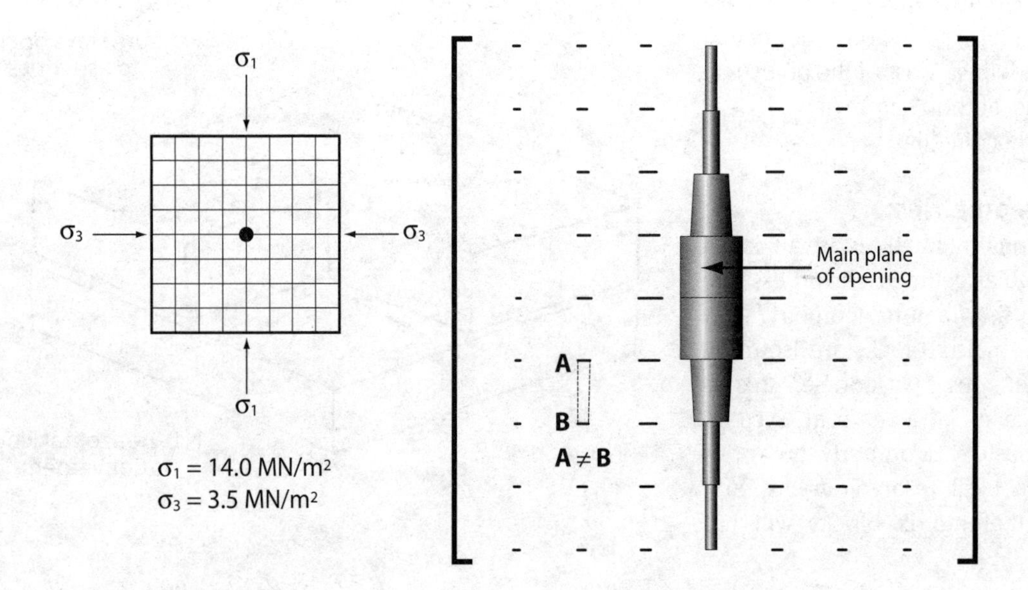

Fig. 18–16 Modeling of Discrete Joints

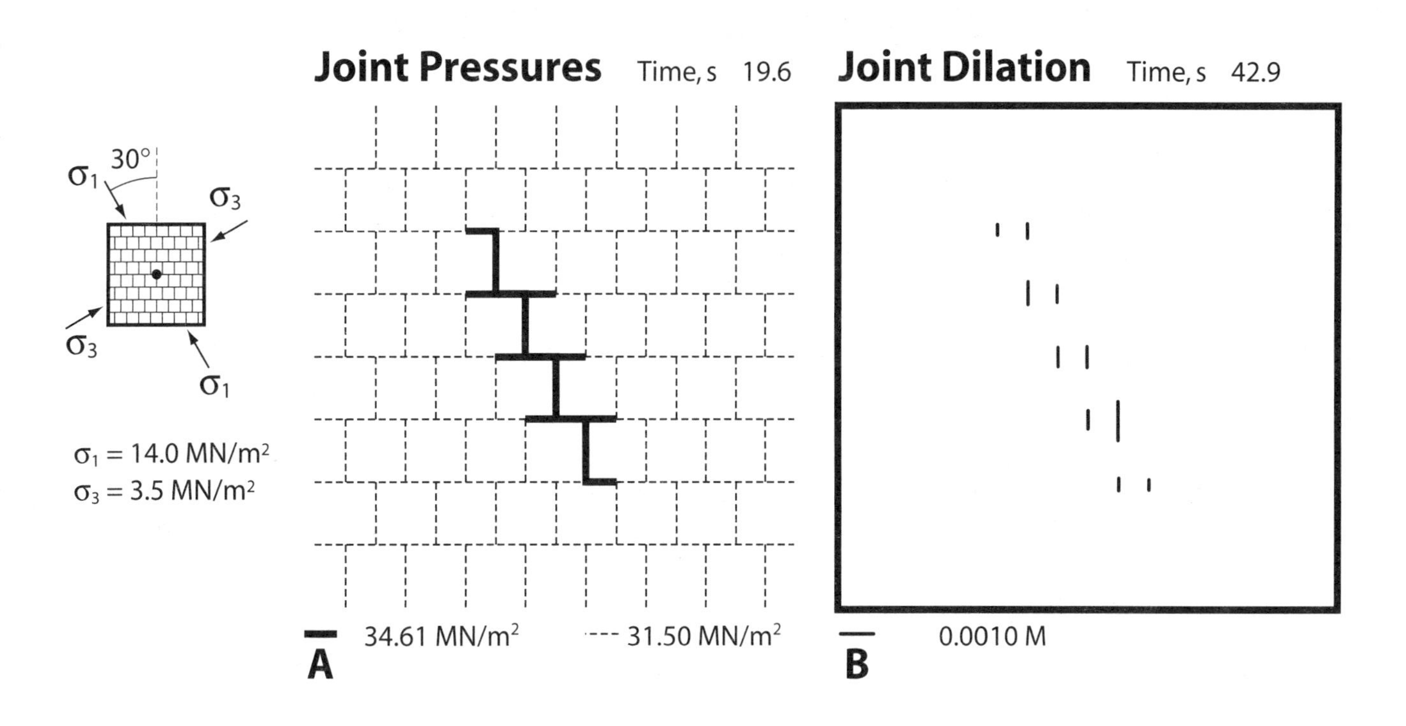

Fig. 18–17 Modeling of Discrete Joints

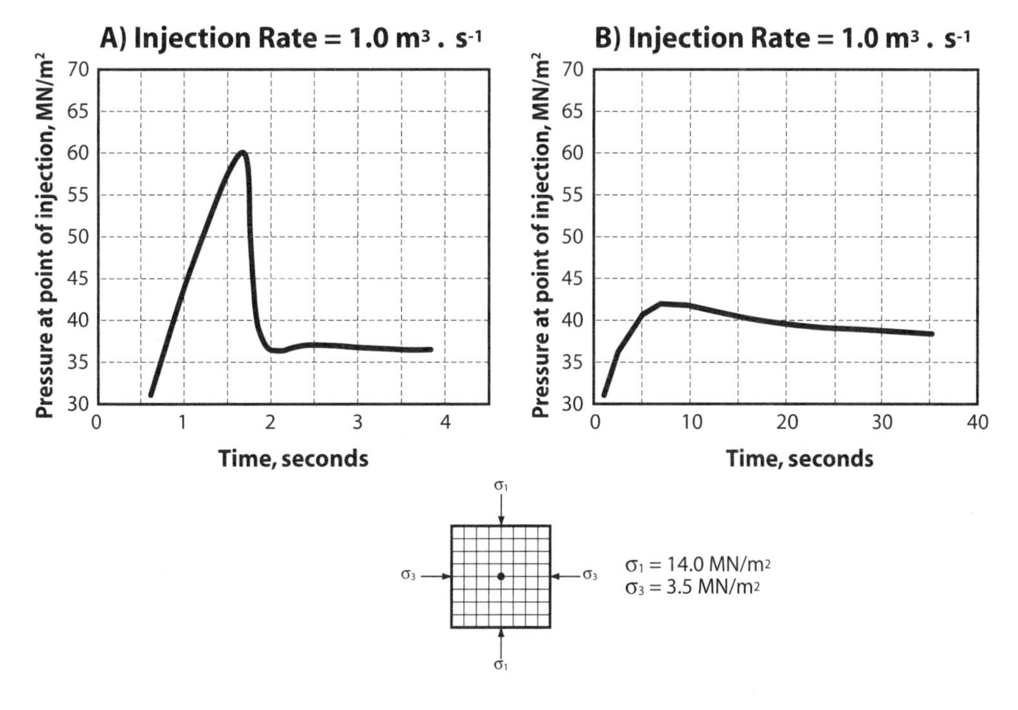

Fig. 18–18 Modeling Discrete Joints—Breakdown Pressures

Resolving Consistency Problems

From the preceding, it is clear that the use of dual porosity type simulators has several pitfalls. Nevertheless, it is a practical reality that our clients or supervisors require some kind of answer. A major component of working on naturally fractured reservoirs is determining the dual porosity approach to use.

To put this in more concrete terms, a typical study outline might look something like this:

1. Geology
2. Production performance analysis
3. Data screening
4. Build simulation datafile
5. Perform history match
6. Tune model
7. Predictions
8. Report

Typically, history matching comprises about one-third of the study and is the largest single effort that is done. It is fair to say that history matching is generally the most challenging.

For a fractured reservoir engineering study, the outline will change, in general terms, as follows:

1. Geology
2. Detailed fracture descriptions—normally core logging
3. Study of origin of fractures
4. Production performance analysis
5. Statistical analysis of initial productivities
6. Detailed review of pressure buildups—computer model matching of observed type curves and derivatives
7. Review of production logging—particularly flow surveys
8. Dual porosity study—do type modeling of groups of matrix blocks
9. Select best model
10. Build full-field model
11. Perform history match
12. Tune model
13. Predictions
14. Report

In some ways, this is just a more detailed version of the first outline. However, these additional steps are not done except for this kind of study.

To look at this a different way, here are some of the additional issues that need to be examined:

1. Geological description is very important. Fortunately, this is one issue that seems to be largely agreed upon.
2. There is more than one type of model that can be used:
   a. Single porosity models can accurately represent fractured reservoirs under the right conditions.
   b. Discrete element modeling.
   c. Use of a Saidi-type columnar model.
   d. Specialty fracture models; these are often stochastically based.
   e. Dual porosity modeling has a number of sub choices:
      i. Straight dual continuum
      ii. Dual permeability systems (interblock transfer)
      iii. Nested matrix solutions (such as MINC).

   These techniques are listed in order of preference. Perhaps, more importantly, the type of model used should be tested for the reservoir conditions at hand with a detailed fine grid matrix and fracture model. This should be within the reach of most modern high-performance PCs.
3. Consistency should be maintained, particularly with pressure transient techniques. The dual continuum solutions will result most likely in interpretations that are not unique or necessarily applicable at all.
4. Use of semianalytical techniques, such as that due to Aronofsky et al. and subsequent modifications, may still have some application.

A few comments are in order on these topics as outlined in the following paragraphs.

Geology

In the author's opinion, there are alternate explanations for the development of fractures and how they physically manifest. This is outlined in the following section. Other controls should also be mentioned. Structural environment is an important control. The author has divided this between extensional and compressive environments. With time, different structural styles will be identified that result in more detailed differentiation. Within structures, such as imbricate thrust sheets, there are variations in fracturing that correlate with structural position. This seems to be related to stress.

Types of models

- In many situations, a single porosity model will provide the same results as a dual porosity simulator much faster and with fewer computational requirements.
- Discrete element modeling is not widely applied. However, the Norwegian Geological Institute has used it with good success. This has been in the North Sea where the overall environment is extensional. The results are interesting, and they have published an extensive report. Considerably more action occurs in a reservoir predicted by these models than one might suspect. This requires some geomechanical input and some additional fracture data. Such models, although available, are still a specialty item.
- Saidi's approach has some compelling logic and is a realistic alternative to dual continuum modeling. The author has not found this specific type of modeling applicable to Western Canada. However, it seems to have worked well for Iranian-type reservoirs. This involves very large structures, massive carbonates, and an anticlinal structure—related to thrust faulting. These reservoirs have high permeability. Clearly, this type of problem is of economic importance.
- There is much to be learned from stochastic fracture models. Again, this is a specialty technique. The author has not seen a full field study using this methodology. Fracture swarms are observed, and this is likely a useful tool to understand the effects of these patterns. If the swarms are dense enough, it is likely that conventional reservoir simulators can be used to replace these areas with high permeability. It is possible to envisage how the stochastic model leads to developing an understanding of patterns that would be put into a conventional reservoir simulator.
- Finally, there may be some changes to the dual continuum concept, such as nested grid blocks, that enable more sophisticated modeling, as computer performance increases steadily.

Detailed grid and fracture modeling

Careful detailed modeling is required before proceeding with dual porosity models. One obvious question that springs to mind is, *Why would we bother with a dual porosity simulator if it is insensitive to grid-block size?* Restated, if the production performance doesn't vary when we change the fracture spacing, what is the point of using a dual porosity simulator? The author's conclusion is that, in the absence of large fracture spacing, there is no point.

Using a dual porosity simulator doubles computer memory requirements and extends execution times. Why use these resources to no purpose? In a number of instances, the author and others known to him have simply discarded the dual porosity simulator.

Of course, the conditions encountered by the author may not be representative of the conditions that others encounter. In this regard, the author cautions, as usual, buyer beware! Restated, do the tests to determine that dual porosity works as expected.

Experimentation

Not long ago, it was fashionable to evaluate the effect of innovation on companies. A number of books were published on this issue and led to some "newspeak" terms such as "intrapreneuring," "out of the box" thinking and strategizing. The evidence that links corporate success and the ability to innovate is compelling. Actually, creating the required circumstances is more difficult.

These books included a number of key ingredients that had been identified: first, an emphasis on action; second, a tolerance for time spent on "hunches," "bootleg," or simply "interesting" projects; and finally, the recognition that not all endeavors lead to immediate results. The latter more pervasively represents a general attitude that investigation of basic ideas will ultimately lead to useful implementations at some time in the future.

Therefore, if engineers are going to work on naturally fractured reservoirs, it is necessary to experiment with the basic tools. This means experimentation with the different techniques on a comparison problem. Does using the different models give you different results? How sensitive is the model to grid-block dimensions? Do minute changes have large effects? Or, are some significant changes required to appreciably alter output effects?

It could be argued that this is the normal course of doing reservoir studies; however, this is one area where expectations of how the model would work and what actually happened were drastically different. There appear to be more gaps in the literature in this area than with other techniques.

Basic transfer functions—single phase

The author has used the steady state assumptions built into the simulator and used the default shape factor. Clearly, the geometry of real matrix blocks is not square. In fact, the shape factor seems to be considered a history-match parameter. This is one area where the author's investigation was truncated. More sensitivity would be prudent if one elects to use a dual continuum approach.

One might hope this were an area where the analytical solutions derived from pressure transient analysis would provide quantitative guidance. However, this is not the case. The calculation of σ is a characteristic dimension and is a relative rather than absolute variable.

Imbibition capillary pressure

One area where fundamental research appears to be developing is with respect to imbibition capillary pressure. In some ways, the simple experiment initially done by Arps is the essence of what is required. We put oil-saturated rock in a water bath, and the oil and the water interchange places. It is known that the water gets drawn into the smaller pore spaces and the oil is apparently ejected into the larger pore spaces and ultimately out of the rock.

This is almost a form of capillary pressure drive. From this it may be concluded that the distribution of pore size is important and the wettability of the rock is also important.

In reducing this to practice, the Arps experiment does not really represent a matrix block geometry. His model also had only one open face, whereas a real matrix block would be open on many sides. Ideally, the lab experiment would represent the dimensions of the block and allow for gravity segregation and actual phases present. Realistic time comparisons could be rather long. Even for common grid-block dimensions of 3–5 meters (9.8–11.5 feet), this would require a very large laboratory apparatus. The author has not read any work on this, using scaling, that has been reduced to practice.

In addition, there are more than water-oil processes taking place, and a more extensive suite of experiments would be required to fully analyze reservoirs with gas caps and water legs (or water injection).

Rock properties for fractures

One other area of debate is the actual properties for the fractures. Some of the simplified analysis has utilized the assumption that the fractures were truly open, like two parallel plates. In this case, the fractures could be represented as vertical equilibrium (straightline) pseudo relative permeability curves and VE (straightline) pseudo capillary pressure curves.

Of course, real fractures are not parallel plates. The fractures are normally held open by irregularities on the surface of the fracture (as discussed earlier, the asperities). Hence, the matrix blocks will contact each other at certain contact points. This means there will be some kind of relative permeability relation and some kind of capillary pressure function.

This data is not commonly found, and this leaves us back with a history-match parameter. This is not necessarily an unrealistic requirement; however, there are limits to the degrees of freedom that allow us to make predictions, and this is relatively basic data. Most people assume VE curves and do a sensitivity. This is not very elegant, but there are not any other real choices.

Multiphase transfer functions

The author has not used these transfer functions and cannot offer direct experience. However, here are a few suggestions:

- Examine the papers on dual porosity formulation. There is a capillary pressure relationship for the matrix. It is important to realize that this is a rock property and really implies that the grid block is thinner than the thickness of the transition zone. If the reservoir matrix blocks are quite thick, on the same order or thicker than the transition zone, a pseudo capillary pressure relation is required. This is no different from a single porosity simulation.
- There are transfer functions based on the Aronofsky-type assumptions. The method has some limitations, and the authors have been commendably frank with these problems.[23]

- There seems to be relatively little lab data upon which the functions can be based experimentally. The same data, which has been published, is therefore used by many of the authors. There is some data quality screening available on these papers.
- Modifications have been proposed for the transfer functions. To the best of the author's knowledge, they have not been incorporated in commercial simulator code.
- Some recent papers indicate convection to be significant. Most of this work relates to the giant Cantarell complex in Mexico.

It is hoped that more material will appear on the issue of transfer functions in the future.

Consistency with pressure transient techniques and interpretations

Sabet in his book *Well Test Analysis* discusses two circular (concentric) regions in the chapter, "Testing of Naturally Fractured Reservoirs." The basis was a paper by Adams et al. for which he developed an analytical model to match observed performance. The author has seen many tests that look like this and has a slightly different explanation, which is outlined in the last section of this chapter. Adams' technique is practical and is definitely observable.[24]

Warren and Root's paper was developed theoretically and did not include practical examples. This is not too surprising, since the Warren and Root solution has some unlikely assumptions. The classic Horner Plot for this solution consists of two parallel lines, but, as outlined previously, is virtually never observed. Part of this is related to the fact that the presumed character occurs very early and would usually be obscured by wellbore storage. However, it is possible to use downhole shut-in techniques and still these effects are not noted.

Author's Model—"Poddy" Reservoir

The author has worked on a number of fractured reservoirs in Western Canada that exhibit a number of common characteristics. Evaluating this type of pool is normally complex and presents a number of difficulties. These pools also exhibit distinct types of pressure transient response, which do not correspond to an accepted dual porosity description.

A number of different approaches used to characterize the reservoirs are described, or more particularly, the fracture systems in the Western Canadian pools studied by the author. In the following, the methodology used for these different techniques is outlined as well as the results obtained.

Lithology

Lithology has a large influence on the degree of fracturing that can be expected. This is demonstrated in Figure 18–19, which depicts rock ductility versus depth for various lithologies.[25] Quartzite and dolomites have very low ductility and are prone to brittle failure or fracturing. Limestones exhibit high ductility and are the least likely lithology to fracture. Sandstones exhibit intermediate behavior.

Core pictures

One of the best ways to get an immediate understanding of the character of a reservoir is to look at the core. Take pictures extensively. Invariably, there is much discussion in meetings about what the core might look like. The conjecture can go on forever. The use of pictures allows one to communicate the results quickly and with a higher degree of confidence and acceptance. An example of a picture is shown in Figure 18–20.[26] A drawing that shows how these pictures were taken is shown in Figure 18–21.

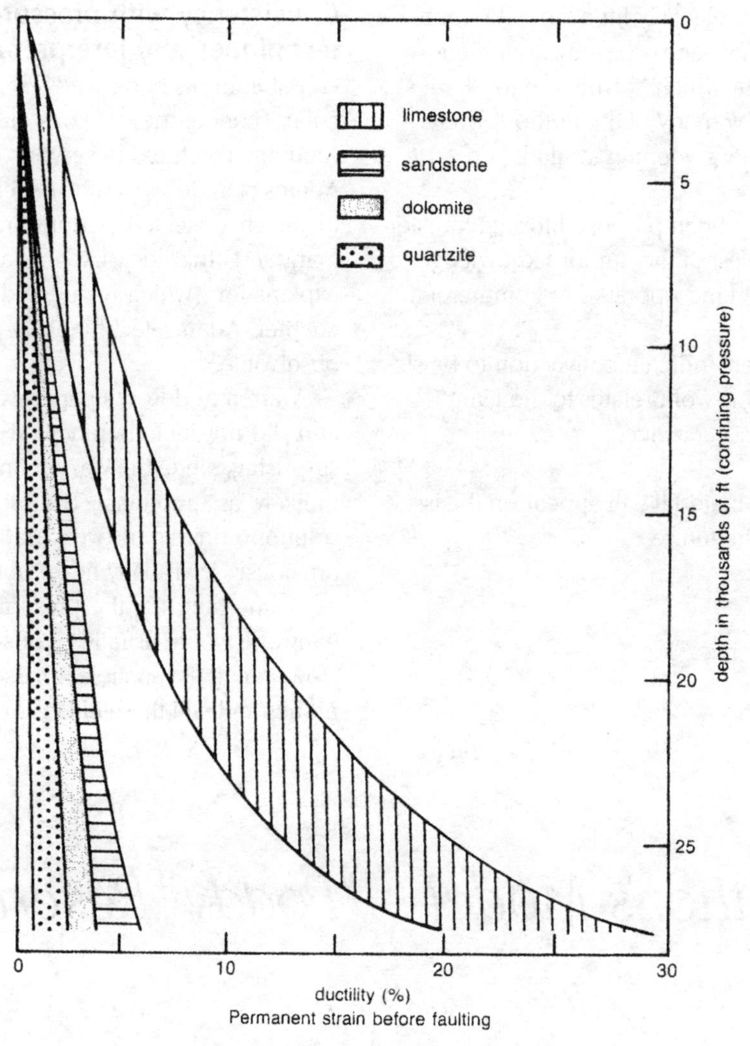

Fig. 18–19 Rock Ductility Chart Various Lithologies

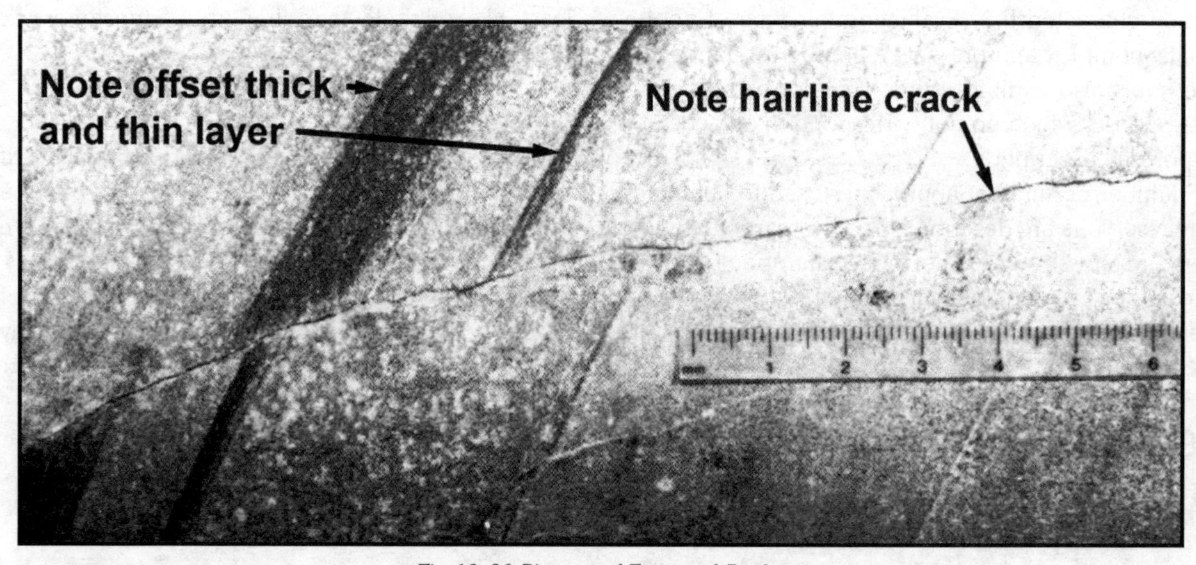

Fig. 18–20 Pictures of Fractured Rocks

Core logging

ARE conducted a major reservoir simulation, and Figure 18–22 shows a core log of the reservoir. Note that most of the fractures observed are on a relatively small scale, approximately a hairline in width, and appear to have a fairly extensive network. There are normally no signs of large displacements. It was noted that the fractures were found only in limited vertical sections of the core.

Conventional core analysis

Obtaining most of the fundamental information requires a basis of observation, which, for most reservoirs, starts with core. Information can be derived from the following conventional core analysis data and plots:

- Core permeability versus core porosity—all data
- Core permeability versus core porosity—sorted by lithology
- Vertical permeability versus horizontal permeability
- k_{90} plotted versus k_{max}

Fractured reservoirs do not show the typical straight-line relation on core permeability versus core porosity (semilog) plots. Typically, lower porosity rock is more prone to fracturing. Fractured reservoirs also tend to have higher anisotropy, which is seen as large variations in k_{90} versus k_{max}.

Example of conventional core analysis from central foothills of Alberta

In Western Canada, there is a typical core permeability versus core porosity relationship. The following diagrams and points are based on an example from the central foothills:

- Core permeability versus core porosity is shown in Figure 18–23. The data shows tremendous scatter, which is typical of a fractured reservoir. There are 762 data points, which have an average air permeability of 22.39 mD and an average core porosity of 4.11%. At first glance, the data looks completely random. However, careful examination shows there is a triangular distribution. As outlined earlier, lower porosity core is more prone to fracturing. The lower bound of the triangle approximates matrix properties.
- Fractured samples were then isolated, as shown in Figure 18–24. These 564 points comprise about ⅔ of the total sample set. Note that the average porosity drops from 4.11% for all samples to 3.65%. At the same time, the permeability increases from 22.39 mD for all samples to 29.15 mD for fractured samples only.

Nonfoothills core

It is interesting to compare the previous results with a nonfoothills reservoir. The Brazeau River Elkton/Shunda is located about 50 km to the east of the preceding example. The pool is located on a subcrop edge and is a type of fractured reservoir. However, the fracturing is derived from a different source—karst development. Core permeability versus core porosity from this field is shown in Figure 18–25. It is immediately obvious that the data does not resemble the shotgun blast or triangular appearance of the foothills reservoir.

The average permeability and porosity are 56.54 mD and 9.25% (574 points). The foothills reservoir core averaged 22.39 mD with a porosity of 4.11%. It is interesting to note that, at an average porosity of 9.25%, the nonfoothills core has an average permeability of only 0.3 mD to air at surface conditions.

Wabamun (D-1)

Another reservoir that shows this type of core permeability versus core porosity has been plotted for a Wabamun D-1 reservoir as shown in Figure 18–26. The core data shows a classic triangular distribution of core permeability versus core porosity. In the author's opinion, this is diagnostic of fracturing derived from foothills-type structural deformation.

Borehole imaging

ARE has had some experience with borehole imaging. HEF Consulting indicates that the best fractures on logs do not always correspond to the best production. In very tight rock, which is highly resistive, small-scale fractures will stand out clearly, although they have small apertures. In HEF's experience, areas that appeared foggy had a more complete network of small-scale fractures and were more likely to correspond to high-productivity zones.

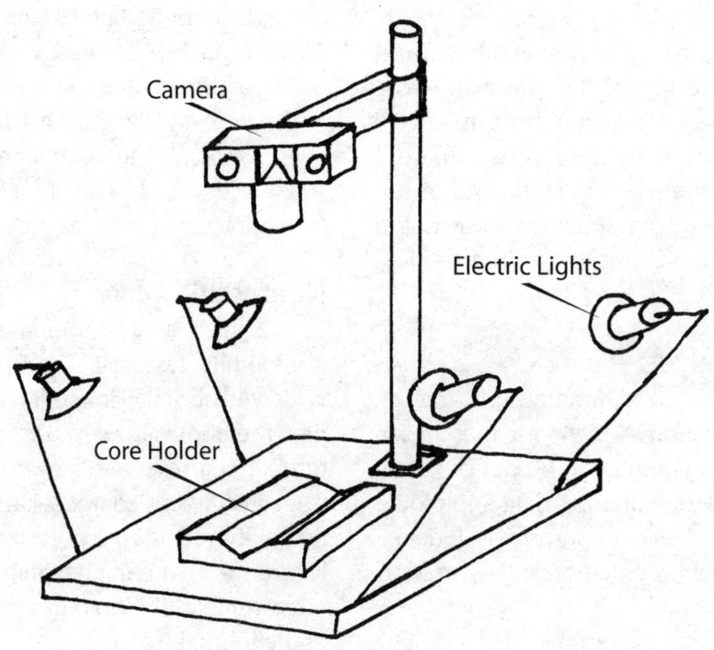

Note: Cloth over core holder of contrasting color helps

Fig. 18–21 Apparatus for Core Pictures

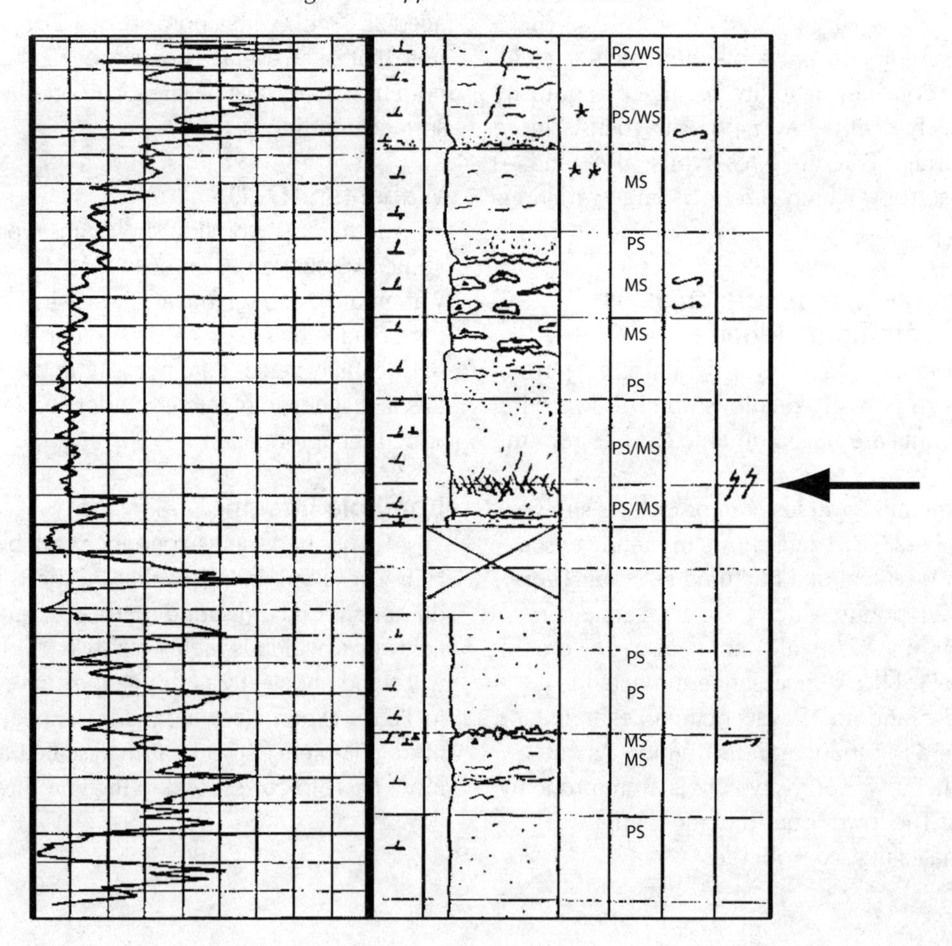

Fig. 18–22 Core Log—Fractured Reservoir

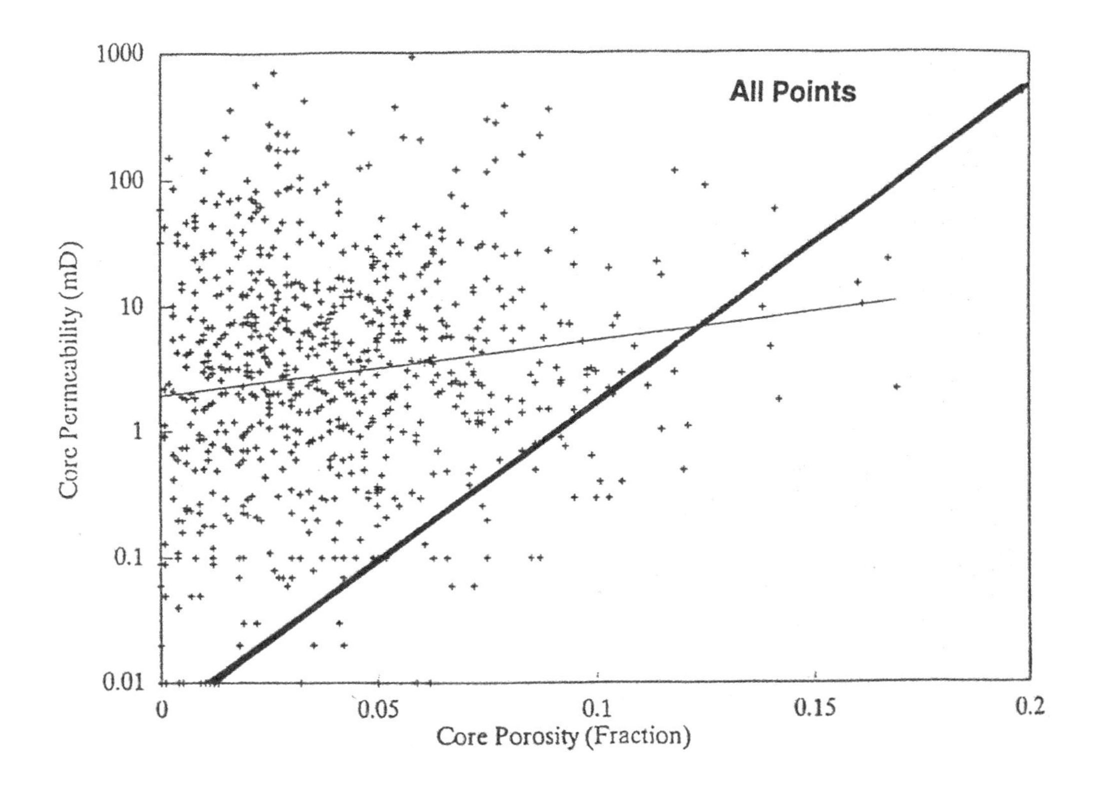

Fig. 18–23 Core Porosity vs. Core Permeability—All Samples

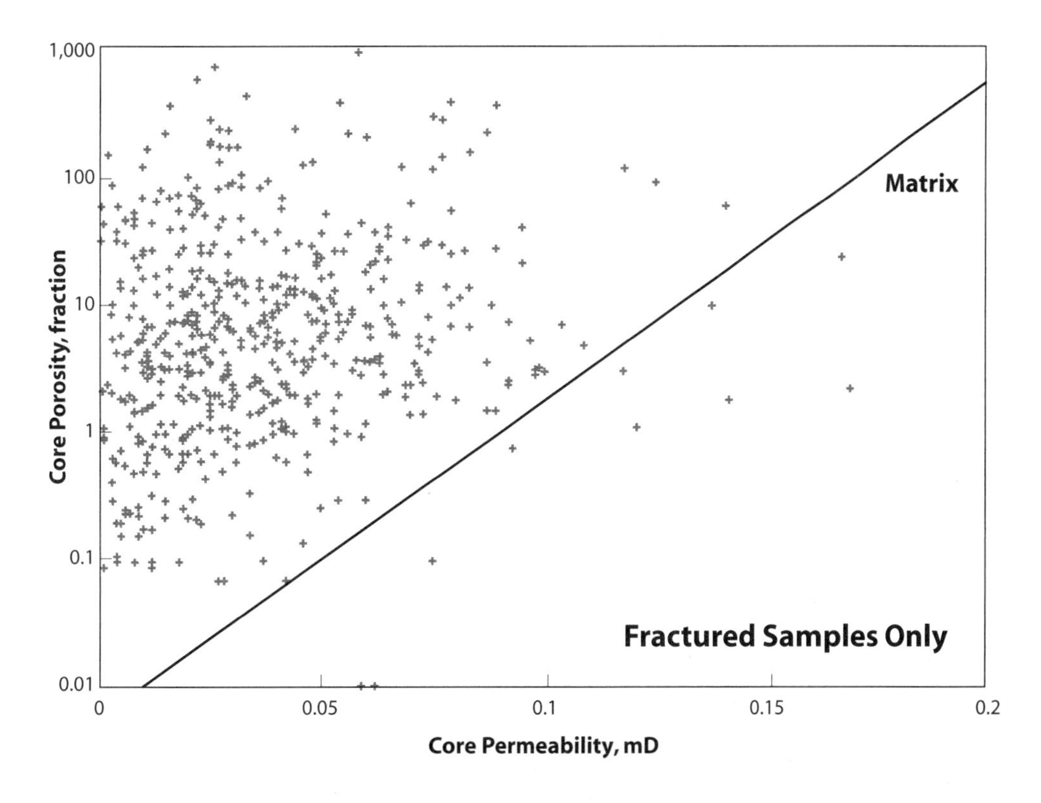

Fig. 18–24 Core Porosity vs. Core Permeability—Fractured Samples Only

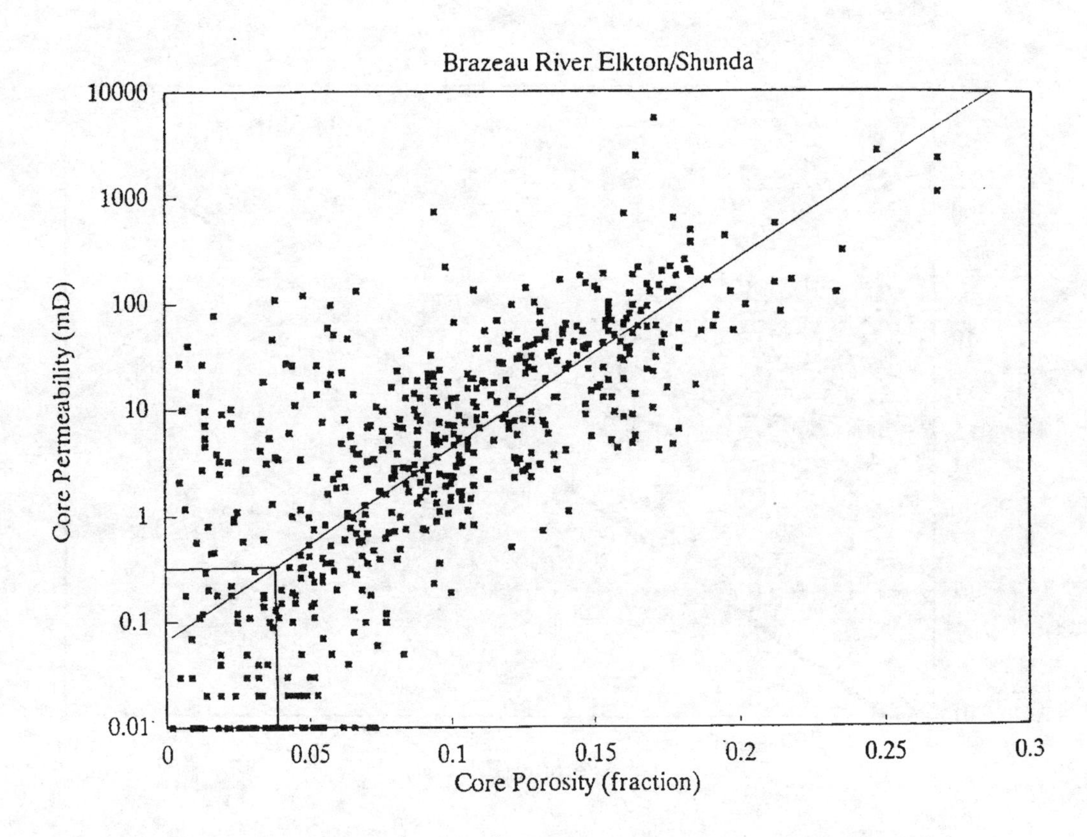

Fig. 18–25 Core Porosity vs. Core Permeability—Same Formation, Nonfractured

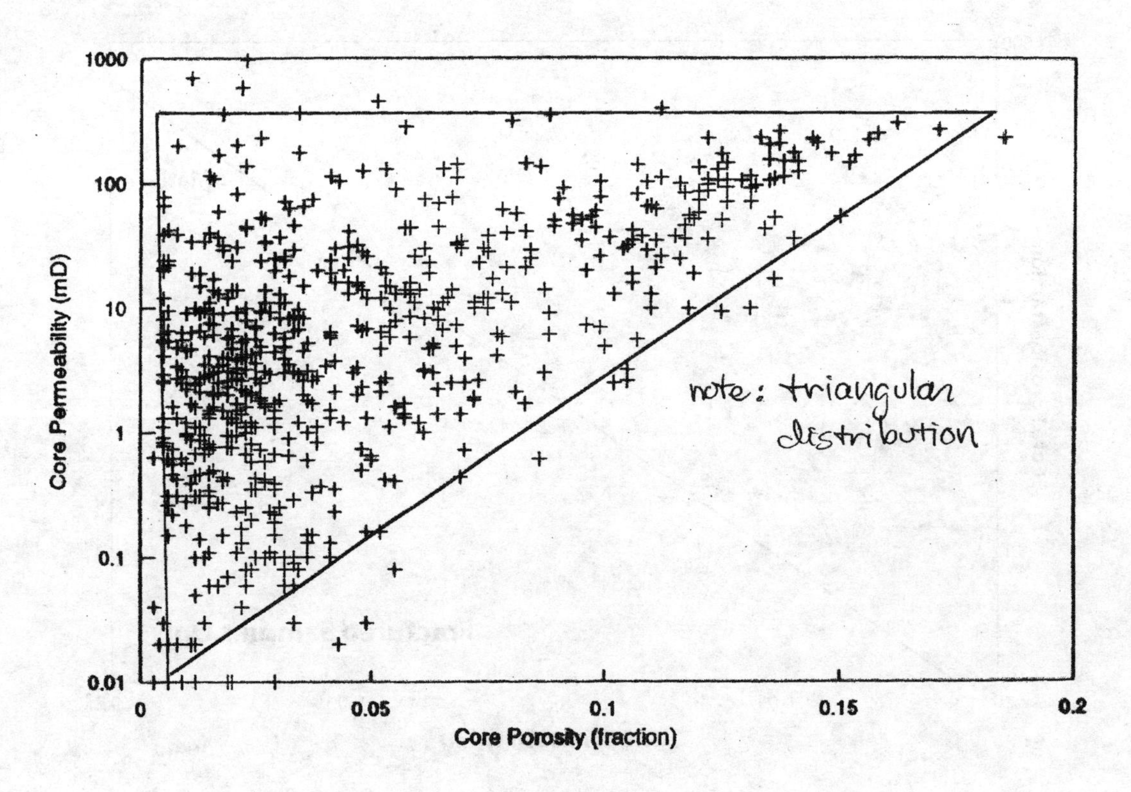

Fig. 18–26 Core Porosity vs. Core Permeability—Second Fractured Reservoir

Structural style effect

Figure 18–27 shows a cross section through a Western Canadian sandstone reservoir. As part of an economic evaluation, the author plotted well AOF and type curve reservoir interpretation. This diagram shows the basic building blocks of structural style in Western Canada, which is a thrust fault. Well performance is strongly affected, depending on where in the structure wells are completed:

- Along the top of the overthrust sheet leading edge, well deliverabilities are very high. Type curves for this reservoir were typically single porosity radial responses.
- Underneath the thrust sheet and adjacent to the shear zone, long slivers of reservoir rock are dragged up or broken off. In this area, a number of bilinear test results were obtained.
- Further ahead, a series of faults are of a much smaller scale than the thrust sheet. In these areas, type curves indicate systems with concentric circular regions of permeabilities.
- Behind the thrust sheet on the back limb, well tests indicated hydraulic fractures and mixed medium deliverabilities.

One thing was immediately obvious. There was good permeability across the front of the thrust sheet.

This is not the only structural model that exists. Such variations are well-known in two other areas. The first example from the literature is from Iran, where there are large anticlinal structures. This is shown in Figure 18–28.[27] The second example from the literature is a rollover from the Bullmoose-Sukunka trend in northeastern British Columbia shown in Figure 18–29.[28] In the anticlinal structure, maximum fracturing is associated with the downwarp side of the flanks and the end of the structure. There are other examples. An oilfield in Northern Alberta is associated with a buried meteorite impact structure. Impact structures have distinctive features as shown in Figure 18–30.[29] Salt diapirs are very common in the United States Gulf Coast and cause different structures as well.

Effect of fracture pattern

Part of the effect of fracture pattern is discussed in the section on pressure transient analysis. This effect of fracture patterns is obvious on pressure tests. Most of the initial work in this area relates to numerical flow modeling or the use of electrical models. This work utilizes fracture systems that have been created on a statistical basis, such as those shown in Figure 18–31(a).[30] These networks can be modeled numerically with an applied pressure gradi-

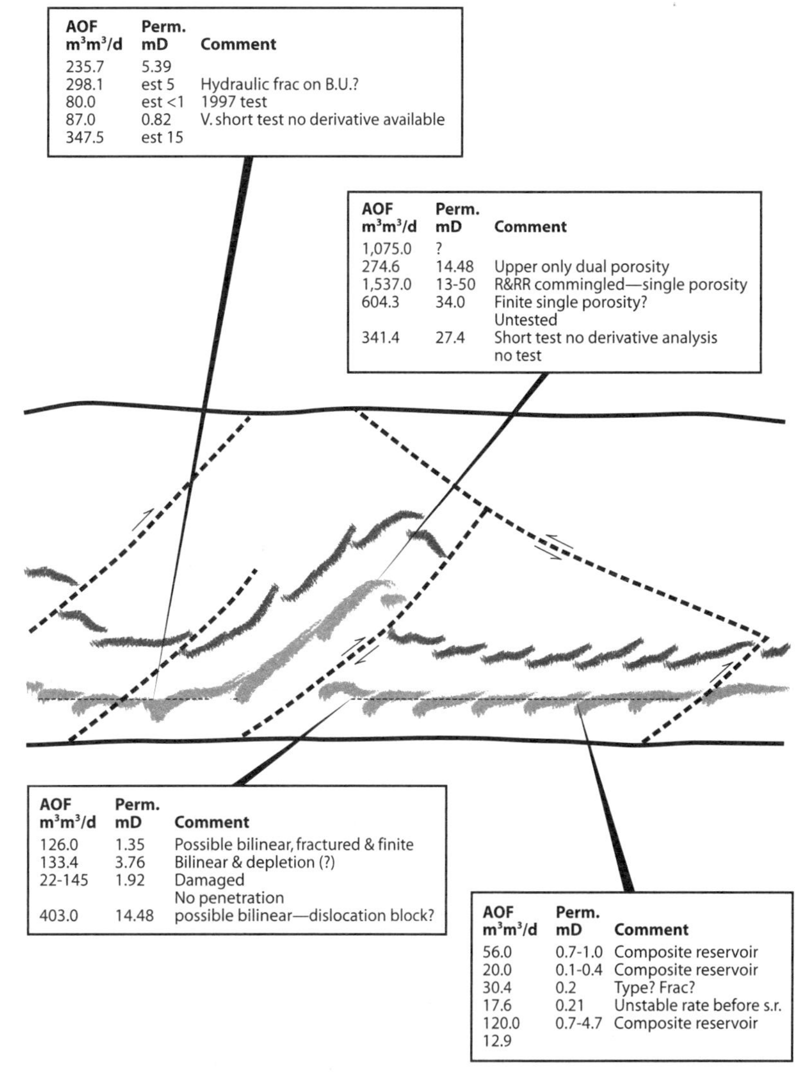

Fig. 18–27 Well Performance Variation with Structure—Western Canada

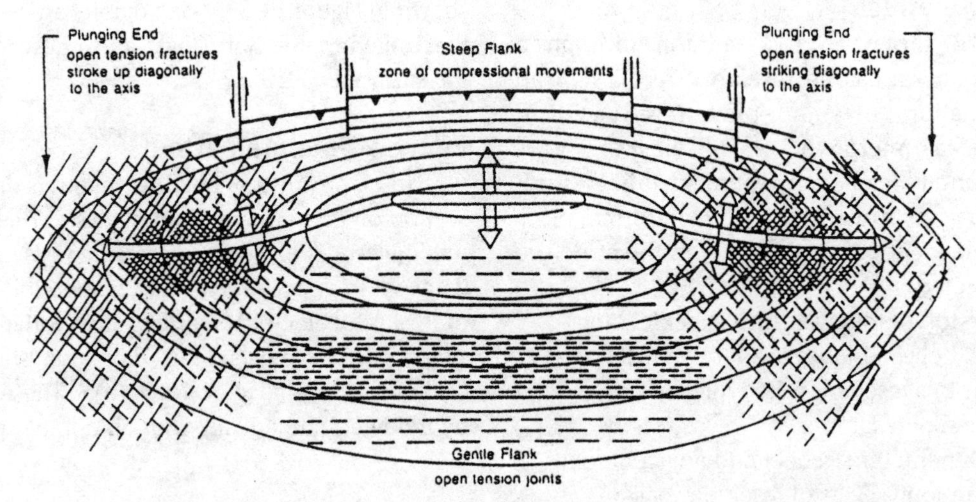

Fig. 18–28 Variation in Fracture Pattern with Position—Iranian Reservoirs

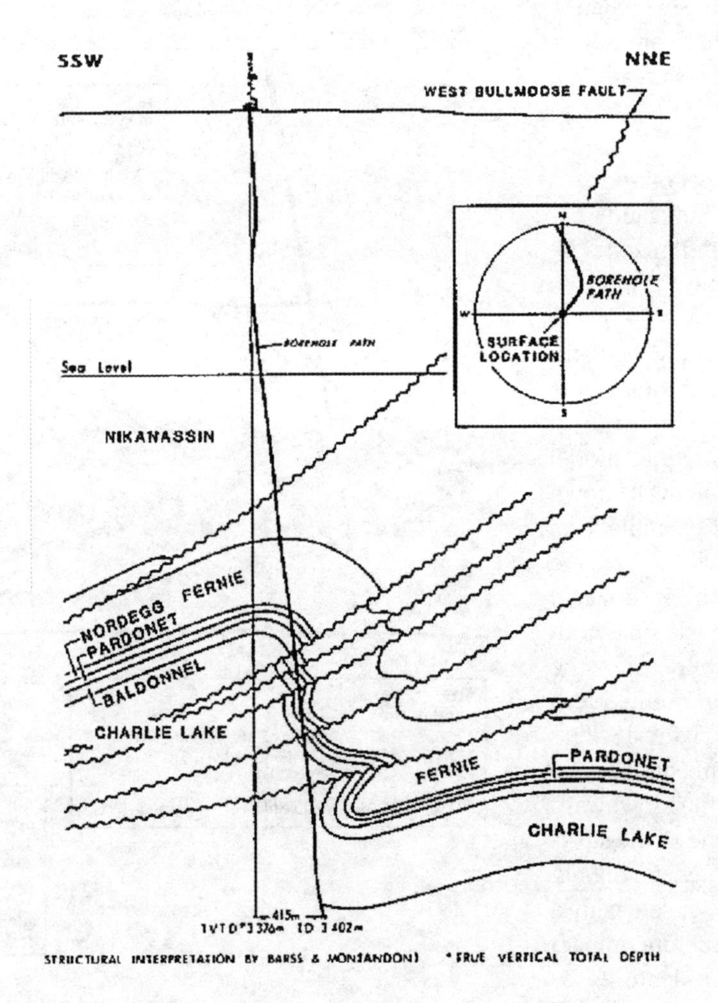

Fig. 18–29 Structure of Bullmoose Sukunka Trend

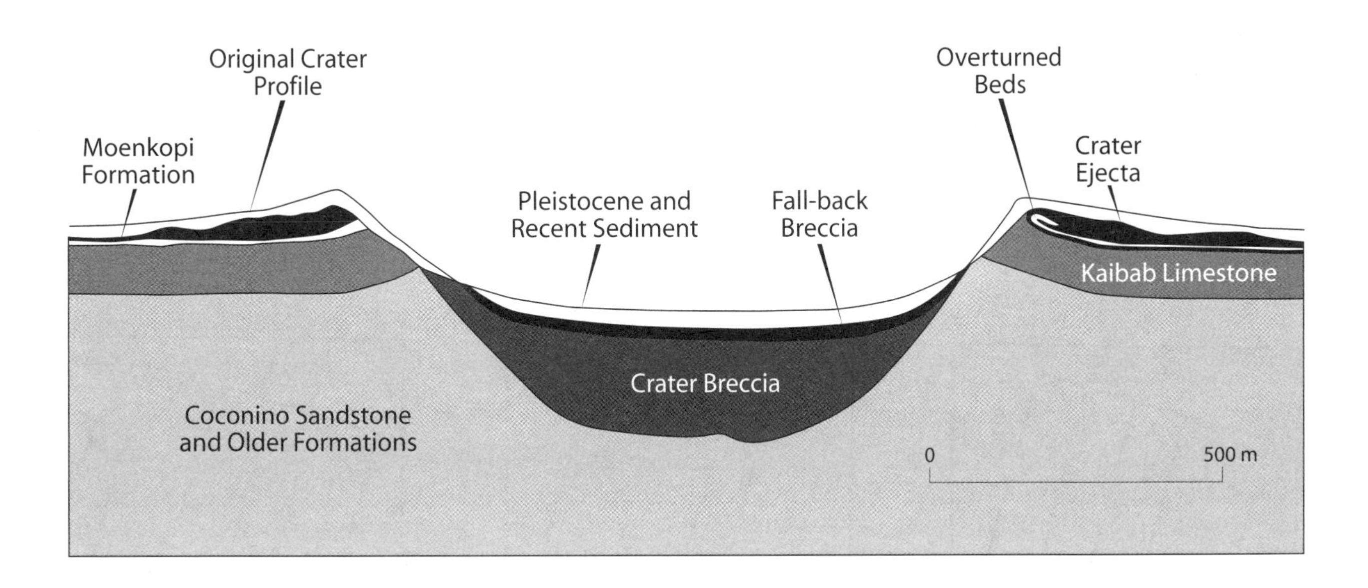

Fig. 18–30 Impact Structure

ent and effective permeabilities calculated as shown in Figure 18–31(c). Note that the orientation of the potential gradient is changed.

A particularly interesting result is that permeability is controlled by fracture connectivity as shown in Figure 18–31(b). At first glance, it is tempting to think that permeability would be highest along the trend of the fractures; however, this is not the case.

Classic fracture set descriptions

Price was responsible for much of the early work on fractured reservoir description as shown in Figure 18–32.[31] He found that many of the fractures associated with anticlines are conjugate shears. This has led to the suggestion that there would be good permeability downdip on thrust sheets. This would be oriented with the general long direction of the fractures.

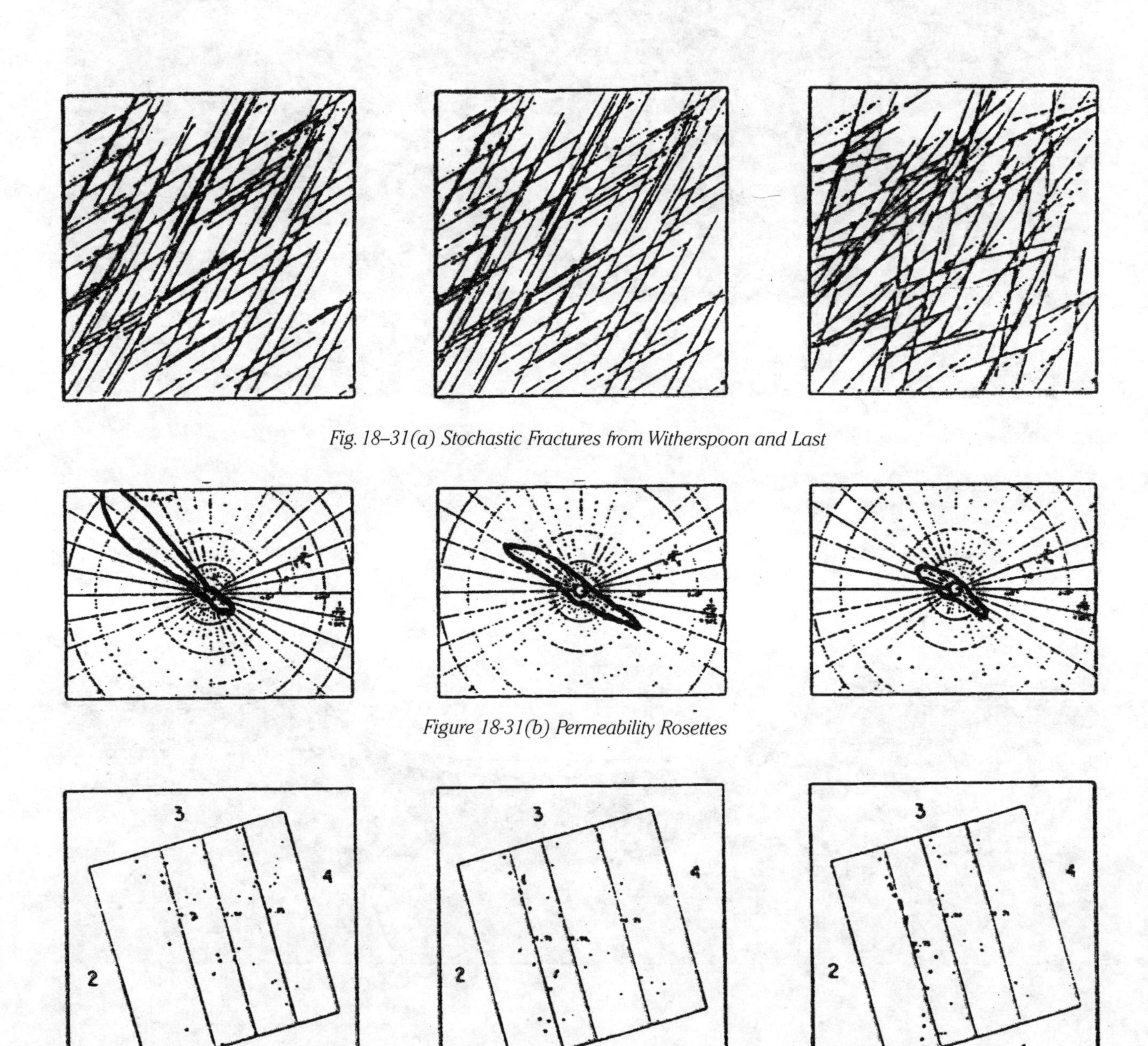

Fig. 18–31(a) Stochastic Fractures from Witherspoon and Last

Figure 18-31(b) Permeability Rosettes

Fig. 18–31(c) Applied Pressure Gradients (Note these are rotated)

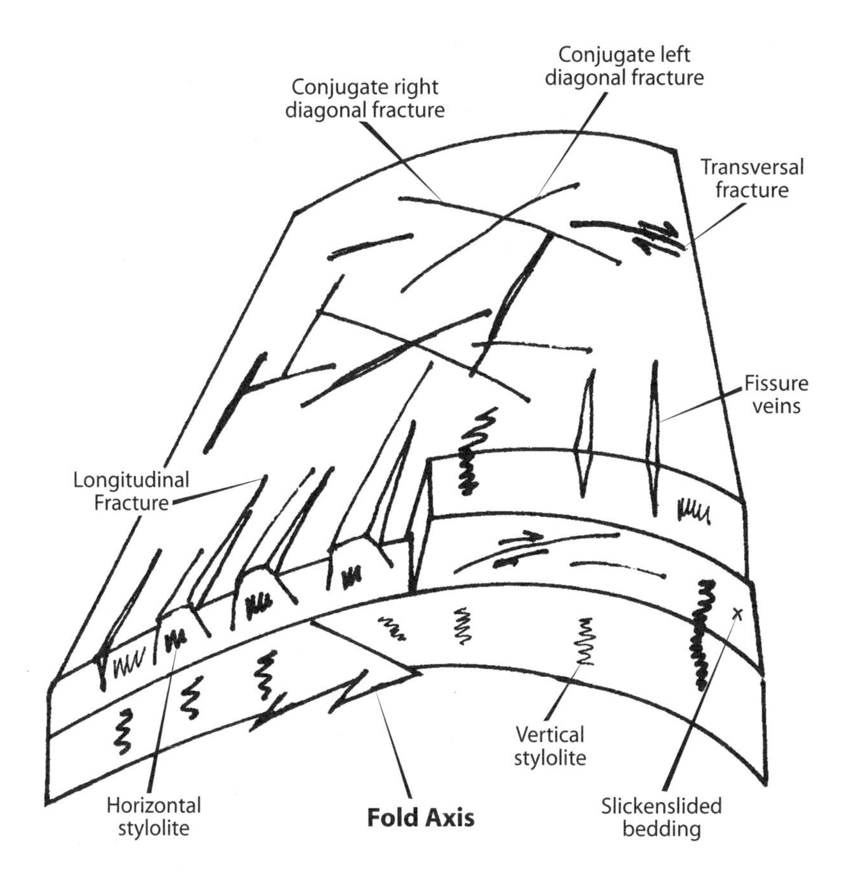

Fig. 18–32 Conjugate Fractures—Classic Fracture Pattern

Note that there are longitudinal fractures shown in his diagram. However, most petroleum reservoirs are located well below surface, which is a compressive environment. The longitudinal fractures are unlikely to be completely open.

Based on the production and material balance analysis of many of these reservoirs, data has always indicated good communication along the leading edges. The data in the figures described previously resolves this problem; fracture connectivity governs the directional permeability, which would be oriented along the top of the thrust sheets—particularly in the rollovers.

Therefore, connectivity governs permeability rather than fracture orientation, based on small-scale modeling and practical large-scale experience. Scale is also an important consideration.

Physical failure of rock

Figure 18–33 shows how lab failures appear. Note the fine scale of the fracturing and the pattern of fractures that develops. This looks very similar to the cores retrieved from a number of projects ARE has worked on.

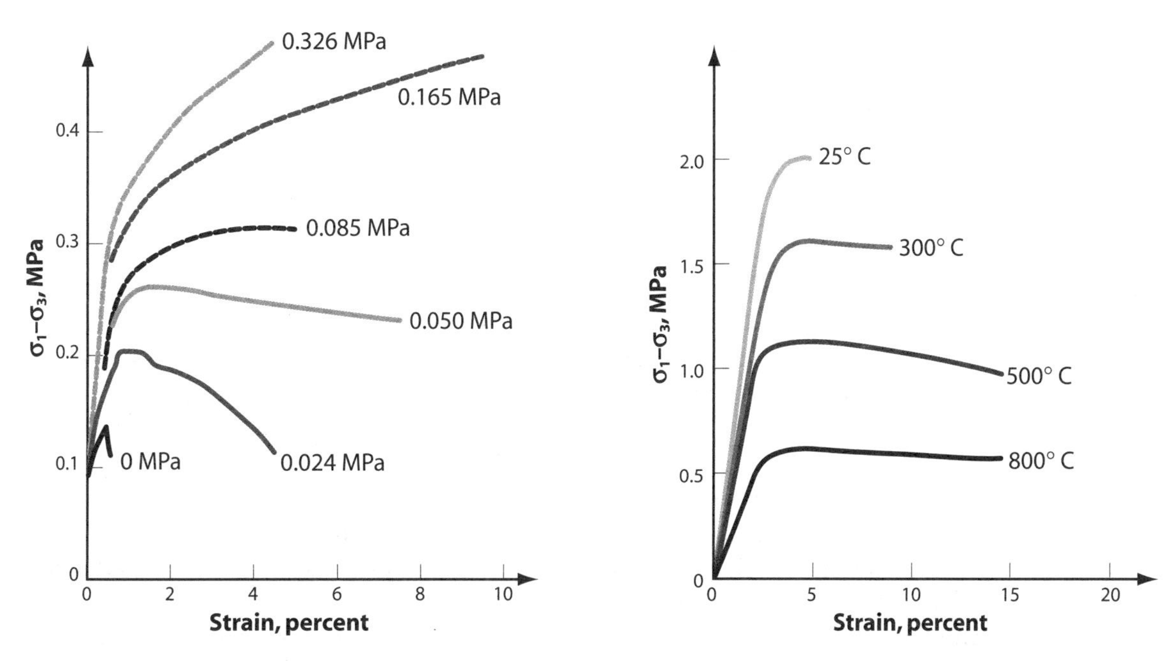

Fig. 18–33 Example Stress-Strain Curves for Rock at Various Confining Pressures and Temperatures

Figure 18–34 was taken from a rock mechanics textbook. These figures show a significant change in physical behavior from 25–800° C (77–1,472 °F). This graph indicates that temperature will have a noticeable effect on the physical behavior of rocks.

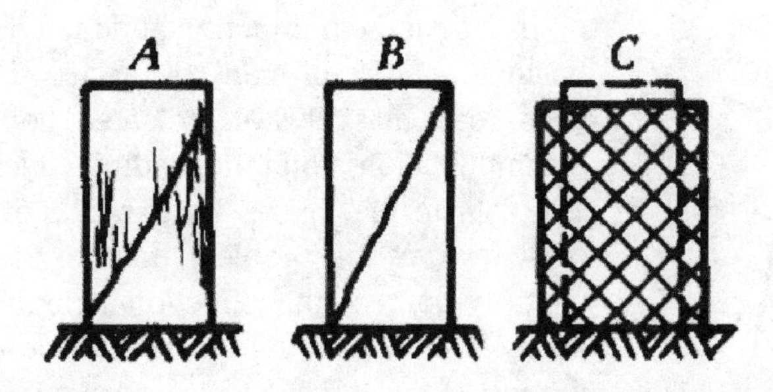

Fig. 18–34 Fracture Patterns on Failed Samples

Some of the reservoirs in western Canada that the author has worked on have temperatures of about 103° C (217° F). However, this is the current temperature, which is lower than would have existed with paleo burial depths. Determining the expected fractures must be based on paleo temperatures when the structures were formed. Therefore, since mountain building takes place during long time periods, the physical properties of the rock may not have been uniform.

Stress sensitivity of fracture conductivity

Permeability reductions as high as 60% are common when net overburden (NOB) is applied to core samples. This is a result of closing of microfractures and changes to the pore fabric. This was identified in Muskat's classic work.[32] Further work was done by Fatt and Davis, as well as Wilhelmi and Somerton in "Simultaneous Measurement of Pore and Elastic Properties of Rocks under Triaxial Stress Conditions."[33, 34] This work pertains mostly to rocks that are relatively competent; i.e., nonfractured reservoirs.

The conductivity of joints is greatly affected by the state of stress in the reservoir. To date, relatively little work appears to have been done in Canada and the United States on this subject. The best examples that the author has found come from the North Sea. One example is shown in Figure 18–35, which relates normal stress (i.e., applied across the fracture) to the fracture closure and conductivity. Note that the fracture closes and conductivity is reduced with increased normal stress.

The best joint description is based on the Barton Bandis joint behavior model. Determining the overall conductivity of a joint depends on the roughness of the joint, the aperture, and whether the joint is filled with gouge—caused by grinding while rock material moves. Figure 18–36 shows the fracture conductivity, dilation, and shear stress with shear displacement.[35]

- At a critical value, the joint starts to slip. Immediately after, due to grinding of asperities, the shear stress remains essentially constant with further displacement.
- In the second diagram, the aperture increases in width (dilates) with continued displacement.
- As dilation continues, the conductivity of the joint continues to increase.

Restated, the conductivity of a fractured rock will depend on the stress in the reservoir as well as rock displacement history.

Shear direction also has an important effect. Data supporting this is shown in Figure 18–37. The experiment by Makurat shows an increase in aperture and conductivity with forward movement. However, if the movement is reversed, then the final aperture is smaller, due to flattened asperities, and the final conductivity is lower when total displacement returns to zero.[36]

Modeling implications

From this it may be concluded that we really should be modeling with stress-dependent permeability in naturally fractured reservoirs. For the most part, this is not included in current models.

Tectonic environment

It is apparent that fracture properties are stress dependent and vary with structural location. Frequently, both of these factors are largely controlled by plate tectonics. Probably the most striking difference the author has observed seems to fall in two different basin types. Some basins are predominantly extensional. This seems to fit the United States Gulf Coast area and parts of the North Sea. Some areas are predominantly compressional. This includes the foothills of the Western Canadian Sedimentary Basin, the eastern slopes of Columbia, Bolivia, Ecuador, and Peru, as well as the reservoirs on the north shore of Cuba. Because of these variations, very different reservoir and fracture behaviors are likely. Overall, it would seem, at least superficially, that extensional environments are more likely to be favorable.

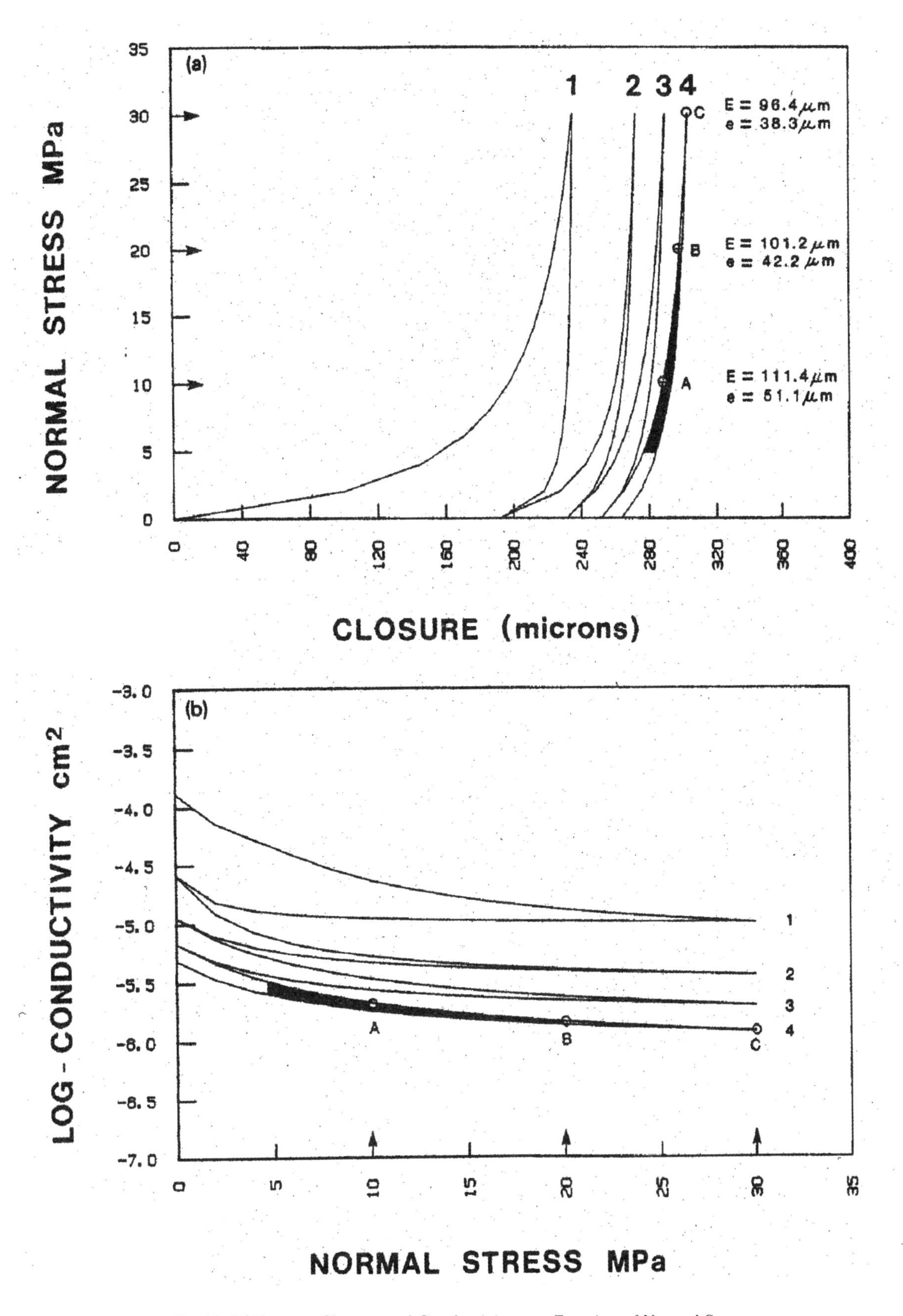

Fig. 18–35 Fracture Closure and Conductivity as a Function of Normal Stress

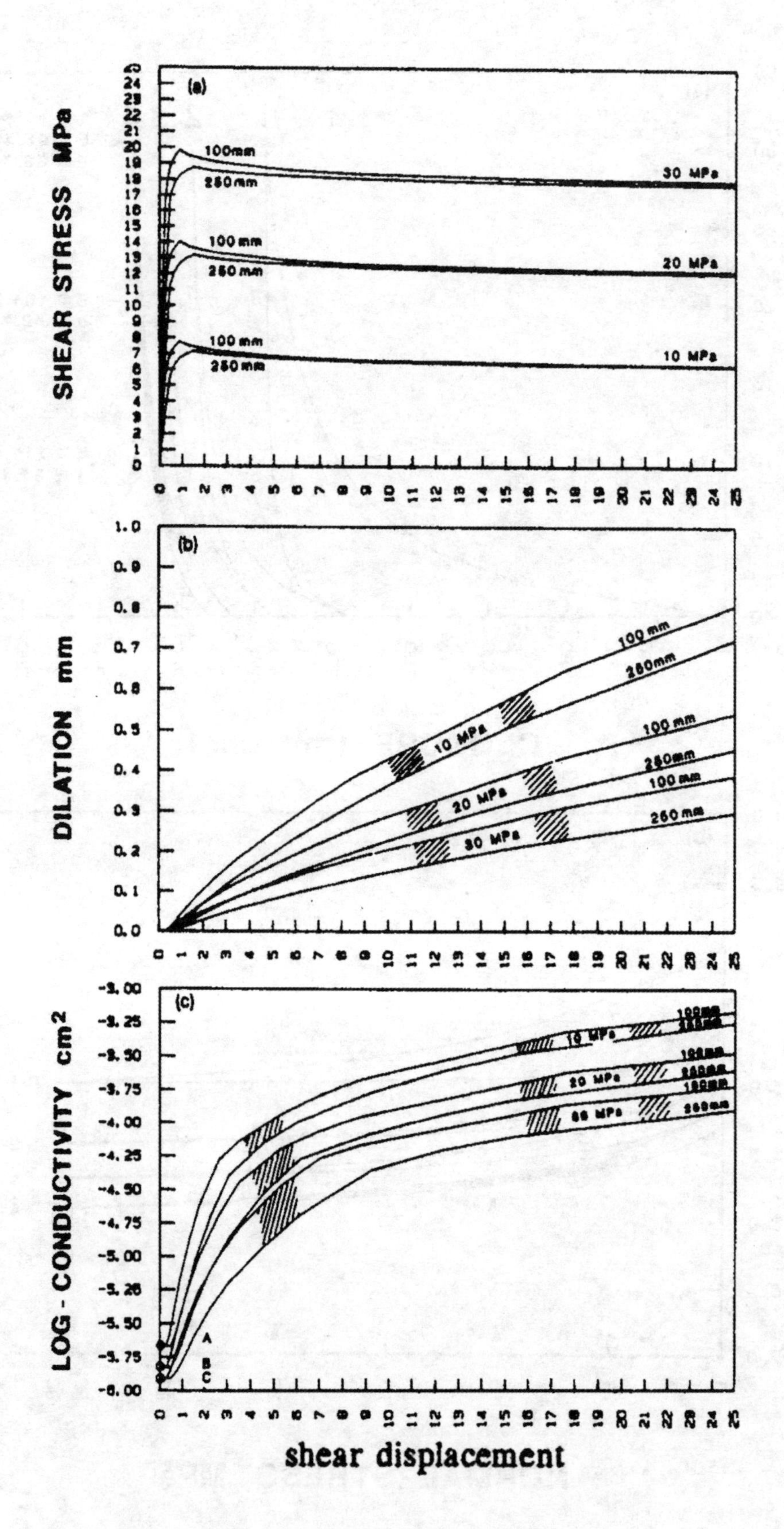

Fig. 18–36 Dilation and Conductivity with Shear Displacement

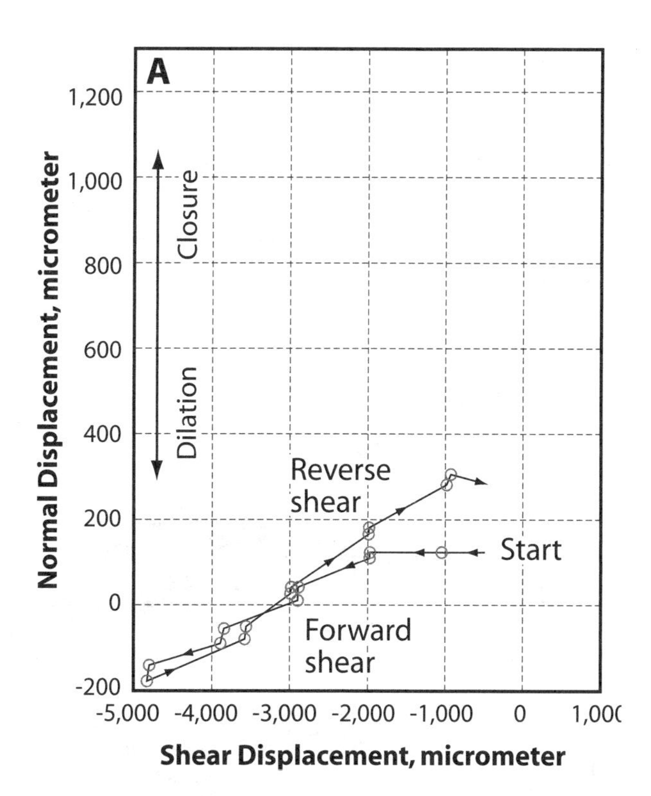

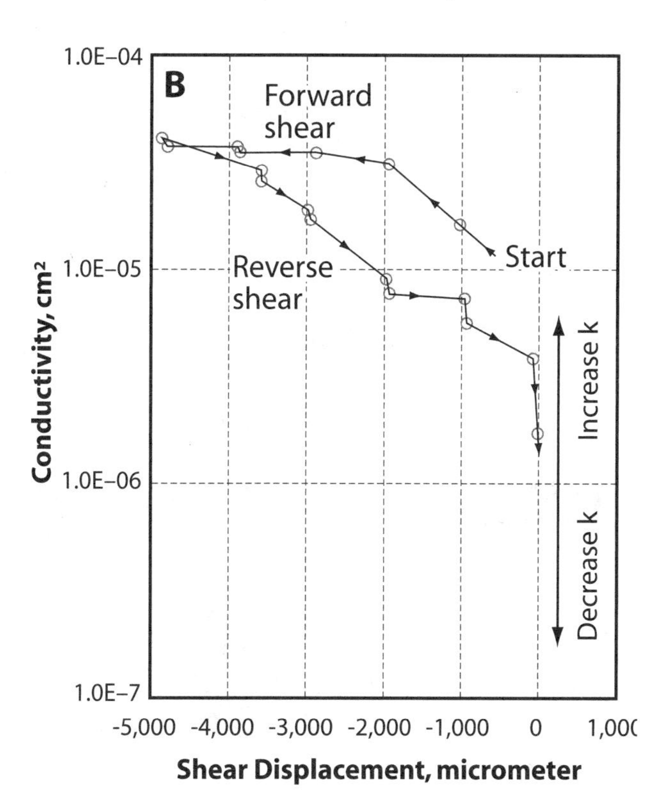

Fig. 18–37 Vertical Displacement and Conductivity with Shear Displacement

Production logging

On one central Alberta foothills reservoir, a production spinner survey was run. The interpretation of the log coincided with a presentation to working interest partners of the pressure transient analysis. Two models were presented, a composite (concentric rings of permeability) reservoir and a poddy (isolated pods of distinct porosity and permeability) model.

Based on limited entry from three thin zones identified on the spinner survey, the more sophisticated, less uniform poddy interpretation was chosen instead of a concentric permeability model. Furthermore, the latter model gave an excellent match of long-term production performance.

Poddy model

The general idea of this model is shown with a geostatistical realization in Figure 18–38.

Rocks do not have uniform mechanical properties. In fact, they are strongly influenced by original bedding. It is logical that the rock would not crack uniformly but rather in beds of stiffer rocks—often low porosity. This is reminiscent of a geostatistical representation for porosity.

Monte Carlo

One clarification must be made. This is *not* a purely random process; therefore, it is not possible to use a Monte Carlo technique to generate this type of characterization. It involves a constant standard deviation (either a sill or a nugget effect). This approach totally ignores the spatial information. An example of entirely random geologic description is shown in Figure 18–39.

Grid construction

The poddy reservoir model is initialized using a dry gas fluid description. Output from the simulation was entered into a well test package. The type curve from this simulation is shown in Figure 18–40. The overall shapes obtained are similar to that expected from a hydraulically fractured well and are very similar to the expected models described earlier. A number of more detailed examples are outlined in the paper describing this study.[37]

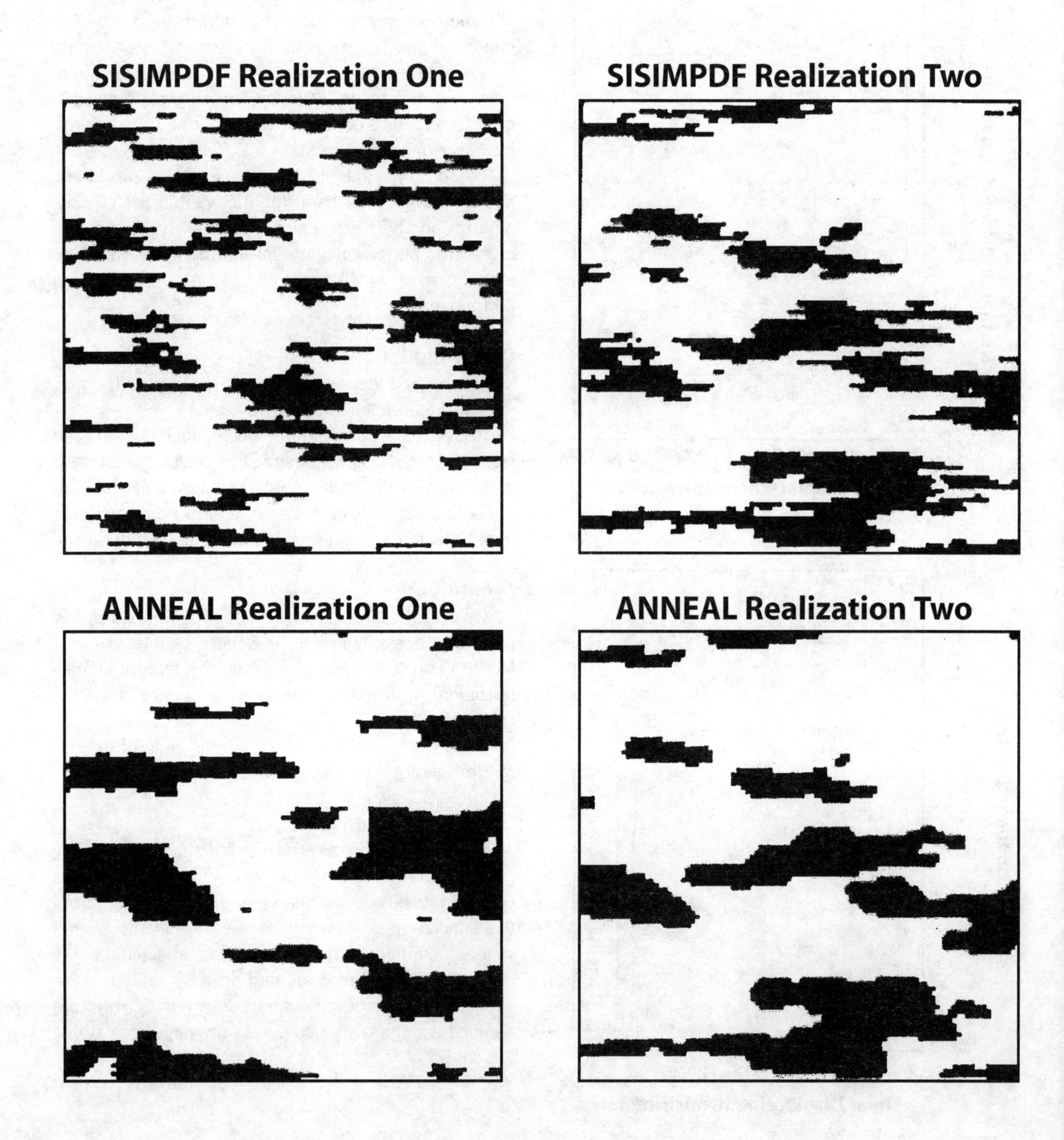

Fig. 18–38 Geostatistical Suggestion of "Poddy" Reservoirs

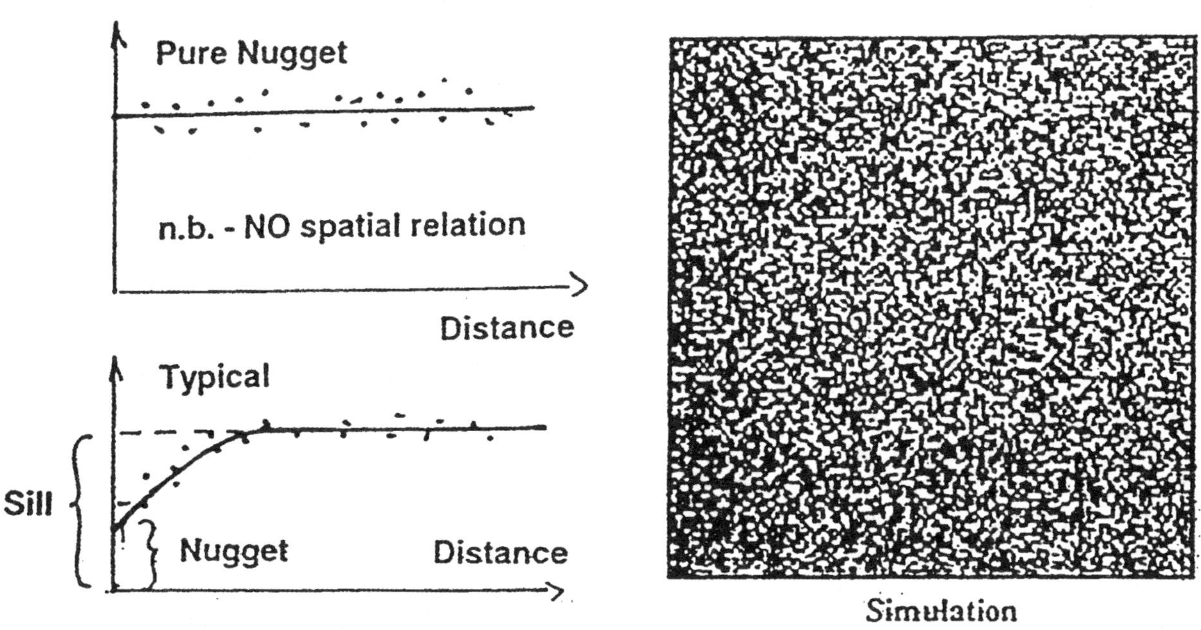

Fig. 18–39 Monte Carlo Type Modeling

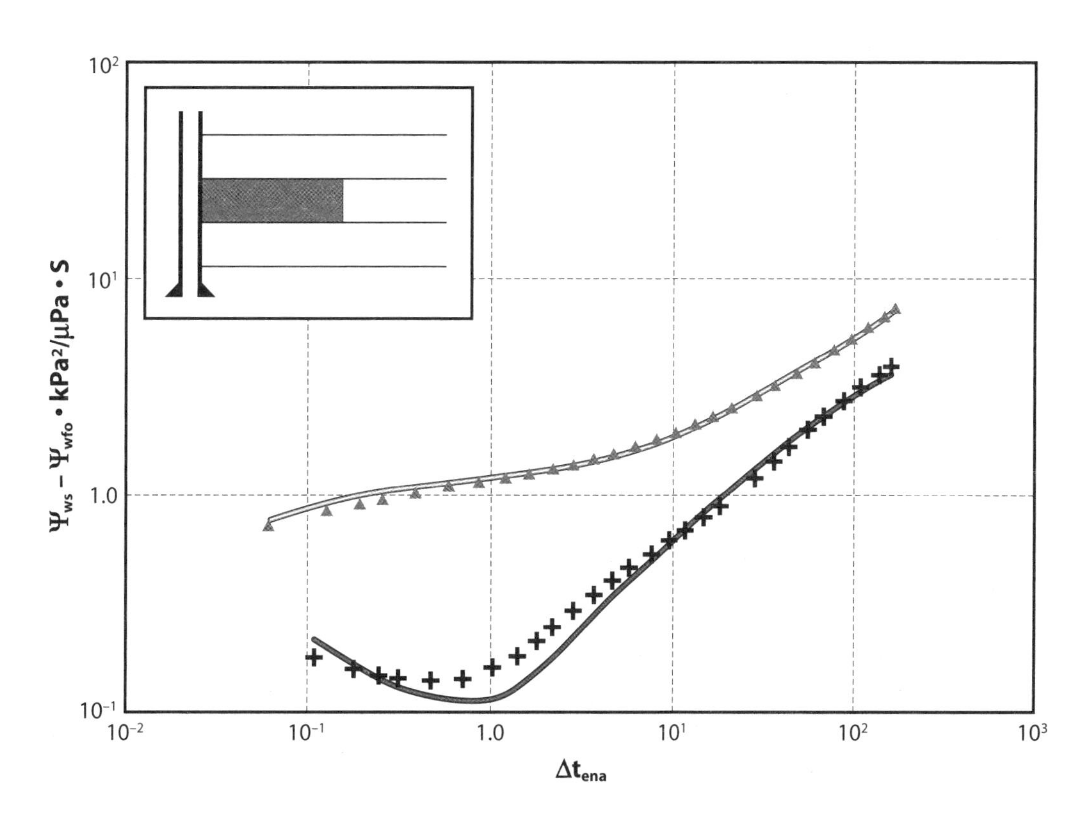

Fig. 18–40 Pressure Transient Response- Type Curve—"Poddy" Reservoir with Dry Gas

Poddy reservoir model with condensate dropout

An example is shown for a gas condensate system as well. This reservoir has a very high CVD dropout, on the order of 40%. Figure 18–41 shows that the condensate drops out at the interface between the high-permeability pod and the surrounding lower permeability matrix at 30 days.

Note that there is an element of feathering in permeability and porosity at the edges of the pods. A gradual change has been approximated in the model by a ring of intermediate permeability at 1.0 mD at the outer edge of the pod.

Limitations of pressure transient interpretations presented

In all of the cases examined, the gas properties within the well test package were calculated based on molar average P_cs and T_cs to derive pseudo critical pressures and temperatures, respectively. The fluid in this reservoir is near critical, and estimates of properties that are calculated based on standard correlations will have larger-than-normal errors. Finally, pressure transient solutions are based on single-phase assumptions, which are obviously not applicable. The interpretations from the well test package will be approximate.

Production profiles

Examining the production profiles of these pod representations shows a decline behavior consistent with that displayed as shown in Figure 18–42. Initially, there is a rapid drop-off in well productivity as the pod is depleted. The rate then declines somewhat steeply for about 9 to 18 months. Following this, there is a long-term exponential decline at slow rates of decline.

GORs

On this same figure, the GOR on this type of system shows an unexpected behavior. The outline has a hump in which the GOR increases after the flowing bottomhole pressure drops below the dew point. Eventually it drops to levels slightly above the initial levels.

The condensate is forced out of the gas at the interface of the pod and primarily deposits in the low-permeability surrounding matrix. With lower amounts of liquid produced, the GOR increases. In time, the liquid saturation in the high-permeability pod increases. When the pod fills up to S_{or}, the GOR drops back toward initial levels.

This can be quite misleading and could be interpreted as a bubblepoint system if a homogeneous reservoir is incorrectly assumed.

"Pod" relative permeability

Sensitivities were run using gas-oil relative permeability curves and stick (gravity segregation or miscible gas-oil relative permeability) curves. The results of these runs did not change the trends described previously.

Pressure buildup response with condensate

Figure 18–43 shows the pressure transient response with condensate. This curve is virtually indistinguishable from the earlier curves derived for dry gas. Note that homogeneous radial systems have a notable character on the derivative curve. The pod masks this effect.

Implications of poddy reservoir characterization

The pod moves the relative permeability impairment away from the well. Therefore, it has less effect on productivity. Most of the condensate dropout is at the interface between the low-permeability and the high-permeability regions. The condensate also drops out inside a much larger rock volume than for a homogeneous radial system. Since the condensate is held up within the tight matrix, condensate gas ratios are reduced, i.e., GLRs are increased. Less condensate will be recovered from the well.

If one assumes a radial homogeneous system, this is an apparent paradox. The condensate is stripped (must be dropping out), yet the productivity is not impaired (by k_{rg} reduction from dropout) to the degree expected.

With lower CVD dropout levels, it is unlikely that the GLR would ever drop. See the works of Carlson and Myer or Yadavelli and Jones regarding hydraulically fractured wells (which were discussed in the previous chapter).[38, 39]

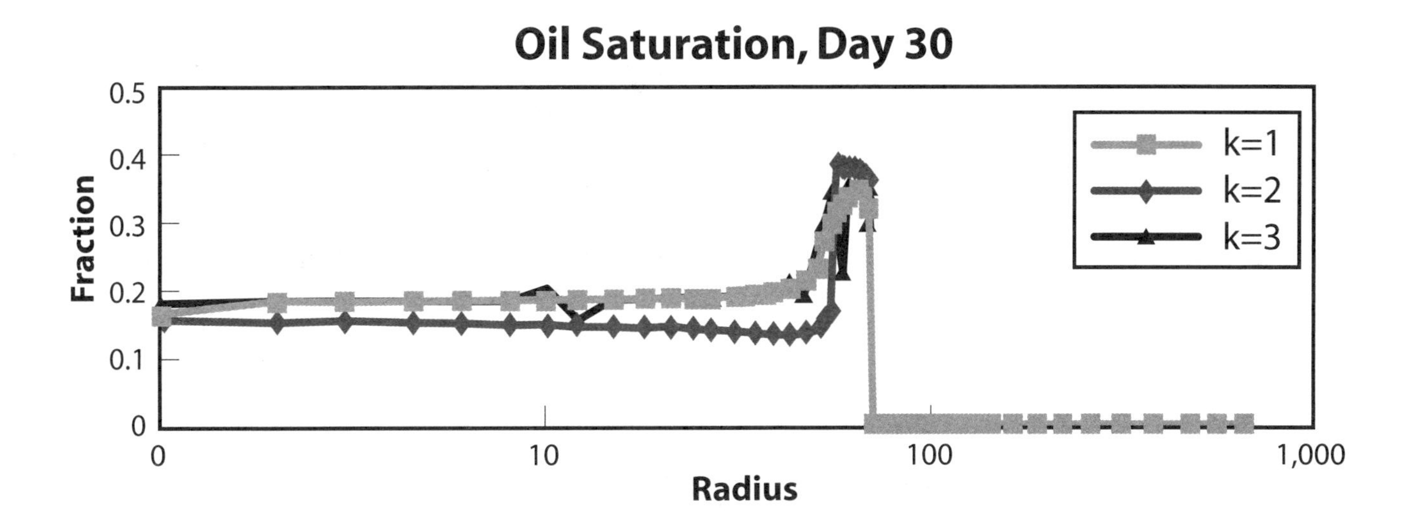

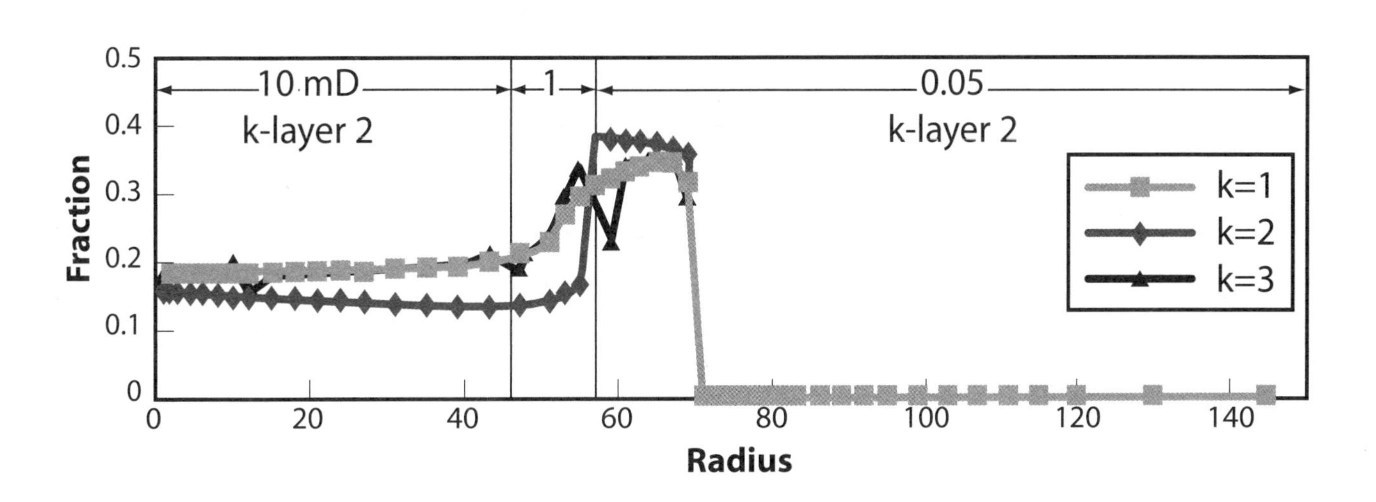

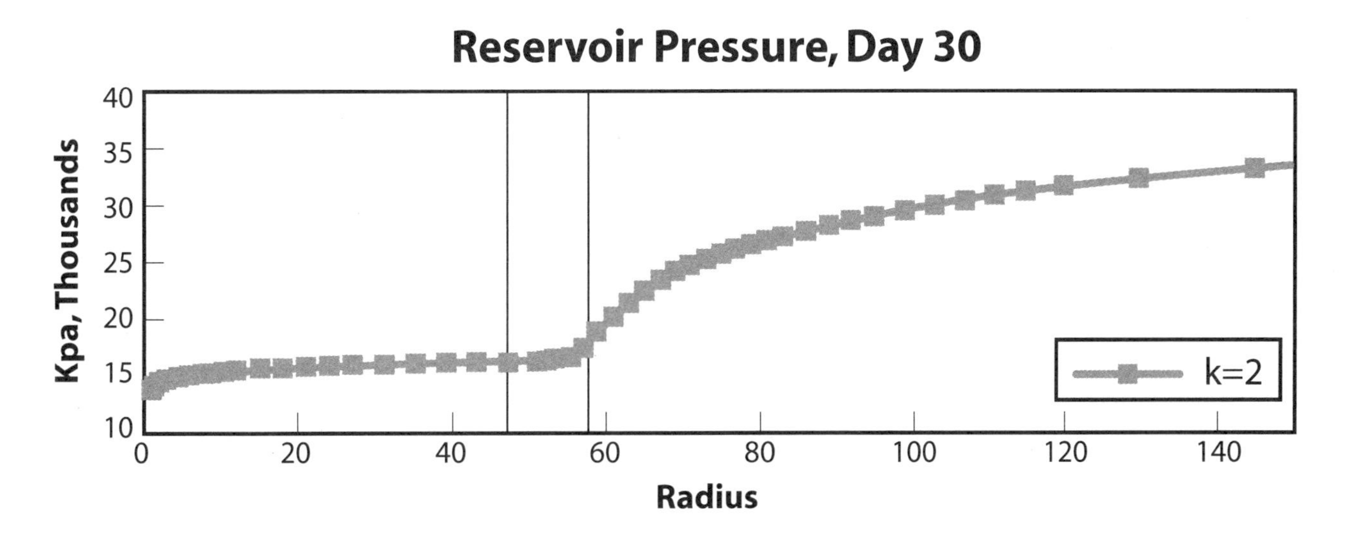

Fig. 18–41 Dropout of Condensate around Reservoir Pods

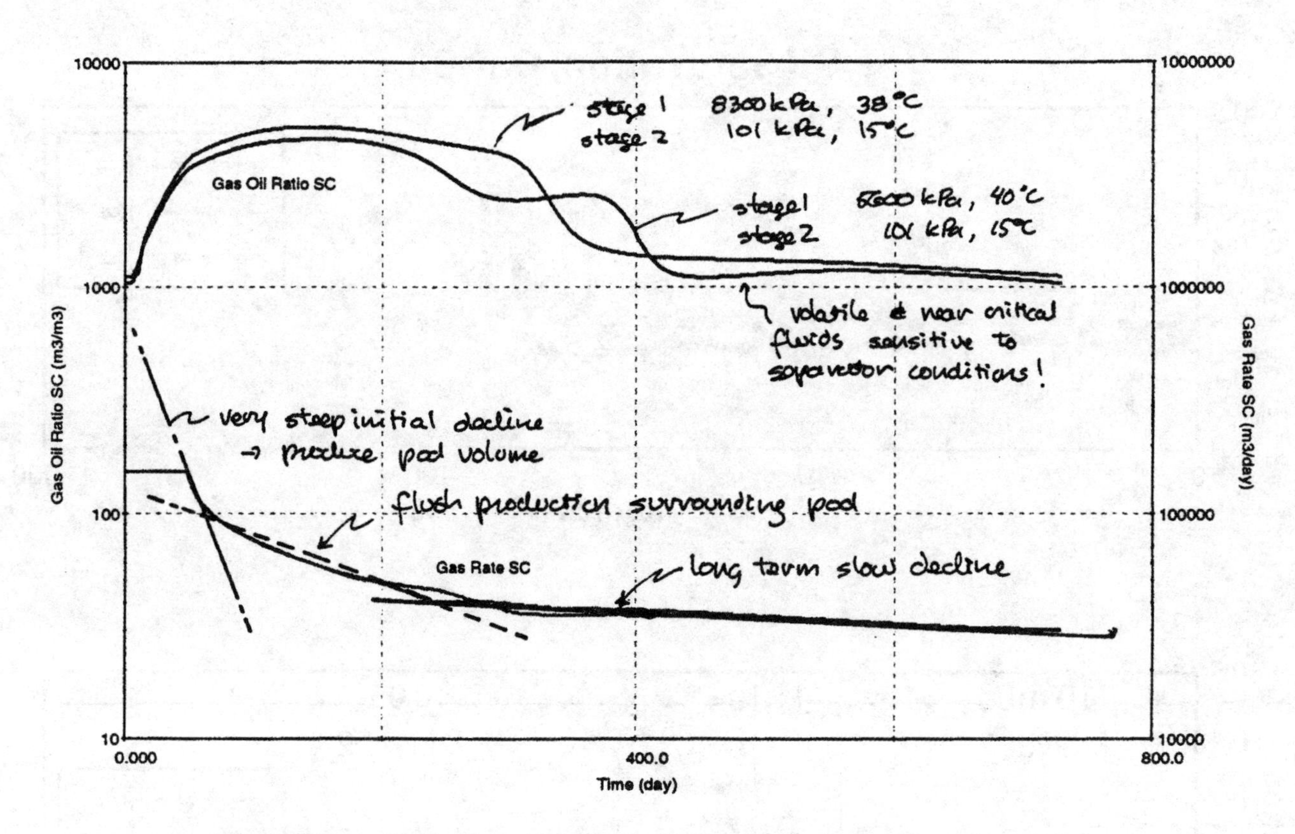

Fig. 18–42 Production Profile and GOR Profile for Rich Gas Condensate Reservoir with Pods

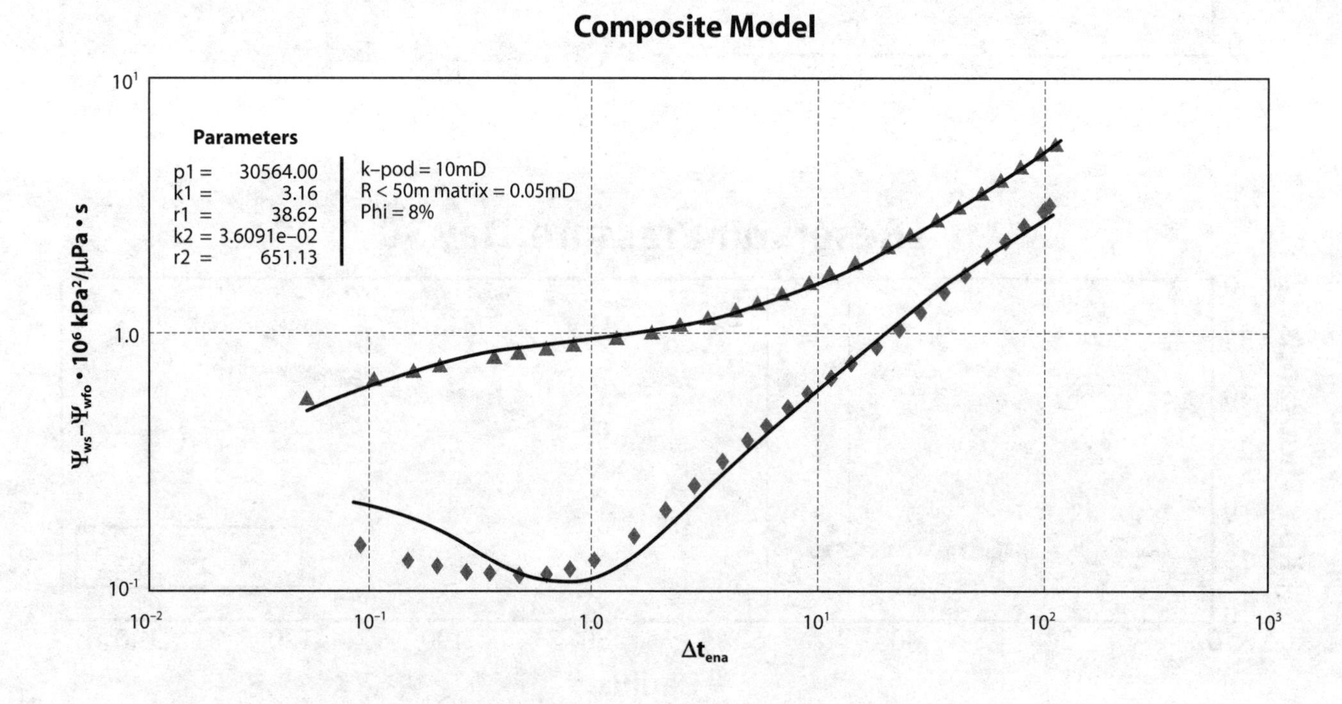

Fig. 18–43 Pressure Transient Response—Type Curve—"Poddy" Reservoir, Rich Gas Condensate

Dolomitization

The cause of dolomitization is not known with certainty. Increased dolomitization typically increases porosity and permeability. The process on one of the reservoirs that ARE studied was thought to be related to four general types of development:

- Early diagenetic, matrix selective, facies-controlled dolomitization
- Unconformity-related dolomitization
- Fault-related dolomitization
- Hydrothermal dolomitization

In particular, mechanisms 2 and 3 were thought to be important in the reservoir under study.

Characteristic well test response

Perhaps the fashion of developing a type curve to depict characteristic well test response has been most popular with well test analysis. The logical step at this point was to evaluate what the buildup test responses looked like. Four type curves were available for a-50-K, c-71-L, d-25-D, and d-36-D. Two examples are shown in Figures 18–44 and 18–45. All these wells appear to have been hydraulically fractured, although no such treatments were used in this field. This is consistent with the author's experience in the central foothills of Alberta and other areas.

Poddy reservoir development

The natural question is, *What type of well test response would be obtained from intensely fractured beds of limited areal extent?* The author calls this model poddy fracture development. Based on the work done by Witherspoon et al., it would seem likely that the pods may have a significant directional permeability characteristic.[40]

Of course, no analytical model has yet been developed for such a pattern. A reservoir simulator was a logical method of generating type curves. The output from the model would be input into a well test analysis package to enable patterns to be identified in a manner consistent with well test interpretation.

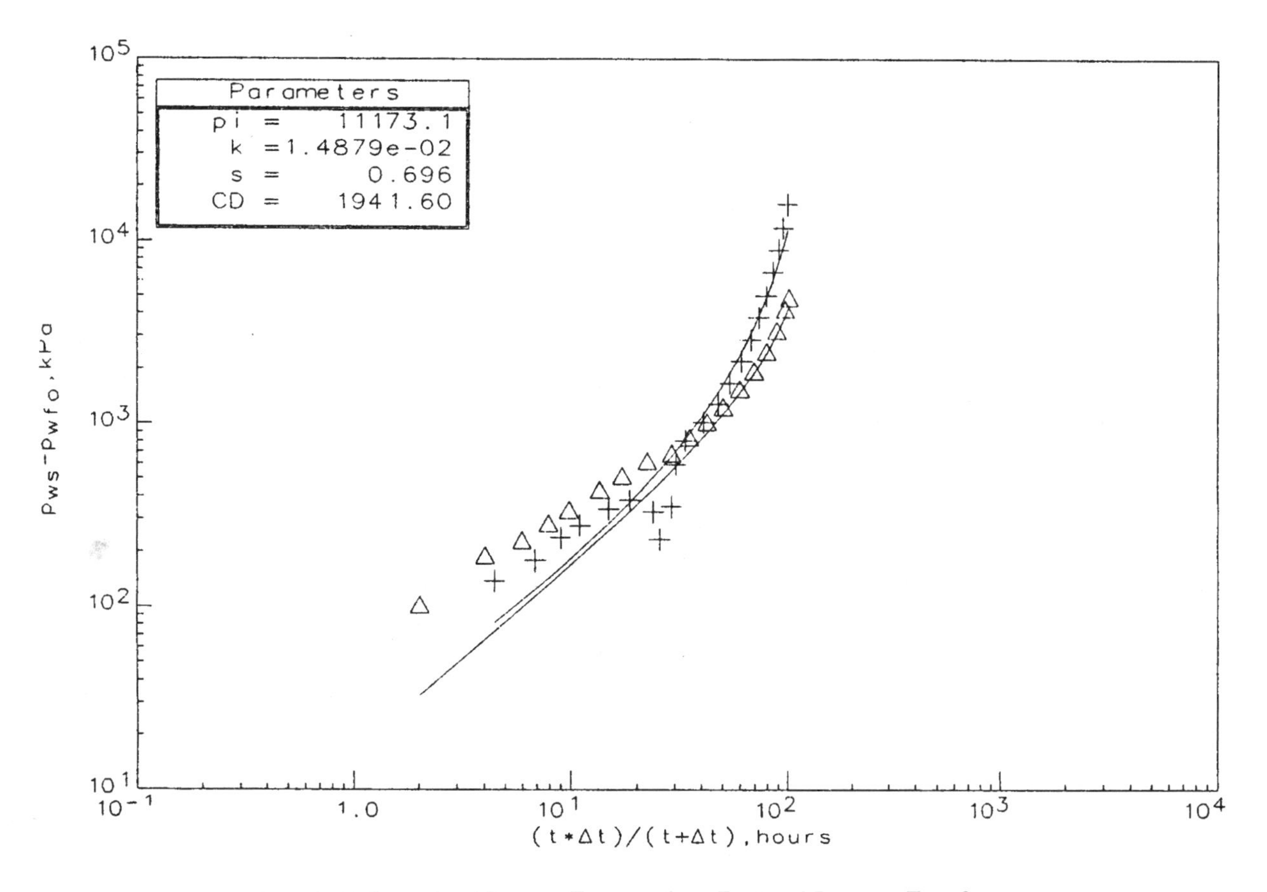

Fig. 18–44 Example of Pressure Transient from Fractured Reservoir-Type Curve

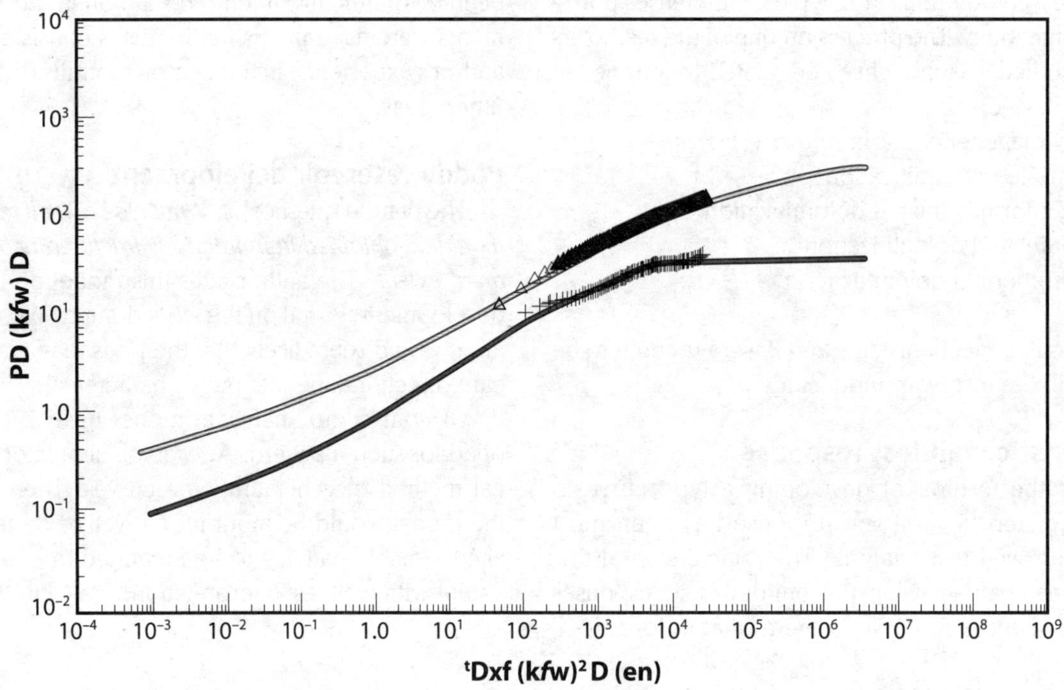

Fig. 18–45 Another Example of Pressure Transient from Fractured Reservoir—Type Curve

Uniform reservoir

A simulator grid was developed and uniform reservoir properties assigned. With this, the simulator well test package could be tuned and errors evaluated. Some noise was evident, and there was some error in derived properties. The type curve overestimated permeability by 13%, and the Horner plot underestimated permeability by 17% for a radial homogeneous system. Overall, such an error would be considered relatively minor in field terms. Note that no wellbore storage has been used.

Next, directional permeabilities were evaluated. In order to save space, these plots are not shown; however, the average permeability was overestimated by 10.2% on the type curve and underestimated by 4.2% on the Horner plot (a systematic trend?).

Pod model

A pod model, with no directional permeability, was developed as shown in Figure 18–46. Note that the pod does not extend very far laterally. The results are shown in Figures 18–47 and 18–48. These results are startling. First, the shape of the curves is very similar to that observed on the actual well tests. Second, calculated reservoir properties are well below the true reservoir properties.

A compound pod was developed as shown in Figure 18–49. The results of the well test analysis are shown in Figures 18–50 and 18–51. In this case, there is considerable apparent noise that is related to complex boundaries and changes in pod properties.

Finally, a last quality-control run was made with a multilayered system. Overall, the levels of error were low and the character of the type curves generated followed expectations from analytical solutions.

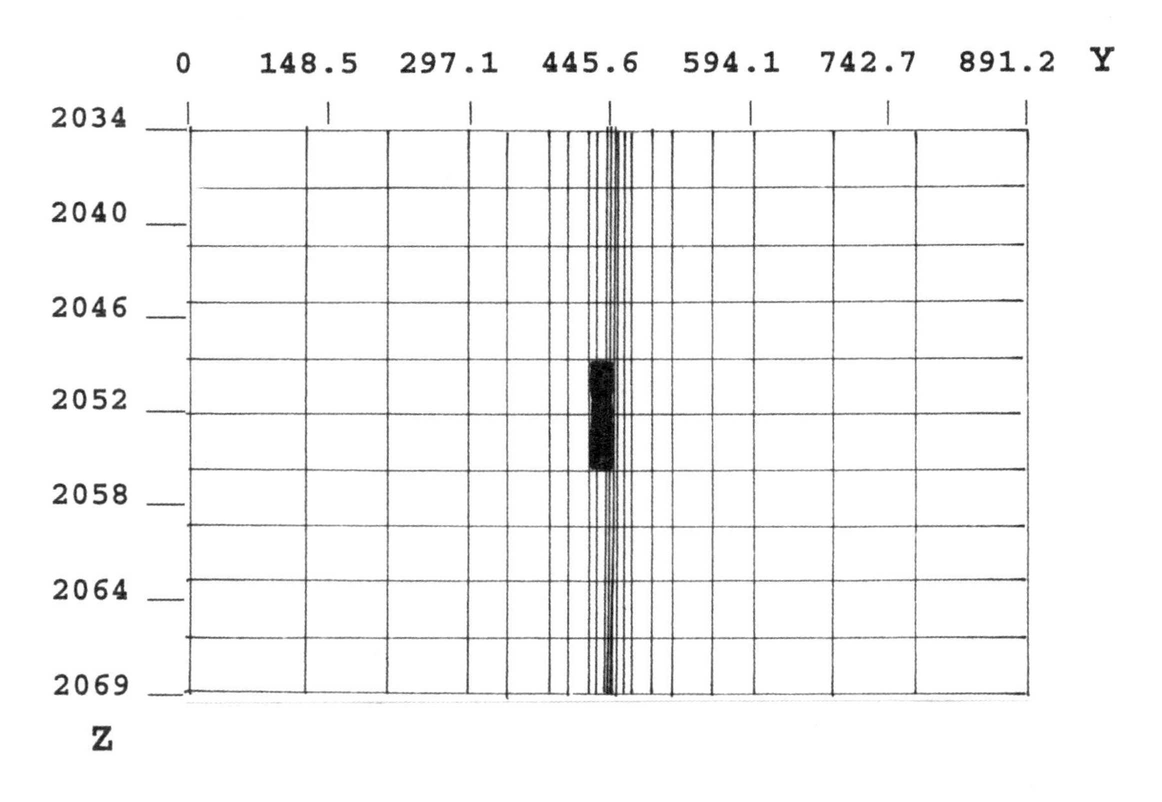

Fig. 18–46 Cross Section of Simulation Grid for "Poddy" Model

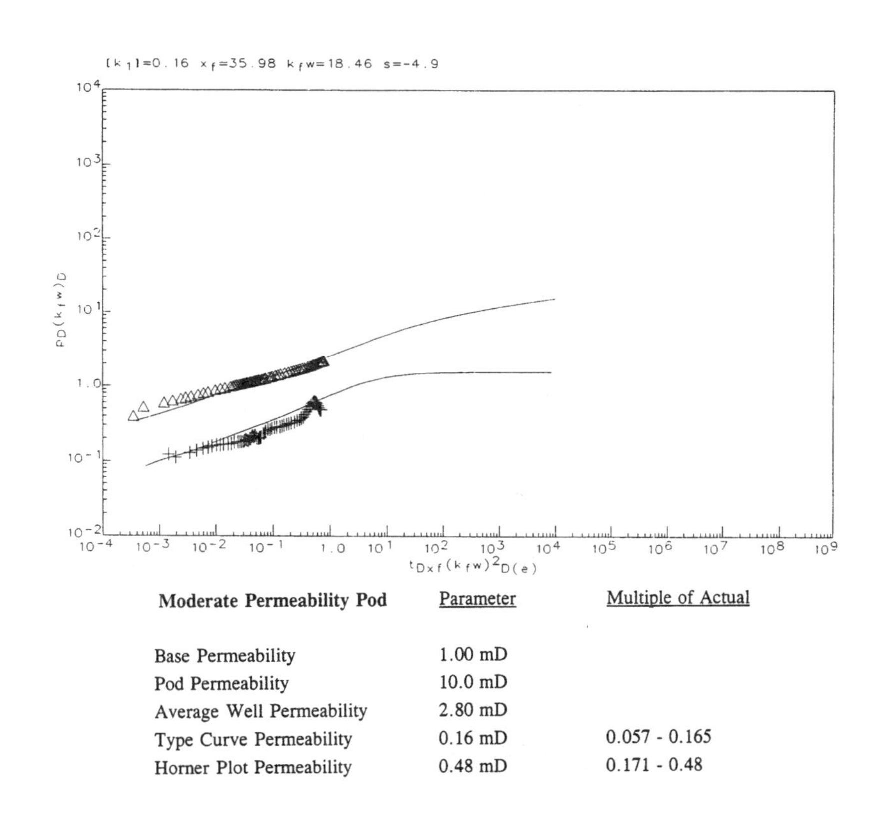

Moderate Permeability Pod	Parameter	Multiple of Actual
Base Permeability	1.00 mD	
Pod Permeability	10.0 mD	
Average Well Permeability	2.80 mD	
Type Curve Permeability	0.16 mD	0.057 - 0.165
Horner Plot Permeability	0.48 mD	0.171 - 0.48

Fig. 18–47 Simulation Prediction of Performance with "Poddy" Model-Type Curve

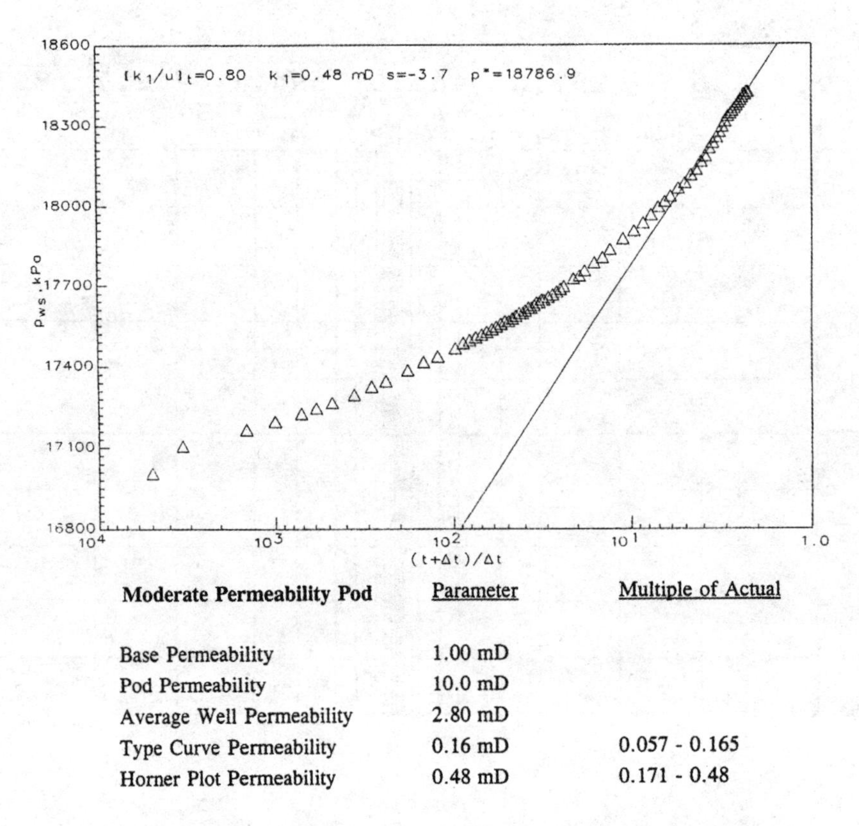

Moderate Permeability Pod	Parameter	Multiple of Actual
Base Permeability	1.00 mD	
Pod Permeability	10.0 mD	
Average Well Permeability	2.80 mD	
Type Curve Permeability	0.16 mD	0.057 - 0.165
Horner Plot Permeability	0.48 mD	0.171 - 0.48

Fig. 18–48 Simulation Prediction of Performance with "Poddy" Model—Horner Plot

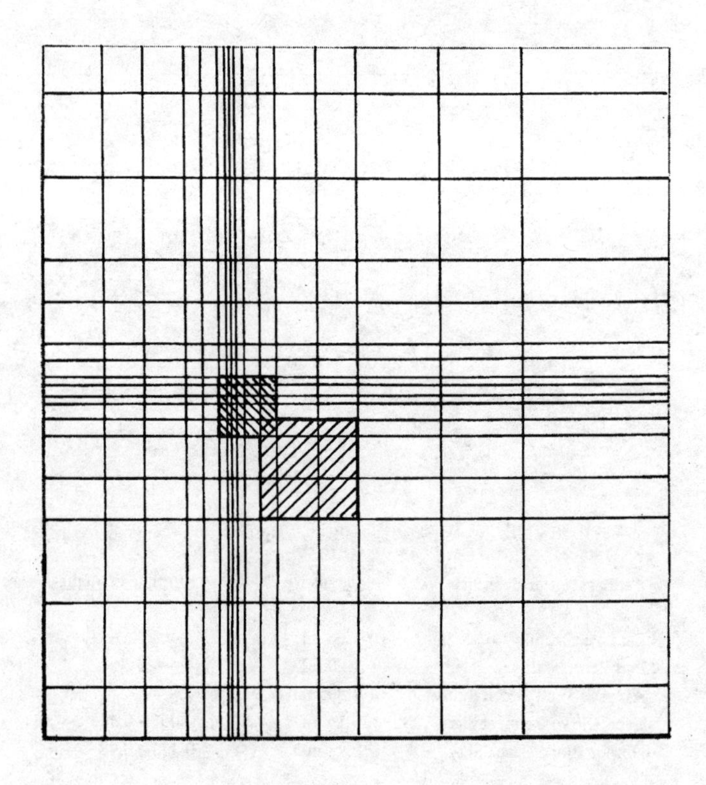

Fig. 18–49 Grid Layout for Complex Pod Model

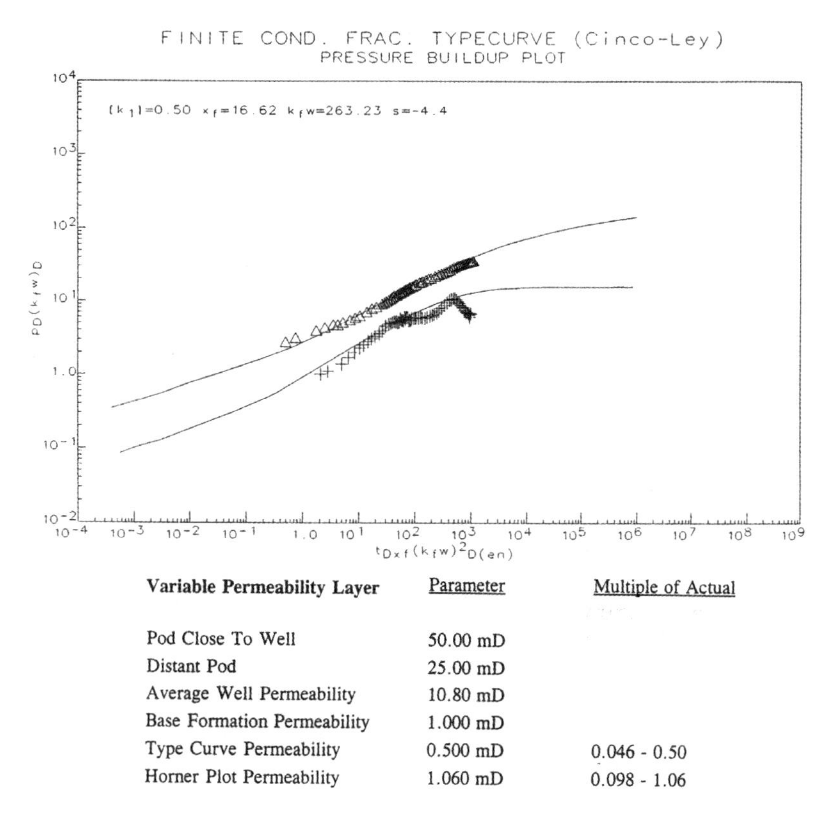

Variable Permeability Layer	Parameter	Multiple of Actual
Pod Close To Well	50.00 mD	
Distant Pod	25.00 mD	
Average Well Permeability	10.80 mD	
Base Formation Permeability	1.000 mD	
Type Curve Permeability	0.500 mD	0.046 - 0.50
Horner Plot Permeability	1.060 mD	0.098 - 1.06

Fig. 18–50 Simulation Prediction of Performance with "Complex Pod" Model-Type Curve

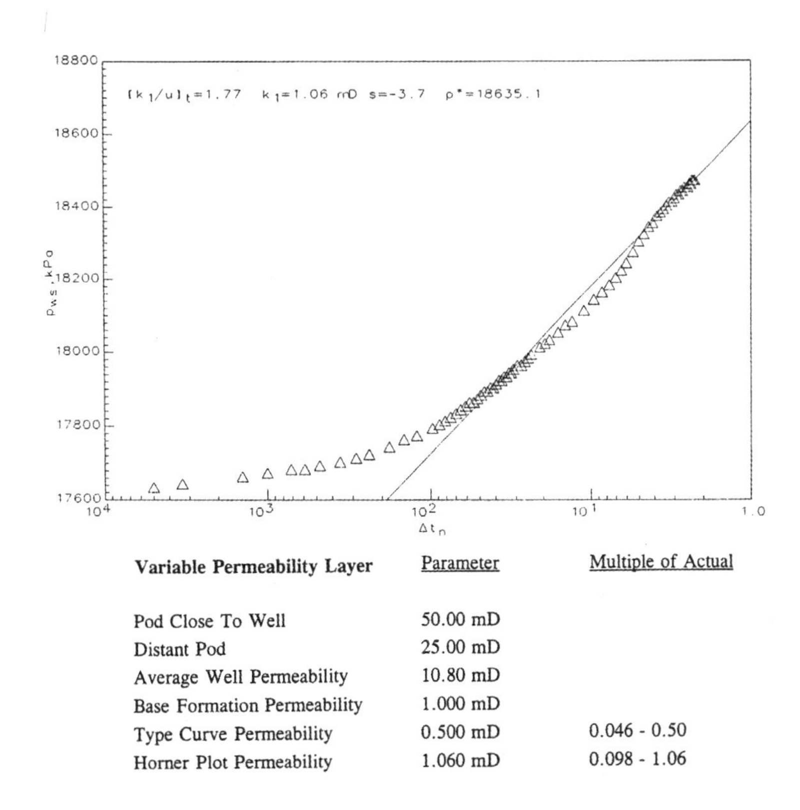

Variable Permeability Layer	Parameter	Multiple of Actual
Pod Close To Well	50.00 mD	
Distant Pod	25.00 mD	
Average Well Permeability	10.80 mD	
Base Formation Permeability	1.000 mD	
Type Curve Permeability	0.500 mD	0.046 - 0.50
Horner Plot Permeability	1.060 mD	0.098 - 1.06

Fig. 18–51 Simulation Prediction of Performance with "Complex Pod" Model—Horner Plot

Concentric systems

Often, it is possible to get matches of poddy reservoirs based on concentric reservoir systems. Under the assumptions of this less-sophisticated, more uniform system, it appears that most wells drilled always encountered an area of good reservoir development—a logical conundrum. This is the model suggested years ago by Adams and more recently by Grant.[41] As discussed earlier, this poddy nature can also be represented geostatistically. Any well that encounters a pod will have a negative skin.

Reservoir characterization

The pods, as shown in the preceding diagrams, are of limited areal extent. Well test analysis indicates that well productivity is controlled strongly by near-wellbore conditions, i.e., the pods. They are ubiquitous. The approach used was to assign an average permeability, the pods being of sufficiently small scale that a salt-and-pepper approach can be used on a field scale or areal simulation basis.

Unlike the Warren and Root model, this view allows wells to be centered in apparently improved permeability, even if it is only on a patchwork basis. It is consistent with the variable performance witnessed in fractured reservoirs in general. Finally, it relies on typical geological heterogeneity with respect to lithology, layering, and rock properties.

Conclusions

- A systematic approach is important in fractured reservoirs, which are more complex than single porosity reservoirs.
- Fractured reservoirs in Western Canada show a number of distinctive characteristics, such as core permeability versus core porosity and pressure transient response.
- To date, the author has not found reservoirs in Western Canada that behave anything like Warren and Root sugar cube models. (Some fault boundaries have been encountered.)
- Reservoir conditions required to get Warren and Root type behavior are extremely restrictive, to the point where such behavior should be extremely rare.
- The use of a dual continuum approach causes serious inconsistencies for length scales typical of Western Canadian reservoirs.
- A poddy type reservoir model is put forth. Such a model could also be described geostatistically.
- The poddy model is consistent with pressure tests analyzed and with empirical observations.
- The poddy model has interesting implications for gas condensate systems. Condensate is trapped in the formation where formation permeability changes. The author believes that this reservoir characterization will help explain the behavior of some foothills reservoirs, such as Waterton.
- The poddy type of reservoir description should be expanded to a 3-D geostatistical representation.

References

1 van Golf-Racht, T.D., *Fundamentals of Fractured Reservoir Engineering, Developments in Petroleum Science 12*, Elsevier, 1982.

2 Aguilera, R., *Naturally Fractured Reservoirs*, 2nd ed., PennWell Publishing Co., 1995.

3 Saidi, A., *Reservoir Engineering of Fractured Reservoirs—Fundamental and Practical Aspects*, Total Edition Presse, 1987.

4 Nelson, *Geologic Analysis of Naturally Fractured Reservoirs*, 2nd ed., Butterworth-Heinemann, 2001.

5 Odeh, A.S., "Unsteady-State Behavior of Naturally Fractured Reservoirs," *Society of Petroleum Engineers Journal*, 1965.

6 Sonier, "Discussion of Improved Calculations for Viscous and Gravity Displacement in Matrix Blocks in Dual-Porosity Simulators," *Journal of Petroleum Technology*, June 1988.

7 Gilman, J.R., and H. Kazemi, "Authors' Reply to Discussion of Improved Calculations for Viscous and Gravity Displacement in Matrix Blocks in Dual Porosity Simulators," *Journal of Petroleum Technology*, June 1988.

8 Sabathier, J.C., "Discussion of Improved Calculations for Viscous and Gravity Displacement in Matrix Blocks in Dual Porosity Simulators," *Journal of Petroleum Technology*, June 1988.

9 Saidi, A., "Discussion of Improved Calculations for Viscous and Gravity Displacement in Matrix Blocks in Dual Porosity Simulators," *Journal of Petroleum Technology*, June 1988.

10 "Further Discussion of Improved Calculations for Viscous and Gravity Displacement in Matrix Blocks in Dual Porosity Simulators," *Journal of Petroleum Technology*, December 1989.

11 Gilman, J.R., and H. Kazemi, "Author's Reply to Further Discussion of Improved Calculations for Viscous and Gravity Displacement in Matrix Blocks in Dual-Porosity Simulators," *Journal of Petroleum Technology*, January 1990.

12 Warren and Root, "The Behavior of Naturally Fractured Reservoirs," *Society of Petroleum Engineers Journal* (September): 245–255, 1963.

13 Bourdet, D., J.A. Ayoub, and Y. M. Pirard, "Use of Pressure Derivative in Well Test Interpretation," SPE 12777, Society of Petroleum Engineers, 1984.

14 Bidaux, P., et al., "Analysis of Pressure and Rate Transient Data From Wells in Multi-layered Reservoirs: Theory and Application," SPE 24679, Society of Petroleum Engineers, 1992.

15 Sabet, M.A., *Well Test Analysis*, Gulf Publishing, 1991.

16 Streltsova, T.D., *Well Testing in Heterogeneous Formations, an Exxon Monograph*, John Wiley and Sons, 1988.

17 DaPrat, G., *Well Test Analysis for Fractured Reservoir Evaluation*, Elsevier, 1990.

18 Lai, C.-H., and C.-W. Liu, "Accuracy Examination of the Multiple Interacting Continua Approximation," *Proceedings*, National Science Council, 2001.

19 Aguilera, R. "Graphical Solution of Imbibition Equations Used To Predict Oil Recovery by Water Influx in Naturally Fractured Reservoirs," *Journal of Petroleum Technology*, 1975.

20 Aronofsky, J.S., L. Masse, and S.G. Natanson, "A Model for the Mechanism of Oil Recovery from the Porous Matrix Due to Water Invasion in Fractured Reservoirs," *Transactions of AIME 17*, Vol. 213, 1958.

21 Kazemi, H., "A Fully-Implicit, Three-Dimensional, Two-Phase, Control Volume Finite Element Model for the Simulation of Naturally Fractured Reservoirs," SPE 36279, Society of Petroleum Engineers, 1996.

22 Last, J.C., and T.V. Harper, "Response of Fractured Rock Subject to Fluid Injection—Part 1, Development of a Numerical Model," *Tectonophysics*, 1990.

23 Kazemi, H., J.R. Gilman, and A.M. Elsharkawy, "Analytical and Numerical Solution of Oil Recovery From Fractured Reservoirs With Empirical Transfer Functions," SPE 19849, *SPE Reservoir Evaluation and Engineering* (May): 219–227, 1992.

24 Adams, A.R., H.J. Ramey, Jr., and R.J. Burgess, "Gas Well Testing in a Fractured Carbonate Reservoir," *Transactions of AIME*, Vol. 243, 1968.

25 Handin, J.W., et al., "Experimental Deformation of Sedimentary Rocks under Confining Pore Pressure Test," *AAPG Bulletin*, Vol. 47, no. 5, pp. 717–755, 1963.

26 Kulander, B.R., S.L. Dean, and B.J. Ward, Jr., Fractured Core Analysis: Interpretation, Logging, and Use of Natural and Induced Fractures in Core, *AAPG Methods in Exploration Series No. 8*, AAPG, 1990.

27 Martin, G.H., "Petrofabric Studies May Find Fracture-Porosity Reservoirs," *World Oil*, pp. 52–53, February 1963.

28 Wade, R.P., and K. Aziz, "Stimulating the Triassic Carbonates in the Foothills Gas Trend of Northeast British Columbia," *Journal of Canadian Petroleum Technology*, Oct.–Dec. 1981.

29 Suppe, J., *Principles of Structural Geology*, Prentice Hall, 1985.

30 Long, J.C.S., et al., "Porous Media Equivalents for Networks of Discontinuous Fractures," *Water Resources Research*, Vol. 18, no. 3, pp. 645–658, June 1982.

31 Price, N.J., "The Tectonic Significance of Mesoscopic Subfacies in the Southern Rocky Mountains of Alberta and British Columbia," *Canadian Journal of Earth Science*, Vol. 4, no. 1, pp. 30–70, 1967. Diagram after Leroy, G., Cours de Geologie de Production IFP, Ref. 24, 429.

32 Muskat, *Physical Principles of Oil Production*, McGraw-Hill, 1949.

33 Fatt and Davis, "Reduction in Permeability with Overburden Pressure," *Transactions of AIME*: 329, 1952.

34 Wilhelmi, B., and W.H. Somerton, "Simultaneous Measurement of Pore and Elastic Properties of Rocks under Triaxial Stress Conditions," *Society of Petroleum Engineers Journal*. 1967.

35 Barton, N., S. Bandis, and K. Kakhtar, "Strength Deformation and Conductivity Coupling of Rock Joints, Intl. Journal of Rock Mechanics and Mining Science," *Geomech. Abstr.* 22, No. 3, pp. 131–140.

36 Makurat, et al., "Joint Conductivity Variation Due to Normal and Shear Deformation," *Proceedings* International Symposium on Rock Joints, 1990.

37 Carlson, M.R., et al., "Optimization of the Blueberry Debolt Oil Pools: Significant Production Increases for a Mature Field," *Journal of Canadian Petroleum Technology Special Edition*, Vol. 38, no. 13, 1999

38 Carlson, M.R., and J.G. Myer, "The Effects of Retrograde Liquid Condensation on Single Well Productivity Determined Via Direct (Compositional) Modelling of a Hydraulic Fracture in a Low Permeability Reservoir," Denver, Colorado: SPE Rocky Mountain Regional and Low Permeability Reservoirs Symposium, 1995.

39 Yadavelli, S.K., and J.R. Jones, "Interpretation of Pressure Transient Data From Hydraulically Fractured Gas Condensate Wells," SPE 36556, Denver, Colorado: SPE Annual Technical Conference and Exhibition, October 6–9, 1996.

40 Schlueter, E.M., et al., "The Fractal Dimension of Pores in Sedimentary Rocks and Its Influence on Permeability," *Engineering Geology*, Vol. 48, 1997.

41 Grant, J.R., "Predicting the Long-Term Recovery From Naturally Fractured Reservoirs Utilizing Production Test Data and a Conventional Reservoir Simulator," SPE 35597, Society of Petroleum Engineers, 1996.

19

Thermal Reservoir Simulation

Introduction

Thermal simulation is probably the most challenging aspect of reservoir simulation at present. From a numerical simulation point of view, the inclusion of thermal effects involves a number of different issues from conventional reservoir engineering. The author does not believe a good overview exists in the literature presenting the current state of the art. Here is the approach taken by the author:

1. Provide a brief description of the various processes that exist today and compare the differences and common aspects:
   a. Primary production with sand
   b. Steamflooding
   c. Cyclic steam stimulation (CSS)
   d. Steam-assisted gravity drainage (SAGD)
2. Describe the aspects of thermal simulation that are not common to the simulations described earlier in this book:
   a. Formulation
   b. Steam properties
   c. Heat losses
   d. Thermal conductivities
   e. Geomechanics
   f. Equilibrium ratios
   g. Temperature-dependent viscosities
   h. Differences in relative permeability
   i. Detailed grids
   j. Nine-point difference schemes
3. Numerical difficulties with thermal simulation:
   a. Steam injection well instabilities
   b. Countercurrent flow
   c. Slabbing
   d. Averaging thermal properties
4. Detailed process discussion:
   a. Convection
   b. Solution gas
   c. Foamy oil flow
   d. Thermal expansion

   e. Hydraulic fracturing
   f. Shearing
   g. Dilation
5. Discuss typical issues that exist in bitumen-prone areas:
   a. Channel environment
   b. Shale plugs
   c. Inclined heterolithic strata (IHS)
   d. Interbedding
   e. Discontinuities
   f. Water above and below
   g. Gas above and below
6. Describe a number of design issues that are handled through simulation:
   a. Operating conditions
   b. Subcool well controls
   c. Well length
   d. Well spacing
   e. Vertical spacing between horizontal injector and producer wells
   f. Startup
   g. Artificial lift
   h. Design tradeoffs
   i. Global optimization

State of technology

Thermal reservoir simulation is still changing rapidly primarily due to changes in the modeling of SAGD. This is also the topic of most interest in the current literature. Steamflooding and cyclic steam stimulation will be addressed here but in considerably less detail than SAGD. Conversely, although there is a great deal of material presented here on SAGD, it has been somewhat difficult to assemble a description of SAGD in a coherent fashion, due to the rapid evolution of this technology. The first successful demonstration of the SAGD process was about 20 years ago. Undoubtedly, there will be considerable demand for SAGD in the near future, particularly in Canada, Venezuela, China, and the United States.

Brief Description of Main Processes

Primary recovery

Provided the oil viscosity is sufficiently low, primary production is possible in heavy oil reservoirs. This utilizes the classical solution gas drive mechanism. Although this is not a thermal recovery technique, it has been included for two reasons. Historically, great efforts have been made to minimize sand production by the use of liners and sand screens. Optimization often concerned minimizing the number of sand cleanouts that were required to keep a well on production using conventional sucker-rod pumps. With the development of the progressive cavity pump, it is now possible to produce sand and oil simultaneously. Empirical evidence clearly shows that much greater production is obtained if sand production is allowed.

This has led to new problems: wells produced in this manner develop large piles of sand, which must be disposed of or utilized in an environmentally acceptable manner.

The reason for these higher production rates has been the subject of considerable research. The net results of this research indicated that wormholes, or to be more technically precise, shear zones, develop. There is a considerable permeability enhancement in the sheared zone. This process is similar to the mechanics of the sand in the reservoir with thermal processes. Hence, this type of production will be analyzed using similar technology—simulators.

Steam flooding, cyclic steam stimulation (CSS), and steam-assisted gravity drainage (SAGD)

In conventional heavy oil, steam is used as a displacing medium. The basic predictive technique is based on classic Buckley-Leverett frontal advance theory, much like a waterflood. The early work on this topic involved determining how much heat would be lost down the well to overburden and underburden and was aimed at getting the steam to the steam front, where frontal advance could take place. This reservoir mechanism relies on some movement of the oil behind the front. The general process is shown diagrammatically in Figure 19–1. This technique is used in a number of places. Much of the development of steamflooding took place in the Kern River field of California. Note that the key to the process is delivering heat to the formation via convection; although conduction is ever present and must be accounted for to properly characterize the process, conduction is a very slow and generally ineffective means of heating. On the other hand, many Kern River operators have capitalized on the 20+ years of conduction hot plate heating—especially from a lower zone since steam overrides—by subsequently recompleting and carrying out thermal operations in adjacent, especially upper, zones.

In some reservoirs, the oil is not sufficiently mobile to allow it to move behind the front. Often, such very heavy, very viscous oil is referred to as bitumen. This precludes direct displacement. The process relies on obtaining injectivity into the cold bitumen with steam and hot water and, in the process, convectively heating the reservoir, and then producing the heated fluids back. This process is repeated for a number of cycles, hence the name, cyclic steam stimulation. In some places, this process is known as huff and puff. The key to the process is getting the steam into the formation despite the bitumen's immobility. This is done by a mixture of geomechanical processes, which include hydraulically fracturing/deforming the reservoir and disturbing the sand grain structure of what are usually unconsolidated sands. This is shown in Figure 19–2. This technique was largely developed in the Cold Lake area of northeastern Alberta. From a reservoir modeling perspective, predicting performance proved very difficult. The major developments required were the direct modeling of the hydraulic fracturing/deformation process and accounting for changes in the sand permeability and porosity.

Cyclic steam stimulation proved useable in a significant portion of the tar sands, mostly in the Clearwater formation, which is relatively shallow and has initial oil viscosities on the order of 100,000 cp. The flagship of this technology is Esso's Cold Lake project, which began commercial operation in 1985. Subsequently, it did not prove to be useful in the deeper, more viscous McMurray formation (initial oil viscosities on the order of 1,000,000 cp). At this time, horizontal wells were becoming a reality. Dr. Roger Butler is widely credited with the development of the SAGD process.

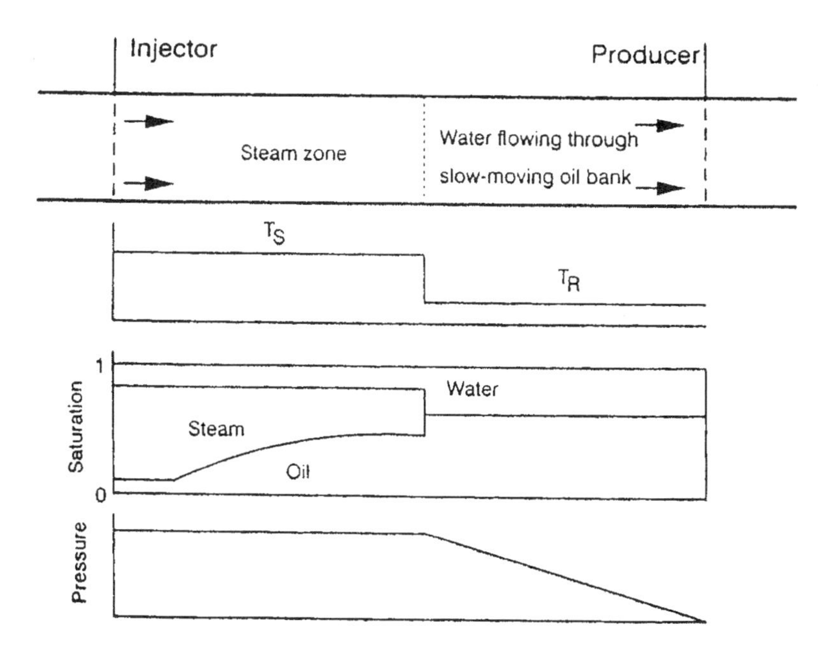

Diagram showing the Distribution of Temperature, Pressure, and Saturations in a Hypothetical One-Dimensional Steamflood

Fig. 19–1 Distribution of Temperature, Pressure, and Saturations for Idealized Steamflood

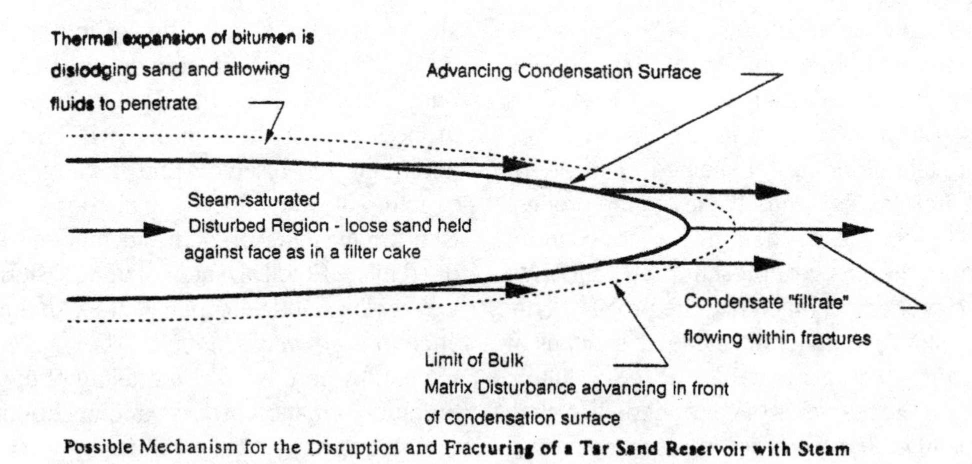

Fig. 19–2 Mechanism Associated with Fracturing Tar Sands, Cyclic Steam Stimulation (CSS)

The concept behind SAGD is relatively simple. Two horizontal wells are drilled in the formation—one directly above the other. Communication is developed between the two wells via circulation. Heat conduction "melts" (deviscosifies) the bitumen between the two wells. When communication is developed, steam is injected into the top well and melted production is gathered in the bottom well. This is shown in Figure 19–3.

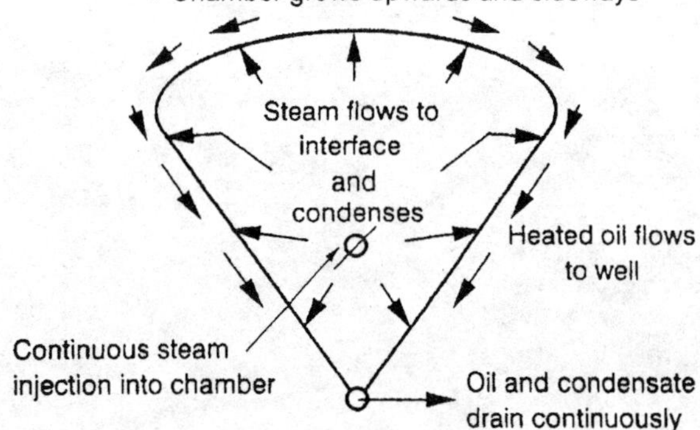

Fig. 19–3 Basic Description of SAGD Process

The original implementation was done by Esso in a test with a horizontal well and a vertical well. The success of the method was not apparent immediately, and Esso discontinued the idea and the associated research. A paper was written on this test.

The underground test facility (UTF)

The Alberta government, through its research organization the Alberta Oil Sands Technology and Research Authority (AOSTRA), was convinced to try a more detailed investigation of the SAGD concept in an underground test facility (UTF). The results are shown diagrammatically in Figure 19–4.

This was a very ambitious project, with wells drilled horizontally from mine shafts and utilizing extensive instrumentation. The results can only be described as a runaway success. Four different phases have been implemented at this site. The project was turned over to a commercial operator—originally the Gibson Petroleum Corporation.

Subsequent analysis of the original Esso test, in light of the successful UTF experiment, indicates that the experiment was, in fact, quite successful with the benefit of hindsight.

Public data

At present, only select information is available in the public domain with respect to SAGD. The UTF results have been available to those who participated in the project or who subsequently joined. In a small number of pilot projects, the data is still confidential for a defined period. The first commercial schemes should have public results soon, which will expand the amount of information available on this topic.

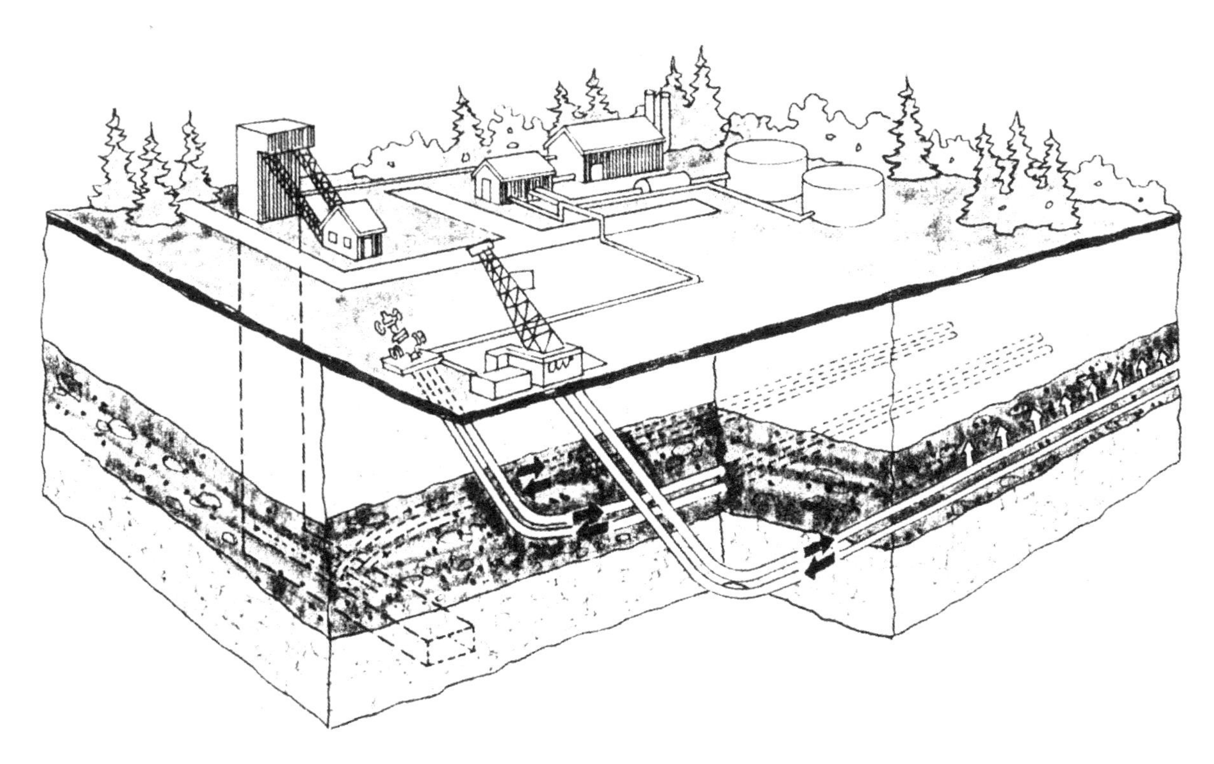

Fig. 19–4 UTF Project

Unique Aspects of Thermal Simulation

Formulation

Thermal reservoir simulators are similar to conventional black oil simulators in that they are based on the volumetric material (mass) balance (sometimes called the continuity equation) and Darcy's Law. In the case of a thermal simulator, the model also conserves energy.

Other references

Much of the current steamflood and steam stimulation technologies are described quite well in existing texts. This material will not be repeated in detail here. Butler has prepared a considerable amount of material in his textbook on the analytical development of SAGD. While this material was pivotal in developing that technology, it is the author's belief that numerical simulation, which much more completely and accurately incorporates multidimensional effects, variable reservoir description, and other important dynamics including geomechanics, will prove to be the ultimate design methodology.

Steam properties

The properties of steam are well-known. The key points relevant to simulation are:

- the position of the vapor-liquid line and the critical point
- the amount of energy required to heat steam
- the latent heat of vaporization

The vapor liquid line is shown in Figure 19–5.[1] Dry, superheated steam is not used in reservoir processes. Saturated steam is used, and, as a result, pressure and temperature are intimately related. It is interesting that the amount of heat required to generate steam is not a strong function of temperature and pressure. This is also indicated in Figure 19–6. The amount of heat that is required to change liquid into vapor (water into steam) changes with temperature as shown on the same diagram. Note that the heat of vaporization, or latent heat, varies about 15% over most expected SAGD operating conditions.

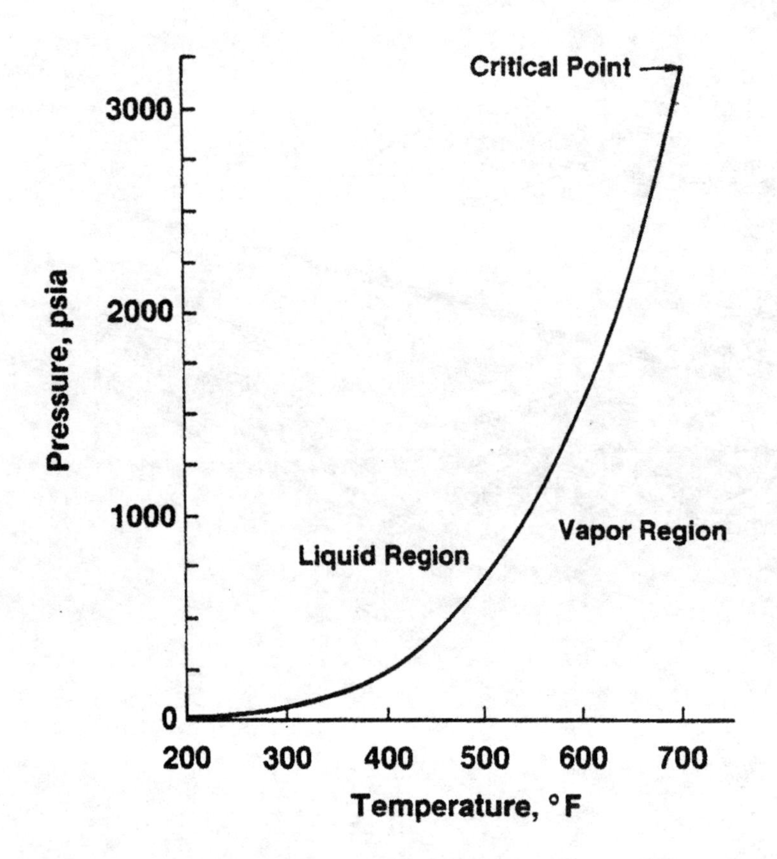

Fig. 19–5 Phase Diagram for Steam

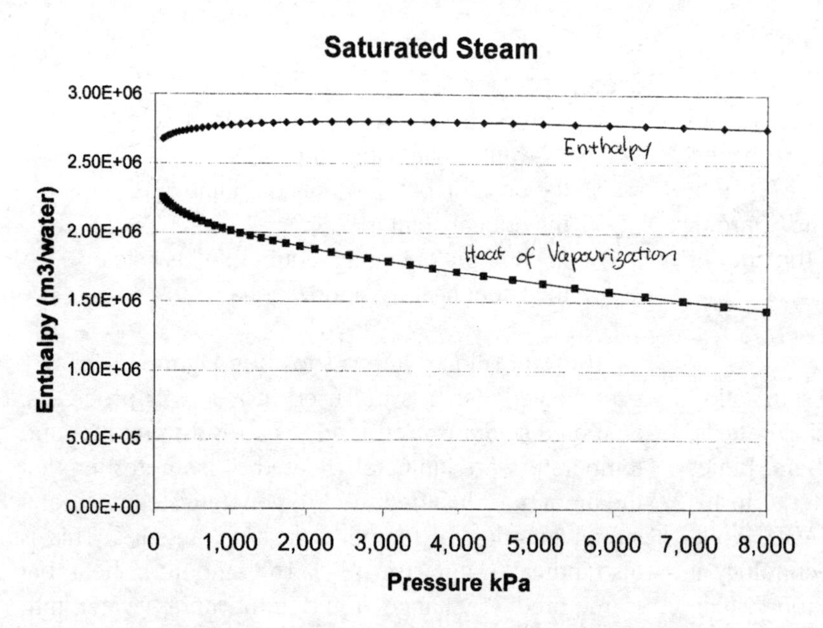

Fig. 19–6 Enthalpy and Heat of Vaporization for Steam

Since this data is obviously of critical importance, reservoir simulators typically have steam properties built in. This knowledge is needed for setting well controls.

Heat losses

Steam injection results in considerable loss of heat down the wellbore and to the formation overburden and underburden. The classic texts on thermal recovery deal with this issue in detail. Losses from around the well are calculated analytically as well as the overburden and underburden heat losses.

Thermal reservoir simulators are designed to handle these losses. Following the normal practice, these calculations are done analytically outside of the reservoir grid. This requires a number of lines of input that give the required parameters for these heat loss calculations. Normally, the wellbore, overburden, and underburden are calculated separately and have separate input.

Thermal conductivities

As with conventional reservoir simulation, finding accurate input is critical. Thermal conductivities will vary from area to area. Potential sources of data include the *Canadian Heavy Oil Association Reservoir Handbook*, the *AOSTRA Technical Handbook on Oil Sands, Bitumens, and Heavy Oils*, as well as the appendices to Roger Butler's book, *Thermal Recovery of Oil and Bitumen*.[2, 3, 4] Typically, the SPE monographs have considerable useful data in the appendices, and there is a *Thermal Recovery* monograph by Michael Prats.[5] The latter is oriented to steam flooding and consequently has less data than subsequent publications. Finally, Thomas Boberg's *Exxon Monograph on Thermal Methods of Oil Recovery* provides an excellent general background with very useful data in its appendices.[6]

Thermal conductivity is a function of the matrix and of the material filling the pore volume. In addition, it can be a function of temperature. The calculation of thermal conductivity can be done arithmetically or by correlations. Two correlations are used:

- correlation of Anand, Somerton, and Gomaa[7]
- correlation of Somerton, Keese, and Chu[8]

Heat capacity

As with thermal conductivity, the different components of the reservoir within a grid block have different heat capacities. The rock matrix has a different heat capacity from that of water, steam, gas, and/or oil present in the pores. The total heat capacity can be calculated by adding the capacities volumetrically.

Enthalpies

Enthalpy ($H \equiv U + PV$) is used in heat calculations within the simulator. Enthalpy is determined by a polynomial approximation. Most programs have defaults for common components; however, it is possible to input custom components or change the correlation. Note that the correlations for the vapor and liquid phases are entered separately.

Heat of vaporization

It is possible to enter the heat of vaporization from correlations. Two correlations include Watson's correlation as shown in Equation 19.1 and the correlation of Reidal shown in Equation 19.2.[9]

$$\frac{\Delta H_n / T_n}{R} = \frac{1.092(\ln P_c - 1.013)}{0.930 - T_{r_n}} \qquad (19.1)$$

$$\frac{\Delta H_1}{\Delta H_2} = \left(\frac{1 - T_{v2}}{1 - T_{r1}}\right)^{0.38} \qquad (19.2)$$

Enthalpies and heat of vaporization are related. The heat of vaporization is equal to the difference between the enthalpy in the vapor phase less the enthalpy in the liquid phase. Therefore, the system can be overdetermined. The manner in which the data is entered will control how the calculation is actually made.

Solution gas

Most bitumen contains relatively little solution gas. As a result, most preliminary SAGD modeling did not include solution gas in the formulation of SAGD reservoir simulations. It is fair to say that solution gas recovery in heavy oils is relatively low. The latter is a large function of oil viscosity. The lower viscosity heavy oils will have higher primary recovery factors. This is shown in Figure 19–7.

Data is available that provides some heavy oil solution gas heat capacities. The purpose of this work was aimed at providing background information for fire flooding.[10]

A number of other important implications with respect to solution gas include:

- In cold bitumens, it is not clear if significant gas would migrate through the high-viscosity bitumen.
- In conventional heavy oil recovery, G. E. Smith proposed the concept of foamy oil to explain some of the observed viscosity characteristics. Considerable research has followed on foamy oil flow.[11]

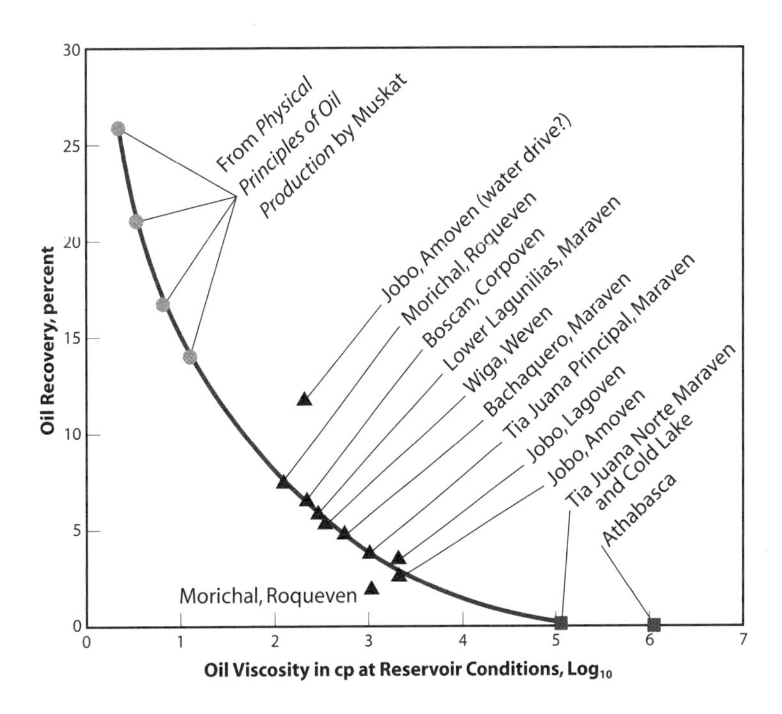

Fig. 19–7 Viscosity of Different Reservoirs at Reservoir Temperature and Effect on Recovery

- Some of this research does not support the existence of foamy oil but does suggest that emulsions are formed. The new interpretation has been obtained as the result of using large pore networks in detailed laboratory experiments.
- For the SAGD process, gas evolution affects thermal conductivity outside of the steam chamber, and the evolved gas collects at points where it may affect the loss of heat to the overburden (methane has a low thermal conductivity). It is possible to introduce other noncondensible gases, such as nitrogen, to accomplish this same effect.

Some of the earlier modeling of the SAGD process indicated that better history-match results were obtained if solution gas was not included. The implications were that the solution gas was not detected in the steam chamber and insufficient methane gas was present to indicate that gas was being evolved from the surrounding bitumen.

Equilibrium ratios

Although it is possible to model reservoirs on a chemical species basis using compositional simulation and an equation of state, this is too time-consuming for thermal simulation. The relative amounts of the various species (such as methane, propane, butane, pentane, etc.) in each phase (for instance gas or liquid) are therefore determined using equilibrium ratios or *k* values.

The manner in which the *k* values are determined varies considerably. One method is to use a sophisticated form of a table lookup based on temperature, pressure, and extensive laboratory data. This can be very accurate. There are simplified estimates of equilibrium ratios based on thermodynamics. Alternatively, a well-defined procedure is outlined in the GPSA data book that is based on temperature, pressure, and composition utilizing convergence pressures.

The methodology utilized in the thermal models that ARE has used are based on a lower level form of inputting equilibrium ratios.

Temperature-dependent viscosities

The whole point of thermal recovery is to obtain the reduction of bitumen viscosity at higher temperatures. It is no surprise that this needs to be input in a table of viscosity versus pressure. The viscosity of heavy oils varies so much that experimental data is required.

Having said this, the data that is usually obtained does not cover the entire range of temperatures and pressures. The viscosity of heavy oils is most strongly dependent on temperature, as shown in Figure 19–8. Therefore, the normal practice is to use a linear extrapolation on a log-log plot to determine viscosity.

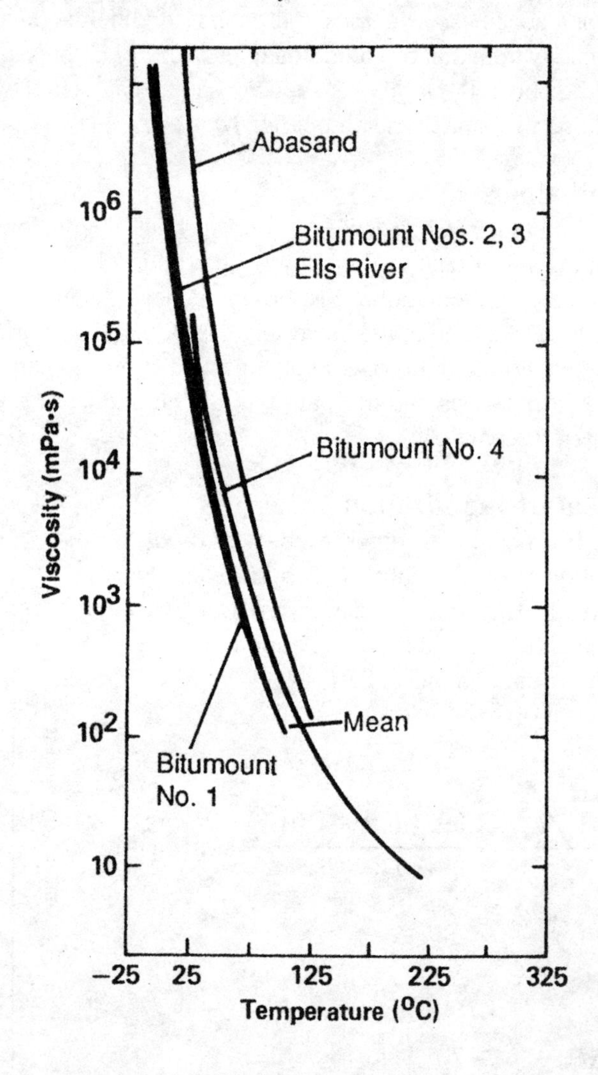

Fig. 19–8 Viscosity of Athabasca Bitumen from Different Areas

Data input, as with PVT and relative permeability data, utilizes a table lookup function. The user must input a table of temperature and viscosity.

High-temperature viscosities

At higher temperatures, for example with the Athabasca bitumen, the last points can deviate from a straight line, as shown in Figure 19–9.

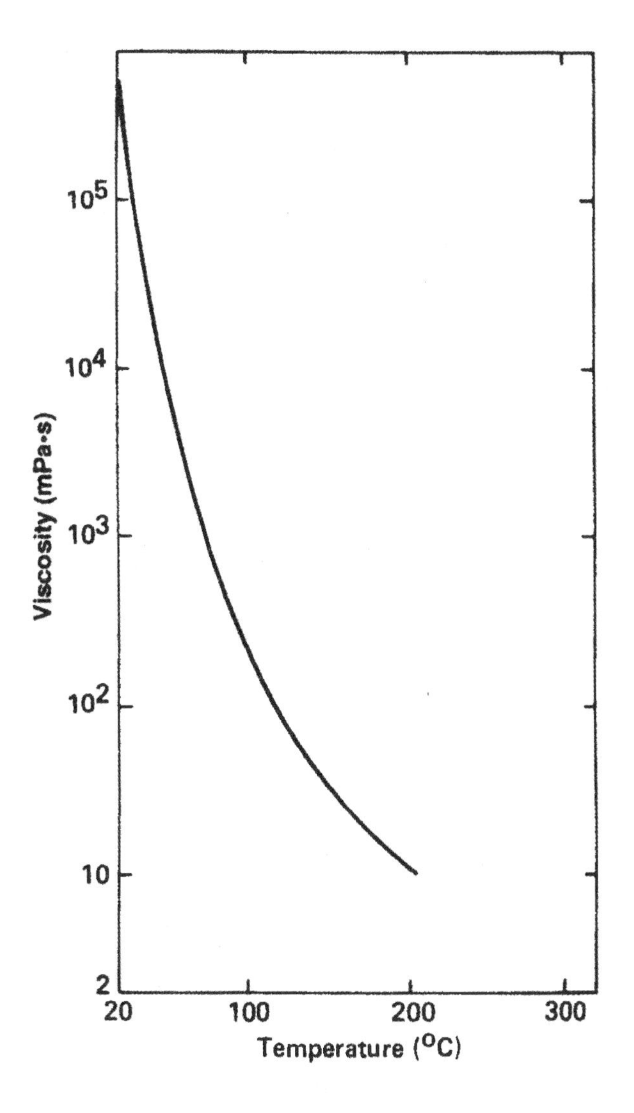

Fig. 19–9 Mean Viscosity of Athabasca Bitumen

This suggests that the trends may always be curved or, perhaps, below 100 cp cold viscosity, this reduction is attenuated with further temperature increases. Still, the majority of viscosities are measured assuming the straight-line relation, and it is common to see measurements of viscosities at three different temperatures to confirm and utilize the (perhaps incorrect) straightline relationship.

Field viscosity data

Figure 19–10 shows some samples that ARE assembled from northeastern Alberta. One of the wells showed properties that seemed to fall into two different groupings. Some layers fit the first grouping, while others fit the second grouping. Surprisingly, the two groupings seemed to occur otherwise randomly from different sands.

Note that great variation exists in viscosity relationships. It has become common to use the viscosity relationship from the UTF in many SAGD simulations. While this is good data, its applicability is unlikely ubiquitous. Then again, the variations from this relationship may be entirely related to oxidation.

Viscosity sampling

The more prudent will realize immediately that testing the entire temperature range is a very good idea before designing a project requiring hundreds of millions of dollars of capital. (The author has a copy of an application to the Alberta government for such a project where the operator did not have lease-specific data.) In truth, this is more difficult than one might think.

Since the bitumen doesn't flow, it is not possible to put the well on production and simply measure viscosity. The best way seems to be extracting the bitumen from core. This is a destructive process and means the bitumen must be taken from the core either by heating or chemical extraction with a solvent.

Oxidation

It is known that the viscosity of bitumen varies with the degree of oxidation. An example for Athabasca bitumen samples is shown in Figure 19–11 (from page 166 of the AOSTRA *Technical Handbook on Oil Sands, Bitumens, and Heavy Oils*).[3] Therefore, it is important to analyze the samples quickly and take the appropriate precautions to minimize oxidation. This makes interpretation of existing viscosity data much more exciting, since it seems that the details of sampling and sample handling often are not included in lab reports.

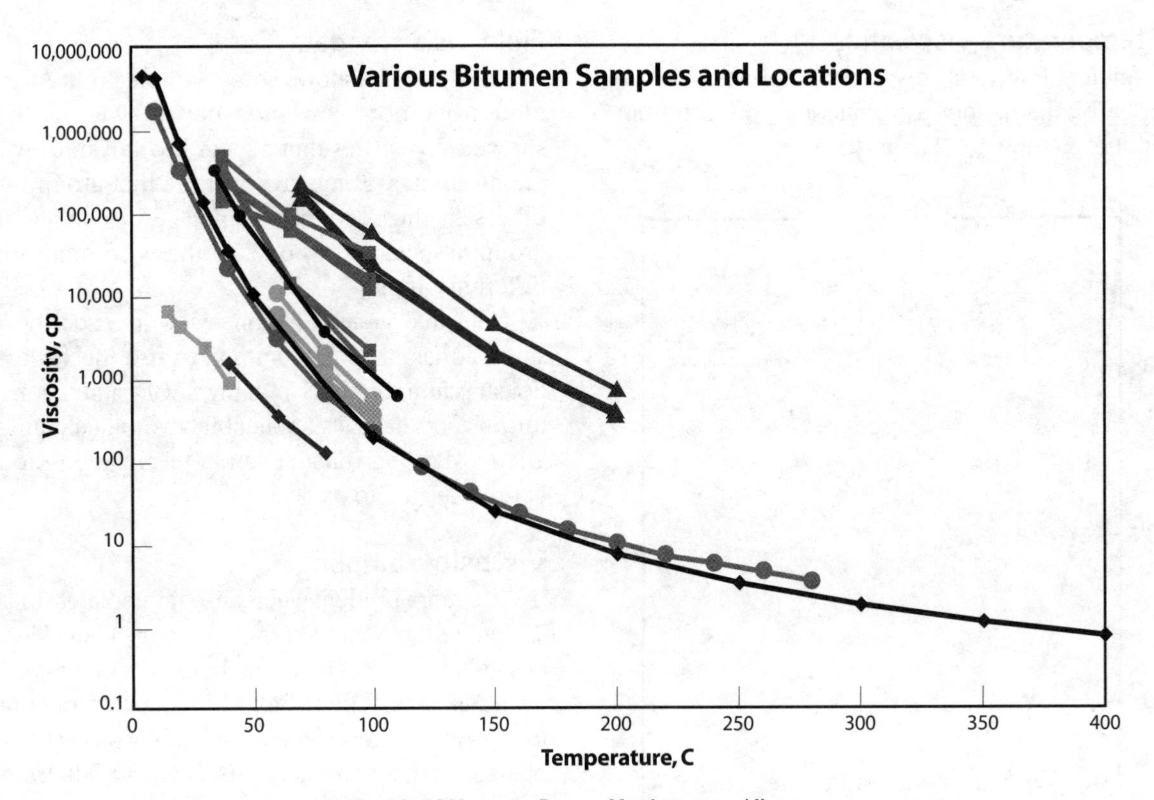

Fig. 19–10 Viscosity Data—Northeastern Alberta

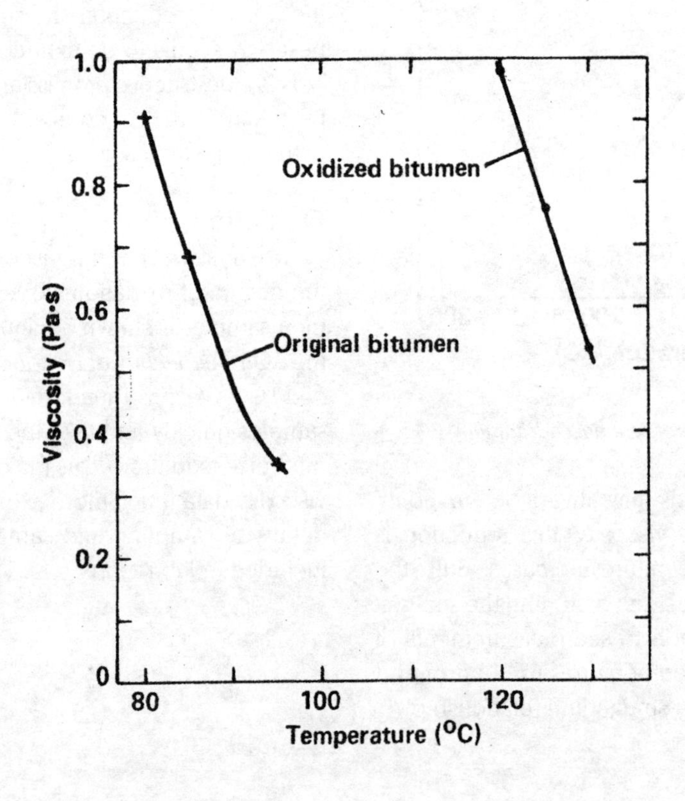

Fig. 19–11 Variation of Viscosity with Oxidation

Start-up heating for SAGD

As outlined earlier, the initial step in the SAGD process is achieving communication between the producer and injector. This is normally accomplished by circulating steam in both the producer and injector. Experience has shown the direction of circulation to be important. More heating at the heel results from circulating down casing first and then up the tubing (which is landed at the end of the horizontal section). Circulating down the tubing first gets the hot fluid to the end of the well. The cooled steam or water outside the tubing, in the annulus, is warmed by the hotter fluids passing through the tubing. The second circulation methodology gives a more even distribution of heat.

Some experience also indicates that benefit is derived from continuing the heating for a period after breakthrough is obtained. During this period, minimal steam or hot water is produced from the injector because the steam chamber is still quite small. Some steam was required in the past to lift the fluids up the wellbore. Continued injection of steam provides gas lift until the chamber is fully developed.

The heating phase seems to last somewhere between 30 and 90 days. Usually, a pressure differential is applied to encourage flow from the injector to the producer. High pressures are used during this stage normally.

From a modeling perspective, this was originally implemented as a radial heat loss calculation. This is input as a heat flux at the outside of the well. The grid takes over outside the edge of the well. An example calculation of this is shown in Figure 19–12. Subsequently, some reservoir simulators have built-in routines to do this calculation. In this case, the appropriate input must be entered.

Subcool operating constraint for SAGD

One of the more bizarre changes from conventional reservoir engineering well controls is the use of subcool. This sounds much like a form of teenage vernacular for *not so good* or, maybe *cool beyond belief*, rather than a technical term. In simplified terms, the production of steam from the lower well is wasteful. Produced steam does not condense inside the formation and release its heat; instead, significant heat of vaporization is carried out of the formation where it cannot be effectively utilized. Since the saturation pressure of steam varies with temperature, what would be a constant bottomhole pressure control must vary. This can be done by specifying a temperature versus pressure relationship. In fact, this is what a steam trap control really does.

In practical terms, it is difficult to measure pressures in a wellbore. There is sufficient vibration and fluctuations from flow and phase effects that stable readings are difficult, if not impossible. On the other hand, stable temperature readings are much easier to obtain. As a result, the well control is expressed as the temperature differential below the steam saturation line. The oil industry now has a new term, subcool, to rival the many pseudo properties. Apparently, the simulator uses a bottomhole pressure. This operating constraint is the subject of considerable interest because it affects the efficiency of the SAGD process. A value of 5° Celsius is typical. This topic will be discussed in more detail later.

Permeability

As described earlier, permeability is normally determined from core data using a semilog plot of core porosity versus log core permeability. This may be linked to a geostatistical description of facies.

Core from unconsolidated sands containing bitumen causes a number of unique difficulties. This has been well documented since 1984, when Dusseault described the difficulties in obtaining undamaged core from unconsolidated oil sands reservoirs.[12]

The problems can be described briefly as follows and are shown in Figure 19–13.

- Since the sands are not cemented, they are prone to mechanical deformation. In this case, the core has rotational friction, which will cause shear dilation. The sand grains ride up on each other.
- Stress is released from the core during the coring process. This causes outward expansion.
- Like conventional core, the core is subject to mud filtrate invasion due to the circulation of drilling mud while coring. However, because the uncemented sands expand and "create" more pore space, a much higher degree of invasion occurs than in conventional core.
- The bitumen is sufficiently viscous that one would not expect significant bitumen displacement.
- Solution gas exsolution will cause significant deformation. The solution gas content of bitumen is quite low, normally under 4 m^3/m^3. However, in relation to the sample volume, this is potentially a large expansion. Note that this corresponds to an oil shrinkage of only 1–2%.

Numerical Example of Well Bore Heat Loss Calculation

Steam is injected into a reservoir having a depth of 460 m using a well with a 17.8 cm (7 in.) diameter casing. The conditions are as follows:

Reservoir Temperature:	*10° C*
Steam Pressure:	*10 MPa*
Steam Quality:	*70%*
Injection Rate:	*160 m³ d⁻¹*

Assume that the thermal conductivity of the overburden is 1.7 W m⁻¹ °C⁻¹ and that its volumetric heat capacity is 2410 kJ m⁻³ °C⁻¹. Neglect the effect of the cement around the casing.

a. Calculate the heat injection rate in megawatts.

b. Assuming that the steam is injected directly down the casing, calculate the heat loss rate in megawatts after 1, 10, 100, and 1,000 days of injection. Express this loss as a percentage of the heat input.

c. Plot the temperature as a function of distance from the casing surface for each of the times in (b).

d. Calculate the heat loss rate in watts per square meter for each of the times in (b) and compare to the heat loss that would be expected for a flat surface at the same steam temperature.

e. Assume that the steam is injected into a 7.3 cm (2⅞ in.)-outside-diameter tubing and that this is isolated from the casing by a thermal packer at the bottom so that the annulus is filled with air at atmospheric pressure. Assume that the emissivity of the facing tubing and casing surfaces is 0.8. The internal diameter of the casing may be taken as 16.5 cm (6.5 in.). For these conditions repeat the calculations in parts (b) and (d) and compare the answers using a bar chart. Calculate the heat-transfer coefficients for a casing temperature of 250° C and assume that they do not vary with time. The thermal conductivity of air in the annulus may be taken as 0.0145 W m⁻¹ °C⁻¹.

f. Using the heat flows determined in (e), calculate the casing temperature for each time and make separate improved estimates of the heat-transfer coefficient U. Repeat the heat-flow calculations of (e) and repeat until consistent values are obtained. Revise the bar chart produced in (e) and also plot a graph showing the casing temperature as a function of the steaming time (use a log scale for time).

Solution

$L = 460$ m	$R_2 = 7/2$ in. = 0.0889 m
$T_R = 10°$ C	$T_S = 311°$ C
Steam Quality = 70%	Injection Rate = 160 m³d⁻¹ = 160,000 kg d⁻¹
$K = 1.7$ W m⁻¹ °C	Vol Heat Cap = 2,410 kJ m⁻³ °C⁻¹

a. Heat-Injection Rate

From Steam Tables:

Vapor Enthalpy	=	2,724.7 kJ kg⁻¹
Liquid Enthalpy	=	1,407.6 kJ kg⁻¹
Heat in 70% Quality Steam	=	2,724.7 x 0.7 + 1,407.6 x 0.3
	=	2,329.57 kJ kg⁻¹ above 0° C
Heat Above T_R	=	2,329.57 – 4.2 x 10 = 2,287.6 kJ kg⁻¹
Injection Rate	=	160,000 x 2,287.6 = 366 x 10⁶ kJ d⁻¹
	=	366 x 10⁶ x 1,000/(24 x 3,600 x 10⁶)
	=	4.24 MW

b. Heat Loss With Injection Down the Casing

Days of Injection	1	10	100	1,000
Dimensionless Time ($\alpha t/R_w^2$)[1]	7.71	77.1	771	7710
In(Dimensionless Time)	2.04	4.35	6.65	8.95
Heat Loss in Megawatts from (2.62)	0.75	0.44	0.29	0.21
Heat Loss as Percent of Input	17.5	10.4	6.8	5.0

[1] $\alpha = 1.7/(2{,}410 \times 1{,}000) = 7.05 \times 10^{-7}\ m^2s^{-1} = 0.060946\ m^2d^{-1}$

c. Dimensionless Temperatures

log (R/R_w)	R/R_w	Dimentionless Temperatures $t^* = 7.71$	$t^* = 77.1$	$t^* = 771$
0.00	1.00	1.00	1.00	1.00
0.25	1.78	0.66	0.80	0.85
0.50	3.16	0.33	0.59	0.70
0.75	5.62	0.09	0.39	0.55
1.00	10.00	0.00	0.18	0.39
1.25	17.78	0.00	0.05	0.25
1.50	31.62	0.00	0.00	0.14
1.75	56.23	0.00	0.00	0.04
2.00	100.00	0.00	0.00	0.00

d. Calculation of Heat Loss Rate

External Area of Well Casing = $2\pi R_w L = 256.9$ m²

Days of Injection	1	10	100	1,000
Heat Loss W m–2				
From Well	2,919	1,726	1,144	810
Flat Surface (2.22)	1,169	370	117	37

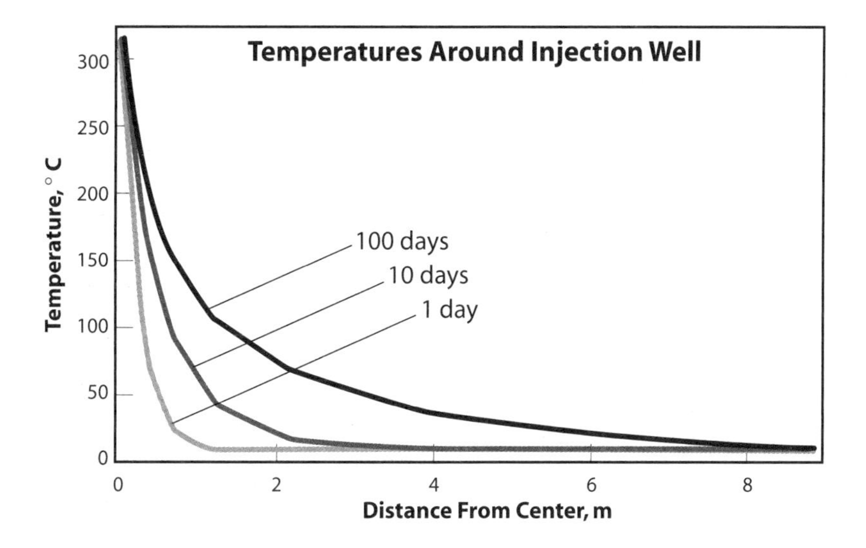

e. Use Equation 2.80 to calculate h_R.

$$\sigma = 5.669 \times 10^{-8}\ \text{W m}^{-1}\ \text{K}^{-4}$$
$$T_i = 273 + 311 = 584\ \text{K}$$
$$T_o = 273 + 250 = 523\ \text{K}$$

Emissivity for Both Surfaces equals 0.8.

$$\frac{R_i}{R_o} = \frac{2.875}{6.5} = 0.442; \qquad R_i = 0.442 \times 3.25 \times 0.0254 = 0.0365\ \text{m}$$

$$h_R = 28.4\ \text{W m}^{-2}\ {}^{\circ}\text{C}^{-1}$$

Use Equations 2.84 to calculate h_C.

$$\delta = \frac{0.5(6.5 - 2.875)2.54}{100} = 0.0460\ \text{m}$$

Average gas temperature = 553.5 K

N_{Ra} factor from Figure 2–17 = 7.5 x 106

$$N_{Ra} = 44{,}600; \qquad K_{eff}/K_g = 0.13 N_{Ra}^{0.25} = 1.89$$

$$h_c = \frac{1.89\ K}{R_i \ln(R_o/R_i)} = 0.9$$

$$U = h_R + h_C = 29.3\ \text{W m}^{-2}\ {}^{\circ}\text{C}^{-1}$$

$$R_w = 0.01812\ \text{m}$$

Calculation of Loss Rate

Time in Days	1	10	100	1,000
Dimensionless Time	186	1,860	18,600	186,000
ln(Dimensionless Time)	5.22	7.53	9.83	12.13
Heat Loss in Megawatts	0.38	0.36	0.18	0.13

Fig. 19–12 Heat Loss Calculation Around Wellbore

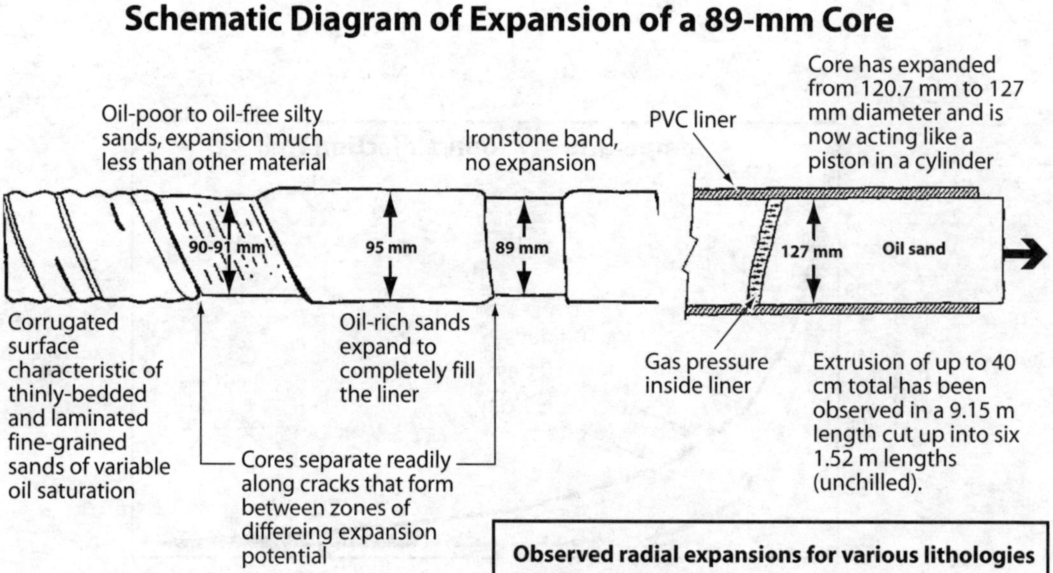

Schematic Diagram of Expansion of a 89-mm Core

Oil-poor to oil-free silty sands, expansion much less than other material
Ironstone band, no expansion
PVC liner
Core has expanded from 120.7 mm to 127 mm diameter and is now acting like a piston in a cylinder
90-91 mm
95 mm
89 mm
127 mm
Oil sand
Corrugated surface characteristic of thinly-bedded and laminated fine-grained sands of variable oil saturation
Oil-rich sands expand to completely fill the liner
Cores separate readily along cracks that form between zones of differeing expansion potential
Gas pressure inside liner
Extrusion of up to 40 cm total has been observed in a 9.15 m length cut up into six 1.52 m lengths (unchilled).

Observed radial expansions for various lithologies	
Ironstone concretions	89 mm (no expansion)
Basal clays and clayey silts	89-91 mm
Oil-poor to oil-free silty sands	90-93 mm
Fine-grained oil-rich sand	91-95 mm
Coarse-grained oil-rich sand	94-95 mm

Fig. 19–13 Problems of Dilate Core

In Canada and Venezuela, the expansion of bitumen sands normally results in core extruding from the core barrel at surface. It is also common for oil sands recovery to exceed 100% due the linear expansion.

Most oil sands core barrels use white PVC pipe to maintain the physical integrity of the samples. These sands can be remolded in your hands.

These difficulties cause a host of problems:

- Core porosities are overstated. Most core is analyzed using the summation of fluids methods in which the water volumes and bitumen volumes are determined by solvent extraction and then used to calculate porosity. These porosities are higher than those determined from logs due to filtrate invasion.
- As outlined previously, gas expansion is a major expansion process. This is not measured by the summation of fluid method. As a result, porosities taken from Boyle's law porosimeters are normally 3% higher. Most permeability samples are measured with Boyle's law porosimeters.
- The water saturations from oil sands cores are too high by virtue of filtrate invasion. This can be proven by comparison to log-calculated saturations as well as capillary pressure tests. Note that water saturations are known from the oil sands mining side of the business. In fact, the recovery process used relies on the fact that the sand grains are surrounded by water.

Some apparent contradictions exist in the literature on this topic. The geotechnical literature has concentrated on gas exsolution; there are passing references to the imbibition of drilling mud filtrate. The logging literature has identified expansion in general and its effect on total porosity, which is calculated by summation of fluid calculations involving water and tar. There is passing reference to solution gas exsolution in the petroleum literature, which is regarded as insignificant since heavy oils have low solution gas-oil ratios.

This is not the end of it. The geotechnical engineers appear to assume that gas exsolution involves only positive pressure displacement from the core. The bitumen is assumed to be immobile. While oil is displaceable in the conventional oil industry, it is well-known that water is imbibed into conventional oil cores despite vastly more solution gas that is expelled from light oils.

Strangely, the geotechnical literature does not always discuss mechanical rotational shearing in the specialized core barrels. Although the liners do not rotate, core bits do rotate. Yet, the geotechnical engineers do all their sample preparation for laboratory analysis on cores frozen to -18° Celsius to prevent mechanical rearrangement of the sand grains. They conclude that freezing the formation ahead for coring is potentially valuable, ergo mechanical deformation must occur.

Pore volume compressibility and stress relief is cited in much of the conventional oil literature, despite the fact that pore volume compressibilities are much lower. Offsetting this are much larger reductions in overburden pressure. The pore volume compressibilities of unconsolidated sands are much higher, in some cases 10 times higher or more. Thus, some fluids should be sucked into the core.

The generation of gas pathways is an opportunity for water to imbibe. Normally, gas segregates by gravity in wellbores rising.

Porosities from cores obtained from Gulf Coast unconsolidated sands are known to be optimistic. The latter porosities are also determined from the summation of fluids method, since the cores have no mechanical strength.

Heavy oil sands analysis

The following section on core analysis for bitumens digresses somewhat from permeability. Summation of fluids is rarely used in the conventional side of the business. The process involves putting small samples of the oil sands in a Dean Stark apparatus, which is shown in Figure 19–14.

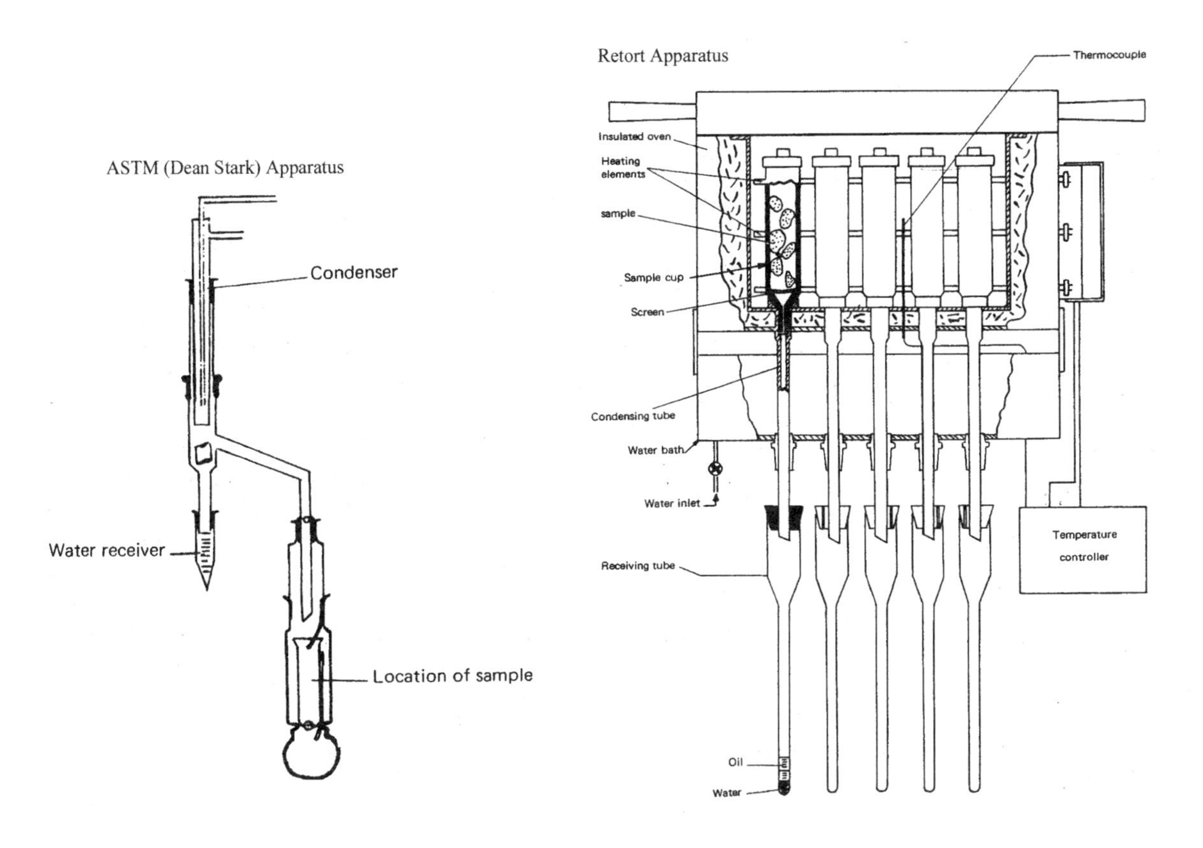

Fig. 19–14 Dean Stark Apparatus

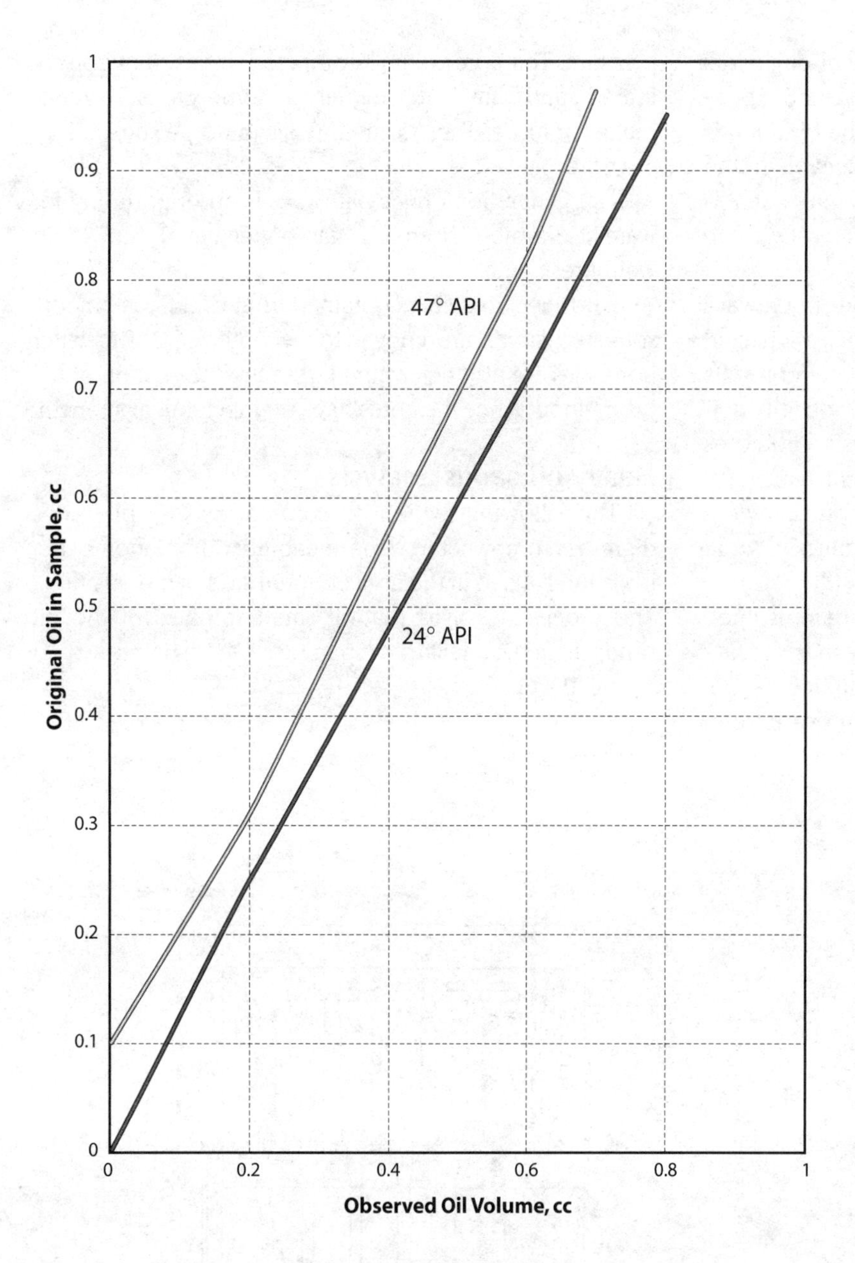

Fig. 19–15 Coking Corrections

The volume and weight of the total sample is measured first. This can be done by a variety of methods. The oil and water are extracted by using toluene as a solvent as well as heat. This is done with an electric heater, for safety reasons. The water is condensed and measured in a graduated container. Note that this measurement is made at a specific temperature. At high temperatures, water is driven off from clays and other minerals. Since the specific gravity of water and the bitumen are known, the volume of oil can be calculated by subtraction.

For additional money, the volume of oil can also be measured. Common sense indicates that this should result in a tarry mess left in the sample and that the volume of oil recovered from the condenser will be less than the volume of bitumen that was originally in the sample. This is, in fact, the case. The volume of oil is adjusted by cooking known volumes of the bitumen and toluene and determining a correction factor. An example of such correction factors is shown in Figure 19–15.[13]

Some of the descriptions indicate that the total volume of the sample is determined by immersion in mercury, and this will compensate for the gas-filled portion of samples. Normally mercury immersion destroys samples. Some references seen by the author say this does not present a problem and the mercury can be recovered in the extraction process. Mercury vapor is highly poisonous—at least according to the spy novels. The labs visited by the author usually have water/glycol mixtures on top of the mercury collection trays to stop vaporization.

Determining correct saturations from logs and calibrating to core

Woodhouse describes the correct method of calibrating logs and core.[14] As is a common theme in this book, much of the problems with reservoir simulation have little do with the actual simulator. The problem, more often than not, is getting the correct input data.

The basic assumption is simple. Relatively little bitumen is displaced from the core due to its high viscosity. The author questions whether this is completely true; some little bubble of evolved gas is going to cause a spurt of bitumen. The oil actually shrinks with gas exsolution, so it will have shrunk about 1–2%. In practical terms, these are tolerable errors. The log calculations then are tuned to weight-percent bitumen.

The errors in core analysis saturations do not affect the correct calculation of original oil in place. The increased porosity compensates for the filtrate invasion in cores. This does become significant, however, in thermal recovery calculations—i.e., simulations—where the extra connate water must be heated as part of the process. Therefore, it is important to get the actual porosities and water saturations correct.

One other significant item should be mentioned. As shown in Figure 19–16, uncemented formations typically have lower values of *m*, the cementation exponent in the Archie equation.[15] It is also known that *n*, the saturation exponent, normally follows *m*. This can be confirmed in a number of ways. A log analyst can use a Pickett plot on a water-bearing zone to empirically determine *m*. In addition, it is possible to have the sands tested under overburden conditions.

The following are some actual field results determined by experiment on cores obtained from the Hangingstone area. Examples of this data are shown in Figure 19–17 and 19–18. The porosity and saturation exponents from the complete series of tests were tabulated in Table 19–1.

The tests were conducted at overburden pressure. Since these cores do not fully recover from dilations, the lower water zones from the adjoining oil sands leases were analyzed by ARE using a Pickett plot. An *in situ* value of 1.35 was determined for *m* for this area. This value is consistent with the lower range of the data in Table 19–1. The following calculation was based directly on the core data. Note that *n* cannot be determined from a Pickett plot.

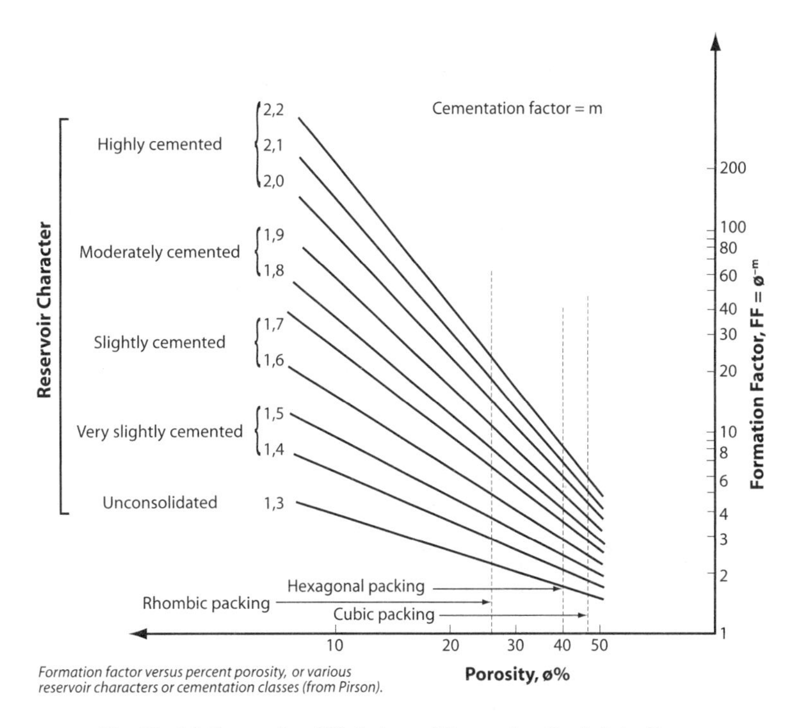

Fig. 19–16 Generalized Variation of Formation Resistivity Factor with Cementation

Table 19–1 Cementation (*m*) and Saturation (*n*) Exponent Values from Various Athabasca McMurray Formation Sand Special Core Tests

Well	m	n
5-13-84-11W4M	1.45	1.81
4-15-84-12W4M	1.47	1.70
15-20-84-11W4M	1.35	1.81
14-29-84-11W4M	1.36	1.70
13-27-84-11W4M	1.38	1.83
Average	1.40	1.77

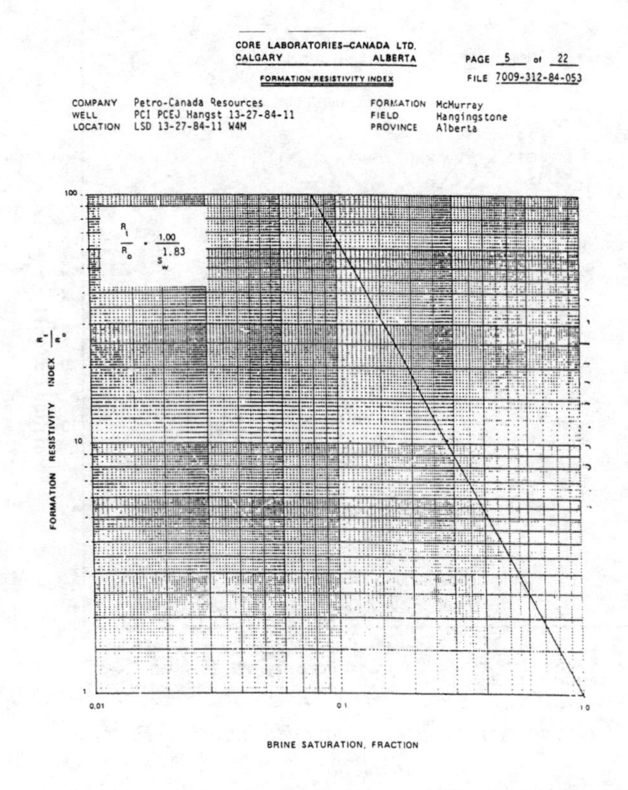

Fig. 19–17 Formation Resistivity Factor m*—Actual Lab Data*

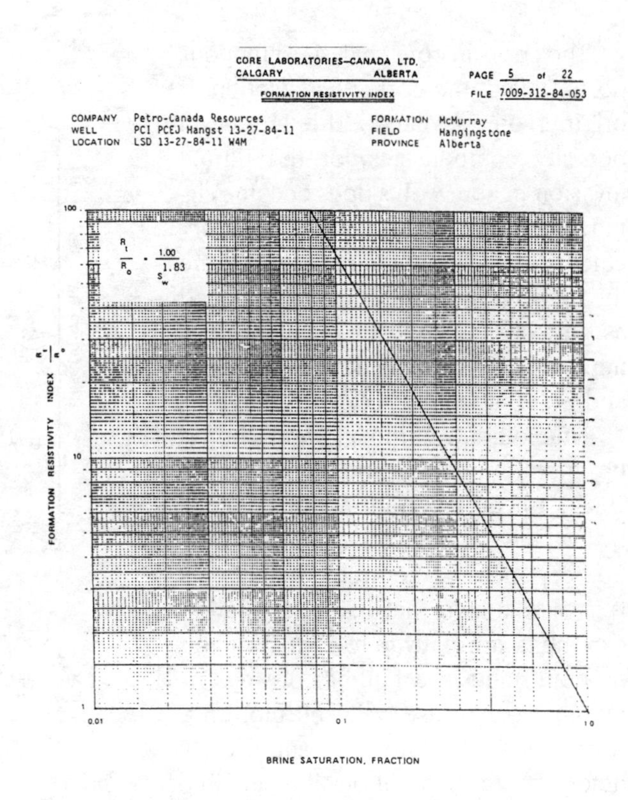

Fig. 19–18 Formation Resistivity Factor n*—Actual Lab Data*

$$S_w^{\,n} = a \times Rw \,/\, \varphi^m \times Rt \qquad (19.2a)$$

$$S_w^{\,2.00} = (1.0 \times 0.853) \,/\, ((0.30)^{2.00} \times 105.31) \qquad (19.2b)$$

$$S_w = 30.0\%$$

$$S_w^{\,1.77} = (1.0 \times 0.853) \,/\, ((0.30)^{1.40} \times 105.31) \qquad (19.2c)$$

$$S_w = 17.1\%$$

Therefore, if $m = n = 2$ were incorrectly assumed, then the S_w would have been 30% instead of 17.1%. Using the revised values of m and n will significantly reduce the calculated *in situ* water saturations.

Air permeabilities

The standard design process for SAGD was developed on the UTF projects. The predictions of performance were based on core permeabilities derived directly from air permeabilities on dilated cores. From the viewpoint of a conventional reservoir and simulation engineer, this approach is wrong on the face of it. Interestingly, one of the statements from an earlier paper (SPE 21529) is most telling:

> *Prior to Phase A operations, in-house opinions of what the bulk effective vertical permeability of the formation would prove to be ranged over two orders of magnitude, resulting in correspondingly varied estimates of economic viability. The central objective of Phase A test was to resolve this uncertainty by direct measurement of the SAGD rate in a minimal volume of reservoir.*[16]

Given these comments, it is indeed fortunate that the pilot went forward. The substantive contingent of geotechnical engineers and the extensive geomechanical instrumentation suggests that the use of higher than undisturbed liquid permeabilities was not accidental. Note that substantial experience in the area of cyclic steam injection also supports allowance for the effects of dilation on permeability.

Liquid permeabilities

There is experimental data available that shows the difference between liquid and air permeabilities. Again, this data is from Hangingstone. This suite of data is unusually complete. It includes permeability for cleaned and uncleaned samples and both air and liquid permeability data. The liquid permeabilities were determined at net overburden pressure (NOB). A simulated overburden pressure of 5,000 kPa was used.

There are two factors to consider in evaluating the effects of NOB:

- Normally, a horizontally drilled plug is placed vertically in a triaxial apparatus. The effective pressure applied surrounds the entire sample, even in what would be the vertical sides of the core plug. As a result, NOB tests typically provide stresses that are higher than that which would actually exist in the ground. Consequently, NOB tests tend to report lower permeability than they should.
- Most unconsolidated samples do not behave elastically. Hence, not all of the unloading effects are replicated by reloading the sample. Typically, the grains do not return to as compact a form as they would have been *in situ*.

To some degree, these two factors may cancel out; however, it is not easy to estimate which factor dominates. Thus, some care (educated guess) must be used in interpreting these results.

The experimental results are shown on Figure 19–19. In total, 25 plugs were tested for liquid permeability. The associated air permeabilities are also shown on the plot. The center of the grouping of the dark diamonds is located in the same position as seen on previous plots of air core permeability versus core porosity. The light squares clearly show the difference in permeability measured with liquid and with overburden pressure.

The ratio of k_{air}/k_{liquid} has been plotted on Figure 19–20. As can be seen from the histogram, the average correction factor will be in the range of 0.25 to 0.30. The arithmetic average was calculated to be 0.249. This is a useful indicator of how much correction should be applied to core air permeabilities to derive *in situ* reservoir permeabilities.

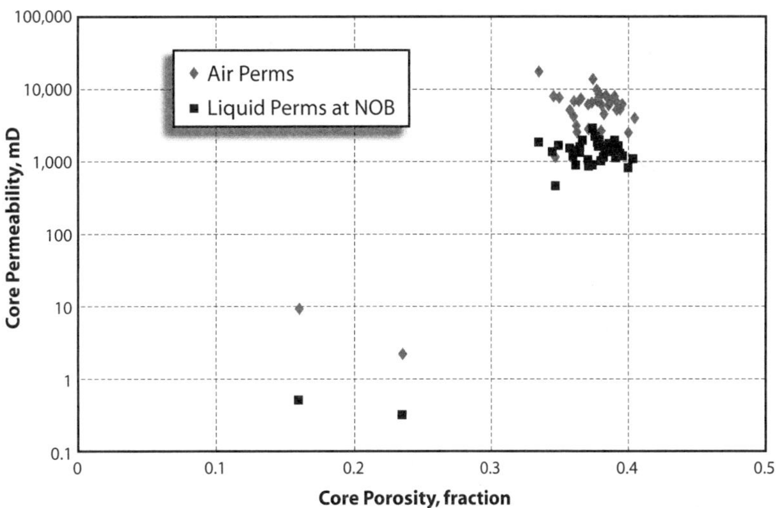

Fig. 19–19 Net Overburden (NOB) Liquid Permeability as a Fraction of Air Permeability

Documented history matches (UTF Phase A and B) did not utilize such corrections. It may be concluded that there must be some significant mechanism occurring in the reservoir that causes air permeabilities to be representative in SAGD modeling.[17]

Geomechanical effects

The justification for the beneficial effects of geomechanics is based on:

- extensive documentation in the literature
- laboratory results
- numerical simulations
- field observations

This combination must lead to the conclusion that oil sands reservoirs are being affected by geomechanical effects. A background explanation will be made of geomechanics in the following.

Geomechanical literature

Geomechanics is not an area that is currently included in standard petroleum engineering programs. This is a classic area where cross-disciplinary understanding is critical, much like the link between reservoir simulation and geology. Some university programs have combined these two disciplines. The following list follows up on earlier references to literature on the subject and summarizes a portion of what is available. This material concentrates on simulations.

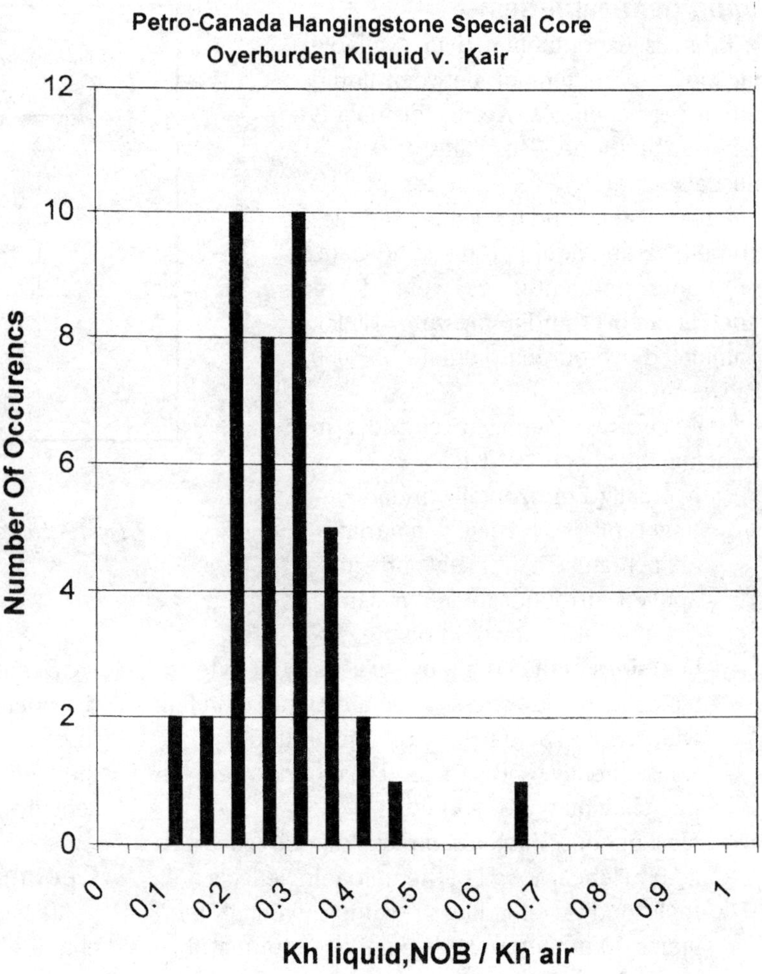

- Espinoza and Mirabel (1988) used geomechanics to simulate compaction and subsidence associated with heavy oil reservoirs in Venezuela. There is considerable expertise within PDVSA with regard to geomechanics. Unfortunately, only a small part of the Spanish literature is translated.[18]
- Abou-Kassem and Farouq Ali (1985) recognized the need for a coupled reservoir/geomechanics model.[19] Later, Tortike and Farouq Ali (1991) described their attempts to predict numerically the failure of oil sands due to steam-induced stresses. They constructed a 3-D steam-water-oil-gas model incorporating geomechanics. Plastic strains and volumetric strains were found to increase the permeability and porosity of the material. Tortike, Farouq Ali et al. (1994) used their model to simulate cyclic steam stimulation.[20]
- A model for oil sands production was developed by Wan et al. (1989) to describe the effective stress-strain behavior of oil sands. Although the model was developed for casing/formation interaction studies, it had applicability to other geomechanics problems since it incorporated thermal effects and strain softening with dilation at low confining stresses.[21]
- Vaziri (1986) developed another model, which could be calibrated with existing geomechanical laboratory tests.[22]
- Beattie et al. (1991) and Denbina et al. (1991) utilized a porosity dilation model coupled to a permeability enhancement algorithm to implicitly model prescribed deformation regions (simulating dynamic horizontal fractures) in cyclic steam stimulation (CSS) history matches of Esso's Cold Lake wells. These first-time successes at Cold Lake history matching were largely attributed to the approach of using a thermal reservoir simulator implicitly coupled to a deformational model (relative permeability hysteresis was also employed). This follows from the mechanistic observation from these studies that formation compaction is by far the dominant mechanism supplying drive energy to the early cycles of the CSS process; solution gas drive was the most important of the remaining mechanisms. Significantly, gravity drainage accounted for little of the oil produced in the first two cycles, but increased in importance subsequently.[23,24]
- Chalaturnyk and Scott (1991) modeled the geomechanics of Phase A of the AOSTRA UTF pilot project. Thermal values from a reservoir simulator were fed into a geomechanical model, which used the temperatures to calculate an equivalent single-phase hydraulic conductivity. They found that, for multiwell patterns, zones of

Fig. 19–20 Ratio of $Kh_{liquid\ NOB}/Kh_{air}$ *Hangingstone Samples*

failure within the cold reservoir might occur between the well pairs, rather than near an injector or producer well.[25]

- Settari et al. (1992) examined the geomechanical aspects of fluid injection into oil sands for the PCEJ cyclic steam stimulation project. Dilation in the fracture failure region was identified as one of the critical design parameters. The result was a combined laboratory and simulation study to determine the compressibility and dilation values for numerical simulation.[26]
- Singhal et al. (1998) discussed the effect of formation dilation during steam injection if some threshold pressure was exceeded in the formation. During production, the deformation was not entirely reversed. Effective permeabilities were increased, enabling larger-than-expected injectivity and productivity.[27]
- Satriana et al. (1998) modeled an Indonesian steamflood and demonstrated that oil recovery can be accelerated due to increased injectivity because of dilation. A coupled reservoir/geomechanics simulator was used. Results showed permeability enhancement caused by reduced effective stress for all injection cycles. Field measurements from tiltmeters were used to back calculate increases in reservoir porosity, and these had increased from 33% to a porosity of 38%. A history match showed the existence of permeability enhancement due to the reduced effective stress and limited fracturing in later injection cycles.[28]
- Yuan et al. (1999) developed a model for wormhole propagation in cold production from heavy oil reservoirs. These self-eroding cavity channels in poorly consolidated sandstones are responsible for the unusually high productivity of wells. Daily production rates are generally proportional to the cumulative sand production, indicating the wormholes continuously propagate farther into the reservoir.[29]
- Dusseault et al. (1994) addressed the geomechanics of the wormhole phenomenon, as well as the growth of a remolded zone of dilated material around the wellbore. Both resulted in sand production. They recognized the coupling of sand geomechanics with fluid flow models as being at the forefront of engineering in the oil industry.[30]
- Shen, C. (2000) of Imperial Oil conducted a numerical investigation of SAGD process using a single horizontal well. He included a dilated zone around a horizontal well with a reduced capillary pressure (more on this issue later) and enhanced permeability. A parametric sensitivity was done on different geomechanical properties. Oil rates were found to be five times greater than for the base case without any dilation.[31]
- Ito, Ichikawa, and Hirata (2000) examined the SAGD steam chamber growth at the UTF Phase B and Hangingstone Phase I projects.[32]
- Ito et al. did further work at Hangingstone, such as resolving anomalous thermocouple readings by including a high-permeability zone laterally from the injection well. JACOS has used high steam injection pressures of 5,000 kPa with the intention of causing the oil sands to fail in shear. Steam chamber rise rates were significantly higher at Hangingstone than at the UTF. Early growth was reported to be 1.0 m/day as compared to 0.1 m/day at the UTF. The authors interpreted this to be due to geomechanical effects. Ito et al. (2002) concluded that elevated operating pressures, as high as the fracture pressure, appear to be required to achieve feasible growth of the SAGD steam chamber.[33]
- McLellan et al. (2000) used geomechanics to study the caprock integrity of the SAGD process, rather than the SAGD process itself. Here, a coupled reservoir/geomechanical simulation was done to examine breakthrough by steam through a caprock intersected by a weak discontinuity.[34]
- Wong and Li (2000) developed a stress-dependent model for permeability changes in oil sands due to shear dilation. Their model differs from other approaches in that they predict the permeability changes in one direction under continuous shearing. The result is an anisotropic permeability change.[35]
- Settari, Walters, and Behie (2000) described their geomechanical reservoir modeling of diverse problems such as injection into oil sands, compaction problems in the North Sea, and brine disposal. They concluded that geomechanical modeling was feasible on a full field scale.[36]
- Denbina et al. (2001) concluded that geomechanical effects were essential in matching foamy oil behavior, using permeability enhancements

(made functionally dependent on local pressure drop in the simulator) to provide a realistic method to model wormhole growth. They also found that suppressing gas relative permeability—to simulate gas entrainment—was essential in abating pressure decline and maintaining a high overall system compressibility.[37]

Geomechanical behavior

Oils sands are a frictional material; i.e., they derive their mechanical strength from the frictional resistance of the sand grains. Because they are uncemented by calcareous or siliceous adhesions at sand grain contact points, they are extremely weak when unsupported by a confining stress. The bitumen, being a highly viscous fluid, is unable to provide any strength other than under rapid loading.

Coffee test (confining stress)

ARE uses an effective demonstration during presentations. A vacuum packed package of ground coffee is used. The grounds are a granular material and the vacuum packing provides a confining stress of atmospheric pressure. Vacuum packed coffee is generally packaged in a form similar to a brick. Most people can stand on one of these bricks. There will be some deformation with a person standing on it; however, if you take a pair of scissors, or a knife, and puncture the package, the "brick" will immediately collapse into a loose pile of ground coffee that cannot support any weight without deforming completely. This is the effect of confining stress.

Effective stress

Since the oil sands are a frictional material, its strength is highly dependent upon the effective confining stress. Effective stress is the portion of the total stress in excess of the fluid pressure.

$$\sigma' = \sigma - p \qquad (19.3)$$

where

σ' = effective stress
σ = total stress
p = fluid pressure.

The general concept is shown in Figure 19–21. Stress is a vector quantity, therefore for the three principal stresses, σ_1, σ_2, σ_3, there are three effective stresses, σ'_1, σ'_2, σ'_3. Strength and stiffness are functions of the effective confining stress, σ'_3. As the effective confining stress increases, the strength and stiffness also increase.

Soil mechanics

The content of these comments is related directly to the response of oil sands to loading and relies on concepts developed in soil mechanics. Since this field of study is unique to civil and geological engineering, a brief description has been provided.

Interlocked structure

The fundamental control of permeability is the structure of the sand. In typical bitumen deposits, the sands are fine grained and the grains of sand are deposited with the long axes of the grains lying horizontal. In many cases, the structure of the sand has been the result of deep burial in the past, rather than glaciation, and results in compaction and some diagenesis. The net result is that the sand structure consists of a matrix of interlocked grains.

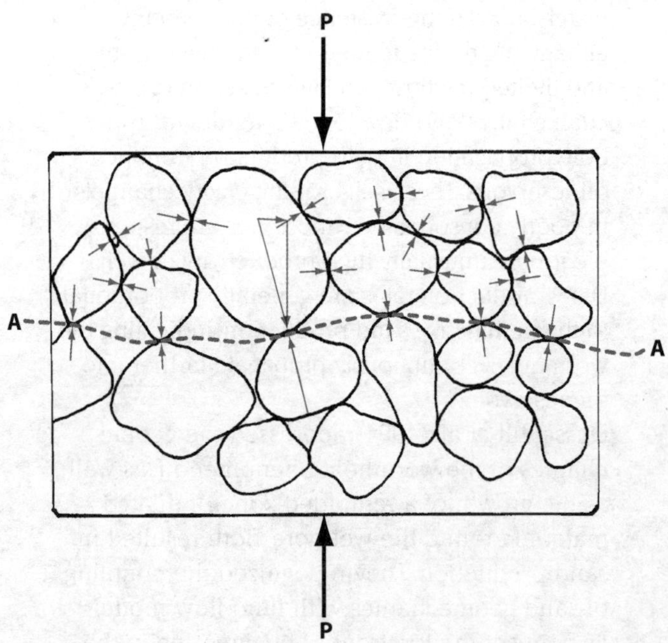

Fig. 19–21 Concept of Effective Stress

When subject to loading, the possible intergranular behavior includes:

- grain override
- grain shearing
- grain rotation
- grain compaction/expansion

Because of this granular structure, these noncemented (unconsolidated) sands do not have any substantive tensile strength. Further, determining the bulk properties of granular material requires that the original structure of the grains not be disturbed. This requires that samples be carefully obtained and handled.

To date, no reliable methodology has been developed to reproduce the same sand grain structure once it has been disturbed.

Material testing

Methods were developed in civil engineering to quantify the strength of unconsolidated samples. The two methods developed were the direct shear apparatus and the triaxial cell. Basic depictions of these two apparatuses are shown in Figures 19–22 and 19–23.[38]

Drained and undrained tests

These tests can be run in two fundamentally different ways. In the first method, pore fluids are allowed to escape, which is termed a drained test. In the second method, the pore fluids are not allowed to dissipate. This is termed an undrained test. This has a profound effect on the behavior of the tested materials. In an undrained test, pore pressures will initially support the majority of the incremental load. In a drained test, sufficient time is allowed for the pore pressures to dissipate during loading. The speed of drainage is a function of permeability.

Typical responses

The results from these types of tests are shown in Figure 19–24 on the upper left side. The test is controlled by strain, and the loads are read from a strain cell. Otherwise, samples will explode if controlled by load when the material softens. Note that Δl represents strain, which is horizontal for the first apparatus and vertical (axial) for the second apparatus. In a similar fashion, the Δh represents either vertical dilation or circumferential dilation for the two different apparatuses respectively. In the generalized figures, a clear distinction is made between loose and dense sands. The dense sands show a peak behavior, after which their load-bearing capacity is reduced. Loose sands do not show a loss in peak strength. Before complete shear failure can take place, the interlocking of sand grains in dense sand must be overcome in addition to the frictional resistance at the point of contact. After a peak stress is reached, at a low value of shear displacement, the degree of interlocking decreases, and the shear stress necessary to continue shear displacement is reduced. This is termed strain softening.

Dilation

The lower diagram on the left side shows the changes in volume, which result from shearing and dilation. The structure of dense sands expands, which has the effect of increasing porosity and permeability. The loose sands contract with strain, which has the effect of lowering porosity and permeability.

Failure envelope

Multiple tests can be combined to determine a failure envelope, as shown on the right side of the diagram. This is termed a Mohr-Coulomb failure envelope. This material is also presented in Figure 19–25.

Alternate representation of failure envelope

Mohr's circle can be used to define stress as a combination of shear and normal stresses for any plane. By using the properties of this relationship, it is possible to represent the state of stress with a single point. The construction is shown Figure 19–26.

The left or y-axis is p' and the lower or x-axis is the average stress. By using this construction, a stress condition can be described with a point rather than with the complete Mohr's circle. This technique makes it easy to plot the stress history (path) of samples or points in a reservoir simulation.

Stress path testing

The testing techniques described previously are usually conducted with a constant confining pressure. It is possible to conduct testing under different stress paths by varying the confining pressure by computer control. In this manner, the stress history of an element in a heavy oil reservoir can be approximated. This type of triaxial testing was used in the Touhidi-Baghini thesis (1998).[39] In these cases, the vertical stress (σ_1') divided by the confining stress (σ_3') can be plotted against axial strain. The maximum value of σ_1'/σ_3' is then considered to be the point of failure.

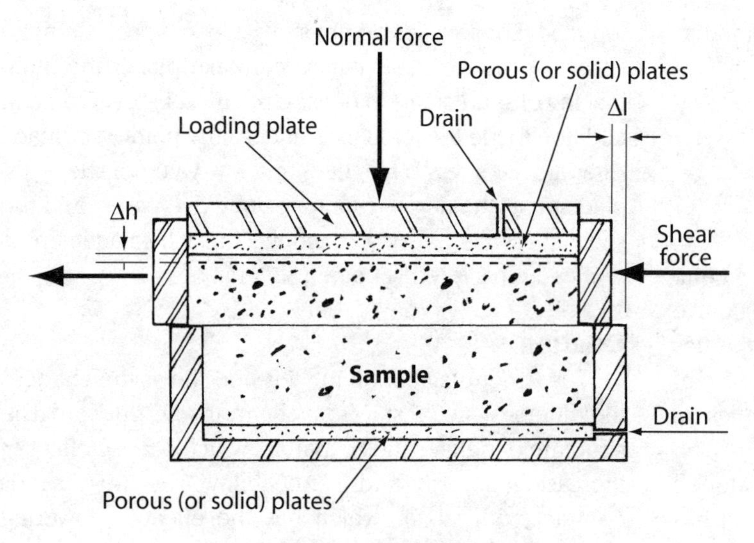

Fig. 19–22 Direct Shear Stress Apparatus

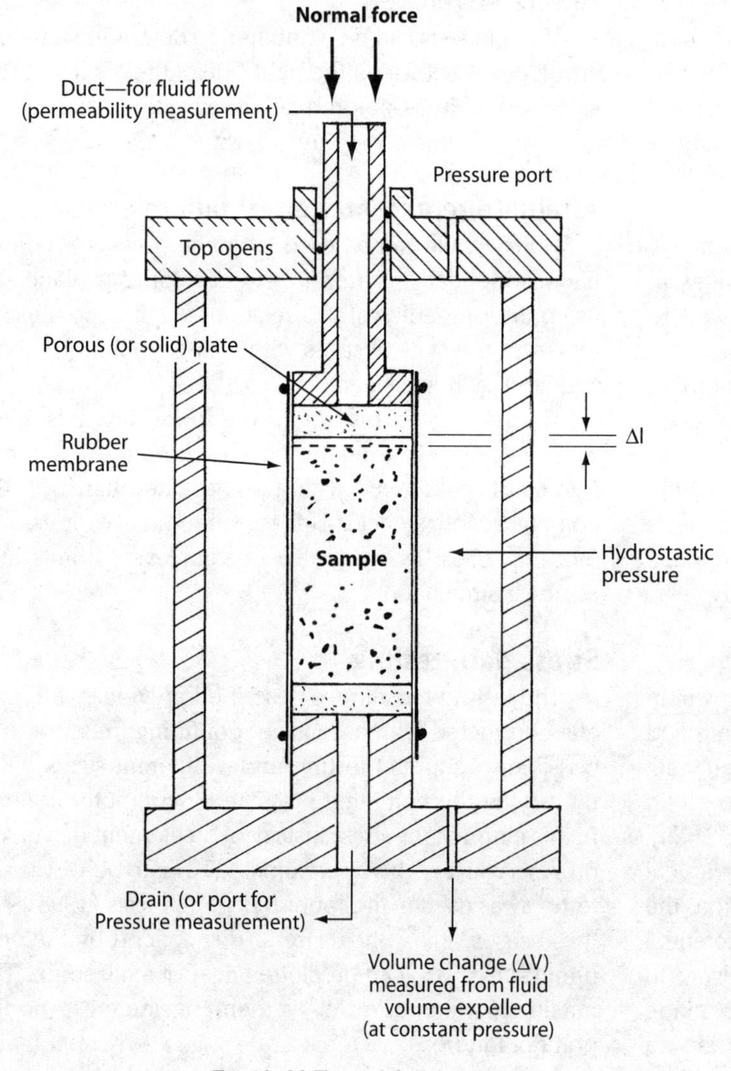

Fig. 19–23 Triaxial Cell Apparatus

Constitutive models of sand behavior

One of the major problems readily apparent from the previous description is how to describe the physical behavior of these sands. The use of the Mohr-Coulomb model to determine the failure point has been discussed. In fact, there are a number of different ways of describing how sands change. One model is called the elastic-plastic model; another is the hyperbolic model. These two are the most common methods used; there are, in fact, a number of others, which will not be discussed in this text. These methods concentrate on bulk deformation of the sand and how much load can be carried.

However, for reservoir engineering purposes we are interested in a little more. How does one tie the loads and the changing effective stresses (as a result of changing reservoir pressures) together with porosity and permeability? This is not easy. There are a number of ways to do this. One will be discussed in the following.

A triaxial test apparatus can be modified to simultaneously measure fluid flow through the failing specimen. In this manner, permeability can be related to stresses and axial and volumetric strains. Obtaining samples that are undisturbed can be very difficult.

Touhidi-Baghini (1998) and Scott (1997) studied the change in absolute permeability of oil sands while specimens were undergoing shearing. They tested a sample of nonbituminous McMurray Formation sands excavated from a natural river outcrop. Specimens were obtained parallel and perpendicular to bedding. An example of the permeability increase, relative to the initial absolute permeability, is given in Figures 19–27 and 19–28.[39, 40] At volumetric strains of 4%, roughly corresponding to the peak strengths of the samples, the vertical permeabilities increased by 100%. Some specimens had an increase in absolute permeability of

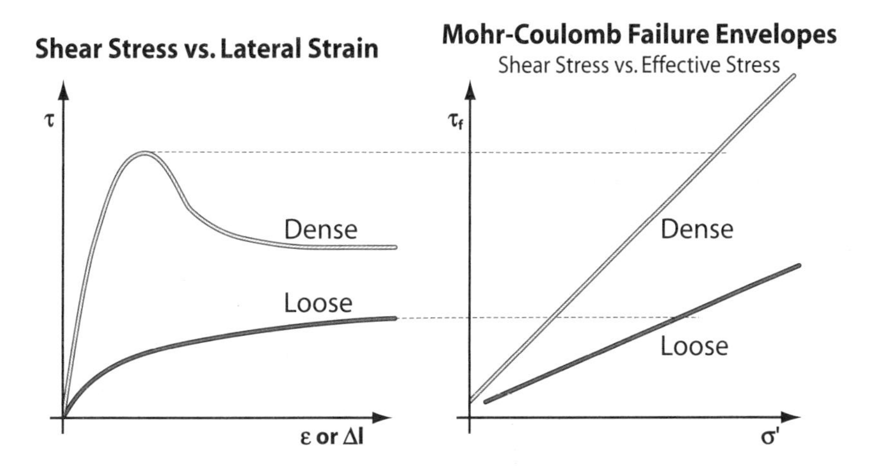

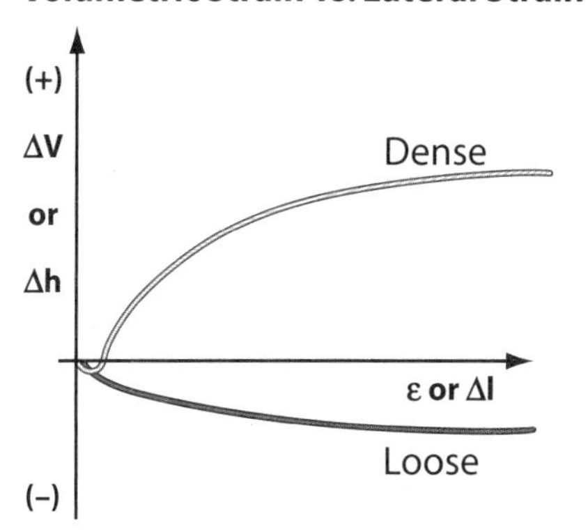

Fig. 19–24 Shear Test Results

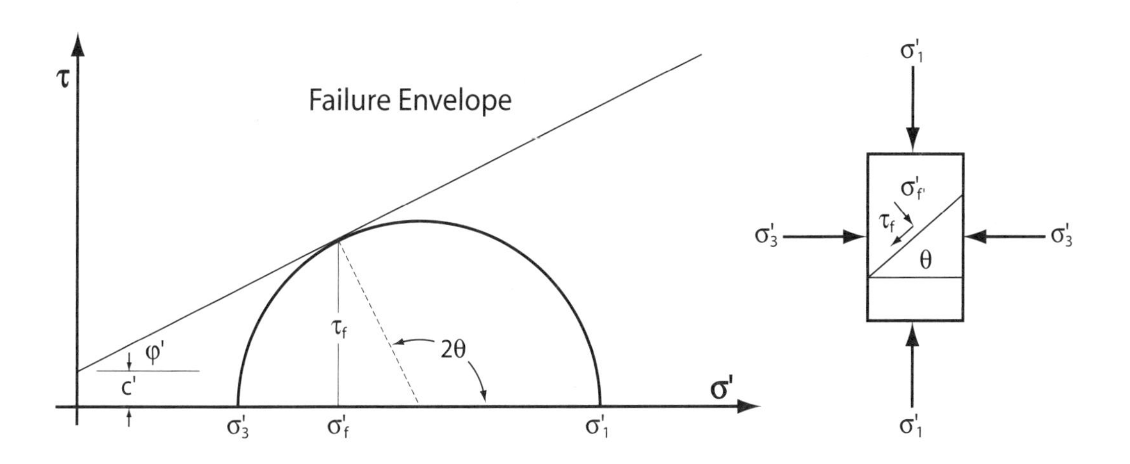

Fig. 19–25 Mohr-Coulomb Envelope

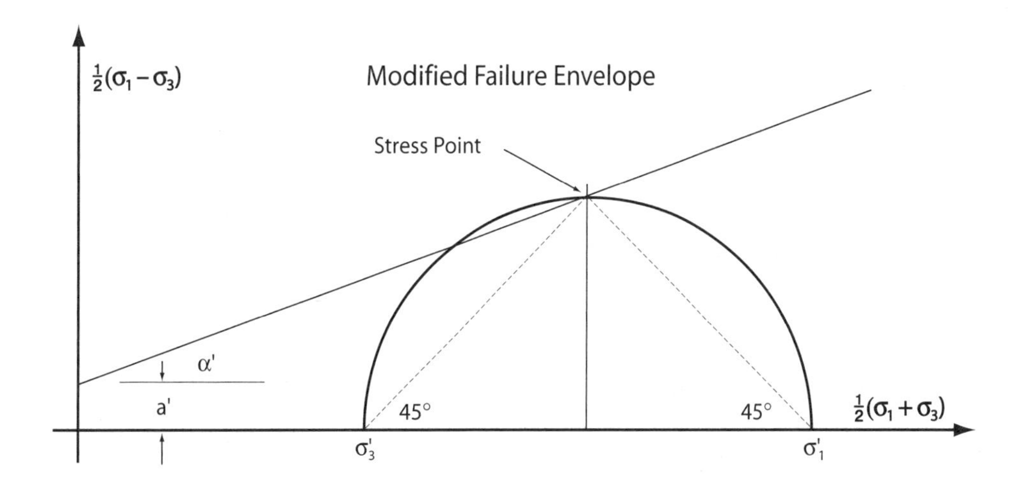

Fig. 19–26 Alternative Representation of Mohr-Coulomb Failure Envelope

one order of magnitude during the test. Permeability increases were attributed to dilation and shearing. Dilation resulted in a pore volume increase and widened flow paths, and shearing increased pore connectivity.

The horizontal/vertical permeability anisotropy was found to be a factor of 1.4. For a 2% volumetric strain, the vertical permeability increased by 40%, and the horizontal permeability increased by 20%. This increase in vertical permeability has extremely significant implications for the advance of the steam front in SAGD.

Due to the unique sampling process used, this is probably the most accurate data available on how permeability varies with shearing.

The previous data considered the case where the bitumen is heated and may flow. However, shearing can occur in the cold bitumen. In this case the bitumen remains immobile. The change in the volume of sand and porosity still causes changes in fluid flow properties.

According to the data of Oldakowski, which is shown in Figure 19–29, the permeability to water is also known to increase with shearing.[41] This data differs from the previous data in that different stress paths were used and these tests were performed at *in situ* cold reservoir conditions. *The increase in water permeability is substantive with shearing.* This leads naturally into a discussion of relative permeability.

Relative permeability

There is considerable debate in the literature regarding relative permeability curves for bitumen reservoirs. There are a number of relevant issues:

- whether the shape of the oil curve is convex or concave
- whether the curves are temperature sensitive
- whether the curves are water or oil wet, which is reflected in the level of the water relative permeability

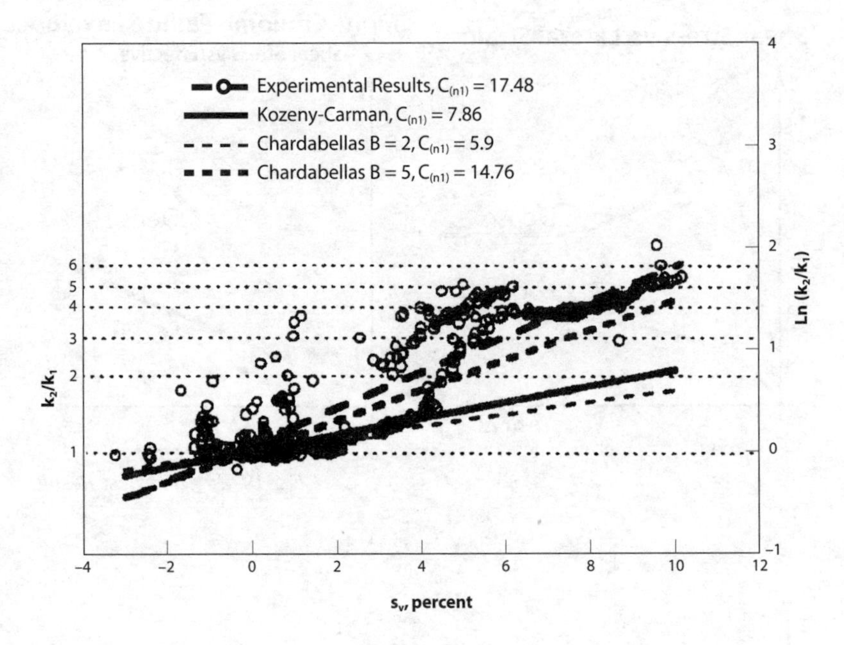

Fig. 19–27 Effect of Shear on Absolute Vertical Permeability

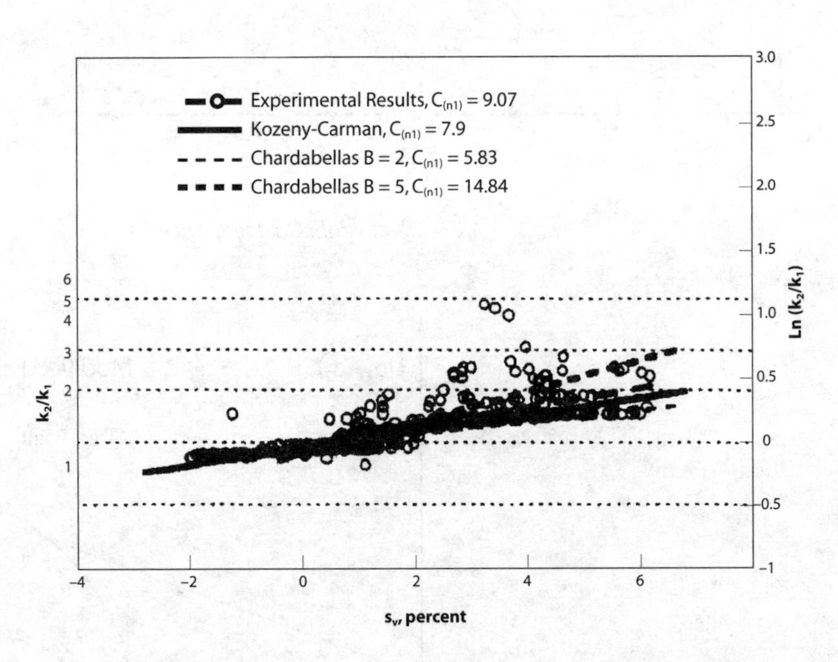

Fig. 19–28 Effect of Shear on Absolute Horizontal Permeability

at the residual oil saturation end-point as well as the value of the irreducible water saturation

- whether there are hysteresis effects associated with changes from drainage to imbibition

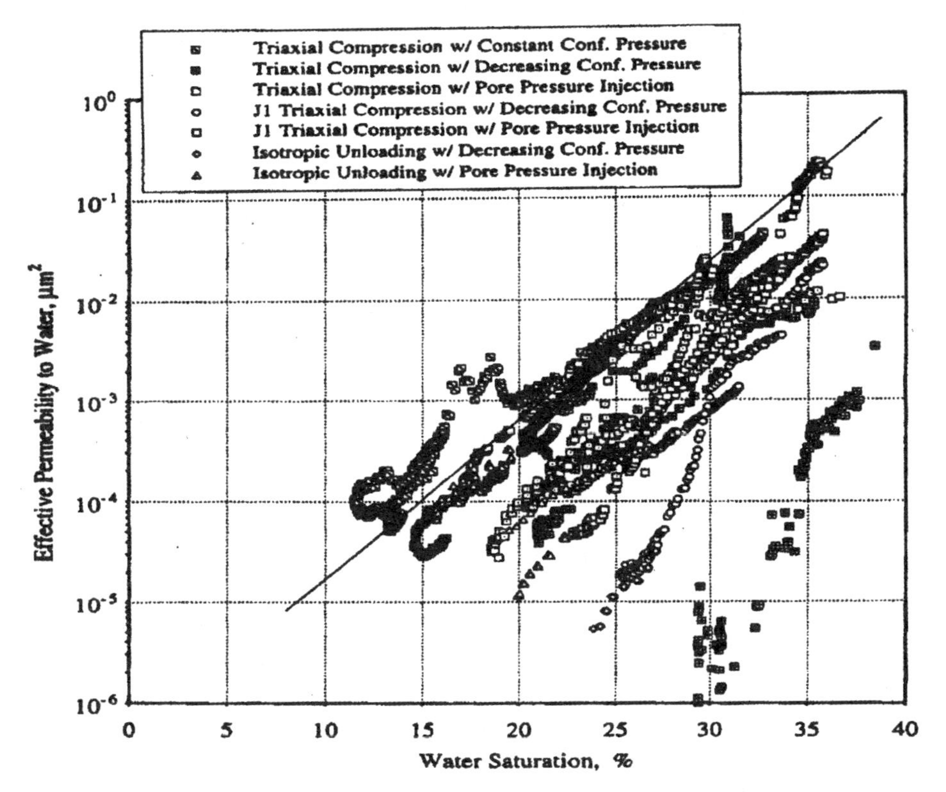

Fig. 19–29 Effect of Shear on Water Permeability—Oldakowski

Requirements for accurate relative permeability data

Obtaining accurate data is very difficult and expensive. There are a number of fundamental problems:

- The tests must be run at high temperature. Hong states that the most important area for performance prediction is the high-pressure gas-oil relative permeability. For accurate measurement of residual oil saturation to steam, both inlet and outlet must be maintained at wet steam conditions. This is certainly consistent with the observations from the Hangingstone hot water/steam flood test.
- For unsteady state tests, maintaining such conditions does not represent a long period and temperature drift is more easily controlled. However, the procedure for steady state tests takes considerable time and maintaining consistent conditions is more difficult. Therefore, most tests are done as unsteady state tests. Displacement (unsteady) tests do not measure relative permeability directly and require considerable interpretation.
- Some of these interpretation methods are graphical, which can be quite sensitive to make correctly and can create substantial error. More recently, there is a trend to use numerical simulation to history match relative permeabilities. Unfortunately, it is necessary to assume the shape of the curves in this process, which is one of the unknowns. In fact, this methodology results in multiple solutions. Steady state data is substantially superior, but it is expensive. Such tests probably represent less than one-third of the available data.
- Capillary pressure end effects can distort relative permeability curves. There are two solutions to this. First, use longer core samples, and second, use high displacement rates. Normally, the high rates required are significantly above reservoir displacement rates, which can lead to unrepresentative data and to viscous fingering. The later is a significant issue for heavy oils.
- Wettability is also an important issue. Cores need to be cut with bland drilling mud systems. Further, when dealing with noncemented cores, great care must be taken to ensure that disturbance is minimized and that the loading conditions will be replicated. This includes overburden loading and effective stress changes.
- The chemistry of the oil, particularly the polarity of the components, can also have a strong effect on relative permeability. Good quality live samples—i.e., with solution gas entrained—are required.

It is impossible to solve all of these conditions at once, although one can design around the majority of them. Such a program is very expensive and rarely done.

Three-phase relative permeability

Typically, three-phase relative permeability has been calculated using a correlation. The correlation is a variant of those developed by Stone. Although a standard practice, the correlations have been assembled based on a limited number of available data sets. None of these included heavy oil samples. Abou-Kassen and Farouq Ali make the following comments:

> *Three-phase relative permeabilities are the Achilles heel of reservoir simulation. This is more so for thermal simulation. Relative permeabilities measured in the laboratory are questionable in the light of instability theory (viscous fingering). Frequently, the relative permeabilities become a history match parameter. The laboratory value of end points are still of value. In thermal simulation, the temperature dependence of relative permeability should be taken into account. The same can be said of hysteresis. In many cases, two-phase relative permeability data is used in some variation of Stone's model to obtain three-phase relative permeabilities. Myhill has referred to the problem of obtaining the low oil saturations behind the steam front using Stone's model. Once the end points are obtained, it is easy to use the Naar-Wygal-Henderson equations. Kaeraie gives a modified version of these, suggesting that one way of including temperature dependence is to make the end points in the equations temperature dependent. Relative permeability hysteresis should be included in the simulation of cyclic steam stimulations. This can be accomplished in several ways; a simple approach was discussed by Bang. Sato has discussed the role of relative permeabilities on thermal simulation.*[19]

Again, the low saturations behind the steam front suggest that the gas- (including steam) bitumen relative permeability set is critical to residual saturations.

Laboratory scale modeling

Butler and Chow used straightline pseudo relative permeability curves to history-match some lab experiments. Their conclusions indicated that SAGD simulation was insensitive to the choice of relative permeability curves. They referenced Dake in the development of these curves.

The key assumption in the use of the VE technique is that the height of the capillary pressure transition zone is much smaller than the vertical dimension of the grid block. Although most capillary pressure lab data overestimates the thickness of transition zones, most logs indicate transitions that are at least one meter thick. Since the capillary transition zone is the same size as the grid blocks (particularly for a lab experiment), it was inappropriate to use segregated relative permeability curves. Dake outlines the use of pseudo relative permeability curve development with finite capillary pressure transition zones.

Many of the water bitumen relations are quite straight in comparison to conventional light oil relative permeability curves. In this case, the history match likely worked simply because the curves closely resembled what the lab curves would have looked like—had they been obtained.

Sasaki, Akibayashi, Yazawa, Doan, and Farouq Ali experimented with different relative permeability curves to match some of their scaled laboratory models. As shown in Figure 19–30, they found that the shape of the steam chamber was strongly affected by the choice of relative permeability curves. Their work utilized a number of different assumptions. They strongly disagreed with Butler and Chow's conclusions that SAGD is insensitive to the choice of relative permeability curves. The varying shape of the chambers would make this point obvious. However, their best match utilized straightline curves. Again, this may represent rock properties.[42]

It is unfortunate that the latter authors did not separately investigate the effects on the gas-bitumen and the water-bitumen curves.

For conventional light oil modeling, using straightline curves will often work, is an accurate representation of reservoir physics, and can be substantiated by detailed modeling with multiple layers. These curves are rate sensitive and often do not apply in predictions. At higher rates, segregation may not occur to the same degree, and the capillary pressure transition often is not accounted for in this process. It may be concluded that conditions for using VE should be proven before they are used for predictions.

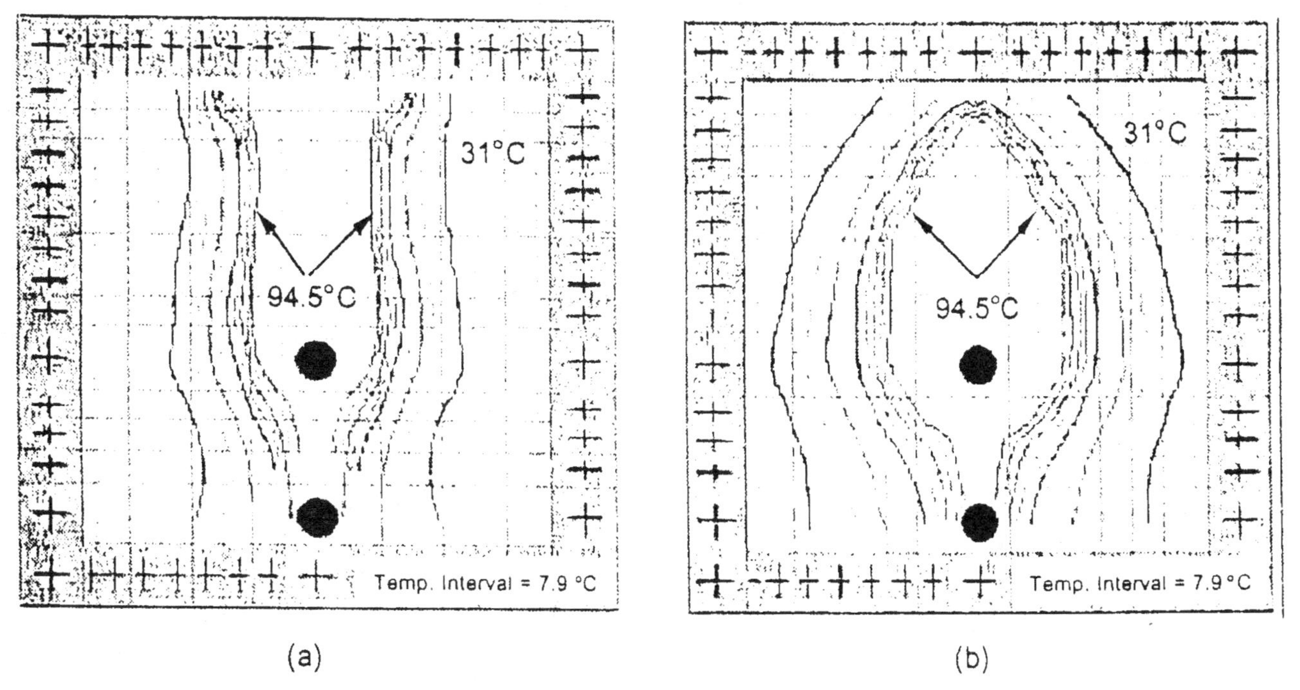

Numerical simulation temperature contours (at τ= 550 min)
with linear relative permeability functions
(a) with zero endpoint saturations, (b) with non-zero endpoint saturations

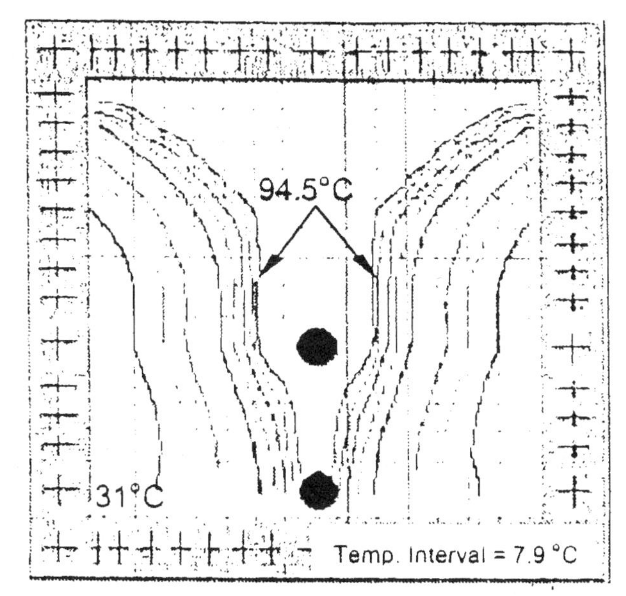

Numerical simulation temperature contours(at t = 550 min)
with non-linear relative permeability functions

Fig. 19–30 Numerical Simulation Results with Different-Shaped Relative Permeability Curves

Importance of relative permeability data

Shen investigated the effects of capillary pressure, undulating horizontal wellbores, and the rates that would be obtained. In this study, the relative permeability curves used are as shown in Figure 19–31. This paper also discusses the importance of countercurrent drainage and capillary imbibition pressure thresholds. Earlier work by Edmunds indicated that sufficiently high capillary imbibition pressures could prevent the SAGD process.

Earlier work by Adegbesan utilized straightline relative permeabilities.

Kisman and Yeung modeled the Burnt Lake Oil Sands Lease. In this case, there was no experimental relative permeability or wettability data. They indicated that the UTF simulation was not particularly sensitive to relative permeability curves. However, they did conduct sensitivities to relative permeability curve end-points. The data they used is shown in Figure 19–32.[43] They performed some different sensitivities with different shaped curves. The extent of all of the modifications is not clear. However, the use of Stone's relative permeability correlations was predicated on the k_{ro} at the irreducible water saturation being equal to 1. Otherwise, unpredictable values may occur. Overall, the effects of their various runs indicate changes from worst to best case of only 10%. However, they do not seem to have changed the shape of the curves greatly.

It should be noted that Kisman and Yeung used a very low gas-bitumen residual oil saturation.

In the final analysis, it may be that geomechanics, numerical dispersion, and heterogeneity have more effect than the shape of the relative permeability curves. In ARE's experience, one thing stands out—the water relative permeability end-point is increased significantly. In conventional reservoir simulation, this is usually the effect of layering. Pseudo relative permeability curves are easy to generate for history matching. Unfortunately, they are not transferable from one reservoir simulation to another and are of limited to no use for predictions. It is always better to model the heterogeneities directly in the model with quantitative reservoir characterization.

It should also be pointed out that, in his textbook, Butler notes that he has observed fingering at the top of steam chambers, and this displacement is very likely to

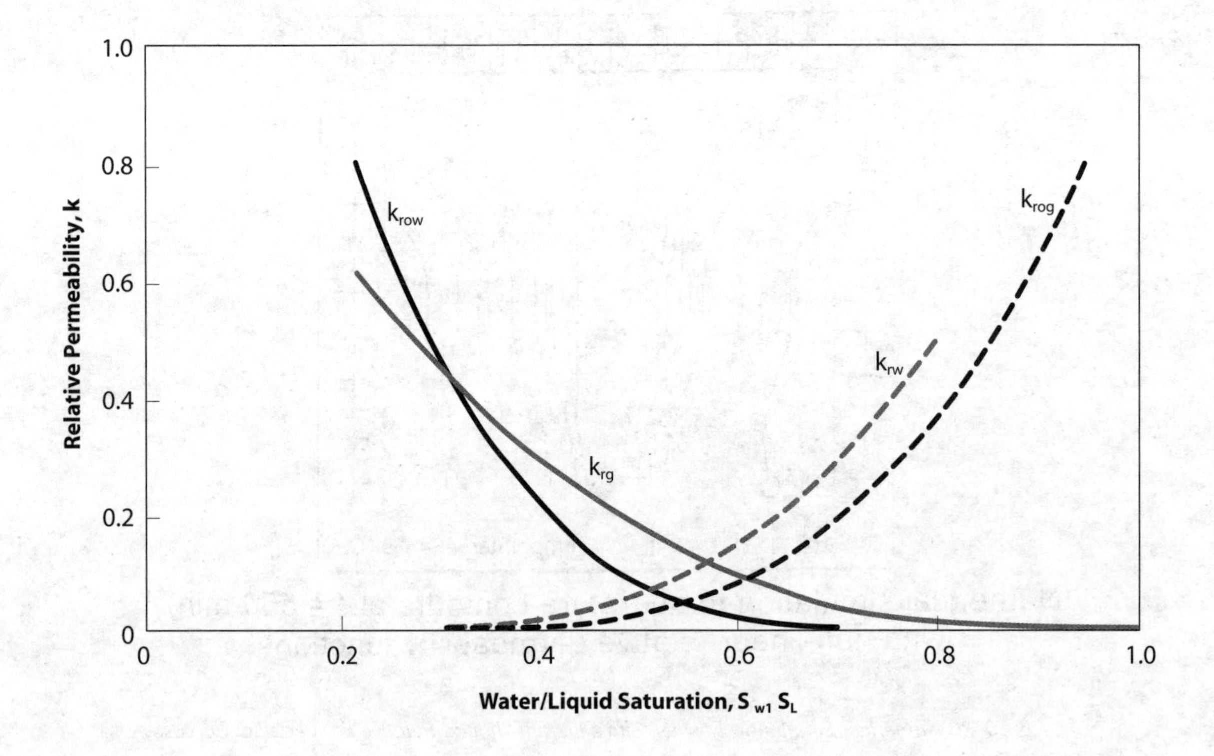

Fig. 19–31 Relative Permeability Data Used by Shen

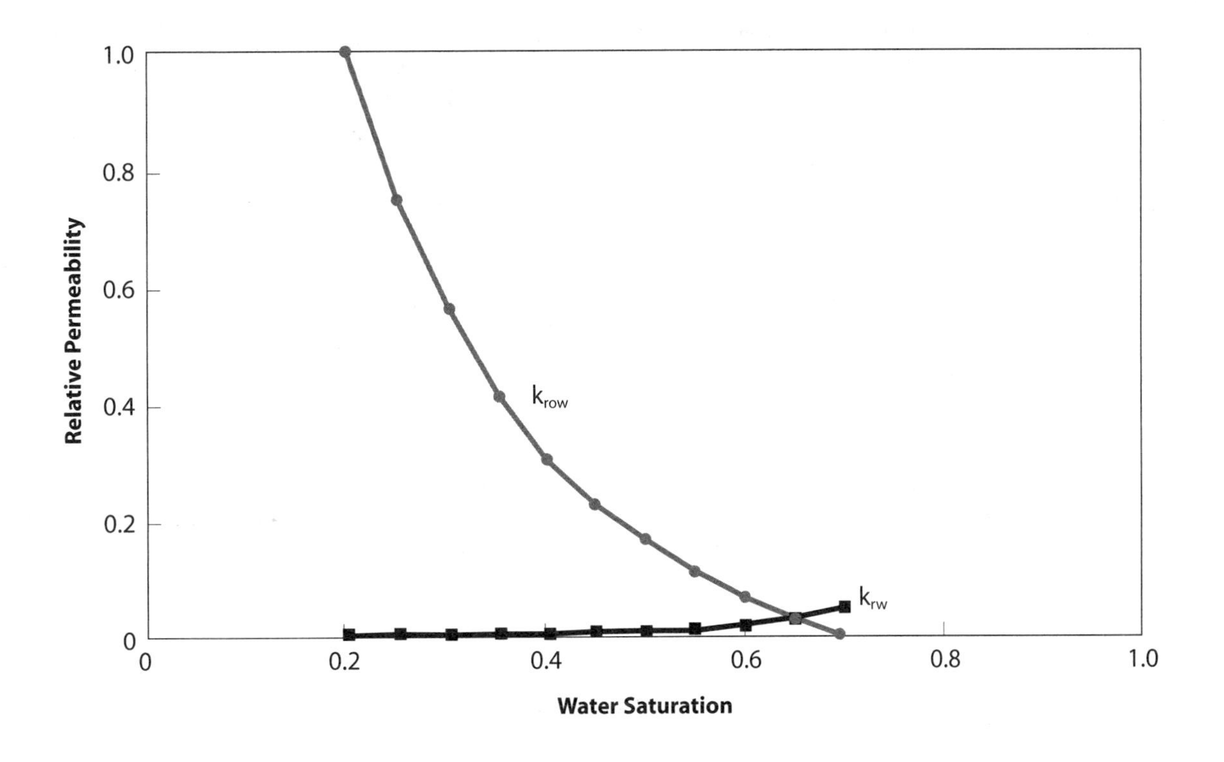

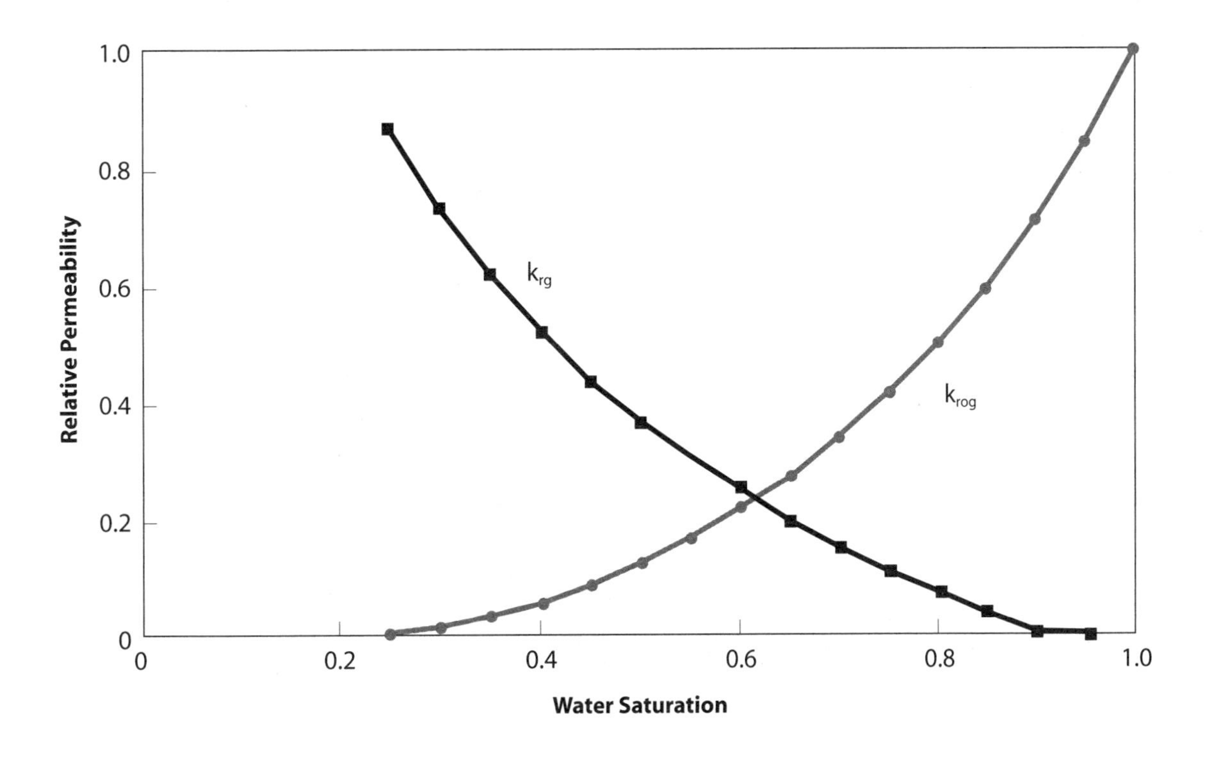

Fig. 19–32 Relative Permeability Curves Used by Kisman and Yeung

be unstable. This would be reflected in relative permeability curves that are history matched. The sides of the chamber appear to be inherently stable.

Relative permeability data available

There is a range of data outlined in a number of textbooks on classic heavy oil reservoir engineering.

1. IFP shows a number of curves, all of which are water-wet samples. Most of these are for cemented sandstones. These curves are shown in Figure 19–33.[44]
2. Exxon (Boberg) showed some curves used as demonstration examples and analyzes the sensitivity to different curves. These are shown in Figure 19–34. Exxon's data also includes gas-oil relative permeability curves.[6]
3. *Gravdrain* (Butler's textbook) showed curves for cemented samples, as shown in Figure 19–35.[5,45] Butler also showed some changes in relative permeability related to temperature. Note that not all researchers agree on this point. It is quite possible that some samples are temperature sensitive while others are not.
4. Maini and Okazawa showed relative permeability curves for silica sand that was packed into a core holder. These are shown in Figure 19–36. In this case, they inferred temperature sensitivity. However, these were unsteady state tests. The endpoint saturations were also forced to a common value, which may not be true. They discussed many of the limitations of history-matched relative permeability curves.[46]
5. Polikar, Puttagunta, DeCastro, and Farouq Ali also investigated the shape of relative permeability curves in "Relative Permeability Curves for Bitumen and Water in Oil Sand Systems." Their research used the steady state methodology and a noncemented sample. They found convex oil relative permeability curves as shown in Figure 19–37. Again, note the high S_{or} for the water-oil curve without any steam present. They compared

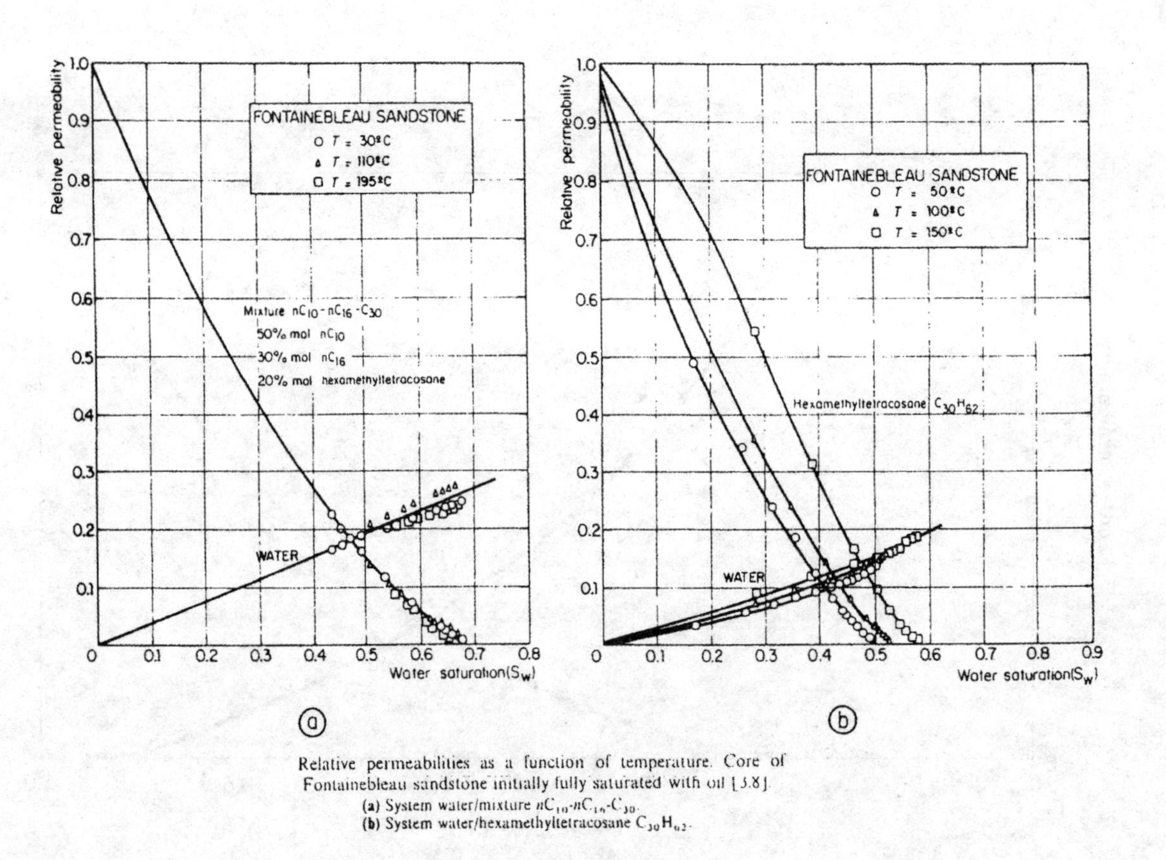

Relative permeabilities as a function of temperature. Core of Fontainebleau sandstone initially fully saturated with oil [38].
(a) System water/mixture nC_{10}-nC_{16}-C_{30}.
(b) System water/hexamethyltetracosane $C_{30}H_{62}$.

Fig. 19–33 (a, b) IFP Heavy Oil Relative Permeability Data

Effect of Temperature on Absolute Permeability

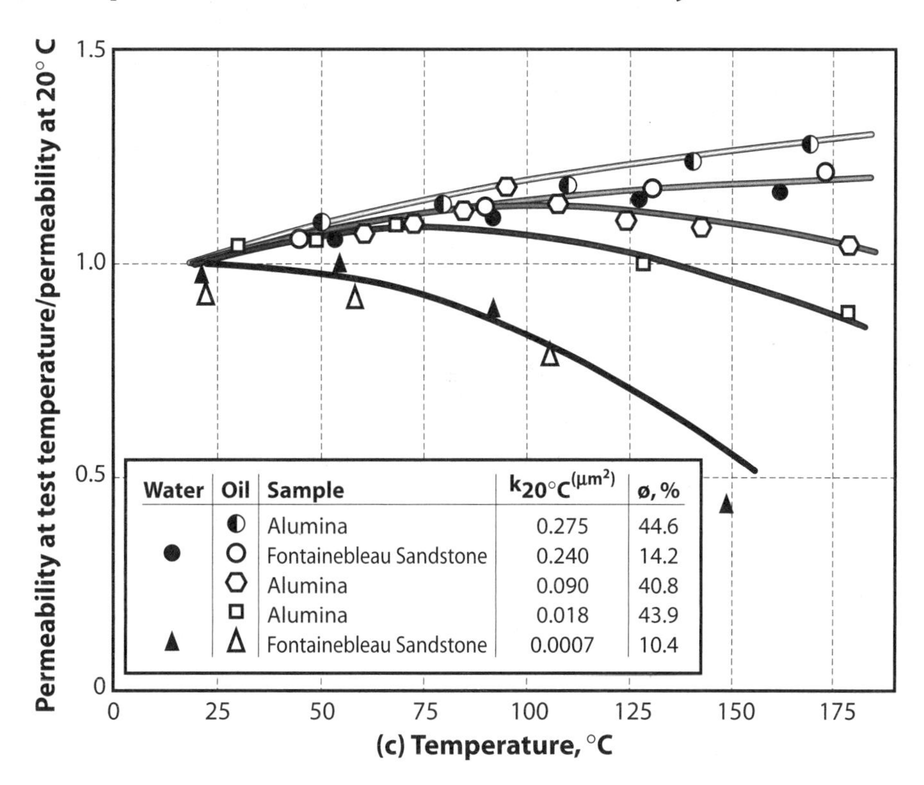

Water	Oil	Sample	$k_{20°C}$ (µm²)	ø, %
	◐	Alumina	0.275	44.6
●	○	Fontainebleau Sandstone	0.240	14.2
	⬡	Alumina	0.090	40.8
	□	Alumina	0.018	43.9
▲	△	Fontainebleau Sandstone	0.0007	10.4

Effect of the Effective Stress on Porosity and Permeability
at Two Different Temperatures

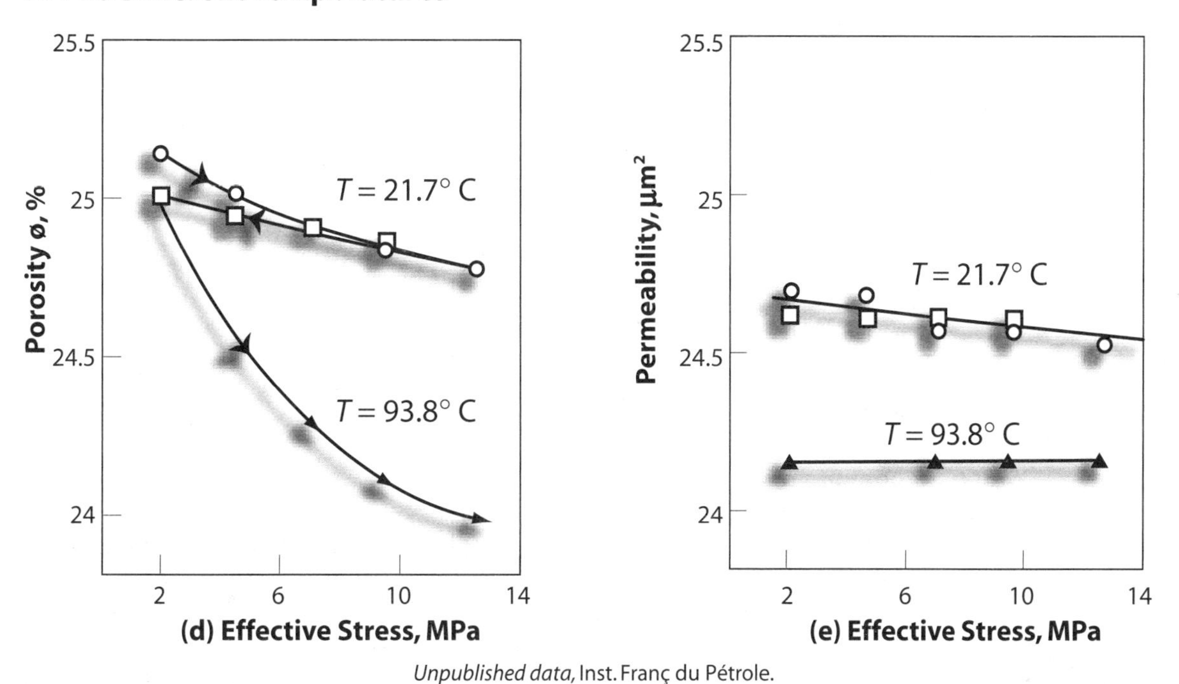

Unpublished data, Inst. Franç du Pétrole.

Fig. 19–33 (c–e) Heavy Oil Relative Permeability Data (Subsidiary Effects)

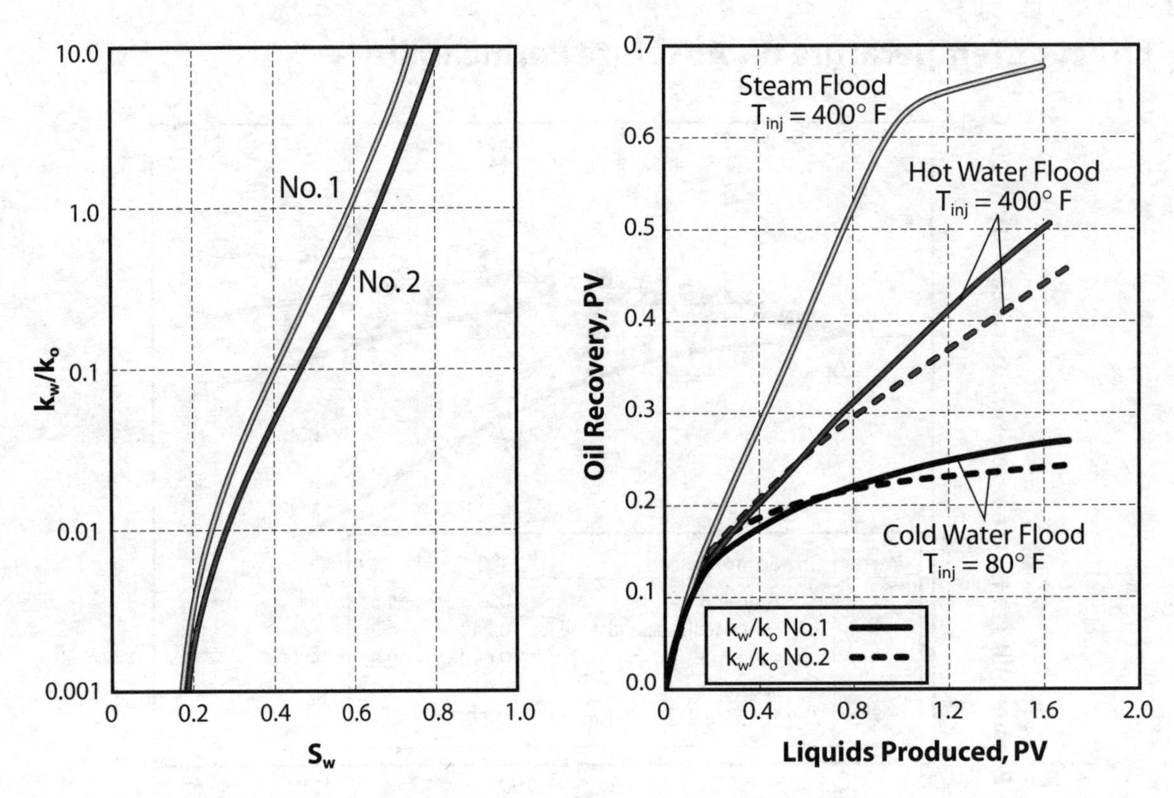

Fig. 19–34 Relative Permeability Data of Exxon for Heavy Oil and Effect on Process

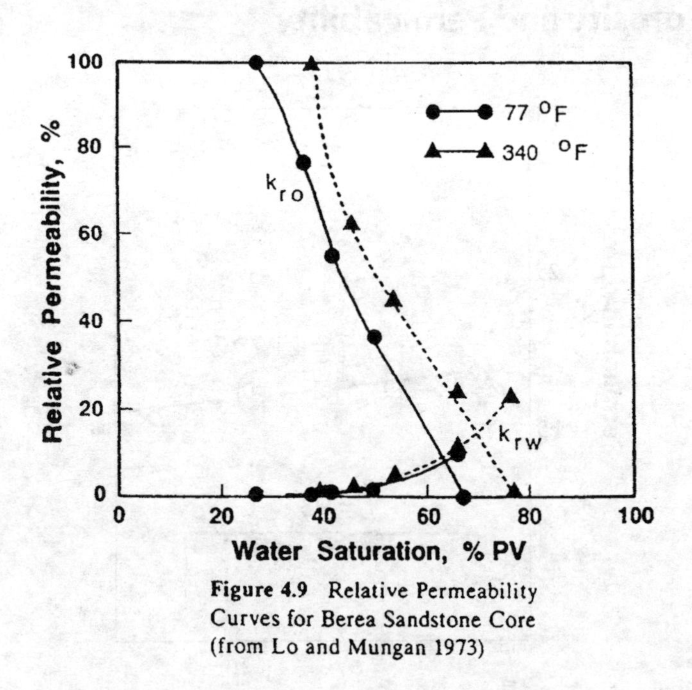

Fig. 19–35 Relative Permeability Used by Butler

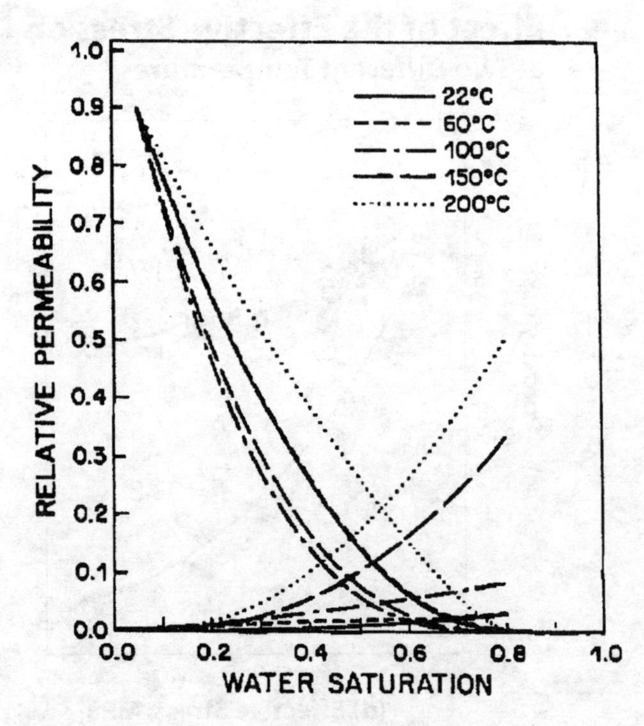

Fig. 19–36 Temperature-Dependent Relative Permeability From Maini and Okazawa

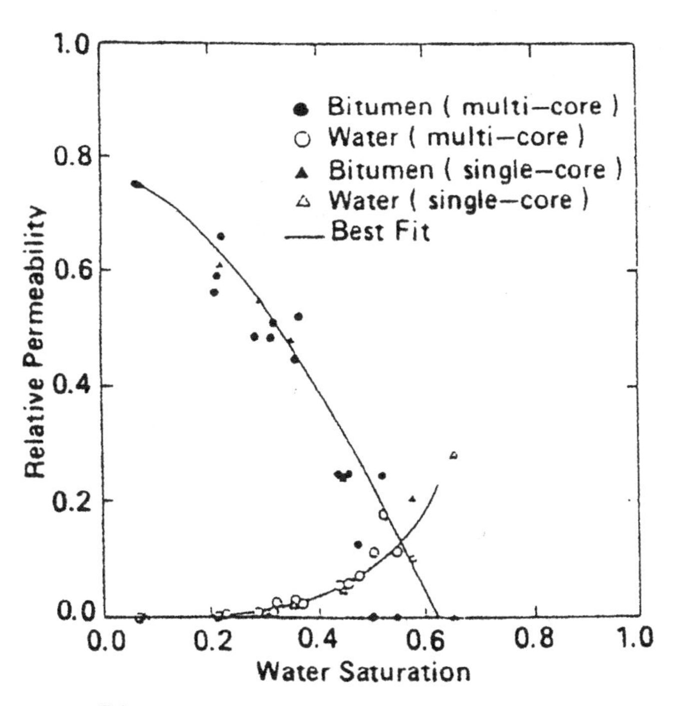

Bitumen-water relative permeability curves

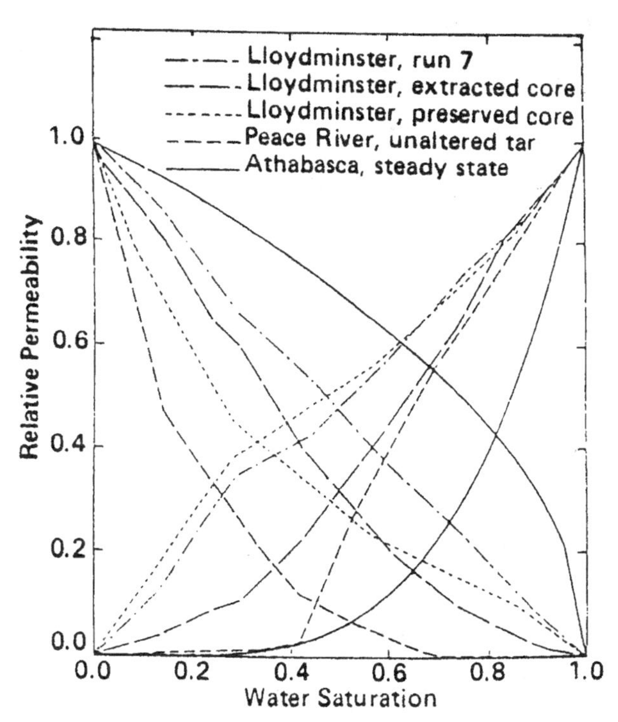

Normalized heavy oil-water relative permeability curves from experiments

Fig. 19–37 Experimental Curves From Polikar, Puttagunta, DeCastro, and Farouq Ali

the shape derived for a number of other different experiments. This underscores the degree of variation found in tests.[47]

6. Donnelly used the curves shown in Figure 19–38 for the Hilda Lake SAGD project.[48] His predictions were outstanding using this data. The initial water saturations were very high, apparently with no negative effect on oil production. His S_{or} on the gas-liquid relative permeability was very low, which, as described previously, is critical. His k_{ro} curve drops to zero at a S_{or} of 30%, and his water relative permeability curve extends beyond the S_{or}. This is somewhat unusual, as most k_{rw} curves would end at S_{or}. However, it would allow for simultaneous steam and water movements. The end of the k_{rw} curve appears to coincide with the likely end-point of the gas-liquid relative permeability curve.

Unfortunately, only two sets of the previous data included gas-bitumen relative permeability curves.

Hangingstone steam flood test

Earlier in the chapter, temperature dependence of relative permeability endpoints was discussed. Figure 19–39 shows results with these effects from the Hangingstone area of Alberta. This water flood/steam flood test dates from 1984. It was performed on a fresh state core plug as well as a companion sample. The companion sample was used to derive conventional reservoir properties, such as initial oil and water saturations.

The fresh state sample was placed in a hydrostatic core holder in a temperature-controlled oven. A temperature of 100° Celsius was used with a net overburden pressure of 600 kPa. This would be equivalent to a depth of approximately 300 meters.

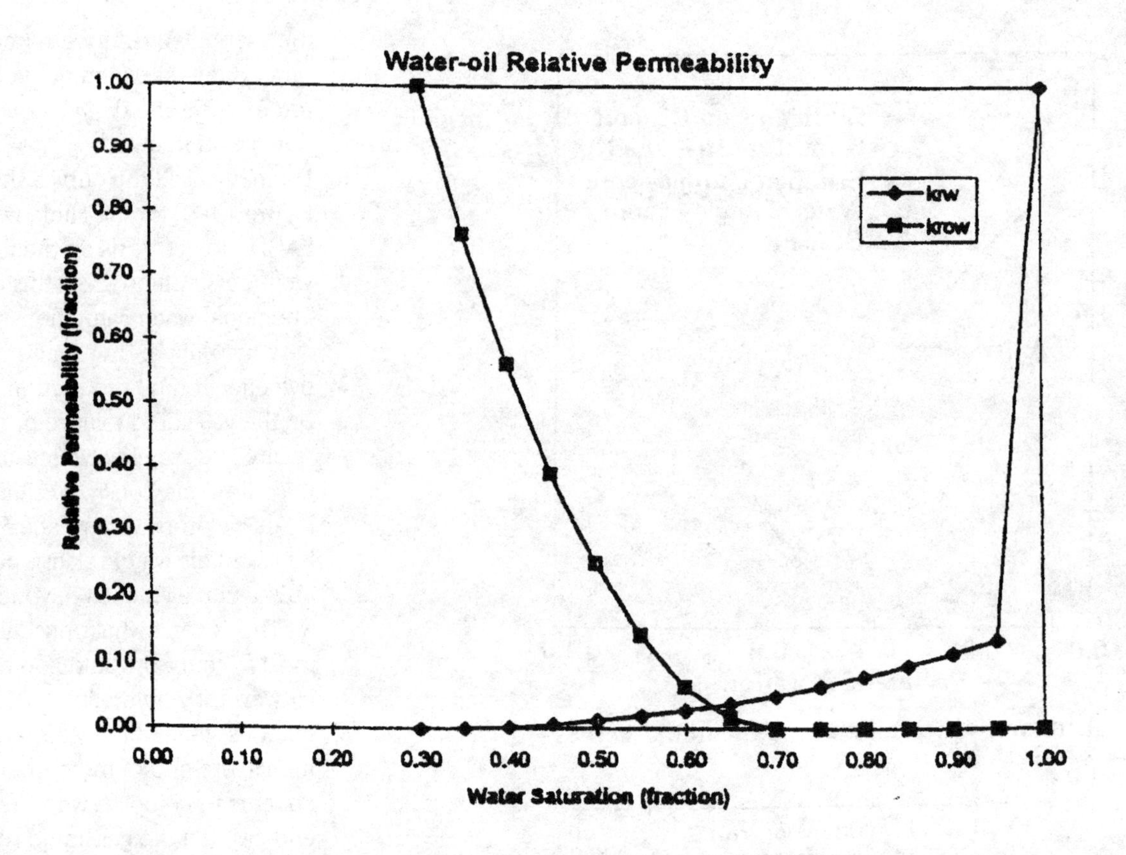

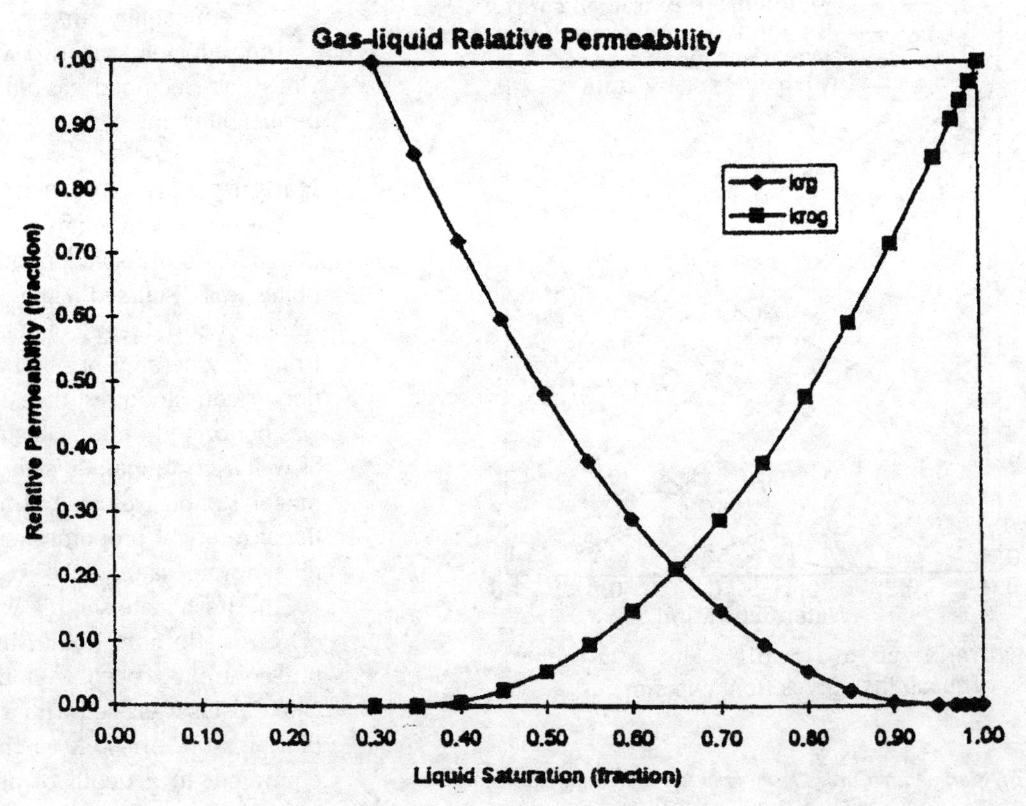

Fig. 19–38 Relative Permeability Curves used by Donnelly for Hilda Lake

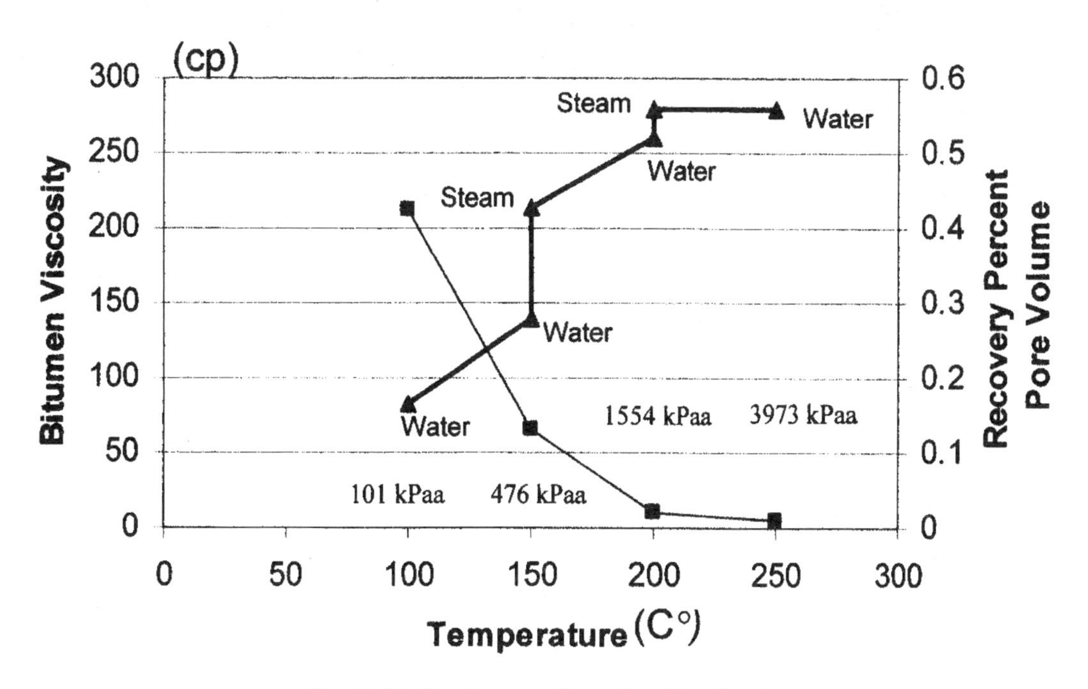

Fig. 19–39 Hangingstone Steamflood Test Results

The sample was flooded with hot water at 100° Celsius and 4,900 kPa to residual oil, and the permeability was determined. The oven temperature was then raised to 150° Celsius, the sample was allowed to equilibrate, and was again flooded with hot water at 4,900 kPa. The pressure was dropped to 450 kPa, and the sample was flooded with steam. The sample was flooded until oil production ceased and the permeability to steam was determined. Alternating water and steam was injected similarly at 200° Celsius with pressures of 4,900 kPa for steam and 1,500 kPa for water. This process was repeated once more at 250° Celsius with injection occurring at 4,900 kPa for water and 3,900 kPa for steam. The results are outlined as follows:

Sample	k_{air}	Porosity	S_{wi}	S_{oi}	S_{or}
P-6	5,860 mD	34.8 %			23.5
P-6a	6,190 mD	36.7 %	12.0	79.3	

The residual saturations were determined by Dean Stark extraction. Following this, air permeabilities, porosity, and grain densities were determined. Finally, the sample was resaturated with water and a liquid permeability determined.

Note that the companion sample saturations sum to 91.3% and not to 100%. With the summation of fluid method, only water and oil are measured. The likely source of the difference is air that had been imbibed in the core during unloading or gas that has been liberated from the bitumen. The sample on which the flooding was done was restressed, and this air or gas would most likely have been compressed out of the core. This makes the original *in situ* S_w 13.1%, the S_o 86.8%, and the S_{or} 25.7%.

Detailed analysis of recovery

It is interesting to analyze closely the recovery associated with the individual steps in the previous process. Recall that there was an initial displacement with hot water, followed by three stages at different temperatures in which hot water was first injected, followed by steam. Recovery data is summarized in Table 19–2.

The water sweeps an increment of recovery of 20.8% of the OOIP. At this temperature, Athabasca bitumen has a viscosity of more than 200 cp. When the temperature is increased, the viscosity drops to about 60 cp, and the water is able to displace an additional 14.4% of the OOIP. At this point, flooding with steam at the same temperature adds another 18.6%. Steam dramatically increases

Table 19–2 Summary of Water Steamflooding Test Results—Hangingstone Athabasca McMurray Sands

Temperature	Phase	Pressure	Incremental Recovery	Cumulative Recovery	Percent Recovery	Permeability
℃		*kPaa*	*% Pore Volume*	*% Pore Volume*	*OOIP*	*mD*
100	water	4,900	0.1650	0.165	20.8	70.4
150	water	4,900	0.1150	0.279	35.2	29.4
150	steam	450	0.1480	0.427	53.8	314.0
200	water	4,900	0.0920	0.520	65.6	30.1
200	steam	1,500	0.0380	0.558	70.4	321.0
250	water	4,900	0.0005	0.558	70.4	33.2
250	steam	3,900	0.0000	0.558	70.4	465.0

recovery. Increasing the temperature and flooding with water adds an additional 11.8% of OOIP. Switching to steam adds only a small increment of 4.8%.

A number of additional observations can be made. It seems that higher temperature steam adds to recovery up to 200° Celsius or 1,500 kPa. Conversely, running a steam chamber below 1,500 kPa will decrease recovery. One consideration in SAGD design is heat losses. Lower temperatures will reduce heat losses. However, this evidently comes at the expense of recovery. Temperatures above 200° Celsius and 3,900 kPa do not seem to add to recovery from a core flood point of view.

Another interesting observation is that the steam permeability increases from 321 mD to 465 mD when flooding is changed from 200 to 250° Celsius. Since there is no increment in recovery, it would seem that there should be no change in oil saturation. This is likely an effective stress effect. The intergranular pressures are reduced, with the higher pore pressures leading to lower effective stresses and higher permeability. It is also clear that there is a large change in permeability between the steam and hot water. Note that permeability is independent of viscosity, and, therefore, this would have to be a relative permeability effect. This would also indicate that ultimate recovery is more strongly affected by gas-bitumen relative permeability than water-bitumen relative permeability.

History match relative permeability curves

Much of the historical design procedure for SAGD projects was based on the work conducted at the UTF. These curves have become widely used. Originally, the relative permeability relations used by AOSTRA were derived on what AOSTRA has described as general experience. The author questioned a number of the people directly involved, and it does not appear that experimental data was available.

The history-match curves are, in fact, nothing more than pseudo relative permeability curves, a topic that has been discussed extensively in this book. As such, these UTF curves represent a number of different things including:

- numerical dispersion
- layering/heterogeneity
- viscous fingering
- capillary pressure transition zones
- production rates
- completion configurations
- geomechanical effects
- all in addition to the fundamental properties of the rock

Of course, many of these factors are unique to the UTF project. Therefore, the use of these curves provides little assurance that they are applicable elsewhere. This does not imply that the methodology used was incorrect. In fact, this was the limit of technology as it existed at the time.

A conference paper by Polikar, Puttagunta, DeCastro and Farouq Ali summarized a number of relative permeability curves used in simulation. The curves are shown in Figure 19–40.[49] Note that there is considerable variation in the curves. The degree of history matching and the grid-block sizes are not outlined. However, it does serve to show the variation in curves used. It seems that some degree of skepticism is indeed warranted.

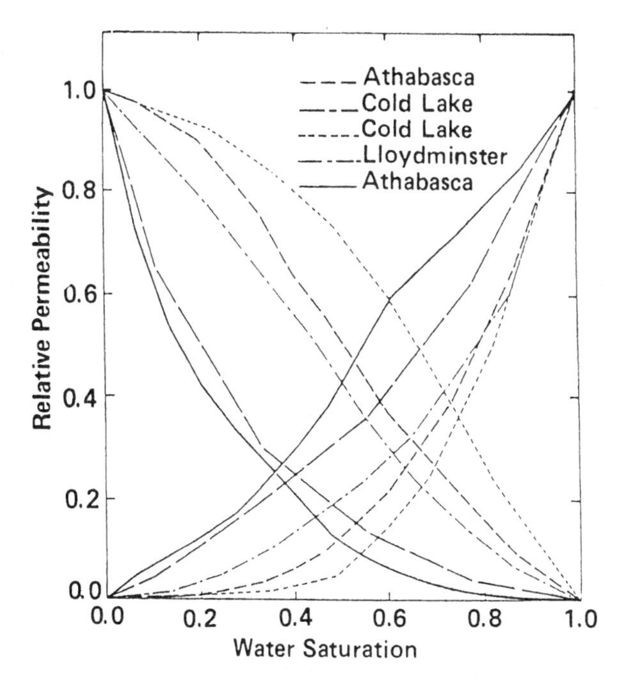

Fig. 19–40 Normalized Relative Permeability Curves Used in Simulation—Various Sources

Choosing relative permeability curves

A number of difficulties have been outlined previously. As practicing reservoir engineers, a decision must be made, for better or for worse. If a unique set of data cannot be identified, then a range of sensitivities should be performed. It should also be noted that, despite many of these uncertainties, thermal reservoir simulation has proven to be capable of predicting the performance of SAGD projects. This is well-documented in the literature.

Relative permeability end-point—connate water

Faced with the determination of the connate water saturation end-point, ARE summarized a variety of different data sources as shown in the following:

Hangingstone (an offset reservoir)	
P_c Data	8–18%, avg. 13%
Hot Water/Steam Flood at Hangingstone	13%
UTF (Offsetting Reservoir) History Match	15%
Polikar et al. (Technical Paper) Clean Sand	6%
Log Analysis	11%
Maini and Okazawa (Paper) Clean Sand	5%
Bennion et al. Preserved Core	
(Offset) Reservoir	3 and 5.8%

Most connate water saturations on cemented rocks are higher, normally in the range of 25%. Although such curves have been discussed, they will not represent accurately oil sands conditions of connate water saturations. The most accurate data in our view is log analysis prepared for the actual lease under study. ARE has rounded up the log-derived S_{wc} slightly to 15%, which is similar to the majority of existing models.

Relative permeability end-point—residual oil saturation

Polikar, DeCastro, Puttagunta, and Farouq Ali made an extensive study of noncemented sand in "Effect of Temperature on Bitumen-Water End-Point Relative Permeabilities and Saturations." They found little temperature effect. Note that the residual oil saturations in these tests, which do not include steam, were relatively high—averaging 46%.

Maini and Okazawa calculated all of their end-points at a residual oil saturation of 20%. Experimental end-points were not shown in their results.

Bennion, Sarioglu, et al. performed tests on stacked, preserved core material. In general, ARE has found that careful screening of stacked data in comparison to solid samples indicates core stacks are not reliable. The problem is the multiple capillary end effects that exist at sample junctures. This problem is exacerbated if the samples do not mesh exactly inside the core holder. Noncemented cores are less likely to cause this problem because the material will mold. Their results indicate water-bitumen residual saturations of 31.1% and 34.6%, respectively.[50]

Further guidance on this matter is provided by Farouq Ali/Butler (via previous Butler reference) as shown in Figure 19–41. Note that the bitumen viscosity of the Athabasca deposit is about 1.7 million cp at an initial reservoir temperature of 8–10° Celsius. This falls slightly off the diagram and to the left. There is no correlation between original reservoir bitumen viscosity at original reservoir temperature and residual oil saturation. There is a range or belt of values across the chart. Note that this is for a variety of steamfloods. Therefore, this would represent the effects of various degrees of water-oil and gas-oil relative permeability.

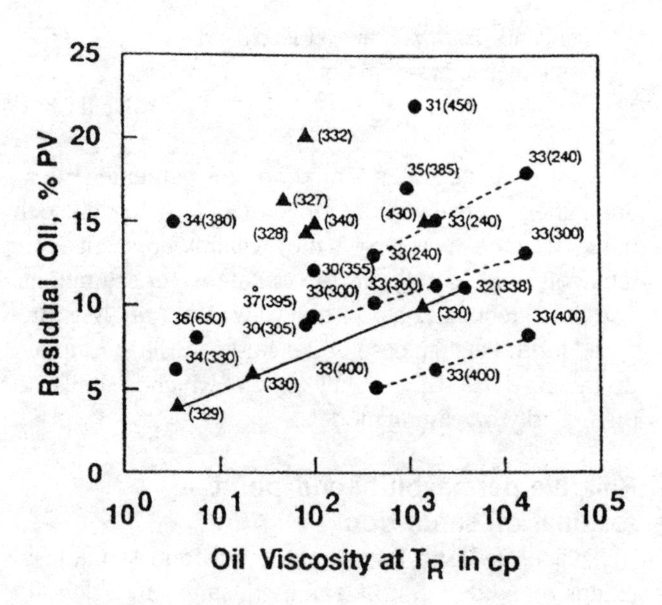

Fig. 19–41 Empirical Data of Residual Oil Saturation

Butler has presented calculations that indicate the residual oil saturations are lower for the SAGD process than for conventional steam flooding. He estimates a typical steam chamber to have a residual oil saturation of 0.10 at 2,000 days.

As outlined from the water flood/steam flood sample and other data, it is likely that the end-point saturation for gas-bitumen and water-bitumen are quite different. There are other considerations here:

- The Stone II phase relative permeability correlation was designed with a consistent S_{or}.
- In the SAGD process, fluid is free to drain vertically, unlike the lab tests to which frontal advance calculations have been calibrated. In fact, the process probably takes place across a thin zone rather than at a sharp interface.
- This displacement zone is probably of the order of one meter or less. This is less than the thickness of a single grid block. Therefore, the process will not be modeled using the one-meter grid blocks selected. The relative permeability curves have to account for this entire zone.

Accordingly, in past simulations, ARE has assigned a single value for residual oil saturation for both the oil-water and gas-oil relative permeabilities.

Steam-water relative permeability and capillary pressure

Steam-water relative permeability is more complex than oil-water permeability in that there can be mass exchange between the two phases. Experimental investigations of this have been done by Horne and indicate that low interfacial tensions and/or mass exchange do occur.[51] A case can be made for comparison with gas-condensate relative permeability where liquids drop out of the gas, and this alters the formation of relative permeability curves. This topic was discussed earlier. Thus far, there does not appear to have been much research in this area.

The results of Horne contradict some earlier heavy oil steamflood testing and have also been applied or developed for geothermal research. These issues seem logical and, like the temperature dependence of relative permeability, there may not be uniform or consistent results.

Geomechanical effects on mudstones

Geomechanical effects can be expected to occur in mudstones. As steam condenses on the underside of a mudstone, the hot water is free to penetrate the mudstones. This is not analogous to trapping gas. Gas accumulations are trapped by capillary imbibition pressures, not low permeability. Steam would be a nonwetting phase. In fact, as outlined in Bachu and Undershultz, mudstones and shales act as aquitards and are not barriers to wetting phase flow—i.e., water.

Since the water can penetrate mudstones, their permeability will also increase due to pore pressure effects as well as shear failure. These effects will be more pronounced where shear forces and dilation are the strongest, which normally would be near the injector, producer, and in the area directly above the well pair.

There is also hot plate heating above mudstones, which propagates by conduction. This reduces the viscosity of the oil above the mudstone and eventually leads to bypassing of the barrier as the oil becomes mobile.

Probably the most convincing evidence of this was demonstrated in the AOSTRA UTF Phase A, where a mudstone existed between the injector and the producer. Communication was established and this, in ARE's opinion, represents structural changes to the mudstone fabric resulting from geomechanical effects.

Numerical Difficulties with Thermal Simulation

Numerical difficulties with reservoir simulation for black oil models are relatively rare. This is not true, at least yet, for thermal simulations. The following are trouble spots the author has encountered.

Grid-block requirements

Thermal simulations are affected by the choice of grid size, as are conventional reservoir simulations. However, since the mechanism is somewhat different, SAGD simulations require much smaller grids than are typical for other processes. Based on sensitivities conducted by ARE, an optimal grid size is approximately 1 meter horizontally (*j* direction) by 1 meter vertically (*k* direction). In the cases described in the following, all simulations were conducted with one 500-meter grid block along the horizontal well (*j* direction).

There is a noticeable amount of error associated with this grid spacing. The accuracy of output would be on the order of ±10%. Figures 19–42, 19–43, and 19–44 show a grid sensitivity.

Nine-point difference schemes

Due to the high mobility of steam and the contrast with the low mobility bitumen, most thermal reservoir simulators include nine-point difference schemes to control grid orientation effects. The latter was discussed in chapter 2 of this book.

The implementations on the simulators used by the author permit nine-point differencing in only one plane. Fortunately, the plane can be chosen as either *ik* or *jk*.

On some of the simulations, this seemed to have no effect on the execution time. One might expect the more complex difference scheme to increase run time. On the runs that the author examined carefully, the results appeared to be essentially the same. Interestingly, the largest difference seemed to be in improved material balance, which was of significant benefit.

While the runs under discussion can hardly be called an exhaustive review, it does seem that the nine-point differencing was beneficial. The results that were obtained are shown in Figures 19–45, 19–46, and 19–47. The differences in material balance are shown in Table 19–3.

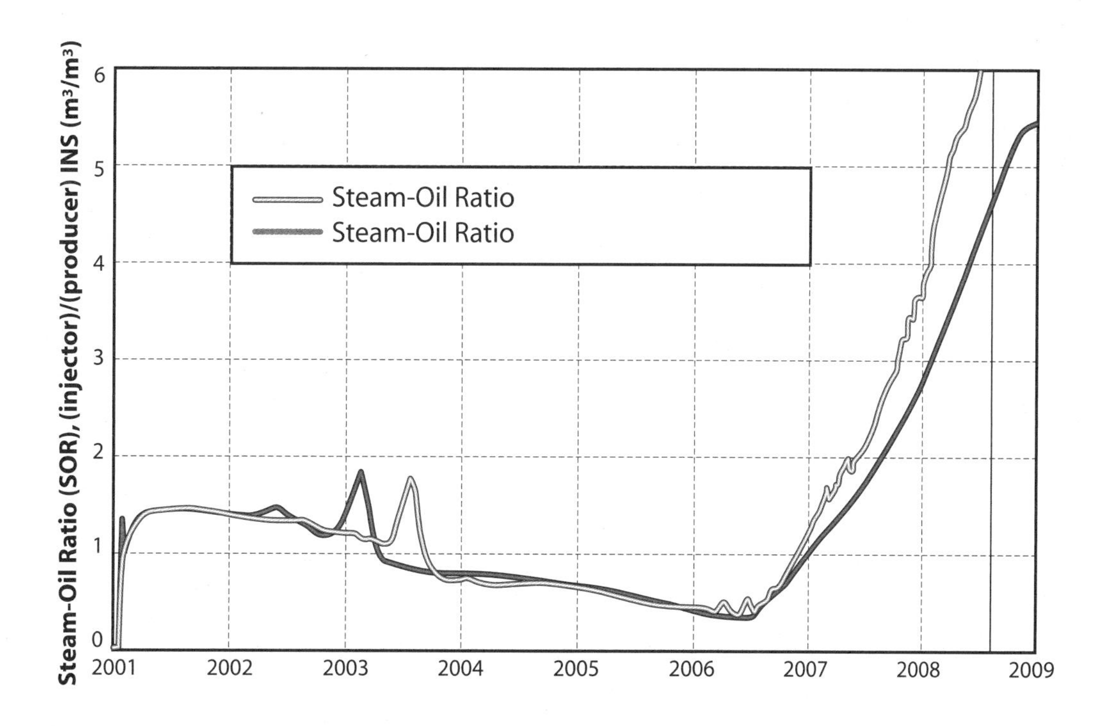

Fig. 19–42 Grid Sensitivity Steam–Oil Ratio (SOR)

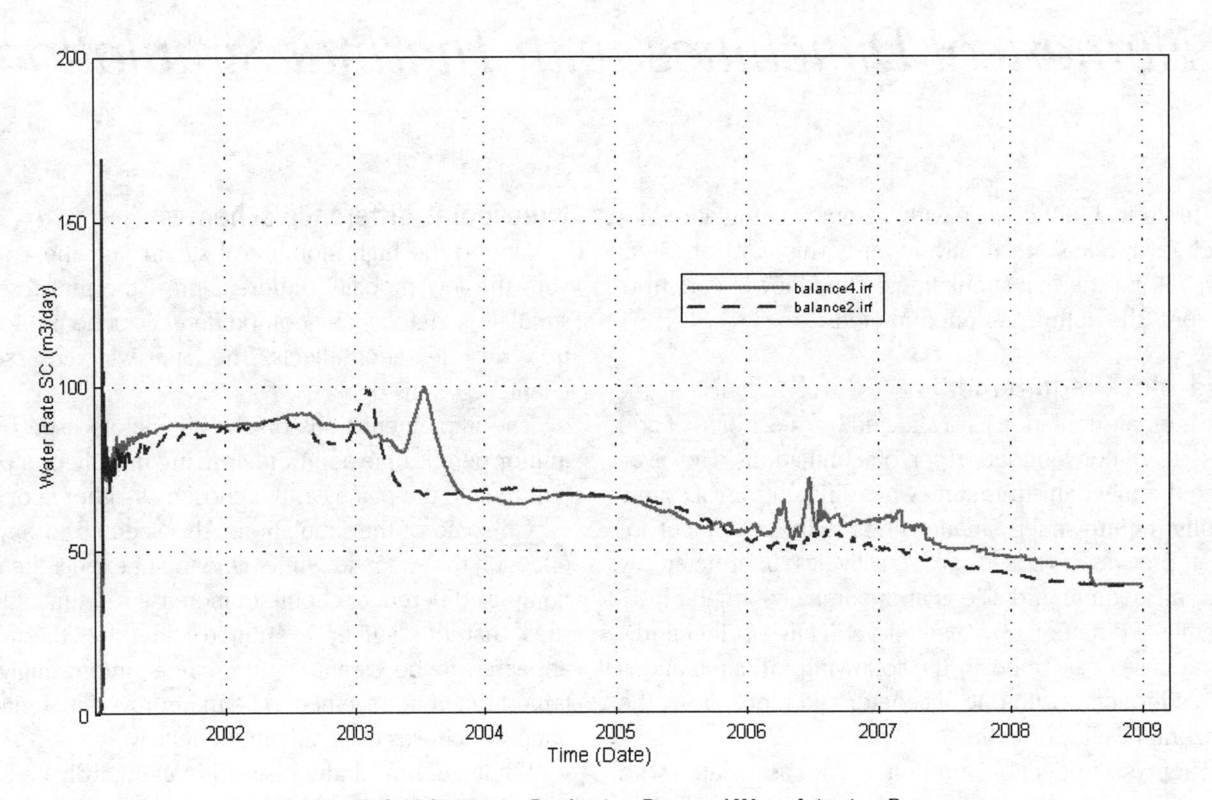

Fig. 19–43 Grid Sensitivity Production Rate and Water Injection Rate

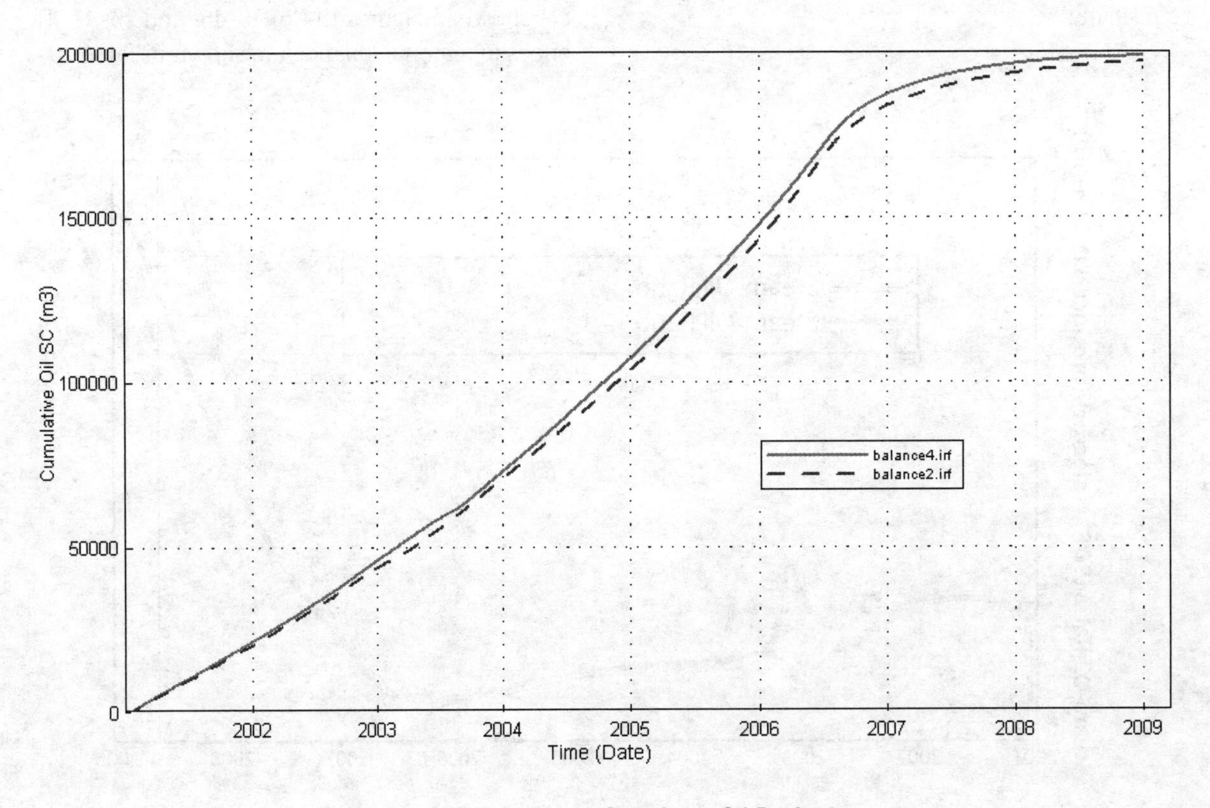

Fig. 19–44 Grid Sensitivity Cumulative Oil Production

Material balances

Obtaining low material balance errors is much more difficult in thermal simulation. Normal black oil simulations can usually obtain material balance errors of significantly less that one-half a percent and often in hundredths of a percent. The thermal simulations that involved gas and water legs often involved material balance errors in the 1–2% range. The simulation vendor felt this was within technical norms. The author was able to achieve better results with just bitumen and water but no overlying gas. Thermal simulations generally are not as stable and have more difficulty in achieving convergence. Thermal shock fronts move convectively, unlike pressure shocks, which move diffusively. Generally speaking, thermal simulations require smaller timesteps than normal black oil simulations.

Thermal simulators also provide an energy balance. ARE has generally found that the energy balance generally correlates with the material balance—i.e., whatever the level of accuracy reflected in the material balance will correspond to about that level in the energy balance. The energy balance should also be checked along with the material balance. It also provides some interesting insight into the overall efficiency of the SAGD process.

The majority of the cost in SAGD is related to producing steam. The material balance can be used directly in the calculation of efficiency. Not all produced heat can be recovered, however, improved efficiency in recovering heat will significantly reduce steam generating costs.

Longer run times

Most SAGD simulations are run with 2-D cross-sectional models. Considerably more computation is required with thermal models than with black oil models. As outlined previously, grid-block sizes have to be quite small. Normally, with conventional reservoir simulations, larger grid blocks can be used away from the wells. The nature of the SAGD mechanism does not allow large grid blocks to be used. In SAGD, the steam chamber front progresses by melting a grid block's oil, thereby precluding the use of large grid blocks. Some minor changes can be made by moving from 1- to 1.5-meter grid blocks. Based on the sensitivities run by the author, this is not recommended.

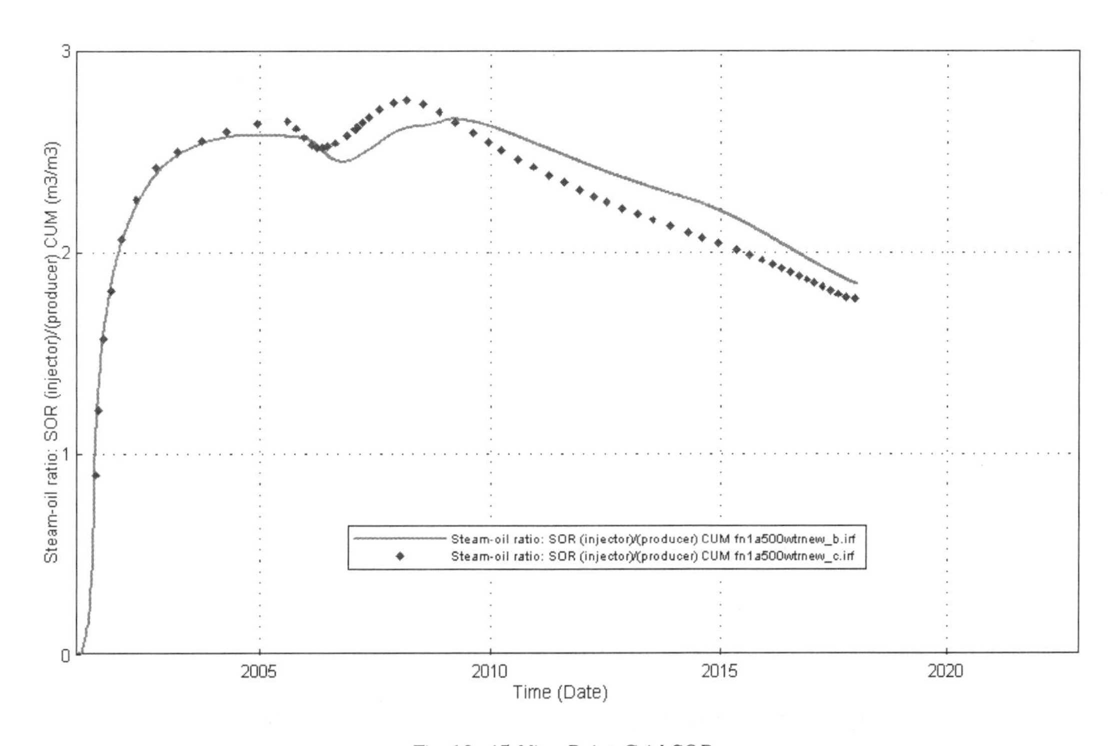

Fig. 19–45 Nine-Point Grid SOR

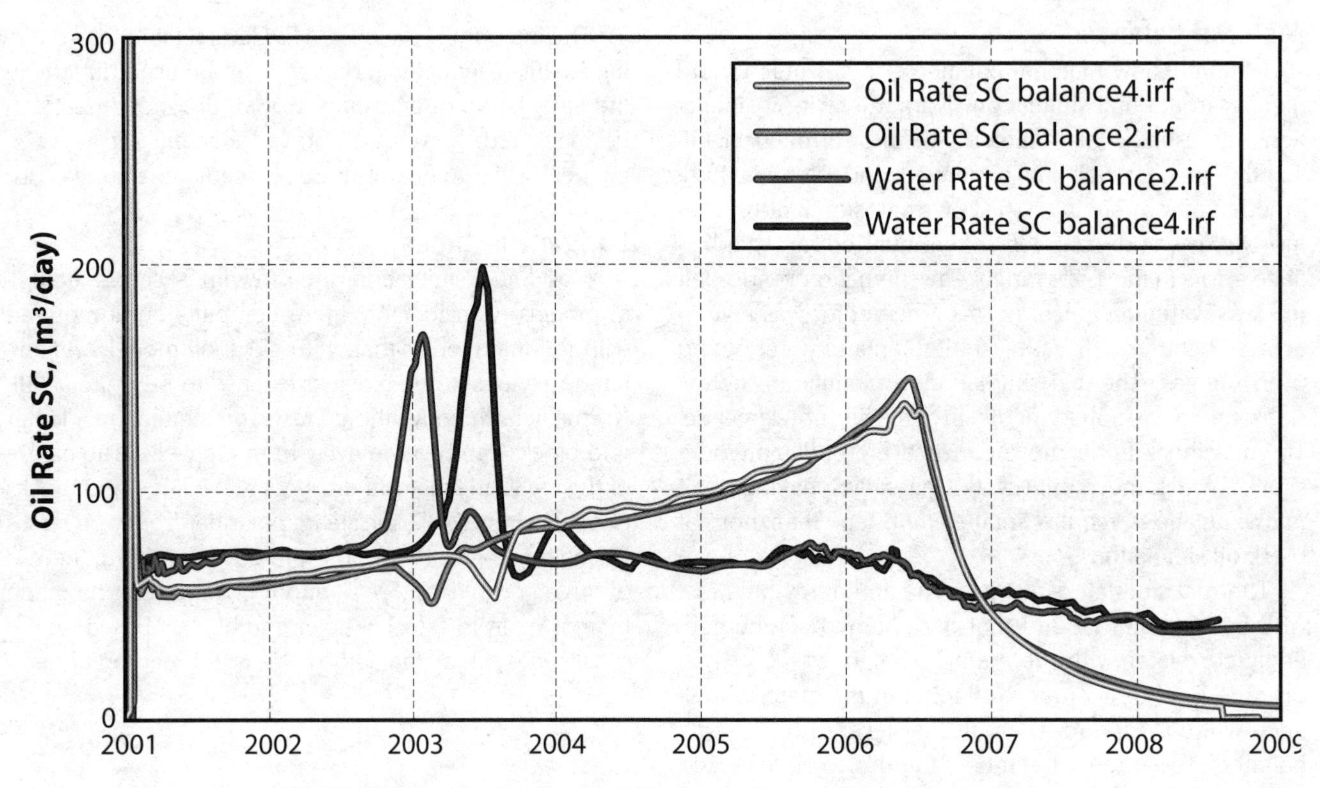

Fig. 19–46 Nine-point Grid Sensitivity Production Rate and Water Injection Rate

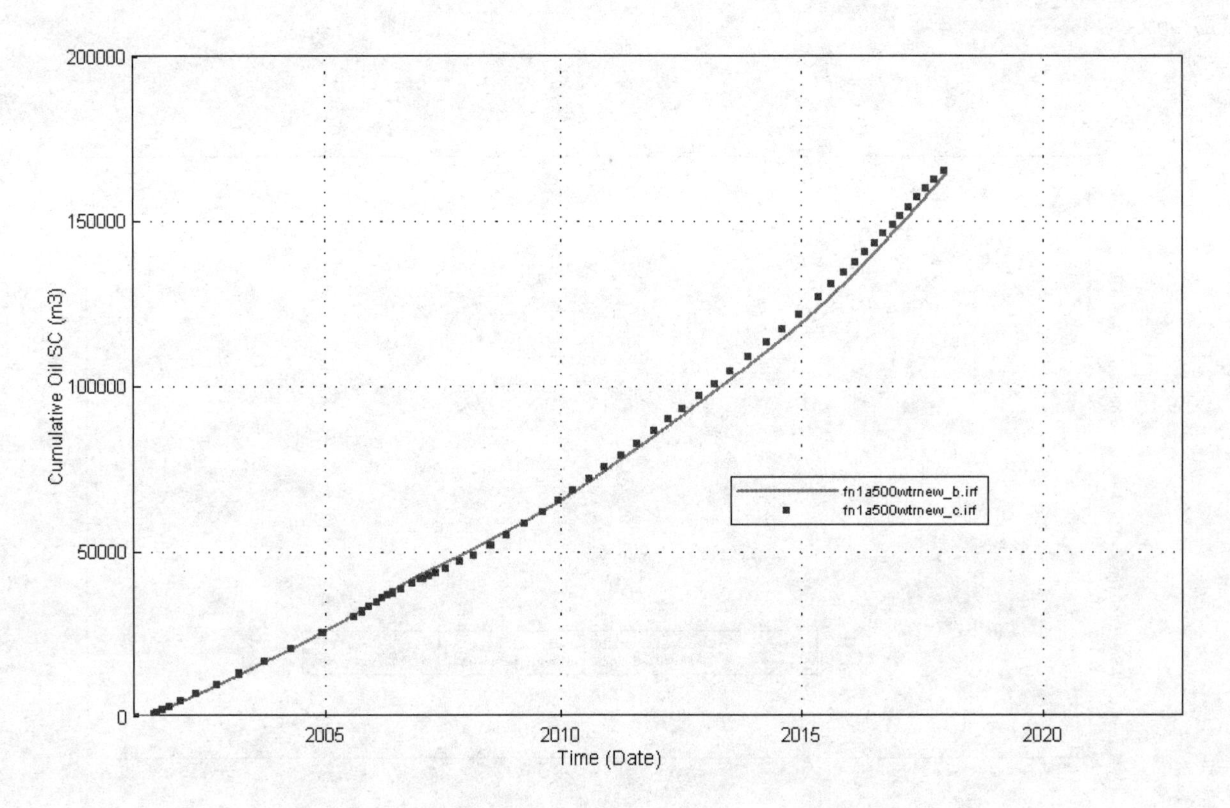

Fig. 19–47 Nine-point Grid Sensitivity Cumulative Oil Production

Table 19–3 Material Balance Statistics from Grid Differencing (Nine-Point vs. Five-Point Sensitivity)

Without Nine-Point Discretization							
		Initial Total	Current Total	Net Inj/Prod	Net Heat Lost	Error	% Error
Water	(gmol)	4.38E+09	3.07E+09	-1.29E+09		-2.20E+07	-0.2522
Oil	(gmol)	4.67E+08	1.41E+08	-3.34E+08		8.14E+06	0.8636
CH4	(gmol)	5.56E+06	2.63E+06	0.00E+00		-2.93E+06	-35.8378
Energy	(J)	-4.29E+13	3.07E+14	5.49E+14	-1.89E+14	-1.06E+13	-0.9777

With Nine-Point Discretization							
		Initial Total	Current Total	Net Inj/Prod	Net Heat Lost	Error	% Error
Water	(gmol)	4.38E+09	2.98E+09	-1.39E+09		-1.50E+07	-0.1716
Oil	(gmol)	4.67E+08	1.38E+08	-3.37E+08		7.51E+06	0.7978
CH4	(gmol)	5.56E+06	5.92E+06	0.00E+00		3.59E+05	3.1256
Energy	(J)	-4.29E+13	2.99E+14	5.28E+14	-1.78E+14	-8.42E+12	-0.8027

Steam injection well instabilities

For the injection well, there are often stability problems. Steam has a viscosity more like a gas—in the range of 0.015 cp. The viscosity of steam and water are shown in Figure 19–48. There are very high permeabilities used in the unconsolidated sands, particularly when accounting for geomechanical effects. The net effect of this is that when the steam injection rate is controlled, the solution is very sensitive to the well block pressure. This often leads to instabilities.

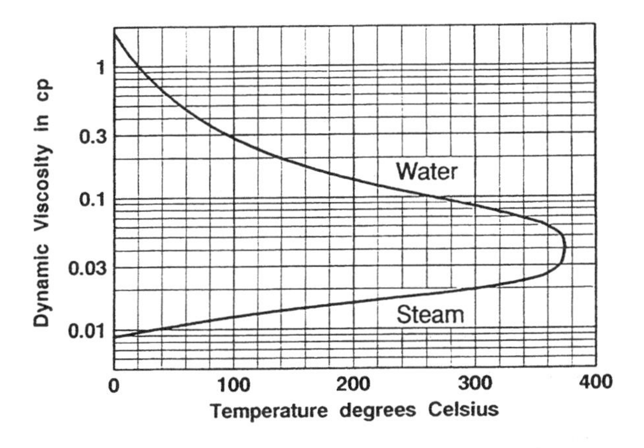

Fig. 19–48 Viscosity of Steam and Water

There are a number of options available. It is possible to tighten the tolerances on the pressure solution and, in some cases, the pressure solution for the well equation alone. Sometimes this seems to work, and sometimes it merely increases run times.

One of the simulator developers simply changes the well index to reduce the instabilities. This significantly decreased run times and improved the material balance, but it may not be desirable if the well will go on a pressure control later. In this case, the injectivity of the well may be unrealistically reduced.

Having said this, most horizontal sections are not completely open. Screens are placed across roughly one-quarter to one-half of the horizontal well section. Realistically, conventional reservoir experience indicates skins are real. The author has not found any available data where the skin has been measured.

The author has experienced stability problems on horizontal wells in conventional oil reservoirs where there were pressure solution problems on the producers. In this case, the production was not affected by quite large reductions in the well index. The limiting factor seemed to be the amount of oil that could converge on that part of the reservoir. It is possible that the surrounding formation will control the rate at which steam can be injected.

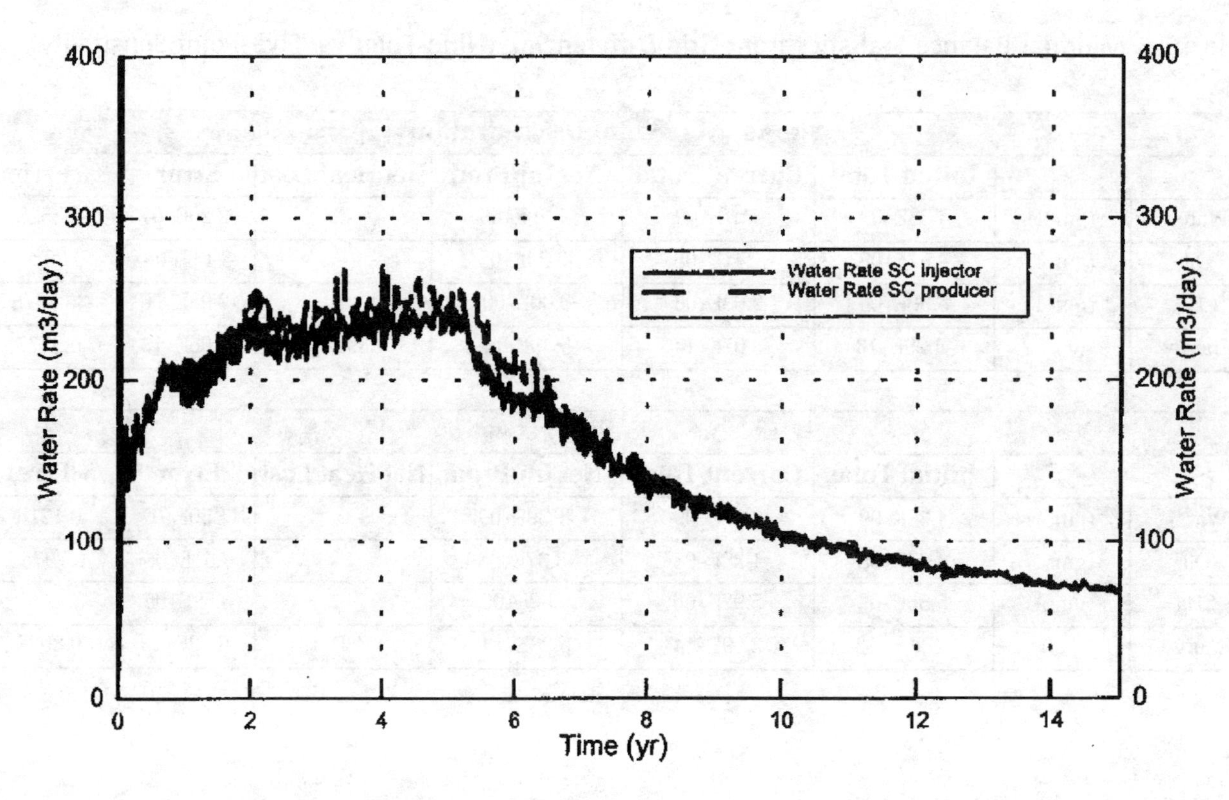

Fig. 19–49 Stability Difficulties with Steam Injection

This problem is very easy to spot on a horizontal reservoir simulation. A prime example is shown in Figure 19–49. If nothing else, the results are not visually pleasing. It is probable that a smaller grid block would not solve this problem, although the author has not tested this. Slightly larger grid blocks might help.

Countercurrent flow

In a number of runs, the author had difficulties when there were gas and water over the bitumen. The objective of these runs was to match the pressure at the top of the steam chamber with the pressure in the overlying zones. Thus, the low-density steam was rising while the higher density water was falling. In addition, there were phase changes with the water as it was contacted by the steam.

This can lead to numerical problems and very slow execution times. It appears that the countercurrent flow caused difficulties for the reservoir simulator. The problem appears to disappear when the fluids enter the steam chamber, which, if the volume is large, can quench the steam chamber, or when the steam pressure is high enough to cause displacement of the steam into the overlying zones.

Phase shifts

In one of the simulators used by the author, a control is available that limits the number of times the phases can change from successive iterations. The description for the use of this control states:

> *When the reservoir conditions are at saturated values and the reservoir fluids contain very volatile components, or the total heat capacity of a grid block is small, it may happen that a very small change in pressure or temperature will cause phase appearance or disappearance in a grid block in almost every iteration during a timestep.*

A keyword can be used to limit the number of phase changes to a specified number. This may cause the numerical performance to improve, but it can come at the expense of material balance errors.

Heat capacities and thermal conductivities

When STARS, a product of CMG, was developed simultaneously with SAGD, there was a bug in the calculation of heat transfer. The simulator identified the grid block with the highest temperature, a form of upstream weighting. The transfer of heat was then calculated using the thermal conductivity of this upstream block to the adjoining blocks. In some circumstances, this led to oscillatory behavior and convergence problems when contrasting thermal conductivities existed in adjacent grid blocks.

For this reason, one will find that older data sets, which are used to develop subsequent simulations, often have identical thermal capacities and conductivities for the rock matrix, gas, and water. This step was designed to avoid instabilities that might occur. The author has found these common properties in a number of data sets.

Since that time, the calculation has been changed and now uses a harmonic average thermal conductivity between grid blocks. In the author's experience, the use of averaged thermal properties seemed to effect no noticeable change in results; hence, it is likely that this practice is no longer required.

Slabbing

Very often in the early period of the oil production, there is an anomalous spike of oil production. This occurs in most runs and seems to coincide with the period immediately after the upward plume of steam hits the top of the reservoir.

This appears to represent a numerical effect. The oil in the stack of grid blocks adjacent to the steam plume seems to melt at once, which leads to the short-term oil production rate spike shown in Figure 19–50. The author has termed this slabbing, because the whole slab of grid blocks appears to melt at once or in a short-spaced series.

This effect has also been noted by at least one other author, with a similar explanation.

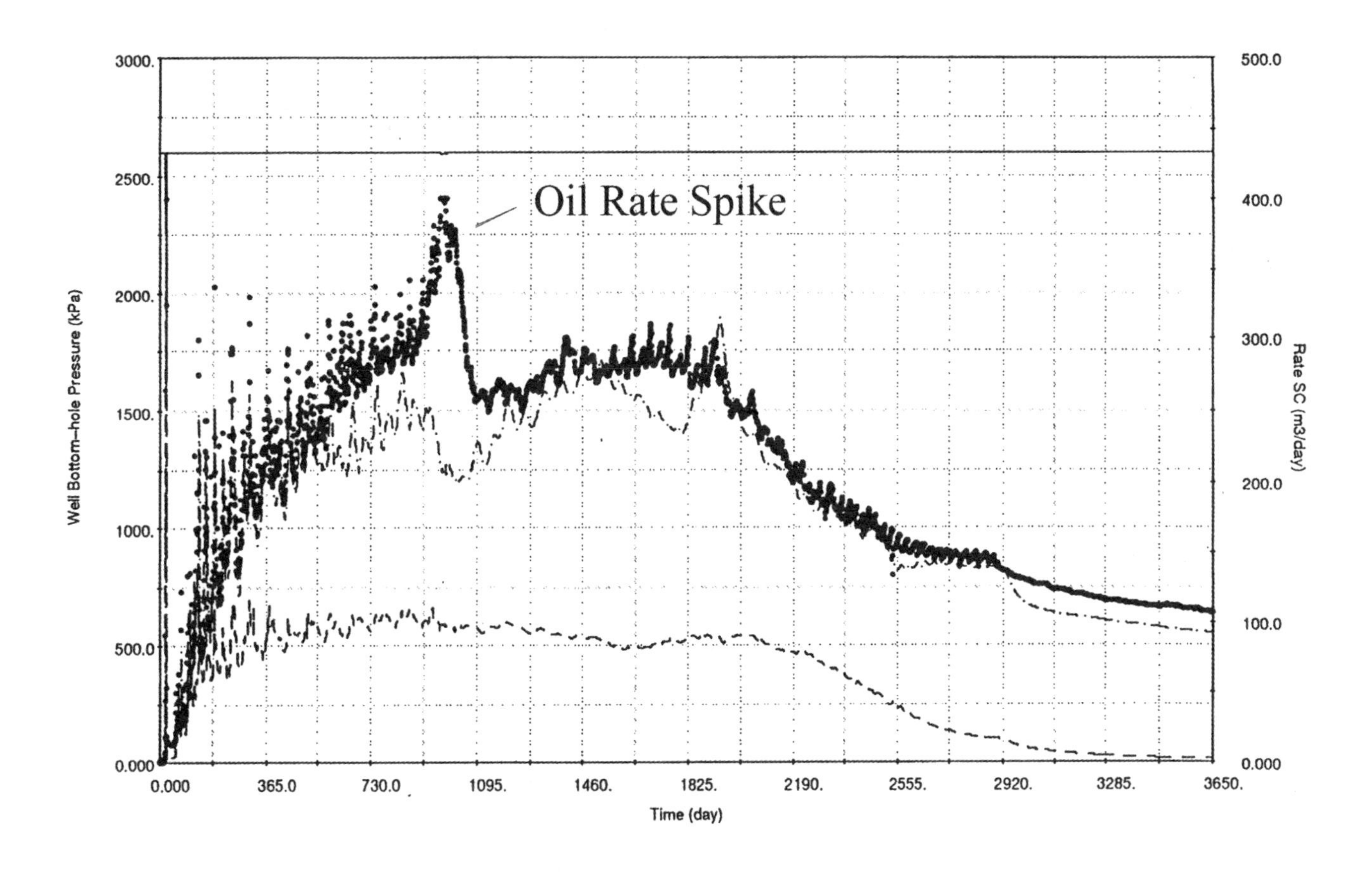

Fig. 19–50 Slabbing—Oil Rate Spike

Detailed SAGD Process Description

Steamflood front mechanism

Extensive work on these issues has been done with respect to both water and steam flooding, with good matches obtained between analytical and laboratory experiments. An example of this is shown in Figure 19–51.[52]

Historically, steamflood work has concentrated on the interface between the oil and the bitumen. It is based on the assumption that the steam will condense at the cold bitumen front. This process would be dominated by water-oil relative permeability curves. Behind this region is saturated steam, and this process will be affected by gas-bitumen relative permeability. It is likely that the process in a SAGD chamber involves recovery both at the interface and in the steam chamber during a period of time.

Convection

In the strictest sense, convection means the transfer of heat by the motion of fluids that have a heat capacity. In classic steam flooding, the movement of steam behind the front is a convective process.

The context for this discussion about convection concerns the movement of steam inside a SAGD chamber. This was clearly envisaged by Butler, who, in his theoretical developments, shows material moving down the interior of the chamber as shown in Figure 19–52.

This is observed in reservoir simulation. It was first identified in the literature by Ito et al.[53] They show a series of diagrams that illustrate a progressive change in the depth of the mobile material and a gradation of saturations. These diagrams are shown in Figures 19–53, 19–54, 19–55, 19–56, 19–57, and 19–58.

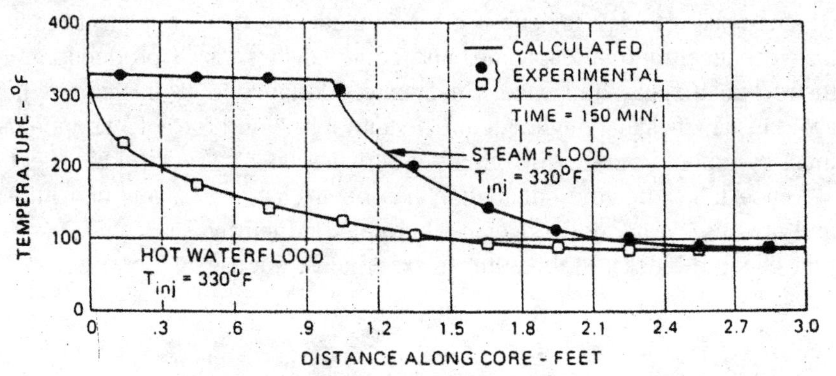

Comparison of calculated and experimental temperature profiles for laboratory model

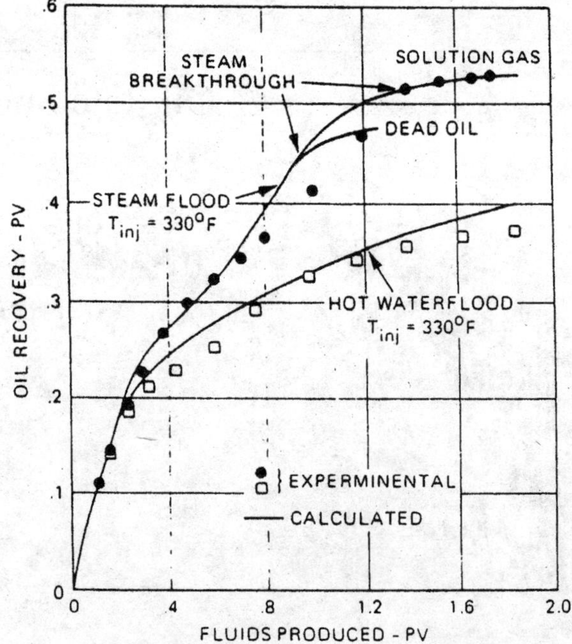

Comparison of calculated and experimental oil recovery curves for laboratory model

Fig. 19–51 Steam Flooding Advance

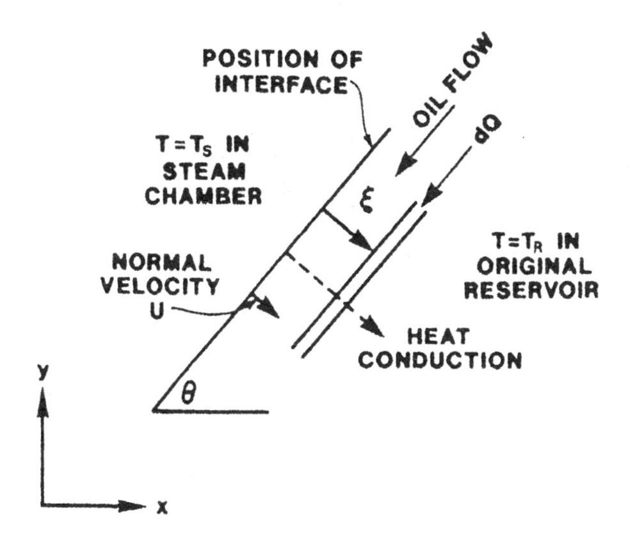

Fig. 19–52 SAGD Theory Developed by Butler

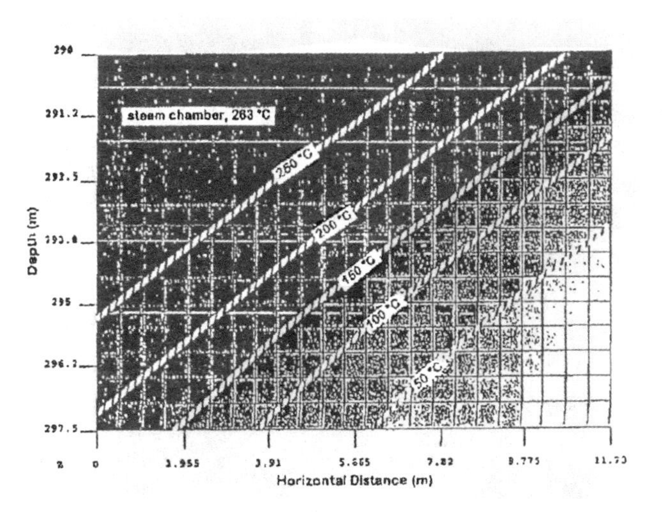

Fig. 19–54 Detailed Isotherms of SAGD Simulation

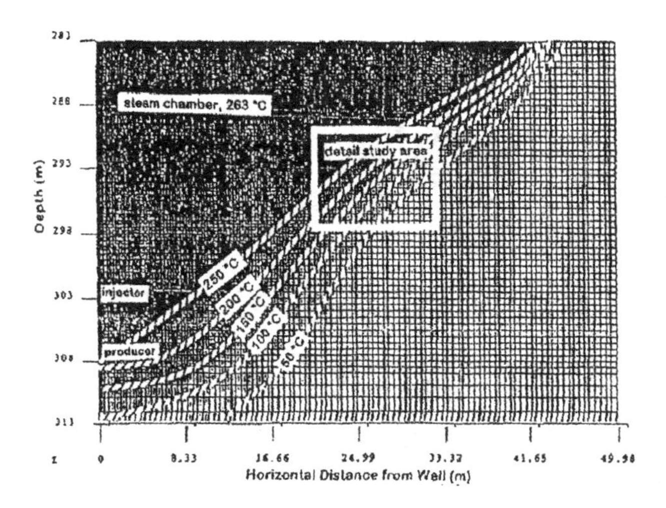

Fig. 19–53 Simulation Results Adjacent to Steam Chamber

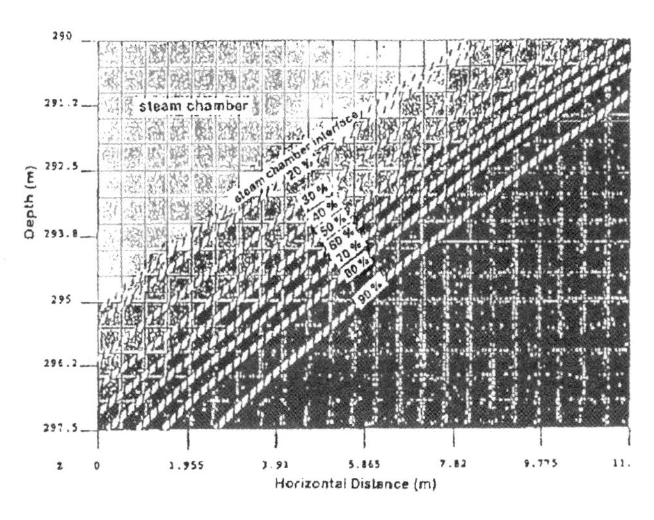

Fig. 19–55 Detailed Contours of Oil Saturation at Edge of SAGD Chamber

This seems to have caused a small furor. The author sees no conflict between the theoretical development and the results of simulation. The equation developed by Butler is shown as follows:

$$q = \sqrt{\frac{2\phi \Delta S_o k_o g \alpha h}{m v_s}} \qquad (19.4)$$

Details of this gradation in saturations are not specifically highlighted in Butler's theory, which uses some clever integration and assumed properties. The assumption in question is oil permeability—i.e., the product of k_{ro} and absolute permeability. He did account for the variation in the viscosity of the oil with temperature gradients in the mobile zone. He solved for oil flow, since he was principally interested in predicting production rates. Water flow in the mobile zone did not appear to be of

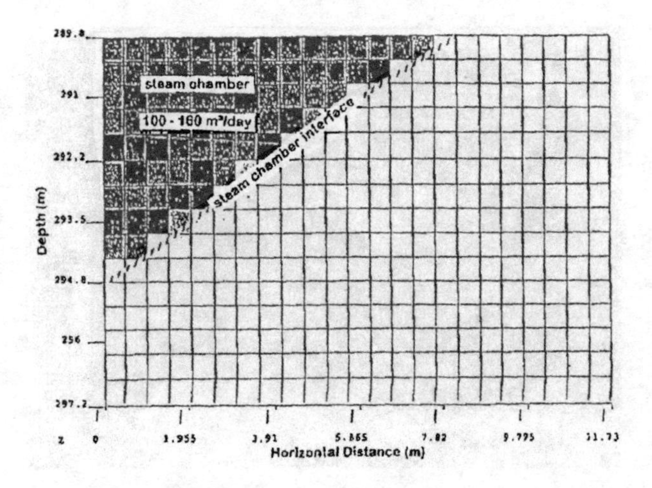

Fig. 19–56 Detailed Steam Flow Contours at Edge of SAGD Chamber

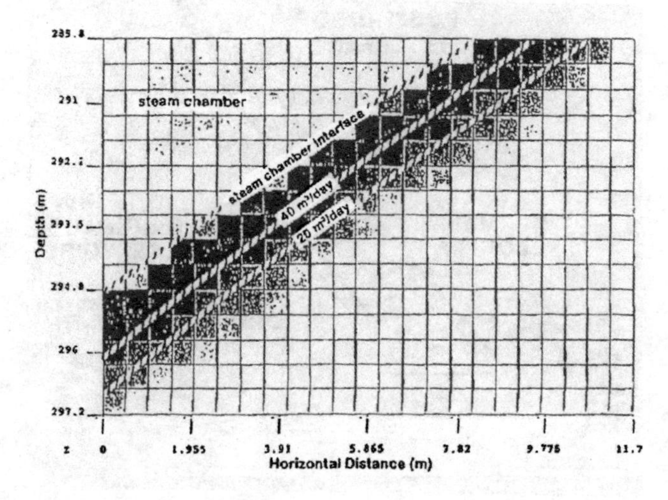

Fig. 19–57 Detailed Water Flow Contours at Edge of SAGD Chamber

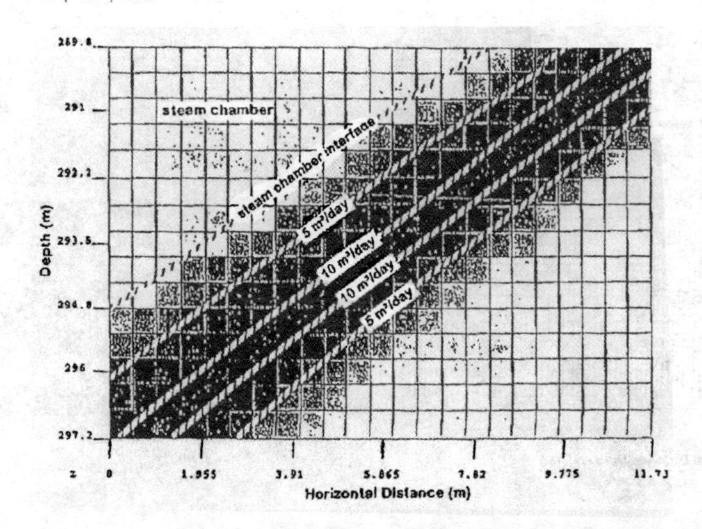

Fig. 19–58 Detailed Oil-Flow Contours at Edge of SAGD Chamber

particular interest, and he presents no calculations. He must have been aware that water (steam condensate) was flowing, but, normally, water is not sold.

The results have been interpreted to indicate that the rate is independent of the shape of the interface. We know, for simulation, that SAGD recovery efficiency is heavily affected by rates and operating conditions. This also affects steam chamber shape, which will be discussed in more detail later in this chapter. There were, of course, some justifiable simplifications made in the development of the theory.

The author has discussed some of the simulation models directly with Dr. Butler. One of the problems he observed was that there is a discrepancy between the rates and permeabilities predicted from his theory and those predicted by simulations. Numerical errors in the simulations were suggested as a possible source of this discrepancy. Having done a number of grid sensitivities, the author does not believe the problems are numerical but concurs with the comment about permeability. The author's explanation is somewhat different in that geomechanics indicates large permeability changes near the wellbore and more moderate changes away from the wellbore. Allowing for some variation in relative permeabilities, the two methods likely produce substantially similar results.

Operating conditions

Operating conditions provide some interesting implications. Referring to Figure 19–59, the steam goes in the upper horizontal injector, condenses at the interface, and then is produced out of the reservoir from the lower horizontal producer. Evidently, a certain proportion of the heat in the reservoir is simply produced and does no useful work. The amount of heat that is transferred to the cold bitumen is going to depend on the temperature difference between the steam and the cold bitumen and the amount of heat that is swept away through the producer. More heat should be transferred with slower side-chamber velocities and higher temperatures. Conversely, a thinner zone would require less material to travel through. Let us not forget that the objective of production is largely to produce as much as possible, as fast as possible.

The heat that is produced from the producing steam is not totally wasted. The heat can be (and is) recovered in heat exchangers. This changes optimization. A better measure of efficiency than the cumulative steam oil ratio (CSOR) would be: *(heat in – recovered heat out) divided by cumulative oil production.*

Solution gas

Most early simulations done in cyclic steam were done with no solution gas in the oil. It has been argued that, at low solution gas oil contents of 4 m^3/m^3 (23 scf/bbl) or less, solution gas drive will be minimal. Further, it is likely that cold bitumen is sufficiently viscous that gas bubbles will not form a continuous gas saturation in the reservoir and lead to gas relative permeability.

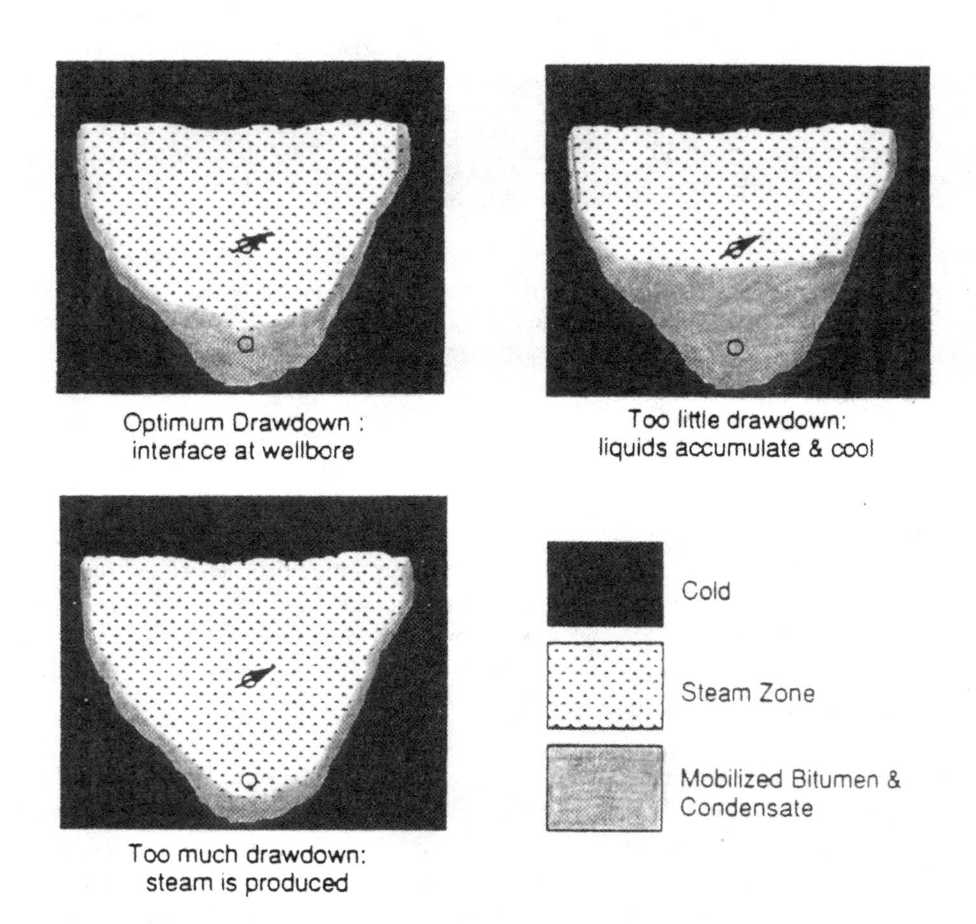

Fig. 19–59 Heat Loss Due to Production

Layers of gas are found in some heavy oil reservoirs, and this has suggested a number of things:

- The layers may indicate sealing barriers within the reservoir.
- The gas should segregate.

Gas sealing, as outlined earlier in this book, is controlled by capillary imbibition barrier pressures. This also varies for wetting and nonwetting phases. Consequently, even though there are thin traps of gas within an oil zone, this does not imply that there are seals to steam or that increases in reservoir pressure due to steam injection won't cause the gas to leak.

From the perspective of segregation, geological time is evidently available for the *in situ* conditions. The author has heard that solid residues from heavy oil wells will remold if left in a beaker for several days. So, while bitumen can be near solid, it is not static on the scale of several days. Clearly, reservoir temperature would affect how quickly segregation would occur. To the author's knowledge, there are no tests on this issue on the pore scale level.

Having said this, reservoir simulation model results are affected by the inclusion of solution gas. A number of these issues with heavy oil relative permeability were outlined earlier. This issue remains unresolved.

Foamy oil flow

Foamy oil was originally proposed by Dr. Gerry Smith. His proposal was based on pressure transient test interpretations of heavy oil producers and was based on a logical argument that there must be some viscosity reduction to account for interpreted permeabilities. This is not a simple issue and includes the effect of worm-hole propagation and the attendant increases in formation permeability.

It has also spawned a great deal of fundamental research on oil flow in reservoirs. Some of this work has included tests with glass micromodels to view what is happening on the pore scale. The most recent research has indicated that foaming does not appear to be occurring in glass micromodels.

In the author's opinion, the permeability is changing more than the liquid viscosity, and this is leading to the high permeabilities on primary heavy oil production. However, more research is required to categorically prove permeability increase over viscosity decrease (foam flow). It appears that the foamy oil research is more mature than the geomechanical analysis.

Thermal expansion

Thermal expansion occurs when the oil sands are heated. There are a number of implications to this:

- The fluids often have a higher coefficient of thermal expansion than the solids (sand grains). In cold bitumen, which has a very high viscosity, this pressure will not be able to dissipate laterally and vertically. To some extent, the sand will explode. Butler analyzed expected pore pressure increases and dissipation time in a paper entitled, "The Expansion of Tar Sands During Thermal Recovery."[54]
- If the steam chamber expands, the expansion and stresses must be absorbed vertically and laterally. Thus, the cold sand between chambers will be squeezed horizontally. At the same time, the vertical expansion of the steam chamber will cause the overburden to be "jacked up" to some extent, which will decrease vertical stresses within the steam chambers. During the early periods of heating, there will be concentric expansion surrounding the injection and production well for a SAGD pattern.

In effect, the differential expansion of oil sands fluids and solids will reduce the effective stress or reduce sand grain contact pressures. This will increase permeability somewhat, but it will enhance the tendency of the oil sands to shear. The differential expansion between the cold and hot areas will cause shear stresses, and this will facilitate shearing. The latter greatly increases permeability.

In the literature, there are examples of stress path changes with time. Timing is very important. The effective stress decrease originally occurs at temperatures before the bitumen becomes mobile. The stresses from expansion are absorbed ahead of the steam chamber in the cold bitumen. Therefore, shearing should occur in the cold bitumen and within the steam chamber and cold bitumen transition.

Shearing

The general concept of shearing is shown in Figure 19–60. In essence, the stresses induced by the steam chamber (both fluid pressures and thermal expansion) cause the oil sands to fail. When this occurs, the permeability is enhanced as described earlier.

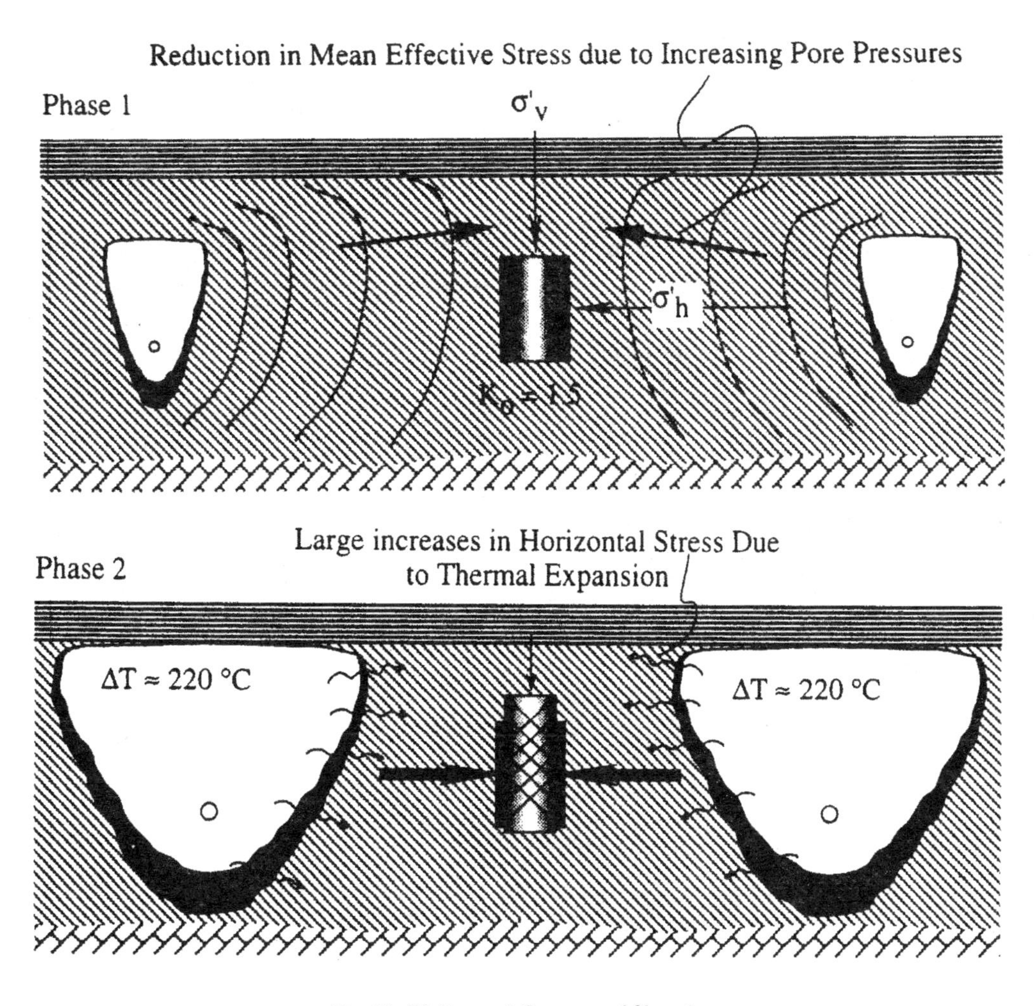

Fig. 19–60 General Concept of Shearing

The Phase A pilot at AOSTRA's UTF project was extensively instrumented. Piezometer readings within the Phase A pilot clearly demonstrate that pressures were being elevated to near-steam pressures well in advance of the steam chamber, which is indicated by the presence of a temperature increase. Another example is available from the JACOS Hangingstone projects. This data is indicative of fluid mobility within the cold reservoir outside of the steam chamber. These elevated pressures reduce the effective stresses within the cold oil sands, which encourages the tendency toward shear failure.

Required permeability changes

Based on the early UTF modeling, it would seem that very high permeability increases and porosity changes are required throughout the reservoir. However, if one considers a steam chamber, reservoir permeability has the most effect in the immediate vicinity of the well where there is a flow concentration. This is potentially significant. Large dilation effects may not be required throughout the entire reservoir to achieve production increases. It also suggests the possibility of stimulation.

ARE conducted a sensitivity by comparing two cases:

- a reservoir of high uniform permeabilities of 7,500 mD horizontally and 2,500 mD vertically (nearwelltarget.dat)
- a reservoir of undilated permeabilities of 1,775 mD horizontally and 1,509 mD vertically (nearwellbase.dat)

A third run was tuned to match the production profiles and cumulative recovery. The results are shown on Figures 19–61, 19–62, 19–63, and 19–64. Indeed, the permeability changes required are far less drastic than is suggested by modeling air permeabilities for the entire reservoir.

The maximum increase in permeability is a factor of 2 in the near-wellbore region. It is appropriate to consider the effects of drilling and completion also. The well is drilled to a larger diameter than the liner. Most projects use either slots or a screen to prevent sand production; therefore, the sand falls onto the slots or screens. In this case, not all of the dilation comes from reservoir engineering effects. This occurs directly adjacent to the well where it will have the most impact.

Away from the wellbore, a 30% increase in permeability was required. The modeling showed that no reasonable near wellbore alteration in permeability could account for the production increase required. From this, ARE concluded that geomechanical effects must have occurred at, or ahead of, the steam chamber interface. The results are unlikely to be unique. The changes in permeability are unlikely to be as abrupt as was history matched. A transition from various levels of permeability could be incorporated. However, there would be no basis to determine which of these solutions would be the best. These results should be viewed as approximate. Ideally, laboratory measurements could be made of Athabasca oil sands to directly determine these effects. In practice, obtaining undisturbed cores is both expensive and difficult.

With an *in situ* porosity of 28% and an index of disturbance of about 30%, the near-well pair porosity would increase to about 36.4%. Based on pressure transient analysis the undisturbed permeability of the formation is about 2,995 mD. An increase in permeability by a factor of 2, due to disturbance, indicates the near-well permeability required is about 5,850 mD. This is within 20% of the average air permeability derived from core. In this context, the previous history-match permeabilities are logical.

Figures 19–65, 19–66, and 19–67 show the permeability distributions and the shape of the steam chamber derived.

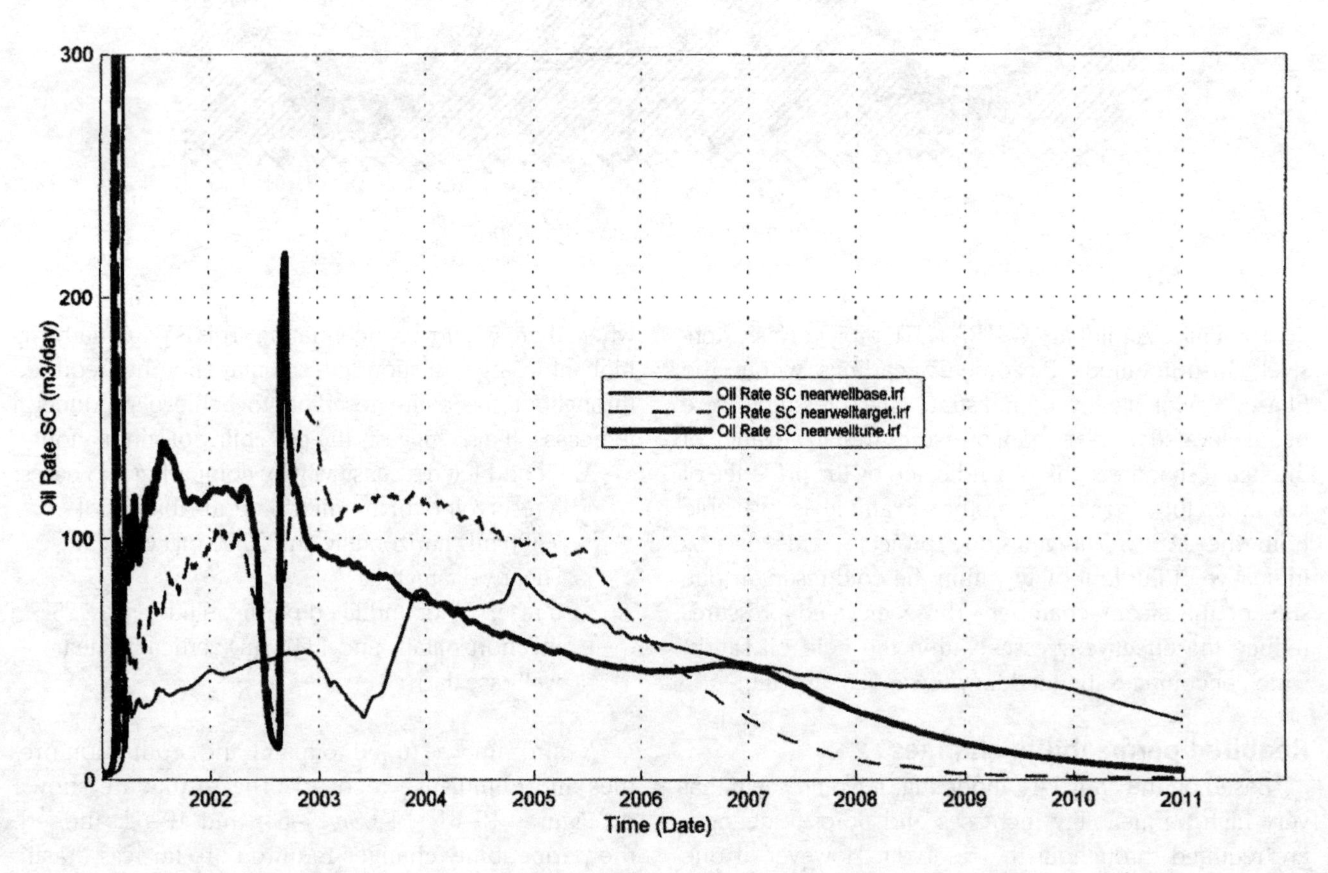

Fig. 19–61 Near-Well Permeability Enhancement—Oil Rates

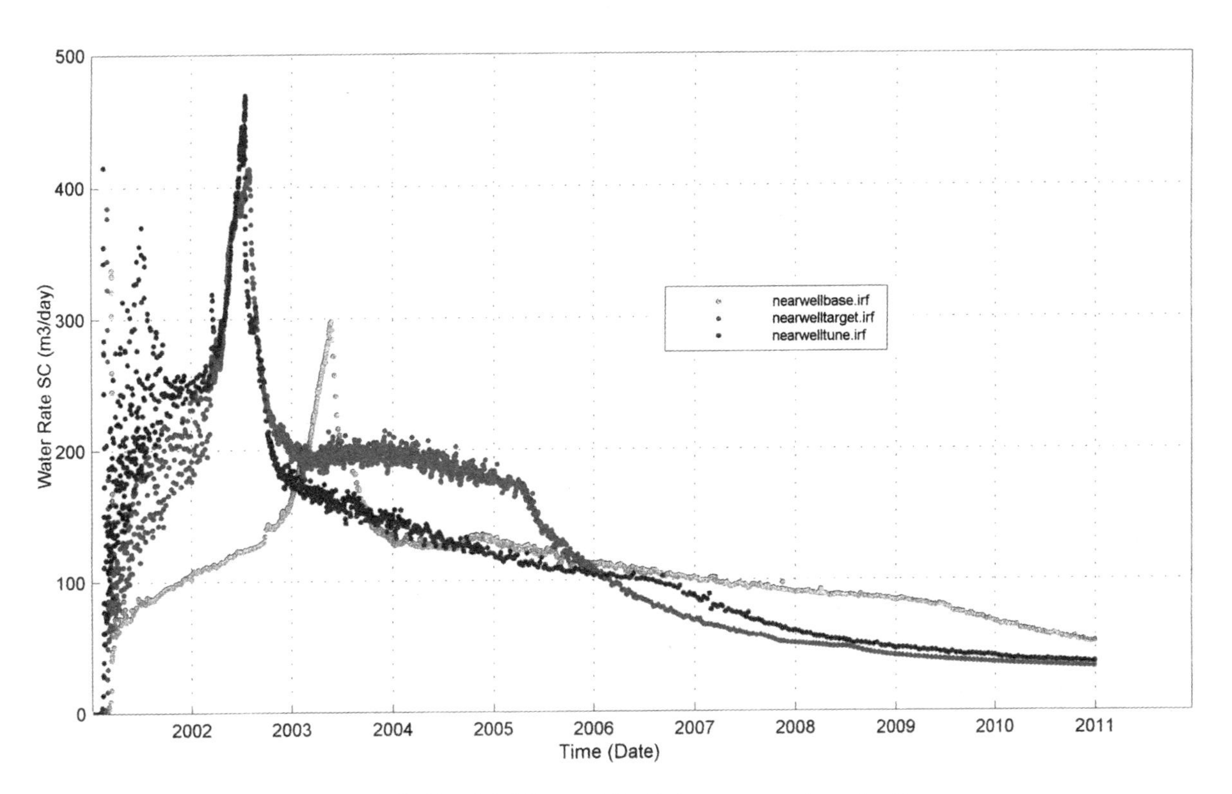

Fig. 19–62 Near-Well Permeability Enhancement—Injection Rates

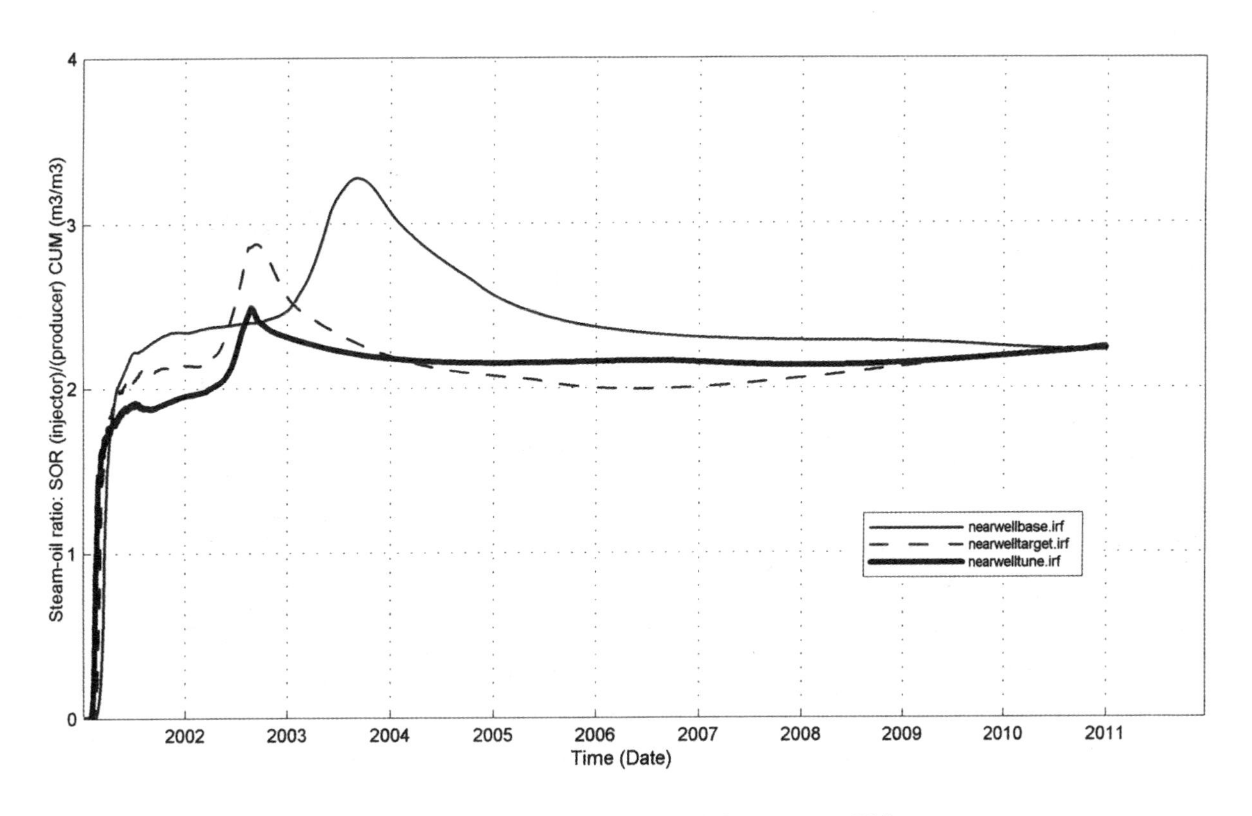

Fig. 19–63 Near-Well Permeability Enhancement—SOR

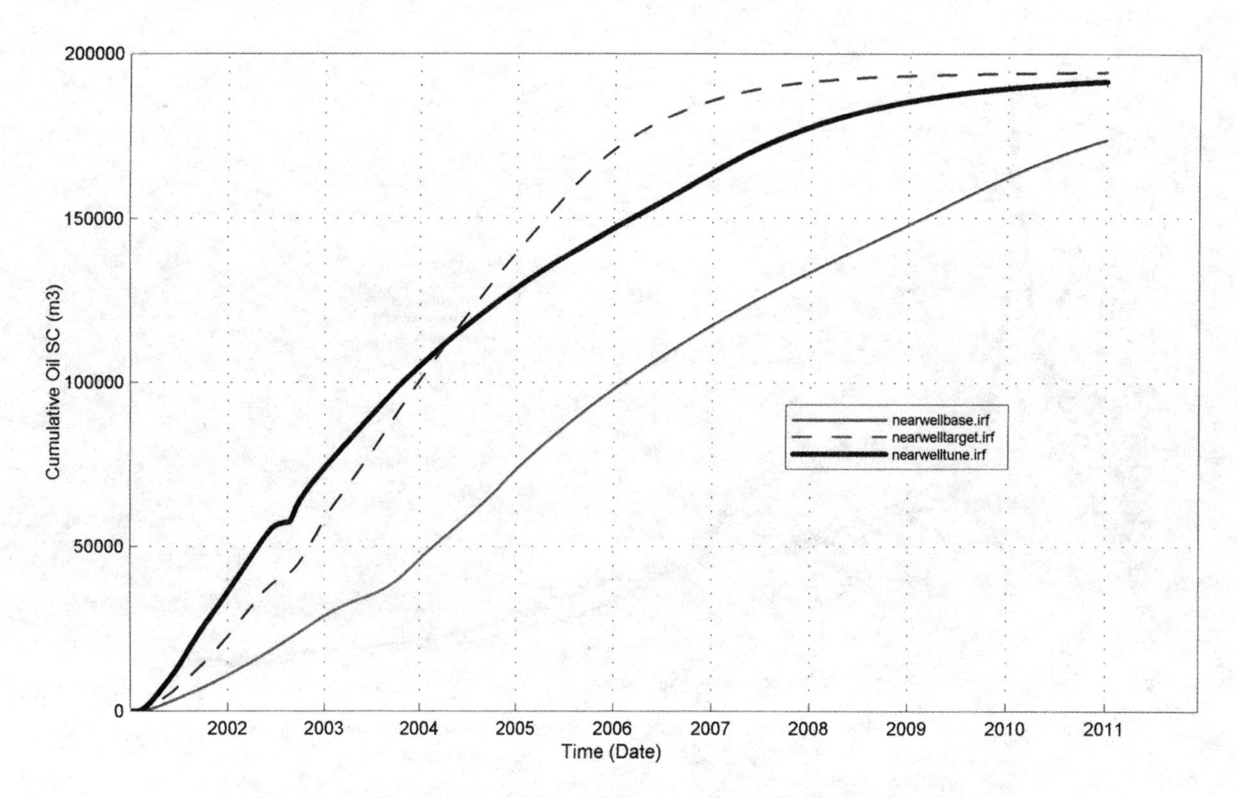

Fig. 19–64 Near-Well Permeability Enhancement—Cumulative Oil Production

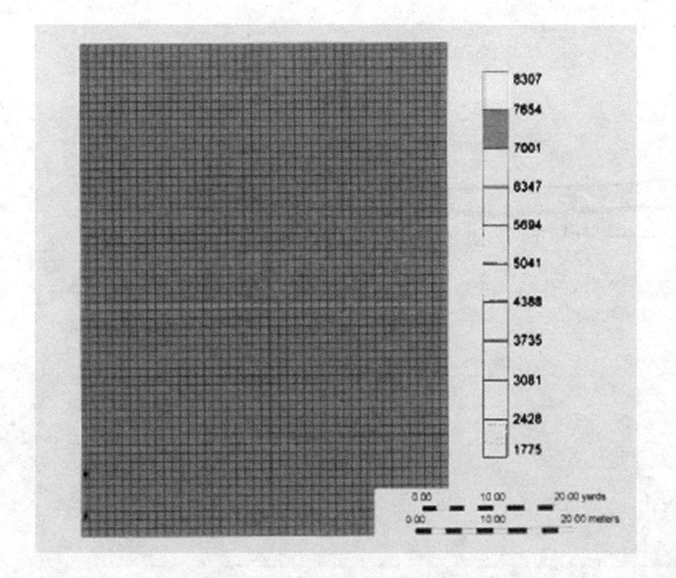

Fig. 19–65 Permeability Distribution in Uniform High-Permeability Reservoir

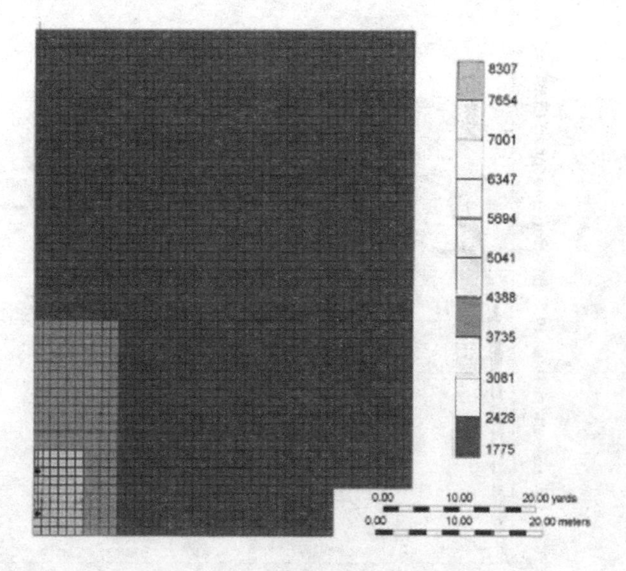

Fig. 19–66 Permeability Distribution to Match Use of Air Permeabilities

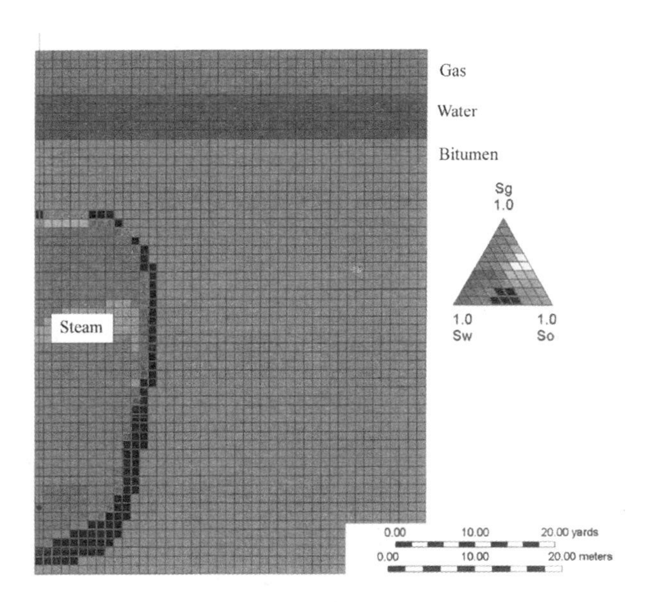

Fig. 19–67 Ternary Diagram of Steam Chamber with Localized Permeability Enhancement

Hydraulic fracturing

The possibility of using hydraulic fracturing for SAGD has been debated. The concept can be divided into two possibilities:

- the use of hydraulic fracturing as a start-up technique—i.e., to establish communication between the injector and producer
- the use of hydraulic fracturing as a general injection technique

At present, the concept of injecting above fracture pressure is not popular primarily due to concerns about containment above and below the pay zones and possible distortions to the steam chamber. It seems to have worked in cyclic steam stimulation.

The first possibility could shorten the start-up time dramatically. Thermal convection is a slow process, and reducing the 3 to 6 month startup time would be economically advantageous. The AEUB has allowed this to proceed on AEC's Foster Creek project. In the author's opinion, this is a realistic application; however, time will tell.

Typical Issues in Bitumen-Prone Areas

Reservoir characterization

Invariably, distribution of permeability and porosity is a major issue in reservoir simulation. The two largest areas of bitumen deposition are the northeastern section of Alberta and the Orinoco belt of Venezuela. The best pay is usually found in areas with stacked fluvial channels. This is true of the Kern River field in California, also.

Channel environment

Fluvial environments are characterized by rapid lateral changes. Experience has shown that channels have characteristic dimensions. For instance, as shown in Figure 19–68, it is possible to estimate the length of river meanders based on modern geomorphologic studies.[55,56] For the internal structure of fluvial channels, a typical width to thickness ratio is 10:1 (Kupfersberger and Deutsch).[57]

Channels or, more correctly, point bars have characteristic cross sections. An example of this is shown in Figure 19–69.[58]

There are a number of key features. Note that the point bar accretes on the inside of the bends and grows to the outside of the bends as shown in Figure 19–70.[59] If the river goes through an annual cycle or has periodic flooding, then there will be deposition of different energies and particle sizes. In the final stage, when the channel is buried, the watercourse that is shown on the right of Figure 19–71 can be filled with either permeable or impermeable material.[60] This is known as an abandonment plug.

A number of discontinuities are present, which are usually muds or shales. To summarize, there are three kinds of mud relevant to production:

- Interbedded muds
- Inclined heterolithic strata (IHS)
- Abandonment plugs

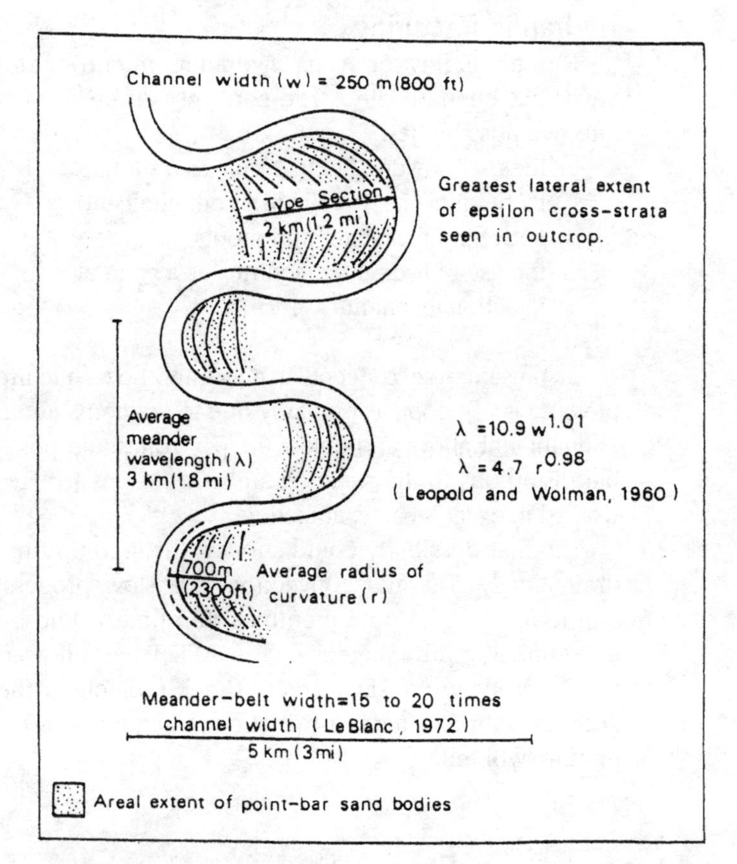

Fig. 9. Schematic representation of large scale point bar deposits based on channel width and meander wave length (from Flach and Mossop, 1985).

Fig. 19–68 Characteristic Dimensions of Channels

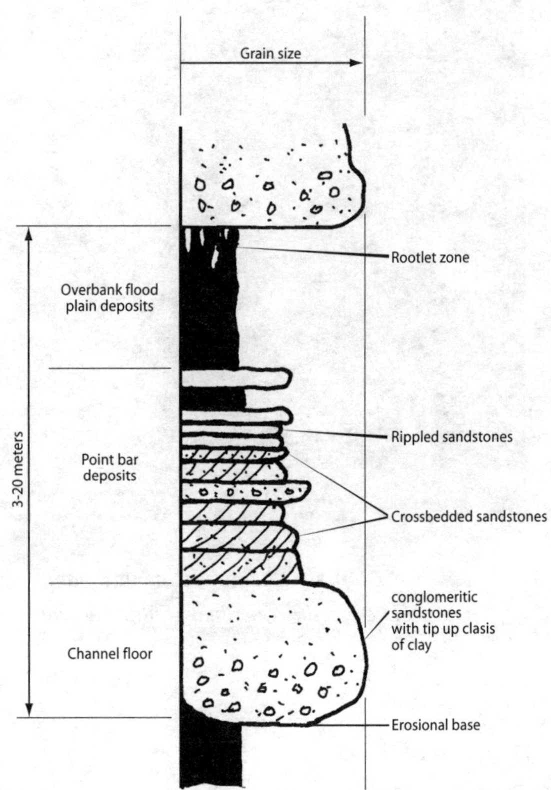

Fig. 19–69 Characteristic Cross Section of Fluvial Channels

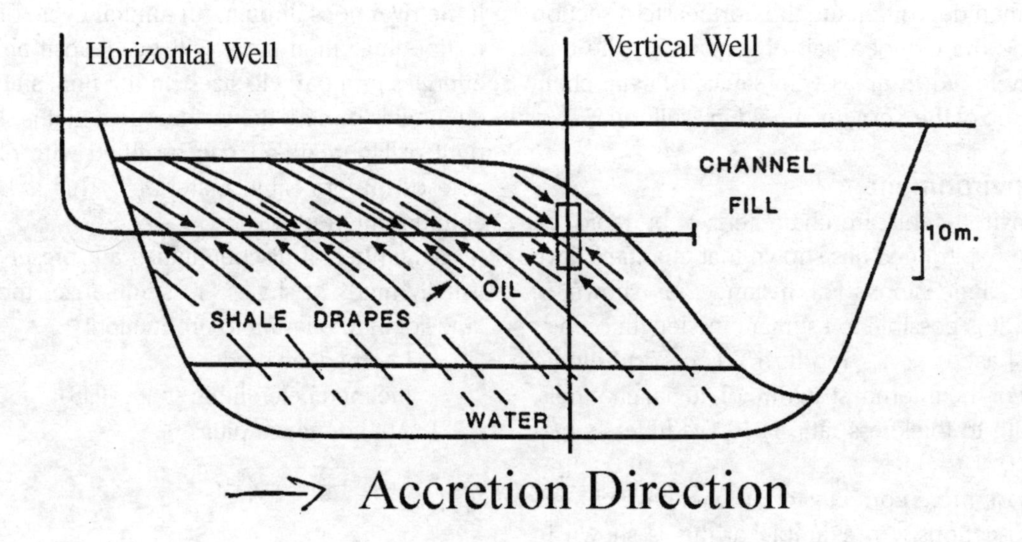

Fig. 19–70 Development of Point-Bar Deposits

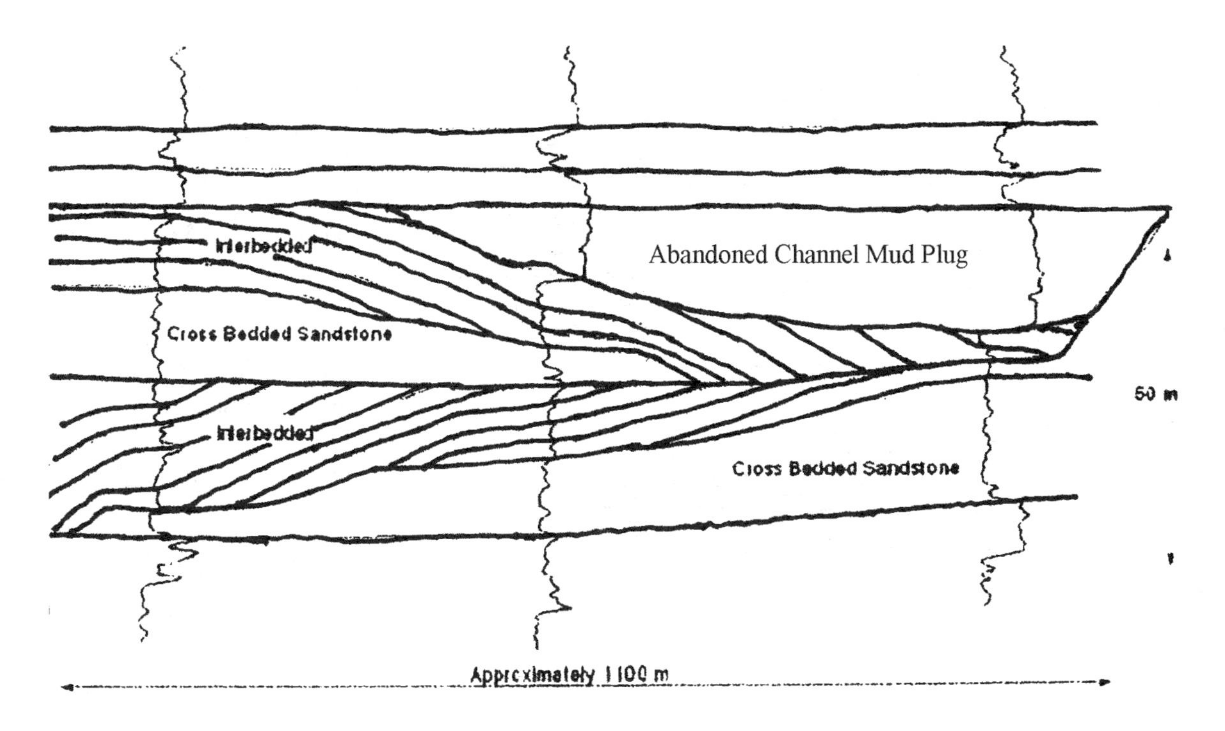

Fig. 19–71 Abandonment Plug—Point-Bar Deposits

Geostatistics

Geostatistics is a highly appropriate tool for analyzing the distribution of these different facies; however, such descriptions must be properly tuned. This involves extremely detailed description of rock properties and geometrical distribution. From the point of view of geometrical distributions, ARE has not been able to find other analogues to the McMurray in the Athabasca area.

The work of Strobl et al. provides some insight, in particular with respect to IHS; however, this is insufficient for the purposes of quantitative description.[61] More detailed outcrop studies could provide useful base data. There are a number of river outcrops, and data is available from several surface mining operations. It will take years of research to produce quantitative results.

There have been some attempts to use data from Prudhoe Bay and the Tillamook estuary in Oregon; however, all of these have significantly different morphology and scales. The use of inappropriate analogues does not provide any useful insight.

Fractal geostatistics were used for the UTF sites, as described in Mukherjee et al. Due to computational limitations, the results of a number of individual pattern models were concatenated. This is shown in Figures 19–72 and 19–73.[62]

Figure 19–74 shows a method in which the distribution of mudstones was broken into smaller scale models. Adding the results of these individual runs permits one to perform a detailed large-scale model. Note that width of the element of symmetry is 35.0 meters. This corresponds to a total pattern width of 70 meters.

Discontinuous barriers in SAGD

In conventional oil reservoir engineering as well as simulation, the effect of these interbeds has been quantitatively calculated. The first step was to determine statistically the shale distribution as discussed in chapter 4. The major reservoir performance effect is to reduce vertical permeability due to increased flow path lengths. This is shown in Figure 19–75. Note that these calculations are usually done with single-phase flow.

Returning to the previous model prepared by Muhkerjee, the fractal model results in a much less continuous reservoir description. This is important for SAGD simulation since continuous shales form barriers to steam rise, whereas a series of smaller shales with gaps in between will present a minimal disruption to steam rise. This is shown in Figure 19–76.[63]

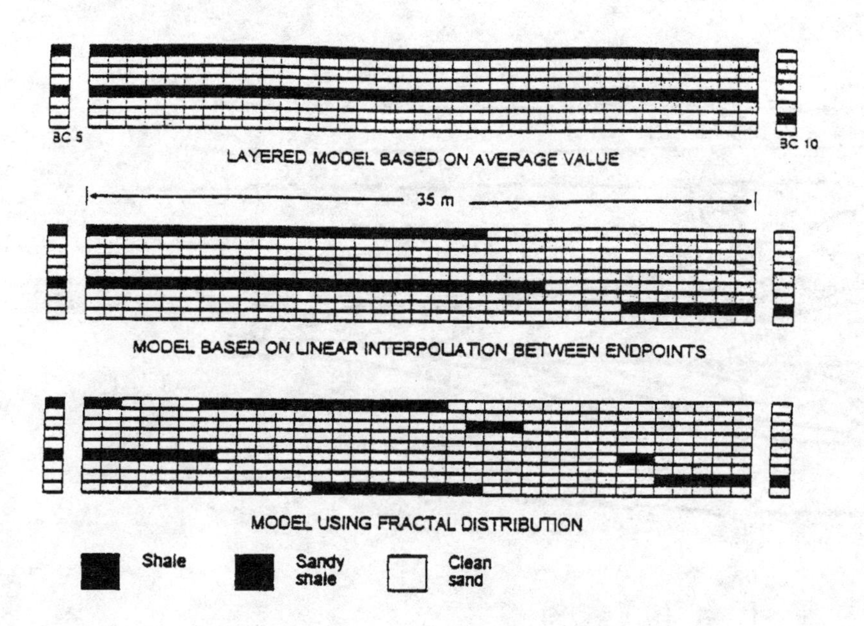

Fig. 19–72 Distribution of Barriers in UTF Modeling—Vertical Cross Sections

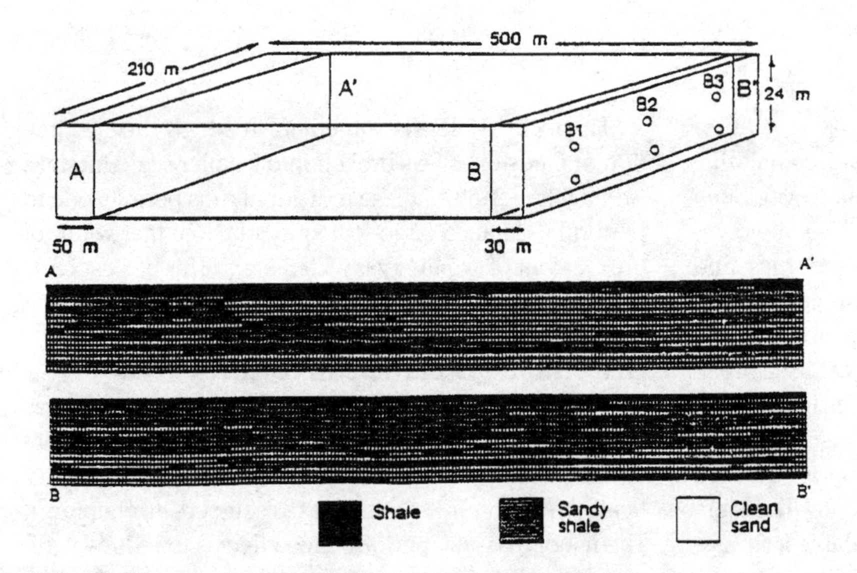

Fig. 19–73 Distribution of Barriers in UTF Modeling—Arrangement of Cross Sections

It may be concluded that the effects of barriers in SAGD involve fundamentally different processes and physics than in conventional reservoir engineering. This issue has been approached from an experimental perspective also.

Lab testing of mudstone gap size on SAGD performance

Yang and Butler did an extensive laboratory study on this issue using scaled lab experiments. They examined a number of interesting geometries.[64]

The first involved the use of a plastic barrier that extended from immediately adjacent to the well all the way across the pattern. This is shown on Figure 19–77. The results showed that, as long as the heat could transfer upward and there was a small gap, the bitumen would be drained. By analogy, a very small gap in the mudstones is all that is required to provide good vertical communication.

The second case involved the use of a plastic barrier that extended from the well all the way across the pattern with only a gap at the edge of the pattern. The results of this experiment are shown on Figure 19–78. In essence, the steam migrated along the bottom of the plastic barrier and then grew upwards from the end. Following this, the bitumen started to drain away from the well on top of the barrier and then reversed backward toward the well underneath the barrier.

Both of these experiments are strong indications that, while discontinuous mudstones may delay oil recovery, they do not impair ultimate recovery. It also establishes that even a small gap near the wells will have a major effect. A gap that occurs most of the way across the pattern will eventually provide drainage.

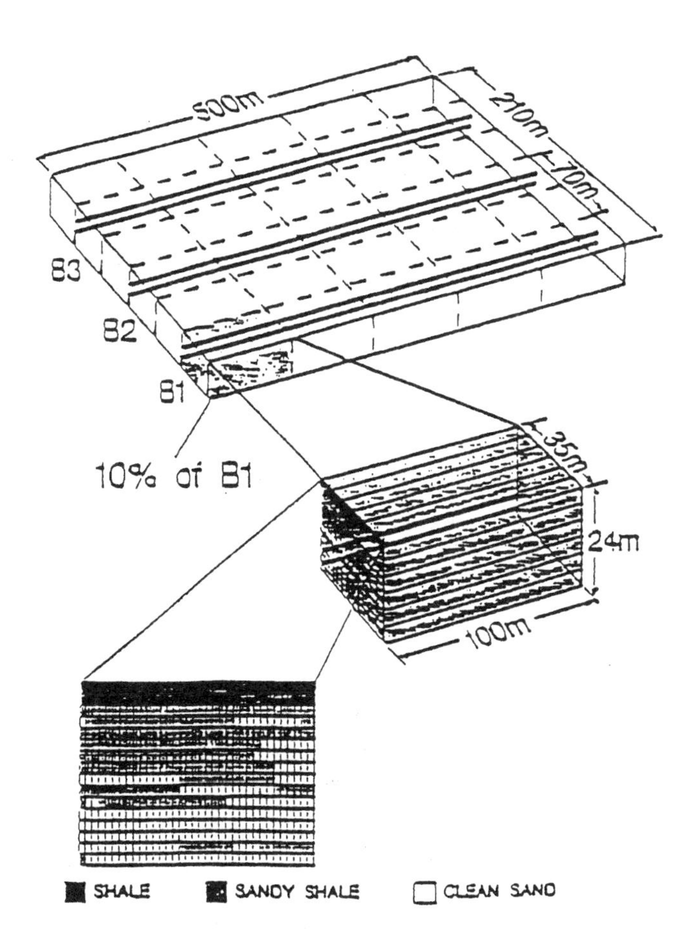

Fig. 19–74 Method of Combining Results From Multiple Models

Index properties of images

Section	Plane	n	*Axial ratio*	θ_m (°)	*V.M.* (%)	p
Vertical	Z - X	742	0.693	1	18.8	4×10^{-12}
Vertical	Z - X	815	0.698	2	17.7	8×10^{-12}
Vertical	Z - Y	872	0.686	-10.2	20.1	5×10^{-16}
Vertical	Z - Y	914	0.679	6.6	24.7	7×10^{-25}
Horizontal	Y - X	756	0.706	39	7.6	10^{-2}
Horizontal	Y - X	616	0.694	-15.5	15.3	5×10^{-7}

n = total number of particles,
θ_m = vector mean direction (Degrees),
V.M. = vector magnitude (%),
p = probability level of significance.

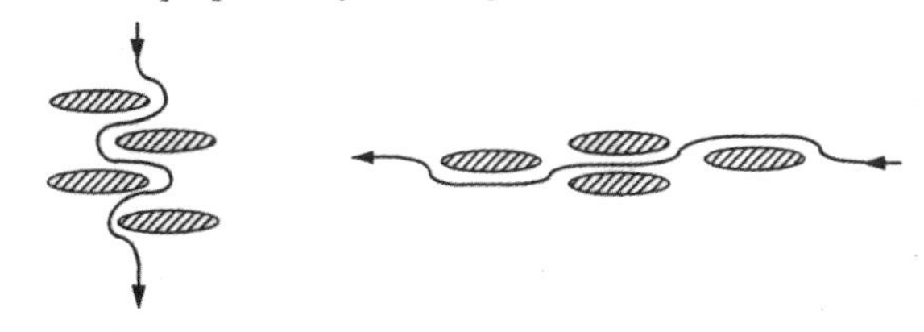

Flow paths perpendicular and parallel to grain axes

Fig. 19–75 Increased Flow Path Lengths—Single Phase

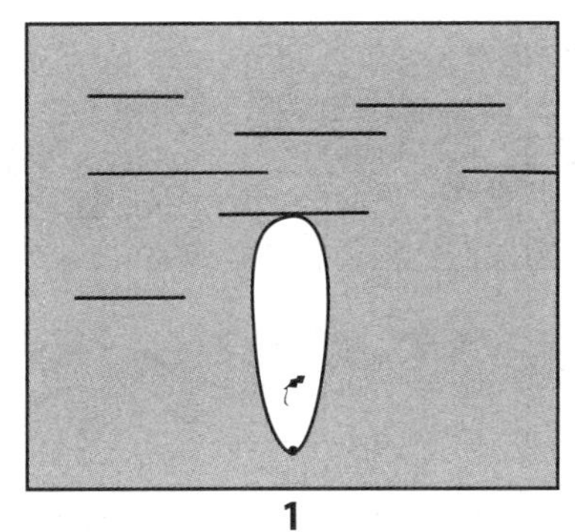

1
Initial rise to first shale contact

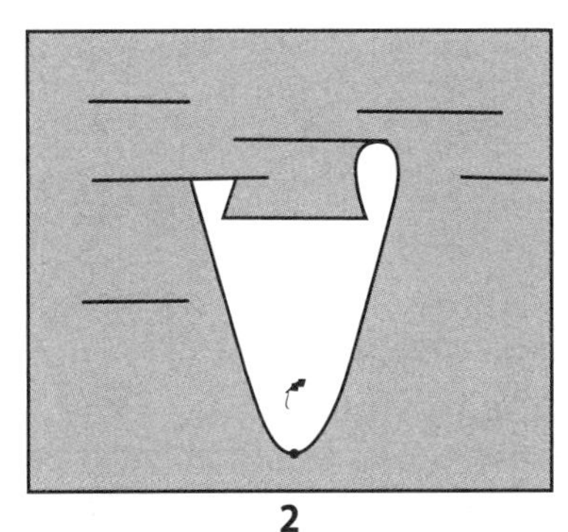

2
Spreading around first shale

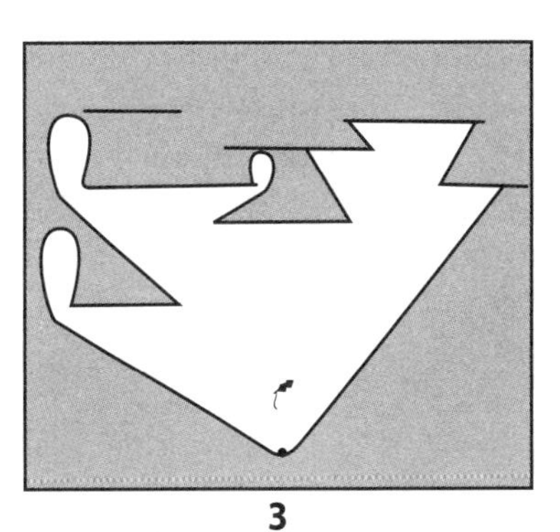

3
Multiple rise and slope development

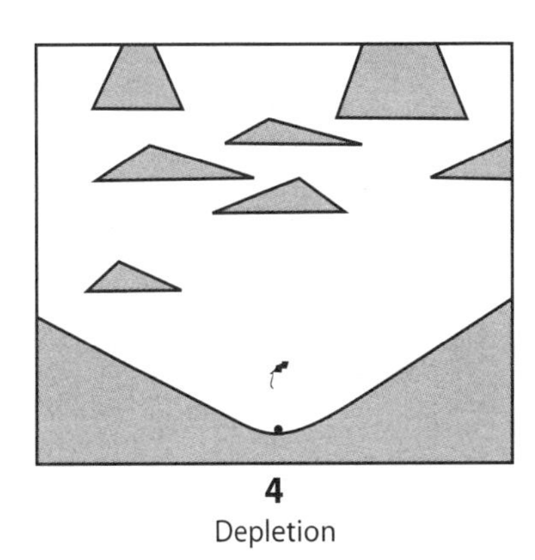

4
Depletion

Fig. 19–76 Growth of Steam Chamber around Barrier

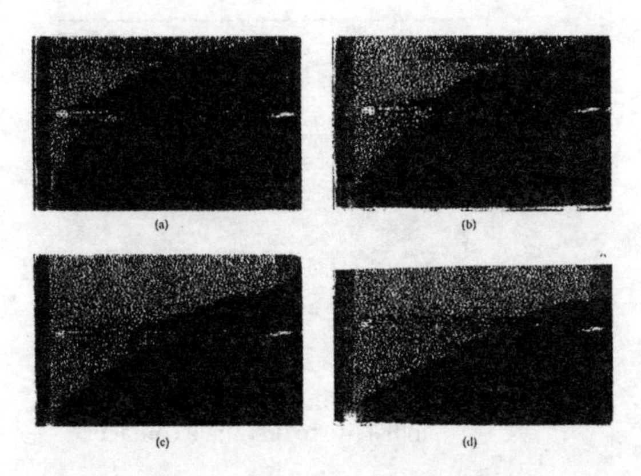

Fig. 19–77 Physical Flow Model of Yang and Butler

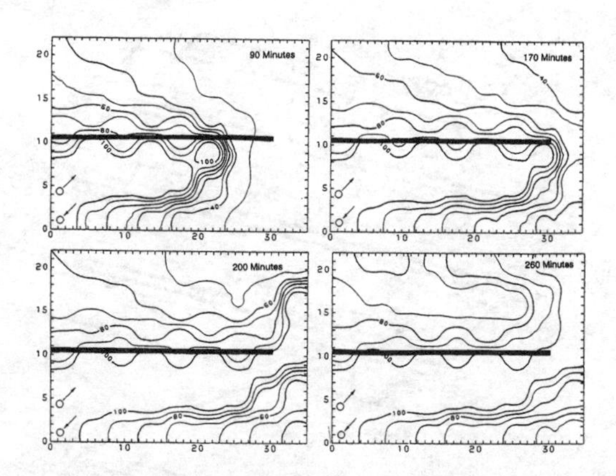

Fig. 19–78 Cross Section of Isotherms and Flow Vectors around Lab Plastic Barrier

Reservoir simulation of UTF and Hangingstone

More recently, Ito, Ichikawa, and Hirata have shown that steam will detour around low-permeability areas. This is shown in Figure 19–79, which illustrates the results from reservoir simulation and a vertical array of thermocouples. Further results are shown in Figure 19–80.[65]

Ito et al. state further that low-permeability zones and such phenomena were seen in seven out of nine observation wells in the UTF Phase B area, which comprised three injector/producer pairs. The diameter of the detours was calculated to be less than 10 meters in most cases. Based on this rate, 77.5% of wells will have some form of barrier in them.

Since the fundamental geological environment of the McMurray sands is similar, it can reasonably be extrapolated that the majority of bitumen wells will have some kind of barrier present. Large sections of pay, without any barriers whatsoever, are the exception for SAGD development.

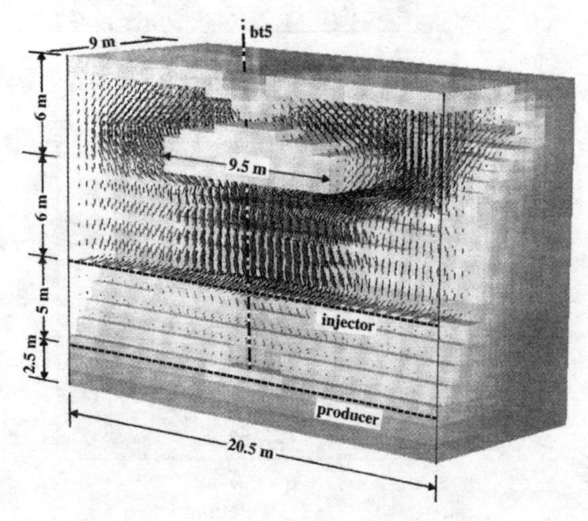

Fig. 19–79 Growth of Steam Chamber and Steam Flow around a Barrier

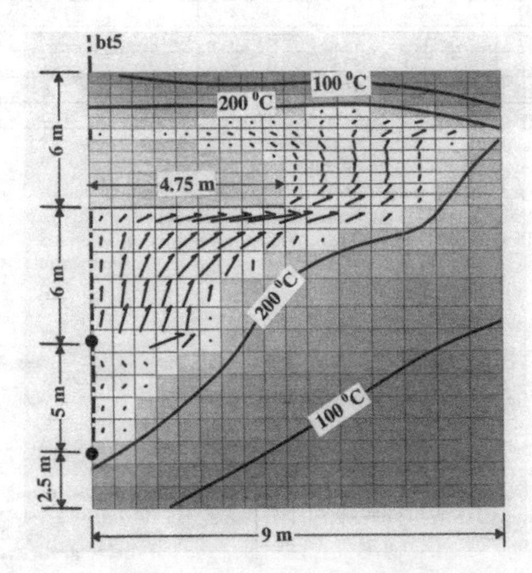

Fig. 19–80 Simulation Cross Section of Isotherms and Flow Vectors around Barrier

Kern River shales

Similar experience has been demonstrated in the Kern River field. Figure 19–81 depicts the shale distribution determined using detailed descriptions for this field.[66] Key conclusions from this study are:

- Detailed heterogeneity modeling is required and can result in good performance matches.
- Discontinuous shales allow significant oil drainage from upper to lower sands. As a result, lower sands have higher apparent recoveries and upper sands have lower apparent recoveries.
- Small pattern element or single sand models are inadequate to predict recovery.

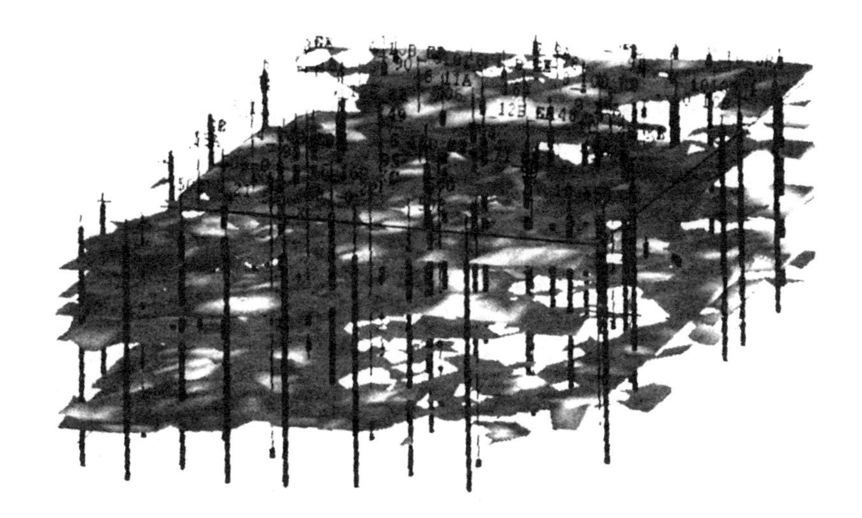

Fig. 19–81 Stochastic Distribution of Shales in Kern River

Properties of mudstones

The shales found in the unconsolidated sediments of northeastern Alberta are not shales in the context of conventional reservoir engineering. In fact, the properties of shales vary considerably.

For instance, as a junior engineer, the author found many of the log analysis examples from the Gulf Coast did not apply to Alberta. Gulf Coast shales are much cleaner and typically have resistivities of 2–3 ohms. In the Western Canadian Sedimentary Basin, shales typically are more a mix of silt and clay and have resistivities that are typically higher, usually 8–10 ohms. One trick in log analysis is to use the resistivity of the overlying or underlying shales to estimate shale resistivity. The inherent assumption is that the clay in the sand is the same as the shales above and below.

The question then becomes, *Is the shale in the sandstones pure clay or a mix of clay and silt?* A strong argument can be made that the pore sizes are sufficiently small that only the clay portion of the overlying sands and shales is really in the reservoir rock. Therefore, shale adjustments should be done with clean clay resistivities of 2 to 3 ohm-meters instead of 8 to 10 ohm-meters. The converse argument is that the clays are intrinsically different—i.e., there are many clay minerals with different conductivities—and that the local shale, really clay and silt, should still be used.

The truth is somewhere in between. The clays are different, but the shales above and below are not clean clays. The trick that can be safely used in the Gulf Coast needs to be applied with caution elsewhere.

The mudstones in uncemented formations are quite different from pure clays. They are, in fact, a mixture of clay, silt, and sand. There is considerably more sand than is typical of most shales and, as a result, the properties of the mudstones are not typical of shales. They have considerably higher permeability and porosity. In this regard, there is a substantial deviation in the pattern of reservoir performance that one comes to expect from conventional reservoir experience.

This cannot be determined by looking at logs. It requires detailed geological descriptions. Thin sections have proved to be very useful in understanding the behavior of these rocks.

Layering

The effect of discontinuous barriers has been discussed in some detail. Like almost all other reservoirs, the fluvial channel environment also has layering and interbedding.

ARE performed some studies using a description based on logs from an actual well. The model included interbedded sands and silts immediately above the injector and midway through the section. The largest effect was to cause the shape of the steam chamber to widen. This is the opposite of the upside-down triangle predicted by Butler's theory.

Interbedding

Interbedding is a series of alternating mudstones and sands. This reservoir is clearly of lower quality than the massive cross-bedded sands that are present in channels. Experience at the UTF indicates that the steam chamber slows in height growth in these sections. This was summarized in an overview context by Farouq Ali.[67]

Shape of the SAGD steam chamber

To date, the most comprehensive data regarding the actual shape of the chambers comes from the UTF site. The plot shown in Figure 19–82 does not show the classic upside-down triangle predicted by Butler from theory that is demonstrated in many lab experiments.[68] The theory is not wrong per se, but actual reservoirs are rarely homogeneous.

The author has dubbed this the manta ray shape, since the chamber undulates somewhat and is pointed at the edges. There has been some intense discussion at ARE as to whether this shape was caused by geomechanical effects. Overall, based on ARE's geomechanical runs, geomechanics do not appear to be responsible for the shape of the chamber. In the author's opinion, the shape of the chamber is controlled mostly by geological effects such as layering and discontinuities.

ARE did a number of sensitivities with the shape of the chamber. In the first case, we input alternating layers with permeability increased by a factor of 2 and decreased (divided) by a factor of 2. The porosity was left constant. This is shown in Figure 19–83.

Fig. 19–83 Alternating Layer Sensitivity

In the second case, we input a gradation of permeabilities as shown in Figure 19–84. This was expected to give the manta ray shape. The actual effect was to change the shape of the chamber into the shape of the superman crest with the truncated upper edges of the upside-down triangle rather than a manta ray.

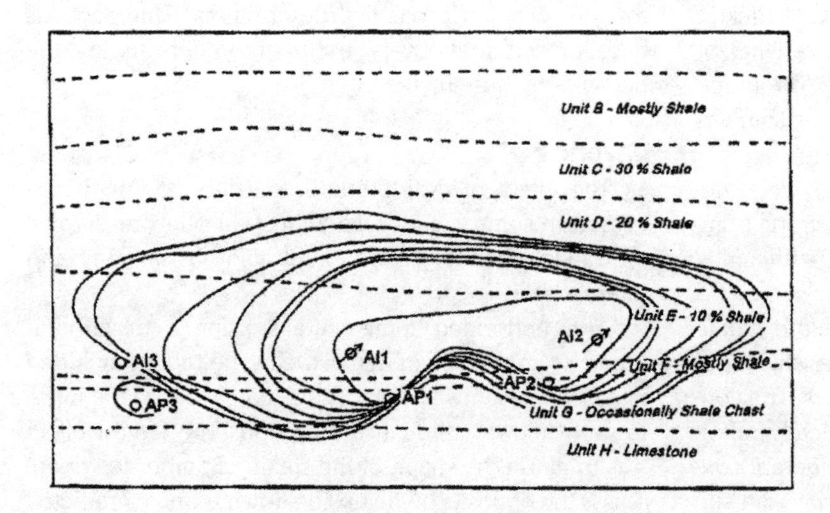

Fig. 19–82 Shape of Steam Chamber from the UTF

In the third case, we input a series of discontinuous barriers. The pattern is too dense and too closely spaced to be realistic. The shape generated certainly indicates an erratic development as shown in Figure 19–85.

In the final analysis, it seems likely that a combination of Case Two with gradational permeability changes, Case Three with some discontinuities, and the Layered Case with interbedded layers near the horizontal wells (which produced the right-side-up triangle) could be combined to match the shape observed in the UTF Phase A. Note that there was an interbedded layer in this pilot area.

It will be interesting to see further results as they become available.

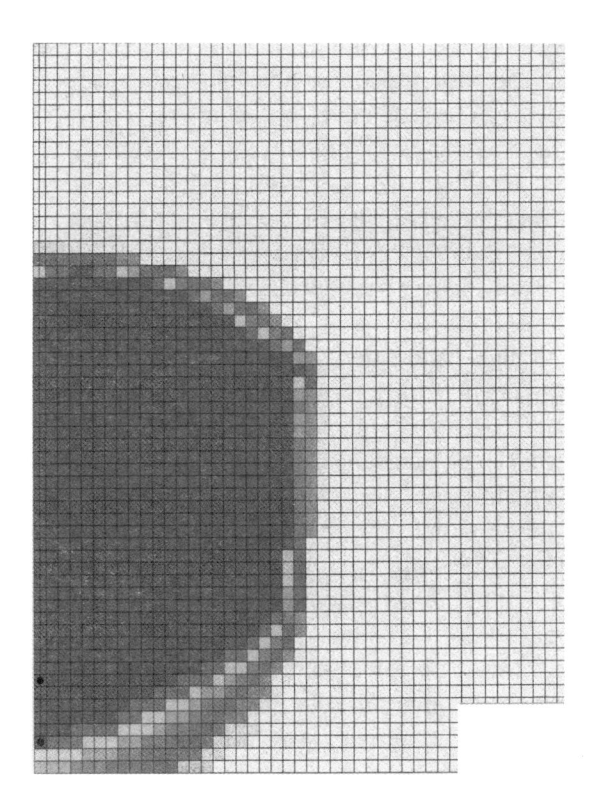

Fig. 19–84 Gradation of Permeabilities

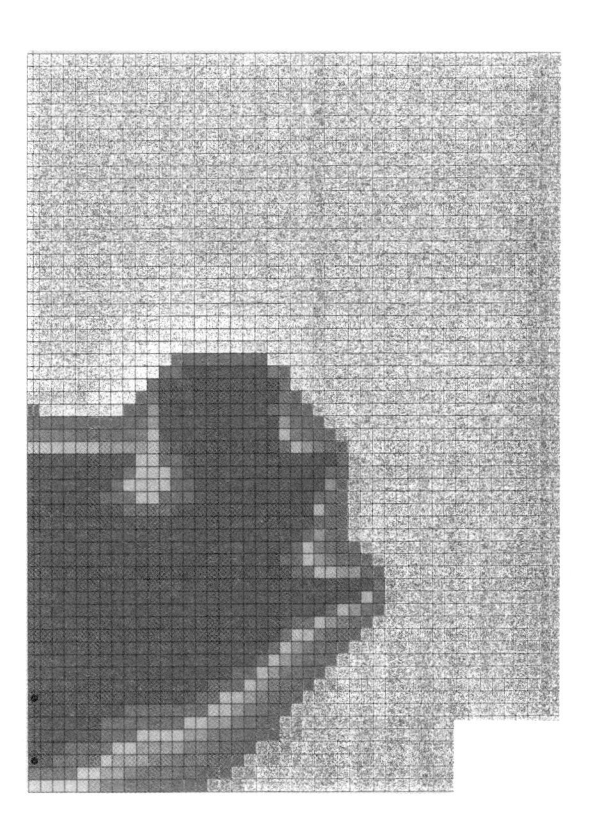

Fig. 19–85 Discontinuous Barriers

Tracking the development of the steam chamber

It is easiest to track the overall development of the steam chamber by monitoring the progress of isotherms. This change is quite dramatic. However, there is some conduction away from the chamber and a gradation of temperatures. Therefore, the change of colors on the output plots does not indicate the precise location of the steam chamber interface.

The ternary diagram indicates where the steam chamber itself is located. The ternary diagram is a color-coded triangular plot, which indicates whether the fluids in the formation are water, oil, or gas. The ternary diagram cannot indicate if the gas is steam or hydrocarbon (methane).

The pressure plot shows a general increase in pressure within the steam chamber. However, pressure increases precede the steam chamber. The pressure increase in the reservoir is caused by a number of occurrences. The oil expands with temperature, as does the matrix. If the amount of thermal expansion is different for the matrix and pore fluid, then the pore pressure can either increase or decrease. The reservoir fluids can move to equalize these pressures. Until high temperatures are reached, the bitumen viscosity is very high. As a result, these pore pressure increases dissipate slowly. Pressure increases can occur from stress increases also.

Water above and below

Generally speaking, water below is not thought to be a problem in SAGD. Some vertical standoff is required, but this is in the order of 1 to 3 meters. This may leave a small basement of oil that is unrecoverable.

Top water can also affect recovery. When the steam chamber rises vertically into the water zone, the water will segregate by gravity. The water zone also absorbs a great deal of heat. A tiny skiff of water on top of an oil column will not be significant. Therefore, it is the relative amount of water and oil pay that is significant. A general rule of thumb is that up to one-sixth of the total pay can be water filled without noticeable effects on production performance. At proportions larger than one-sixth water, it is a matter of economics.

Gas above and below

Gas below may seem counterintuitive. In fact, the gas is not truly below—i.e., below the oil pay—but rather distributed in layers through the pay. As discussed previously, thin layers of gas do not necessarily indicate a barrier to steam or higher injection pressures, unless the mudstones are quite thick. Gas layers are more common in the western part of the McMurray bitumen trend.

Gas above the oil pay also affects performance. So far, most of the efforts regarding this issue have been directed to estimating the effects of gas cap pressure reduction caused by gas production. This has resulted in some of the bitterest controversy ever seen in the Canadian industry. The Alberta Provincial Government has held three hearings on this to date: the General Bitumen Inquiry, the Surmont Hearing, and the Chard-Leismer hearing. No decision has been rendered on the last hearing. The first two are available on the Internet at the AEUB website. They are in a .pdf file format and are free. Rather than try to paraphrase these comprehensive reports, the author recommends that individuals read them directly.

Implications of thief zone

From a SAGD design perspective, the thief zone represents a serious technical problem. The steam chamber must be run at lower temperatures and pressures to prevent the loss of steam. Steam loss cannot be prevented entirely. This can represent a real loss in efficiency, which translates directly into the potential economics of the project.

Thief zone

The top of the Wabiskaw/Upper McMurray sands are thought to be in communication on a regional basis. This can be represented in a number of ways. One way is to use a dummy well to remove high-pressure fluids. Another way is to extend the grid with increasingly large grid blocks. Since there is no steam chamber advance into bitumen in water and gas zones, larger grid blocks can be used.

ARE has prepared a number of such sensitivities using both methods. The initial sensitivities were done with a half well pair pattern model. Early runs with the sensitivities indicated high steam-oil ratios and high heat losses for a single pattern. These results were, however, not realistic. Since the actual projects are implemented in pads with multiple wells, modeling was changed to include a 2½-pattern arrangement. On a proportionate basis, steam losses dropped considerably and reduced steam-oil ratios. The author recommends that thief zone losses be modeled, as best one can, with elements of symmetries typical of actual development.

The production of gas has introduced some significant pressure gradients in the overlying upper sands. Restated, the thief zone pressures are unlikely to be constant overlying a potential SAGD development. If this is the case, there may be influx from the boundaries adjacent to higher pressured sands and losses at the boundaries adjacent to the lower pressured overlying sands. The combined steam chamber from multiple patterns may be expected to equalize due to the much lower viscosity of steam than water.

One potential advantage of modeling the reservoir away from the pattern is that overburden heat losses can be accounted for. We used null grid blocks for the reservoir outside of the pattern—with an extra margin. The underburden heat losses in the overlying thief zone away from the margin were not calculated. In the end, we felt that all of the heat was lost through the dummy well, and this was probably a more conservative assumption. In reality, a new pattern might be built eventually next to the existing pattern, and some of this heat might be recovered.

This suggests that there may be considerable merit to attempting to block the permeability in the regional sands adjacent to a SAGD project. Experience in conventional oil and gas suggests that implementing this will not be easy.

Field thief-zone losses

Field data is available from one project at Kearl Lake in Alberta. This was a pilot project where there was an overlying gas zone. The pressure losses here reduced the efficiency of the steam injection. Natural gas injection was used in a circle around the pilot well to repressurize the area. Pattern containment is a classic pilot project problem with conventional waterflooding. Since the patterns do not repeat themselves to provide containment, it is well-known that pilot responses typically underestimate field scale waterflood implementation. The increased pressures that occurred due to gas injection did improve production response; however, a considerable amount of injected gas was required.

From an operational perspective, it will be difficult to evaluate the exact operating pressure without knowing the thief-zone pressures. We have not found any way to detect breakthrough easily from the simulations that we have done. Gulf has indicated, in a government application, that they feel that water cuts will indicate when this occurs. Given the problems with relative permeability determination, outlined previously, the author is skeptical. Experience will help resolve this issue; unfortunately, such information is not in the public domain yet.

SAGD Design Issues

Criteria for evaluating performance

The major criteria for evaluating SAGD are:

- **Breakthrough Time.** This is the time that it takes for communication to occur between the horizontal injection and production wells. It is a function of input heat as well as the distance between the injector and producer.
- **Production Rates.** This affects economics directly. Higher rates translate into revenue. It is common practice to refer to the maximum rate, because this usually indicates the overall level for the production forecast.
- **Recovery Factor.** This is an indication of the total amount of resource that can be expected to be recovered. However, a high recovery factor does not necessarily translate into the most efficient economic recovery.
- **Steam-Oil Ratio (SOR)/Oil-Steam Ratio (OSR).** This is an indication of the efficiency with which bitumen is recovered. It can be specified on an instantaneous and cumulative basis. Normally the cumulative SOR is used. It inversely describes the overall efficiency (i.e., a low SOR indicates an efficient SAGD process). Usually, the instantaneous SOR is quite high at the beginning of the steam injection process. This is inherent in the physics of the process. However, toward the end of the steam injection process, the instantaneous SOR can be an indication of when the project should be terminated economically.
- **Rise Rate.** Many people use rise rate as an indication of SAGD chamber performance. This is normally expressed in cm/day or in./day. The author regards this as potentially misleading. Chamber shapes are not necessarily the classic upside-down triangular, and discontinuities, even of significant dimensions, can affect rise rate without really reflecting whether the SAGD process is working. However, given a reasonably homogeneous reservoir, this can be a useful indicator.

In the near future, the author believes that these criteria will be changed somewhat. This will be discussed under global optimization.

Well length

Once the turn has been made in a horizontal well, the cost of a horizontal well does not appear to vary greatly with length. This strongly suggests the use of longer wells. Friction pressure losses in the steam injection well can be significant. Optimal chamber development suggests that the steam chamber should be as uniform as possible. Detailed well models have been developed for modeling the flow in horizontal SAGD wells. The general result is that large casing sizes are used for SAGD wells. Typically, 177.8 mm (7 in.) casing seems to be the most popular. In some cases, 244.5 mm (9⅝ in.) casing is used. This suggests that there will be a point beyond which longer wells are not advantageous.

Since 3-D modeling is so time intensive, the author has seen relatively little work in the literature on this issue. Intuitively, one would expect that as heterogeneity increased, steam chamber problems would increase. There is not yet much practical experience with this issue. We do know that the fluvial environments have rapid changes in reservoir properties and, in the author's estimation, this will favor the use of shorter rather than longer wells.

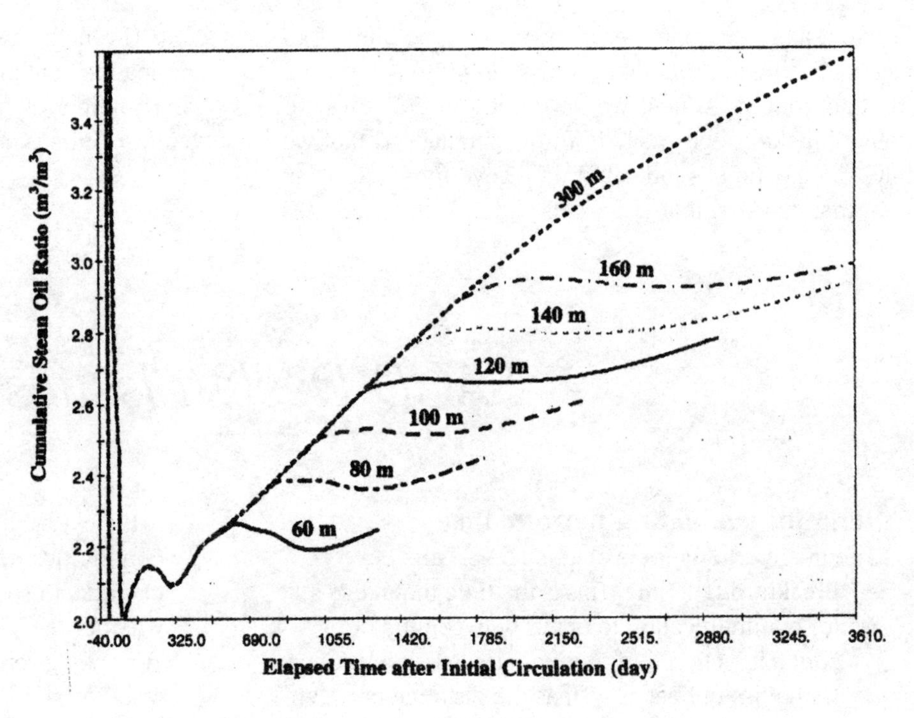

Fig. 19–86 Effect of Horizontal Well Spacing—Cumulative SOR

There are practical considerations in the use of longer horizontal wells. The chance of workover failures and completion failures will increase with length. Undoubtedly, one of the more important aspects in the horizontal section is physical friction. Screens associated with slotted liners are physically delicate.

It appears that the best well length is not known yet. The longest wells proposed seem to be about 1,000 meters (3,250 feet), and the shortest are about 500 meters (1,500 feet). From an economic perspective, ignoring operational problems and heterogeneity, it would seem that the longer wells would be more economical. Based on experience in the conventional side of the business, the author has used short well lengths in simulations. The average seems to be about 750 meters (2,500 feet).

Well spacing

More work has been done on the spacing between wells. The best summary on this issue is the work of Ito, Ichikawa, and Hirata, which is shown in Figure 19–86.

This work indicates clearly that there is a link between recovery and well spacing. This will depend on the height of the formation also. It seems that most designs for commercial projects are in the range of 70 to 110 meters between wells, with 90 being typical. The author suggests that this is an appropriate step for designing a field project in the detailed design stages.

Vertical spacing between producer and injector

The effect of vertical spacing has also been studied. This was examined by Edmunds and Gittins and is shown in Figure 19–87. With larger vertical spacing, more time is required to initiate communication between the injector and the producer. Ultimately this is an issue of timing and changes in rates.[69]

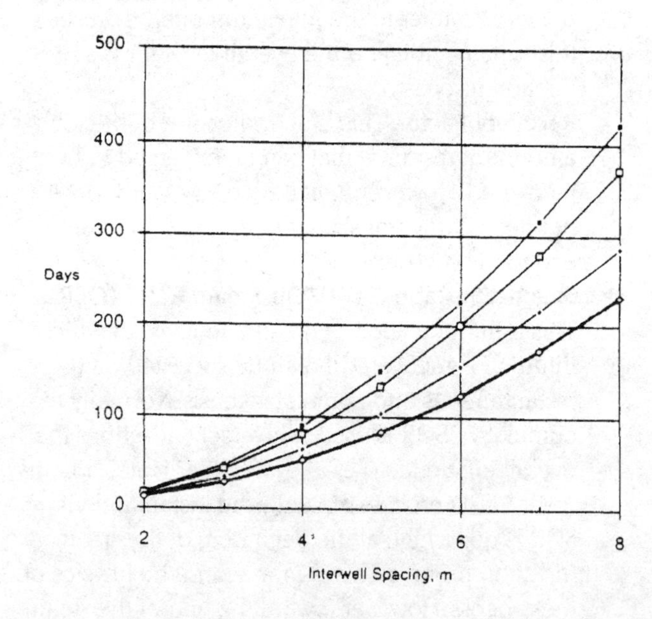

Fig. 19–87 SAGD Start-up Time vs. Spacing and Potential Difference

SAGD start-up

Most SAGD simulations, as previously discussed, have been performed using 2-D cross sections. Actual wells are never spaced evenly apart. In the past, detailed studies have been conducted using the actual directional surveys from the wells. True 3-D models are difficult to run since the longitudinal grid blocks are normally quite large; they should be of the same order of size as the horizontal dimension. Most of the comments in the literature are quite oblique on this point. The concern seems to be to develop the chamber in a manner that is as even as possible.

Given that most conventional reservoirs demonstrate permeability variations, as do oil sands reservoirs, the author's expectation would be that steam chamber growth would be more strongly dominated by reservoir factors than the distance between wells.

The author has an example where long distances were used, and there is a section that apparently does not develop because of increased well spacing. This is a hypothetical example. The concern with this result is that the block is so wide that lateral heating of the grid-block volume is not possible. This simulation, which is shown in Figure 19–88, would not, in the author's opinion, actually represent what would occur in the reservoir.

To more correctly model along the length of the well, a finer grid should be used as shown in Figure 19–89. The steam chambers now coalesce along the axis, consistent with cross-sectional modeling.

The data of Strobl et al. from the UTF site shown in Figure 19–90 supports the view that reservoir properties affect chamber shape.[70] This is related to the IHS discussed earlier and the rip up breccia in the left corner.

The IHS, which occurs on a fine scale, is difficult to model. This would not be a problem if permeability were input as a true tensor, as discussed in chapter 2.

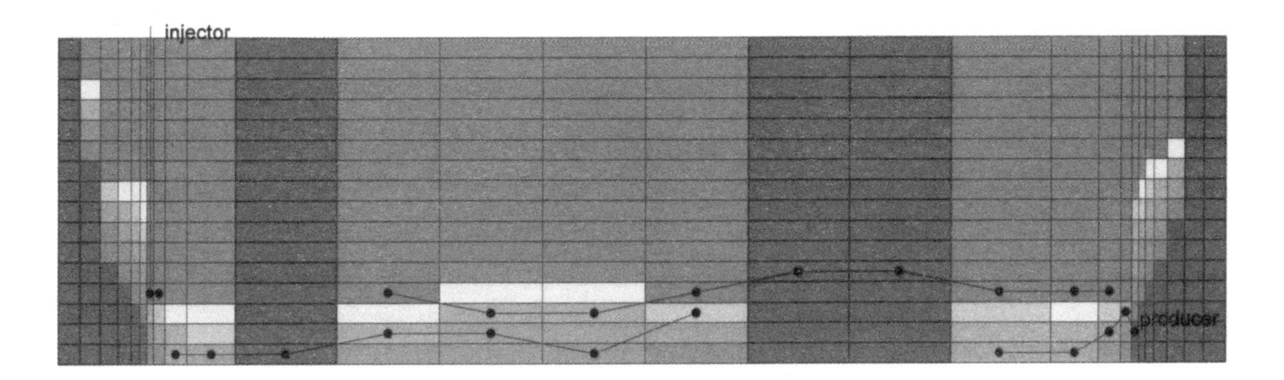

Fig. 19–88 3-D Optimization Using a Coarse Grid along Horizontal Well Axis

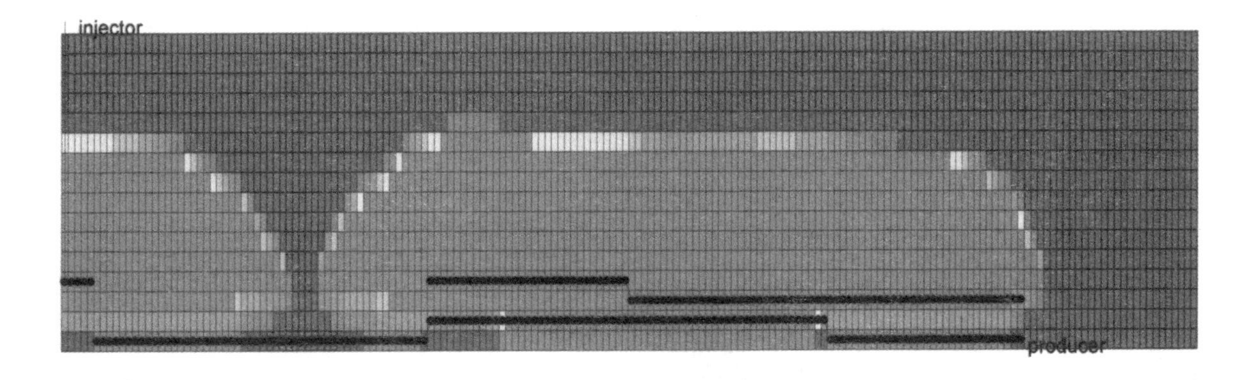

Fig. 19–89 3-D Optimization Using a Fine Grid along Horizontal Well Axis

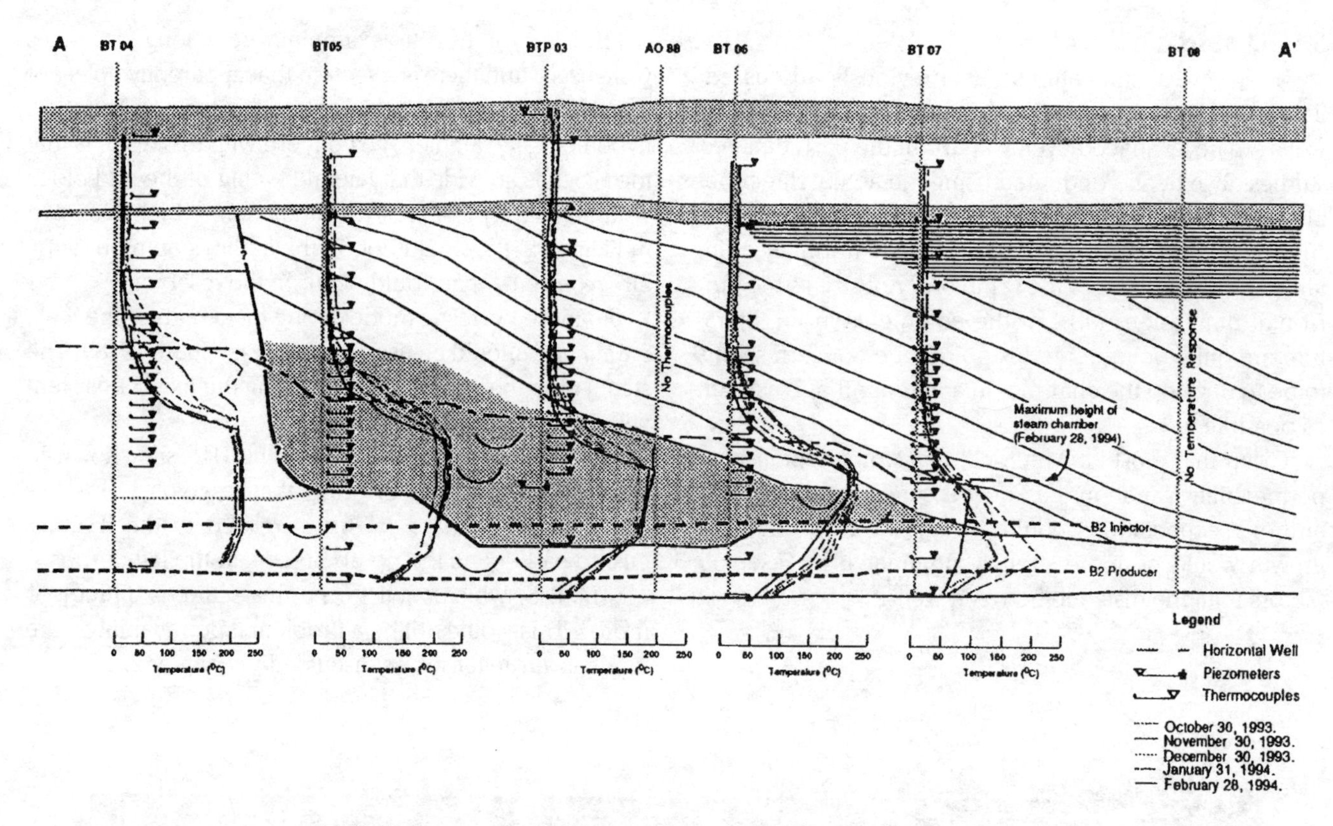

Fig. 19–90 Temperature Profiles and Effects of IHS Measured in UTF Reservoir

Artificial lift

Provided sufficient pressure is available, the SAGD process, once started, will provide steam lift naturally. The flow of these fluids to surface is not straightforward. Figure 19–91 shows a plot from the UTF Phase A and documents pressure and temperature conditions in the riser.[71] Recall that the UTF Phase A wells were actually located in a mine, and the fluids did not return to surface as they would in a regular well.

Edmunds points out that they did not operate the well where there was intermittent flow and that these conditions are quite undesirable. Certainly, the intermittent production seems to have interrupted the surface facilities. Slugging is hardly new to the oil business; this phenomenon is observed on many conventional wells. Still, some idea of how big and how frequent the slugs will be is quite helpful. Sometimes these details are not transmitted to facility design people, who may design facilities for steady state conditions.[72]

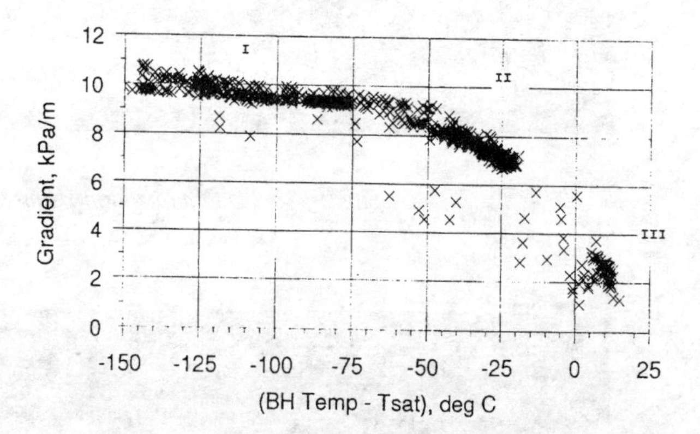

Fig. 19–91 Vertical Riser (from Mine) to Surface vs. Inlet Subcooling

One of the implications of Edmunds' paper seems to be that intermittent production is inherently bad. This is certainly true if the surface facilities were not designed to handle it. Conversely, Donnelly describes the intentional use of intermittent flow for the Hilda Lake project.[73]

Wellbore hydraulics programs have been designed to calculate these effects, which have been called geysering. The analogy is to the geyser in Yellowstone National Park in the United States. Donnelly compares the lift at Hilda Lake to a coffee percolator (Perco-Lift?). Some of the results from the Edmunds paper are shown in Figure 19–92.

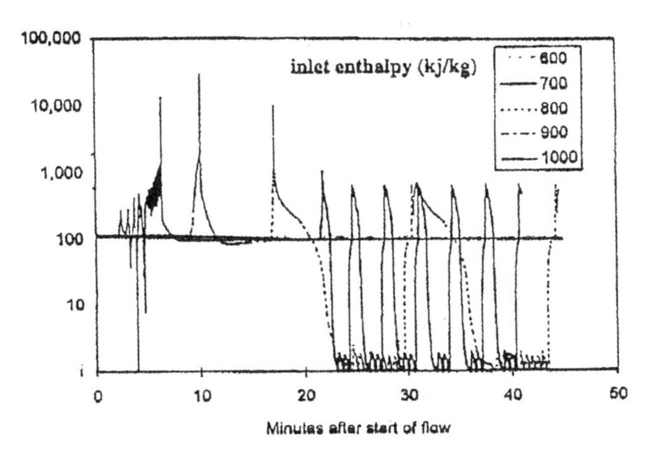

Fig. 19–92 Effect of Steam Inlet Enthalpy on Vertical Flow in Well

Indeed, lift does occur on an intermittent basis. This program has been linked with a reservoir simulator to solve both intermittent wellbore and reservoir problems simultaneously. Butler has presented calculations using steady state correlations. This information may also be used to determine lift from SAGD wells. Butler's assumptions are somewhat different.[74]

The lift of oil can be aided by steam injection during the early stages. No extra facilities are required since all of the required equipment is in place for the initial start-up phase. Some companies plan to use natural gas as a form of gas lift.

The use of gas lift can be expanded to lower pressures by staging the lift. This process, called E-lift, is shown in Figure 19–93 and was originally developed by Dr. Ken Kisman.

Conventional sucker-rod pumps are limited by stuffing box rubbers and limited rate capacity. The rates from most SAGD wells typically make electric submersible pumps (ESPs) more appropriate. However, they are limited in the temperature range across which they can operate. Most people seem to feel that 170° Celsius or about 800 kPa (340° Fahrenheit or 116 psi) is the limit of current technology. ESPs also have limited sand production capacities. This may change in the future.

ESPs are more costly to operate and are more difficult to put in the wellbore. At present, there is limited power available in northeastern Alberta. ESPs consume large amounts of electricity.

The cost and viability of artificial lift is a subject of interest from an optimization perspective. Early optimization analyses have concentrated on minimizing steam-oil ratios. These attempts are quite preliminary and will be discussed later.

Geomechanical effects

The following information is required for geomechanical effects:

- *in situ* stresses
- formation strengths
- permeability variation

In situ stresses

In situ stresses have been determined for some areas. There are a number of ways of doing this. The stresses can be estimated from overburden calculations. In very general terms (ignoring detailed tectonic stress analysis) the largest stresses are horizontal at depths above 175 meters. In moderate depths, from 175 to 575 meters, the

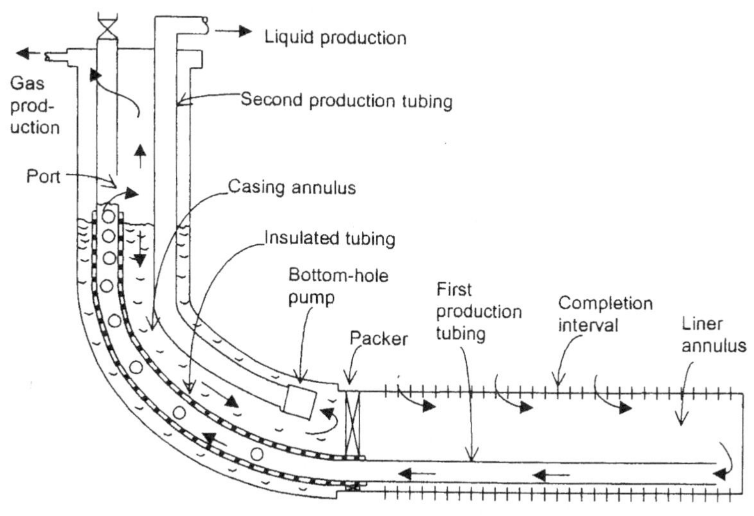

Fig. 19–93 E-Lift Production Scheme

horizontal and vertical stresses are approximately equal. At deeper depths, below 575 meters, the largest stresses will be vertical.

It is possible to use mini-fracs to determine the minimum stresses experimentally. This technique has been used in the conventional side of the business for some time. These tests and data are used for hydraulic fracture treatment design. It is also possible to estimate formation permeability with such tests.

These tests were applied at the UTF site, and the results were published. The UTF site is quite shallow and, as a result, it is expected that maximum stresses would be horizontal.

It should be pointed out that horizontal stresses are often not uniform. There is a directional preference, and this promotes failure.

This type of test and data interpretation is usually done by specialist geotechnical engineers.

Formation strengths

The degree of dilation that occurs is strongly dependent on the physical properties of the reservoir sands. This type of testing is not routine yet. Most work has been done in the university environment. Specialist geotechnical companies build and operate testing equipment.

The most important aspect is getting samples that are not disturbed. This was discussed from a reservoir engineering perspective earlier in this book. Dusseault (1977) examined the effect of sampling disturbance on specimen properties. Geomechanical properties such as shear strength, stiffness, compressibility, and permeability are very sensitive to disturbance.[75]

Rajani and Sanchez (1988) attempted to characterize the geomechanical properties of the oil sands within the heavy oil belt in Venezuela. They concede that sample disturbance, as indicated by higher core porosities, had affected their measurements of absolute permeability and hoped that a relationship could be found to apply similar results to *in situ* conditions. Chalaturnyk and Scott (1992) described how reservoir properties could be obtained from a geomechanical laboratory.[76] Reservoir engineers were cautioned to use the appropriate stress paths so that the laboratory tests would suit their particular recovery process. Vasquez et al. (1999) conducted mechanical and thermal tests on unconsolidated sands for the Tia Juana field heavy oil SAGD project in Venezuela.[77] Core disturbance, when coring appears to have resulted in disturbed samples, is indicated by higher than expected core porosities. The resultant friction angles of 25° to 27° appear to be low. This was attributed to sample disturbance.

Getting good data is very difficult, and many sophisticated attempts result in failure. Therefore, acquiring good data requires some persistence. From a practical perspective, the author has had to rely on the best data from offsets, which are often a considerable distance from the actual field where work is being conducted.

- Scott et al. (1994) described their laboratory results on Cold Lake oil sands. The volume and permeability changes measured resulted from changes in the effective stresses, shear stress, and temperatures. In these cases, it is necessary to borrow from others' experience.[78]
- Oldakowski (1994) studied the stress-induced permeability changes of Athabasca oil sands at reservoir temperatures (~8°C) and noted that, although absolute permeabilities were in the order of 1,000 to 5,000 mD, the effective permeabilities to the water phase were 0.01 mD. It is difficult to displace water around an immobile bitumen phase. Shear stresses resulted in distortion and dilation, which resulted in an increase in absolute permeability if the confining stress was below a critical level. Oldakowski found that dilatancy was pronounced at failure and postfailure and that it was a function of both the confining stress and the original porosity of the specimens.[79]
- Chalaturnyk (1996) studied the UTF Phase A pilot project. His laboratory program examined thermal volume change, thermal conductivity, compressibility, stress-strain and strength behavior, gas evolution, and the composition and properties of oil sands, shale, and limestone. There was an error made in the loading procedure for the samples, which caused some disturbance effects. Accordingly, he did not use his own dilation data. Instead, he used the data of Oldakowski.[80]
- The best data to date seems to be the data of Touhidi-Baghini. These samples were taken from the McMurray in a river outcrop. In this area, the McMurray is not bitumen saturated. This data includes a comprehensive suite of tests. It will be discussed in more detail in the following section.[81]

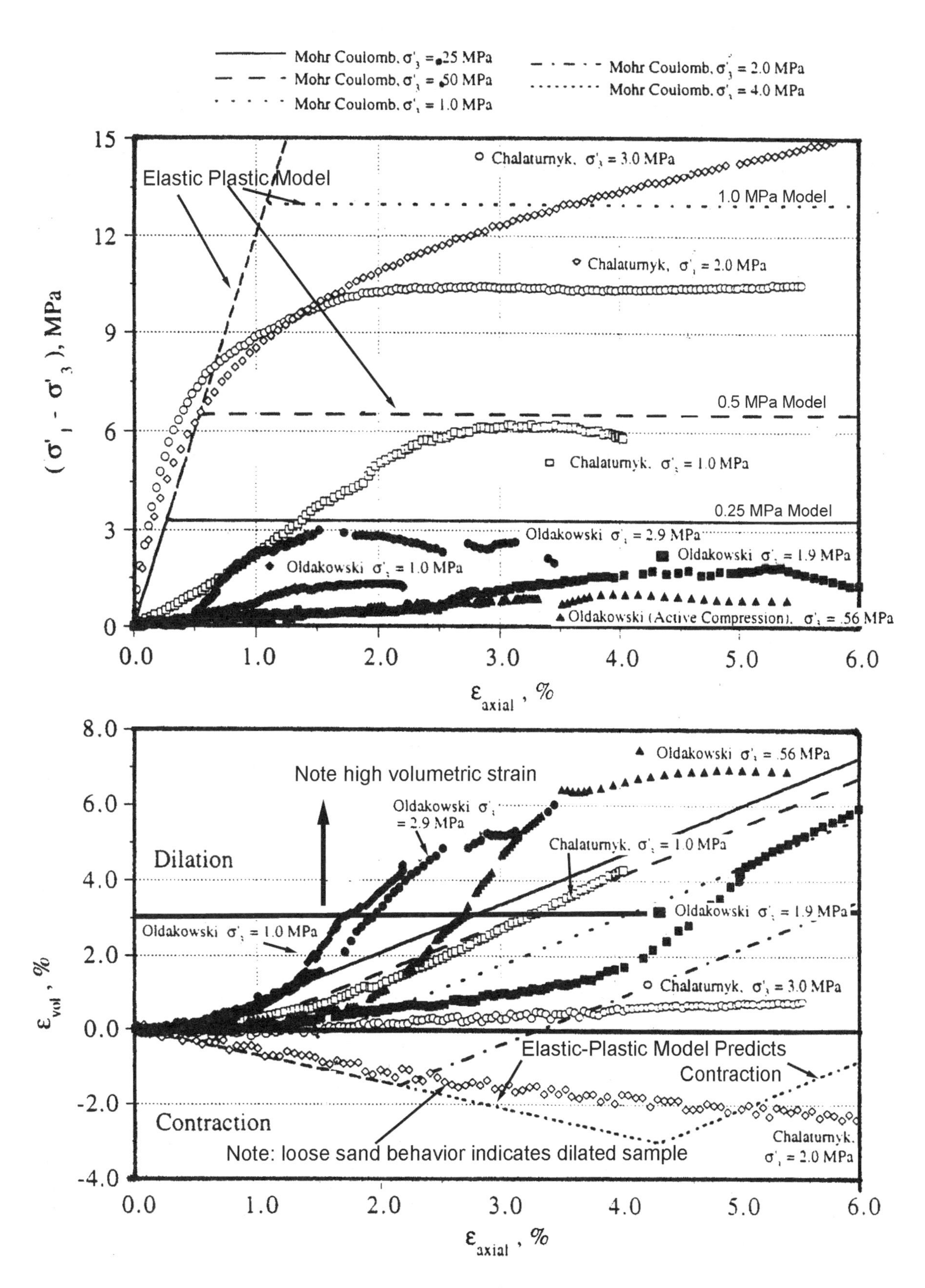

Fig. 19–94 Deviatoric Stress vs. Axial Strain—Experimental Data

Failure model and permeability changes

Different failure models, which describe the manner in which the oil sands react to changing pore pressure and physical loads, can be used. Most researchers (Agar, 1984; Kosar, 1989) have concluded that oil sands are generally nonlinear and exhibit strain softening behavior. For oil sands, ARE has utilized a hyperbolic model.[82, 83]

The next most popular model is the elasto-plastic model. This represents a compromise between matching real behavior and keeping the model as simple as possible. Selection of a perfectly elastic plastic model results in an underestimation of the shear-induced volume changes occurring during the SAGD process. Figure 19–94 compares experimental and idealized stress strain relationships for oil sands.

The primary reason for the elasto-plastic model's inability to represent adequately the dilatant volume changes is related to the method used in computing volumetric strains. Plastic volume change is only computed when an element is in a state of yield. Below yield, the model only calculates elastic volume change. The stress paths illustrated in Figure 19–95 show that elements along the left boundary reach the peak failure envelope only briefly. They then undergo an increase in the mean normal effective stress, causing a return to an elastic state and no further calculation of plastic volume changes. In reality, oil sands are a strain-softening material. They would continue to deform and soften once the peak failure envelope was reached.

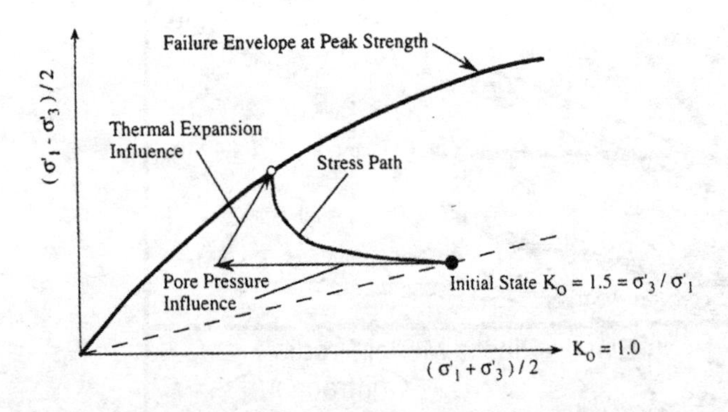

Fig. 19–95 Demonstration of Stress Path

Modifications to the strength parameters can be made to force the model to yield much more quickly and initiate shear-induced volume changes. While this is a realistic alternative, it causes the stresses and strains in the reservoir to be unrealistic.

Reservoir simulation and geomechanics

The most advanced geomechanical/reservoir simulators would solve the stress, strain, and fluid mobility functions simultaneously. This would be a truly coupled analysis. Attempts have been made to develop such programs. However, the matrices are very large and apparently difficult to solve. Coupled solutions involving single-phase fluids are available; however, these solutions are of limited interest to a reservoir engineer. The results have been applied largely in the disciplines of hydrogeology and geotechnical engineering.

It is possible to use a simpler form of geomechanics. The CMG STARS simulator does this by associating a predetermined dilation and its associated increases in porosity and permeability with an increase in pore pressure. This is a distinct improvement compared to the omission of geomechanical effects. It has a major shortcoming in that it assumes, *a priori*, that all the reservoir rock will fail and dilate once a critical pore pressure is attained. Implicit in this assumption is that the total earth stresses are such that the increase in pore pressure will necessarily result in dilation. No account is taken of the actual stress state of the rock, nor can it account for the changes in stresses due to thermal expansion or a readjustment of stress.

Analyses performed by ARE using the simplistic geomechanics facilities within STARS have provided quite satisfactory results. However, this is not thought to be a unique solution and may be purely coincidental.

Chalaturnyk (1996) used an alternate methodology. His analysis consisted of using results from the STARS thermal reservoir simulator to obtain temperatures and pressures, which were then used as input for a geomechanical stress-strain simulator. The results of his geomechanical model are therefore not used to modify the reservoir simulation.[80]

To be fair, Chalaturnyk met his objective of demonstrating that geomechanical effects should be accounted for. It was shown that displacements within the reservoir are capable of significantly influencing reservoir properties. Vertical strains of 2.5%, horizontal strains of 0.3%, and volumetric strains of 2.5% were found to increase the absolute permeabilities by 30% to 40% during SAGD operations. Chalaturnyk and Li used the same methodology to calculate geomechanical effects at discrete timesteps.[84] However, these methodologies are not, in the author's opinion, sufficient for design purposes.

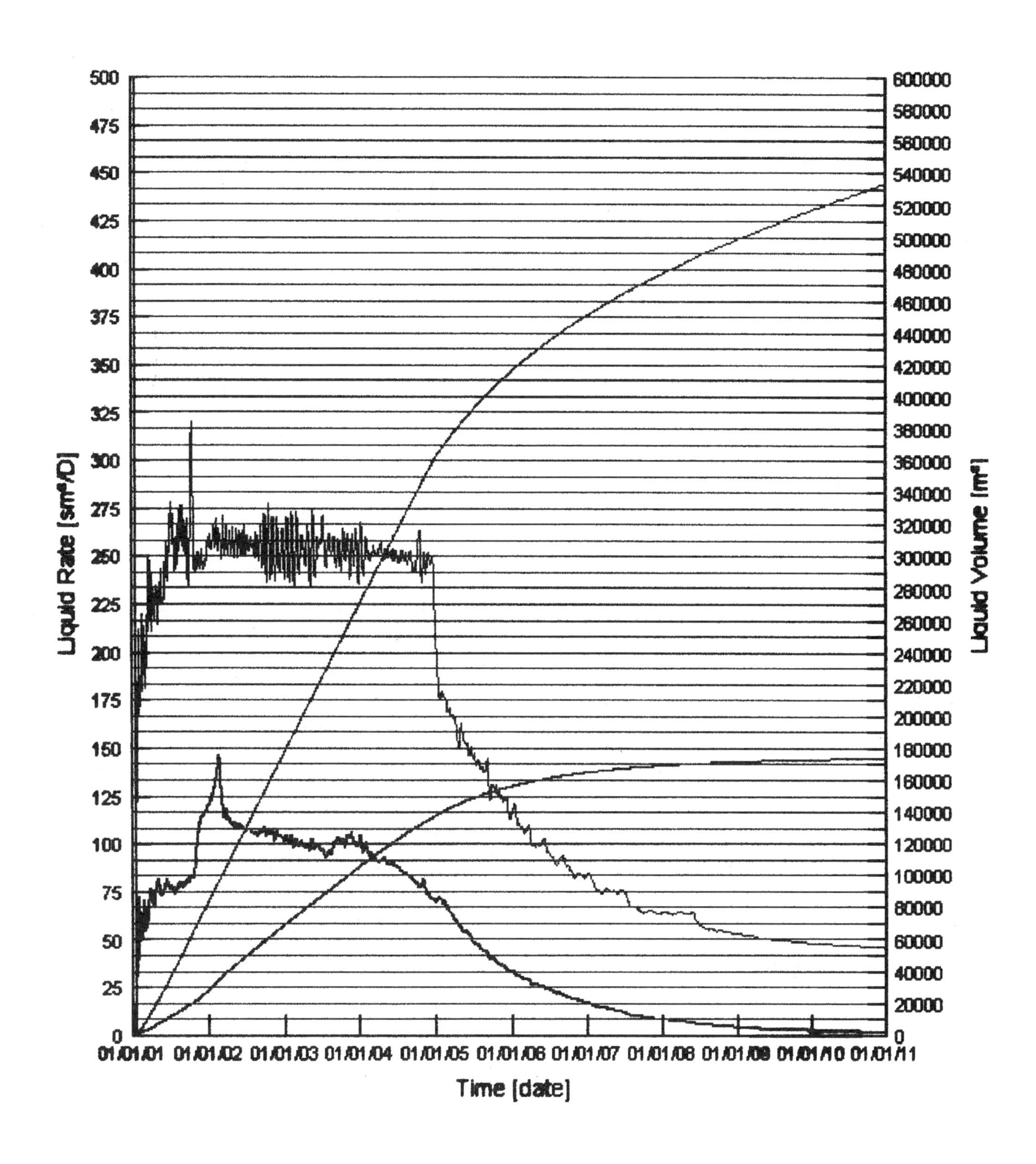

Fig. 19–96 Geomechanics Simulation—Oil Rates

Another method to avoid the computationally onerous requirements of a true simultaneous solution is to solve the stress-strain equations with one program and the conventional reservoir simulation equations with another program. The output from each is used to modify the operation of the other. As an example, the temperatures and pressures from the reservoir simulator would be input into the stress analysis program to obtain updated effective stresses. Similarly, the volumetric strains from the stress analysis package would be used to update porosities and permeabilities in the reservoir simulator.

When done in sufficiently small timesteps, this method results in realistic predictions of geomechanical reservoir behavior. Note that this type of twinned (or "leapfrog") analysis is correctly called an uncoupled or partially coupled analysis because the solution of stress and flow equations are done separately.

Fung, Buchanan, and Wan (1994) described their numerical model (CMG's STARS simulator) for the solution of poro-elasto-plasticity and multiphase thermal flow in oil sands reservoirs. The stress analysis was done with a finite element program, which was uncoupled with the thermal reservoir model. Formation properties are updated at the end of timesteps. This is somewhat similar to the change in saturations with the IMPES method. The authors conducted a series of simulations, with and without geomechanical effects, and concluded that the geomechanical effects were significant. These tests were done for cyclic steam stimulation. During injection, the reduction in effective stresses caused shear failure around their well, with the increased porosity enhancing the permeability and injectivity. Volume and permeability increases throughout the reservoir resulted in a larger heated volume and total recovery than for the case where geomechanics was ignored. The linkage between permeability is done through the porosity, which changes with dilation.[85]

This program also allows modifications of relative permeability relations with dilation. This is a useful feature, although the author has not come across any data that deals with this topic directly. The objective of this feature is the expectation that sheared zones would have relative permeability curves more similar to those of a fractured zone, i.e., linear relative permeability curves. The work of Oldakowski certainly shows that the permeability of water changes in cold bitumen with shearing.

There is another reservoir simulator (Settari and Walters) that uses essentially the same process. The model differs in that a hyperbolic description of failure is available, in addition to the poro-elastic-plastic model. The permeability changes are handled differently. Increased permeability is linked through volumetric strain and Rowe's dilatancy theory. This is somewhat more convenient for varying porosities and ties in more directly with geomechanical testing. The results are similar.[86]

Example results using geomechanics

The results of a run with geomechanics are shown in Figure 19–96. Here the advantages of running at a higher operating pressure are obvious.

The formation has undergone some volumetric strain at and above the wells as shown in Figure 19–97. The associated increase in porosity is shown in Figure 19–98. Figure 19–99 shows the permeability increase in the vertical direction.

Fig. 19–97 Volumetric Strain—Geomechanics Run

In time, most of the reservoir achieves a transmissibility multiplier of 1.5 to 2.0, with higher values at the wells. This transmissibility increase is due to the dilation of the oil sands due to shearing under low confining stresses.

Fig. 19–98 Porosity—Geomechanics Run

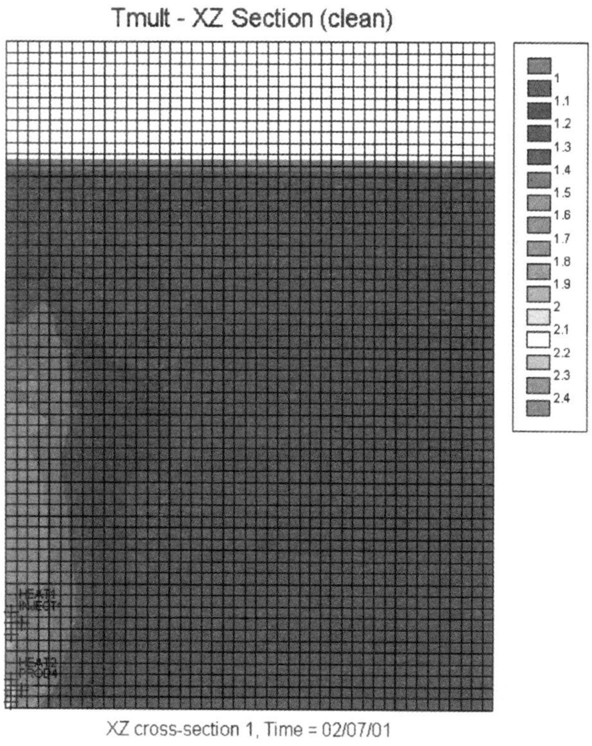

Fig. 19–99 Vertical Permeability Multiplier

Other geomechanics studies of operating conditions

The most effective means of ensuring the failure and dilation of oil sands is to operate the steam chamber at or near the preexisting minimum total stress. Thus, the effective stress and the strength of the oil sands falls towards zero. A typical Athabasca oil sands reservoir has an anisotropic stress state before steaming. Therefore, raising injection pressure will ensure failure.

The effect of different operating pressures on the failure of oil sands is described in Chalaturnyk and Li (2001). The authors examine shallow, medium depth, and deep reservoirs (150-m, 265-m, 742-m depths, respectively) which are comparable depths to those of the Dover (UTF), Surmont, and Senlac projects respectively. Three injection pressures were studied for each case, corresponding to multiples of the virgin reservoir pressure. In their study, none of the effective stresses was close to zero. Two virgin stress states were studied: a horizontal stress equal to the vertical stress and a horizontal stress 50% greater than the vertical stress.

In their examples, operating a deep reservoir at 15,000 kPa increased the absolute permeability by 26%, while an operating pressure of 5,000 kPa had no appreciable benefit. In general, operating at the lowest injection pressures did not result in failure. This is because the benefits of the higher pressure were offset by increases in thermal stresses, which were too localized at early times to allow for vertical stress relief; i.e., the stresses increased isotropically, which resulted in higher oil sands strengths. The beneficial effect of failure was most pronounced for the case with the anisotropic stress state, as expected.

None of their examples studied the case of operating the steam chamber near overburden pressures. This would certainly result in shear failure and dilation of oil sands. Note also that their modeling uses a different method of including geomechanics. They did not use a closely coupled model, and they have used an elastoplastic failure. Both of these factors will lead to results that are understated.

SAGD monitoring programs

At present, SAGD is undergoing significant development. In the recent past, instrumentation of pilots has provided critical data on steam chamber development. While such monitoring may prove to be of limited use in the future, the current state of the art suggests there are many benefits to monitoring. Accordingly, the author recommends comprehensive monitoring. The considerable experience on this matter is briefly summarized in the following.

Laing et al. (1988) described the instrumentation program at the Phase A pilot at AOSTRA's UTF project.[87] A comprehensive monitoring scheme was subsequently installed at the UTF Phase B Pilot (Collins, 1994). The design of this instrumentation was based on the observed behavior from the Phase A instrumentation.[88]

Gronseth (1989) monitored a cyclic steam injection operation in the Clearwater Formation.[89] Surface vertical uplifts near the injection wells exceeded 100 mm, and extensions within the reservoir exceeded 160 mm. Lateral displacements away from the injection wells were 60 mm. Towson and Khallad (1991) described the surface deformation monuments for the PCEJ steam stimulation project.[90] They reported a maximum uplift of 35 mm within a concentric pattern. Periodically, horizontal stresses were measured, and these were found to increase in time. Bilak et al. (1988) described a rigorous method of positioning surface heave monuments. These displacements were the result of dilation of the oil sands.[91]

Components to consider include:

- **Piezometers.** These have been influential in explaining the mechanisms that take place in the reservoir. Such instrumentation has also been used to evaluate the effects of pressure depletion from overlying gas zones. This has also been used by Petro-Canada, Rio-Alto, and Conoco (Gulf Canada). Piezometer design should include some duplication of instrumentation to monitor drift. Vibrating wire and bubble piezometers are recommended.
- **Thermocouples.** The advance of the steam chamber can be determined using temperature. This can indicate the effect of baffles, barriers, or variations in reservoir quality.
- **Inclinometers.** High-resolution gyroscopic surveys are periodically run to measure the lateral displacement of the reservoir and casing.
- **Extensometers.** It is possible to anchor several points in the reservoir and then, by using a steel bar of known properties, determine any shortening or expansion between these two points. These bars are made of Invar and are corrected for changes in temperature. Such monitoring allows one to estimate the degree of volumetric expansion from geomechanics, which is directly related to permeability increases.
- **Surface Monuments.** These are surveyed using conventional technology. Significant heaves can occur. This has proven to be an environmental problem in the Cold Lake area where heaves of the order of one ft. have substantially altered drainage patterns (affecting lakes).

SAGD surface facilities

There are substantial surface facilities associated with SAGD projects. This is the largest component of capital expenditure and will have a major impact on project design. It seems that, in general, it is cheaper to pipe steam than to use distributed generation facilities.

Unlike cyclic steam operations, SAGD operations use 100% quality steam. This is achieved by using wellhead separators rather than superheating the steam at the plant. Generation of steam appears to be most efficient at a steam quality of about 80%.

Interestingly, the change in enthalpy to obtain steam at different temperatures is relatively flat. Thus, the amount of fuel to generate steam should not vary greatly for different temperatures and pressures. This is also the largest operating cost.

The design of piping is temperature dependent. Higher pressures require heavier piping but also result in smaller diameters to move the same mass of steam. Natural gas pipelines are usually run at about 1,000 psi, since this gives minimum volumes and results in a good tradeoff between pressure (pipe thickness) and physical volumes (pipe diameter). In the author's opinion, this will also be true of steam, although the pressure may be different. Steam piping also requires insulation.

Separation of bitumen from water is normally temperature dependent. This is shown in Figure 19–100.[92] Specific temperatures are required and separators may therefore need additional heating. Some operators have opted to use diluent to change the bitumen density. This involves a real cost, and diluent recovery can be possible at some stage.

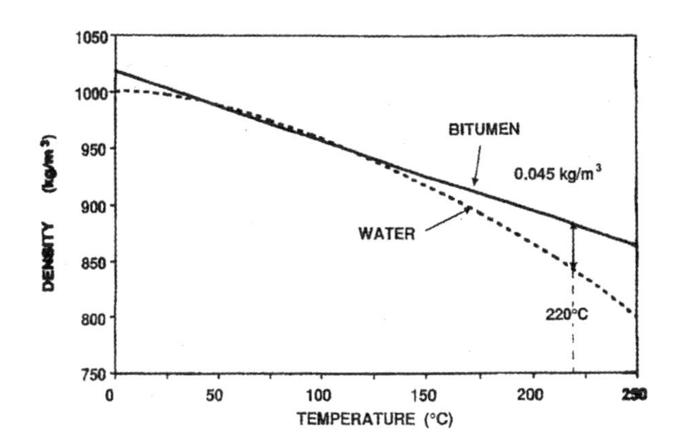

Fig. 19–100 Density Comparison, Water vs. Bitumen

Heat exchangers are used to recover heat from produced fluids; this will affect the calculation of efficiency.

This density reversal may well occur in the reservoir also, depending on reservoir temperature.

SAGD optimization

There have been some early attempts at optimization. Most of these have been aimed at lowering steam oil ratios (SORs). Unfortunately, these optimizations have ignored a great many relevant points:

- The requirement of high injection pressures to obtain high permeabilities (Ito et al. have done preliminary work considering these effects.)
- Surface separation requirements
- Artificial lift costs
- Thief-zone losses
- Actual facility design criteria for piping
- Operating conditions
- Well spacing
- Well lengths

To date, there is no example of extensive optimization in the literature. The conventional side of the business has resorted to nodal analysis for optimization. This seems like a logical methodology to be applied to SAGD operations as well.

Future SAGD Developments

Coinjection

- One possibility for increasing the pressure in the steam chamber, without losing the benefit of large latent heats at lower operating pressures, is to take advantage of partial pressures. By maintaining a condensable or noncondensible gas in the steam chamber, other than steam, the benefits of lower operating temperatures and higher latent heat capacities can be maintained. Multiple gases could be injected to take advantage of the various dew points of the constituent solvents.
- Butler (1999) and Jiang et al. (2000) described the SAGP process, in which a small amount of a noncondensible gas (NCG) is added to the steam injection stream. This reduces the steam requirements by creating a gas blanket at the top of the reservoir. Ultimate recoveries are the same as for SAGD.[93,94,95]
- Nasr et al. (1998) of the ARC discussed SAGD operating strategies and addressed the potential for solvent and gaseous additives in steam to improve SOR and reduce water requirements. Laboratory experiments of naphtha in steam found that the addition of naphtha was beneficial. Higher naphtha concentrations reduced the operating temperature due to the partial pressure effects. Coinjection of naphtha and steam improved the SOR and accelerated SAGD initialization.[96]
- Ito et al. (2001) examined the effect of methane injection on the SAGD process for JACOS. They found that the injection during the early stages of SAGD reduced the ultimate oil recovery. However, injection of methane in the later stages results in an improved SOR and vertical conformance. Partial pressure effects were explicitly recognized as being able to reduce the condensation temperature.[97]

Alternatives to SAGD

There are other alternatives to SAGD available as a recovery process. Again, high operating pressures will be necessary to take advantage of the geomechanical aspects, especially since thermal strains will be absent for the ambient temperature process.

- Palmgren and Edmunds (1995), of IFP and CS Resources, respectively, examined using high-temperature naphtha to replace steam in the SAGD process, in a process labeled naphtha assisted gravity drainage (NAGD). Naphtha is commonly the diluent used for pumping and pipeline transport of heavy oil; therefore, the injection of naphtha vapor combines the thermal process with the diluent mechanism. The authors found that reducing the injection pressure from 1,600 kPa to 1,000 kPa reduced the oil production rate by 85%. Lower permeabilities were found to result in higher levels of naphtha retained in the reservoir due to the increased resistance to flow in the reservoir. The economics for NAGD were comparable to those for SAGD.[98]
- Butler et al. (1991) describe the vapor extraction (VAPEX) process, in which a mixture of a noncondensible gas and a volatile solvent such as propane or butane are injected at ambient temperatures. The solvent dissolves in the bitumen, thus reducing its viscosity to allow production. One of the key controls is how the solvent diffuses into the oil. Geomechanical expansion of the pore space and permeability increases would enhance this process. The pore space is occupied by the noncondensible gas, and the solvent is recovered from the produced fluids.[99]

Summary

Thermal simulation is very challenging. It is one area where computation remains a significant issue. Fortunately, it appears the capabilities of computers continue to improve dramatically.

Thermal simulation involves some major paradigm shifts from conventional reservoir simulation in the following areas.

- There is considerably more input required. The properties of steam and other thermal parameters must be added to the input stream.
- The process mechanisms for cyclic steam stimulation and SAGD are considerably different from those found in conventional reservoir engineering.
- Geomechanics is extremely important in many thermal applications.
- Reservoir simulation, with the appropriate input data, has successfully predicted thermal project performance.

Much of the experience, from a thermal simulation perspective, is common to conventional reservoir engineering and conventional simulation:

- Geological description and heterogeneity remain critical.
- The techniques used to evaluate heterogeneity are common with conventional reservoir simulation. Field experience, reservoir simulations, and laboratory tests show that discontinuous barriers are very common and do not represent a barrier to successful implementation of SAGD.

Conventional simulations, ignoring geomechanics, are successful only because they rely on core permeabilities taken from highly disturbed, dilated specimens. By doing so, these analyses implicitly include geomechanical effects in their results.

Only small discontinuities in barriers are required for steam to effectively drain bitumen. Mudstone breaks, which are inherently discontinuous in a fluvial environment, do not significantly alter production performance.

There are many facets of reservoir engineering related to the SAGD process that are still under active development. This is an emerging technology. The design process is not routine yet. Thermal simulation has proven to be sufficiently accurate to predict production performance for a number of projects. While there have been some engineering failures for SAGD projects, to date the majority have been successful.

References

1 Hong, K.C., *Steamflood Reservoir Management*, Thermal Enhanced Oil Recovery, PennWell Publishing Co., 1994.

2 The Canadian Heavy Oil Association, *Reservoir Handbook*, 1991.

3 Hepler, L.G. and C. His, eds., *AOSTRA Technical Handbook on Oil Sands, Bitumens, and Heavy Oils*, Alberta Oil Sands Technology and Research Authority, 1989.

4 Butler, R.M., *Thermal Recovery of Oil and Bitumen*, Prentice Hall, 1991.

5 Prats, M., *Thermal Recovery,* SPE Monograph, Vol. 7, Henry L. Doherty Series, Society of Petroleum Engineers, 1982.

6 Boberg, T.C., *Thermal Methods of Oil Recovery, An Exxon Monograph*, John Wiley and Sons, 1988.

7 Anand, J., W.H. Somerton, and E. Gomaa, "Predicting Thermal Conductivities of Formations from Other Known Properties," *Society of Petroleum Engineers Journal*, 1973.

8 Somerton, W.H., J.A. Keese, and S. L. Chu, "Thermal Behavior of Unconsolidated Oil Sands," SPE 4506, Society of Petroleum Engineers, 1974.

9 Perry, J.H., et al., *Perry's Chemical Engineers' Handbook*, 5th ed., McGraw-Hill, 1977.

10 Mehrotra, A.K., and W.Y. Svrcek, "Correlations for Properties of Bitumen Saturated with CO2, CH4, and N2, and Experiments with Combustion Gas Mixtures," *Journal of Canadian Petroleum Technology*, 1982.

11 Smith, G.E., "Fluid Flow and Sand Production in Heavy Oil Reservoirs under Solution Gas Drive," *SPE Production and Facilities*, 1988.

12 Dusseault, M.B., "Sample Disturbance in Athabasca Oil Sand," *Journal of Canadian Petroleum Technology*, 1980.

13 After API, RP40, *Recommended Practice for Core Analysis procedure,* 1st ed., August 1960.

14 Woodhouse, R., "Discussion of How Much Core-Sample Variance Should a Well-Log Model Reproduce?" *SPE Reservoir Evaluation and Engineering*, Society of Petroleum Engineers, 2000.

15 Monicard, R.P., *Properties of Reservoir Rocks: Core Analysis*, Editions Technip, 1980.

16 Edmunds, N.R., et al., "Review of the Phase A Steam-Assisted Gravity Drainage Test: An Underground Test Facility," SPE 21529, Society of Petroleum Engineers, 1991.

17 Mukherjee, N.J., N.R. Edmunds, and S.D. Gittins, "Impact and Mitigation of Certain Geological and Process Factors in the Application of SAGD at AOSTRA's UTF," Paper HWC94-51, Petroleum Society, 1994.

18 Espinoza, C.E., "A New Formulation for Numerical Simulation of Compaction, Sensitivity Studies for Steam Injection," SPE 12246, Society of Petroleum Engineers, 1983.

19 Abou-Kassem, J.H., and S.M. Farouq Ali, "Appraisal of Steamflood Models," SPE 13947, Society of Petroleum Engineers, 1985.

20 Pooladi-Darvis, M., S.M. Farouq Ali, and W.S. Tortike, "Steam Heating of Fractured Formations Containing Heavy Oil—An Analysis of the Basic Premises," Paper 94-64, Petroleum Society, 1994.

21 Wan, R., et al., "A Constitutive Model for the Effective Stress-Strain Behavior of Oil Sands," Paper 89-40-66, Petroleum Society, 1989.

22 Vaziri, H.H., "Mechanics of Fluid and Sand Production from Oil Sand Reservoirs," 86-37-75 Petroleum Society, 1986.

23 Beattie, C.I., T.C. Boberg, and G.S. McNab, "Reservoir Simulation of Cyclic Steam Stimulation in the Cold Lake Oil Sands," *Society of Petroleum Engineers Journal*, 1991.

24 Denbina, E.S., T.C. Boberg, and M.B. Rotter, "Evaluation of Key Reservoir Drive Mechanisms in the Early Cycles of Steam Stimulation at Cold Lake," *Society of Petroleum Engineers Journal*, 1991.

25 Chalaturnyk, R.J., J.D. Scott, and G. Yang, "Geomechanical Modeling of Phase A at the AOSTRA Underground Test Facility," Petroleum Society 91-7, 1991.

26 Settari, A., Y. Ito, and K.N. Jha, "Coupling of a Fracture Mechanics Model and a Thermal Reservoir Simulator for Tar Sands," *Journal of Canadian Petroleum Technology*, 1992.

27 Singhal, A. K., et al., "Screening and Design Criteria for Steam Assisted Gravity Drainage (SAGD) Projects," SPE 50410, Society of Petroleum Engineers, 1998.

28 Satriana, D., et al., "Effects of Geomechanics on Streamflood in Shallow Rindu Zones, Duri Field, Indonesia," Proceedings of the 7th International Conference on Heavy Crude and Tar Sands, Beijing, China, Paper no. 1998, 120, Oct. 27-30, 1998.

29 Yuan, J.-Y., B. Tremblay, and A. Babchin, "A Wormhole Network Model of Cold Production in Heavy Oil," SPE 54097, Society of Petroleum Engineers, 1999.

30 Geilikman, M., M.B. Dusseault, and F.A. Dullien, "Sand Production and Yield Propagation around Wellbores," Paper 94-89, Petroleum Society, 1994.

31 Shen, C., "Numerical Investigation of SAGD Process Using a Single Horizontal Well," *Journal of Canadian Petroleum Technology*, 2000.

32 Ito, Y., M. Ichikawa, and T. Hirata, "The Growth of the Steam Chamber during the Early Period of the UTF Phase B and Hangingstone Phase 1 Projects," Canadian International Petroleum Conference, June 4–8, 2000.

33 Ito, Y., T. Hirata, and M. Ichikawa, "The Effect of Operating Pressure on the Growth of the Steam Chamber Detected at the Hangingstone SAGD Project," Paper 2002-286, Petroleum Society's Canadian International Petroleum Conference, June 11–13, 2002.

34 McLellan, P., R. Read, and K. Gillen, "Assessing Caprock Integrity for Steam Assisted Gravity Drainage (SAGD) Projects in Heavy Oil Reservoirs," Paper, Calgary: 4th International Conference on Horizontal Well Technology, November 6–8, 2000.

35 Wong, R.C.K., and Y. Li., "A Deformation-Dependent Model for Permeability Changes in Oil Sand due to Shear Dilation," Paper 2000-22, Petroleum Society's Canadian International Petroleum Conference, June 6–8, 2000.

36 Settari, A., D.A. Walters, and G.A. Behie, "Use of Coupled Reservoir and Geomechanical Modeling for Integrated Reservoir Analysis and Management," Paper 2000-78, Petroleum Society's Canadian International Petroleum Conference, June 6–8, 2000.

37 Denbina, E. S., et al., "Modeling Cold Production for Heavy Oil Reservoirs," *Journal of Canadian Petroleum Technology*, 2001.

38 Craig, R. F., *Soil Mechanics*, Van Norstrand Reinhold, 1976.

39 Touhidi-Baghini, A., *Absolute Permeability of McMurray Formation Oil Sands at Low Confining Stresses*, Department of Civil and Environmental Engineering, University of Alberta, Ph.D. Dissertation, 1998.

40 Touhidi-Baghini, A., and J.D. Scott, "Absolute Permeability Changes of Oil Sand during Shear," 51st Canadian Geotechnical Conference, 1997, pp. 729–736.

41 Oldakowski, K., *Stress-Induced Permeability Changes of Athabasca Oil Sands*, Master of Science Thesis, Dept. of Civil Engineering, Univ. of Alberta, Edmonton, Alberta, 1994.

42 Sasaki, K., et al., "Numerical and Experimental Modeling of the Steam-Assisted Gravity Drainage (SAGD) Process," Paper 99-21, Petroleum Society, 1999.

43 Kisman, K.E., and K.C. Yeung, "Numerical Study of the SAGD Process in the Burnt Lake Oil Sands Lease," SPE 30276, International Heavy Oil Symposium, Calgary, Alberta, Canada, June 19–21, 1995.

44 Burger, J., P. Sourieau, and M. Combarnous, *Thermal Methods of Oil Recovery*, Editions Technip, 1985.

45 Lo H.Y. and N. Mungan, "Effect of Temperature on Water-Oil Relative Permeabilities in Oil-Wet and Water-Wet Systems," SPE 4505, 1973.

46 Maini, B.B., and T. Okazawa, "Effects of Temperature on Heavy Oil-Water Relative Permeability of Sand," *Journal of Canadian Petroleum Technology*, May–June, 1987.

47 Polikar, M., et al., "Relative Permeability Curves for Bitumen and Oil Sand Systems," *Journal of Canadian Petroleum Technology,* Vol. 28, no. 1, 89-01-09 (January-February), 1985.

48 Donnelly, J.K., "Application of Steam Assisted Gravity Drainage to Cold Lake," CIM Paper 97-192.

49 Polikar, M., et al., "Relative Permeability Curves for Bitumen and Oil Sand Systems," CIM Paper 89-01-09, Conference Paper (diagram not in JCPT), 37th ATM of the Petroleum Society of CIM.

50 Bennion, D.B, et al., "Steady State Bitumen-Water Relative Permeability Measurements at Elevated Temperatures in Consolidated Porous Media," *PetSocCIM 93-25*, PetSocCIM Annual Technical Meeting, May 9–12, 1983.

51 Horne, R.N., et al., "Steam-Water Relative Permeability," Proceedings, World Geothermal Congress 2000, Kyusha-Tohoku, Japan, May 28–June 10, 2000.

52 Weinstein, H.G., J.A. Wheeler, and E.G. Woods, "Numerical Model for Steam Stimulations," *Society of Petroleum Engineers Journal*, p. 65, February 1977.

53 Ito, Y. and S. Suzuki, "Numerical Simulation of the SAGD Process in the Hangingstone Oil Sands Reservoir," CIM 96-57, 47th ATM of the Petroleum Society of CIM, June 10–12, 1996.

54 Butler, R.M., "The Expansion of Tar Sands During Thermal Recovery," *Journal of Canadian Petroleum Technology*, 1986.

55 Flach, P.D., and G.D. Mossop. Depositional Environments of Lower Cretaceous McMurray Formation, Athabasca Oil Sands, Alberta. *AAPG Bulletin*, Vol. 69, no. 8, pp. 1195–1207, 1985; and Miall, A.D. *The Geology of Fluvial Deposits*, Springer Verlag, 1996.

56 Figure after Leopold, L.B., and M.G. Wolman, "River Floodplains: Some Observations on Their Formation," USGS Paper 282C, 1957.

57 Kupfersberger, H., and C.V. Deutsch, "Methodology for Integrating Analogue Geologic Data in 3-D Variogram Modeling," *AAPG Bulletin*, Vol. 83, no. 8, (August): 1262–1278, 1999.

58 After Pettijohn, F.J., *Sedimentary Rocks,* 3rd ed., Harper and Row, New York, 628 p, 1975.

59 Aguilera, R., et al., *Horizontal Wells, Formation, Evaluation, Drilling and Production, Including Heavy Oil Recovery*, Gulf Publishing, 1991.

60 Suggett, J., and S. Youn, "Christina Lake Thermal Project," SPE/PS-CIM International Conference on Horizontal Well Technology, November 6–8, 2000.

61 Strobl, R.S., et al., "Application of Outcrop Analogues and Detailed Reservoir Characterization to the AOSTRA Underground Test Facility, McMurray Formation, Northeastern Alberta," Memoir 18 Canadian Society of Petroleum Geologists, 1997.

62 Mukherjee, N., et al., "A Comparison of Field versus Forecast Performance for Phase B of the UTF Project in the Athabasca Oil Sands," Houston, Texas: 6th UNITAR International Conference on Heavy Crude and Tar Sands, February 1995.

63 Edmunds, N.R., J.A. Haston, and D.A. Best, "Analysis and Implementation of the Steam Assisted Gravity Drainage Process at the AOSTRA UTF," Paper no. 125, The Fourth UNITAR/UNDP International Conference on Heavy Crude and Tar Sands, Vol. 4, In Situ Recovery, Edmonton Alberta, August 7–12, 1988.

64 Yang, G., and R.M. Butler, "Effects of Reservoir Heterogeneities on Heavy Oil Recovery by Steam-Assisted Gravity Drainage," *Journal of Canadian Petroleum Technology*, 1992.

65 Ito, Y., M. Ichikawa, and T. Hirata, "The Growth of the Steam Chamber during the Early Period of the UTF Phase B and Hangingstone Phase 1 Projects," Canadian International Petroleum Conference, June 4–8, 2000.

66 Williams, L.L., W.S. Fong, and M. Kumar, "Effects of Discontinuous on Multizone Steamflood Performance in the Kern River Field," SPEREE, Vol. 4, no. 5, October 2001.

67 Farouq Ali, S.M., "Is There Life After SAGD?" *Journal of Canadian Petroleum Technology,* Distinguished Authors Series, 1997.

68 Ito, Y., and S. Suzuki, "Numerical Simulation of the SAGD Process in the Hangingstone Oil Sands Reservoir," CIM 96-57, 47th ATM of the Petroleum Society of CIM, June 10–2, 1996.

69 Edmunds, N.R. and S.D. Gittins, "Effective Application of Steam Assisted Gravity Drainage of Bitumen to Long Horizontal Well Pairs," *Journal of Canadian Petroleum Technology*, 1993.

70 Strobl, R.S., et al., "Application of Outcrop Analogues and Detailed Reservoir Characterization to the AOSTRA Underground Test Facility, McMurray Formation, North Eastern Alberta," eds. Pemberton, S.G. and D.P. James, *Petroleum Geology of the Cretaceous Mannville Group, Western Canada,* CSPG Memoir 18, 1997.

71 Edmunds, N.R., and W.K. Good, "The Nature and Control of Geyser Phenomena in Thermal Production Risers," *Journal of Canadian Petroleum Technology*, Vol. 35, no. 4, 1996.

72 Edmunds, N.R., "On the Difficult Birth of SAGD" *Journal of Canadian Petroleum Technology,* Distinguished Authors Series, 1999.

73 Donnelly, J.K., "Who Invented Gravity?" *Journal of Canadian Petroleum Technology,* Distinguished Authors Series, 1998.

74 Butler, R.M., *Horizontal Wells for the Recovery of Oil, Gas and Bitumen*, Petroleum Society Monograph, No. 2, 1994.

75 Dusseault, M.B., "The Geotechnical Characteristics of the Athabasca Oil Sands," Ph.D. thesis, Department of Civil Engineering, University of Alberta, Edmonton, Alberta, 1977.

76 Chalaturnyk, R.J., and J.D. Scott, "Evaluation of Reservoir Properties from Geomechanical Tests," *Journal of Canadian Petroleum Technology*, 1992.

77 Vasquez, H.A.R., et al., "Mechanical and Thermal Properties of Unconsolidated Sands and Its Applications to the Heavy Oil SAGD Project in the Tia Juana Field, Venezuela," SPE 54009, Caracas, Venezuela: SPE Latin American and Caribbean Petroleum Engineering Conference, April 21–23, 1999.

78 Scott, J.D., D. Adhikary, and S.A. Proskin, "Volume and Permeability Changes Associated with Steam Stimulation in an Oil Sands Reservoir," *Journal of Canadian Petroleum Technology*, 1994.

79 Oldakowski, K., "Stress Induced Permeability Changes of Athabasca Oil Sands," Master of Science thesis, Department of Civil Engineering, University of Alberta, Edmonton, Alberta, 1994.

80 Chalaturnyk, R.J., "Geomechanics of the Steam Assisted Gravity Drainage Process in Heavy Oil Reservoirs," Ph.D. thesis, Department of Civil Engineering, University of Alberta, Edmonton, Alberta, 1996.

81 Touhidi-Baghini, A., "Absolute Permeability of Mcmurray Formation Oil Sands at Low Confining Stresses," Ph.D. thesis, Department of Civil Engineering, University of Alberta, 1998.

82 Agar, J.G., "Geotechnical Behavior of Oil Sands at Elevated Temperatures and Pressures," Ph.D. thesis, Department of Civil Engineering, University of Alberta, Edmonton, Alberta, 1984.

83 Kosar, K.M., and K. Been, "Large Scale Laboratory Fracturing Test in Oil Sands," Paper 89-40-83, Petroleum Society, 1989.

84 Chalaturnyk, R.J., and P. Li, "When Is It Important to Consider Geomechanics in SAGD Operations?" Paper no. 2001-46, Calgary, Alberta: Petroleum Society of CIM's Canadian International Petroleum Conference, June 12–14, 2001.

85 Fung, L.S.-K., L. Buchanan, and R.G. Wan, "Coupled Geomechanical-Thermal Simulation for Deforming Heavy-Oil Reservoirs," *Journal of Canadian Petroleum Technology,* Vol. 33, no. 4, (April): 22–28. 1994.

86 Settari, A., D.A. Walters, and G.A. Behie, "Use of Coupled Reservoir and Geomechanical Modeling for Integrated Reservoir Analysis and Management," Paper no. 2000-78, Calgary, Alberta: Petroleum Society of CIM's Canadian International Petroleum Conference, June 4–8, 2000.

87 Laing, J.M., et al., "Geotechnical Instrumentation of the AOSTRA Mine-Assisted Underground Steaming Trial," Proceedings Paper 112, 4th UNITAR/UNDP International Conference on Heavy Crude and Tar Sands, 1988.

88 Collins, P.M., "Design of the Monitoring Program for AOSTRA's Underground Test Facility, Phase B Pilot," Proceedings Paper 91-87, Banff, Alberta: Joint CIM/AOSTRA Technical Conference, April 1994.

89 Gronseth, J.M., "Geomechanics Monitoring of Cyclic Steam Stimulation Operations in the Clearwater Formation," Proceedings Paper No. 34, Calgary, Alberta: CIM District 4 Meeting (October), 1989.

90 Towson, D., and A. Khallad, "The PCEJ Steam Stimulation Project," (preprint) Paper no. 91-108, Banff, Alberta: Petroleum Society of CIM/AOSTRA Technical Conference, April 21–24, 1991.

91 Bilak, R.A., L. Rothenberg, and M.B. Dusseault, "Use of Surface Displacements to Monitor EOR Projects," Proceedings 5th UNITAR/UNDP International Conference on Heavy Crude and Tar Sands: 267–277, 1988.

92 Kovalsky, J.A., and P. Spargo, "High Temperature and Pressure Separation of Bitumen and Water at AOSTRA's Underground Test Facility (UTF) Pilot," Challenges and Innovations, Heavy Oil and Oilsands Technical Symposium, 1991.

93 Butler, R.M., "The Steam and Gas Push (SAGP)," *Journal of Canadian Petroleum Technology*, 1999.

94 Jiang, Q., R.M. Butler, and C.-T. Yee, "The Steam and Gas Push (SAGP)—2: Mechanism Analysis and Physical Model Testing," *Journal of Canadian Petroleum Technology*, 2000.

95 Jiang, Q., R.M. Butler, and C.-T. Yee, "The Steam and Gas Push (SAGP)—3: Recent Theoretical Developments Laboratory Results," *Journal of Canadian Petroleum Technology*, 2000.

96 Nasr, T.N., H. Golbeck, G. Korpany, and G. E. Pierce, "SAGD operating strategies," *Proceedings SPE 50411*, Calgary, Alberta: SPE/Petroleum Society of CIM International Conference on Horizontal Well Technology, . November 1–4, 1998.

97 Ito, Y., M. Ichikawa, and T. Hirata, "The Effect of Gas Injection on Oil Recovery during SAGD Projects," *Journal of Canadian Petroleum Technology*, Vol. 40, no. 1, (January): 20–23, 2001.

98 Palmgren, C., and N.R. Edmunds, "High Temperature Naptha (sic) to Replace Steam in the SAGD Process," Proceedings SPE 30294, Calgary, Alberta: International Heavy Oil Symposium, June 19–21, 1995.

99 Butler, R.M., and I.J. Mokrys, "A New Process (VAPEX) for Recovering Heavy Oils Using Hot Water and Hydrocarbon Vapour," *Journal of Canadian Petroleum Technology*, 1991.

20

Problem Sets

Introduction

This chapter deals mostly with problem sets that the author has created for use in industry short courses. They have been used many times, and potential bugs should be resolved. The data sets have been prepared for both Computer Modelling Group and Geoquest software. Originally, there was going to be a copy of all the data sets included in the back of the book. However, the book was getting somewhat long and it would have taken a considerable period of time for readers to reproduce the data sets manually. For these reasons, the data sets will, instead, be located on a website at http://www.applied-reservoir-engineering.com. The files can be downloaded from this website free of charge. This will also allow for future revisions, additional exercises, and perhaps different data set formats.

There are also some final thoughts on reservoir simulation at the end of the chapter.

Grid Problem Set

This problem set is designed to illustrate the effects of grid density on the accuracy of solutions obtained from numerical simulation. This is a simple example that utilizes a five-spot waterflood pattern, perhaps one of the most common situations that one may encounter. Pay particular attention to the water cut performance as the grid becomes progressively finer.

This is a relatively simple problem set. The second objective is to allow the reader to become familiar with a simulation data set, since no simulation experience is assumed.

Element of symmetry

Only a portion of a complete five-spot has been modeled. This saves on the amount of computing time and memory required. The basic grid for the first run is shown in Figure 20–1.

The element of symmetry has also been extended to the edge of the model. The simulator assumes that the wells are located in the center of the grid block. This degree of accuracy allows simulator results to be compared against analytical solutions, although this is not covered in this chapter.

Two changes are required to effect comparison with analytical solutions. First, the permeability of the edge blocks must be cut in half, since the outside portion does not exist. In addition, the porosity of the edge blocks must be halved also, to give the correct total pore volume. Note also that, by applying the 50% factor to permeabilities and porosities on all four edges, the corner blocks will get this treatment twice. Hence, the porosity and permeability in the corner blocks will be 25% of their original values.

Changes required

Required changes have been highlighted in the following section. Note that only a few numbers need to be changed in the data set.

Analysis

Although this problem set may seem simplified, it contains some important implications. A number of people use a minimum of two grid blocks between wells as a rule of thumb for grid definition. The author suggests that this is not a good criterion, as explained in chapter 10.

The simulator provides comparable results on recovery factors; however, it is apparent that very different results are obtained for water breakthrough times. Often, this is an important factor in determining pore volumes during history matching.

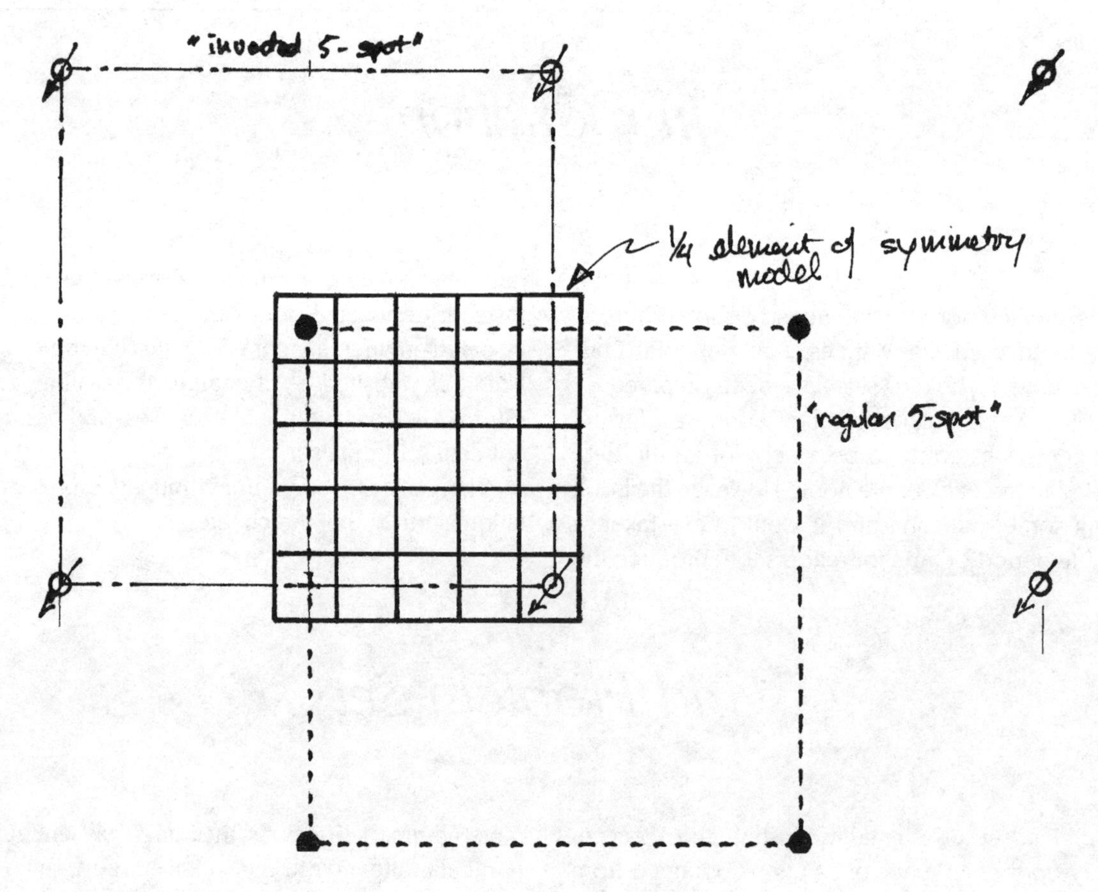

Fig. 20–1 Idealized Element of Symmetry

Coning Problem Set

This problem set has a number of interesting twists. It highlights the differences in behavior between gas and water coning. The final solution is somewhat counterintuitive. It is a dramatic example of why datum pressure should be used to analyze flow in the reservoir. It is easy to move ahead blindly using straight pressure rather than actual flow potential (datum pressure). Recall that potential includes the appropriate hydrostatic head—i.e., the gravity term—with the respective phase pressures.

Constraints

The objective of this problem set is to maximize oil recovery on an offshore platform. There is a water cut restriction due to limitations with the topside facilities. The water cut must be kept below 20%. Further, the topside facilities are limited to a GOR of 750 m^3/m^3. The only variables that can be changed are the location of the perforations and the bottomhole flowing pressure. These items have been highlighted in the data set.

Description

The physical reservoir properties are shown in Figure 20–2. The reservoir has been divided into layers. Note that the formation permeability and porosity vary vertically with depth. There is also a gas-oil and a water-oil contact, which are located at a depth of 1,500 and 1,359 meters, respectively.

As with the earlier data sets, the fluid properties and rock properties, such as capillary pressure and relative permeability remain the same. Note also that depths and pressures are the same, or very similar, throughout all of the problem sets.

Process

The author recommends using a graphical approach to understanding the problem. This means evaluating the flow in the reservoir. Plot production and then plot both pressure and datum pressure. What is dominating the reservoir performance?

Grid

Note that a radial grid has been used. Most simulators have this capability. Radial models can be very efficient for single-well problems.

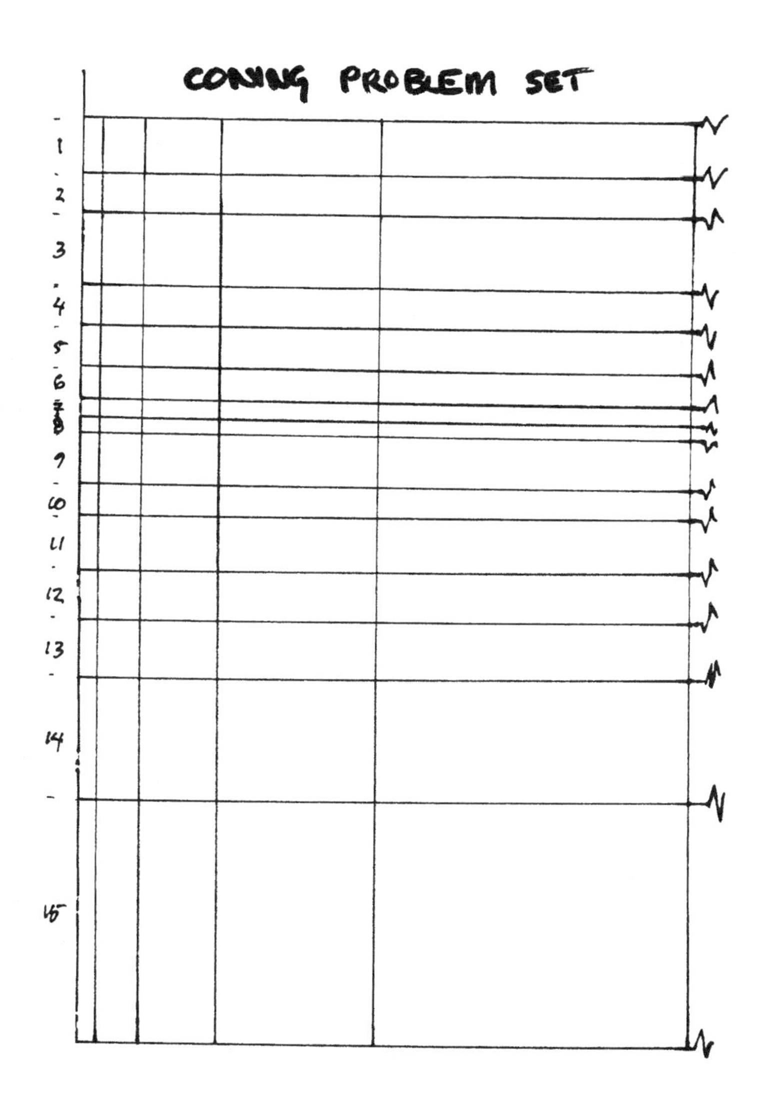

Fig. 20–2 Coning Model Grid

Results

The results of your various runs should be tabulated as shown in Table 20–1.

In addition, make a description of the reservoir physics. What are the relative coning susceptibilities of gas versus water?

Analysis of coning problem

A production plot shows that the peak GOR production does not occur at the end of the run. Therefore, this constraint must be plotted against time.

The reservoir flow developed from raw pressure is very misleading when compared to the potential or pressure datum approach. Clearly, the high-permeability layers act as a conduit. Gas is also more prone to coning than water. Despite the larger density difference, the gas has a significantly lower viscosity. The latter dominates the propensity to cone gas ahead of water. In fact, in some cases it is possible to increase recovery by completing wells in the water leg and not the oil section (as indicated in chapter 16 in the section dealing with water and gas coning, this is a technique known as "reverse coning").

Reservoir simulation is usually associated with studying areal movements of fluid and secondary recovery processes—"big picture" issues. This problem set demonstrates that it also can be useful for completion design.

Table 20–1 Blank Table for Recording Completion Sensitivity Results

Case	Cumulative Recovery	Layers Perforated	Bottomhole Pressure	Maximum GOR	Maximum Water Cut
	E3M3	k_n:k_{n+x}	kPa	M3/M3	Percent
Base					
1					
2					
3					
4					
5					
6					
7					
8					

Cross-Sectional Study Problem Set

This is the most difficult exercise, since pseudo relative permeability is the most difficult concept in the text. In the author's opinion, pseudos are widely misunderstood. In fact, although this text covers relative permeability and pseudo relative permeability concepts in chapters 7 and 8, respectively, there's nothing like putting these concepts into practice to really get a good feel for them. The author has found that it is easy to look at the formulae and gain a general idea from a textbook; it is another matter entirely to actually implement pseudo relative permeability correctly in a model. Aside from the basic experience, it provides another valuable insight. It is possible to duplicate the results of a detailed simulation with accuracy using pseudo relative permeability. This exercise will give the reader an instinctive feel for what can be done.

Process

A detailed six-layer run is provided to start. This is a reservoir engineering problem.

1. First, understand the reservoir from a qualitative perspective. This means understanding the structure, the layering, and the type of fluids in place.
2. Next, run the program provided.
3. Transfer the production results to graphical format. While a graphical postprocessor can be used, readers are advised to plot the results by hand. It gives one more time to think about what is happening in the reservoir. Using a postprocessor may see to be more exciting or clever; however, it can lead to shortcuts in analysis.
4. Observe how the saturations change in the reservoir. What is actually happening? A graphical postprocessor greatly speeds up this process. Note that these sections also can be plotted by hand, and this is appropriate at times.
5. Write a short description of the relevant reservoir mechanisms.
6. Develop water-oil rock pseudo relative permeabilities. This is done on a grid column by grid column basis. A specific example of how to do this is outlined in the following section. In the author's experience, translating the theory to practice in this area is the most difficult.
7. Simplify the model from a six-layer model to a single-layer model. First, an average permeability and average porosity must be generated. Second, an average capillary pressure curve is developed. In this simplified example, each layer has the same S_w and P_c curve. This would rarely be the case in actual practice. A set of rock pseudo relative permeability curves needs to be developed, i.e., k_{rw} and k_{row}. Were simplifications required in this process? How certain is the reader of the general applicability?
8. Run the new single-layer model. Plot the results and compare them to the first model run with six layers. How do the results compare?
9. What happens if the model is run without the pseudo relative permeability curves? Recall that each of the layers actually utilized only one unique (lab) relative permeability relationship. If this run doesn't match the six-layer model, how much is it off? How severe is the layering? Can a Dykstra-Parsons ratio be calculated, and how severe is this compared to most reservoirs? What does this tell us about the relative permeability curves that we actually use or history match in reservoir simulation?
10. Finally, develop pseudo relative permeability curves for the wells. This takes into account local wellbore effects. Note that different curves may be required for all three wells.
11. If one were completing a study, one more step would be required, which is to transfer all this information into a single-layer reservoir model. What other issues would have to be resolved?

Cross-sectional analysis

In this problem set, an apparently simple problem set actually requires a great deal of detailed analysis for correct interpretation. The author recommends that, to truly understand this material, more is required than simply reading or reviewing the solution that follows.

Saturations show that the water fingers up the high-permeability streak in the middle of the zone, as shown in Figure 20–3.

Gas breakout occurs at the top of the structure. This gives a high GOR in the top well. The production performance for the wells is shown in Figure 20–4.

The wells show increasing water production, which is related to flank water influx.

The pressure distribution in the reservoir is shown in Figure 20–5. The average pressure drops in the first five years. After this, the aquifer maintains the pressure at approximately 7,400 kPa.

The production performance for the field is shown in Figure 20–6. Oil rates, water cuts, and GOR trends are indicated.

It may be concluded from this that a formation or rock relative permeability characterization will not be sufficient. The effects of partial completion must also be included.

Less use of pseudo relative permeability is made today than in the past. However, on large simulations, this may still be a practical necessity. The generation of these curves is time intensive. Utilities are available to generate these curves. Note that shifting the end-points on the pseudo relative permeability is often sufficient to achieve the required result.

This is by no means the end of the problem. After this, it would be necessary to include these curves in an areal simulation. If there is cross-flow into the cross section, the history match generated may not be representative. This has occurred in practice on studies performed by the author.

This problem set also demonstrates another benefit that can be derived from using a cross-sectional study. The problem has been reduced in complexity, which allows one to concentrate on understanding the basic mechanisms in the reservoir.

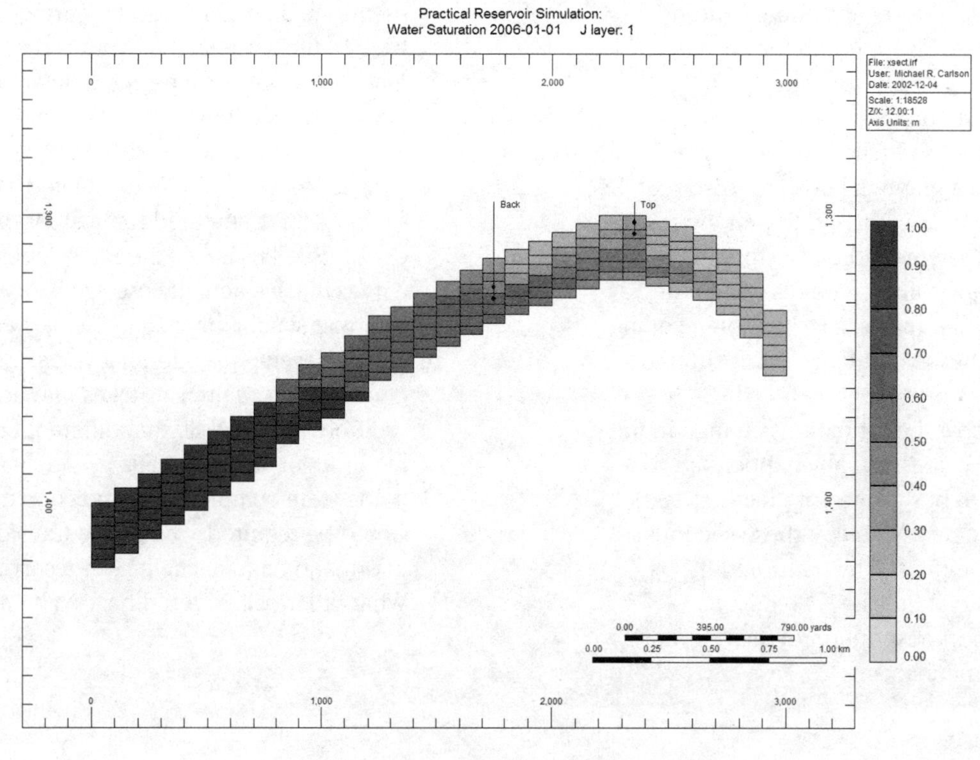

Fig. 20–3 Cross-Sectional Model—Viscous Fingering

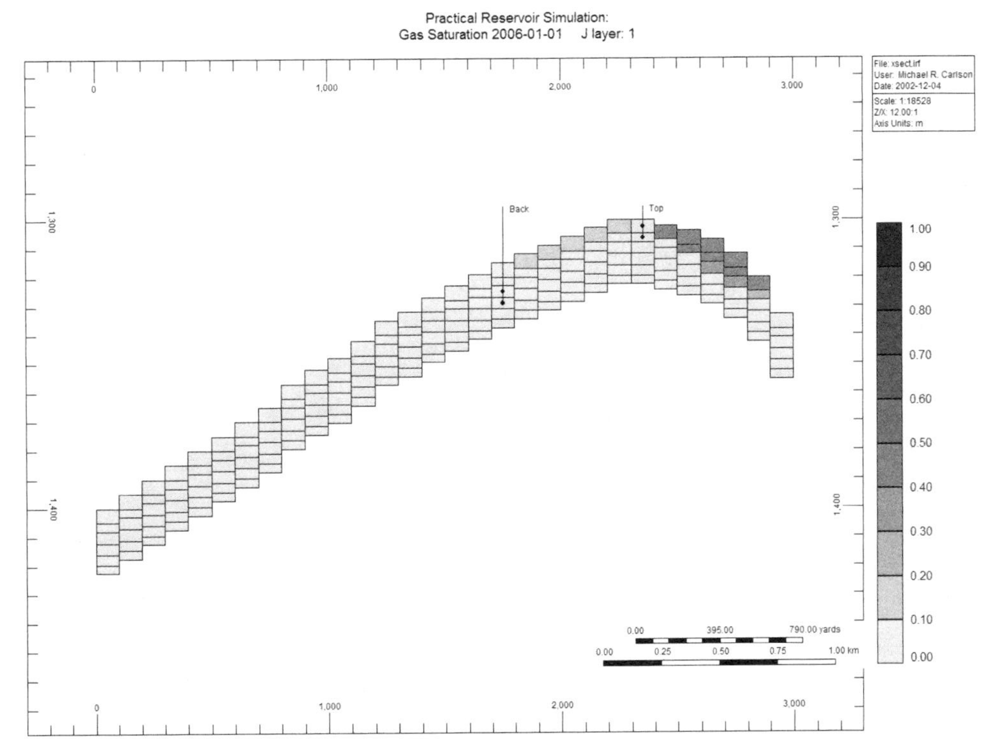

Fig. 20–4 Cross-Sectional Model—Gas Breakout

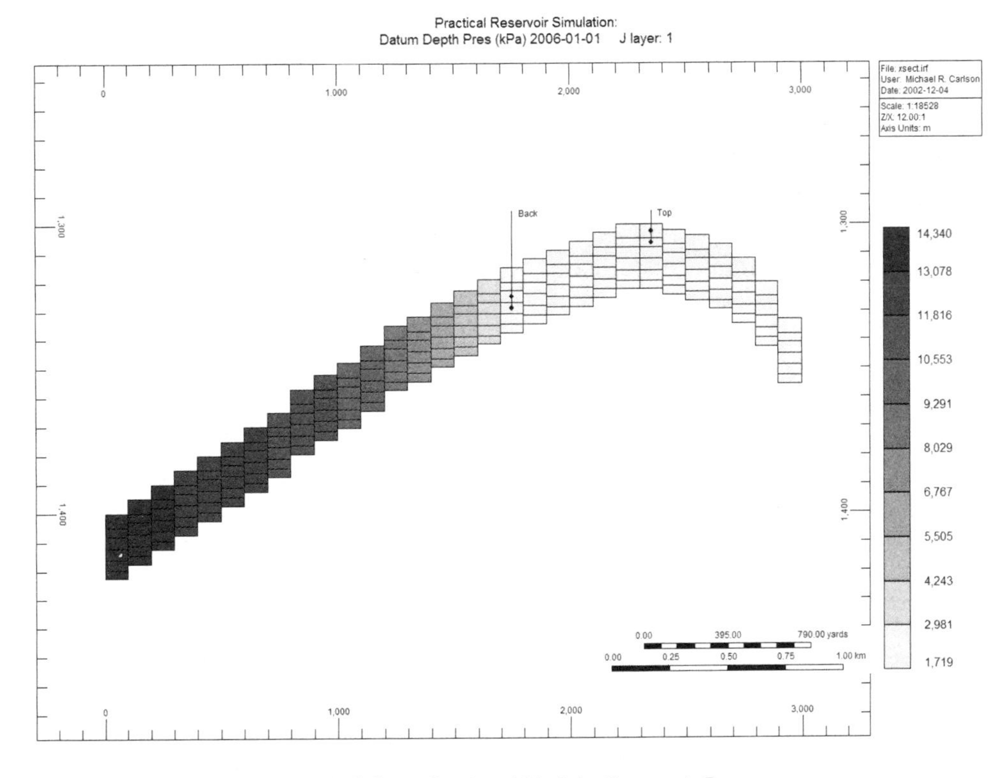

Fig. 20–5 Cross-Sectional Model—Reservoir Pressure

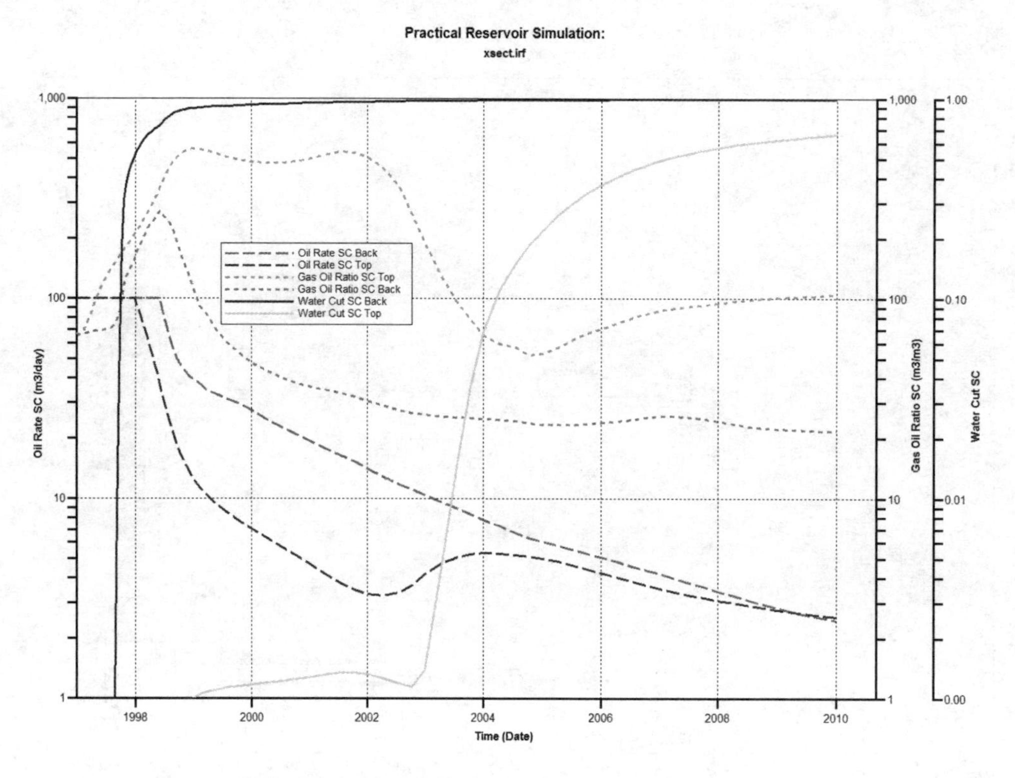

Fig. 20–6 Original Reservoir Model with Six Layers

If the reader were actually doing this study for an employer or a client, what would be written up regarding the risks and assumptions of the results? The pseudo curves are also rate dependent. How sensitive are the curves to areal variations in rate? One of the major reasons for implementing simulation is waterflooding. Government conservation regulations historically allowed one to produce at higher rates when the project was recognized (i.e., accredited). Typically, the permitted rates were doubled. Do the relative permeability curves used apply to the new rates? This would mean changing the relative permeability curves on the water injection sensitivities. In some cases, the reservoir simulator will not allow one to change the relative permeability curves during the run section and is usually set in the initial conditions.

In a more modern context, 3-D simulations are the norm today. However, many people formed their perception of simulation results when the use of these dangerous shortcuts were necessary. Note that upscaling can make use of pseudo relative permeability so these issues are far from dead. Most upscaling is developed assuming single-phase flow. In the author's opinion, many of the current "discoveries" in upscaling were previously found with pseudo relative permeability.

Areal Problem Set

This problem set is the author's favorite. It is not a trite exercise, and is based on some direct experiences in Halfway (formation name) reservoirs in northwestern Alberta and northeastern British Columbia. The objectives are outlined in the following:

1. Demonstrate the extent to which outside influences affect the development outcome of the reservoir. This makes the exercise very interesting in that it approximates the actual development or exploitation process.
2. Provide input on a cross-discipline basis. Successful simulations are strongly influenced by geological input and impact development decisions. The outcome can have a strong influence on future locations or may provide valuable insight into what is fundamentally a geological focus. An example of this is the extension of a pool. Simulation input and results often influence production engineering issues, such as well completions.
3. Keep a business focus. Historically, many people involved in reservoir simulation have been the more technically oriented engineers. Very often, management really wants economic implications to be considered at the technical level. This may be as simple as including the economics in a format that can be used readily in an economics model.
4. Get the geological model right. This is the most difficult part of the simulation. This example is very stylized in that the input was created in an orderly fashion. Having said this, the model has some elements of actual reservoir complexity. Although this is substantially less than reality, the problems of geological description are apparent.

To reiterate, this exercise affords an ideal opportunity to make mistakes. The results are not reported to anyone, and there is no real money on the line.

Outline

The general idea is this. Assume that you are operating a small oil company and that you are allocated a certain amount of money from which to obtain land and drill. A reservoir model has been built, and well results are obtained when money is spent. In addition, you can derive revenue from production with which to drill further wells or purchase land. You will be provided with a generalized play and some seismic data. The final objective is to devise a geological map and develop a reservoir simulation model.

Financial

You company is awarded 5,000 K$ to start. Oil prices are $24.00 per barrel, and the royalty rate is a flat 20%. The gas price is $2.50 per scf with a 15% royalty rate. Operating costs are $2,500/well/month, regardless of whether it is an oil or gas well.

Geological model

The basic geological concept is shown in Figure 20–7 (heavy, grey lines are lag shales). The sands are a combination of shoreface and channel sands. In the problem set area, the sea was located towards the bottom of the maps and/or grids when this reservoir was being deposited. Geophysics were run with the results shown in Figure 20–8. This is an amplitude variation map and is somewhat representative of porosity and sand thickness development.

Land sales

When this exercise is used as part of a course, land is sold via auction in the style adopted by the western provincial governments. This transforms this problem set into a competitive game. It is therefore possible to have one person assume the role of game master and others can competitively model/develop the reservoir.

The land sale is very bureaucratic. Part of this reflects the realities of dealing with government, and part of it is pure administration. Bids must be provided on quarter section by section data using the forms provided and blue ink! The first time the author attempted to do this exercise, several hours were required to determine the results from the first land sale. The form in Table 20–2 was

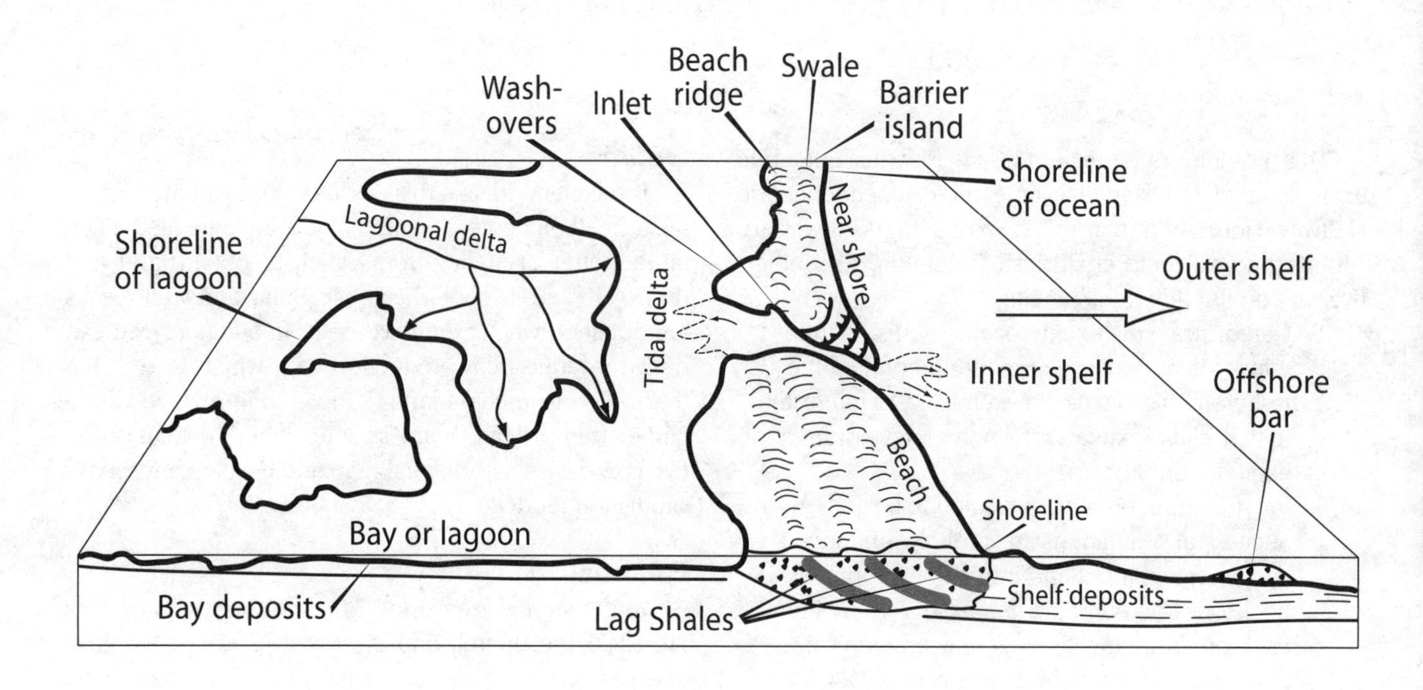

Fig. 20–7 Geological Model—Areal Problem Set

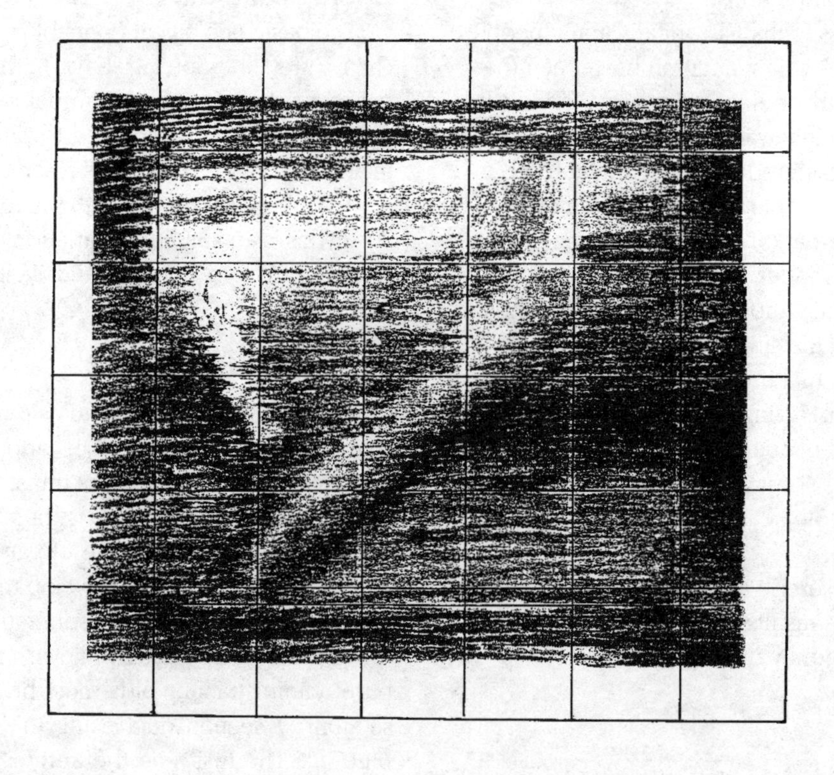

Fig. 20–8 Geophysical Amplitude Survey Results, Areal Problem Set

Table 20–2 Official Government Form for Submitting Land Bids (Read instructions carefully and literally)

Offer to Purchase Land From Government					
(use red ink)					
Section Number	________________				
LSDs (circle)		SE	SW	NE	NW
Amount	$ ________________				
Signed	________________				
Company	________________				

adopted so that each parcel of land could be rapidly sorted during lunch. This greatly facilitated the progress of the game.

Note that land is sold by the quarter section. The available leases are shown in Figure 20–9. Pay careful attention to the following section on permitted drilling and production spacing locations.

Well drilling and completion costs

Wells cost an even 1,000 K$ and completions 250 K$. Well drilling locations must be specified exactly on the grid and shown diagrammatically. It is also possible to obtain core data at the cost of 50 K$ per well.

Drilling locations and production requirements

Wells must be drilled in the specified target areas shown in Figure 20–10. Although wells may be drilled in any of the target areas, oil wells can only be produced from the oil targets and gas wells can only be produced from the gas targets. There is an area of overlap between these targets. These four areas within a section are the safest place to drill.

The government (gamemaster) may allow you to produce out of target upon special application, provided no other operators have a bona fide objection. The government is conscious of maximizing royalties. Such applications normally involve costs.

Caveat emptor

Bad things happen in real life, and this exercise is no exception. First, dry holes are quite possible and AEUB regulations are generally in effect. You may get oil, gas, or water. You may drill wells with no reservoir development or that have poor permeability.

Reservoir engineering data

All reservoir engineering data is provided in this exercise, as it was in the other examples. This has been done to concentrate on the geological model. The core porosity versus permeability is shown in Figure 20–11 for both the channel and beach sand facies. A base map is also provided for mapping as shown in Figure 20–12. A number of maps will be required and photocopying the various maps is suggested by the author.

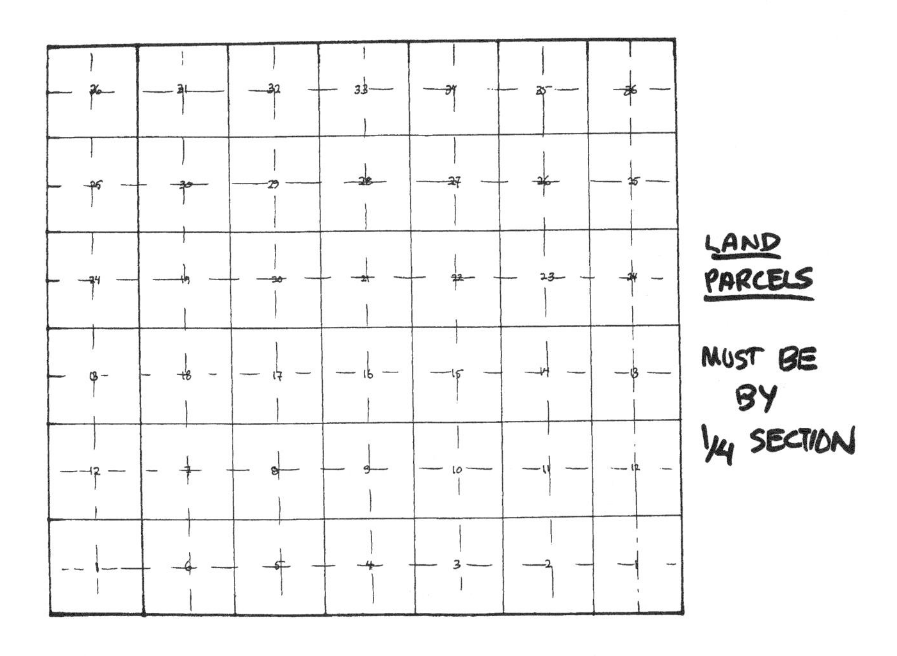

Fig. 20–9 Land Parcels Available for Sale

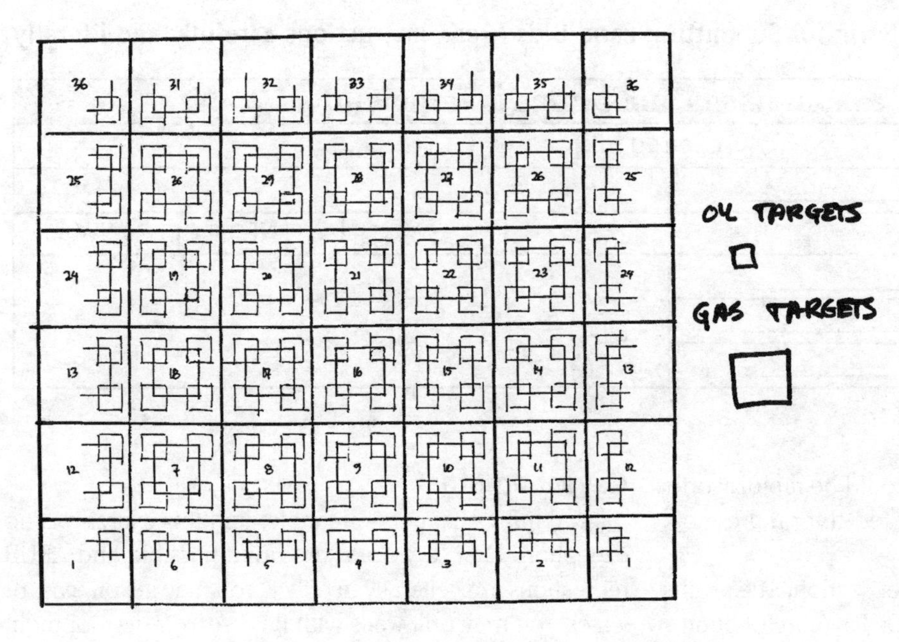

Fig. 20–10 Allowable Drilling Locations

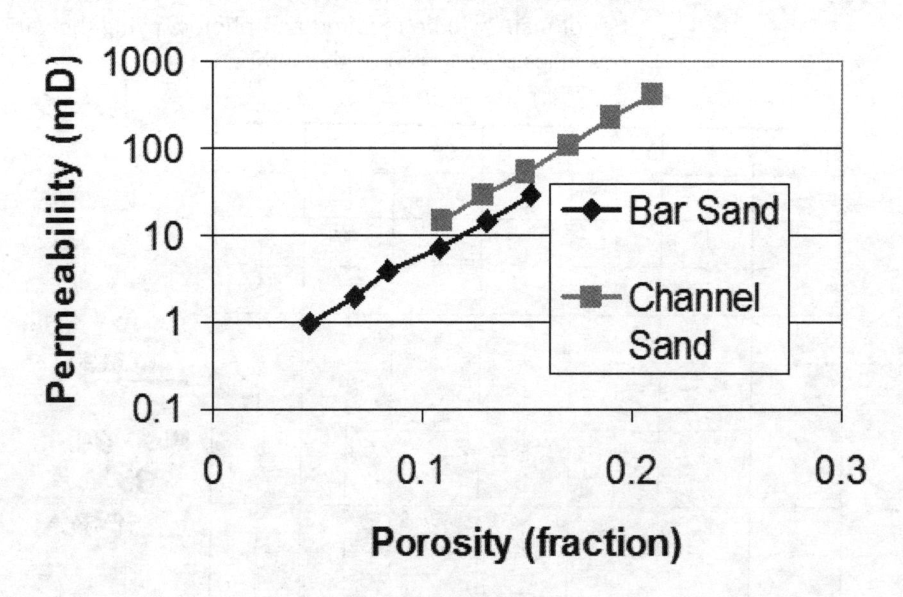

Fig. 20–11 Core Data for Areal Data Set

Well results (confidentiality)

When a well is drilled, the logs and production results are held confidential from your competitors for a period of one year. During a five-day course, this equates to one day. After that, a copy of your logs will be distributed to your competitors. This means that, at any given time, every group in the class will be dealing with a different set of data.

Analysis

This is a nasty problem. To date, the author has not heard of anyone who has solved it and correctly determined the geology. Note in particular that there are three overlapping sands. This problem is discussed in chapter 18 in the section dealing with structural style effect. The extent of the channels has not been completely identified. Moreover, this problem set is highly idealized.

The contacts are the same across all three sands. They are, in effect, in partial communication by virtue of the intersecting channels; this is known as juxtaposition. The geological model was based on actual experience and is representative of an area in Northern Alberta. In the author's opinion, it is impossible to solve this problem without a geological model.

The problem set is like real life in that a partial set of information is available. Unlike a real reservoir, a correct solution is known. This is an interesting control possibility. Can we actually deter-

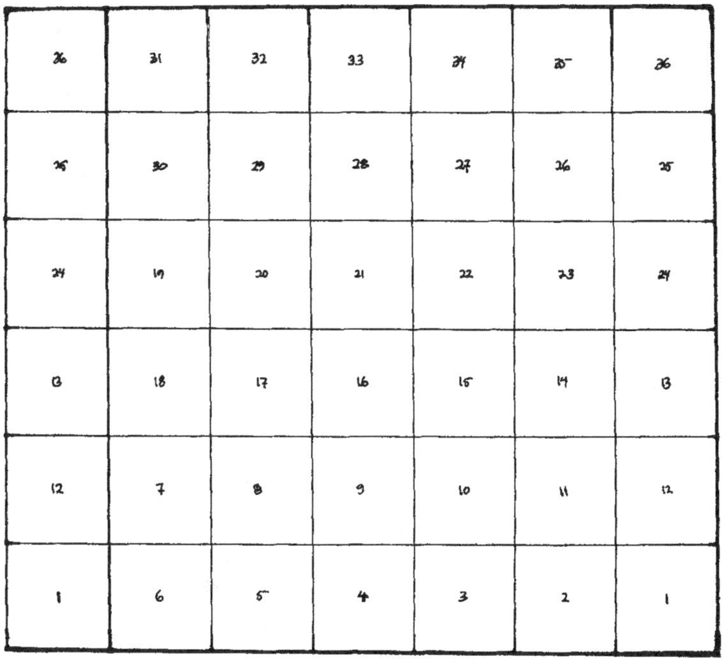

36	31	32	33	34	35	36
25	30	29	28	27	26	25
24	19	20	21	22	23	24
13	18	17	16	15	14	13
12	7	8	9	10	11	12
1	6	5	4	3	2	1

Fig. 20–12 Base Map of Areal Problem Set Study Area

mine the right answer with simulation? The author doubts that it is possible to arrive at a correct, or more to the point, a unique answer.

At least one group taking the course went broke drilling in the gas cap. They were shut in by the oil operators, which is consistent with Alberta oil and gas conservation laws. One of the participants worked for the AEUB. He was none too pleased! Other groups went broke drilling in the water leg or in low-permeability areas. This is an interesting aspect as well. Winning the game is not just a matter of technical skill. One must assess risk and take appropriate hedging action. The situation is competitive.

One other issue worth investigation is *What is the optimal recovery scheme?* Some areas are low permeability that do not make for economical wells. The proper location in channels would indicate intuitively a better development plan.

Each time the author has observed others attempt to solve this problem, there have been different developments. This is an artifact of history and chance. People interpret the seismic differently and use different profiles for drilling. It would be interesting to investigate how much variance would occur in recovery or net present value (NPV) based on different development scenarios. One branch of statistical analysis allows us to evaluate relative probabilities by combining a grouping of experts' analysis. This would be an interesting investigation also.

During one course offered by the author, there was sufficient time for the groups to prepare a report. The production forecasts for the three different groups were entirely different! This was used for course evaluation. The scope of the report was to recommend a development plan based on mapping and reservoir simulation. This course was attended by several complete exploitation teams.

Procedure

The author has worked to simplify the procedure so that it is possible to enter the geological input more quickly without having to resort to a contour package. This, like real life, is a major part of simulation input.

Beyond This Book

In this book, the author hopes to have changed the reader's attitude towards simulation. Getting good results is a matter of putting together many different elements and disciplines. Simulation technique is not enough. In the author's opinion, the biggest hang-up usually comes from not having a good visualization or sufficient geology. Ideally, the techniques and details of running the simulator are second nature, and the engineer can concentrate on getting the correct input. This is not realistic, in particular for the first few simulations. Cheat (fairly) by reading geological papers on reservoirs similar to the one being modeled, by reviewing technical papers on offset reservoirs, by reading all government applications, by using a systematic approach, by careful data organization, and by setting aside time to specifically review global issues. It is not easy, and it takes time to do it correctly.

Technique

Many techniques have been developed in the field of reservoir engineering. It is not possible to remember all of the information that is available. Even Albert Einstein is said to have looked up his own number in the telephone book. The person who asked for his number was so surprised that they asked him why he had not committed it to memory. His reply was to the effect that he had enough important things (theoretical physics) to remember. He was happy simply to know where to find the rest.

Develop a set of reference materials. At a minimum, a practicing reservoir engineer should have one or two of the classic reservoir engineering textbooks and several of the SPE and Petroleum Society of CIM monographs.

Continuous development

Some areas of reservoir engineering are still progressing rapidly. Such areas include pressure transient analysis, compositional behavior, naturally fractured reservoirs, and reservoir characterization with geostatistics. These areas require that one read some of the professional journals such as the *Journal of Canadian Petroleum Technology* (*JCPT*), *Journal of Petroleum Technology* (*JPT*), *Society of Petroleum Engineering Journal* (*SPEJ*), and *SPE Reservoir Evaluation & Engineering* (*SPEREE*). At one time, operating companies supported membership in a reasonable number of professional societies. This is no longer the case. The author recommends that at least an engineer belong to APEGGA, the Petroleum Society of CIM, and the Society of Petroleum Engineers.

Contacts

Finally, friends may be able to help. This will not always be possible since they may be working for competitors. The company may have a few gurus in some areas of technical reservoir engineering, geology, or simulation. These people are amazing. Very often, they remember details of wells drilled more than 20 years earlier and will recall why certain decisions were made. There are not as many of them as there used to be, but a few are known to the author.

An engineer can hire a consultant to critique the approach. Some consultants are reluctant to do this since they want to do the work. On the other hand, one very good log analyst prefers to advise consultants. A Canadian heavy oil expert often comes on site for a day and reviews projects or data. Apparently, his suggestions can keep a number of people busy for a considerable period, even when the project is at a mature stage. Many consultants, including the author, do not object to providing advice as long as they are paid for the time expended.

Conclusion

It is hoped the reader found the material in this book to be useful. Please contact the author if you find any errors (there are some undoubtedly). Good luck with simulation studies that you do, have to review, or that you commission others to do.

Appendix A

Sour Samples

Introduction

This section has been included as an appendix since it does not have universal application. However, when working with sour fluids, this type of information is very important. Sour fluids are relatively common in western Canada.

Surface sampling—recombination

This is the standard procedure when saturated reservoir fluids are being sampled. The path of a small volume of sour retrograde condensate gas (the bubble) will be traced to illustrate potential changes that can occur. As the bubble flows from the distant part of the formation, part of the bubble condenses on the way to the well. Liquid buildup occurs around the well, the amount and timing of which depend on layering and well geometry. Bubbles entering the well contain part gas and part movable liquid. As the bubble travels up the wellbore, the pressure and temperature drop, resulting in the bubble shrinking and more and more hydrocarbon liquid being formed.

In some cases, the sulphur will start to precipitate on the walls of the tubing. If this is a new well, it is just being tested prior to putting the well on production; the H_2S will start to react with the tubulars and part of the gas will be adsorbed between the grains of the metal. Near the surface, some methanol may be added to prevent the well from freezing off due to hydrate formation, a common problem for sour wells. At surface, the gas bubble goes through a choke, which causes a rapid depressurization and further cooling. At this point, the bubble enters a high-pressure separator, where the gas is separated into separate bubbles of liquid and gas. Within the separator, it is intended that chemical equilibrium be achieved.

Limitations to equilibrium include slugging from the wellbore, and sulphur compounds can be relatively slow in attaining equilibrium. It is from such a nonequilibrium condition that most separator liquid and gas samples are taken. The high-pressure separator GOR needs to be recorded in order to accurately recombine the samples in the lab.

Sour gas storage

The effect of sample age was earlier identified as an unknown. The issue that immediately springs to mind is the effect of corrosion. Metal corrosion in sour gas environments is an electrochemical process, which consists of anodes and cathodes. Metal corrosion can be divided into two basic reactions: oxidation and reduction. Oxidation represents the corrosion of the metal that can occur at the anode. Reduction is the reaction that occurs at the cathode. In an acidic solution, which is often found in the oilfield, the following reactions are common.

$$Fe \longrightarrow Fe2^{+} + 2e^{-} \text{ [oxidation]} \quad (A.1)$$

$$2H^{+} + 2e^{-} \longrightarrow H_2 \text{ (gas) [reduction]} \quad (A.2)$$

The oxidation and reduction reactions are dependent on each other; therefore, whatever affects one reaction affects the other. In other words, if the reduction reaction is increased, the oxidation reaction, or corrosion, also increases.

From a PVT point of view, the generation of iron sulphide would reduce the molar volume of hydrogen sulphide. This molar volume would be offset by the hydrogen gas generated. The latter is a very light gas with a low critical temperature and pressure. Hydrogen would be expected to have a strong effect on phase behavior. Significant corrosion (FeS) might be evidenced by reduced H_2S content as well as increased hydrogen content in the produced gas. One would expect that determining sample stability would revolve around comparing sample cylinder weights from before sampling and after discharge, opening cylinders and attempting to weigh precipitated FeS, and monitoring for the amount of H_2 generated. However, observing weight differences in sample cylinders is very difficult, because they are quite heavy in relation to the weight of sulphur content.

Rhodes' work dramatically shows that very small amounts of water in sample cylinders accelerate corrosion. Most reservoir samples are saturated with water vapor at reservoir temperature, so condensation is inevitable when the samples are at the surface. At the very least, samples should be saturated at the high-pressure separator conditions.[1]

Influence of containers

The most comprehensive paper on this was by Price and Cromer. They used a GC designed specifically to detect a range of sulphur species, which include CS_2, SO_2, methane thiol, ethane thiol, ethyl methyl sulphide, and a host of other compounds. Their methodology consisted of adding 100 ppm of each of the following components: hydrogen sulphide, carbonyl sulphide, ethyl mercaptan, dimethyl disulphide, and ethyl methyl sulphide into a natural gas sample. Composition changes were monitored for a period of time using a series of different containers.[2]

The results were extremely interesting. Some component concentrations declined while new products appeared. Cromer and Price identified a number of potential reactions that could result in these various products. The authors concluded that silanized glass was the best container.

The author's interpretation is somewhat different and suggests implications that are more serious. First, since the concentrations eventually stabilize, it may be concluded that there was a chemical equilibrium. In general, chemical equilibria are temperature and pressure sensitive. If, for example, a perfect bottomhole sample were transferred to a cell, it would be necessary for the cell to stabilize in a range of up to 200 hours, approximately one week, just to reestablish system equilibrium. A GC analysis of a room-temperature sample cylinder would give a completely different composition than the reequilibrated sample in a PVT cell. Of course, the chief objective is to characterize reservoir fluids. Therefore, determining sulphur compound concentrations from a sample container stored at room temperature could be very misleading.

Although this work was very helpful, it does not appear to have been continued to higher levels of H_2S concentrations. The mix of the five sulphur compounds does not likely represent an actual gas sample, which would probably have a larger variety of species and a different distribution of concentrations.

Sulphur chemistry

Preliminary research on sulphur chemistry was done using overview texts. The results of this investigation were very enlightening. First, sulphur vapor is comprised of a number of different molecules ranging from S_1 to S_8. At low temperature and pressure, the vapor is green (S_8); at intermediate temperatures and pressure, it is deep cherry red (S_3 and S_4); and at high temperatures and pressures, it is dark violet (S_2). Liquid sulphur also shows a variety of species, and work has been done on the equilibrium of various lengths of chains and ring sizes. In particular, liquid sulphur contains polysulphides as well as dissolved H_2S gas.

It seems likely that determining the critical temperature and pressure could be quite difficult for sulphur since the proportion of different elemental species (S_8 or S_6, etc.) changes with temperature and pressure. This could be solved by using yet another pseudo property. The pseudo critical sulphur temperature and pressure could be made a function of temperature and pressure. Such a technique could adjust for the proportion of sulphur species. At the actual critical point, the pseudo and actual properties would converge. This pseudo property could be used, or back calculated from, EOS calculations to predict phase envelopes and other PVT processes such as a CVD or CCE lab test.

Chemical equilibria

A summary of various reactions was found. In the liquid state, there is an equilibrium between H_2S and polysulphides (H_xS_y). This does not appear to relate to the gas phase based on the material presented, although there may be an indirect effect via sulphur vapor-sulphur liquid equilibrium.[3, 4, 5]

The reaction of sulphur and alkanes is discussed. In particular, methane with sulphur resulting in CS_2, H_2S, and H_2 is described, but at temperatures well above reservoir temperatures. Reaction kinetics and observations indicate that the most likely mechanism involves liquid (molten) sulphur on the wall of the reaction vessel. Molten sulphur contains many free radicals. The latter could easily extract hydrogen from methane to form various products. The rates of reaction were also strongly affected by increased surface area in the reaction vessel.[6, 7, 8, 9]

Other research material details reactions, again at high temperatures, that involve sulphur and olefins. Gas phase reactions are documented for sulphur and olefins; however, the elemental sulphur was generated by photolysis of COS. The carbon monoxide concentration, which is a by-product of COS disassociation, changes and was involved in the reaction. Therefore, the data is not directly applicable to oil and gas problems.

Overall, there are many potential reactions. The effect of surfaces is consistent with Cromer and Price's observations. The formation of H_2S was catalyzed by carbonate surfaces. If these surfaces are necessary to complete the reaction mechanism, thought should be given to including such surfaces in a PVT cell, while attempting to reestablish equilibria at reservoir temperature and pressure.

Used sample containers

Informal discussions with lab personnel indicated that used pressure cylinders, which had contained sour gas, could reduce the amount of metallic degradation that could be expected. The most common explanation is that the H_2S is adsorbed into the cracks between metal grains. It would seem logical that sample cylinders could be presoaked with sour gases in various concentrations so that containers and samples could be matched. This could be done potentially through an industry consortium.

Given the preceding analysis, it is natural to wonder if perhaps the real advantage is that the catalyzing effect of the sample surface is somehow inhibited. Perhaps a sulphur-phobic surface would produce more stability than attempts with chemically inert surfaces such as Teflon. Polishing the interior of sample containers might also inhibit unwanted reactions if surface area strongly affects the reaction rate. Sample containers are normally evacuated immediately before sampling. This would probably decrease the advantages gained from presoaking, if absorption is in fact the underlying mechanism.

Sulphur solubility

Further investigation indicated what has long been known—i.e., elemental sulphur is dissolved in sour gases. More recently, this work has been extended. This has been driven by some extremely sour reservoirs, such as those found in the Bearberry area of Alberta. Although the equilibrium and mechanism of this reaction are discussed in some of the sulphur chemistry texts, the more recent work does not appear to have approached the problem from this direction.

Roberts and Hyne have applied a technique developed by Ziger and Eckert that incorporates van der Waal's EOS with aspects of solubility parameter theory to formulate a semiempirical correlation for the solubility of solids in fluid phases. This analysis was based on S_8. The material was not extrapolated to the prediction of phase behavior.[10]

Potential effect of sulphur vapor

The elemental sulphur content of the gas could conceivably have a very significant effect on the phase behavior. ARE was able to obtain two estimates of the critical temperature and pressure. The first was 1,038° C (1,311 °K or 1,900° F) and 116 bars or 11,760 kPa (1,705 psi). The second was 1,041° C (1,314 °K or 1,906 °F) and 207.0 bars or 20,970 kPa (3,042 psi). The critical pressures differ significantly.

Such high values would have the effect of shifting the phase diagram far to the right. The possible saturated sulphur solubility was extrapolated from the data of Davis, Lau, and Hyne.[11] At about 90° C and 27,000 kPa, the solubility was estimated to be 0.7 g/m^3. The molecular weight of sulphur is 32.066 and, when converted to a molar concentration, amounts to approximately 0.052 mole percent. The data of Kennedy and Wieland places the sulphur content at 7.32 lbs sulphur/mmscfd (0.12 g/m^3) in a mixture of 75.5:24.5 mole percent ratio of methane to hydrogen sulphide.[12] The latter corresponds to 0.009 mole percent sulphur. Such low contents would seem unlikely to have much effect on the phase behavior.

Davis et al. used a Hastelloy C-276 high nickel bar stock to prevent corrosion. Kennedy and Wieland do not identify the composition of their sulphur equilibrium vessel; however, it does contain 40 and 80 mesh screens—which are likely to be wire. Davis et al. overheated the sulphur until it was molten for a 16-hour period. It was then equilibrated at the target temperature for at least 32 hours. Therefore, the equilibrium should be approached from an oversaturated condition.

However, as discussed earlier, the container surface appears to be involved in the equilibrium mechanism. Cromer and Price also show the equilibrium to be affected by storage cylinder surfaces and quite long equilibrium times, up to 200 hours. Geochemistry indicates that the generation of H_2S may be catalyzed by carbonates, indicating that the reverse reaction (generating sulphur) would be similarly affected. Finally, these tests were generated with synthetic gases. There is data to suggest that there are a considerable number of subsidiary sulphur compounds involved in simultaneous equilibria. Therefore, the author concludes that the tests conducted to date may not reflect actual reservoir conditions.

Nonreactive cells

Robinson describes a new PVT apparatus that solves mercury interaction problems. The sample container utilized was made from an artificial sapphire and is clear and nonreactive. A piston with tapered edges is utilized that allows low liquid dropout volumes to be more accurately measured. There are some limitations to this apparatus. At present, it has a lower pressure rating than many steel cells and can only handle very small sample volumes.[13]

Vagtborg solved the problem of mercury interaction in 1954 by using a separating disk with Teflon seals for a highly sour oil sample.[14]

Complex phase behavior

Heng-Joo Ng et al. report very interesting results using the previously described cell. The fluid tested was a mixture of methane, carbon dioxide, and hydrogen sulphide with 49.88, 9.87, and 40.22 mole percent concentrations, respectively. The resultant phase diagram is shown in Figure A–1. The system has three different critical points. Note the upturn at low temperatures and the lack of a unique critical point. These problems were present with the EOS characterization. However, it is difficult to differentiate between mathematical problems and chemistry problems. Note that the primary objective of the research conducted was for use in gas processing facilities at low temperatures.[15]

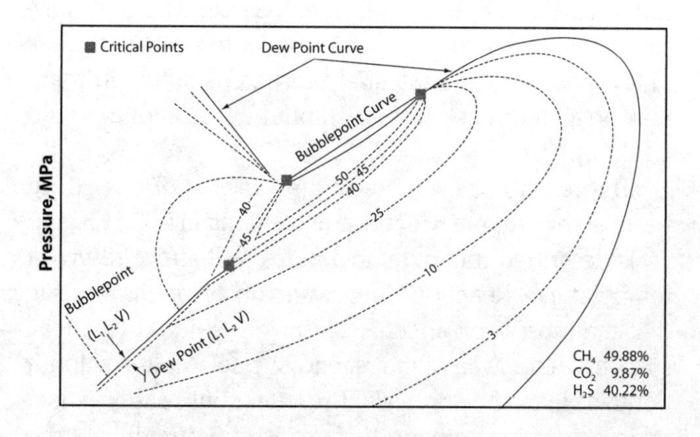

Fig. A–1 Phase Diagram with Multiple Critical Points for a Sour Gas System (after Heng-Joo Ng et al.)

References

1 Rhodes, H.L., "Determination of Hydrogen Sulphide Content in Natural Gas, Evaluation of Containers for Preparation of Calibration Standards, and Sample Collection Procedures." Report of Investigation 8394, U.S. Department of the Interior, Bureau of Mines, 1979.

2 Price, J.G.W. and D.K. Cromer, "Influence of Containers on Sour Gas Samples," *Petroleum Reservoir International*, March 1980.

3 Carlson M.R., and B. Cawston, "Obtaining PVT Data for Very Sour Retrograde Condensate Gas and Volatile Oil Reservoirs; A Multi-Disciplinary Approach," SPE 35653, Calgary, Alberta, Canada: Gas Technology Conference (Apr. 20–May 1), 1996.

4 Hunt, J.M., *Petroleum Geochemistry and Geology*, W.H. Freeman and Company, 1995.

5 Tissot, B.P., and D.H. Welte, *Petroleum Formation and Occurrence*, 2nd ed., Springer Verlag, 1978.

6 Nickless, G., ed., *Inorganic Sulphur Chemistry*, Elsevier Publishing Co., 1968.

7 Tuller, W.T., *The Sulphur Data Book*, McGraw-Hill, 1954.

8 Tobolsky, A.V., *The Chemistry of Sulphides*, John Wiley and Sons, 1968.

9 Senning, A., *Sulphur in Organic and Inorganic Chemistry*, New York: Marcel Dekker Inc., 1972.

10 Roberts, B.E., and J.B. Hyne, "Correlation of the Solubility of Sulphur in H2S and Sour Gas Mixtures," Alberta Research Limited Bulletin, 1993.

11 Davis, P.M., C.S.C. Lau, and J.B. Hyne, "Data on the Solubility of Sulphur in Sour Gases," *Alberta Research Council Limited Bulletin*, 1992.

12 Kennedy, H.T., and D.B. Wieland, "Equilibrium in the Methane – Carbon Dioxide – Hydrogen Sulphide – Sulphur System," *Transactions of AIME*, Vol. 219, 1960.

13 Robinson, D.B., "Experimental Methods for Measurement of Phase Equilibria at High Pressures," *Pure and Applied Chemistry*, Vol. 65, no. 5: 971–976, 1993.

14 Vagtborg, H., Jr., "Equilibrium Vaporization Ratios for a Reservoir Fluid Containing a High Concentration of Hydrogen Sulphide," *Transactions of AIME*, Vol. 201, 1954.

15 Ng, Heng-Joo, D.B. Robinson, and A-D Leu, "Critical Phenomena in a Mixture of Methane, Carbon Dioxide and Hydrogen Sulphide," *Fluid Phase Equilibria, 19*, Elsevier Science Publishers: 273–286, 1985.

Appendix B

Study Review Procedure

Introduction

Often, reservoir engineers are provided with a numerical simulation study as part of the technical support for a property. The author has developed a procedure that will aid in assessing the applicability of simulation results based on experience in a corporate reserves group, as an evaluation consultant, as an exploitation engineer, and as a simulation engineer/consultant. The procedure consists of:

- Applying consistency checks
- Identifying the critical issues
- Evaluating the simulation technique
- Reviewing the simulation report

This procedure has proven to be time effective. Simulation results may complement the reservoir engineering process, but they cannot replace an experienced engineer's judgment. In particular, risk and economic limits must be assessed in the application of most reservoir engineering calculations.

This procedure will be outlined after a short discussion of real-life studies.

Real-life studies

In one study that was performed by the author, the client had stipulated the assumptions prior to the start of the work. Soon it became obvious that some of these assumptions were not at all reasonable; however, the model was not changed when these disparities were pointed out. In the end, it led to one of the worst studies in the author's experience.

The point is twofold. First, for the most part we impute an expectation that people, which includes clients, bosses, companies, and ourselves, will behave logically as a whole. In fact, this is far from reality. While we can speculate as to causes and motives, it is usually impossible to tell the exact reasons why some things occur. For the most part, the most common causes turn out to be a combination of oversights, miscommunications, clerical errors, differing opinions, and simple ignorance (one cannot know everything).

On this basis, it is logical that studies that have significant financial impact should be checked thoroughly. Engineers who work in civil engineering are often shocked by the lack of checking that occurs in the oil industry. Certainly, much of the work done by the author as a junior engineer in the oil industry was never checked in detail. Perhaps the reason that this persists is that reservoir engineering decisions rarely put individual lives at risk. On the other hand, large sums of money are at stake. The emphasis on protection of life is probably a good thing.

Reserves evaluation is one area where checking occurs, although this is not universally true.

The second point is that studies that fail are not always the simulation engineer's fault. Since bosses and clients rarely like to be proven wrong (who does?), normally these faults are not documented, or they are phrased in innocuous terms (so as to preserve one's career). Excessively polite warnings or qualifications are actually VERY LARGE RED FLAGS and should be considered with *extreme* care.

In one instance, bizarre study results were put out with no comments at all. In this particular case, the author knew the individual, an engineer who was exceptionally competent. A polite phone call resulted in revealing the story behind the bizarre approach and results. This situation, like the one described previously, was the result of specific client directions. Therefore, if something looks strange, try to talk privately with the individual who did the work. Most technical engineers are eager to disown anything they know to be wrong.

Applying Consistency Checks

Perform simple numerical checks

The following checks have proven to be useful in the past:

- Match original oil in place (OOIP) derived from the model to those calculated by volumetrics and material balance. Note that often the OOIP will be changed in the process of history matching. Are the changes reasonable?
- What percentages of the OOIP and original gas in place have been produced? Does the model match the actual production that has occurred?
- Calculate the recovery factors for all subsidiary products. Are these numbers reasonable?
- Check the water saturations in the model versus the results from log analysis.

These checks, and any others that one may think are appropriate, will identify a number of potential errors. Some examples observed include:

- Sensitivity cases can be mixed up. One such error found by the author involved a gas injection sensitivity result that was mixed in with a waterflood sensitivity. This was found by checking the solution gas recovery as a percentage of OGIP. Simulators create enormous amounts of output; clerical errors are inevitable.
- Changes to the OOIP in history matching may have been excessive. In one review, a 40% increase had been made in the downdip portion of the reservoir. Subsequent analysis indicated that this effect could have been represented more reasonably by an active aquifer.
- Check metric conversions. The built-in metric conversions within a simulator often use slightly different conversion factors. In rare instances, there may be a mistake in the numerical model. More commonly, a model will be run in Imperial units, and the results will be converted manually for the report.

Plot history match and main prediction together

The majority of simulation engineers prefer linear rate versus time plots to semilog rate versus time plots. Invariably, evaluations engineers prefer the latter, since they commonly use decline analysis. In order for one's experience base to have its full impact, the results of the study should be plotted on semilog rate versus time plots.

The predictive phase and the history-match phase of a study are usually performed at different times and using separate computer runs. Very often, the history-match plot and the prediction are not plotted together. However, a combined plot can be very informative.

In one case, the production rate at the end of the history match and the start of the prediction were not identical—production occurred at different rates. In this case, our critique of a simulation caused a client to rethink his approach to optimizing production. The client has subsequently achieved highly successful results with a different, more realistic, and less-costly optimization program.

Specific problems that the author has observed include:

- Unrealistic production forecasts. In one case that the author reviewed, modeling predicted production response in a reimplemented waterflood, predicted response exceeded the original (successful) waterflood peak production.
- Simulations that have not been properly tuned. The initial production in one case was too high, which led to optimistic recovery predictions.
- Inappropriate model physics. Projected profiles, when compared to offset projects, showed different production responses. This indicated that critical reservoir mechanisms had not been accounted for.

Plot individual well history matches and predictions together.

As in the previous section, plotting the prediction and history match on consistent plots for individual wells is recommended. Items to watch for include:

- Tuning of individual wells. Although the overall pool performance may match, the author has seen simulations where the best wells had 25% production jumps—at the expense of minor producers. The resultant prediction was skewed optimistically.
- Systematic trends in mismatches. For example, an active water leg will result in wells watering out in relation to their dip position.
- How boundaries have been handled. If the model is not a closed system, such as a window study or where only a part of the reservoir has been modeled, there will likely be a buffer area that is extremely difficult to match.

After completing the checks listed previously, the overall purpose of the simulation should be examined before evaluating the simulation technique and proceeding to a detailed review of the report.

Identifying Critical Issues

Common critical issues

A number of different situations exist where a simulation is normally encountered:

- A property that has been on primary production that an operating company has decided to waterflood is probably the most common issue. They are looking for waterflood reserves based on the results of a simulation.
- A well drilled recently in a new field that has either a gas-oil or water-oil contact in close proximity to the completed interval represents another common issue. A simulation has been provided to support a recovery factor as well as GOR or water cut trends.
- An operating company has a mature field, which they have modeled to optimize production. Small increases in overall recovery factors may represent major increases in the remaining reserves. It can be difficult to distinguish incremental reserves from production acceleration.
- A large gas field has been developed that exhibits retrograde liquid condensation. A study has been used to determine the amount of NGL and gas recovery.
- A large miscible flood has been installed, and the operating company has performed a simulation. Since limited production history exists, a simulation is often the best method of engineering analysis available to estimate recovery.

The nature of numerical models introduces some modeling requirements that are necessary to achieve the objectives of such studies. The following attempts to present some of the techniques of reservoir simulation and how these specific requirements arise.

Evaluating Simulation Technique

Tailoring to critical issue

Using the concepts developed previously, check the most important issues. Some examples might be:

- For a waterflood implemented after a period of primary production, one of the most critical aspects of model design is characterizing geological layering or heterogeneity. This can be done either with detailed layering built into a model or via pseudo relative permeability curves.
- For a well drilled into a pool with a water or gas contact, the critical modeling issue likely will be accounting properly for near-wellbore effects. This would usually involve a detailed coning study, defining layers, covering the correct range of production rates, and including the effects of hysteresis.
- For a mature field optimization, the most critical aspect of the study likely would be how to build a balanced model. This could require preliminary coning and cross-sectional studies that must be integrated later into an areal model. Integration is the major issue.
- For a retrograde gas-condensate reservoir, the study likely will require a compositional simulator. Proper PVT characterization will comprise a major portion of the study.
- For a miscible flood, a number of critical issues are possible. The same simulator may not address all of the issues effectively. If miscibility is the critical issue, a compositional simulator is in order. If maximizing sweep efficiency is the critical concern, then a pseudo miscible simulator is probably the most effective solution.

Note that other factors such as well spacing, laboratory relative permeability, PVT properties, injectivity, and timing of wells are still important.

Any numerical model, depending on its design, will take into account various factors to varying degrees. Tradeoffs are usually required in the design of a model. A study may have been designed to address a specific isolated objective. However, the objectives that the simulation engineer has designed his model around and one's perception of the critical issues must coincide. Otherwise, although the simulation may have met its objectives, it may not be of direct use in the broader reserves picture.

Check that grid can model reservoir flow physically

Reducing the number of grid blocks can be taken too far. The following example occurred in a study proposal that was submitted to the author. The objective was to determine horizontal well potential in a tight sandstone reservoir. Production in vertical wells was not economical due to low rates and GOR penalties. The study was launched to determine if a horizontal well would increase productivity sufficiently to enable economic production and if GORs would be low enough to avoid penalties.

Flow nets are a useful tool to help visualize the flow in the reservoir. Since the reservoir is tight (less than 1 mD), localized gas saturations would be expected to arise. A fine grid made with small grid blocks and thin layers would be required around the well. An element of symmetry could be utilized through the center of the well to reduce computation.

The proposal for the situation described previously recommended a very coarse grid of three layers with the wellbore centered in blocks with 100-ft. horizontal dimensions. It is important to realize that, although the simulator would still have run, these large grid blocks could not correctly represent the physics in the reservoir. The model would still produce results, despite the fact that inaccurate results would be generated. The author did not consider the proposal favorably.

Many wells have been hydraulically fractured. Usually models are run and minor adjustments are made via pseudo well relative permeability curves. Wells that have been fraced will show slower GOR increases than unfracced wells. This will not always produce correct results if the combined propped half-length of the fractures is greater than 25% of the interwell distance. Under these conditions, the deviation from a radial flow pattern is too severe to be represented radially in conjunction with pseudo well curves.

Studies by the author have shown that waterflood response in a reservoir that has large fracs is not as pronounced as predicted by radial flow models. The approach to such a simulation must be changed—a type well study with the fracture modeled directly is more appropriate. Grids are not always implemented correctly. Unfortunately, physically incompatible results will not automatically cause the simulator to crash. In such cases, the model will calculate meaningless results with great precision, following the maxim, *garbage in, garbage out* (GIGO).

Check implementation of wells

The wellbore equation used in grid blocks is for steady state flow. Hence, buildups cannot be modeled, unless a coning-type model has been built. The author has encountered studies where buildups were matched using coarse grids.

When fluid contacts exist in the reservoir, one of the more sophisticated techniques described earlier must be used. The use of pseudo well relative permeabilities has a number of pitfalls. These problems include rate dependence, a common occurrence with the implementation of secondary recovery, which can easily result in distorted rate predictions. While the use of pseudo well relative permeabilities was once an absolute necessity, there are now many circumstances where the use of well pseudos can be avoided.

Check layering (heterogeneity) for EOR

The effects of layering are perhaps the most important design criterion for enhanced oil recovery. Coincidentally, single-layer areal studies are common. Choosing appropriate layers requires detailed work. The most comprehensive approaches will feature:

- A detailed description of the layers by a geologist. Different physical reservoir properties will usually correlate to different stratigraphic units.
- A detailed petrophysical study in which the porosity, permeability, and rock lithology have been correlated.
- A computerized log analysis of all the wells. Computerized log analyses that allow more consistent calculations and more sophisticated correlations are also possible using a computer.

- Statistical layering techniques developed by Testerman, which provide a quantitative basis for determining layers.
- If none of these is available, a range of typical heterogeneities can be evaluated by using a Hearn-type pseudo relative permeability technique in conjunction with a Dykstra-Parsons type coefficient.

At least one of these must be used, particularly if only primary production has been history matched. Water saturations during primary production will remain close to the connate water saturation. Therefore, almost none of the real relative permeability curves will have been traversed. Under these circumstances, the effects of layering and heterogeneity will not be accounted for.

If no layering calculations have been made, the resulting waterflood prediction may be seriously optimistic. Under these circumstances, the quality of the history match on primary production will not indicate the degree of accuracy of the waterflood prediction.

If a significant period of waterflooding were history matched, it is likely that the history-matched pseudo relative permeability accounts for the effects of layering and heterogeneity. For this reason, a waterflood feasibility simulation study often will recommend that the study be updated after water breakthrough has occurred. In this way, an improved optimization strategy can be developed.

Check for thorough and up-to-date geology

The geological phase of data input is the most involved. Maps of net pay, porosity, and permeability must be developed and entered into the simulator. In most cases, the geology should be updated immediately before the study. Using outdated geology, without all the wells and available information, has proven to be a waste of effort. Conversely, in the early stages of development, a complete reservoir description is not possible; hence, the results of a study performed early will be incomplete. Correct geological interpretation also affects such parameters as layering, degree of fracturing, and the probability of water influx. A comprehensive geological study is required before simulation. As a general guideline, the geological segment should comprise about the same amount of time and cost as the reservoir simulation study. The specifics of evaluating geological studies goes well beyond the scope of this text. However, most engineers should have seen enough geological studies to be aware of the depth of study.

In the author's opinion, most simulations that fail do so, directly or indirectly, because of inadequate geological descriptions.

Check for proper initialization

Check how the simulator has been initialized. The use of pseudo relative permeabilities creates the need for modifications to capillary pressure. Some programs implement these changes automatically and others do not. Note that a model will likely run with incorrect data. If the saturations have been specified manually, check to see that a run has been made with no well production. This ensures that the pool is in gravity equilibrium before starting the history match.

Check history-matching technique and changes

This will likely be the largest single proportion of work for a study. Check first for an organized approach. This is not fixed since the requirements vary from study to study. For instance, a coning study with detailed layering and lab data should not require development of pseudo relative permeabilities.

A history-match log outlining the various changes for individual runs is a positive sign. This indicates a high degree of organization on behalf of the simulation engineer. In addition, check that the process and the changes made are consistent. A table containing all of the runs made, execution times, etc. is an efficient way of summarizing this data.

Check for numerical errors

The solution of a reservoir simulator involves numerical techniques and the answers are only approximate. It is not possible to calculate the degree of error analytically. To quantify errors, it is necessary to rerun the problem with successively finer grids. When the differences are small, compared to the accuracy of the input data, an adequate solution has been obtained. The presence of such a sensitivity is normally a very positive sign.

Check for prediction consistency

The prediction scenarios must be analyzed for consistency. A few suggested techniques are as follows:

- Using a photocopy of the pool map, shade in projected waterflood sweeps. If the zone is thick, try this method on a cross section.
- Look at the pressure history. Frequently rapid injection is required to fill voidage, after which water injection rates can be reduced. Rapid pressurization will improve the speed of response and economics. However, the injection rates must remain realistic.
- Check that the date for implementation of injection is realistic. The speed of waterflood response is negatively affected strongly by the amount of solution gas that has broken out. It takes time to drill wells, build water injection plants, and obtain government approvals. If a study were predicated on injection starting five years prior to the time the facilities actually could be installed, suspect that the true response will not be as quick. In smaller pools, considerably less than five years of production could be significant.

Since most simulation results may be used in economic runs, the predictive cases in the report should not only be plotted but should be provided in tables on a yearly basis.

Simulation in the Overall Reservoir Life Cycle

The final element of the procedure is a detailed review of the simulation study report. If the report has not been kept up-to-date as the study progressed, there is a good chance that important decisions were not recorded. Report writing also falls at the period where people try to correct budget overruns, resulting in an incomplete explanation of the study. It is possible that a report will not be documented sufficiently to make a useful assessment of the results.

Check data preparation

A model can only be as good as the input data. Therefore, input data should be checked meticulously. The assumptions should be reasonable. As indicated earlier, the author feels simulation reports should include a copy of the final data input deck. This is easily imported into word processing packages. With appropriate comments and a reduced text size available on almost all laser printers, there is no reason why there should not be a permanent record of the input file within a report.

An explanation should be included in this section of the report explaining the basic methodology or approach to the problem. For waterfloods, check the relative permeability end-point saturations.

History-match plots

A complete set of history-match plots should be included either in the report or as an appendix. One item the author finds very helpful on history-match plots is a small key map. With this, it is possible to mentally correlate history-match results for a particular well with its location in the unit or pool.

Almost every numerical modeling study includes a history match. Surprisingly, in all of the literature that the author has reviewed, there have been no firm guidelines to determine the adequacy of a history match. This is probably because conceptual implementation is as important as the quality of matches on individual wells.

Assuming, for the time being, that the study has been properly designed, it is rare for all of the wells in an areal simulation to match historical data perfectly.

The degree to which individual wells match can vary considerably. For this reason, it can be difficult to decide if the match is satisfactory overall. The following approach is suggested:

- Make a list of all of the wells and then assign each well a score from 1 to 10 for:
  a. overall history match quality
  b. the importance of the well to the pool
- Multiply factor *a* times *b* and sum the products for all of the wells. Divide by the sum of the *b* factors. This can be done easily using a spreadsheet.

The author does not have a standard numerical threshold. However, after using this method, conclusions can be made regarding the overall history-match quality.

Check that all of the production has been obtained on a cumulative basis. The base production should be almost 100% accurate. If not, then a well has probably been turned off by the simulator's bottomhole pressure routine, and the history-match base production will not be correct. In addition, the cumulative amounts of production on secondary products should be evaluated. Although history-match trends may look good, the cumulative volumes can be off by quite a bit with rapidly rising water cuts. Realistically, expect the values to fall within ±10% for the majority of values and 15–20% on the exceptions.

For an areal study, look for a summary diagram that has the total pore volume and permeability changes displayed. Simulators tend to be more sensitive to pore volume changes than permeability changes. A pore volume change of 15% should not cause concern, nor should a doubling of permeability. Larger changes may occur, but should be the exception, not the rule. If there are extreme changes or ridiculous faults that are necessary to obtain a history match, then it is likely that there was a problem in conceptual implementation.

Look through computer output

Ask for a copy of the final computer outputs and timestep logs. This may be the only method of finding numerical instabilities.

Stability problems occur more frequently with gas. Due to its low density and low viscosity, this phase is the most likely to move faster than the simulator can handle. This will manifest itself usually as negative values of gas saturations in the output arrays. Once again, expect the problems to coincide with well blocks. These problems are far more likely to exist on one of the older IMPES-type simulators. The more modern simulators use adaptive implicit techniques. The switch from IMPES to implicit can be based on either threshold values (fast) or stability calculations (slower), so this type of error can still occur. Contrary to popular belief, these instability errors will not show up on the material balance. The author has found these instabilities in actual studies.

In summary, numerical errors are very real. A grid sensitivity study, although rare in areal simulations, should be taken as a positive sign if it is in the report. It should be performed with all of the phases that will eventually be present in the field. Some relatively simple tests on a five-spot waterflood pattern will show that simulation results are less sensitive to recovery than to water breakthrough timing. A 7 × 7 grid, of a quarter element of symmetry for a five-spot waterflood pattern, would not be an unusual requirement to match water breakthrough times.

Coarser grids are commonly used, and history is matched via water breakthroughs. Suspect that the history matched pseudo well/reservoir relative permeability curves compensate for dispersion, in addition to well completion, coning, or layering effects.

Conclusions

A systematic procedure to evaluate a simulation is presented. The use of some relatively straightforward commonsense checks will usually screen out major flaws in a simulation study.

- Simulation is not a substitute for commonsense reservoir engineering. The results of a simulation study may complement the evaluation process; however, it cannot replace all of the elements that should be considered.
- To effectively model a reservoir, the simulation engineer must understand the critical depletion mechanisms in the first place. To use the results of a simulation study, one's opinion of the critical components must agree with the way in which the study was performed.
- The details of simulation techniques are important. This is particularly true for the more involved miscible and compositional simulations. Do not hesitate to dig into these issues.
- Be particularly wary when dynamic pseudo relative permeability functions have been used and when waterflood performance has been predicted based on a primary production history match.

Some of the checks described are designed to find simple clerical errors. Due to the large amount of data involved in simulations, these are more common than one might expect. This needs to be checked.

The author has encountered statements in some reserves monographs that state "the quality of a history match is an indication of whether simulation results are appropriate for use in reserves calculations." This overlooks the fundamentals of simulation technique and is therefore dangerously simplistic and outright misleading. Buyer Beware!

Index

A

B

C

D

E

F

G

H

I

J

K

L

M

N

O

P

Q

R

S

T

U

V

W

Z

Notes

Notes

Notes

Notes